# Basic College Mathematics with Early Integers

## SECOND EDITION

## Marvin L. Bittinger

*Indiana University Purdue University Indianapolis*

## Judith A. Penna

*Indiana University Purdue University Indianapolis*

**PEARSON**

Addison
Wesley

Boston  Columbus  Indianapolis  New York  San Francisco  Upper Saddle River
Amsterdam  Cape Town  Dubai  London  Madrid  Milan  Munich  Paris  Montréal  Toronto
Delhi  Mexico City  São Paulo  Sydney  Hong Kong  Seoul  Singapore  Taipei  Tokyo

| | |
|---|---|
| *Editorial Director* | Christine Hoag |
| *Editor in Chief* | Maureen O'Connor |
| *Executive Editor* | Cathy Cantin |
| *Executive Project Manager* | Kari Heen |
| *Associate Editors* | Joanna Doxey and Christine Whitlock |
| *Editorial Assistant* | Jonathan Wooding |
| *Production Manager* | Ron Hampton |
| *Composition* | Pre-Press PMG |
| *Production Services* | Sally Lifland/Lifland et al., Bookmakers |
| *Editorial Services* | Martha K. Morong/Quadrata, Inc. |
| *Art Editor and Photo Researcher* | Geri Davis/The Davis Group, Inc. |
| *Senior Media Producer* | Ceci Fleming |
| *Associate Producer* | Nathaniel Koven |
| *Content Development Managers* | Rebecca Williams (MathXL), Mary Durnwald (TestGen) |
| *Executive Marketing Manager* | Michelle Renda |
| *Marketing Assistants* | Margaret Wheeler and Alicia Frankel |
| *Prepress Supervisor* | Caroline Fell |
| *Manufacturing Manager* | Evelyn Beaton |
| *Senior Manufacturing Buyer* | Carol Melville |
| *Senior Media Buyer* | Ginny Michaud |
| *Text Designer* | Geri Davis/The Davis Group, Inc. |
| *Senior Designer/Cover Design* | Beth Paquin |
| *Cover Photograph* | © Photoshot Holdings, Ltd./Alamy |

**Photo Credits**

Photo credits appear on page C-1.

Library of Congress Cataloging-in-Publication Data
Bittinger, Marvin L.
  Basic college mathematics with early integers. — 2nd ed. /
by Marvin L. Bittinger, Judith A. Penna.
      p. cm.
  Includes index.
1. Arithmetic—Textbooks.    I.  Penna, Judith A.
II.  Bittinger, Marvin L., Basic mathematics.
III.  Title.
  QA152.3.B528 2011
  510—dc22                                                               2009020862

3 4 5 6 7 8 9 10—CRK—14 13 12 11

© 2011, 2007. Pearson Education, Inc.

**Addison-Wesley**
is an imprint of

www.pearsonhighered.com

ISBN-13: 978-0-321-61341-7
ISBN-10: 0-321-61341-4

# Contents

# Index of Applications

# Authors' Note to Students

Welcome to *Basic College Mathematics with Early Integers*. Having a solid grasp of the mathematical skills taught in this book will enrich your life in many ways, both personally and professionally, including increasing your earning power and enabling you to make wise decisions about your personal finances.

As we wrote this text, we were guided by the desire to do everything possible to help you learn its concepts and skills. The material in this book has been developed and refined with feedback from users of the previous edition so that you can benefit from their class-tested strategies for success. Regardless of your past experiences in mathematics courses, we encourage you to consider this course as a fresh start and to approach it with a positive attitude.

One of the most important things you can do to ensure your success in this course is to allow enough time for it. This includes time spent in class and time spent out of class studying and doing homework. To help you derive the greatest benefit from this textbook, from your study time, and from the many other learning resources available to you, we have included an organizer card at the front of the book. This card serves as a handy reference for contact information for your instructor, fellow students, and campus learning resources, as well as a weekly planner. It also includes a list of the Study Tips that appear throughout the text. You might find it helpful to read all of these tips as you begin your course work.

Knowing that your time is both valuable and limited, we have designed this objective-based text to help you learn quickly and efficiently. You are led through the development of each concept, then presented with one or more examples of the corresponding skills, and finally given the opportunity to use these skills by doing the interactive margin exercises that appear on the page beside the examples. For quick assessment of your understanding, you can check your answers with the answers placed at the bottom of the page. This innovative feature, along with illustrations designed to help you visualize mathematical concepts and the extensive exercise sets keyed to section objectives, gives you the support and reinforcement you need to be successful in your math course.

To help apply and retain your knowledge, take advantage of the new Skill to Review exercises when they appear at the beginning of a section and the comprehensive mid-chapter reviews, summary and reviews, and cumulative reviews. Read through the list of supplementary material available to students that appears in the preface to make sure you get the most out of your learning experience, and investigate other learning resources that may be available to you.

Give yourself the best opportunity to succeed by spending the time required to learn. We hope you enjoy learning this material and that you will find it of benefit.

Best wishes for success!
Marv Bittinger
Judy Penna

**Related Bittinger Paperback Titles**

- Bittinger: *Fundamental College Mathematics,* 5th Edition
- Bittinger: *Basic College Mathematics ,* 11th Edition
- Bittinger: *Introductory Algebra,* 11th Edition
- Bittinger: *Intermediate Algebra,* 11th Edition
- Bittinger/Beecher: *Introductory and Intermediate Algebra,* 4th Edition

**Accuracy**
Students rely on accurate textbooks, and our users value the Bittinger reputation for accuracy. All Bittinger titles go through an exhaustive checking process to ensure accuracy in the problem sets, mathematical art, and accompanying supplements.

# Preface

## New in This Edition

To maximize retention of the concepts and skills presented, six highly effective review features are included in this Second Edition. Student success is increased when review is integrated throughout each chapter.

### Six Types of Integrated Review

**Skill to Review** exercises, found at the beginning of most sections, link to a section objective. These exercises offer a just-in-time review of a previously presented skill that relates to new material in the section. For convenient studying, section and objective references are followed by two practice exercises for immediate review and reinforcement. Exercise answers are given at the bottom of the page for immediate feedback.

**Skill Maintenance Exercises**, found in each exercise set, review concepts from other sections in the text to prepare students for their final examination. Section and objective references appear next to each Skill Maintenance exercise. All Skill Maintenance answers are given at the back of the book.

A **Mid-Chapter Review** reinforces understanding of the mathematical concepts and skills just covered before students move on to new material. Section and objective references are included. Exercise types include Concept Reinforcement, Guided Solutions, Mixed Review, and Understanding Through Discussion and Writing. Answers to all exercises in the Mid-Chapter Review are given at the back of the book.

The **Chapter Summary and Review** at the end of each chapter is expanded to provide more comprehensive in-text practice and review.

- **Key Terms, Properties, and Formulas** are highlighted, with page references for convenient review.
- **Concept Reinforcement** offers true/false questions to enhance students' understanding of mathematical concepts.
- **Important Concepts** are listed by section objectives, followed by *worked-out examples* for reference and review and *similar practice exercises* for students to solve, in this *new* feature.
- **Review Exercises**, including Synthesis exercises and two new multiple-choice exercises, are organized by objective and cover the whole chapter.
- **Understanding Through Discussion and Writing** exercises strengthen understanding by giving students a chance to express their thoughts in spoken or written form.

Section and objective references for all exercises are included. Answers to all exercises in the Summary and Review are given at the back of the book.

**Chapter Tests**, including Synthesis questions and a new multiple-choice question, allow students to review and test their comprehension of chapter skills prior to taking an instructor's exam. Answers to all questions in the Chapter Tests are given at the back of the book. Section and objective references for each question are included with the answers.

A Cumulative Review after every chapter starting with Chapter 3 revisits skills and concepts from all preceding chapters to help students recall previously learned material and prepare for exams. Answers to all Cumulative Review exercises are coded by section and objective at the back of the book to help students identify areas where additional practice is needed.

## Other New Elements

A new design enhances the Bittinger guided-learning approach. Margin exercises are now located next to examples for easier navigation, and answers for those exercises are given at the bottom of the page for immediate feedback.

Content changes include streamlined coverage in Sections 1.3 and 1.5 and also in 3.4, where two objectives are combined. Coverage of rounding and estimating, formerly found in Section 1.4, is now presented in Section 1.6, where new material on estimating quotients has been added. Material relating to credit card interest rates has moved from Section 7.8 to Section 7.7 in this edition. In addition, we have added over 200 new examples and over 1350 new exercises.

# Hallmark Features

**Revised!**   The **Bittinger Student Organizer** card at the front of the text helps students keep track of important contacts and dates and provides a weekly planner to help schedule time for classes, studying, and homework. A helpful list of study tips found in each chapter is also included.

**New!**   **Chapter Openers** feature motivating real-world applications that are revisited later in the chapters. This feature engages students and prepares them for the upcoming chapter material. (See pages 1, 189, and 257.)

**New!**   **Real-Data Applications** encourage students to see and interpret the mathematics that appears every day in the world around them. (See pages 75, 342, and 421.) Over 700 applications are new to this edition, and many are drawn from the fields of business and economics, life and physical sciences, social sciences, medicine, and areas of general interest such as sports and daily life.

**Study Tips** appear throughout the text to give students pointers on how to develop good study habits as they progress through the course, encouraging them to get involved in the learning process. (See pages 9, 153, and 259.) For easy reference, a list of Study Tips by chapter, section, and page number is included in the Bittinger Student Organizer.

**Caution Boxes** are found at relevant points throughout the text. The heading "*Caution!*" alerts students to coverage of a common misconception or an error often made in performing a particular mathematics operation or skill. (See pages 160, 230, and 264.)

**Revised!**   Optional **Calculator Corners** are located where appropriate throughout the text. These streamlined Calculator Corners are written to represent current calculators. A calculator icon indicates exercises suitable for calculator use. (See pages 234, 270, and 399.)

## Immediate Practice and Assessment in Each Section

OBJECTIVES ➞ SKILL TO REVIEW ➞ EXPOSITION ➞ EXAMPLES WITH DETAILED ANNOTATIONS AND VISUAL ART PIECES ➞ MARGIN EXERCISES ➞ EXERCISE SETS

**Objective Boxes** begin each section. A boxed list of objectives is keyed by letter not only to section subheadings, but also to the section exercise sets and the Mid-Chapter Review and the Summary and Review exercises, as well as to the answers to the questions in the Chapter Tests and Cumulative Reviews. This correlation enables students to easily find appropriate review material if they need help with a particular exercise or skill at the objective level. (See pages 128, 204, and 258.)

**New!** **Skill to Review** exercises, found at the beginning of most sections, link to a section objective and offer students a just-in-time review of a previously presented skill that relates to new material in the section. For convenient studying, objective references are followed by two practice exercises for immediate review and reinforcement. Answers to these exercises are given at the bottom of the page for immediate feedback. (See pages 19, 278, and 407.)

**Revised!** **Annotated Examples** provide annotations and color highlighting to lead students through the structured steps of the examples. The level of detail in these annotations is a significant reason for students' success with this book. This edition contains over 200 new examples. (See pages 166, 232, and 364.)

**Revised!** The **art and photo program** is designed to help students visualize mathematical concepts and real-data applications. Many applications include source lines and feature graphs and drawings similar to those students see in the media. Color is used consistently in geometric shapes and fractional parts to convey meaning. For example, a red border emphasizes perimeter, a black outline with a yellow fill emphasizes area, three-dimensional figures illustrating volume are blue, and fractional parts are shown in purple. (See pages 59, 143, 224, 308, 432, and 577).

**Revised!** **Margin Exercises**, now located next to examples for easier navigation, accompany examples throughout the text and give students the opportunity to work similar problems for immediate practice and reinforcement of the concept just learned. Answers are now available at the bottom of the page. (See pages 137, 302, and 400.)

## Exercise Sets

To give students ample opportunity to practice what they have learned, each section is followed by an extensive exercise set *keyed by letter to the section objectives* for easy review and remediation. In addition, students also have the opportunity to synthesize the objectives from the current section with those from preceding sections. **For Extra Help** icons, shown at the beginning of each exercise set, indicate supplementary learning resources that students may need.

- **Skill Maintenance Exercises**, found in each exercise set, review concepts from other sections in the text to prepare students for their final examination. Section and objective codes appear next to each Skill Maintenance exercise for easy reference. All Skill Maintenance answers are given at the back of the book. (See pages 156, 211, and 398.)
- **Vocabulary Reinforcement Exercises** provide an integrated review of key terms that students must know to communicate effectively in the language of mathematics. These appear once per chapter in the Skill Maintenance portion of an exercise set. (See pages 34, 179, and 313.)
- **Synthesis Exercises** help build critical-thinking skills by requiring students to use what they know to synthesize, or combine, learning objectives from the current section with those from previous sections. These are available in nearly every exercise set. (See pages 79, 196, and 381.)

## Mid-Chapter Review

**New!** A **Mid-Chapter Review** gives students the opportunity to reinforce their understanding of the mathematical skills and concepts just covered before they move on to new material. Section and objective references are included for convenient studying, and answers to all the Mid-Chapter Review exercises are given at the back of the book. The types of exercises are as follows:

- **Concept Reinforcement** are true/false questions that enhance students' understanding of mathematical concepts. True/false exercises are also found in the Summary and Review at the end of the chapter. (See pages 35, 164, and 297.)
- **Guided Solutions** present worked-out problems with blanks for students to fill in the correct expressions to complete the solution. (See pages 35, 164, and 297.)

- **Mixed Review** provides free-response exercises, similar to those in the preceding sections in the chapter, reinforcing mastery of skills and concepts. (See pages 35, 164, and 297.)
- **Understanding Through Discussion and Writing** lets students demonstrate their understanding of mathematical concepts by expressing their thoughts in spoken or written form. This type of exercise is also found in each Chapter Summary and Review. (See pages 36, 165, and 298.)

## Matching Feature

**Translating for Success** problem sets give extra practice with the important "Translate" step of the process for solving applied problems. After translating each of ten problems into its appropriate equation or inequality, students are asked to choose from fifteen possible translations, encouraging them to comprehend the problem before matching. (See pages 176, 321, and 368.)

## End-of-Chapter Material

**Revised!**   The **Chapter Summary and Review** at the end of each chapter is expanded to provide more comprehensive in-text practice and review. Section and objective references and answers to all the Chapter Summary and Review exercises are included in the text. (See pages 80, 180, and 329.)

- **Key Terms, Properties, and Formulas** are highlighted, with page references for convenient review. (See pages 80, 180, and 329.)
- **Concept Reinforcement** offers true/false questions to enhance student understanding of mathematical concepts. (See pages 80, 180, and 329.)
- **New! Important Concepts** are listed by section objectives, followed by *a worked-out example* for reference and review and *a similar practice exercise* for students to solve. (See pages 80–82, 180–182, and 329–331.)
- **Review Exercises,** including Synthesis exercises and two new multiple-choice exercises, covering the entire chapter are organized by objective. (See pages 82–84, 182–184, and 331–333.)
- **Understanding Through Discussion and Writing** exercises strengthen understanding by giving students a chance to express their thoughts in spoken or written form. (See pages 84, 184, and 333.)

**Chapter Tests**, including Synthesis questions and a new multiple-choice question, allow students to review and test their comprehension of chapter skills prior to taking an instructor's exam. Answers to all questions in the Chapter Test are given at the back of the book. Section and objective references for each question are included with the answers. (See pages 185, 387, and 459.)

**New!**   **A Cumulative Review** now follows every chapter starting with Chapter 3; this review revisits skills and concepts from all preceding chapters to help students recall previously learned material and prepare for exams. Answers to all Cumulative Review exercises are coded by section and objective at the back of the book to help students identify areas where additional practice is needed. (See pages 255, 337, and 389.)

# For Extra Help

## Student Supplements

**New!** **Worksheets for Classroom or Lab Practice**
(ISBN: 978-0-321-61354-7)

These classroom- and lab-friendly workbooks offer a list of learning objectives, vocabulary practice problems, and extra practice exercises with ample work space for every section of the text.

**Student's Solutions Manual** (ISBN: 978-0-321-60544-3)
By Judith Penna

Contains completely worked-out annotated solutions for all the odd-numbered exercises in the section-level exercise sets. Also includes fully worked-out annotated solutions for all the exercises (odd- and even-numbered) in the Mid-Chapter Reviews, the Summary and Reviews, the Chapter Tests, and the Cumulative Reviews.

**Chapter Test Prep Videos**

Chapter Tests can serve as practice tests to help you study. Watch instructors work through step-by-step solutions to all the Chapter Test exercises from the textbook. Chapter Test Prep videos are available on YouTube (search using Bittinger-BasicMathEI) and in MyMathLab. They are also included on the Video Resources on DVD described below and available for purchase at www.MyPearsonStore.com.

**Video Resources on DVD Featuring Chapter Test Prep Videos**
(ISBN: 978-0-321-61355-4)

- Complete set of lectures covering every objective of every section in the textbook
- Complete set of Chapter Test Prep videos (see above)
- All videos include optional English and Spanish subtitles.
- Ideal for distance learning or supplemental instruction
- DVD-ROM format for student use at home or on campus

**InterAct Math Tutorial Website** (www.interactmath.com)

Get practice and tutorial help online! This interactive tutorial website provides algorithmically generated practice exercises that correlate directly to the exercises in the textbook. Students can rework an exercise as many times as they like with new values each time for unlimited practice and mastery. Every exercise is accompanied by an interactive guided solution that provides helpful feedback for incorrect answers, and students can also view a worked-out sample problem that leads them through an exercise similar to the one they're working on.

**MathXL® Tutorials on CD** (ISBN: 978-0-321-60951-9)

This interactive tutorial CD-ROM provides algorithmically generated practice exercises that are correlated at the objective level to the exercises in the textbook. Every practice exercise is accompanied by an example and a guided solution designed to involve students in the solution process. Selected exercises also include a video clip to help students visualize concepts. The software provides helpful feedback for incorrect answers and can generate printed summaries of students' progress.

## Instructor Supplements

**Annotated Instructor's Edition** (ISBN: 978-0-321-56171-8)

Includes answers to all exercises printed in blue on the same page as the exercises. Also includes the student answer section, for easy reference.

**Instructor's Solutions Manual** (ISBN: 978-0-321-60543-6)
By Judith Penna

Contains brief solutions to the even-numbered exercises in the exercise sets. Also includes fully worked-out annotated solutions for all the exercises (odd- and even-numbered) in the Mid-Chapter Reviews, the Summary and Reviews, the Chapter Tests, and the Cumulative Reviews.

**Printed Test Forms** (ISBN: 978-0-321-60949-6)
By Laurie Hurley

- Contains one diagnostic test and one pretest for each chapter, plus two cumulative tests per chapter, beginning with Chapter 2.
- **New!** Includes two versions of a short mid-chapter quiz.
- Provides eight test forms for every chapter and eight test forms for the final exam.
- For the chapter tests, four free-response tests are modeled after the chapter tests in the main text, two tests are designed for 50-minute class periods and organized so that each objective in the chapter is covered on one of the tests, and two tests consist of multiple-choice questions. Chapter tests also include more challenging Synthesis questions.
- For the final exam, three test forms are organized by chapter, three forms are organized by question type, and two forms are multiple-choice tests.
- Also includes Extra Practice Exercises for select sections.

**Instructor's Resource Manual**
(ISBN: 978-0-321-61353-0)

- Features resources and teaching tips designed to help both new and adjunct faculty with course preparation and classroom management.
- **New!** Includes a mini-lecture for each section of the text with objectives, key examples, and teaching tips.
- Additional resources include general first-time advice, sample syllabi, teaching tips, collaborative learning activities, correlation guide from First Edition to Second Edition, video index, and transparency masters.

# Additional Media Supplements

**MyMathLab** **MyMathLab® Online Course (access code required)**

MyMathLab is a series of text-specific, easily customizable online courses for Pearson Education's textbooks in mathematics and statistics. Powered by CourseCompass™ (our online teaching and learning environment) and MathXL® (our online homework, tutorial, and assessment system), MyMathLab gives instructors the tools they need to deliver all or a portion of their course online, whether their students are in a lab setting or working from home. MyMathLab provides a rich and flexible set of course materials, featuring free-response exercises that are algorithmically generated for unlimited practice and mastery. Students can also use online tools, such as video lectures, animations, interactive math games, and a multimedia textbook, to independently improve their understanding and performance. Instructors can use MyMathLab's homework and test managers to select and assign online exercises correlated directly to the textbook, and they can also create and assign their own online exercises and import TestGen tests for added flexibility. MyMathLab's online gradebook—designed specifically for mathematics and statistics—automatically tracks students' homework and test results and gives the instructor control over how to calculate final grades. Instructors can also add offline (paper-and-pencil) grades to the gradebook. MyMathLab also includes access to the **Pearson Tutor Center** (www.pearsontutorservices.com). The Tutor Center is staffed by qualified mathematics instructors who provide textbook-specific tutoring for students via toll-free phone, fax, email, and interactive Web sessions. MyMathLab is available to qualified adopters. For more information, visit our website at www.mymathlab.com or contact your sales representative.

**MathXL** **MathXL® Online Course (access code required)**

MathXL® is a powerful online homework, tutorial, and assessment system that accompanies Pearson Education's textbooks in mathematics or statistics.

With MathXL, instructors can

- create, edit, and assign online homework and tests using algorithmically generated exercises correlated at the objective level to the textbook.
- create and assign their own online exercises and import TestGen tests for added flexibility.
- maintain records of all student work tracked in MathXL's online gradebook.

With MathXL, students can

- take chapter tests in MathXL and receive personalized study plans based on their test results.
- use the study plan to link directly to tutorial exercises for the objectives they need to study and retest.
- access supplemental animations and video clips directly from selected exercises.

MathXL is available to qualified adopters. For more information, visit our website at www.mathxl.com, or contact your Pearson sales representative.

**TestGen®** (www.pearsoned.com/testgen) enables instructors to build, edit, and print tests using a computerized bank of questions developed to cover all the objectives of the text. TestGen is algorithmically based, allowing instructors to create multiple but equivalent versions of the same question or test with the click of a button. Instructors can also modify test bank questions or add new questions. The software and test bank are available for download from Pearson Education's online catalog.

**PowerPoint® Lecture Slides** present key concepts and definitions from the text. Slides are available to download from within MyMathLab and from Pearson Education's online catalog.

**Pearson Math Adjunct Support Center** (http://www.pearsontutorservices.com/math-adjunct. html) is staffed by qualified instructors with more than 100 years of combined experience at both the community college and university levels. Assistance is provided for faculty in the following areas: suggested syllabus consultation, tips on using materials packed with your book, book-specific content assistance, and teaching suggestions, including advice on classroom strategies.

# Acknowledgments

Our deepest appreciation to all of you who helped to shape this edition by reviewing and spending time with us on your campuses. In particular, we would like to thank the following reviewers of *Basic College Mathematics*, 11th Edition, and *Basic College Mathematics with Early Integers*, 2nd Edition:

Kimberly Caldwell, *Volunteer State Community College*
Amelia Canavan, *Century College*
Alison Cater, *California State University, Hayward*
Julane B. Crabtrere, *Johnson County Community College*
Kelly Groginski, *Onondaga Community College*
Karen Hale, *Onondaga Community College*
Cassandra V. Johnson, *Robeson Community College*
Joe Jordan, *John Tyler Community College*
David Kedrowski, *Mid Michigan Community College*
Lynette J. King, *Gadsten State Community College*
Edward Kinley, *Pennsylvania Culinary Institute*
Dottie Lapre, *Nash Community College*
Edith Lester, *Volunteer State Community College*
Lynn Marecek, *Santa Ana College*
Shauna Mullins, *Murray State University*
Ruela Overby, *Richland College*
Brian Peterman, *Century College*
Nimisha Raval, *Central Georgia Technical College–Macon Campus*
Ronald Rushing, *Southwest Georgia Technical College*
Robert Thompson, *Lane Community College*
Stephen Toner, *Victory Valley Community College*
S. Paige Wood, *Kilgore College*

Endless hours of hard work by Martha Morong, Sally Lifland, Jane Hoover, Geri Davis, and Judy Beecher have led to a finished product of which we are immensely proud. Other strong support has come from Laurie Hurley for the Printed Test Bank and for accuracy checking, along with accuracy checker Holly Martinez and proofreader Patty LaGree. Michelle Lanosga assisted with applications research. We also wish to recognize Tom Atwater and Margaret Donlan, who wrote video scripts, and Barbara Johnson, who presented video lessons.

In addition, a number of people at Pearson have contributed in special ways to the development and production of this textbook including the Developmental Math team: Vice President, Executive Director of Development Carol Trueheart, Senior Development Editor Dawn Nuttall, Production Manager Ron Hampton, Senior Designer Beth Paquin, Associate Editors Joanna Doxey and Christine Whitlock, Editorial Assistant Jonathan Wooding, Associate Media Producer Nathaniel Koven, and Media Producer Ceci Fleming. Executive Editor Cathy Cantin and Executive Marketing Manager Michelle Renda encouraged our vision and provided marketing insight. Kari Heen, Executive Project Manager, deserves special recognition for overseeing every phase of the project and keeping it moving.

# Whole Numbers

# Real-World Application

When construction is completed on the slender 150-story Chicago Spire, it will be the tallest building in the United States, rising to a height of 2000 ft. The height of Chicago's Willis Tower, formerly known as the Sears Tower, is 1450 ft. How much taller will the Chicago Spire be?

*Sources: Chicago Historical Society; USA Today research*

*This problem appears as Exercise 1 in Section 1.8.*

# 1.1 Standard Notation

## OBJECTIVES

**a** Give the meaning of digits in standard notation.

**b** Convert from standard notation to expanded notation.

**c** Convert between standard notation and word names.

### TO THE STUDENT

At the front of the text, you will find the Bittinger Student Organizer card. This pullout card will help you keep track of important dates and useful contact information. You can also use it to plan time for class, study, work, and relaxation. By managing your time wisely, you will provide yourself the best possible opportunity to be successful in this course.

We study mathematics in order to be able to solve problems. In this section, we study how numbers are named. We begin with the concept of place value.

## a Place Value

Consider the numbers in the following table.

**Three Most Populous Countries in the World**

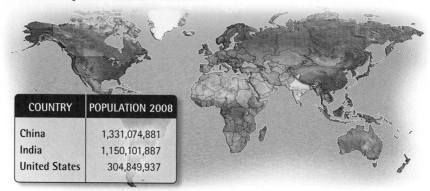

| COUNTRY | POPULATION 2008 |
|---------|-----------------|
| China | 1,331,074,881 |
| India | 1,150,101,887 |
| United States | 304,849,937 |

SOURCE: msnu.edu

A **digit** is a number 0, 1, 2, 3, 4, 5, 6, 7, 8, or 9 that names a place-value location. For large numbers, digits are separated by commas into groups of three, called **periods**. Each period has a name: *ones, thousands, millions, billions, trillions,* and so on. To understand the population of India in the table above, we can use a **place-value chart**, as shown below.

| PLACE-VALUE CHART | | | | | | | | | | | | | | |
|---|---|---|---|---|---|---|---|---|---|---|---|---|---|---|
| Periods → | Trillions | | | Billions | | | Millions | | | Thousands | | | Ones | | |
| | | | | | | 1 | 1 | 5 | 0 | 1 | 0 | 1 | 8 | 8 | 7 |
| | Hundreds | Tens | Ones | Hundreds | Tens | Ones | Hundreds | Tens | Ones | Hundreds | Tens | Ones | Hundreds | Tens | Ones |

1 billion    150 millions    101 thousands    887 ones

**EXAMPLES** In each of the following numbers, what does the digit 8 mean?

1. 278,342      8 thousands
2. 872,342      8 hundred thousands
3. 28,343,399,223      8 billions
4. 1,023,850      8 hundreds
5. 98,413,099      8 millions
6. 6328      8 ones

Do Margin Exercises 1–6 (in the margin at left).

---

What does the digit 2 mean in each number?

1. 526,555

2. 265,789

3. 42,789,654

4. 24,789,654

5. 8924

6. 5,643,201

*Answers*

1. 2 ten thousands    2. 2 hundred thousands
3. 2 millions    4. 2 ten millions    5. 2 tens
6. 2 hundreds

**EXAMPLE 7** *Hurricane Relief.* Private donations for relief for Hurricanes Katrina and Rita, which struck the Gulf Coast of the United States in 2005, totaled $3,378,185,879. What does each digit name?

**Source:** The Center on Philanthropy at Indiana University

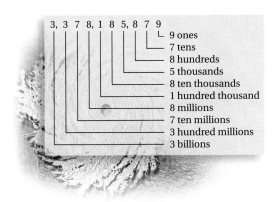

3, 3 7 8, 1 8 5, 8 7 9
- 9 ones
- 7 tens
- 8 hundreds
- 5 thousands
- 8 ten thousands
- 1 hundred thousand
- 8 millions
- 7 ten millions
- 3 hundred millions
- 3 billions

Do Exercise 7.

Do Exercise 7.

## b Converting from Standard Notation to Expanded Notation

To answer questions such as "How many?", "How much?", and "How tall?", we often use whole numbers. The set, or collection, of **whole numbers** is

0, 1, 2, 3, 4, 5, 6, 7, 8, 9, 10, 11, 12, . . . .

The set goes on indefinitely. There is no largest whole number, and the smallest whole number is 0. Each whole number can be named using various notations. The set 1, 2, 3, 4, 5, . . . , without 0, is called the set of **natural numbers**.

Consider the data in the table below showing the Advanced Placement exams taken most frequently by the class of 2007.

| EXAM | NUMBER TAKEN |
|---|---|
| U.S. History | 333,561 |
| English Language and Composition | 282,230 |
| Calculus AB | 211,693 |
| Biology | 144,796 |

SOURCE: The College Board

The number of Biology exams taken was 144,796. This number is expressed in **standard notation**. We write **expanded notation** for 144,796 as follows:

144,796 = 1 hundred thousand + 4 ten thousands
+ 4 thousands + 7 hundreds
+ 9 tens + 6 ones.

**7. Federal Payroll.** In December 2006, the payroll for civilian employees of the federal government was $13,896,346,626. What does each digit name?

**Source:** U.S. Census Bureau

*Answer*

**7.** 1 ten billion; 3 billions; 8 hundred millions; 9 ten millions; 6 millions; 3 hundred thousands; 4 ten thousands; 6 thousands; 6 hundreds; 2 tens; 6 ones

**Write expanded notation.**

**8.** 1895

**9.** 333,561, the number of Advanced Placement U.S. History exams taken by the class of 2007

**10.** 1670 ft, the height of the Taipei 101 Tower in Taiwan

**11.** 2718 mi, the length of the Congo River in Africa

**12.** 104,094 square miles, the area of Colorado

**EXAMPLE 8** Write expanded notation for 1815 ft, the height of the CN Tower in Toronto, Canada.

1815 = 1 thousand + 8 hundreds + 1 ten + 5 ones

**EXAMPLE 9** Write expanded notation for 3400.

3400 = 3 thousands + 4 hundreds + 0 tens + 0 ones,   or

3 thousands + 4 hundreds

**EXAMPLE 10** Write expanded notation for 211,693, the number of Advanced Placement Calculus exams taken by the class of 2007.

211,693 = 2 hundred thousands + 1 ten thousand
        + 1 thousand + 6 hundreds + 9 tens + 3 ones

Do Exercises 8–12.

## c Converting Between Standard Notation and Word Names

We often use **word names** for numbers. When we pronounce a number, we are speaking its word name. Russia won 72 medals in the 2008 Summer Olympics in Beijing, China. A word name for 72 is "seventy-two." Word names for some two-digit numbers like 36, 51, and 72 use hyphens. Others like 17 use only one word, "seventeen."

**2008 Summer Olympics Medal Count**

| COUNTRY | GOLD | SILVER | BRONZE | TOTAL |
|---|---|---|---|---|
| United States of America | 36 | 38 | 36 | 110 |
| People's Republic of China | 51 | 21 | 28 | 100 |
| Russia | 23 | 21 | 28 | 72 |
| Great Britain | 19 | 13 | 15 | 47 |
| Australia | 14 | 15 | 17 | 46 |

SOURCE: beijing2008.cn

*Answers*

**8.** 1 thousand + 8 hundreds + 9 tens + 5 ones
**9.** 3 hundred thousands + 3 ten thousands + 3 thousands + 5 hundreds + 6 tens + 1 one
**10.** 1 thousand + 6 hundreds + 7 tens + 0 ones, or 1 thousand + 6 hundreds + 7 tens
**11.** 2 thousands + 7 hundreds + 1 ten + 8 ones
**12.** 1 hundred thousand + 0 ten thousands + 4 thousands + 0 hundreds + 9 tens + 4 ones, or 1 hundred thousand + 4 thousands + 9 tens + 4 ones

**EXAMPLES** Write a word name.

**11.** 36, the number of gold medals won by the United States

Thirty-six

**12.** 15, the number of silver medals won by Australia

Fifteen

**13.** 100, the total number of medals won by the People's Republic of China

One hundred

Do Exercises 13–15.

For word names for larger numbers, we begin at the left with the largest period. The number named in the period is followed by the name of the period; then a comma is written and the next number and period are named. Note that the name of the ones period is not included in the word name for a whole number.

**EXAMPLE 14** Write a word name for 46,605,314,732.

Forty-six **billion,**

six hundred five **million,**

three hundred fourteen **thousand,**

seven hundred thirty-two

The word "and" *should not* appear in word names for whole numbers. Although we commonly hear such expressions as "two hundred *and* one," the use of "and" is not, strictly speaking, correct in word names for whole numbers. For decimal notation, it is appropriate to use "and" for the decimal point. For example, 317.4 is read as "three hundred seventeen *and* four tenths."

Do Exercises 16–19.

**EXAMPLE 15** Write standard notation.

Five hundred six **million,**

three hundred forty-five **thousand,**

two hundred twelve

Standard notation is 506,345,212.

Do Exercise 20.

*Answers*

**13.** Forty-six **14.** Thirteen **15.** Twenty-eight **16.** Two hundred four **17.** Fifty-two thousand, one hundred seventy-seven **18.** One million, eight hundred seventy-nine thousand, two hundred four **19.** Six billion, six hundred seventy-seven million, six hundred two thousand, two hundred ninety-two **20.** 213,105,329

**a** What does the digit 5 mean in each number?

**1.** 235,888    **2.** 253,777    **3.** 1,488,526    **4.** 500,736

*Movie Receipts.* Box-office receipts on the opening weekend of *Shrek the Third* were $121,629,270.

**Source:** Box Office Mojo

What digit names the number of:

**5.** thousands?    **6.** ten millions?

**7.** tens?    **8.** hundred thousands?

 **b** Write expanded notation.

**9.** 5702    **10.** 3097    **11.** 93,986    **12.** 38,453

*Stair-Climbing Races.* The figure below shows the number of stairs in four buildings in which stair-climbing races are held. In Exercises 13–16, write expanded notation for the number of stairs in each race.

**Stair-Climbing Races**

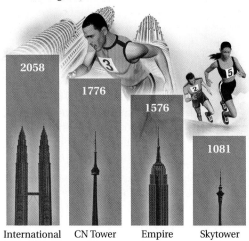

| International Towerthon, Kuala Lumpur, Malaysia | CN Tower Stair Climb, Toronto, Ontario, Canada | Empire State Building Run-Up, New York | Skytower Vertical Challenge, Auckland, New Zealand |

SOURCE: towerrunning.com

**13.** 2058 steps in the International Towerthon, Kuala Lumpur, Malaysia

**14.** 1776 steps in the CN Tower Stair Climb, Toronto, Ontario, Canada

**15.** 1576 steps in the Empire State Building Run-Up, New York City, New York

**16.** 1081 steps in the Skytower Vertical Challenge, Auckland, New Zealand

*Population Projections.* The table below shows the projected population in 2050 for six countries. In Exercises 17–22, write expanded notation for the population of the given country.

**PROJECTED POPULATION IN 2050**

| COUNTRY | POPULATION |
| --- | --- |
| China | 1,424,161,948 |
| Hong Kong | 6,172,725 |
| Japan | 99,886,568 |
| Monaco | 32,964 |
| Surinam | 617,249 |
| United States | 420,080,587 |

SOURCES: International Programs Center, U.S. Census Bureau, U.S. Department of Commerce

**17.** 1,424,161,948 in China

**18.** 6,172,725 in Hong Kong

**19.** 99,886,568 in Japan

**20.** 32,964 in Monaco

**21.** 617,249 in Surinam

**22.** 420,080,587 in the United States

 Write a word name.

**23.** 85

**24.** 48

**25.** 88,000

**26.** 45,987

**27.** 123,765

**28.** 111,013

**29.** 7,754,211,577

**30.** 43,550,651,808

**31.** *Wilderness Areas.* In 2009, in the most sweeping land-protection law passed in fifteen years, all development was banned on huge swaths of land in nine states. The largest was in California, where 700,634 acres were newly designated as wilderness. Write a word name for 700,634.

**Source:** The Wilderness Society

**32.** *Busiest Airport.* In 2006, the world's busiest airport, Hartsfield in Atlanta, had 84,846,639 passengers. Write a word name for 84,846,639.

**Source:** Airports Council International World Headquarters, Geneva, Switzerland

**33.** *Auto Racing.* Helio Castroneves, winner of the 2009 Indianapolis 500 auto race, won a record prize of $3,048,005. Write a word name for 3,048,005.
Source: ESPN

**34.** *College Football.* The 2008 Rose Bowl game was attended by 93,923 fans. Write a word name for 93,923.
Source: Fox Sports

Write standard notation.

**35.** Two million, two hundred thirty-three thousand, eight hundred twelve

**36.** Three hundred fifty-four thousand, seven hundred two

**37.** Eight billion

**38.** Seven hundred million

**39.** Fifty thousand, three hundred twenty-four

**40.** Twenty-six billion

**41.** Six hundred thirty-two thousand, eight hundred ninety-six

**42.** Seventeen thousand, one hundred twelve

Write standard notation for the number in each sentence.

**43.** *Ice Cream Purchases.* Americans buy one billion, six hundred million gallons of ice cream and frozen desserts each year.
Source: International Dairy Foods Association

**44.** *Learning a Language.* There are two hundred million Chinese children studying English.
Source: U.S. Department of Education

**45.** *Pacific Ocean.* The area of the Pacific Ocean is sixty-four million, one hundred eighty-six thousand square miles.

**46.** The average distance from the sun to Neptune is two billion, seven hundred ninety-three million miles.

# Synthesis

*To the student and the instructor*: The Synthesis exercises found at the end of every exercise set challenge students to combine concepts or skills studied in that section or in preceding parts of the text. Exercises marked with a ▦ symbol are meant to be solved using a calculator.

**47.** How many whole numbers between 100 and 400 contain the digit 2 in their standard notation?

**48.** ▦ What is the largest number that you can name on your calculator? How many digits does that number have? How many periods?

# 1.2 Addition

## a Addition of Whole Numbers

Addition of whole numbers corresponds to combining or putting things together.

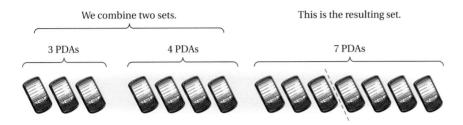

We combine two sets.     This is the resulting set.

3 PDAs     4 PDAs     7 PDAs

We say that the **sum** of 3 and 4 is 7. The numbers added are called **addends**. The addition that corresponds to the figure above is

$$3 + 4 = 7.$$

Addend   Addend   Sum

To add whole numbers, we add the ones digits first, then the tens, then the hundreds, then the thousands, and so on.

**EXAMPLE 1**   Add: 6878 + 4995.

Place values are lined up in columns.

$$
\begin{array}{r}
\overset{\phantom{6}\phantom{8}\overset{1}{\phantom{7}}\phantom{8}}{6\;8\;7\;8} \\
+\;4\;9\;9\;5 \\
\hline
3
\end{array}
$$

Add ones. We get 13 ones, or 1 ten + 3 ones. Write 3 in the ones column and 1 above the tens. This is called *carrying*, or *regrouping*.

$$
\begin{array}{r}
\overset{\phantom{6}\overset{1}{\phantom{8}}\overset{1}{\phantom{7}}\phantom{8}}{6\;8\;7\;8} \\
+\;4\;9\;9\;5 \\
\hline
7\;3
\end{array}
$$

Add tens. We get 17 tens, so we have 10 tens + 7 tens. This is also 1 hundred + 7 tens. Write 7 in the tens column and 1 above the hundreds.

$$
\begin{array}{r}
\overset{\overset{1}{\phantom{6}}\overset{1}{\phantom{8}}\overset{1}{\phantom{7}}\phantom{8}}{6\;8\;7\;8} \\
+\;4\;9\;9\;5 \\
\hline
8\;7\;3
\end{array}
$$

Add hundreds. We get 18 hundreds, or 10 hundreds + 8 hundreds. This is also 1 thousand + 8 hundreds. Write 8 in the hundreds column and 1 above the thousands.

$$
\begin{array}{r}
\overset{\overset{1}{\phantom{6}}\overset{1}{\phantom{8}}\overset{1}{\phantom{7}}\phantom{8}}{6\;8\;7\;8} \\
+\;4\;9\;9\;5 \\
\hline
1\;1\;8\;7\;3
\end{array}
$$

Add thousands. We get 11 thousands.

We show you these steps for explanation. You need write only this.

$$
\begin{array}{r}
\overset{1\;1\;1}{6\;8\;7\;8} \\
+\;4\;9\;9\;5 \\
\hline
1\;1\;8\;7\;3
\end{array}
$$

Addends

Sum

## OBJECTIVES

**a** Add whole numbers.

**b** Use addition in finding perimeter.

## STUDY TIPS

### USING THIS TEXTBOOK

*One of the most important ways in which to improve your math study skills is to learn the proper use of the textbook. Here we highlight a few points that we consider most helpful.*

**Be sure to note the symbols a, b, c, and so on, that correspond to the objectives you are to master in each section.**   The first time you see them is in the margin at the beginning of the section; the second time is in the subheadings of each section; and the third time is in the exercise set for the section. You will also find symbols like [1.1a] or [1.2c] next to the skill maintenance exercises in each exercise set, in the mid-chapter review, and in the review exercises at the end of the chapter, as well as in the answers to the chapter tests and the cumulative reviews. These objective symbols allow you to refer to the appropriate place in the text when you need to review a topic.

**EXAMPLE 2**   Add: $391 + 276 + 789 + 498$.

$$
\begin{array}{r}
3\ \overset{2}{9}\ 1 \\
2\ 7\ 6 \\
7\ 8\ 9 \\
+\ 4\ 9\ 8 \\
\hline
4
\end{array}
$$

Add ones. We get 24, so we have 2 tens + 4 ones. Write 4 in the ones column and 2 above the tens.

$$
\begin{array}{r}
\overset{3}{3}\ \overset{2}{9}\ 1 \\
2\ 7\ 6 \\
7\ 8\ 9 \\
+\ 4\ 9\ 8 \\
\hline
5\ 4
\end{array}
$$

Add tens. We get 35 tens, so we have 30 tens + 5 tens. This is also 3 hundreds + 5 tens. Write 5 in the tens column and 3 above the hundreds.

$$
\begin{array}{r}
\overset{3}{3}\ \overset{2}{9}\ 1 \\
2\ 7\ 6 \\
7\ 8\ 9 \\
+\ 4\ 9\ 8 \\
\hline
1\ 9\ 5\ 4
\end{array}
$$

Add hundreds. We get 19 hundreds.

Do Exercises 1–4.

Add.

**1.** $6203 + 3542$

**2.**
$$
\begin{array}{r}
7\ 9\ 6\ 8 \\
+\ 5\ 4\ 9\ 7 \\
\hline
\end{array}
$$

**3.**
$$
\begin{array}{r}
9\ 8\ 0\ 4 \\
+\ 6\ 3\ 7\ 8 \\
\hline
\end{array}
$$

**4.**
$$
\begin{array}{r}
1\ 9\ 3\ 2 \\
6\ 7\ 2\ 3 \\
9\ 8\ 7\ 8 \\
+\ 8\ 9\ 4\ 1 \\
\hline
\end{array}
$$

How do we perform an addition of three numbers, like $2 + 3 + 6$? We could do it by adding 3 and 6, and then 2. We can show this with parentheses:

$$2 + (3 + 6) = 2 + 9 = 11. \qquad \text{Parentheses tell what to do first.}$$

We could also add 2 and 3, and then 6:

$$(2 + 3) + 6 = 5 + 6 = 11.$$

Either way the result is 11. It does not matter how we group the numbers. This illustrates the **associative law of addition**, $a + (b + c) = (a + b) + c$. We can also add whole numbers in any order. That is, $2 + 3 = 3 + 2$. This illustrates the **commutative law of addition**, $a + b = b + a$. Together, the commutative and associative laws tell us that to add more than two numbers, we can use any order and grouping we wish. Adding 0 to a number does not change the number: $a + 0 = 0 + a = a$. That is, $6 + 0 = 0 + 6 = 6$, or $198 + 0 = 0 + 198 = 198$. We say that 0 is the **additive identity**.

*Answers*

**1.** 9745   **2.** 13,465   **3.** 16,182
**4.** 27,474

---

**Calculator Corner**

**Adding Whole Numbers**   This is the first of a series of *optional* discussions on using a calculator. A calculator is *not* a requirement for this textbook. Check with your instructor about whether you are allowed to use a calculator in the course.

There are many kinds of calculators and different instructions for their usage. We have included instructions here for a low-cost calculator. Be sure to consult your user's manual as well.

To add whole numbers on a calculator, we use the $\boxed{+}$ and $\boxed{=}$ keys. For example, to find $314 + 259 + 478$, we press $\boxed{3}\ \boxed{1}\ \boxed{4}\ \boxed{+}\ \boxed{2}\ \boxed{5}$ $\boxed{9}\ \boxed{+}\ \boxed{4}\ \boxed{7}\ \boxed{8}\ \boxed{=}$. The display reads $\boxed{\phantom{xxx}1051\phantom{xxx}}$, so $314 + 259 + 478 = 1051$.

**Exercises:**   Use a calculator to find each sum.

**1.** $73 + 48$

**2.** $925 + 677$

**3.** $826 + 415 + 691$

**4.** $253 + 490 + 121$

## (b) Finding Perimeter

Addition can be used when finding perimeter.

> **PERIMETER**
>
> The distance around an object is its **perimeter**.

**EXAMPLE 3** Find the perimeter of the figure.

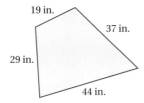

We add the lengths of the sides:

Perimeter = 29 in. + 19 in. + 37 in. + 44 in.

We carry out the addition as follows.

$$\begin{array}{r} \overset{2}{2}\,9 \\ 1\,9 \\ 3\,7 \\ +\ 4\,4 \\ \hline 1\,2\,9 \end{array}$$

The perimeter of the figure is 129 in.

> Do Exercises 5 and 6.

**EXAMPLE 4** Find the perimeter of the octagonal (eight-sided) resort swimming pool.

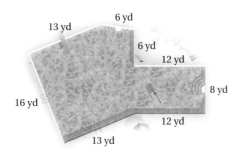

Perimeter = 13 yd + 6 yd + 6 yd + 12 yd + 8 yd
    + 12 yd + 13 yd + 16 yd

When we carry out the addition, we get 86. Thus, the perimeter of the pool is 86 yd.

> Do Exercise 7.

Find the perimeter of each figure.

**5.**

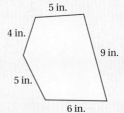

**6.**

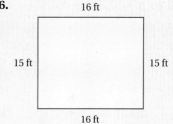

Solve.

**7. Index Cards.** Two standard sizes for index cards are 3 in. by 5 in. and 5 in. by 8 in. Find the perimeter of each type of card.

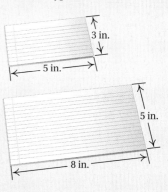

*Answers*

**5.** 29 in.   **6.** 62 ft   **7.** 16 in.; 26 in.

**a** Add.

**1.** 3 6 4
 + 2 3

**2.** 1 5 2 1
 + 3 4 8

**3.** 8 6
 + 7 8

**4.** 7 3
 + 6 9

**5.** 1 7 1 6
 + 3 4 8 2

**6.** 7 5 0 3
 + 2 6 8 3

**7.** 9 9
 + 1

**8.** 9 9 9
 + 1 1

**9.** 8113 + 390

**10.** 271 + 3338

**11.** 356 + 4910

**12.** 280 + 34,902

**13.** 3870 + 92 + 7 + 497

**14.** 10,120 + 12,989 + 5738

**15.** 4 8 2 5
 + 1 7 8 3

**16.** 3 6 5 4
 + 2 7 0 0

**17.** 2 3,4 4 3
 + 1 0,9 8 9

**18.** 4 5,8 7 9
 + 2 1,7 8 6

**19.** 7 7,5 4 3
 + 2 3,7 6 7

**20.** 9 9,9 9 9
 + 1 1 2

**21.** 4 5
 2 5
 3 6
 4 4
 + 8 0

**22.** 3 8
 2 7
 3 2
 1 4
 + 7 6

**23.** 1 2,0 7 0
 2,9 5 4
 + 3,4 0 0

**24.** 4 2,4 8 7
 8 3,1 4 1
 + 3 6,7 1 2

**25.** 4 8 3 5
 7 2 9
 9 2 0 4
 8 9 8 6
 + 7 9 3 1

**26.** 9 8 9
 5 6 6
 8 3 4
 9 2 0
 + 7 0 3

**b** Find the perimeter of each figure.

**27.**
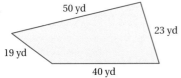
50 yd

23 yd

19 yd

40 yd

**28.**

14 mi

13 mi

8 mi

22 mi

10 mi

47 mi

**29.**
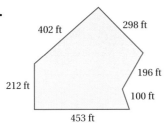
402 ft

298 ft

196 ft

212 ft

100 ft

453 ft

**30.**
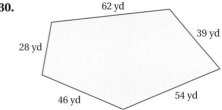
62 yd

39 yd

28 yd

46 yd

54 yd

**31.** Find the perimeter of a standard hockey rink.

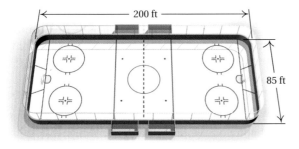
200 ft

85 ft

**32.** In Major League Baseball, how far does a batter travel in circling the bases when a home run has been hit?

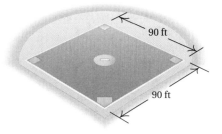
90 ft

90 ft

## Skill Maintenance

The exercises that follow begin an important feature called *Skill Maintenance exercises*. These exercises provide an ongoing review of topics previously covered in the book. You will see them in virtually every exercise set. It has been found that this kind of continuing review can significantly improve your performance on a final examination.

**33.** What does the digit 8 mean in 486,205? [1.1a]

**34.** The population of the world is projected to be 9,346,399,468 in 2050. Write a word name for 9,346,399,468. [1.1c]
**Source:** U.S. Census Bureau

## Synthesis

**35.** A fast way to add all the numbers from 1 to 10 inclusive is to pair 1 with 9, 2 with 8, and so on. Use a similar approach to add all numbers from 1 to 100 inclusive.

# 1.3 Subtraction

## OBJECTIVE

**a** Subtract whole numbers.

## STUDY TIPS

**USING THIS TEXTBOOK**

**Read and study each step of each example.** The examples include important side comments that explain each step. These examples and annotations have been carefully chosen so that you will be fully prepared to do the exercises.

**Stop and do the margin exercises as you study a section.** This gives you immediate reinforcement of each concept as it is introduced and is one of the most effective ways to master the mathematical skills in this text. Don't deprive yourself of this benefit!

**Note the icons listed at the top of each exercise set.** These refer to the distinctive multimedia study aids that accompany the book.

## **a** Subtraction of Whole Numbers

Subtraction is finding the difference of two numbers. Suppose you pick 6 pints of blueberries and give your neighbor 2 pints.

The subtraction that represents this is

$$6 \quad - \quad 2 \quad = \quad 4.$$

Minuend   Subtrahend   Difference

The **minuend** is the number from which another number is being subtracted. The **subtrahend** is the number being subtracted. The **difference** is the result of subtracting the subtrahend from the minuend.

In the subtraction above, note that the difference, 4, is the number we add to 2 to get 6. This illustrates the relationship between addition and subtraction and leads us to the following definition of subtraction.

---
**SUBTRACTION**

The difference $a - b$ is that unique whole number $c$ for which $a = c + b$.
---

We see that $6 - 2 = 4$ because $4 + 2 = 6$.

To subtract whole numbers, we subtract the ones digits first, then the tens digits, then the hundreds, then the thousands, and so on.

**EXAMPLE 1** Subtract: $9768 - 4320$.

```
  9 7 6 8      Subtract
- 4 3 2 0      ones.
─────────
        8
```

```
  9 7 6 8      Subtract
- 4 3 2 0      tens.
─────────
      4 8
```

```
  9 7 6 8      Subtract
- 4 3 2 0      hundreds.
─────────
    4 4 8
```

> We show these steps for explanation. You need write only this.

```
  9 7 6 8      Subtract              9 7 6 8
- 4 3 2 0      thousands.          - 4 3 2 0
─────────                          ─────────
5 4 4 8                            5 4 4 8
```

Because subtraction is defined in terms of addition, we can use addition to *check* subtraction.

*Subtraction:*

```
    9  7  6  8
 -  4  3  2  0
 ─────────────
    5  4  4  8
```

?

*Check by Addition:*

```
    5  4  4  8
 +  4  3  2  0
 ─────────────
    9  7  6  8
```

Do Exercise 1.

**EXAMPLE 2** Subtract: $348 - 165$.

We have

$$
\begin{array}{rl}
3 \text{ hundreds } + 4 \text{ tens } + 8 \text{ ones} = & 2 \text{ hundreds } + 14 \text{ tens } + 8 \text{ ones} \\
- 1 \text{ hundred } \;\; - 6 \text{ tens} - 5 \text{ ones} = & -1 \text{ hundred } \;\; - \;\; 6 \text{ tens} - 5 \text{ ones} \\
\hline
= & 1 \text{ hundred } + 8 \text{ tens } + 3 \text{ ones} \\
= & 183.
\end{array}
$$

First, we subtract the ones.

```
    3  4  8
 -  1  6  5
 ──────────
          3
```
Subtract ones.

We cannot subtract the tens because there is no whole number that when added to 6 gives 4. To complete the subtraction, we must *borrow* 1 hundred from 3 hundreds and regroup it with the 4 tens. Then we can do the subtraction 14 tens − 6 tens = 8 tens.

```
   2  14
   3̶  4̶  8
 - 1  6  5
 ──────────
          3
```
Borrow one hundred. That is, 1 hundred = 10 tens, and 10 tens + 4 tens = 14 tens. Write 2 above the hundreds column and 14 above the tens.

```
   2  14
   3̶  4̶  8
 - 1  6  5
 ──────────
   1  8  3
```
Subtract tens; subtract hundreds.

This is what you should write.

```
   2  14
   3̶  4̶  8
 - 1  6  5
 ──────────
   1  8  3
```

*Check:*

```
      1
   1  8  3
 + 1  6  5
 ──────────
   3  4  8
```

The answer checks because this is the top number in the subtraction.

---

**1.** Subtract. Check by adding.
```
   7  8  9  3
 - 4  0  9  2
```

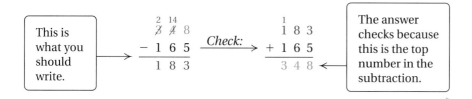

**Calculator Corner**

**Subtracting Whole Numbers** To subtract whole numbers on a calculator, we use the $\boxed{-}$ and $\boxed{=}$ keys. For example, to find $63 - 47$, we press $\boxed{6}\,\boxed{3}\,\boxed{-}\,\boxed{4}\,\boxed{7}\,\boxed{=}$. The calculator displays $\boxed{16}$, so $63 - 47 = 16$. We can check this result by adding the subtrahend, 47, and the difference, 16. To do this, we press $\boxed{1}\,\boxed{6}\,\boxed{+}\,\boxed{4}\,\boxed{7}\,\boxed{=}$. The sum is the minuend, 63, so the subtraction is correct.

**Exercises:** Use a calculator to perform each subtraction. Check by adding.

**1.** $57 - 29$

**2.** $81 - 34$

**3.** $145 - 78$

**4.** $612 - 493$

**5.**
```
   4  9  7  6
 - 2  8  4  8
```

**6.**
```
   1  2 , 4  0  6
 -    9  8  1  3
```

*Answer*
1. 3801

**EXAMPLE 3** Subtract: 6246 − 1879.

$$
\begin{array}{r}
\phantom{-}6\ 2\ \overset{3\ 16}{\cancel{4}\,\cancel{6}} \\
-\ 1\ 8\ 7\ 9 \\
\hline
7
\end{array}
$$

We cannot subtract 9 ones from 6 ones, but we can subtract 9 ones from 16 ones. We borrow 1 ten to get 16 ones.

$$
\begin{array}{r}
\phantom{-}6\ \overset{1\ \ \ 13}{2\ \cancel{4}\,\cancel{6}} \\
-\ 1\ 8\ 7\ 9 \\
\hline
6\ 7
\end{array}
$$

We cannot subtract 7 tens from 3 tens, but we can subtract 7 tens from 13 tens. We borrow 1 hundred to get 13 tens.

$$
\begin{array}{r}
\overset{11\ 13}{6\ \cancel{2}\ \cancel{4}\,\cancel{6}} \\
-\ 1\ 8\ 7\ 9 \\
\hline
4\ 3\ 6\ 7
\end{array}
$$

We cannot subtract 8 hundreds from 1 hundred, but we can subtract 8 hundreds from 11 hundreds. We borrow 1 thousand to get 11 hundreds. Finally, we subtract the thousands.

This is what you should write.

$$
\begin{array}{r}
\overset{11\ 13}{\cancel{6}\ \cancel{2}\ \cancel{4}\,\cancel{6}} \\
-\ 1\ 8\ 7\ 9 \\
\hline
4\ 3\ 6\ 7
\end{array}
$$

*Check:*

$$
\begin{array}{r}
4\ 3\ 6\ 7 \\
+\ 1\ 8\ 7\ 9 \\
\hline
6\ 2\ 4\ 6
\end{array}
$$

The answer checks because this is the top number in the subtraction.

Subtract. Check by adding.

**2.**
$$
\begin{array}{r}
8\ 6\ 8\ 6 \\
-\ 2\ 3\ 5\ 8 \\
\hline
\end{array}
$$

**3.**
$$
\begin{array}{r}
7\ 1\ 4\ 5 \\
-\ 2\ 3\ 9\ 8 \\
\hline
\end{array}
$$

Do Exercises 2 and 3.

**EXAMPLE 4** Subtract: 902 − 477.

$$
\begin{array}{r}
\overset{8\ \ 9\ \ 12}{\cancel{9}\,\cancel{0}\ 2} \\
-\ 4\ 7\ 7 \\
\hline
4\ 2\ 5
\end{array}
$$

We cannot subtract 7 ones from 2 ones. We have 9 hundreds, or 90 tens. We borrow 1 ten to get 12 ones. We then have 89 tens.

Do Exercises 4 and 5.

Subtract.

**4.**
$$
\begin{array}{r}
7\ 0 \\
-\ 1\ 4 \\
\hline
\end{array}
$$

**5.**
$$
\begin{array}{r}
5\ 0\ 3 \\
-\ 2\ 9\ 8 \\
\hline
\end{array}
$$

**EXAMPLE 5** Subtract: 8003 − 3667.

$$
\begin{array}{r}
\overset{7\ \ 9\ \ 9\ \ 13}{\cancel{8}\,\cancel{0}\,\cancel{0}\ \cancel{3}} \\
-\ 3\ 6\ 6\ 7 \\
\hline
4\ 3\ 3\ 6
\end{array}
$$

We have 8 thousands, or 800 tens. We borrow 1 ten to get 13 ones. We then have 799 tens.

**EXAMPLES**

**6.** Subtract: 6000 − 3762.

$$
\begin{array}{r}
\overset{5\ \ 9\ \ 9\ \ 10}{\cancel{6}\,\cancel{0}\,\cancel{0}\ \cancel{0}} \\
-\ 3\ 7\ 6\ 2 \\
\hline
2\ 2\ 3\ 8
\end{array}
$$

**7.** Subtract: 6024 − 2968.

$$
\begin{array}{r}
\overset{5\ \ 9\ \ \overset{11}{\cancel{1}}\ \ 14}{\cancel{6}\,\cancel{0}\,\cancel{2}\ \cancel{4}} \\
-\ 2\ 9\ 6\ 8 \\
\hline
3\ 0\ 5\ 6
\end{array}
$$

Do Exercises 6–9.

Subtract.

**6.**
$$
\begin{array}{r}
7\ 0\ 0\ 7 \\
-\ 6\ 3\ 4\ 9 \\
\hline
\end{array}
$$

**7.**
$$
\begin{array}{r}
6\ 0\ 0\ 0 \\
-\ 3\ 1\ 4\ 9 \\
\hline
\end{array}
$$

**8.**
$$
\begin{array}{r}
9\ 0\ 3\ 5 \\
-\ 7\ 4\ 8\ 9 \\
\hline
\end{array}
$$

**9.**
$$
\begin{array}{r}
2\ 0\ 0\ 1 \\
-\ \ \ \ 1\ 2\ 4 \\
\hline
\end{array}
$$

*Answers*

**2.** 6328  **3.** 4747  **4.** 56  **5.** 205
**6.** 658  **7.** 2851  **8.** 1546  **9.** 1877

**1.3** Exercise Set

For Extra Help
**MyMathLab**

*Math XL*
PRACTICE

WATCH

DOWNLOAD

READ

REVIEW

**a** Subtract. Check by adding.

1.  $\begin{array}{r} 6\ 5 \\ -\ 2\ 1 \\ \hline \end{array}$

2.  $\begin{array}{r} 8\ 7 \\ -\ 3\ 4 \\ \hline \end{array}$

3.  $\begin{array}{r} 8\ 6\ 6 \\ -\ 3\ 3\ 3 \\ \hline \end{array}$

4.  $\begin{array}{r} 5\ 2\ 6 \\ -\ 3\ 2\ 3 \\ \hline \end{array}$

5. $86 - 47$

6. $73 - 28$

7. $51 - 37$

8. $64 - 19$

9.  $\begin{array}{r} 5\ 6\ 3 \\ -\ 1\ 9\ 4 \\ \hline \end{array}$

10. $\begin{array}{r} 7\ 9\ 5 \\ -\ 3\ 9\ 8 \\ \hline \end{array}$

11. $\begin{array}{r} 3\ 9\ 1 \\ -\ 3\ 6\ 5 \\ \hline \end{array}$

12. $\begin{array}{r} 3\ 1\ 6 \\ -\ 2\ 4\ 7 \\ \hline \end{array}$

13. $981 - 747$

14. $887 - 698$

15. $683 - 266$

16. $342 - 217$

17. $\begin{array}{r} 7\ 7\ 6\ 9 \\ -\ 2\ 3\ 8\ 7 \\ \hline \end{array}$

18. $\begin{array}{r} 6\ 4\ 3\ 1 \\ -\ 2\ 8\ 9\ 6 \\ \hline \end{array}$

19. $\begin{array}{r} 4\ 5\ 1\ 2 \\ -\ 1\ 7\ 3\ 4 \\ \hline \end{array}$

20. $\begin{array}{r} 8\ 3\ 6\ 4 \\ -\ 5\ 3\ 7\ 5 \\ \hline \end{array}$

21. $5318 - 2249$

22. $9241 - 5643$

23. $3947 - 2858$

24. $7583 - 3641$

25. $\begin{array}{r} 1\ 2,6\ 4\ 7 \\ -\ \ \ 4,8\ 9\ 9 \\ \hline \end{array}$

26. $\begin{array}{r} 1\ 6,2\ 2\ 2 \\ -\ \ \ 5,8\ 8\ 8 \\ \hline \end{array}$

27. $\begin{array}{r} 5\ 1,3\ 4\ 2 \\ -\ 4\ 7,1\ 9\ 8 \\ \hline \end{array}$

28. $\begin{array}{r} 3\ 2,1\ 9\ 4 \\ -\ 2\ 9,2\ 3\ 6 \\ \hline \end{array}$

29. $\begin{array}{r} 8\ 0 \\ -\ 2\ 4 \\ \hline \end{array}$

30. $\begin{array}{r} 9\ 0 \\ -\ 7\ 8 \\ \hline \end{array}$

31. $\begin{array}{r} 6\ 9\ 0 \\ -\ 2\ 3\ 6 \\ \hline \end{array}$

32. $\begin{array}{r} 8\ 0\ 3 \\ -\ 4\ 1\ 8 \\ \hline \end{array}$

**33.**
```
   7 6 4 0
 - 3 8 0 9
```

**34.**
```
   5 2 8 0
 - 3 0 9 1
```

**35.**
```
   6 8 0 8
 - 3 0 5 9
```

**36.**
```
   9 4 0 5
 -   2 5 8
```

**37.**
```
   2 3 0 0
 -   1 0 9
```

**38.**
```
   7 5 0 0
 - 3 6 0 4
```

**39.**
```
   6 0 0 7
 - 1 5 8 9
```

**40.**
```
   8 0 0 3
 -   5 9 9
```

**41.** $90,237 - 47,209$

**42.** $84,703 - 298$

**43.** $101,734 - 5760$

**44.** $15,017 - 7809$

**45.**
```
   7 0 0 0
 - 2 7 9 4
```

**46.**
```
   8 0 0 1
 - 6 5 4 3
```

**47.**
```
   3 9,0 0 0
 - 3 7,6 9 5
```

**48.**
```
   1 7,0 0 0
 - 1 1,5 9 8
```

**49.** $10,008 - 19$

**50.** $40,006 - 147$

**51.** $50,001 - 1984$

**52.** $30,004 - 6749$

## Skill Maintenance

Add.   [1.2a]

**53.**
```
   9 4 6
 +   7 8
```

**54.**
```
   9 0 7 8
 + 3 6 5 4
```

**55.**
```
   5 7,8 7 7
 + 3 2,4 0 6
```

**56.**
```
   8 0 0 4
   6 7 8 9
   7 7 2 0
 + 6 8 5 1
```

**57.** $567 + 778$

**58.** $901 + 23$

**59.** $12,885 + 9807$

**60.** $9909 + 1011$

**61.** Write a word name for 6,375,602.   [1.1c]

**62.** What does the digit 7 mean in 6,375,602?   [1.1a]

## Synthesis

**63.** Fill in the missing digits to make the subtraction true:
$9,\boxed{\phantom{0}}48,621 - 2,097,\boxed{\phantom{0}}81 = 7,251,140.$

**64.** ▦ Subtract:  $3,928,124 - 1,098,947.$

# 1.4 Multiplication

## a) Multiplication of Whole Numbers

### Repeated Addition

The multiplication $3 \times 5$ corresponds to this repeated addition.

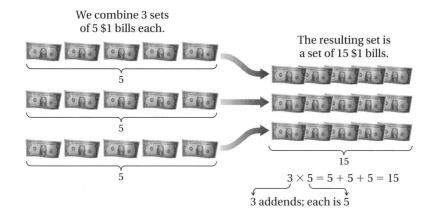

We combine 3 sets of 5 $1 bills each.

The resulting set is a set of 15 $1 bills.

$3 \times 5 = 5 + 5 + 5 = 15$

3 addends; each is 5

The numbers that we multiply are called **factors**. The result of the multiplication is called a **product**.

$$3 \quad \times \quad 5 \quad = \quad 15$$

Factor   Factor   Product

### Rectangular Arrays

Multiplications can also be thought of as rectangular arrays. Each of the following corresponds to the multiplication $3 \times 5$.

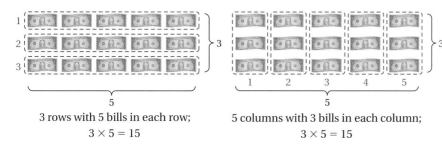

3 rows with 5 bills in each row;
$3 \times 5 = 15$

5 columns with 3 bills in each column;
$3 \times 5 = 15$

When you write a multiplication corresponding to a real-world situation, you should think of either a rectangular array or repeated addition. In some cases, it may help to think both ways.

We have used an "$\times$" to denote multiplication. A dot " $\cdot$ " is also commonly used. (Use of the dot is attributed to the German mathematician Gottfried Wilhelm von Leibniz in 1698.) Parentheses are also used to denote multiplication. For example,

$$3 \times 5 = 3 \cdot 5 = (3)(5) = 3(5) = 15.$$

*Answers*
*Skill to Review:*
1. 903   2. 6454

The product of 0 and any whole number is 0: $0 \cdot a = a \cdot 0 = 0$. For example, $0 \cdot 3 = 3 \cdot 0 = 0$. Multiplying a number by 1 does not change the number: $1 \cdot a = a \cdot 1 = a$. For example, $1 \cdot 3 = 3 \cdot 1 = 3$. We say that 1 is the **multiplicative identity**.

**EXAMPLE 1** Multiply: $5 \times 734$.

We have

$$
\begin{array}{r}
7\ 3\ 4 \\
\times\ \ \ \ \ 5 \\
\hline
2\ 0 \\
1\ 5\ 0 \\
3\ 5\ 0\ 0 \\
\hline
3\ 6\ 7\ 0 \\
\end{array}
$$

← Multiply the 4 ones by 5: $5 \times 4 = 20$.
← Multiply the 3 tens by 5: $5 \times 30 = 150$.
← Multiply the 7 hundreds by 5: $5 \times 700 = 3500$.
← Add.

Instead of writing each product on a separate line, we can use a shorter form.

$$
\begin{array}{r}
\overset{2}{}\ \ \\
7\ 3\ 4 \\
\times\ \ \ \ \ 5 \\
\hline
0 \\
\end{array}
$$

Multiply the ones by 5: $5 \cdot (4\ \text{ones}) = 20\ \text{ones} = 2\ \text{tens} + 0\ \text{ones}$. Write 0 in the ones column and 2 above the tens.

$$
\begin{array}{r}
\overset{1}{}\ \overset{2}{}\ \\
7\ 3\ 4 \\
\times\ \ \ \ \ 5 \\
\hline
7\ 0 \\
\end{array}
$$

Multiply the 3 tens by 5 and add 2 tens: $5 \cdot (3\ \text{tens}) = 15\ \text{tens}$, $15\ \text{tens} + 2\ \text{tens} = 17\ \text{tens} = 1\ \text{hundred} + 7\ \text{tens}$. Write 7 in the tens column and 1 above the hundreds.

$$
\begin{array}{r}
\overset{1}{}\ \overset{2}{}\ \\
7\ 3\ 4 \\
\times\ \ \ \ \ 5 \\
\hline
3\ 6\ 7\ 0 \\
\end{array}
$$

Multiply the 7 hundreds by 5 and add 1 hundred: $5 \cdot (7\ \text{hundreds}) = 35\ \text{hundreds}$, $35\ \text{hundreds} + 1\ \text{hundred} = 36\ \text{hundreds}$.

$$
\begin{array}{r}
\overset{1}{}\ \overset{2}{}\ \\
7\ 3\ 4 \\
\times\ \ \ \ \ 5 \\
\hline
3\ 6\ 7\ 0 \\
\end{array}
$$

You should write only this.

Do Exercises 1–4.

Multiplication of whole numbers is based on a property called the **distributive law**. It says that to multiply a number by a sum, $a \cdot (b + c)$, we can multiply each addend by $a$ and then add like this: $(a \cdot b) + (a \cdot c)$. Thus, $a \cdot (b + c) = (a \cdot b) + (a \cdot c)$. For example, consider the following.

$$4 \cdot (2 + 3) = 4 \cdot 5 = 20 \qquad \text{Adding first; then multiplying}$$

$$4 \cdot (2 + 3) = (4 \cdot 2) + (4 \cdot 3) = 8 + 12 = 20 \qquad \text{Multiplying first; then adding}$$

The results are the same, so $4 \cdot (2 + 3) = (4 \cdot 2) + (4 \cdot 3)$.

Multiply.

**1.**
$$
\begin{array}{r}
5\ 8 \\
\times\ \ \ 2 \\
\hline
\end{array}
$$

**2.**
$$
\begin{array}{r}
3\ 7 \\
\times\ \ \ 4 \\
\hline
\end{array}
$$

**3.**
$$
\begin{array}{r}
8\ 2\ 3 \\
\times\ \ \ \ \ 6 \\
\hline
\end{array}
$$

**4.**
$$
\begin{array}{r}
1\ 3\ 4\ 8 \\
\times\ \ \ \ \ \ \ 5 \\
\hline
\end{array}
$$

*Answers*

**1.** 116   **2.** 148   **3.** 4938   **4.** 6740

Let's find the product $51 \times 32$. Since $32 = 2 + 30$, we can think of this product as

$$51 \times 32 = 51 \times (2 + 30) = (51 \times 2) + (51 \times 30).$$

That is, we multiply 51 by 2, then we multiply 51 by 30, and finally we add. We can write our work this way.

```
      5 1
    × 3 2
    ─────
    1 0 2      Multiplying by 2
  1 5 3 0      Multiplying by 30. (We write a 0 and then multiply 51 by 3.)
```

> You may have learned that such a 0 need not be written. You may omit it if you wish. If you do omit it, remember, when multiplying by tens, to start writing the answer in the tens place.

We add to obtain the product.

```
      5 1
    × 3 2
    ─────
    1 0 2
  1 5 3 0
  ───────
  1 6 3 2      Adding to obtain the product
```

**EXAMPLE 2**  Multiply: $457 \times 683$.

```
      5   2
      6 8 3
    × 4 5 7
    ───────
    4 7 8 1      Multiplying 683 by 7
```

```
      4   1
      5   2
      6 8 3
    × 4 5 7
    ───────
    4 7 8 1
  3 4 1 5 0      Multiplying 683 by 50
```

```
        3   1
        4   1
        5   2
        6 8 3
      × 4 5 7
      ─────────
      4 7 8 1
    3 4 1 5 0      Multiplying 683 by 400. (We write
  2 7 3 2 0 0  ←   00 and then multiply 683 by 4.)
  ───────────
  3 1 2,1 3 1      Adding
```

Do Exercises 5–8.

Multiply.

**5.**
```
    4 5
  × 2 3
```

**6.** $48 \times 63$

**7.**
```
    7 4 6
  ×   6 2
```

**8.** $245 \times 837$

**STUDY TIPS**

**TIME MANAGEMENT**

- **A rule of thumb on study time.** Budget about 2–3 hours for homework and study for every hour you spend in class each week.
- **Scheduling your time.** Make an hour-by-hour schedule of your typical week. Include work, school, home, sleep, study, and leisure times. Try to schedule time for study when you are most alert. Choose a setting that will enable you to focus and concentrate. Plan for success and it will happen!

*Answers*
**5.** 1035   **6.** 3024   **7.** 46,252   **8.** 205,065

1.4  Multiplication   **21**

**Multiply.**

**9.**
$$\begin{array}{r} 4\ 7\ 2 \\ \times\ 3\ 0\ 6 \\ \hline \end{array}$$

**10.** $408 \times 704$

**11.**
$$\begin{array}{r} 2\ 3\ 4\ 4 \\ \times\ 6\ 0\ 0\ 5 \\ \hline \end{array}$$

**12.**
$$\begin{array}{r} 1\ 0\ 0\ 6 \\ \times\ \ \ 7\ 0\ 3 \\ \hline \end{array}$$

**Multiply.**

**13.**
$$\begin{array}{r} 4\ 7\ 2 \\ \times\ 8\ 3\ 0 \\ \hline \end{array}$$

**14.**
$$\begin{array}{r} 2\ 3\ 4\ 4 \\ \times\ 7\ 4\ 0\ 0 \\ \hline \end{array}$$

**15.** $100 \times 562$

**16.** $1000 \times 562$

**17. a)** Find $23 \cdot 47$.

  **b)** Find $47 \cdot 23$.

  **c)** Compare your answers to parts (a) and (b).

**Multiply.**

**18.** $5 \cdot 2 \cdot 4$

**19.** $4 \cdot 2 \cdot 6$

---

**EXAMPLE 3**   Multiply: $306 \times 274$.

Note that $306 = 3$ hundreds $+ 6$ ones.

$$\begin{array}{r} 2\ 7\ 4 \\ \times\ 3\ 0\ 6 \\ \hline 1\ 6\ 4\ 4 \\ 8\ 2\ 2\ 0\ 0 \\ \hline 8\ 3,8\ 4\ 4 \end{array}$$

⌐ Multiplying by 6
⌐ Multiplying by 3 hundreds. (We write 00 and then multiply 274 by 3.)
  Adding

Do Exercises 9–12.

**EXAMPLE 4**   Multiply: $360 \times 274$.

Note that $360 = 3$ hundreds $+ 6$ tens.

$$\begin{array}{r} 2\ 7\ 4 \\ \times\ 3\ 6\ 0 \\ \hline 1\ 6\ 4\ 4\ 0 \\ 8\ 2\ 2\ 0\ 0 \\ \hline 9\ 8,6\ 4\ 0 \end{array}$$

⌐ Multiplying by 6 tens. (We write 0 and then multiply 274 by 6.)
⌐ Multiplying by 3 hundreds. (We write 00 and then multiply 274 by 3.)
  Adding

Do Exercises 13–16.

When we multiply two numbers, we can change the order of the numbers without changing their product. For example, $3 \cdot 6 = 18$ and $6 \cdot 3 = 18$. This illustrates the **commutative law of multiplication:** $a \cdot b = b \cdot a$.

Do Exercise 17.

To multiply three or more numbers, we generally first group them so that we multiply two at a time. Consider $2 \cdot 3 \cdot 4$. We can group these numbers as $2 \cdot (3 \cdot 4)$ or as $(2 \cdot 3) \cdot 4$. The parentheses tell what to do first:

$$2 \cdot (3 \cdot 4) = 2 \cdot (12) = 24.$$   We multiply 3 and 4 and then that product by 2.

We can also multiply 2 and 3 and then that product by 4:

$$(2 \cdot 3) \cdot 4 = (6) \cdot 4 = 24.$$

Either way we get 24. It does not matter how we group the numbers. This illustrates the **associative law of multiplication:** $a \cdot (b \cdot c) = (a \cdot b) \cdot c$.

Do Exercises 18 and 19.

---

*Answers*

**9.** 144,432   **10.** 287,232
**11.** 14,075,720   **12.** 707,218
**13.** 391,760   **14.** 17,345,600
**15.** 56,200   **16.** 562,000
**17.** (a) 1081; (b) 1081; (c) same
**18.** 40   **19.** 48

## (b) Finding Area

The area of a rectangular region can be considered to be the number of square units needed to fill it. Here is a rectangle 3 cm (centimeters) long and 4 cm wide. It takes 12 square centimeters (sq cm) to fill it.

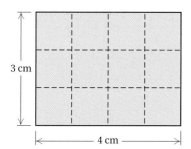

This is a square centimeter (a square unit).

In this case, we have a rectangular array of 3 rows, each of which contains 4 squares. The number of square units is given by 3 · 4, or 12. That is, $A = l \cdot w = 3\,\text{cm} \cdot 4\,\text{cm} = 12\,\text{sq cm}$.

**EXAMPLE 5** *Table Tennis.* Find the area of a standard table tennis table that has dimensions of 9 ft by 5 ft.

If we think of filling the rectangle with square feet, we have a rectangular array. The length $l = 9$ ft and the width $w = 5$ ft. Thus the area $A$ is given by the formula

$$A = l \cdot w = 9\,\text{ft} \cdot 5\,\text{ft} = 45\,\text{sq ft}.$$

Do Exercise 20.

**20. Professional Pool Table.**   The playing area of a pool table used in professional tournaments is 50 in. by 100 in. (There are 6-in. wide rails on the outside that are not included in the playing area.) Determine the playing area.

*Answer*
**20.** 5000 sq in.

**a** Multiply.

**1.**
$$\begin{array}{r} 6\ 5 \\ \times\quad 8 \\ \hline \end{array}$$

**2.**
$$\begin{array}{r} 8\ 7 \\ \times\quad 4 \\ \hline \end{array}$$

**3.**
$$\begin{array}{r} 9\ 4 \\ \times\quad 6 \\ \hline \end{array}$$

**4.**
$$\begin{array}{r} 7\ 6 \\ \times\quad 9 \\ \hline \end{array}$$

**5.** $3 \cdot 509$

**6.** $7 \cdot 806$

**7.** $7(9229)$

**8.** $4(7867)$

**9.** $90(53)$

**10.** $60(78)$

**11.** $(47)(85)$

**12.** $(34)(87)$

**13.**
$$\begin{array}{r} 8\ 7 \\ \times\ 1\ 0 \\ \hline \end{array}$$

**14.**
$$\begin{array}{r} 2\ 3\ 4\ 0 \\ \times\ 1\ 0\ 0\ 0 \\ \hline \end{array}$$

**15.**
$$\begin{array}{r} 9\ 6 \\ \times\ 2\ 0 \\ \hline \end{array}$$

**16.**
$$\begin{array}{r} 8\ 0\ 0 \\ \times\ 7\ 0\ 0 \\ \hline \end{array}$$

**17.**
$$\begin{array}{r} 6\ 4\ 3 \\ \times\quad 7\ 2 \\ \hline \end{array}$$

**18.**
$$\begin{array}{r} 7\ 7\ 7 \\ \times\quad 7\ 7 \\ \hline \end{array}$$

**19.**
$$\begin{array}{r} 4\ 4\ 4 \\ \times\quad 3\ 3 \\ \hline \end{array}$$

**20.**
$$\begin{array}{r} 5\ 4\ 9 \\ \times\quad 8\ 8 \\ \hline \end{array}$$

**21.**
$$\begin{array}{r} 5\ 6\ 4 \\ \times\ 4\ 5\ 8 \\ \hline \end{array}$$

**22.**
$$\begin{array}{r} 4\ 3\ 2 \\ \times\ 3\ 7\ 5 \\ \hline \end{array}$$

**23.**
$$\begin{array}{r} 8\ 5\ 3 \\ \times\ 9\ 3\ 6 \\ \hline \end{array}$$

**24.**
$$\begin{array}{r} 3\ 4\ 6 \\ \times\ 6\ 5\ 9 \\ \hline \end{array}$$

**25.**
$$\begin{array}{r} 6\ 4\ 2\ 8 \\ \times\ 3\ 2\ 2\ 4 \\ \hline \end{array}$$

**26.**
$$\begin{array}{r} 8\ 9\ 2\ 8 \\ \times\ 3\ 1\ 7\ 2 \\ \hline \end{array}$$

**27.**
$$\begin{array}{r} 3\ 4\ 8\ 2 \\ \times\quad 1\ 0\ 4 \\ \hline \end{array}$$

**28.**
$$\begin{array}{r} 6\ 4\ 0\ 8 \\ \times\ 6\ 0\ 6\ 4 \\ \hline \end{array}$$

**29.**
$$\begin{array}{r} 8\ 7\ 6 \\ \times\ 3\ 4\ 5 \\ \hline \end{array}$$

**30.**
$$\begin{array}{r} 3\ 5\ 5 \\ \times\ 2\ 9\ 9 \\ \hline \end{array}$$

**31.**
$$\begin{array}{r} 7\ 8\ 8\ 9 \\ \times\ 6\ 2\ 2\ 4 \\ \hline \end{array}$$

**32.**
$$\begin{array}{r} 6\ 5\ 2\ 1 \\ \times\ 3\ 4\ 4\ 9 \\ \hline \end{array}$$

**33.**
```
    5 6 0 8
  × 4 5 0 0
```

**34.**
```
    4 5 0 6
  × 7 8 0 0
```

**35.**
```
    5 0 0 6
  × 4 0 0 8
```

**36.**
```
    6 0 0 9
  × 2 0 0 3
```

**b**    Find the area of each region.

**37.**

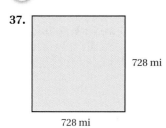

728 mi

728 mi

**38.**

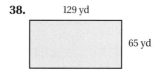

129 yd

65 yd

**39.** Find the area of the region formed by the base lines on a Major League Baseball diamond.

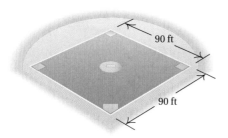

90 ft

90 ft

**40.** Find the area of a standard-sized hockey rink.

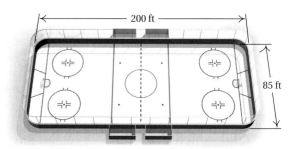

200 ft

85 ft

## Skill Maintenance

Add.   [1.2a]

**41.**
```
    4 9 0 8
    5 6 6 7
  + 2 1 1 0
```

**42.**
```
    9 8 7 6
      8 7 6
        7 6
  +       6
```

**43.**
```
    3 4 0 , 7 9 8
  +   8 6 , 6 7 9
```

**44.**
```
    8 8 , 7 7 7
  + 2 2 , 3 3 3
```

Subtract.   [1.3a]

**45.**
```
    4 9 0 8
  − 3 6 6 7
```

**46.**
```
    9 8 7 6
  −   9 8 7
```

**47.**
```
    3 4 0 , 7 9 8
  −   8 6 , 6 7 9
```

**48.**
```
    8 8 , 7 7 7
  − 2 2 , 3 3 3
```

## Synthesis

**49.** 🖩 An 18-story office building is box-shaped. Each floor measures 172 ft by 84 ft with a 20-ft by 35-ft rectangular area lost to an elevator and a stairwell. How much area is available as office space?

# Division

## OBJECTIVE

**a** Divide whole numbers.

**SKILL TO REVIEW**
Objective 1.3a:
Subtract whole numbers.

Subtract.

**1.**  5 6 4
   − 3 9 7

**2.**  7 0 3 5
   − 2 9 4 4

## a  Division of Whole Numbers

### Repeated Subtraction

Division of whole numbers applies to two kinds of situations. The first is repeated subtraction. Suppose we have 20 doughnuts, and we want to find out how many sets of 5 there are. One way to do this is to repeatedly subtract sets of 5 as follows.

20 doughnuts

How many sets of 5 doughnuts each?

Since there are 4 sets of 5 doughnuts each, we have

$$20 \div 5 = 4.$$

Dividend  Divisor  Quotient

The division $20 \div 5$ is read "20 divided by 5." The **dividend** is 20, the **divisor** is 5, and the **quotient** is 4. We divide the *dividend* by the *divisor* to get the *quotient*.

We can also express the division $20 \div 5 = 4$ as

$$\frac{20}{5} = 4 \quad \text{or} \quad 5\overline{)20}.$$

### Rectangular Arrays

We can also think of division in terms of rectangular arrays. Consider again the 20 doughnuts and division by 5. We can arrange the doughnuts in a rectangular array with 5 rows and ask, "How many are in each row?"

We can also consider a rectangular array with 5 doughnuts in each column and ask, "How many columns are there?" The answer is still 4.

In each case, we are asking, "What do we multiply 5 by in order to get 20?"

Missing factor             Quotient

$$5 \cdot \square = 20 \qquad 20 \div 5 = \square$$

This leads us to the following definition of division.

*Answers*

*Skill to Review:*
**1.** 167  **2.** 4091

## DIVISION

The quotient $a \div b$, where $b \neq 0$, is that unique number $c$ for which $a = b \cdot c$.

This definition shows the relation between division and multiplication. We see, for instance, that

$20 \div 5 = 4$   because   $20 = 5 \cdot 4$.

This relation allows us to use multiplication to check division.

**EXAMPLE 1**   Divide.  Check by multiplying.

**a)**  $16 \div 8$
**b)**  $\dfrac{36}{4}$
**c)**  $7\overline{)56}$

We do so as follows.

**a)**  $16 \div 8 = 2$     *Check*: $8 \cdot 2 = 16$.

**b)**  $\dfrac{36}{4} = 9$     *Check*: $4 \cdot 9 = 36$.

**c)**  $7\overline{)56}^{\,8}$     *Check*: $7 \cdot 8 = 56$.

Do Exercises 1–3.

Divide. Check by multiplying.
**1.** $9\overline{)45}$

**2.** $27 \div 3$

**3.** $\dfrac{48}{6}$

Let's consider some basic properties of division.

## DIVIDING BY 1

Any number divided by 1 is that same number:   $a \div 1 = \dfrac{a}{1} = a$.

For example, $6 \div 1 = 6$ and $\dfrac{15}{1} = 15$.

## DIVIDING A NUMBER BY ITSELF

Any nonzero number divided by itself is 1:   $a \div a = \dfrac{a}{a} = 1$,   $a \neq 0$.

For example, $7 \div 7 = 1$ and $\dfrac{22}{22} = 1$.

## DIVIDENDS OF 0

Zero divided by any nonzero number is 0:   $0 \div a = \dfrac{0}{a} = 0$,   $a \neq 0$.

For example, $0 \div 14 = 0$ and $\dfrac{0}{3} = 0$.

## EXCLUDING DIVISION BY 0

Division by 0 is not defined: $a \div 0$, or $\frac{a}{0}$, is **not defined**.

For example, $16 \div 0$, or $\frac{16}{0}$, is not defined.

Why can't we divide by 0? Suppose the number 4 could be divided by 0. Then if $\square$ were the answer, we would have

$$4 \div 0 = \square,$$

and since 0 times any number is 0, we would have

$$4 = \square \cdot 0 = 0. \qquad \text{False!}$$

Thus, the only possible number that could be divided by 0 would be 0 itself. But such a division would give us any number we wish. For instance,

$$
\left.
\begin{array}{ll}
0 \div 0 = 8 & \text{because} \quad 0 = 8 \cdot 0; \\
0 \div 0 = 3 & \text{because} \quad 0 = 3 \cdot 0; \\
0 \div 0 = 7 & \text{because} \quad 0 = 7 \cdot 0.
\end{array}
\right\} \quad \text{All true!}
$$

We avoid the preceding difficulties by agreeing to exclude division by 0.

Do Exercises 4–7.

Divide, if possible. If not possible, write "not defined."

**4.** $\dfrac{9}{9}$

**5.** $5 \div 0$

**6.** $0 \div 20$

**7.** $\dfrac{8}{1}$

### Division with a Remainder

Suppose we have 22 cans of soda and want to pack them in cartons of 6 cans each. We could fill 3 cartons and have 4 cans left over.

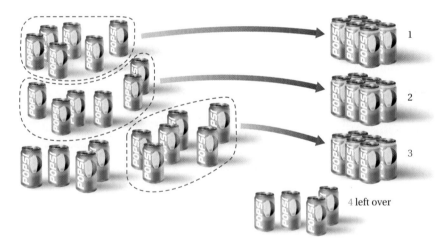

We can think of this as the following division. The leftover cans are the **remainder**.

$$
\begin{array}{r}
3 \; \leftarrow \text{Quotient} \\
6\,\overline{)2\;2} \\
\underline{1\;8} \\
4 \; \leftarrow \text{Remainder}
\end{array}
$$

*Answers*

**4.** 1  **5.** Not defined
**6.** 0  **7.** 8

We express the result as

$$22 \div 6 = 3\,\text{R}\,4.$$

Dividend  Divisor  Quotient  Remainder

Note that

Quotient · Divisor + Remainder = Dividend.

Thus we have

$$3 \cdot 6 = 18 \qquad \text{Quotient · Divisor}$$

and  $18 + 4 = 22.$  Adding the remainder. The result is the dividend.

We now show a procedure for dividing whole numbers.

**EXAMPLE 2**  Divide and check:  $4\overline{)3\ 4\ 5\ 7}$.

First, we try to divide the first digit of the dividend, 3, by the divisor, 4. Since $3 \div 4$ is not a whole number, we consider the first *two* digits of the dividend.

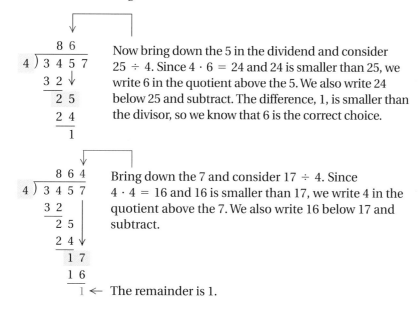

$$
\begin{array}{r}
8 \\
4\,)\overline{3\ 4\ 5\ 7} \\
3\ 2 \\
\hline
2
\end{array}
$$

Since $4 \cdot 8 = 32$ and 32 is smaller than 34, we write an 8 in the quotient above the 4. We also write 32 below 34 and subtract.

What if we had chosen a number other than 8 for the first digit of the quotient? Suppose we had used 7 instead of 8 and subtracted $4 \cdot 7$, or 28, from 34. The result would have been $34 - 28$, or 6. Because 6 is larger than the divisor, 4, we know that there is at least one more factor of 4 in 34, and thus 7 is too small. If we had used 9 instead of 8, then we would have tried to subtract $4 \cdot 9$, or 36, from 34. That difference is not a whole number, so we know 9 is too large. When we subtract, the difference must be smaller than the divisor.

Let's continue dividing.

$$
\begin{array}{r}
8\ 6 \\
4\,)\overline{3\ 4\ 5\ 7} \\
3\ 2\ \downarrow \\
\hline
2\ 5 \\
2\ 4 \\
\hline
1
\end{array}
$$

Now bring down the 5 in the dividend and consider $25 \div 4$. Since $4 \cdot 6 = 24$ and 24 is smaller than 25, we write 6 in the quotient above the 5. We also write 24 below 25 and subtract. The difference, 1, is smaller than the divisor, so we know that 6 is the correct choice.

$$
\begin{array}{r}
8\ 6\ 4 \\
4\,)\overline{3\ 4\ 5\ 7} \\
3\ 2 \\
2\ 5 \\
2\ 4\ \downarrow \\
\hline
1\ 7 \\
1\ 6 \\
\hline
1
\end{array}
$$

Bring down the 7 and consider $17 \div 4$. Since $4 \cdot 4 = 16$ and 16 is smaller than 17, we write 4 in the quotient above the 7. We also write 16 below 17 and subtract.

$\leftarrow$ The remainder is 1.

**Check:**  $864 \cdot 4 = 3456$ and $3456 + 1 = 3457.$

The answer is 864 R 1.

Do Exercises 8–10.

Divide and check.

**8.** $3\overline{)2\ 3\ 9}$

**9.** $5\overline{)5\ 8\ 6\ 4}$

**10.** $6\overline{)3\ 8\ 5\ 5}$

*Answers*
**8.** 79 R 2  **9.** 1172 R 4
**10.** 642 R 3

**EXAMPLE 3**  Divide: 8904 ÷ 42.

Because 42 is close to 40, we think of the divisor as 40 when we make our choices of digits in the quotient.

$$
\begin{array}{r}
2\phantom{00} \\
42\overline{\smash{)}8904} \\
84\phantom{0}\downarrow \\
\hline
50
\end{array}
$$
← *Think:* 89 ÷ 40. We try 2. Multiply 42 · 2 and subtract. Then bring down the 0.

$$
\begin{array}{r}
2\;1\phantom{0} \\
42\overline{\smash{)}8904} \\
84\phantom{0} \\
\hline
5\;0 \\
4\;2\downarrow \\
\hline
8\;4
\end{array}
$$
← *Think:* 50 ÷ 40. We try 1. Multiply 42 · 1 and subtract. Then bring down the 4.

$$
\begin{array}{r}
2\;1\;2 \\
42\overline{\smash{)}8904} \\
84\phantom{00} \\
\hline
5\;0 \\
4\;2 \\
\hline
8\;4 \\
8\;4 \\
\hline
0
\end{array}
$$
← *Think:* 84 ÷ 40. We try 2. Multiply 2 · 42 and subtract.

The remainder is 0, so the answer is 212.

Do Exercises 11 and 12.

-------------------------------------------------- *Caution!* --------------------------------------------------

Be careful to keep the digits lined up correctly when you divide.

--------------------------------------------------------------------------------------------------------------

**Divide.**

11. $45\overline{\smash{)}6030}$

12. $52\overline{\smash{)}3288}$

### Calculator Corner

**Dividing Whole Numbers**  To divide whole numbers on a calculator, we use the ÷ and = keys. For example, to divide 711 by 9, we press

7 1 1 ÷ 9 =. The display reads [ 79 ], so 711 ÷ 9 = 79.

When we enter 453 ÷ 15, the display reads [ 30.2 ]. Note that the result is not a whole number. This tells us that there is a remainder. The number 30.2 is expressed in decimal notation. The symbol "." is called a decimal point. Decimal notation will be studied in Chapter 5. Although it is possible to use the number to the right of the decimal point to find the remainder, we will not do so here.

**Exercises:**  Use a calculator to perform each division.

1. $19\overline{\smash{)}532}$

2. $7\overline{\smash{)}861}$

3. 9367 ÷ 29

4. 12,276 ÷ 341

*Answers*

**11.** 134   **12.** 63 R 12

## Zeros in Quotients

**EXAMPLE 4** Divide: 6341 ÷ 7.

```
      9
  7 ) 6 3 4 1   ←  Think: 63 ÷ 7 = 9. The first digit in the quotient
      6 3 ↓         is 9. We do not write the 0 when we find 63 − 63.
          4         Bring down the 4.
```

```
      9 0
  7 ) 6 3 4 1
      6 3 ↓
          4 1   ←  Think: 4 ÷ 7. If we subtract a group of 7's, such
                   as 7, 14, 21, etc., from 4, we do not get a whole
                   number, so the next digit in the quotient is 0.
                   Bring down the 1.
```

```
      9 0 5
  7 ) 6 3 4 1
      6 3
          4 1   ←  Think: 41 ÷ 7. We try 5. Multiply 7 · 5 and
          3 5      subtract.
            6   ←  The remainder is 6.
```

The answer is 905 R 6.

Do Exercises 13 and 14.

**EXAMPLE 5** Divide: 8169 ÷ 34.

Because 34 is close to 30, we think of the divisor as 30 when we make our choices of digits in the quotient.

```
        2
  3 4 ) 8 1 6 9   ←  Think: 81 ÷ 30. We try 2. Multiply 34 · 2 and subtract.
        6 8 ↓        Then bring down the 6.
        1 3 6
```

```
        2 4
  3 4 ) 8 1 6 9
        6 8
        1 3 6   ←  Think: 136 ÷ 30. We try 4. Multiply 34 · 4 and
        1 3 6 ↓    subtract. The difference is 0, so we do not write it.
              9    Bring down the 9.
```

```
        2 4 0
  3 4 ) 8 1 6 9
        6 8
        1 3 6
        1 3 6
              9 ←  Think: 9 ÷ 34. If we subtract a group of 34's, such as
              0    34 or 68, from 9, we do not get a whole number, so the
              9 ←  last digit in the quotient is 0.
                   The remainder is 9.
```

The answer is 240 R 9.

Do Exercises 15 and 16.

Divide.

**13.** 6 ) 4 8 4 6

**14.** 7 ) 7 6 1 6

Divide.

**15.** 2 7 ) 9 7 2 4

**16.** 5 6 ) 4 4 , 8 4 7

Answers
**13.** 807 R 4  **14.** 1088
**15.** 360 R 4  **16.** 800 R 47

**a**    Divide, if possible. If not possible, write "not defined."

**1.** $72 \div 6$

**2.** $54 \div 9$

**3.** $\dfrac{23}{23}$

**4.** $\dfrac{37}{37}$

**5.** $22 \div 1$

**6.** $\dfrac{56}{1}$

**7.** $\dfrac{0}{7}$

**8.** $\dfrac{0}{32}$

**9.** $\dfrac{16}{0}$

**10.** $74 \div 0$

**11.** $\dfrac{48}{8}$

**12.** $\dfrac{20}{4}$

Divide.

**13.** $277 \div 5$

**14.** $699 \div 3$

**15.** $864 \div 8$

**16.** $869 \div 8$

**17.** $4 \overline{)\ 1\ 2\ 2\ 8}$

**18.** $3 \overline{)\ 2\ 1\ 2\ 4}$

**19.** $6 \overline{)\ 4\ 5\ 2\ 1}$

**20.** $9 \overline{)\ 9\ 1\ 1\ 0}$

**21.** $297 \div 4$

**22.** $389 \div 2$

**23.** $738 \div 8$

**24.** $881 \div 6$

**25.** $5 \overline{)\ 8\ 5\ 1\ 5}$

**26.** $3 \overline{)\ 6\ 0\ 2\ 7}$

**27.** $9 \overline{)\ 8\ 8\ 8\ 8}$

**28.** $8 \overline{)\ 4\ 1\ 3\ 9}$

**29.** $127{,}000 \div 10$

**30.** $127{,}000 \div 100$

**31.** $127{,}000 \div 1000$

**32.** $4260 \div 10$

**33.** $70 \overline{)3692}$

**34.** $20 \overline{)5798}$

**35.** $30 \overline{)875}$

**36.** $40 \overline{)987}$

**37.** $852 \div 21$

**38.** $942 \div 23$

**39.** $85 \overline{)7672}$

**40.** $54 \overline{)2729}$

**41.** $111 \overline{)3219}$

**42.** $102 \overline{)5612}$

**43.** $8 \overline{)843}$

**44.** $7 \overline{)749}$

**45.** $5 \overline{)8047}$

**46.** $9 \overline{)7273}$

**47.** $5 \overline{)5036}$

**48.** $7 \overline{)7074}$

**49.** $1058 \div 46$

**50.** $7242 \div 24$

**51.** $3425 \div 32$

**52.** $48 \overline{)4899}$

**53.** $24 \overline{)8880}$

**54.** $36 \overline{)7563}$

**55.** $28 \overline{)17,067}$

**56.** $36 \overline{)28,929}$

**57.** $80 \overline{)24,320}$

**58.** $90 \overline{)88,560}$

**59.** $285 \overline{)999,999}$

**60.** $3\ 0\ 6\ \overline{)\ 8\ 8\ 8,8\ 8\ 8}$      **61.** $4\ 5\ 6\ \overline{)\ 3,6\ 7\ 9,9\ 2\ 0}$      **62.** $8\ 0\ 3\ \overline{)\ 5,6\ 2\ 2,6\ 0\ 6}$

## Skill Maintenance

In each of Exercises 63–70, fill in the blank with the correct term from the given list. Some of the choices may not be used and some may be used more than once.

**63.** The distance around an object is its _____. [1.2b]

**64.** The _____ is the number from which another number is being subtracted. [1.3a]

**65.** For large numbers, _____ are separated by commas into groups of three, called _____. [1.1a]

**66.** In the sentence $28 \div 7 = 4$, the _____ is 28. [1.5a]

**67.** In the sentence $10 \times 1000 = 10,000$, 10 and 1000 are called _____ and 10,000 is called the _____. [1.4a]

**68.** The number 0 is called the _____ identity. [1.2a]

**69.** The sentence $3 \times (6 \times 2) = (3 \times 6) \times 2$ illustrates the _____ law of multiplication. [1.4a]

**70.** We can use the following statement to check division:
quotient · _____ + _____ = _____. [1.5a]

associative
commutative
addends
factors
area
perimeter
minuend
subtrahend
product
digits
periods
additive
multiplicative
dividend
quotient
remainder
divisor

## Synthesis

**71.** Complete the following table.

| $a$ | $b$ | $a \cdot b$ | $a + b$ |
|---|---|---|---|
|  | 68 | 3672 |  |
| 84 |  |  | 117 |
|  |  | 32 | 12 |

**72.** Find a pair of factors whose product is 36 and:
  **a)** whose sum is 13.
  **b)** whose difference is 0.
  **c)** whose sum is 20.
  **d)** whose difference is 9.

**73.** A group of 1231 college students is going to take buses for a field trip. Each bus can hold 42 students. How many buses are needed?

**74.** ▦ Fill in the missing digits to make the equation true:

$$34,584,132 \div 76\square = 4\square,386.$$

# Mid-Chapter Review

## Concept Reinforcement

Determine whether each statement is true or false.

————————— **1.** If $a - b = c$, then $b = a + c$.  [1.3a]

————————— **2.** We can think of the multiplication $4 \times 3$ as a rectangular array containing 4 rows with 3 items in each row.  [1.4a]

————————— **3.** We can think of the multiplication $4 \times 3$ as a rectangular array containing 3 columns with 4 items in each column.  [1.4a]

————————— **4.** The product of two whole numbers is always greater than either of the factors.  [1.4a]

————————— **5.** Zero divided by any nonzero number is 0.  [1.5a]

————————— **6.** Any number divided by 1 is the number 1.  [1.5a]

## Guided Solutions

Fill in each blank with the number that creates a correct statement or solution.

**7.** Write a word name for 95,406,237.  [1.1c]

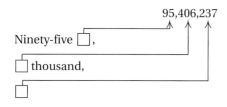

**8.** Subtract: $604 - 497$.  [1.3a]

$$
\begin{array}{ccc}
\square & \square & \square \\
6 & \cancel{0} & \cancel{4} \\
-\,4 & 9 & 7 \\
\hline
\square & \square & \square
\end{array}
$$

## Mixed Review

In each of the following numbers what does the digit 6 mean?  [1.1a]

**9.** 2698

**10.** 61,204

**11.** 146,237

**12.** 586

Consider the number 306,458,129. What digit names the number of:  [1.1a]

**13.** tens

**14.** millions

**15.** ten thousands

**16.** hundreds

Write expanded notation.  [1.1b]

**17.** 5602

**18.** 69,345

Write a word name.  [1.1c]

**19.** 136

**20.** 64,325

Write standard notation.  [1.1c]

**21.** Three hundred eight thousand, seven hundred sixteen

**22.** Four million, five hundred sixty-seven thousand, two hundred sixteen

Add. [1.2a]

23.
```
    3 1 6
  + 4 8 2
```

24.
```
    5 9 3
  + 4 3 7
```

25.
```
    2 6 3 8
  + 5 2 8 4
```

26.
```
    4 6 1 7
    2 4 3 6
  +   4 8 1
```

Subtract. [1.3a]

27.
```
    7 8 6
  - 3 2 1
```

28.
```
    6 2 4
  - 2 8 5
```

29.
```
    3 6 0 2
  - 1 7 4 8
```

30.
```
    5 0 0 4
  -   6 7 6
```

Multiply. [1.4a]

31.
```
    3 6
  ×  6
```

32.
```
    5 6 7
  ×  2 8
```

33.
```
    4 0 7
  × 3 2 5
```

34.
```
    9 4 3 5
  ×   6 0 2
```

Divide. [1.5a]

35. 4 ) 1 0 1 2

36. 3 8 ) 4 2 6 1

37. 6 0 ) 1 3 9 9

38. 5 6 ) 8 0 9 5

39. Find the perimeter of the figure.   [1.2b]

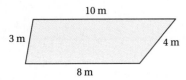

40. Find the area of the region.   [1.4b]

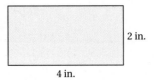

## Understanding Through Discussion and Writing

*To the student and the instructor:* The Discussion and Writing exercises are meant to be answered with one or more sentences. They can be discussed and answered collaboratively by the entire class or by small groups.

41. Explain in your own words what the associative law of addition means.   [1.2a]

42. Is subtraction commutative? That is, is there a commutative law of subtraction? Why or why not?   [1.3a]

43. Describe a situation that corresponds to each multiplication: 4 · $150;   $4 · 150.   [1.4a]

44. Suppose a student asserts that "0 ÷ 0 = 0 because nothing divided by nothing is nothing." Devise an explanation to persuade the student that the assertion is false.   [1.5a]

# 1.6 Rounding and Estimating; Order

## a Rounding

We round numbers in various situations when we do not need an exact answer. For example, we might round to see if we are being charged the correct amount in a store. We might also round to check if an answer to a problem is reasonable or to check a calculation done by hand or on a calculator.

To understand how to round, we first look at some examples using the number line. The number line displays numbers at equally spaced intervals.

**EXAMPLE 1**  Round 47 to the nearest ten.

47 is between 40 and 50. Since 47 is closer to 50, we round up to 50.

**EXAMPLE 2**  Round 42 to the nearest ten.

42 is between 40 and 50. Since 42 is closer to 40, we round down to 40.

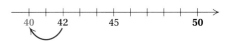

**EXAMPLE 3**  Round 45 to the nearest ten.

45 is halfway between 40 and 50. We could round 45 down to 40 or up to 50. We agree to round up to 50.

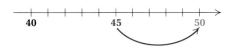

> When a number is halfway between rounding numbers, round up.

Do Margin Exercises 1–7.

We round whole numbers according to the following rule.

### ROUNDING WHOLE NUMBERS

To round to a certain place:

a)  Locate the digit in that place.

b)  Consider the next digit to the right.

c)  If the digit to the right is 5 or higher, round up. If the digit to the right is 4 or lower, round down.

d)  Change all digits to the right of the rounding location to zeros.

### OBJECTIVES

**a** Round to the nearest ten, hundred, or thousand.

**b** Estimate sums, differences, products, and quotients by rounding.

**c** Use < or > for ☐ to write a true sentence in a situation like 6 ☐ 10.

**SKILL TO REVIEW**

Objective 1.1a: Give the meaning of digits in standard notation.

In the number 145,627 what digit names the number of:

1. Tens?
2. Thousands?

Round to the nearest ten.

1. 37

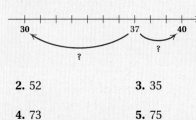

2. 52          3. 35

4. 73          5. 75

6. 88          7. 64

*Answers*

*Skill to Review:*
1. 2    2. 5

*Margin Exercises:*
1. 40    2. 50    3. 40    4. 70    5. 80
6. 90    7. 60

**EXAMPLE 4**  Round 6485 to the nearest ten.

a) Locate the digit in the tens place, 8.

   6  4  8  5
        ↑

b) Consider the next digit to the right, 5.

   6  4  8  5
           ↑

c) Since that digit, 5, is 5 or higher, round 8 tens up to 9 tens.

d) Change all digits to the right of the tens digit to zeros.

   6  4  9  0  ← This is the answer.

> Do Exercises 8–11.

**Round to the nearest ten.**

**8.** 137        **9.** 473

**10.** 235       **11.** 285

**EXAMPLE 5**  Round 6485 to the nearest hundred.

a) Locate the digit in the hundreds place, 4.

   6  4  8  5
     ↑

b) Consider the next digit to the right, 8.

   6  4  8  5
        ↑

c) Since that digit, 8, is 5 or higher, round 4 hundreds up to 5 hundreds.

d) Change all digits to the right of hundreds to zeros.

   6  5  0  0  ← This is the answer.

> Do Exercises 12–15.

**Round to the nearest hundred.**

**12.** 641        **13.** 759

**14.** 1871      **15.** 9325

**EXAMPLE 6**  Round 6485 to the nearest thousand.

a) Locate the digit in the thousands place, 6.

   6  4  8  5
   ↑

b) Consider the next digit to the right, 4.

   6  4  8  5
     ↑

c) Since that digit, 4, is 4 or lower, round down, meaning that 6 thousands stays as 6 thousands.

d) Change all digits to the right of thousands to zeros.

   6 0 0 0  ← This is the answer.

> Do Exercises 16–19.

**Round to the nearest thousand.**

**16.** 7896      **17.** 8459

**18.** 19,343     **19.** 68,500

---------------------------- *Caution!* ----------------------------

7000 is not a correct answer to Example 6. It is incorrect to round from the ones digit over, as follows:

6485 → 6490 → 6500 → 7000.

Note that 6485 is closer to 6000 than it is to 7000.

*Answers*

**8.** 140   **9.** 470   **10.** 240   **11.** 290
**12.** 600   **13.** 800   **14.** 1900   **15.** 9300
**16.** 8000   **17.** 8000   **18.** 19,000
**19.** 69,000

Sometimes rounding involves changing more than one digit in a number.

**EXAMPLE 7**  Round 78,595 to the nearest ten.

a)  Locate the digit in the tens place, 9.

  7 8,5 9 5
       ↑

b)  Consider the next digit to the right, 5.

  7 8,5 9 5
         ↑

c)  Since that digit, 5, is 5 or higher, round 9 tens to 10 tens. To carry this out, we think of 10 tens as 1 hundred + 0 tens and increase the hundreds digit by 1, to get 6 hundreds + 0 tens. We then write 6 in the hundreds place and 0 in the tens place.

d)  Change the digit to the right of the tens digit to zero.

  7 8,6 0 0 ← This is the answer.

Note that if we round this number to the nearest hundred, we get the same answer.

> Do Exercises 20 and 21.

## b) Estimating

Estimating can be done in many ways. In general, an estimate made by rounding to the nearest ten is more accurate than one rounded to the nearest hundred, and an estimate rounded to the nearest hundred is more accurate than one rounded to the nearest thousand, and so on.

**EXAMPLE 8**  Estimate this sum by first rounding to the nearest ten:

  78 + 49 + 31 + 85.

We round each number to the nearest ten. Then we add.

```
  7 8        8 0
  4 9        5 0
  3 1        3 0
+ 8 5      + 9 0
          ───────
           2 5 0 ← Estimated answer
```

> Do Exercises 22 and 23.

**EXAMPLE 9**  Estimate the difference by first rounding to the nearest thousand: 9324 − 2849.

We have

```
  9 3 2 4      9 0 0 0
− 2 8 4 9    − 3 0 0 0
            ──────────
             6 0 0 0 ← Estimated answer
```

> Do Exercises 24 and 25.

**20.** Round 48,968 to the nearest ten, hundred, and thousand.

**21.** Round 269,582 to the nearest ten, hundred, and thousand.

**22.** Estimate the sum by first rounding to the nearest ten. Show your work.
```
    7 4
    2 3
    3 5
  + 6 6
```

**23.** Estimate the sum by first rounding to the nearest hundred. Show your work.
```
    6 5 0
    6 8 5
    2 3 8
  + 1 6 8
```

**24.** Estimate the difference by first rounding to the nearest hundred. Show your work.
```
    9 2 8 5
  − 6 7 3 9
```

**25.** Estimate the difference by first rounding to the nearest thousand. Show your work.
```
    2 3,2 7 8
  − 1 1,6 9 8
```

*Answers*

**20.** 48,970; 49,000; 49,000
**21.** 269,580; 269,600; 270,000
**22.** 70 + 20 + 40 + 70 = 200
**23.** 700 + 700 + 200 + 200 = 1800
**24.** 9300 − 6700 = 2600
**25.** 23,000 − 12,000 = 11,000

In the sentence $7 - 5 = 2$, the equals sign indicates that $7 - 5$ is the *same* as 2. When we round to make an estimate, the outcome is rarely the same as the exact result. Thus we cannot use an equals sign when we round. Instead, we use the symbol $\approx$. This symbol means "**is approximately equal to**." In Example 9, for instance, we could write

$$9324 - 2849 \approx 6000.$$

**EXAMPLE 10**  Estimate the following product by first rounding to the nearest ten and then to the nearest hundred: $683 \times 457$.

*Nearest ten*

```
      6 8 0      683 ≈ 680
  ×   4 6 0      457 ≈ 460
    4 0 8 0 0
  2 7 2 0 0 0
  3 1 2,8 0 0
```

*Nearest hundred*

```
      7 0 0      683 ≈ 700
  ×   5 0 0      457 ≈ 500
  3 5 0,0 0 0
```

*Exact*

```
      6 8 3
  ×   4 5 7
      4 7 8 1
    3 4 1 5 0
  2 7 3 2 0 0
  3 1 2,1 3 1
```

We see that rounding to the nearest ten gives a better estimate than rounding to the nearest hundred.

Do Exercise 26.

**EXAMPLE 11**  Estimate the following quotient by first rounding to the nearest ten and then to the nearest hundred: $12{,}238 \div 175$.

*Nearest ten*

```
            6 8
  1 8 0 ) 1 2,2 4 0
          1 0 8 0
          1 4 4 0
          1 4 4 0
                0
```

*Nearest hundred*

```
            6 1
  2 0 0 ) 1 2,2 0 0
          1 2 0 0
            2 0 0
            2 0 0
              0
```

The exact answer is 69 R 163. Again we see that rounding to the nearest ten gives a better estimate than rounding to the nearest hundred.

Do Exercise 27.

26. Estimate the product by first rounding to the nearest ten and then to the nearest hundred. Show your work.

```
    8 3 7
  × 2 4 5
```

27. Estimate the quotient by first rounding to the nearest hundred. Show your work.

$64{,}534 \div 349$

*Answers*

**26.** $840 \times 250 = 210{,}000;$
$800 \times 200 = 160{,}000$
**27.** $64{,}500 \div 300 = 215$

The next two examples show how estimating can be used in making a purchase.

**EXAMPLE 12** *Microwave Ovens.* Ellen manages a small apartment building and is planning to purchase a new over-the-range microwave oven for each of the 12 units in the building. One model that she is considering costs $248. Estimate, by rounding to the nearest ten, the total cost of the purchase.

We have

$$
\begin{array}{r}
2\ 5\ 0 \\
\times\ \ \ \ 1\ 0 \\
\hline
2\ 5\ 0\ 0.
\end{array}
$$

The microwave ovens will cost about $2500.

Do Exercise 28.

**EXAMPLE 13** *Purchasing a New Car.* Jon and Joanna are shopping for a new car. They are considering buying a Chevrolet Cobalt Sedan LT 1LT. The base price of the car is $16,470. A 4-speed automatic transmission package can be added to this, as well as several other options, as shown in the chart below. Jon and Joanna want to stay within a budget of $20,000.

Estimate, by rounding to the nearest hundred, the cost of the Cobalt with the automatic transmission package and all other options and determine whether this will fit within their budget.

| CHEVROLET COBALT SEDAN LT 1LT | PRICE |
|---|---|
| Base price | $16,470 |
| 4-speed automatic transmission | $1,050 |
| Sport accessory package | $1,715 |
| Cruise control | $275 |
| Interior trim package | $505 |
| 4-wheel anti-lock brakes | $400 |
| Splash guards | $180 |

SOURCE: chevrolet.com

First, we list the base price of the car and then the cost of each of the options. We then round each number to the nearest hundred and add.

$$
\begin{array}{r}
1\ 6,4\ 7\ 0 \\
1\ 0\ 5\ 0 \\
1\ 7\ 1\ 5 \\
2\ 7\ 5 \\
5\ 0\ 5 \\
4\ 0\ 0 \\
+\ \ \ \ \ 1\ 8\ 0 \\
\end{array}
\qquad
\begin{array}{r}
1\ 6,5\ 0\ 0 \\
1\ 1\ 0\ 0 \\
1\ 7\ 0\ 0 \\
3\ 0\ 0 \\
5\ 0\ 0 \\
4\ 0\ 0 \\
+\ \ \ \ \ 2\ 0\ 0 \\
\hline
2\ 0,7\ 0\ 0 \ \leftarrow \text{Estimated cost}
\end{array}
$$

The estimated cost is $20,700. This exceeds Jon and Joanna's budget of $20,000, so they will have to forgo at least one option.

Do Exercises 29 and 30.

## (c) Order

We know that 2 is not the same as 5. We express this by the sentence $2 \neq 5$. We also know that 2 is less than 5. We symbolize this by the expression $2 < 5$. We can see this order on the number line: 2 is to the left of 5. The number 0 is the smallest whole number.

---

### ORDER OF WHOLE NUMBERS

For any whole numbers $a$ and $b$:

1. $a < b$ (read "$a$ is less than $b$") is true when $a$ is to the left of $b$ on the number line.
2. $a > b$ (read "$a$ is greater than $b$") is true when $a$ is to the right of $b$ on the number line.

We call $<$ and $>$ **inequality symbols**.

---

**EXAMPLE 14**  Use $<$ or $>$ for ☐ to write a true sentence: 7 ☐ 11.

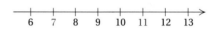

Since 7 is to the left of 11 on the number line, $7 < 11$.

**EXAMPLE 15**  Use $<$ or $>$ for ☐ to write a true sentence: 92 ☐ 87.

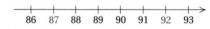

Since 92 is to the right of 87 on the number line, $92 > 87$.

A sentence like $8 + 5 = 13$ is called an **equation**. It is a *true* equation. The equation $4 + 8 = 11$ is a *false* equation. A sentence like $7 < 11$ is called an **inequality**. The sentence $7 < 11$ is a *true* inequality. The sentence $23 > 69$ is a *false* inequality.

> Do Exercises 31–36.

---

Use $<$ or $>$ for ☐ to write a true sentence. Draw the number line if necessary.

**31.** 8 ☐ 12

**32.** 12 ☐ 8

**33.** 76 ☐ 64

**34.** 64 ☐ 76

**35.** 217 ☐ 345

**36.** 345 ☐ 217

---

### STUDY TIPS

#### TEXTBOOK SUPPLEMENTS

Are you aware of all the supplements that exist for this textbook? See the preface for a description of each supplement including the *Student's Solutions Manual* and Video Resources on DVD Featuring Chapter Test Prep Videos.

---

*Answers*

**31.** $<$    **32.** $>$    **33.** $>$    **34.** $<$
**35.** $<$    **36.** $>$

**a** Round to the nearest ten.

**1.** 48 **2.** 532 **3.** 463 **4.** 8945

**5.** 731 **6.** 54 **7.** 895 **8.** 798

Round to the nearest hundred.

**9.** 146 **10.** 874 **11.** 957 **12.** 650

**13.** 9079 **14.** 4645 **15.** 32,839 **16.** 198,402

Round to the nearest thousand.

**17.** 5876 **18.** 4500 **19.** 7500 **20.** 2001

**21.** 45,340 **22.** 735,562 **23.** 373,405 **24.** 6,713,255

**b** Estimate each sum or difference by first rounding to the nearest ten. Show your work.

**25.**
```
   7 8
 + 9 2
```

**26.**
```
   6 2
   9 7
   4 6
 + 8 1
```

**27.**
```
   8 0 7 4
 - 2 3 4 7
```

**28.**
```
   6 7 3
 -   2 8
```

Estimate each sum by first rounding to the nearest ten. State if the given sum seems to be incorrect when compared to the estimate.

**29.**
```
   4 5
   7 7
   2 5
 + 5 6
 ─────
   3 4 3
```

**30.**
```
   4 1
   2 1
   5 5
 + 6 0
 ─────
   1 7 7
```

**31.**
```
   6 2 2
     7 8
     8 1
 + 1 1 1
 ───────
   9 3 2
```

**32.**
```
   8 3 6
   3 7 4
   7 9 4
 + 9 3 8
 ───────
   3 9 4 7
```

Estimate each sum or difference by first rounding to the nearest hundred. Show your work.

**33.**
```
   7 3 4 8
 + 9 2 4 7
```

**34.**
```
   5 6 8
   4 7 2
   9 3 8
 + 4 0 2
```

**35.**
```
   6 8 5 2
 - 1 7 4 8
```

**36.**
```
   9 4 3 8
 - 2 7 8 7
```

Estimate each sum by first rounding to the nearest hundred. State if the given sum seems to be incorrect when compared to the estimate.

37.
```
     2 1 6
        8 4
     7 4 5
  +  5 9 5
  ---------
   1 6 4 0
```

38.
```
       4 8 1
       7 0 2
       6 2 3
  +  1 0 4 3
  -----------
     1 8 4 9
```

39.
```
     7 5 0
     4 2 8
        6 3
  +  2 0 5
  ---------
   1 4 4 6
```

40.
```
     3 2 6
     2 7 5
     7 5 8
  +  9 4 3
  ---------
   2 3 0 2
```

Estimate each sum or difference by first rounding to the nearest thousand. Show your work.

41.
```
     9 6 4 3
     4 8 2 1
     8 9 4 3
  +  7 0 0 4
  -----------
```

42.
```
     7 6 4 8
     9 3 4 8
     7 8 4 2
  +  2 2 2 2
  -----------
```

43.
```
     9 2, 1 4 9
  -  2 2, 5 5 5
  -------------
```

44.
```
     8 4, 8 9 0
  -  1 1, 1 1 0
  -------------
```

Estimate each product by first rounding to the nearest ten. Show your work.

45.
```
       4 5
  ×    6 7
  --------
```

46.
```
       5 1
  ×    7 8
  --------
```

47.
```
       3 4
  ×    2 9
  --------
```

48.
```
       6 3
  ×    5 4
  --------
```

Estimate each product by first rounding to the nearest hundred. Show your work.

49.
```
       8 7 6
  ×    3 4 5
  ----------
```

50.
```
       3 5 5
  ×    2 9 9
  ----------
```

51.
```
       4 3 2
  ×    1 9 9
  ----------
```

52.
```
       7 8 9
  ×    4 3 4
  ----------
```

Estimate each quotient by first rounding to the nearest ten. Show your work.

53. $347 \div 73$

54. $454 \div 87$

55. $8452 \div 46$

56. $1263 \div 29$

Estimate each quotient by first rounding to the nearest hundred. Show your work.

57. $1165 \div 236$

58. $3641 \div 571$

59. $8358 \div 295$

60. $32,854 \div 748$

*Planning a Kitchen.* Perfect Kitchens offers custom kitchen packages with three choices for each of four items: cabinets, countertops, appliances, and flooring. The chart below lists the price for each choice. Customers design their kitchens by making one selection from each group of items.

| CABINETS | TYPE | PRICE |
|---|---|---|
| (a) | Oak | $7450 |
| (b) | Cherry | 8820 |
| (c) | Painted | 9630 |
| **COUNTERTOPS** | **TYPE** | **PRICE** |
| (d) | Laminate | $1595 |
| (e) | Solid surface | 2870 |
| (f) | Granite | 3528 |
| **APPLIANCES** | **PRICE RANGE** | **PRICE** |
| (g) | Low | $1540 |
| (h) | Medium | 3575 |
| (i) | High | 6245 |
| **FLOORING** | **TYPE** | **PRICE** |
| (j) | Vinyl | $625 |
| (k) | Travertine | 985 |
| (l) | Hardwood | 1160 |

**61.** Estimate the cost of remodeling a kitchen with choices (a), (d), (g), and (j) by rounding to the nearest hundred dollars.

**62.** Estimate the cost of a kitchen with choices (c), (f), (i), and (l) by rounding to the nearest hundred dollars.

**63.** Sara and Ben are planning to remodel their kitchen and have a budget of $17,700. Estimate by rounding to the nearest hundred dollars the cost of their kitchen remodeling project if they choose options (b), (e), (i), and (k). Can they afford their choices?

**64.** The Davidsons must make a final decision on the kitchen choices for their new home. The allotted kitchen budget is $16,000. Estimate by rounding to the nearest hundred dollars the kitchen cost if they choose options (a), (f), (h), and (l). Does their budget allotment cover the cost?

**65.** Suppose you are planning a new kitchen and must stay within a budget of $14,500. Decide on the options you would like and estimate the cost by rounding to the nearest hundred dollars. Does your budget support your choices?

**66.** Suppose you are planning a new kitchen and must stay within a budget of $18,500. Decide on the options you would like and estimate the cost by rounding to the nearest hundred dollars. Does your budget support your choices?

**67.** *Company Cars.* A publishing company buys a Honda Accord LX for each of its 112 sales representatives. Each car costs $21,160 with an additional $670 per car in destination charges.

Source: Honda

**a)** Estimate the total cost of the purchase by rounding the cost of each car, the destination charge, and the number of sales representatives to the nearest hundred.

**b)** Estimate the total cost of the purchase by rounding the cost of each car to the nearest thousand and the destination charge and the number of sales representatives to the nearest hundred.

**68.** *Airline Tickets.* A travel club of 176 people decides to fly from Chicago to Seattle. The cost of a round-trip ticket is $535.

**a)** Estimate the total cost of the trip by rounding the cost of the airfare and the number of travelers to the nearest ten.

**b)** Estimate the total cost of the trip by rounding the cost of the airfare and the number of travelers to the nearest hundred.

**69.** *Banquet Attendance.* Tickets to the annual awards banquet for the Riviera Swim Club cost $28 each. Ticket sales for the banquet totaled $2716. Estimate the number of people who attended the banquet by rounding the cost of a ticket to the nearest ten and the total sales to the nearest hundred.

**70.** *School Fundraiser.* For a school fundraiser, Charlotte sells trash bags at a price of $11 per roll. If her sales total $2211, estimate the number of rolls sold by rounding the price per roll to the nearest ten and the total sales to the nearest hundred.

**c**    Use < or > for ☐ to write a true sentence. Draw the number line if necessary.

**71.** 0 ☐ 17

**72.** 32 ☐ 0

**73.** 34 ☐ 12

**74.** 28 ☐ 18

**75.** 1000 ☐ 1001

**76.** 77 ☐ 117

**77.** 133 ☐ 132

**78.** 999 ☐ 997

**79.** 460 ☐ 17

**80.** 345 ☐ 456

**81.** 37 ☐ 11

**82.** 12 ☐ 32

*New Book Titles.* The number of new book titles published in the United States in each of three recent years is shown in the table below. Use this table to do Exercises 83 and 84.

| YEAR | NEW BOOK TITLES |
|------|-----------------|
| 2004 | 190,078 |
| 2005 | 172,000 |
| 2007 | 276,649 |

SOURCE: R. R. Bowker

**83.** Write an inequality to compare the number of new titles published in 2004 and in 2005.

**84.** Write an inequality to compare the number of new titles published in 2005 and in 2007.

**85.** *Wind-Power Capacity.* Wind-power capacity in the United States has increased from 1694 megawatts installed in 2001 to 5249 megawatts installed in 2007. Write an inequality to compare these numbers of megawatts of wind power installed.

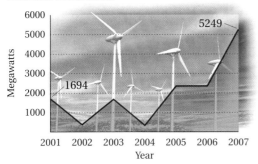

**U. S. Wind-Power Boom**

SOURCE: American Wind Energy Association

**86.** *Life Expectancy.* The life expectancy of a female in the United States in 2015 is predicted to be about 82 yr and that of a male about 76 yr. Write an inequality to compare these life expectancies.

**Life Expectancy in the United States**

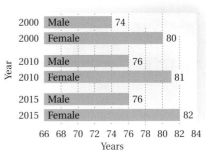

SOURCE: U.S. Census Bureau

# Skill Maintenance

Add. [1.2a]

**87.**
  6 7,7 8 9
+ 1 8,9 6 5

**88.**
  9 0 0 2
+ 4 5 8 7

Subtract. [1.3a]

**89.**
  6 7,7 8 9
− 1 8,9 6 5

**90.**
  9 0 0 2
− 4 5 8 7

Multiply. [1.4a]

**91.**
  4 6
× 3 7

**92.**
  3 0 6
×   5 8

Divide. [1.5a]

**93.** 328 ÷ 6

**94.** 4784 ÷ 23

# Synthesis

**95.–98.** Use a calculator to find the sums and the differences in each of Exercises 41–44. Then compare your answers with those found using estimation. Even when using a calculator it is possible to make an error if you press the wrong buttons, so it is a good idea to check by estimating.

# 1.7

# Solving Equations

## OBJECTIVES

**a** Solve simple equations by trial.

**b** Solve equations like $t + 28 = 54$, $28 \cdot x = 168$, and $98 \cdot 2 = y$.

Find a number that makes each sentence true.

**1.** $8 = 1 + \square$

**2.** $\square + 2 = 7$

**3.** Determine whether 7 is a solution of $\square + 5 = 9$.

**4.** Determine whether 4 is a solution of $\square + 5 = 9$.

Solve by trial.

**5.** $n + 3 = 8$

**6.** $x - 2 = 8$

**7.** $45 \div 9 = y$

**8.** $10 + t = 32$

*Answers*

**1.** 7   **2.** 5   **3.** No   **4.** Yes   **5.** 5
**6.** 10   **7.** 5   **8.** 22

## **a** Solutions by Trial

Let's find a number that we can put in the blank to make this sentence true:

$9 = 3 + \square$.

We are asking "9 is 3 plus what number?" The answer is 6.

$9 = 3 + 6$

Do Exercises 1 and 2.

A sentence with = is called an **equation**. A **solution** of an equation is a number that makes the sentence true. Thus, 6 is a solution of

$9 = 3 + \square$   because   $9 = 3 + 6$ is true.

However, 7 is not a solution of

$9 = 3 + \square$   because   $9 = 3 + 7$ is false.

Do Exercises 3 and 4.

We can use a letter in an equation instead of a blank:

$9 = 3 + n$.

We call $n$ a **variable** because it can represent any number. If a replacement for a variable makes an equation true, it is a **solution** of the equation.

---

**SOLUTIONS OF AN EQUATION**

A **solution of an equation** is a replacement for the variable that makes the equation true. When we find all the solutions, we say that we have **solved** the equation.

---

**EXAMPLE 1**   Solve $y + 12 = 27$ by trial.

We replace $y$ with several numbers.

If we replace $y$ with 13, we get a false equation:   $13 + 12 = 27$.
If we replace $y$ with 14, we get a false equation:   $14 + 12 = 27$.
If we replace $y$ with 15, we get a true equation:   $15 + 12 = 27$.

No other replacement makes the equation true, so the solution is 15.

**EXAMPLES**   Solve.

**2.** $7 + n = 22$
(7 plus what number is 22?)
The solution is 15.

**3.** $63 = 3 \cdot x$
(63 is 3 times what number?)
The solution is 21.

Do Exercises 5–8.

## b Solving Equations

We now begin to develop more efficient ways to solve certain equations. When an equation has a variable alone on one side and a calculation on the other side, we can find the solution by carrying out the calculation.

**EXAMPLE 4**  Solve: $x = 245 \times 34$.

To solve the equation, we carry out the calculation.

$$
\begin{array}{r}
2\ 4\ 5 \\
\times\ \ 3\ 4 \\
\hline
9\ 8\ 0 \\
7\ 3\ 5\ 0 \\
\hline
8\ 3\ 3\ 0
\end{array}
\qquad
\begin{aligned}
x &= 245 \times 34 \\
x &= 8330
\end{aligned}
$$

The solution is 8330.

> Do Exercises 9–12.

Look at the equation

$$x + 12 = 27.$$

We can get $x$ alone by subtracting 12 *on both sides*. Thus,

| | |
|---|---|
| $x + 12 - 12 = 27 - 12$ | Subtracting 12 on both sides |
| $x + 0 = 15$ | Carrying out the subtraction |
| $x = 15.$ | |

---

**SOLVING** $x + a = b$

To solve $x + a = b$, subtract $a$ on both sides.

---

If we can get an equation in a form with the variable alone on one side, we can "see" the solution.

**EXAMPLE 5**  Solve: $t + 28 = 54$.

We have

| | |
|---|---|
| $t + 28 = 54$ | |
| $t + 28 - 28 = 54 - 28$ | Subtracting 28 on both sides |
| $t + 0 = 26$ | |
| $t = 26.$ | |

To check the answer, we substitute 26 for $t$ in the original equation.

Check:
$$
\begin{array}{c}
t + 28 = 54 \\
\hline
26 + 28 \ \overset{?}{\ } \ 54 \\
54 \ \Big| \qquad \text{TRUE} \qquad \text{Since } 54 = 54 \text{ is true, 26 checks.}
\end{array}
$$

The solution is 26.

> Do Exercises 13 and 14.

Solve.

**9.** $346 \times 65 = y$

**10.** $x = 2347 + 6675$

**11.** $4560 \div 8 = t$

**12.** $x = 6007 - 2346$

Solve. Be sure to check.

**13.** $x + 9 = 17$

**14.** $77 = m + 32$

*Answers*

**9.** 22,490   **10.** 9022   **11.** 570   **12.** 3661
**13.** 8   **14.** 45

**EXAMPLE 6**  Solve: $182 = 65 + n$.

We have

$$182 = 65 + n$$
$$182 - 65 = 65 + n - 65 \qquad \text{Subtracting 65 on both sides}$$
$$117 = 0 + n \qquad \text{65 plus } n \text{ minus 65 is } 0 + n.$$
$$117 = n.$$

Check:
$$182 = 65 + n$$
$$182 \; ? \; 65 + 117$$
$$\mid 182 \qquad \text{TRUE}$$

**15.** Solve: $155 = t + 78$. Be sure to check.

The solution is 117.

Do Exercise 15.

**EXAMPLE 7**  Solve: $7381 + x = 8067$.

We have

$$7381 + x = 8067$$
$$7381 + x - 7381 = 8067 - 7381 \qquad \text{Subtracting 7381 on both sides}$$
$$x = 686.$$

Check:
$$7381 + x = 8067$$
$$7381 + 686 \; ? \; 8067$$
$$8067 \mid \qquad \text{TRUE}$$

The solution is 686.

Solve. Be sure to check.

**16.** $4566 + x = 7877$

**17.** $8172 = h + 2058$

Do Exercises 16 and 17.

We now learn to solve equations like $8 \cdot n = 96$. Look at

$$8 \cdot n = 96.$$

We can get $n$ alone by dividing by 8 *on both sides*. Thus,

$$\frac{8 \cdot n}{8} = \frac{96}{8} \qquad \text{Dividing by 8 on both sides}$$
$$n = 12. \qquad \text{8 times } n \text{ divided by 8 is } n.$$

To check the answer, we substitute 12 for $n$ in the original equation.

Check:
$$8 \cdot n = 96$$
$$8 \cdot 12 \; ? \; 96$$
$$96 \mid \qquad \text{TRUE}$$

Since $96 = 96$ is a true equation, 12 is the solution of the equation.

---

**SOLVING** $a \cdot x = b$

To solve $a \cdot x = b$, divide by $a$ on both sides.

---

**EXAMPLE 8**  Solve: $10 \cdot x = 240$.

We have

$$10 \cdot x = 240$$
$$\frac{10 \cdot x}{10} = \frac{240}{10} \qquad \text{Dividing by 10 on both sides}$$
$$x = 24.$$

Check:  
$$10 \cdot x = 240$$
$$\overline{10 \cdot 24 \; ? \; 240}$$
$$240 \; | \qquad \text{TRUE}$$

The solution is 24.

**EXAMPLE 9**  Solve: $5202 = 9 \cdot t$.

We have

$$5202 = 9 \cdot t$$
$$\frac{5202}{9} = \frac{9 \cdot t}{9} \qquad \text{Dividing by 9 on both sides}$$
$$578 = t.$$

Check:  
$$5202 = 9 \cdot t$$
$$\overline{5202 \; ? \; 9 \cdot 578}$$
$$| \; 5202 \qquad \text{TRUE}$$

The solution is 578.

| Do Exercises 18–20.

**EXAMPLE 10**  Solve: $14 \cdot y = 1092$.

We have

$$14 \cdot y = 1092$$
$$\frac{14 \cdot y}{14} = \frac{1092}{14} \qquad \text{Dividing by 14 on both sides}$$
$$y = 78.$$

The check is left to the student. The solution is 78.

**EXAMPLE 11**  Solve: $n \cdot 56 = 4648$.

We have

$$n \cdot 56 = 4648$$
$$\frac{n \cdot 56}{56} = \frac{4648}{56} \qquad \text{Dividing by 56 on both sides}$$
$$n = 83.$$

The check is left to the student. The solution is 83.

| Do Exercises 21 and 22.

Solve. Be sure to check.

**18.** $8 \cdot x = 64$

**19.** $144 = 9 \cdot n$

**20.** $5152 = 8 \cdot t$

Solve. Be sure to check.

**21.** $18 \cdot y = 1728$

**22.** $n \cdot 48 = 4512$

*Answers*

**18.** 8   **19.** 16   **20.** 644   **21.** 96   **22.** 94

**a**  Solve by trial.

**1.** $x + 0 = 14$

**2.** $x - 7 = 18$

**3.** $y \cdot 17 = 0$

**4.** $56 \div m = 7$

**b**  Solve. Be sure to check.

**5.** $x = 12{,}345 + 78{,}555$

**6.** $t = 5678 + 9034$

**7.** $908 - 458 = p$

**8.** $9007 - 5667 = m$

**9.** $16 \cdot 22 = y$

**10.** $34 \cdot 15 = z$

**11.** $t = 125 \div 5$

**12.** $w = 256 \div 16$

**13.** $13 + x = 42$

**14.** $15 + t = 22$

**15.** $12 = 12 + m$

**16.** $16 = t + 16$

**17.** $10 + x = 89$

**18.** $20 + x = 57$

**19.** $61 = 16 + y$

**20.** $53 = 17 + w$

**21.** $3 \cdot x = 24$

**22.** $6 \cdot x = 42$

**23.** $112 = n \cdot 8$

**24.** $162 = 9 \cdot m$

**25.** $3 \cdot m = 96$

**26.** $4 \cdot y = 96$

**27.** $715 = 5 \cdot z$

**28.** $741 = 3 \cdot t$

**29.** $8322 + 9281 = x$

**30.** $9281 - 8322 = y$

**31.** $47 + n = 84$

**32.** $56 + p = 92$

**33.** $45 \cdot 23 = x$

**34.** $23 \cdot 78 = y$

**35.** $x + 78 = 144$

**36.** $z + 67 = 133$

**37.** $6 \cdot p = 1944$

**38.** $4 \cdot w = 3404$

**39.** $5 \cdot x = 3715$

**40.** $9 \cdot x = 1269$

**41.** $x + 214 = 389$

**42.** $x + 221 = 333$

**43.** $567 + x = 902$

**44.** $438 + x = 807$

**45.** $234 \cdot 78 = y$

**46.** $10,534 \div 458 = q$

**47.** $18 \cdot x = 1872$

**48.** $19 \cdot x = 6080$

**49.** $40 \cdot x = 1800$

**50.** $20 \cdot x = 1500$

**51.** $2344 + y = 6400$

**52.** $9281 = 8322 + t$

**53.** $m = 7006 - 4159$

**54.** $n = 3004 - 1745$

**55.** $165 = 11 \cdot n$

**56.** $660 = 12 \cdot n$

**57.** $58 \cdot m = 11,890$

**58.** $233 \cdot x = 22,135$

**59.** $491 - 34 = y$

**60.** $512 - 63 = z$

## Skill Maintenance

Divide.  [1.5a]

**61.** $1283 \div 9$

**62.** $1278 \div 9$

**63.** $1 \ 7 \ ) \ 5 \ 6 \ 7 \ 8$

**64.** $1 \ 7 \ ) \ 5 \ 6 \ 8 \ 9$

Use $>$ or $<$ for ☐ to write a true sentence.  [1.6c]

**65.** $123 \ \square \ 789$

**66.** $342 \ \square \ 339$

**67.** $688 \ \square \ 0$

**68.** $0 \ \square \ 11$

**69.** Round 6,375,602 to the nearest thousand.  [1.6a]

**70.** Round 6,375,602 to the nearest ten.  [1.6a]

## Synthesis

Solve.

**71.** ▦ $23,465 \cdot x = 8,142,355$

**72.** ▦ $48,916 \cdot x = 14,332,388$

# Applications and Problem Solving

**a** Solve applied problems involving addition, subtraction, multiplication, or division of whole numbers.

## (a) A Problem-Solving Strategy

One of the most important ways in which we use mathematics is as a tool in solving problems. To solve a problem, we use the following five-step strategy.

---

**FIVE STEPS FOR PROBLEM SOLVING**

1. **Familiarize** yourself with the problem situation. If the problem is presented in words, this means to read and reread it carefully until you understand what you are being asked to find. Some or all of the following can also be helpful.

   a) Make a drawing, if it makes sense to do so.

   b) Make a written list of the known facts and a list of what you wish to find out.

   c) Assign a letter, or *variable*, to the unknown.

   d) Organize the information in a chart or a table.

   e) Find further information. Look up a formula, consult a reference book or an expert in the field, or do research on the Internet.

   f) Guess or estimate the answer and check your guess or estimate.

2. **Translate** the problem to an equation using the variable.

3. **Solve** the equation.

4. **Check** to see whether your possible solution actually fits the problem situation and is thus really a solution of the problem. Although you may have solved an equation, the solution of the equation might not be a solution of the original problem.

5. **State** the answer clearly using a complete sentence and appropriate units.

---

The first of these five steps, becoming familiar with the problem, is probably the most important. It provides a solid foundation for translating the problem to an equation that represents the situation accurately.

**EXAMPLE 1** *Community Colleges and New Jobs.* Community colleges are playing an increasingly large role in educating America's work force. The numbers of new jobs requiring a community college degree that will be created between 2004 and 2014 are shown in the table on the next page. Find the total number of new jobs for registered nurses, nursing aides and orderlies, and medical assistants.

| JOB | NUMBER |
|---|---|
| Registered nurse | 703,000 |
| Customer service | 471,000 |
| Nursing aide; orderly | 325,000 |
| Heavy-truck driver | 223,000 |
| Maintenance; repair | 202,000 |
| Medical assistant | 202,000 |
| Executive secretary/assistant | 192,000 |
| Sales representative | 187,000 |
| Carpenter | 186,000 |

SOURCES: U.S. Bureau of Labor Statistics;
College Board Center for Innovative Thought

1. **Familiarize.**  First, we assign a letter, or variable, to the number we wish to find. We let $n$ = the total number of new jobs created for registered nurses, nursing aides and orderlies, and medical assistants. Since we are combining numbers, we will add.

2. **Translate.**  We translate to an equation:

| Registered nurses' jobs | plus | Nursing aides' and orderlies' jobs | plus | Medical assistants' jobs | is | Total number of jobs |
|---|---|---|---|---|---|---|
| 703,000 | + | 325,000 | + | 202,000 | = | $n$. |

3. **Solve.**  We solve the equation by carrying out the addition.

$$
\begin{array}{r}
7\,0\,3,0\,0\,0 \\
3\,2\,5,0\,0\,0 \\
+\ 2\,0\,2,0\,0\,0 \\
\hline
1,2\,3\,0,0\,0\,0
\end{array}
$$

$$703{,}000 + 325{,}000 + 202{,}000 = n$$
$$1{,}230{,}000 = n$$

4. **Check.**  We check 1,230,000 in the original problem. There are many ways in which this can be done. For example, we can repeat the calculation. (We leave this to the student.) Another way is to check whether the answer is reasonable. In this case, we would expect the total to be greater than the number of each individual type of new job, and it is. We can also estimate the expected result by rounding. Here we round to the nearest hundred thousand:

$$703{,}000 + 325{,}000 + 202{,}000 \approx 700{,}000 + 300{,}000 + 200{,}000$$
$$= 1{,}200{,}000.$$

Since $1{,}200{,}000 \approx 1{,}230{,}000$, our answer seems reasonable. If the estimate had differed greatly from the possible solution found in step (3), we would suspect that the possible solution is incorrect.

5. **State.**  The total number of new jobs created for registered nurses, nursing aides and orderlies, and medical assistants between 2004 and 2014 is 1,230,000.

Do Exercises 1–3.

Refer to the table above to do Exercises 1–3.

1. Find the total number of new jobs that will be created for customer-service representatives, executive secretaries/assistants, and sales representatives.

2. Find the total number of new jobs that will be created for heavy-truck drivers, maintenance and repair workers, and carpenters.

3. Find the total number of new jobs listed in the table.

*Answers*

1. 850,000 new jobs    2. 611,000 new jobs
3. 2,691,000 new jobs

**EXAMPLE 2** *Checking Account Balance.* The balance in Francisco's checking account is $573. He uses his debit card to buy a juicer that costs $99. Find the new balance in his checking account.

1. **Familiarize.** We first make a drawing or at least visualize the situation. We let $B$ = the new balance in Francisco's checking account. We start with $573 and take away $99.

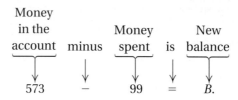

2. **Translate.** We translate to an equation:

$$
\underbrace{573}_{\substack{\text{Money} \\ \text{in the} \\ \text{account}}} \; \underbrace{-}_{\text{minus}} \; \underbrace{99}_{\substack{\text{Money} \\ \text{spent}}} \; \underbrace{=}_{\text{is}} \; \underbrace{B.}_{\substack{\text{New} \\ \text{balance}}}
$$

3. **Solve.** This equation tells us what to do. We subtract.

$$
\begin{array}{r}
\overset{\;\;\;16}{\overset{4\;\;\cancel{6}\;13}{\cancel{5}\;\cancel{7}\;\cancel{3}}} \\
-\;\;9\;9 \\
\hline
4\;7\;4
\end{array}
\qquad
\begin{array}{l}
573 - 99 = B \\
\phantom{5}474 = B
\end{array}
$$

4. **Check.** To check our answer of $474, we can repeat the calculation. We can also note that the answer should be less than the original amount, $573, and it is. Another way to check is to add the money spent, $99, to the new balance, $474: $99 + $474 = $573. We get the original balance, so the answer checks. We can also estimate:

$$\$573 - \$99 \approx \$570 - \$100 = \$470 \approx \$474.$$

This tells us that the answer is reasonable.

5. **State.** The new balance in Francisco's checking account is $474.

Do Exercise 4.

In the real world, problems may not be stated in written words. You must still become familiar with the situation before you can solve the problem.

**EXAMPLE 3** *Travel Distance.* Abigail is driving from Indianapolis to Salt Lake City to attend a family reunion. The distance from Indianapolis to Salt Lake City is 1634 mi. In the first two days, she travels 1154 mi to Denver. How much farther must she travel?

1. **Familiarize.** We first make a drawing or at least visualize the situation. We let $d$ = the remaining distance to Salt Lake City.

**4. Checking Account Balance.**
The balance in Laura's checking account is $457. She uses her debit card to buy a digital-picture frame that costs $49. Find the new balance in her checking account.

*Answer*

4. $408

2. **Translate.** We want to determine how many more miles Abigail must travel. We translate to an equation:

| Distance already traveled | plus | Distance to go | is | Total distance of trip |
|:---:|:---:|:---:|:---:|:---:|
| ↓ | ↓ | ↓ | ↓ | ↓ |
| 1154 | + | $d$ | = | 1634. |

3. **Solve.** To solve the equation, we subtract 1154 on both sides.

$$1154 + d = 1634$$
$$1154 + d - 1154 = 1634 - 1154$$
$$d = 480$$

$$\begin{array}{r} \overset{5\ \ 13}{1\,\cancel{6}\,\cancel{3}\,4} \\ -\ 1\ 1\ 5\ 4 \\ \hline 4\ 8\ 0 \end{array}$$

4. **Check.** We check our answer of 480 mi in the original problem. This number should be less than the total distance, 1634 mi, and it is. We can add the distance traveled, 1154, and the distance left to go, 480: $1154 + 480 = 1634$. We can also estimate:

$$1634 - 1154 \approx 1600 - 1200$$
$$= 400 \approx 480.$$

The answer, 480 mi, checks.

5. **State.** Abigail must travel 480 mi farther to Salt Lake City.

Do Exercise 5.

5. **Reading Assignment.** William has been assigned 234 pages of reading for his history class. He has read 86 pages. How many more pages does he have to read?

**EXAMPLE 4** *Total Cost of Chairs.* What is the total cost of 6 Adirondack chairs if each one costs $169?

1. **Familiarize.** We first make a drawing or at least visualize the situation. We let $C =$ the cost of 6 chairs.

$169    $169    $169    $169    $169    $169

*Answer*

5. 148 pages

**2. Translate.** We translate to an equation:

| Number of chairs | times | Cost of each chair | is | Total cost |
|---|---|---|---|---|
| 6 | × | $169 | = | C. |

**3. Solve.** This sentence tells us what to do. We multiply.

$$\begin{array}{r} 1\ 6\ 9 \\ \times \qquad 6 \\ \hline 1\ 0\ 1\ 4 \end{array}$$

$6 \times 169 = C$

$1014 = C$

**4. Check.** We have an answer, 1014, that is much greater than the cost of any individual chair, which is reasonable. We can repeat our calculation. We can also check by estimating:

$$6 \times 169 \approx 6 \times 170 = 1020 \approx 1014.$$

The answer checks.

**5. State.** The total cost of 6 chairs is $1014.

Do Exercise 6.

**6. Total Cost of Gas Grills.** What is the total cost of 14 gas grills, each with 520 sq in. of total cooking surface, if each one costs $398?

**EXAMPLE 5** *Area of an Oriental Rug.* The dimensions of the oriental rug in the Fosters' front hallway are 42 in. by 66 in. What is the area of the rug?

**1. Familiarize.** We first make a drawing to visualize the situation. We let $A$ = the area of the rug and use the formula for the area of a rectangle, $A = \text{length} \cdot \text{width} = l \cdot w$. Since we usually consider length to be larger than width, we will let $l = 66$ in. and $w = 42$ in.

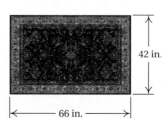

42 in.

66 in.

**2. Translate.** We substitute in the formula:

$$A = l \cdot w = 66 \cdot 42.$$

**3. Solve.** We carry out the multiplication.

$$\begin{array}{r} 6\ 6 \\ \times\ 4\ 2 \\ \hline 1\ 3\ 2 \\ 2\ 6\ 4\ 0 \\ \hline 2\ 7\ 7\ 2 \end{array}$$

$A = 66 \cdot 42$

$A = 2772$

**4. Check.** We can repeat the calculation. We can also round and estimate:

$$66 \times 42 \approx 70 \times 40 = 2800 \approx 2772.$$

The answer checks.

**5. State.** The area of the rug is 2772 sq in.

Do Exercise 7.

**7. Bed Sheets.** The dimensions of a flat sheet for a queen-size bed are 90 in. by 102 in. What is the area of the sheet?

*Answers*

**6.** $5572    **7.** 9180 sq in.

**EXAMPLE 6** *Packages of Paper Towels.* A paper-products company produces 3304 rolls of paper towels. How many 12-roll packages can be filled? How many rolls will be left over?

1. **Familiarize.** We first make a drawing. We let $n$ = the number of 12-roll packages that can be filled. The problem can be considered as repeated subtraction, taking successive sets of 12 rolls and putting them into $n$ packages.

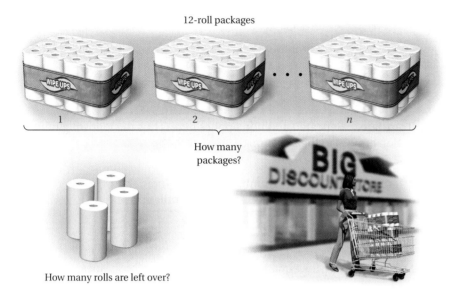

12-roll packages

How many packages?

How many rolls are left over?

2. **Translate.** We translate to an equation:

| Number of rolls | divided by | Number in each package | is | Number of packages |
|:---:|:---:|:---:|:---:|:---:|
| ↓ | ↓ | ↓ | ↓ | ↓ |
| 3304 | ÷ | 12 | = | $n$. |

3. **Solve.** We solve the equation by carrying out the division.

$$
\begin{array}{r}
2\,7\,5 \\
1\,2\,\overline{)\,3\,3\,0\,4} \\
2\,4\phantom{\,0\,4} \\
\hline
9\,0\phantom{\,4} \\
8\,4\phantom{\,4} \\
\hline
6\,4 \\
6\,0 \\
\hline
4
\end{array}
$$

$$3304 \div 12 = n$$
$$275\,\mathrm{R}\,4 = n$$

4. **Check.** We can check by multiplying the number of packages by 12 and adding the remainder, 4:

$$12 \cdot 275 = 3300,$$
$$3300 + 4 = 3304.$$

5. **State.** Thus, 275 twelve-roll packages of paper towels can be filled. There will be 4 rolls left over.

Do Exercise 8.

**8. Packages of Paper Towels.** The paper-products company in Example 6 also produces 6-roll packages. How many 6-roll packages can be filled with 2269 rolls of paper towels? How many rolls will be left over?

*Answer*

**8.** 378 packages with 1 roll left over

**EXAMPLE 7** *Automobile Mileage.* The 2009 Toyota Matrix gets 21 miles to the gallon (mpg) in city driving. How many gallons will it use in 3843 mi of city driving?

Source: Toyota

1. **Familiarize.** We first make a drawing. We let $g$ = the number of gallons of gasoline used in 3843 mi of city driving.

3843 mi to drive

2. **Translate.** Repeated addition applies here. Thus the following multiplication applies to the situation.

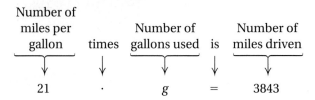

| Number of miles per gallon | times | Number of gallons used | is | Number of miles driven |
|---|---|---|---|---|
| 21 | · | $g$ | = | 3843 |

3. **Solve.** To solve the equation, we divide by 21 on both sides.

$$21 \cdot g = 3843$$
$$\frac{21 \cdot g}{21} = \frac{3843}{21}$$
$$g = 183$$

```
        1 8 3
   2 1 ) 3 8 4 3
        2 1
        ‾‾‾
        1 7 4
        1 6 8
        ‾‾‾‾‾
            6 3
            6 3
            ‾‾‾
              0
```

4. **Check.** To check, we multiply 183 by 21.

```
      1 8 3
   ×    2 1
   ‾‾‾‾‾‾‾‾
      1 8 3
    3 6 6 0
   ‾‾‾‾‾‾‾‾
    3 8 4 3
```

The answer checks.

5. **State.** The Toyota Matrix will use 183 gal of gasoline.

Do Exercise 9.

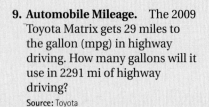

**9. Automobile Mileage.** The 2009 Toyota Matrix gets 29 miles to the gallon (mpg) in highway driving. How many gallons will it use in 2291 mi of highway driving?

Source: Toyota

## Multistep Problems

Sometimes we must use more than one operation to solve a problem, as in the following example.

**EXAMPLE 8** *Weight Loss.* To lose one pound, you must burn about 3500 calories in excess of what you already burn doing your regular daily activities. The chart on the next page shows how long a person must engage in several types of exercise in order to burn 100 calories. For how long would a person have to run at a brisk pace in order to lose one pound?

*Answer*

9. 79 gal

1. **Familiarize.** This is a multistep problem. We begin by visualizing the situation.

| ONE POUND 3500 CALORIES | | | |
|:---:|:---:|:---:|:---:|
| 100 cal 8 min | 100 cal 8 min | ... | 100 cal 8 min |

We will first find how many hundreds are in 3500. This will tell us how many times a person must run for 8 min in order to lose one pound. Then we will multiply to find the total number of minutes required for the weight loss.

We let $x$ = the number of hundreds in 3500.

2. **Translate.** We translate to an equation. Repeated addition applies here, so we will multiply.

$$
\underbrace{100}_{100} \text{ calories} \quad \underbrace{\text{times}}_{\cdot} \quad \underbrace{\text{How many hundreds}}_{x} \quad \underbrace{\text{is}}_{=} \quad \underbrace{3500}_{3500}
$$

3. **Solve.** We divide by 100 on both sides of the equation.

$$100 \cdot x = 3500$$
$$\frac{100 \cdot x}{100} = \frac{3500}{100}$$
$$x = 35$$

$$
\begin{array}{r}
3\,5 \\
1\,0\,0 \overline{)\,3\,5\,0\,0} \\
3\,0\,0 \\
\hline
5\,0\,0 \\
5\,0\,0 \\
\hline
0
\end{array}
$$

Thus, there are 35 hundreds in 3500. We know that running for 8 min will burn 100 calories. This must be done 35 times in order to burn the 3500 calories required to lose one pound. We let $t$ = the time it takes to lose one pound. Thus we have the following.

$$t = 35 \times 8$$
$$t = 280$$

$$
\begin{array}{r}
3\,5 \\
\times \quad 8 \\
\hline
2\,8\,0
\end{array}
$$

4. **Check.** $280 \div 8 = 35$, so there are 35 8's in 280 min, and $35 \cdot 100 = 3500$, the number of calories that must be burned in order to lose one pound. The answer checks.

5. **State.** You must run for 280 min, or 4 hr 40 min, at a brisk pace in order to lose one pound.

To burn 100 calories, you must:

- Run for 8 minutes at a brisk pace, or
- Swim for 2 minutes at a brisk pace, or
- Bicycle for 15 minutes at 9 mph, or
- Do aerobic exercises for 15 minutes, or
- Golf, walking, for 20 minutes, or
- Play tennis, singles, for 11 minutes

Do Exercise 10.

**10. Weight Loss.** Use the information in Example 8 to determine how long an individual must swim at a brisk pace in order to lose one pound.

*Answer*

**10.** 70 min, or 1 hr 10 min

You will find it helpful to look for the words, phrases, and concepts in the table below as you familiarize yourself with applied problems. They will be useful when you translate the problems to equations.

### KEYWORDS, PHRASES, AND CONCEPTS

| ADDITION (+) | SUBTRACTION (−) | MULTIPLICATION (·) | DIVISION (÷) |
|---|---|---|---|
| add | subtract | multiply | divide |
| added to | subtracted from | multiplied by | divided by |
| sum | difference | product | quotient |
| total | minus | times | repeated subtraction |
| plus | less than | of | missing factor |
| more than | decreased by | repeated addition | finding equal quantities |
| increased by | take away | rectangular arrays | |
| | how much more | | |

The following tips will also be helpful in problem solving.

### PROBLEM-SOLVING TIPS

1. Look for patterns when solving problems. Each time you study an example in the text, you may observe a pattern for problems found later in the exercise sets or in other practical situations.

2. When translating in mathematics, consider the dimensions of the variables and constants in the equation. The variables that represent length should all be in the same unit, those that represent money should all be in dollars or all in cents, and so on.

3. Make sure that units appear in the answer whenever appropriate and that you completely answer the original problem.

### STUDY TIPS

#### SOLVING APPLIED PROBLEMS

Don't be discouraged if, at first, you find the exercises in this section to be more challenging than those in earlier sections or if you have had difficulty working applied problems in the past. Your skill will improve with each problem you solve. After you have done your homework for this section, you might want to work extra problems from the text or from the online tutorial that accompanies the text. As you gain experience solving applied problems, you will find yourself becoming comfortable with them.

# Translating for Success

**1.** *Brick-Mason Expense.* A commercial contractor is building 30 two-unit condominiums in a retirement community. The brick-mason expense for each building is $10,860. What is the total cost of bricking the buildings?

**2.** *Heights.* Dean's sons are on the high school basketball team. Their heights are 73 in., 69 in., and 76 in. How much taller is the tallest son than the shortest son?

**3.** *Account Balance.* James has $423 in his checking account. Then he deposits $73 and uses his debit card for purchases of $76 and $69. How much is left in the account?

**4.** *Purchasing Camcorder.* A camcorder is on sale for $423. Jenny has only $69. How much more does she need to buy the camcorder?

**5.** *Purchasing Coffee Makers.* Sara purchases 8 coffee makers for the newly remodeled bed-and-breakfast hotel that she manages. If she pays $52 for each coffee maker, what is the total cost of her purchase?

The goal of these matching questions is to practice step (2), *Translate*, of the five-step problem-solving process. Translate each problem to an equation and select a correct translation from equations A–O.

**A.** $8 \cdot 52 = n$

**B.** $69 \cdot n = 76$

**C.** $73 - 76 - 69 = n$

**D.** $423 + 73 - 76 - 69 = n$

**E.** $30 \cdot 10{,}860 = n$

**F.** $15 \cdot n = 195$

**G.** $69 + n = 423$

**H.** $n = 10{,}860 - 300$

**I.** $n = 423 \div 69$

**J.** $30 \cdot n = 10{,}860$

**K.** $15 \cdot 195 = n$

**L.** $n = 52 - 8$

**M.** $69 + n = 76$

**N.** $15 \div 195 = n$

**O.** $52 + n = 60$

*Answers on page A-2*

**6.** *Hourly Rate.* Miller Auto Repair charges $52 per hour for labor. Jackson Auto Care charges $60 per hour. How much more does Jackson charge than Miller?

**7.** *College Band.* A college band with 195 members marches in a 15-row formation in the homecoming halftime performance. How many members are in each row?

**8.** *Shoe Purchase.* A professional football team purchases 15 pairs of shoes at $195 a pair. What is the total cost of this purchase?

**9.** *Loan Payment.* Kendra's uncle loans her $10,860, interest free, to buy a car. The loan is to be paid off in 30 payments. How much is each payment?

**10.** *College Enrollment.* At the beginning of the fall term, the total enrollment in Lakeview Community College was 10,860. By the end of the first two weeks, 300 students had withdrawn. How many students were then enrolled?

**a** Solve.

*Chicago's Tallest Buildings.* When construction is completed on the slender, 150-story Chicago Spire, it will be the tallest building in the United States, rising to a height of 2000 ft. The figure below shows the heights of the four tallest buildings in Chicago. (The Willis Tower was formerly known as the Sears Tower.) Use this information to do Exercises 1–4.

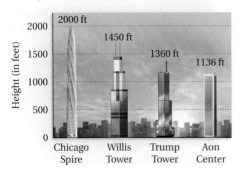

**Chicago's Tallest Buildings**

SOURCES: Chicago Historical Society; *USA Today* research

1. How much taller will the Chicago Spire be than the Willis Tower?

2. How much taller will the Chicago Spire be than the Trump Tower?

3. The Empire State Building in New York is 114 ft taller than the Aon Center. What is the height of the Empire State Building?

4. The Chicago Spire will be 638 ft taller than the Freedom Tower, which is under construction in New York. What is the height of the Freedom Tower?

5. *Caffeine Content.* An 8-oz serving of Red Bull energy drink contains 76 milligrams of caffeine. An 8-oz serving of brewed coffee contains 19 more milligrams of caffeine than the energy drink. How many milligrams of caffeine does the 8-oz serving of coffee contain?
   Source: The Mayo Clinic

6. *Caffeine Content.* Hershey's 6-oz milk chocolate almond bar contains 25 milligrams of caffeine. A 20-oz bottle of Coca-Cola has 32 more milligrams of caffeine than the Hershey bar. How many milligrams of caffeine does the 20-oz bottle of Coca-Cola have?
   Source: *National Geographic*, "Caffeine," by T. R. Reid, January 2005

7. A carpenter drills 216 holes in a rectangular array in a piece of plywood to make a pegboard. There are 12 holes in each row. How many rows are there?

8. Lou arranges 504 entries on a spreadsheet in a rectangular array that has 36 rows. How many entries are in each row?

**9.** *TV Watching.* On average, people age 12 and older spent 1704 hr watching TV in 2008. This is 202 hr more than in 2000. Determine the number of hours people in this age group spent watching TV in 2000.

**Source:** U.S. Census Bureau

**10.** *Drilling Activity.* In 1988, there were 554 rotary rigs drilling for crude oil in the United States. This was 257 more rigs than were active in 2007. Find the number of active rotary oil rigs in 2007.

**Source:** Energy Information Administration

**11.** *Boundaries between Countries.* The boundary between mainland United States and Canada including the Great Lakes is 3987 mi long. The length of the boundary between the United States and Mexico is 1933 mi. How much longer is the Canadian border?

**Source:** U.S. Geological Survey

**12.** *Longest Rivers.* The longest river in the world is the Nile in Egypt at about 4135 mi. The longest river in the United States is the Missouri–Mississippi at about 3860 mi. How much longer is the Nile?

**13.** *Pixels.* A high-definition television (HDTV) screen consists of small rectangular dots called *pixels*. How many pixels are there on a screen that has 1080 rows with 1920 pixels in each row?

Pixel

**14.** *Crossword.* The *USA Today* crossword puzzle is a rectangle containing 15 rows with 15 squares in each row. How many squares does the puzzle have altogether?

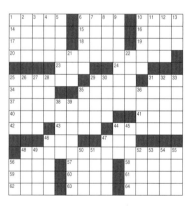

**15.** *Associate's Degrees.* About 697,000 associate's degrees were earned in the United States in 2005. Of these, 429,000 were earned by women. How many men earned associate's degrees in 2005?

**16.** *Bachelor's Degrees.* About 613,000 bachelor's degrees were earned by men in the United States in 2005. This is 215,000 fewer than the number of bachelor's degrees earned by women the same year. How many bachelor's degrees were earned by women in 2005?

**17.** There are 24 hr in a day and 7 days in a week. How many hours are there in a week?

**18.** There are 60 min in an hour and 24 hr in a day. How many minutes are there in a day?

*Housing Costs.* The graph below shows the average monthly rent for a one-bedroom apartment in several cities in June 2008. Use this graph to do Exercises 19–24.

**Average Rent for a One-Bedroom Apartment**

SOURCE: Apartments.com

**19.** How much higher is the average monthly rent in Houston than in Dallas/Fort Worth?

**20.** How much lower is the average monthly rent in Phoenix than in Atlanta?

**21.** Phil, Scott, and Julio plan to rent a one-bedroom apartment in Detroit immediately after graduation, sharing the rent equally. What average monthly rent can each of them expect to pay?

**22.** Maria and her sister Theresa share a one-bedroom apartment in Houston, dividing the average monthly rent equally between them. How much does each pay?

**23.** On average, how much rent would a tenant pay for a one-bedroom apartment in Atlanta during a 12-month period?

**24.** On average, how much rent would a tenant pay for a one-bedroom apartment in Dallas/Fort Worth during a 6-month period?

**25.** *Colonial Population.* Before the establishment of the U.S. Census in 1790, it was estimated that the Colonial population in 1780 was 2,780,400. This was an increase of 2,628,900 from the population in 1680. What was the Colonial population in 1680?

**Source:** *Time Almanac,* 2005

**26.** *Interstate Speed Limits.* The speed limit for passenger cars on interstate highways in rural areas in Montana is 75 mph. This is 10 mph faster than the speed limit for trucks on the same roads. What is the speed limit for trucks?

**27.** *Motorcycle Sales.* Motorcycle sales totaled 710,000 in the United States in 2000. Sales in 2006 exceeded this total by 480,000. How many motorcycles were sold in 2006?

Source: Motorcycle Industry Council

**28.** *Summer Olympics.* There were 43 events in the first modern Olympic games in Athens, Greece, in 1896. There were 259 more events in the 2008 Summer Olympics in Beijing, China. How many events were there in 2008?

Sources: *USA Today* research; beijing2008.cn

**29.** *Yard-Sale Profit.* Ruth made $312 at her yard sale and divided the money equally among her four grandchildren. How much did each child receive?

**30.** *Paper Measures.* A quire of paper consists of 25 sheets, and a ream of paper consists of 500 sheets. How many quires are in a ream?

**31.** *Parking Rates.* The most expensive parking in the United States occurs in midtown New York City, where the average rate is $585 per month. This is $545 per month more than in the city with the least expensive rate, Bakersfield, California. What is the average monthly parking rate in Bakersfield?

Source: Colliers International

**32.** *Trade Balance.* In 2006, foreign visitors spent $422,594,000,000 traveling in the United States, while Americans spent $342,845,000,000 traveling abroad. How much more was spent by visitors to the United States than by Americans traveling abroad?

Source: U.S. Census Bureau

**33.** *Refrigerator Purchase.* Gourmet Deli has a chain of 24 restaurants. It buys a commercial refrigerator for each store at a cost of $1019 each. Determine the total cost of the purchase.

**34.** *Microwave Purchase.* Bridgeway College is constructing new dorms, in which each room has a small kitchen. It buys 96 microwave ovens at $88 each. Determine the total cost of the purchase.

**35.** *"Seinfeld."* A local television station plans to air the 177 episodes of the long-running comedy series "Seinfeld." If the station airs 5 episodes per week, how many full weeks will pass before it must begin re-airing previously shown episodes? How many unaired episodes will be shown the following week before the previously aired episodes are rerun?

**36.** *"Everybody Loves Raymond."* The popular television comedy series "Everybody Loves Raymond" had 208 scripted episodes and 2 additional episodes consisting of clips from previous shows. A local television station plans to air the 208 scripted episodes, showing 5 episodes per week. How many full weeks will pass before it must begin re-airing episodes? How many unaired episodes will be shown the following week before the previously aired episodes are rerun?

**37.** *Automobile Mileage.* The 2008 Hyundai Tucson GLS gets 25 miles to the gallon (mpg) in highway driving. How many gallons will it use in 5900 mi of highway driving?
**Source:** Hyundai

**38.** *Automobile Mileage.* The 2008 Volkswagen Jetta (5 cylinder) gets 21 miles to the gallon (mpg) in city driving. How many gallons will it use in 3465 mi of city driving?
**Source:** Volkswagen of America, Inc.

**39.** *Crossword.* The *Los Angeles Times* crossword puzzle is a rectangle containing 441 squares arranged in 21 rows. How many columns does the puzzle have?

**40.** *Mailing Labels.* A box of mailing labels contains 750 labels on 25 sheets. How many labels are on each sheet?

**41.** *High School Court.* The standard basketball court used by high school players has dimensions of 50 ft by 84 ft.
   **a)** What is its area?
   **b)** What is its perimeter?

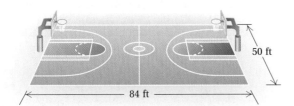

**42.** *College Court.* The standard basketball court used by college players has dimensions of 50 ft by 94 ft.
   **a)** What is its area?
   **b)** What is its perimeter?
   **c)** How much greater is the area of a college court than a high school court? (See Exercise 41.)

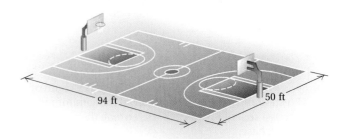

**43.** *Loan Payments.* Dana borrows $5928 for a used car. The loan is to be paid off in 24 equal monthly payments. How much is each payment (excluding interest)?

**44.** *Home Improvement Loan.* The Van Reken family borrows $7824 to build a sunroom on the back of their home. The loan is to be paid off in equal monthly payments of $163 (excluding interest). How many months will it take to pay off the loan?

**45.** *Map Drawing.* A map has a scale of 215 mi to the inch. How far apart *in reality* are two cities that are 3 in. apart on the map? How far apart *on the map* are two cities that, in reality, are 1075 mi apart?

**46.** *Map Drawing.* A map has a scale of 288 mi to the inch. How far apart *on the map* are two cities that, in reality, are 2016 mi apart? How far apart *in reality* are two cities that are 8 in. apart on the map?

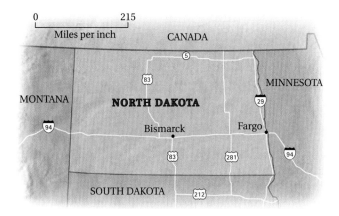

**47.** Copies of this book are usually shipped from the warehouse in cartons containing 24 books each. How many cartons are needed to ship 1344 books?

**48.** The H. J. Heinz Company ships 16-oz bottles of ketchup in cartons containing 12 bottles each. How many cartons are needed to ship 528 bottles of ketchup?

**49.** Elena buys 5 video games at $64 each and pays for them with $10 bills. How many $10 bills does it take?

**50.** Pedro buys 5 video games at $64 each and pays for them with $20 bills. How many $20 bills does it take?

**51.** The balance in Meg's bank account is $568. She uses her debit card for purchases of $46, $87, and $129. Then she deposits $94 in the account after returning a book. How much is left in her account?

**52.** The balance in Dylan's bank account is $749. He uses his debit card for purchases of $34 and $65. Then he makes a deposit of $123 from his paycheck. What is the new balance?

Refer to the information in Example 8 to do Exercises 53–56.

**53.** For how long must you do aerobic exercises in order to lose one pound?

**54.** For how long must you bicycle at 9 mph in order to lose one pound?

**55.** For how long must you play golf, walking, in order to lose one pound?

**56.** For how long must you play tennis singles in order to lose one pound?

**57.** *Seating Configuration.* The seats in the Boeing 737-500 airplanes in United Airlines' North American fleet are configured with 2 rows of 4 seats across in first class and 16 rows of 6 seats across in economy class. Determine the total seating capacity of the plane.
**Source:** United Airlines

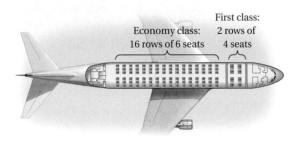

Economy class: 16 rows of 6 seats

First class: 2 rows of 4 seats

**58.** *Seating Configuration.* The seats in the Airbus 320 airplanes in United Airlines' North American fleet are configured with 3 rows of 4 seats across in first class and 21 rows of 6 seats across in economy class. Determine the total seating capacity of the plane.
**Source:** United Airlines

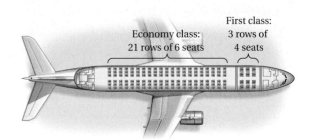

Economy class: 21 rows of 6 seats

First class: 3 rows of 4 seats

**59.** *Bones in the Hands and Feet.* There are 27 bones in each human hand and 26 bones in each human foot. How many bones are there in all in the hands and feet?

**60.** An office for adjunct instructors at a community college has 6 bookshelves, each of which is 3 ft wide. The office is moved to a new location that has dimensions of 16 ft by 21 ft. Is it possible for the bookshelves to be put side by side on the 16-ft wall?

## Skill Maintenance

Round 234,562 to the nearest: [1.6a]

**61.** Hundred.

**62.** Ten.

**63.** Thousand.

Estimate each sum or difference by rounding to the nearest thousand. [1.6b]

**64.** 2783 + 4602 + 5797 + 8111

**65.** 28,430 − 11,977

**66.** 5800 − 2100

**67.** 2100 + 5800

Estimate each product by rounding to the nearest hundred. [1.6b]

**68.** 787 · 363

**69.** 887 · 799

**70.** 10,362 · 4531

## Synthesis

**71.** ▦ *Speed of Light.* Light travels about 186,000 miles per second (mi/sec) in a vacuum such as in outer space. In ice it travels about 142,000 mi/sec, and in glass it travels about 109,000 mi/sec. In 18 sec, how many more miles will light travel in a vacuum than in ice? than in glass?

**72.** Carney Community College has 1200 students. Each instructor teaches 4 classes and each student takes 5 classes. There are 30 students and 1 instructor in each classroom. How many instructors are there at Carney Community College?

# 1.9 Exponential Notation and Order of Operations

## a Writing Exponential Notation

Consider the product $3 \cdot 3 \cdot 3 \cdot 3$. Such products occur often enough that mathematicians have found it convenient to create a shorter notation, called **exponential notation**, for them. For example,

$$\underbrace{3 \cdot 3 \cdot 3 \cdot 3}_{4 \text{ factors}} \text{ is shortened to } 3^{\overset{\displaystyle 4}{\uparrow}}. \leftarrow \text{exponent}$$
$$\qquad\qquad\qquad\qquad \lfloor \text{base}$$

We read exponential notation as follows.

| NOTATION | WORD DESCRIPTION |
|----------|------------------|
| $3^4$ | "three to the fourth power," or "the fourth power of three" |
| $5^3$ | "five-cubed," or "the cube of five," or "five to the third power," or "the third power of five" |
| $7^2$ | "seven squared," or "the square of seven," or "seven to the second power," or "the second power of seven" |

The wording "seven squared" for $7^2$ is derived from the fact that a square with side $s$ has area $A$ given by $A = s^2$.

An expression like $3 \cdot 5^2$ is read "three times five squared," or "three times the square of five."

**EXAMPLE 1**  Write exponential notation for $10 \cdot 10 \cdot 10 \cdot 10 \cdot 10$.

Exponential notation is $10^5$.  5 is the *exponent*.
10 is the *base*.

**EXAMPLE 2**  Write exponential notation for $2 \cdot 2 \cdot 2$.

Exponential notation is $2^3$.

Do Exercises 1–4.

Write exponential notation.
**1.** $5 \cdot 5 \cdot 5 \cdot 5$

**2.** $5 \cdot 5 \cdot 5 \cdot 5 \cdot 5 \cdot 5$

**3.** $10 \cdot 10$

**4.** $10 \cdot 10 \cdot 10 \cdot 10$

*Answers*
**1.** $5^4$   **2.** $5^6$   **3.** $10^2$   **4.** $10^4$

## b Evaluating Exponential Notation

We evaluate exponential notation by rewriting it as a product and then computing the product.

**EXAMPLE 3** Evaluate: $10^3$.

$$10^3 = 10 \cdot 10 \cdot 10 = 1000$$

-------- *Caution!* --------

$10^3$ does not mean $10 \cdot 3$.

**EXAMPLE 4** Evaluate: $5^4$.

$$5^4 = 5 \cdot 5 \cdot 5 \cdot 5 = 625$$

Do Exercises 5–8.

Do Exercises 5–8.

Evaluate.

5. $10^4$    6. $10^2$

7. $8^3$    8. $2^5$

## c Simplifying Expressions

Suppose we have a calculation like the following:

$$3 + 4 \cdot 8.$$

How do we find the answer? Do we add 3 to 4 and then multiply by 8, or do we multiply 4 by 8 and then add 3? In the first case, the answer is 56. In the second, the answer is 35. We agree to compute as in the second case:

$$3 + 4 \cdot 8 = 3 + 32 = 35.$$

Now consider the calculation

$$7 \cdot 14 - (12 + 18).$$

What do the parentheses mean? To deal with these questions, we must make some agreement regarding the order in which we perform operations. The rules are as follows.

| RULES FOR ORDER OF OPERATIONS |
| --- |
| 1. Do all calculations within parentheses ( ), brackets [ ], or braces { } before operations outside. |
| 2. Evaluate all exponential expressions. |
| 3. Do all multiplications and divisions in order from left to right. |
| 4. Do all additions and subtractions in order from left to right. |

It is worth noting that these are the rules that computers and most scientific calculators use to do computations.

**EXAMPLE 5** Simplify: $16 \div 8 \cdot 2$.

There are no parentheses or exponents, so we begin with the third step.

$$16 \div 8 \cdot 2 = 2 \cdot 2$$
$$= 4$$

Doing all multiplications and divisions in order from left to right

### Calculator Corner

**Exponential Notation**

Many calculators have a $y^x$ or $\wedge$ key for raising a base to a power. To find $16^3$, for example, we press $\boxed{1}\ \boxed{6}\ \boxed{y^x}\ \boxed{3}\ \boxed{=}$ or $\boxed{1}\ \boxed{6}\ \boxed{\wedge}\ \boxed{3}\ \boxed{=}$. The result is 4096.

**Exercises:** Use a calculator to find each of the following.

1. $3^5$

2. $5^6$

3. $12^4$

4. $2^{11}$

**EXAMPLE 6** Simplify: $7 \cdot 14 - (12 + 18)$.

$$
\begin{aligned}
7 \cdot 14 - (12 + 18) &= 7 \cdot 14 - 30 && \text{Carrying out operations} \\
& && \text{inside parentheses} \\
&= 98 - 30 && \text{Doing all multiplications} \\
& && \text{and divisions} \\
&= 68 && \text{Doing all additions} \\
& && \text{and subtractions}
\end{aligned}
$$

Do Exercises 9-12.

Do Exercises 9-12.

Simplify.

**9.** $93 - 14 \cdot 3$

**10.** $104 \div 4 + 4$

**11.** $25 \cdot 26 - (56 + 10)$

**12.** $75 \div 5 + (83 - 14)$

**EXAMPLE 7** Simplify and compare: $23 - (10 - 9)$ and $(23 - 10) - 9$.

We have

$$23 - (10 - 9) = 23 - 1 = 22;$$
$$(23 - 10) - 9 = 13 - 9 = 4.$$

We can see that $23 - (10 - 9)$ and $(23 - 10) - 9$ represent different numbers. Thus subtraction is not associative.

Do Exercises 13 and 14.

Simplify and compare.

**13.** $64 \div (32 \div 2)$ and $(64 \div 32) \div 2$

**14.** $(28 + 13) + 11$ and $28 + (13 + 11)$

**EXAMPLE 8** Simplify: $7 \cdot 2 - (12 + 0) \div 3 - (5 - 2)$.

$$
\begin{aligned}
7 \cdot 2 - (12 + 0) &\div 3 - (5 - 2) \\
&= 7 \cdot 2 - 12 \div 3 - 3 && \text{Carrying out operations} \\
& && \text{inside parentheses} \\
&= 14 - 4 - 3 && \text{Doing all multiplications and} \\
& && \text{divisions in order from left to right} \\
&= 10 - 3 && \text{Doing all additions and} \\
&= 7 && \text{subtractions in order from} \\
& && \text{left to right}
\end{aligned}
$$

Do Exercise 15.

**15.** Simplify:

$9 \times 4 - (20 + 4) \div 8 - (6 - 2)$.

**EXAMPLE 9** Simplify: $15 \div 3 \cdot 2 \div (10 - 8)$.

$$
\begin{aligned}
15 \div 3 \cdot 2 &\div (10 - 8) \\
&= 15 \div 3 \cdot 2 \div 2 && \text{Carrying out operations} \\
& && \text{inside parentheses} \\
&= 5 \cdot 2 \div 2 && \text{Doing all multiplications} \\
&= 10 \div 2 && \text{and divisions in order} \\
&= 5 && \text{from left to right}
\end{aligned}
$$

Do Exercises 16-18.

Simplify.

**16.** $5 \cdot 5 \cdot 5 + 26 \cdot 71 - (16 + 25 \cdot 3)$

**17.** $30 \div 5 \cdot 2 + 10 \cdot 20 + 8 \cdot 8 - 23$

**18.** $95 - 2 \cdot 2 \cdot 2 \cdot 5 \div (24 - 4)$

*Answers*

**9.** 51 **10.** 30 **11.** 584 **12.** 84 **13.** 4; 1
**14.** 52; 52 **15.** 29 **16.** 1880 **17.** 253
**18.** 93

**EXAMPLE 10**   Simplify: $4^2 \div (10 - 9 + 1)^3 \cdot 3 - 5$.

$$4^2 \div (10 - 9 + 1)^3 \cdot 3 - 5$$

$$= 4^2 \div (1 + 1)^3 \cdot 3 - 5 \qquad \text{Subtracting inside parentheses}$$

$$= 4^2 \div 2^3 \cdot 3 - 5 \qquad \text{Adding inside parentheses}$$

$$= 16 \div 8 \cdot 3 - 5 \qquad \text{Evaluating exponential expressions}$$

$$\left. \begin{array}{l} = 2 \cdot 3 - 5 \\ = 6 - 5 \end{array} \right\} \qquad \text{Doing all multiplications and divisions in order from left to right}$$

$$= 1 \qquad \text{Subtracting}$$

Simplify.

**19.** $5^3 + 26 \cdot 71 - (16 + 25 \cdot 3)$

**20.** $(1 + 3)^3 + 10 \cdot 20 + 8^2 - 23$

**21.** $81 - 3^2 \cdot 2 \div (12 - 9)$

Do Exercises 19–21.

**EXAMPLE 11**   Simplify: $2^9 \div 2^6 \cdot 2^3$.

$$2^9 \div 2^6 \cdot 2^3 = 512 \div 64 \cdot 8 \qquad \text{There are no parentheses. Evaluating exponential expressions}$$

$$\left. \begin{array}{l} = 8 \cdot 8 \\ = 64 \end{array} \right\} \qquad \text{Doing all multiplications and divisions in order from left to right}$$

**22.** Simplify: $2^3 \cdot 2^8 \div 2^9$.

Do Exercise 22.

## Calculator Corner

**Order of Operations**   To determine whether a calculator is programmed to follow the rules for order of operations, we can enter a simple calculation that requires using those rules. For example, we enter [3] [+] [4] [×] [2] [=]. If the result is 11, we know that the rules for order of operations have been followed. That is, the multiplication $4 \times 2 = 8$ was performed first and then 3 was added to produce a result of 11. If the result is 14, we know that the calculator performs operations as they are entered rather than following the rules for order of operations. That means, in this case, that 3 and 4 were added first to get 7 and then that sum was multiplied by 2 to produce the result of 14. For such calculators, we would have to enter the operations in the order in which we want them performed. In this case, we would press [4] [×] [2] [+] [3] [=].

Many calculators have parenthesis keys that can be used to enter an expression containing parentheses. To enter $5(4 + 3)$, for example, we press [5] [(] [4] [+] [3] [)] [=]. The result is 35.

**Exercises:**   Simplify.

**1.** $84 - 5 \cdot 7$

**2.** $80 + 50 \div 10$

**3.** $3^2 + 9^2 \div 3$

**4.** $4^4 \div 64 - 4$

**5.** $15 \cdot 7 - (23 + 9)$

**6.** $(4 + 3)^2$

*Answers*

**19.** 1880    **20.** 305    **21.** 75    **22.** 4

## Averages

In order to find the average of a set of numbers, we use addition and then division. For example, the average of 2, 3, 6, and 9 is found as follows.

$$\text{Average} = \underbrace{\frac{2 + 3 + 6 + 9}{4}}_{} = \frac{20}{4} = 5 \qquad \begin{array}{l}\text{— The number of addends is } 4.\\[2mm]\text{— Divide by } 4.\end{array}$$

The fraction bar acts as a pair of grouping symbols so

$$\frac{2 + 3 + 6 + 9}{4} \quad \text{is equivalent to} \quad (2 + 3 + 6 + 9) \div 4.$$

Thus we are using order of operations when we compute an average.

---

**AVERAGE**

The **average** of a set of numbers is the sum of the numbers divided by the number of addends.

---

**EXAMPLE 12** *Average Height of Waterfalls.* The heights of the four highest waterfalls in the world are shown in the figure below. Determine the average height of the four falls.

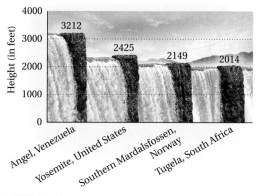

**World's Highest Waterfalls**

SOURCE: World Almanac

The average is given by

$$\frac{3212 + 2425 + 2149 + 2014}{4} = \frac{9800}{4} = 2450.$$

The average height of the world's four highest waterfalls is 2450 ft.

Do Exercise 23.

## d Removing Parentheses within Parentheses

When parentheses occur within parentheses, we can make them different shapes, such as [ ] (also called "brackets") and { } (also called "braces"). All of these have the same meaning. When parentheses occur within parentheses, computations in the innermost ones are to be done first.

**23. Average Number of Career Hits.** The numbers of career hits of five Hall of Fame baseball players are given in the bar graph below. Find the average number of career hits of all five.

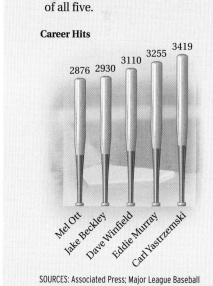

**Career Hits**

SOURCES: Associated Press; Major League Baseball

*Answer*

**23.** 3118 hits

**EXAMPLE 13** Simplify: $[25 - (4 + 3) \cdot 3] \div (11 - 7)$.

$$[25 - (4 + 3) \cdot 3] \div (11 - 7)$$

| | |
|---|---|
| $= [25 - 7 \cdot 3] \div (11 - 7)$ | Doing the calculations in the innermost parentheses first |
| $= [25 - 21] \div (11 - 7)$ | Doing the multiplication in the brackets |
| $= 4 \div 4$ | Subtracting |
| $= 1$ | Dividing |

**EXAMPLE 14** Simplify: $16 \div 2 + \{40 - [13 - (4 + 2)]\}$.

$$16 \div 2 + \{40 - [13 - (4 + 2)]\}$$

| | |
|---|---|
| $= 16 \div 2 + \{40 - [13 - 6]\}$ | Doing the calculations in the innermost parentheses first |
| $= 16 \div 2 + \{40 - 7\}$ | Again, doing the calculations in the innermost brackets |
| $= 16 \div 2 + 33$ | Subtracting inside the braces |
| $= 8 + 33$ | Doing all multiplications and divisions in order from left to right |
| $= 41$ | Adding |

Do Exercises 24 and 25.

Simplify.

**24.** $9 \times 5 + \{6 \div [14 - (5 + 3)]\}$

**25.** $[18 - (2 + 7) \div 3]$
$\quad - (31 - 10 \times 2)$

---

## STUDY TIPS

### PREPARING FOR AND TAKING TESTS

*You are probably ready to begin preparing for your first test. Here are some test-taking study tips.*

- **Make up your own test questions as you study.** After you have done your homework over a particular objective, write one or two questions on your own that you think might be on a test. You will be amazed at the insight this will provide.

- **Do an overall review of the chapter, focusing on the objectives and the examples.** This should be accompanied by a study of any class notes you have taken.

- **Do the exercises in the mid-chapter review and in the end-of-chapter review.** Check your answers at the back of the book. If you have trouble with an exercise, use the objective symbol as a guide to go back and do further study of that objective.

- **Take the chapter test at the end of the chapter.** Check the answers and use the objective symbols at the back of the book as a reference for review.

- **When taking a test, read each question carefully. Try to do all the questions the first time through, but pace yourself.** Answer all the questions, and mark those to recheck if you have time at the end. Very often, your first hunch will be correct.

- **Write your test in a neat and orderly manner.** Doing so will allow you to check your work easily and will also help your instructor follow the steps you took in answering the test questions.

*Answers*

**24.** 46  **25.** 4

**1.9** Exercise Set

For Extra Help

MyMathLab

Math XL
PRACTICE

WATCH

DOWNLOAD

READ

REVIEW

**a** Write exponential notation.

**1.** $3 \cdot 3 \cdot 3 \cdot 3$

**2.** $2 \cdot 2 \cdot 2 \cdot 2 \cdot 2$

**3.** $5 \cdot 5$

**4.** $13 \cdot 13 \cdot 13$

**5.** $7 \cdot 7 \cdot 7 \cdot 7 \cdot 7$

**6.** $9 \cdot 9$

**7.** $10 \cdot 10 \cdot 10$

**8.** $1 \cdot 1 \cdot 1 \cdot 1$

**b** Evaluate.

**9.** $7^2$

**10.** $5^3$

**11.** $9^3$

**12.** $8^2$

**13.** $12^4$

**14.** $10^5$

**15.** $3^5$

**16.** $2^6$

**c** Simplify.

**17.** $12 + (6 + 4)$

**18.** $(12 + 6) + 18$

**19.** $52 - (40 - 8)$

**20.** $(52 - 40) - 8$

**21.** $1000 \div (100 \div 10)$

**22.** $(1000 \div 100) \div 10$

**23.** $(256 \div 64) \div 4$

**24.** $256 \div (64 \div 4)$

**25.** $(2 + 5)^2$

**26.** $2^2 + 5^2$

**27.** $(11 - 8)^2 - (18 - 16)^2$

**28.** $(32 - 27)^3 + (19 + 1)^3$

**29.** $16 \cdot 24 + 50$

**30.** $23 + 18 \cdot 20$

**31.** $83 - 7 \cdot 6$

**32.** $10 \cdot 7 - 4$

**33.** $10 \cdot 10 - 3 \cdot 4$

**34.** $90 - 5 \cdot 5 \cdot 2$

**35.** $4^3 \div 8 - 4$

**36.** $8^2 - 8 \cdot 2$

**37.** $17 \cdot 20 - (17 + 20)$

**38.** $1000 \div 25 - (15 + 5)$

**39.** $6 \cdot 10 - 4 \cdot 10$

**40.** $3 \cdot 8 + 5 \cdot 8$

**41.** $300 \div 5 + 10$

**42.** $144 \div 4 - 2$

**43.** $3 \cdot (2 + 8)^2 - 5 \cdot (4 - 3)^2$

**44.** $7 \cdot (10 - 3)^2 - 2 \cdot (3 + 1)^2$

**45.** $4^2 + 8^2 \div 2^2$

**46.** $6^2 - 3^4 \div 3^3$

**47.** $10^3 - 10 \cdot 6 - (4 + 5 \cdot 6)$

**48.** $7^2 + 20 \cdot 4 - (28 + 9 \cdot 2)$

**49.** $6 \cdot 11 - (7 + 3) \div 5 - (6 - 4)$

**50.** $8 \times 9 - (12 - 8) \div 4 - (10 - 7)$

**51.** $120 - 3^3 \cdot 4 \div (5 \cdot 6 - 6 \cdot 4)$

**52.** $80 - 2^4 \cdot 15 \div (7 \cdot 5 - 45 \div 3)$

**53.** $2^3 \cdot 2^8 \div 2^6$

**54.** $2^7 \div 2^5 \cdot 2^4 \div 2^2$

**55.** Find the average of $64, $97, and $121.

**56.** Find the average of four test grades of 86, 92, 80, and 78.

**57.** Find the average of 320, 128, 276, and 880.

**58.** Find the average of $1025, $775, $2062, $942, and $3721.

**d** Simplify.

**59.** $8 \times 13 + \{42 \div [18 - (6 + 5)]\}$

**60.** $72 \div 6 - \{2 \times [9 - (4 \times 2)]\}$

**61.** $[14 - (3 + 5) \div 2] - [18 \div (8 - 2)]$

**62.** $[92 \times (6 - 4) \div 8] + [7 \times (8 - 3)]$

**63.** $(82 - 14) \times [(10 + 45 \div 5) - (6 \cdot 6 - 5 \cdot 5)]$

**64.** $(18 \div 2) \cdot \{[(9 \cdot 9 - 1) \div 2] - [5 \cdot 20 - (7 \cdot 9 - 2)]\}$

**65.** $4 \times \{(200 - 50 \div 5) - [(35 \div 7) \cdot (35 \div 7) - 4 \times 3]\}$

**66.** $15(23 - 4 \cdot 2)^3 \div (3 \cdot 25)$

**67.** $\{[18 - 2 \cdot 6] - [40 \div (17 - 9)]\} + \{48 - 13 \times 3 + [(50 - 7 \cdot 5) + 2]\}$

**68.** $(19 - 2^4)^5 - (141 \div 47)^2$

## Skill Maintenance

Solve.  [1.7b]

**69.** $x + 341 = 793$

**70.** $4197 + x = 5032$

**71.** $7 \cdot x = 91$

**72.** $1554 = 42 \cdot y$

**73.** $3240 = y + 898$

**74.** $6000 = 1102 + t$

**75.** $25 \cdot t = 625$

**76.** $10{,}000 = 100 \cdot t$

Solve.  [1.8a]

**77.** *Colorado.*  The state of Colorado is roughly the shape of a rectangle that is 273 mi by 382 mi. What is its area?

**78.** On a long four-day trip, a family bought the following amounts of gasoline for their motor home:

23 gallons,     24 gallons,

26 gallons,     25 gallons.

How much gasoline did they buy in all?

## Synthesis

Each of the answers in Exercises 79–81 is incorrect. First find the correct answer. Then place as many parentheses as needed in the expression in order to make the incorrect answer correct.

**79.** $1 + 5 \cdot 4 + 3 = 36$

**80.** $12 \div 4 + 2 \cdot 3 - 2 = 2$

**81.** $12 \div 4 + 2 \cdot 3 - 2 = 4$

**82.** Use one occurrence each of 1, 2, 3, 4, 5, 6, 7, 8, and 9 and any of the symbols $+$, $-$, $\cdot$, $\div$, and $(\ )$ to represent 100.

# Summary and Review

## Key Terms and Properties

digit, p. 2
periods, p. 2
place-value chart, p. 2
whole numbers, p. 3
natural numbers, p. 3
sum, p. 9
addend, p. 9
additive identity, p. 10
perimeter, p. 11

minuend, p. 14
subtrahend, p. 14
difference, p. 14
factor, p. 19
product, p. 19
multiplicative identity, p. 20
dividend, p. 26
divisor, p. 26
quotient, p. 26

remainder, p. 28
rounding, p. 37
equation, p. 42
inequality, p. 42
solution of an equation, p. 48
exponential notation, p. 71
base, p. 71
exponent, p. 71

| | |
|---|---|
| *Associative Law of Addition:* | $a + (b + c) = (a + b) + c$ |
| *Commutative Law of Addition:* | $a + b = b + a$ |
| *Associative Law of Multiplication:* | $a \cdot (b \cdot c) = (a \cdot b) \cdot c$ |
| *Commutative Law of Multiplication:* | $a \cdot b = b \cdot a$ |

## Concept Reinforcement

Determine whether each statement is true or false.

_____ **1.** $a > b$ is true when $a$ is to the right of $b$ on the number line.   [1.6c]

_____ **2.** Any nonzero number divided by itself is 1.   [1.5a]

_____ **3.** For any whole number $a$, $a \div 0 = 0$.   [1.5a]

_____ **4.** Every equation is true.   [1.7a]

_____ **5.** The rules for order of operations tell us to multiply and divide before adding and subtracting.   [1.9c]

_____ **6.** The average of three numbers is the middle number.   [1.9c]

## Important Concepts

**Objective 1.1a**   Give the meaning of digits in standard notation.

**Example**   What does the digit 7 mean in 2,379,465?

2,3 7 9,465

7 means 7 ten thousands.

**Practice Exercise**

**1.** What does the digit 2 mean in 432,079?

**Objective 1.2a**   Add whole numbers.

**Example**   Add: 7368 + 3547.

$$
\begin{array}{r}
\overset{1\ \ 1}{7\ 3\ 6\ 8} \\
+\ 3\ 5\ 4\ 7 \\
\hline
1\ 0,9\ 1\ 5
\end{array}
$$

**Practice Exercise**

**2.** Add: 36,047 + 29,255.

**Objective 1.3a**  Subtract whole numbers.

**Example**  Subtract: 8045 − 2897.

$$
\begin{array}{r}
\overset{7}{\cancel{8}}\ \overset{9}{\cancel{0}}\ \overset{\overset{13}{\cancel{3}}}{\cancel{4}}\ \overset{15}{\cancel{5}} \\
-\ 2\ 8\ 9\ 7 \\
\hline
5\ 1\ 4\ 8
\end{array}
$$

**Practice Exercise**

**3.** Subtract: 4805 − 1568.

---

**Objective 1.4a**  Multiply whole numbers.

**Example**  Multiply: 57 × 315.

$$
\begin{array}{r}
\overset{1}{\phantom{0}}\ \overset{\overset{2}{3}}{\phantom{0}} \\
3\ 1\ 5 \\
\times\ \ \ 5\ 7 \\
\hline
2\ 2\ 0\ 5\ \leftarrow 315 \times 7 \\
1\ 5\ 7\ 5\ 0\ \leftarrow 315 \times 50 \\
\hline
1\ 7{,}9\ 5\ 5
\end{array}
$$

**Practice Exercise**

**4.** Multiply: 329 × 684.

---

**Objective 1.5a**  Divide whole numbers.

**Example**  Divide: 6463 ÷ 26.

$$
\begin{array}{r}
2\ 4\ 8 \\
26\,\overline{)\ 6\ 4\ 6\ 3} \\
5\ 2\phantom{\ 0\ 0} \\
\hline
1\ 2\ 6\phantom{\ 0} \\
1\ 0\ 4\phantom{\ 0} \\
\hline
2\ 2\ 3 \\
2\ 0\ 8 \\
\hline
1\ 5
\end{array}
$$

The answer is 248 R 15.

**Practice Exercise**

**5.** Divide: 8519 ÷ 27.

---

**Objective 1.6a**  Round to the nearest ten, hundred, or thousand.

**Example**  Round 6471 to the nearest hundred.

6 4 7 1

The digit 4 is in the hundreds place. We consider the next digit to the right. Since the digit, 7, is 5 or higher, we round 4 hundreds up to 5 hundreds. Then we change all digits to the right of the hundreds digit to zeros. The answer is 6500.

**Example**  Round to the nearest thousand.

6 4 7 1

The digit 6 is in the thousands place. We consider the next digit to the right. Since the digit, 4, is 4 or lower, we round 6 thousands down, meaning that 6 thousands stays as 6 thousands. Change all digits to the right of the thousands digit to zeros. The answer is 6000.

**Practice Exercises**

**6.** Round 36,468 to the nearest hundred.

**7.** Round 36,468 to the nearest thousand.

**Objective 1.6c** Use < or > for ☐ to write a true sentence in a situation like 6 ☐ 10.

| **Example** Use < or > for ☐ to write a true sentence: 34 ☐ 29. <br> Since 34 is to the right of 29 on the number line, <br>      34 > 29. | **Practice Exercise** <br> **8.** Use < or > for ☐ to write a true sentence: <br>      78 ☐ 81. |
|---|---|

**Objective 1.7b** Solve equations like $t + 28 = 54$, $28 \cdot x = 168$, and $98 \cdot 2 = y$.

| **Example** Solve: $y + 12 = 27$. <br>      $y + 12 = 27$ <br> $y + 12 - 12 = 27 - 12$ <br>      $y + 0 = 15$ <br>        $y = 15$ <br> The solution is 15. | **Practice Exercise** <br> **9.** Solve: $24 \cdot x = 864$. |
|---|---|

**Objective 1.9b** Evaluate exponential notation.

| **Example** Evaluate: $5^4$. <br>      $5^4 = 5 \cdot 5 \cdot 5 \cdot 5 = 625$ | **Practice Exercise** <br> **10.** Evaluate: $6^3$. |
|---|---|

# Review Exercises

The review exercises that follow are for practice. Answers are given at the back of the book. If you miss an exercise, restudy the objective indicated in red next to the exercise or on the direction line that precedes it.

**1.** What does the digit 8 mean in 4,678,952? [1.1a]

**2.** In 13,768,940, what digit tells the number of millions? [1.1a]

Write expanded notation. [1.1b]

**3.** 2793              **4.** 56,078

**5.** 4,007,101

Write a word name. [1.1c]

**6.** 67,819          **7.** 2,781,427

**8.** *Toyoto Sales.* In the first six months of 2008, Toyota Motor Corporation sold 4,817,941 vehicles globally. Write a word name for 4,817,941.
Source: *Indianapolis Star*, July 24, 2008

Write standard notation. [1.1c]

**9.** Four hundred seventy-six thousand, five hundred eighty-eight

**10.** *Subway Ridership.* Ridership on the New York City Subway system totaled one billion, five hundred sixty-three million in 2007.
Source: Metropolitan Transit Authority

Add.  [1.2a]

**11.** $7304 + 6968$

**12.** $27{,}609 + 38{,}415$

**13.** $2703 + 4125 + 6004 + 8956$

**14.**
$$\begin{array}{r} 9\,1{,}4\,2\,6 \\ +\ \ 7{,}4\,9\,5 \\ \hline \end{array}$$

Subtract.  [1.3a]

**15.** $8045 - 2897$

**16.** $9001 - 7312$

**17.** $6003 - 3729$

**18.**
$$\begin{array}{r} 3\,7{,}4\,0\,5 \\ -\ 1\,9{,}6\,4\,8 \\ \hline \end{array}$$

Multiply.  [1.4a]

**19.** $17{,}000 \cdot 300$

**20.** $7846 \cdot 800$

**21.** $726 \cdot 698$

**22.** $587 \cdot 47$

**23.**
$$\begin{array}{r} 8\,3\,0\,5 \\ \times\ \ \ 6\,4\,2 \\ \hline \end{array}$$

Divide.  [1.5a]

**24.** $63 \div 5$

**25.** $80 \div 16$

**26.** $7\,\overline{)\,6\,3\,9\,4}$

**27.** $3073 \div 8$

**28.** $6\,0\,\overline{)\,2\,8\,6}$

**29.** $4266 \div 79$

**30.** $3\,8\,\overline{)\,1\,7{,}1\,7\,6}$

**31.** $1\,4\,\overline{)\,7\,0{,}1\,1\,2}$

**32.** $52{,}668 \div 12$

Round 345,759 to the nearest:  [1.6a]

**33.** Hundred.

**34.** Ten.

**35.** Thousand.

**36.** Hundred thousand.

Use $<$ or $>$ for ☐ to write a true sentence.  [1.6c]

**37.** $67$ ☐ $56$

**38.** $1$ ☐ $23$

Estimate each sum, difference, or product by first rounding to the nearest hundred. Show your work.  [1.6b]

**39.** $41{,}348 + 19{,}749$

**40.** $38{,}652 - 24{,}549$

**41.** $396 \cdot 748$

Solve.  [1.7b]

**42.** $46 \cdot n = 368$

**43.** $47 + x = 92$

**44.** $1 \cdot y = 58$

**45.** $24 = x + 24$

**46.** Write exponential notation: $4 \cdot 4 \cdot 4$.  [1.9a]

Evaluate. [1.9b]

**47.** $10^4$

**48.** $6^2$

Simplify.  [1.9c, d]

**49.** $8 \cdot 6 + 17$

**50.** $10 \cdot 24 - (18 + 2) \div 4 - (9 - 7)$

**51.** $(80 \div 16) \times [(20 - 56 \div 8) + (8 \cdot 8 - 5 \cdot 5)]$

**52.** Find the average of 157, 170, and 168.  [1.9c]

Solve.  [1.8a]

**53.** *Computer Workstation.*  Natasha has $196 and wants to buy a computer workstation for $698. How much more does she need?

**54.** Toni has $406 in her checking account. She is paid $78 for a part-time job and deposits that in her checking account. How much is then in her account?

**55.** *Lincoln-Head Pennies.*  In 1909, the first Lincoln-head pennies were minted. Seventy-three years later, these pennies were first minted with a decreased copper content. In what year was the copper content reduced?

**56.** A beverage company packed 228 cans of soda into 12-can cartons. How many cartons did they fill?

**57.** An apartment builder bought 13 gas stoves at $425 each and 13 refrigerators at $620 each. What was the total cost?

**58.** An apple farmer keeps bees in her orchard to help pollinate the apple blossoms. The bees from an average beehive can pollinate 30 surrounding trees during one growing season. A farmer has 420 trees. How many beehives does she need to pollinate all of them?

Source: Jordan Orchards, Westminster, PA

**59.** *Olympic Trampoline.* Shown below is an Olympic trampoline. Determine the area and the perimeter of the trampoline. [1.2b], [1.4b]

Source: International Trampoline Industry Association, Inc.

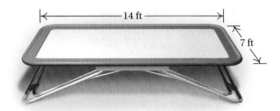

**60.** A chemist has 2753 mL of alcohol. How many 20-mL beakers can be filled? How much will be left over?

**61.** A family budgeted $7825 for food and clothing and $2860 for entertainment. The yearly income of the family was $38,283. How much of this income remained after these two allotments?

**62.** Simplify: $7 + (4 + 3)^2$. [1.9c]
**A.** 32 **B.** 56
**C.** 151 **D.** 196

**63.** Simplify: $7 + 4^2 + 3^2$. [1.9c]
**A.** 32 **B.** 56
**C.** 130 **D.** 196

**64.** $[46 - (4 - 2) \cdot 5] \div 2 + 4$ [1.9d]
**A.** 6 **B.** 20
**C.** 114 **D.** 22

## Synthesis

**65.** ▦ Determine the missing digit $d$. [1.4a]

$$\begin{array}{r} 9\ d \\ \times\quad d\ 2 \\ \hline 8\ 0\ 3\ 6 \end{array}$$

**66.** ▦ Determine the missing digits $a$ and $b$. [1.5a]

$$2\ b\ 1\ \overline{)\ 2\ 3\ 6{,}4\ 2\ 1}^{\,9\ a\ 1}$$

**67.** A mining company estimates that a crew must tunnel 2000 ft into a mountain to reach a deposit of copper ore. Each day, the crew tunnels about 500 ft. Each night, about 200 ft of loose rocks roll back into the tunnel. How many days will it take the mining company to reach the copper deposit? [1.8a]

# Understanding Through Discussion and Writing

**1.** Is subtraction associative? Why or why not? [1.3a]

**2.** Explain how estimating and rounding can be useful when shopping for groceries. [1.6b]

**3.** Write a problem for a classmate to solve. Design the problem so that the solution is "The driver still has 329 mi to travel." [1.8a]

**4.** Consider the expressions $9 - (4 \cdot 2)$ and $(3 \cdot 4)^2$. Are the parentheses necessary in each case? Explain. [1.9c]

CHAPTER
1

Test For Extra Help

Step-by-step test solutions are found on the Chapter Test Prep Videos available via the Video Resources on DVD, in *MyMathLab*, and on You Tube (search "BittingerBasicMathEI" and click on "Channels").

**1.** In the number 546,789, which digit tells the number of hundred thousands?

**2.** Write expanded notation: 8843.

**3.** Write a word name: 38,403,277.

Add.

**4.**
```
  6 8 1 1
+ 3 1 7 8
```

**5.**
```
  4 5,8 8 9
+ 1 7,9 0 2
```

**6.**
```
  1 2 3 9
    8 4 3
    3 0 1
+   7 8 2
```

**7.**
```
  6 2 0 3
+ 4 3 1 2
```

Subtract.

**8.**
```
  7 9 8 3
- 4 3 5 3
```

**9.**
```
  2 9 7 4
- 1 9 3 5
```

**10.**
```
  8 9 0 7
- 2 0 5 9
```

**11.**
```
  2 3,0 6 7
- 1 7,8 9 2
```

Multiply.

**12.**
```
  4 5 6 8
×       9
```

**13.**
```
  8 8 7 6
×   6 0 0
```

**14.**
```
    6 5
×   3 7
```

**15.**
```
    6 7 8
×   7 8 8
```

Divide.

**16.** $15 \div 4$

**17.** $420 \div 6$

**18.** $89 \overline{)8633}$

**19.** $44 \overline{)35,428}$

Round 34,528 to the nearest:

**20.** Thousand.

**21.** Ten.

**22.** Hundred.

Estimate each sum, difference, or product by first rounding to the nearest hundred. Show your work.

**23.**
```
  2 3,6 4 9
+ 5 4,7 4 6
```

**24.**
```
  5 4,7 5 1
- 2 3,6 4 9
```

**25.**
```
    8 2 4
×   4 8 9
```

Use < or > for ☐ to write a true sentence.

**26.** 34 ☐ 17

**27.** 117 ☐ 157

Solve.

**28.** $28 + x = 74$

**29.** $169 \div 13 = n$

**30.** $38 \cdot y = 532$

**31.** $381 = 0 + a$

Solve.

**32.** *Calorie Content.* An 8-oz serving of whole milk contains 146 calories. This is 63 calories more than the number of calories in an 8-oz serving of skim milk. How many calories are in an 8-oz serving of skim milk?
**Source:** *American Journal of Clinical Nutrition*

**33.** A box contains 5000 staples. How many staplers can be filled from the box if each stapler holds 250 staples?

**34.** *Largest States.* The following table lists the five largest states in terms of their land area. Find the total land area of these states.

| STATE | AREA (in square miles) |
|---|---|
| Alaska | 571,951 |
| Texas | 261,797 |
| California | 155,959 |
| Montana | 145,552 |
| New Mexico | 121,356 |

SOURCES: U.S. Department of Commerce; U.S. Census Bureau

**35.** *Pool Tables.* The Bradford™ pool table made by Brunswick Billiards comes in three sizes of playing area, 50 in. by 100 in., 44 in. by 88 in., and 38 in. by 76 in.
**Source:** Brunswick Billiards

  **a)** Determine the perimeter and the playing area of each table.
  **b)** By how much does the area of the largest table exceed the area of the smallest table?

**36.** *Hostess Ding Dongs®.* Hostess packages its Ding Dong snack cakes in 12-packs. How many 12-packs can it fill with 22,231 cakes? How many will be left over?

**37.** *Office Supplies.* Morgan manages the office of a small graphics firm. He buys 3 black inkjet cartridges at $15 each and 2 photo inkjet cartridges at $25 each. How much does the purchase cost?

**38.** Write exponential notation: $12 \cdot 12 \cdot 12 \cdot 12$.

Evaluate.

**39.** $7^3$

**40.** $10^5$

Simplify.

**41.** $35 - 1 \cdot 28 \div 4 + 3$

**42.** $10^2 - 2^2 \div 2$

**43.** $(25 - 15) \div 5$

**44.** $2^4 + 24 \div 12$

**45.** $8 \times \{(20 - 11) \cdot [(12 + 48) \div 6 - (9 - 2)]\}$

**46.** Find the average of 97, 99, 87, and 89.
  **A.** 93      **B.** 124      **C.** 186      **D.** 372

## Synthesis

**47.** An open cardboard container is 8 in. wide, 12 in. long, and 6 in. high. How many square inches of cardboard are used?

**48.** Use trials to find the single-digit number $a$ for which
$$359 - 46 + a \div 3 \times 25 - 7^2 = 339.$$

**49.** Cara spends $229 a month to repay her student loan. If she has already paid $9160 on the 10-yr loan, how many payments remain?

# Integers

## Real-World Application

The tallest mountain in the world, when measured from base to peak, is Mauna Kea (White Mountain) in Hawaii. From its base 19,684 ft below sea level in the Hawaiian Trough, it rises 33,480 ft. What is the elevation of the peak?

*Source: The Guinness Book of Records*

***This problem appears as Exercise 80 in Exercise Set 2.3.***

# 2.1

# The Integers

## OBJECTIVES

**a** State the integer that corresponds to a real-world situation.

**b** Graph integers on the number line.

**c** Determine which of two integers is greater and indicate which, using < or >.

**d** Find the absolute value of an integer.

**SKILL TO REVIEW**

Objective 1.6c: Use < or > for ☐ to write a true sentence in a situation like 6 ☐ 10.

Use < or > to write a true sentence.

**1.** 101 ☐ 99          **2.** 29 ☐ 32

In this section, we introduce the *integers*. To describe integers, we start with the whole numbers, 0, 1, 2, 3, and so on. For each number 1, 2, 3, and so on, we obtain a new number to the left of zero on the number line:

For the number 1, there will be an *opposite* number −1 (negative 1).

For the number 2, there will be an *opposite* number −2 (negative 2).

For the number 3, there will be an *opposite* number −3 (negative 3), and so on.

The **integers** consist of the whole numbers and these new numbers. We picture them on the number line as follows.

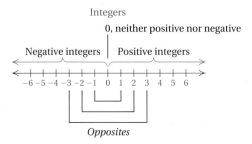

We call the numbers to the left of zero on the number line **negative integers**. The natural numbers are called **positive integers**. Zero is neither positive nor negative. We call −1 and 1 **opposites** of each other. Similarly, −2 and 2 are opposites, −3 and 3 are opposites, −100 and 100 are opposites, and 0 is its own opposite. Opposite pairs of numbers like −3 and 3 are the same distance from 0. The integers extend infinitely on the number line to the left and right of zero.

---

**INTEGERS**

The **integers:** . . . , −5, −4, −3, −2, −1, 0, 1, 2, 3, 4, 5, . . .

---

## a Integers and the Real World

Integers correspond to many real-world problems and situations. The following examples will help you get ready to translate problem situations that involve integers to mathematical language.

**To the student:**

At the front of the text, you will find a Student Organizer card. This pullout card will help you keep track of important dates and useful contact information. You can also use it to plan time for class, study, work, and relaxation. By managing your time wisely, you will provide yourself the best possible opportunity to be successful in this course.

*Answers*

*Skill to Review:*

1. >    2. <

**EXAMPLE 1** Tell which integer corresponds to this situation: The temperature is 4 degrees below zero.

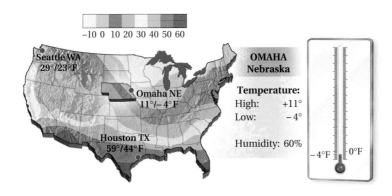

The integer −4 corresponds to the situation. The temperature is −4°.

**EXAMPLE 2** *Elevation.* Tell which integer corresponds to this situation: The shores of California's largest lake, the Salton Sea, are 227 ft below sea level.
**Source:** Salton Sea Authority

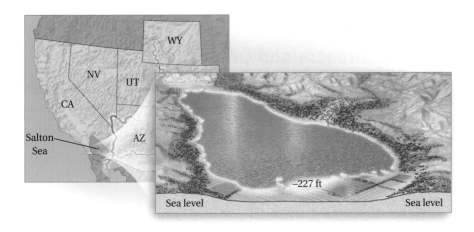

The integer −227 corresponds to the situation. The elevation is −227 ft.

**EXAMPLE 3** *Stock Price Change.* Tell which integers correspond to this situation: Hal owns a stock whose price decreased from $27 per share to $11 per share over a recent time period. He owns another stock whose price increased from $20 per share to $22 per share over the same time period.

The price of the first stock decreased $16 per share. The integer −16 corresponds to the decrease in the value of this stock. The price of the second stock increased $2 per share. The integer 2 represents the increase in the value of the second stock.

Do Exercises 1–5.

Tell which integers correspond to each situation.

1. **Temperature High and Low.** The highest recorded temperature in Nevada is 125°F on June 29, 1994, in Laughlin. The lowest recorded temperature in Nevada is 50°F below zero on January 8, 1937, in San Jacinto.
   **Source:** National Climatic Data Center, Asheville, NC, and Storm Phillips, STORMFAX, INC.

2. **Stock Decrease.** The price of a stock decreased from $41 per share to $38 per share over a recent period.

3. At 10 sec before liftoff, ignition occurs. At 148 sec after liftoff, the first stage is detached from the rocket.

4. The halfback gained 8 yd on first down. The quarterback was sacked for a 5-yd loss on second down.

5. A submarine dove 120 ft, rose 50 ft, and then dove 80 ft.

*Answers*

**1.** 125; −50  **2.** The integer −3 corresponds to the decrease in the value of the stock.
**3.** −10; 148  **4.** 8; −5  **5.** −120; 50; −80

## b) Graph of Integers

To **graph** a number means to find and mark its point on the number line.

**EXAMPLE 4**   Graph: 5.

We locate 5 on the number line and mark its point with a dot.

**EXAMPLE 5**   Graph: 0.

We locate 0 on the number line and mark its point with a dot.

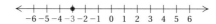

**EXAMPLE 6**   Graph: −3.

We locate −3 on the number line and mark its point with a dot.

Do Exercises 6–8.

## c) The Integers and Order

The integers are named in order on the number line. (See Section 1.6.) For any two numbers on the line, the one to the left is less than the one to the right.

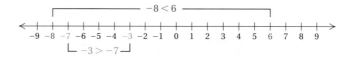

We use the symbol **<** to mean "**is less than.**" The sentence $-8 < 6$ means "$-8$ is less than 6." The symbol **>** means "**is greater than.**" The sentence $-3 > -7$ means "$-3$ is greater than $-7$."

**EXAMPLES**   Use either < or > for ☐ to write a true sentence.

**7.** $-7 \; \square \; 3$         Since $-7$ is to the left of 3, we have $-7 < 3$.

**8.** $6 \; \square \; -12$         Since 6 is to the right of $-12$, we have $6 > -12$.

**9.** $-18 \; \square \; -5$         Since $-18$ is to the left of $-5$, we have $-18 < -5$.

**10.** $25 \; \square \; 23$         Since 25 is to the right of 23, we have $25 > 23$.

**11.** $-8 \; \square \; 0$         Since $-8$ is to the left of 0, we have $-8 < 0$.

**12.** $-11 \; \square \; -13$         Since $-11$ is to the right of $-13$, we have $-11 > -13$.

Do Exercises 9–14.

Graph on the number line.

**6.** $-4$

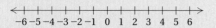

**7.** 1

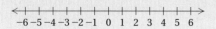

**8.** 4

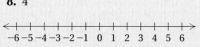

Use either < or > for ☐ to write a true sentence.

**9.** $-3 \; \square \; 7$     **10.** $-8 \; \square \; -5$

**11.** $7 \; \square \; -10$     **12.** $0 \; \square \; -15$

**13.** $79 \; \square \; 63$     **14.** $-9 \; \square \; -8$

*Answers*

**6.**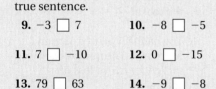

**7.**

**8.**

**9.** <     **10.** <     **11.** >     **12.** >
**13.** >     **14.** <

# d Absolute Value

From the number line, we see that numbers like 4 and −4 are the same distance from zero. Distance is always a nonnegative number. We call the distance of a number from zero the **absolute value** of the number.

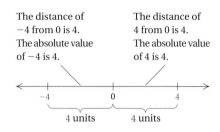

The distance of −4 from 0 is 4.
The absolute value of −4 is 4.

The distance of 4 from 0 is 4.
The absolute value of 4 is 4.

4 units     4 units

> ### ABSOLUTE VALUE
>
> The **absolute value** of a number is its distance from zero on the number line. We use the symbol $|x|$ to represent the absolute value of a number $x$.

**EXAMPLES** Find the absolute value.

**13.** $|-7|$     The distance of −7 from 0 is 7, so $|-7| = 7$.

**14.** $|12|$     The distance of 12 from 0 is 12, so $|12| = 12$.

**15.** $|0|$     The distance of 0 from 0 is 0, so $|0| = 0$.

**16.** $|45| = 45$

**17.** $|-101| = 101$

Do Exercises 15–20.

Find the absolute value.

**15.** $|8|$          **16.** $|0|$

**17.** $|-9|$          **18.** $|-84|$

**19.** $|157|$          **20.** $|-225|$

> To find the absolute value of a number:
>
> 1. If the number is positive or zero, the absolute value is the number itself.
> 2. If the number is negative, make it positive.

------- *Caution!* -------

Absolute value is a distance, so the absolute value of a number is always positive or zero. Absolute value cannot be negative.

*Answers*

**15.** 8    **16.** 0    **17.** 9    **18.** 84
**19.** 157    **20.** 225

**a**    Tell which integers correspond to each situation.

**1.** *Death Valley.*   With an elevation of 282 ft below sea level, Badwater Basin in Death Valley in California has the lowest elevation in the United Sates.
**Source:** Desert USA

**2.** *Pollution Fine.*   The Massey Energy Company, the nation's fourth largest coal producer, was fined $20 million for water pollution in 2008.
**Source:** Environmental Protection Agency

**3.** On Wednesday, the temperature was 24° above zero. On Thursday, it was 2° below zero.

**4.** A student deposited her tax refund of $750 in a savings account. Two weeks later, she withdrew $125 to pay lab fees.

**5.** *Highest and Lowest Temperatures.*   The highest temperature ever created on Earth was 950,000,000°F. The lowest temperature ever created was approximately 460°F below zero.
**Source:** *The Guinness Book of Records,* 2004

**6.** *Extreme Climate.*   Verkhoyansk, a river port in northeast Siberia, has the most extreme climate on the planet. Its average monthly winter temperature is 58.5°F below zero, and its average monthly summer temperature is 56.5°F.
**Source:** *The Guinness Book of Records,* 2004

**7.** In bowling, team A is 34 pins behind team B, and team B is 15 pins ahead of team C.

**8.** During a video game, Sara intercepted a missile worth 20 points, lost a starship worth 150 points, and captured a base worth 300 points.

**9.** The population of Detroit, Michigan, decreased by 39,868 residents from 2000 to 2003.
**Source:** U.S. Census Bureau

**10.** The population of Washington, DC, decreased by 8675 residents from 2000 to 2003.
**Source:** U.S. Census Bureau

**b**    Graph the number on the number line.

**11.** $-5$

**12.** $3$

**13.** $2$

**14.** $-1$

**c** Use either $<$ or $>$ for ☐ to write a true sentence.

**15.** 8 ☐ 0

**16.** 3 ☐ 0

**17.** $-8$ ☐ 3

**18.** 6 ☐ $-6$

**19.** 8 ☐ $-8$

**20.** 0 ☐ $-9$

**21.** $-10$ ☐ $-5$

**22.** $-4$ ☐ $-3$

**23.** $-5$ ☐ $-11$

**24.** $-3$ ☐ $-4$

**25.** $-6$ ☐ $-5$

**26.** $-10$ ☐ $-14$

**27.** $-7$ ☐ 0

**28.** $-3$ ☐ $-2$

**29.** 1 ☐ $-15$

**30.** 2 ☐ $-12$

**d** Find the absolute value.

**31.** $|-3|$

**32.** $|-6|$

**33.** $|18|$

**34.** $|0|$

**35.** $|325|$

**36.** $|-4|$

**37.** $|-29|$

**38.** $|217|$

**39.** $|-300|$

**40.** $|-47|$

**41.** $|53|$

**42.** $|-76|$

## Skill Maintenance

Add. [1.2a]

**43.**　9 1 8 2
　　　$+4\ 3\ 6\ 7$

**44.**　　2 7 8
　　　　$+8\ 2\ 9$

**45.**　3 2,0 4 7
　　　$+1\ 8,5\ 6\ 2$

Subtract. [1.3a]

**46.**　6 5 1
　　　$-4\ 3\ 2$

**47.**　1 2 0 7
　　　$-\ \ 9\ 4\ 8$

**48.**　4 3,0 0 4
　　　$-3\ 4,2\ 2\ 6$

## Synthesis

Use either $<$, $>$, or $=$ for ☐ to write a true sentence.

**49.** $|-5|$ ☐ $|-2|$

**50.** $|4|$ ☐ $|-7|$

**51.** $|-8|$ ☐ $|8|$

**52.** List in order from the least to the greatest: $-6, 7, -10, |-6|, 2^2, |3|, 1^6, -5, 0$.

# 2.2

# Addition of Integers

## OBJECTIVES

**a** Add integers without using the number line.

**b** Find the opposite, or additive inverse, of an integer.

---

**SKILL TO REVIEW**
Objective 1.3a: Subtract whole numbers.

Subtract.

**1.** $14 - 6$          **2.** $5 - 3$

---

Add using the number line.

**1.** $0 + (-3)$

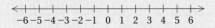

**2.** $1 + (-4)$

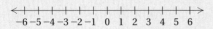

**3.** $-3 + (-2)$

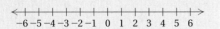

**4.** $-3 + 7$

**5.** $-2 + 2$

**6.** $-1 + (-3)$

---

*Answers*

*Skill to Review:*
**1.** 8     **2.** 2

*Margin Exercises:*
**1.** $-3$     **2.** $-3$     **3.** $-5$     **4.** 4
**5.** 0     **6.** $-4$

---

We now consider addition of integers. First, to gain an understanding, we add using the number line. Then we consider rules for addition.

### Addition on the Number Line

To find $a + b$, we start at 0, then move to $a$, and then move according to $b$.

a) If $b$ is positive, we move from $a$ to the right.

b) If $b$ is negative, we move from $a$ to the left.

c) If $b$ is 0, we stay at $a$.

**EXAMPLE 1** Add: $3 + (-5)$.

We start at 0 and move to 3. Then we move 5 units left since $-5$ is negative.

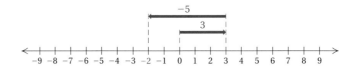

$3 + (-5) = -2$

**EXAMPLE 2** Add: $-4 + (-3)$.

We start at 0 and move to $-4$. Then we move 3 units left since $-3$ is negative.

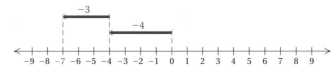

$-4 + (-3) = -7$

**EXAMPLE 3** Add: $-4 + 9$.

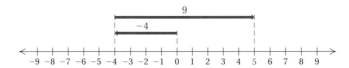

$-4 + 9 = 5$

**EXAMPLE 4** Add: $-6 + 0$.

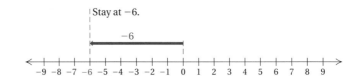

$-6 + 0 = -6$

Do Margin Exercises 1–6.

---

## (a) Addition Without the Number Line

You may have noticed some patterns in the preceding examples. These lead us to rules for adding without using the number line that are more efficient for adding larger or more complicated numbers.

---

### RULES FOR ADDITION OF INTEGERS

1. *Positive numbers*: Add the same as arithmetic numbers. The answer is positive.
2. *Negative numbers*: Add absolute values. The answer is negative.
3. *A positive and a negative number*:
   - If the numbers have the same absolute value, the answer is 0.
   - If the numbers have different absolute values, subtract the smaller absolute value from the larger. Then:
     a) If the positive number has the greater absolute value, the answer is positive.
     b) If the negative number has the greater absolute value, the answer is negative.
4. *One number is zero*: The sum is the other number.

---

Rule 4 is known as the **identity property of 0**. It says that for any number $a$, $a + 0 = a$.

**EXAMPLES** Add without using the number line.

**5.** $-12 + (-7) = -19$     Add the absolute values, 12 and 7, getting 19. Make the answer *negative*, $-19$.

**6.** $-1 + 8 = 7$     The absolute values are 1 and 8. The difference is 7. The positive number has the larger absolute value, so the answer is *positive*, 7.

**7.** $-36 + 21 = -15$     The absolute values are 36 and 21. The difference is 15. The negative number has the larger absolute value, so the answer is *negative*, $-15$.

**8.** $5 + (-5) = 0$     The numbers have the same absolute value. The sum is 0.

**9.** $-7 + 0 = -7$     One number is zero. The sum is the other number, $-7$.

**10.** $-9 + 3 = -6$

**11.** $15 + (-4) = 11$

**12.** $-13 + (-27) = -40$

> Do Exercises 7-20.

Suppose we wish to add several numbers, some positive and some negative, as follows. How can we proceed?

$$15 + (-2) + 7 + 14 + (-5) + (-12)$$

---

Add without using the number line.

**7.** $-5 + (-6)$

**8.** $-9 + (-3)$

**9.** $-4 + 6$

**10.** $-7 + 3$

**11.** $5 + (-7)$

**12.** $-20 + 20$

**13.** $-11 + (-11)$

**14.** $10 + (-7)$

**15.** $0 + (-26)$

**16.** $-6 + 8$

**17.** $-4 + (-5)$

**18.** $-8 + 2$

**19.** $9 + (-7)$

**20.** $-19 + (-11)$

*Answers*

**7.** $-11$    **8.** $-12$    **9.** 2    **10.** $-4$
**11.** $-2$    **12.** 0    **13.** $-22$    **14.** 3
**15.** $-26$    **16.** 2    **17.** $-9$    **18.** $-6$
**19.** 2    **20.** $-30$

The commutative and associative laws hold for integers. Thus we can change grouping and order as we please when adding. For instance, we can group the positive numbers together and the negative numbers together and add them separately. Then we add the two results.

**EXAMPLE 13** Add: $15 + (-2) + 7 + 14 + (-5) + (-12)$.

a) $15 + 7 + 14 = 36$    Adding the positive numbers

b) $-2 + (-5) + (-12) = -19$    Adding the negative numbers

c) $36 + (-19) = 17$    Adding the results of (a) and (b)

We can also add the numbers in any other order we wish, say, from left to right as follows:

$$15 + (-2) + 7 + 14 + (-5) + (-12) = 13 + 7 + 14 + (-5) + (-12)$$
$$= 20 + 14 + (-5) + (-12)$$
$$= 34 + (-5) + (-12)$$
$$= 29 + (-12)$$
$$= 17$$

Do Exercises 21–24.

## b  Opposites, or Additive Inverses

Suppose we add two numbers that are **opposites**, such as 6 and $-6$. The result is 0. When opposites are added, the result is always 0. Opposites are also called **additive inverses**. Every integer has an opposite, or additive inverse.

> **OPPOSITES, OR ADDITIVE INVERSES**
>
> Two numbers whose sum is 0 are called **opposites**, or **additive inverses**, of each other.

**EXAMPLES** Find the opposite, or additive inverse, of each number.

**14.** 34    The opposite of 34 is $-34$ because $34 + (-34) = 0$.

**15.** $-8$    The opposite of $-8$ is 8 because $-8 + 8 = 0$.

**16.** 0    The opposite of 0 is 0 because $0 + 0 = 0$.

Do Exercises 25–30.

To name an opposite, we use the symbol $-$, as follows.

> **SYMBOLIZING OPPOSITES**
>
> The opposite, or additive inverse, of a number $a$ can be named $-a$ (read "the opposite of $a$," or "the additive inverse of $a$").

Note that if we take a number, say 8, and find its opposite, $-8$, and then find the opposite of the result, we will have the original number, 8, again.

Add.

**21.** $(-15) + (-37) + 25 + 42 + (-59) + (-14)$

**22.** $42 + (-81) + (-28) + 24 + 18 + (-31)$

**23.** $-2 + (-10) + 6 + (-7)$

**24.** $-35 + 17 + 14 + (-27) + 31 + (-12)$

Find the opposite, or additive inverse.

**25.** $-4$      **26.** 8

**27.** $-7$      **28.** $-34$

**29.** 0      **30.** 12

*Answers*

**21.** $-58$   **22.** $-56$   **23.** $-13$
**24.** $-12$   **25.** 4   **26.** $-8$   **27.** 7
**28.** 34   **29.** 0   **30.** $-12$

## THE OPPOSITE OF THE OPPOSITE

The opposite of the opposite of a number is the number itself. (The additive inverse of the additive inverse of a number is the number itself.) That is, for any number $a$,

$$-(-a) = a.$$

**EXAMPLE 17** Evaluate $-x$ and $-(-x)$ when $x = 16$.

We replace $x$ in each case with 16.

a)  If $x = 16$, then $-x = -16 = -16$.  The opposite of 16 is $-16$.
b)  If $x = 16$, then $-(-x) = -(-16) = 16$.  The opposite of the opposite of 16 is 16.

**EXAMPLE 18** Evaluate $-x$ and $-(-x)$ when $x = -3$.

We replace $x$ in each case with $-3$.

a)  If $x = -3$, then $-x = -(-3) = 3$.
b)  If $x = -3$, then $-(-x) = -(-(-3)) = -(3) = -3$.

Note that in Example 18 we used an extra set of parentheses to show that we are substituting the negative number $-3$ for $x$. Symbolism like $--x$ is not considered meaningful.

Do Exercises 31–36.

A symbol such as $-8$ is usually read "negative 8." It could be read "the additive inverse of 8," because the additive inverse of 8 is negative 8. It could also be read "the opposite of 8," because the opposite of 8 is $-8$. Thus a symbol like $-8$ can be read in more than one way. A symbol like $-x$, which has a variable, should be read "the opposite of $x$" or "the additive inverse of $x$" and *not* "negative $x$," because we do not know whether $x$ represents a positive number, a negative number, or 0.

We can use the symbolism $-a$ to restate the definition of opposite, or additive inverse.

## OPPOSITES, OR ADDITIVE INVERSES

For any integer $a$, the opposite, or additive inverse, of $a$, denoted $-a$, is such that

$$a + (-a) = (-a) + a = 0.$$

### Signs of Numbers

A negative number is sometimes said to have a "negative sign." A positive number is said to have a "positive sign." When we replace a number with its opposite, we can say that we have "changed its sign."

**EXAMPLES** Change the sign. (Find the opposite.)

**19.** $-3$   $-(-3) = 3$     **20.** $-13$   $-(-13) = 13$
**21.** $0$   $-(0) = 0$       **22.** $14$   $-(14) = -14$

Do Exercises 37–40.

Evaluate $-x$ and $-(-x)$ when:

**31.** $x = 14$.      **32.** $x = 1$.

**33.** $x = -19$.     **34.** $x = -16$.

**35.** $x = 2$.       **36.** $x = -9$.

Find the opposite. (Change the sign.)

**37.** $-4$      **38.** $-21$

**39.** $0$       **40.** $5$

***Answers***

**31.** $-14; 14$   **32.** $-1; 1$   **33.** $19; -19$
**34.** $16; -16$   **35.** $-2; 2$   **36.** $9; -9$
**37.** $4$   **38.** $21$   **39.** $0$   **40.** $-5$

**a** Add. Do not use the number line except as a check.

**1.** $-9 + 2$

**2.** $-5 + 2$

**3.** $-10 + 6$

**4.** $4 + (-3)$

**5.** $-8 + 8$

**6.** $4 + (-4)$

**7.** $-3 + (-5)$

**8.** $-6 + (-8)$

**9.** $-7 + 0$

**10.** $-10 + 0$

**11.** $0 + (-27)$

**12.** $0 + (-36)$

**13.** $17 + (-17)$

**14.** $-20 + 20$

**15.** $-17 + (-25)$

**16.** $-23 + (-14)$

**17.** $18 + (-18)$

**18.** $-13 + 13$

**19.** $-18 + 18$

**20.** $11 + (-11)$

**21.** $8 + (-5)$

**22.** $-7 + 8$

**23.** $-4 + (-5)$

**24.** $10 + (-12)$

**25.** $13 + (-6)$

**26.** $-3 + 14$

**27.** $-25 + 25$

**28.** $40 + (-40)$

**29.** $63 + (-18)$

**30.** $85 + (-65)$

**31.** $-6 + 4$

**32.** $-3 + 2$

**33.** $-2 + (-5)$

**34.** $-7 + (-6)$

**35.** $-22 + 3$

**36.** $35 + (-19)$

**37.** $-5 + (-7) + 6$

**38.** $-10 + (-8) + 3$

**39.** $-4 + 7 + (-4)$

**40.** $-1 + 20 + (-1)$

**41.** $75 + (-14) + (-17) + (-5)$

**42.** $28 + (-44) + 17 + 31 + (-94)$

**43.** $-44 + (-3) + 95 + (-5)$

**44.** $24 + 3 + (-44) + (-8) + 63$

**45.** $98 + (-54) + 113 + (-998) + 44 + (-612) + (-18) + 334$

**46.** $-455 + (-123) + 1026 + (-919) + 213 + 111 + (-874)$

**b** Find the opposite, or additive inverse.

**47.** 24            **48.** −84            **49.** −26            **50.** 36

**51.** 103            **52.** −16            **53.** 46            **54.** −216

Find $-x$ when:

**55.** $x = 9.$            **56.** $x = -26.$            **57.** $x = -14.$            **58.** $x = 52.$

Find $-(-x)$ when:

**59.** $x = -65.$            **60.** $x = 31.$            **61.** $x = 5.$            **62.** $x = -18.$

Change the sign. (Find the opposite.)

**63.** −14            **64.** 33            **65.** 10            **66.** −17

## Skill Maintenance

Multiply.  [1.4a]

**67.** $\begin{array}{r} 1\ 9\ 2 \\ \times\quad 1\ 8 \\ \hline \end{array}$          **68.** $\begin{array}{r} 2\ 6\ 4 \\ \times\ 5\ 1\ 9 \\ \hline \end{array}$          **69.** $\begin{array}{r} 6\ 4\ 0\ 3 \\ \times\quad 7\ 0\ 8 \\ \hline \end{array}$

Divide.  [1.5a]

**70.** $6\ \overline{)\ 2\ 4\ 5\ 1}$          **71.** $5\ 4\ \overline{)\ 6\ 9\ 0\ 4}$          **72.** $4\ 0\ 4\ \overline{)\ 8\ 9{,}6\ 1\ 5}$

Round 641,539 to the nearest:  [1.6a]

**73.** Ten.          **74.** Hundred.          **75.** Thousand.

## Synthesis

**76.** For what integers $x$ is $-x$ positive?          **77.** For what integers $x$ is $-x$ negative?

Add.

**78.** 🖩 $-2{,}706{,}835 + (-645{,}684)$          **79.** 🖩 $-345{,}882 + (-295{,}097)$

Tell whether the sum is positive, negative, or zero.

**80.** If $n = m$ and $n$ is negative, then $-n + (-m)$ is _____ .          **81.** If $n$ is positive and $m$ is negative, then $-n + m$ is _____ .

# 2.3 Subtraction of Integers

## OBJECTIVES

**a** Subtract integers and simplify combinations of additions and subtractions.

**b** Solve applied problems involving addition and subtraction of integers.

## a Subtraction

We now consider subtraction of integers.

> **SUBTRACTION**
>
> The difference $a - b$ is the number $c$ for which $a = b + c$.

Consider, for example, $45 - 17$. *Think*: What number can we add to 17 to get 45? Since $45 = 17 + 28$, we know that $45 - 17 = 28$. Let's consider an example whose answer is a negative number.

**EXAMPLE 1** Subtract: $3 - 7$.

*Think*: What number can we add to 7 to get 3? The number must be negative. Since $7 + (-4) = 3$, we know the number is $-4$: $3 - 7 = -4$. That is, $3 - 7 = -4$ because $7 + (-4) = 3$.

Do Exercises 1–3.

The definition above does not provide the most efficient way to do subtraction. We can develop a faster way to subtract. As a rationale for the faster way, let's compare $3 + 7$ and $3 - 7$ on the number line.

To find $3 + 7$ on the number line, we start at 0, move to 3, and then move 7 units to the *right* since 7 is positive.

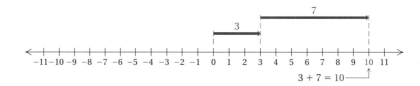

To find $3 - 7$ we do the "opposite" of adding 7: From 3, we move 7 units to the *left* to subtract 7. This is the same as *adding* the opposite of 7, $-7$, to 3.

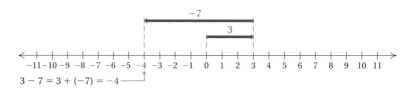

Do Exercises 4–6.

Look for a pattern in the following examples.

| SUBTRACTING | ADDING AN OPPOSITE |
|---|---|
| $5 - 8 = -3$ | $5 + (-8) = -3$ |
| $-6 - 4 = -10$ | $-6 + (-4) = -10$ |
| $-7 - (-10) = 3$ | $-7 + 10 = 3$ |
| $-7 - (-2) = -5$ | $-7 + 2 = -5$ |

Subtract.

**1.** $-6 - 4$

*Think*: What number can be added to 4 to get $-6$:

$$\square + 4 = -6?$$

**2.** $-7 - (-10)$

*Think*: What number can be added to $-10$ to get $-7$:

$$\square + (-10) = -7?$$

**3.** $-7 - (-2)$

*Think*: What number can be added to $-2$ to get $-7$:

$$\square + (-2) = -7?$$

Subtract. Use the number line, doing the "opposite" of addition.

**4.** $-4 - (-3)$

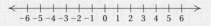

**5.** $-4 - (-6)$

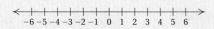

**6.** $5 - 9$

*Answers*

**1.** $-10$ **2.** 3 **3.** $-5$ **4.** $-1$
**5.** 2 **6.** $-4$

Do Exercises 7–10.

Perhaps you have noticed that we can subtract by adding the opposite of the number being subtracted. This can always be done.

> ## SUBTRACTING BY ADDING THE OPPOSITE
>
> For any numbers $a$ and $b$,
> $$a - b = a + (-b).$$
> (To subtract, add the opposite, or additive inverse, of the number being subtracted.)

Complete the addition and compare with the subtraction.

**7.** $4 - 6 = -2$;
$\quad 4 + (-6) = $ _____

**8.** $-3 - 8 = -11$;
$\quad -3 + (-8) = $ _____

**9.** $-5 - (-9) = 4$;
$\quad -5 + 9 = $ _____

**10.** $-5 - (-3) = -2$;
$\quad -5 + 3 = $ _____

This is the method generally used for quick subtraction of integers.

**EXAMPLES**   Subtract.

**2.** $2 - 6 = 2 + (-6) = -4$
> The opposite of 6 is $-6$. We change the subtraction to addition and add the opposite.
> *Check*: $-4 + 6 = 2$.

**3.** $4 - (-9) = 4 + 9 = 13$
> The opposite of $-9$ is 9. We change the subtraction to addition and add the opposite.
> *Check*: $13 + (-9) = 4$.

**4.** $-4 - (-3) = -4 + 3 = -1$
> Adding the opposite.
> *Check*: $-1 + (-3) = -4$.

**5.** $-10 - (-13) = -10 + 13 = 3$
> Adding the opposite.
> *Check*: $3 + (-13) = -10$.

Subtract.

**11.** $2 - 8$

**12.** $-6 - 10$

**13.** $13 - 5$

**14.** $-8 - (-11)$

**15.** $-8 - (-8)$

**16.** $2 - (-5)$

Do Exercises 11–16.

**EXAMPLES**   Subtract by adding the opposite of the number being subtracted.

**6.** $3 - 5$   *Think*: "Three minus five is three plus the opposite of five."
$3 - 5 = 3 + (-5) = -2$

**7.** $-16 - 7$   *Think*: "Negative sixteen minus seven is negative sixteen plus the opposite of seven."
$-16 - 7 = -16 + (-7) = -23$

**8.** $8 - (-9)$   *Think*: "Eight minus negative nine is eight plus the opposite of negative nine."
$8 - (-9) = 8 + 9 = 17$

**9.** $-11 - (-6)$   *Think*: "Negative eleven minus negative six is negative eleven plus the opposite of negative six."
$-11 - (-6) = -11 + 6 = -5$

Do Exercises 17–21 on the following page.

*Answers*

**7.** $-2$   **8.** $-11$   **9.** $4$   **10.** $-2$   **11.** $-6$
**12.** $-16$   **13.** $8$   **14.** $3$   **15.** $0$   **16.** $7$

Subtract by adding the opposite of the number being subtracted.

**17.** $3 - 11$

**18.** $12 - 5$

**19.** $-12 - (-9)$

**20.** $-7 - 10$

**21.** $-4 - (-4)$

Simplify.

**22.** $-6 - (-2) - (-4) - 12 + 3$

**23.** $20 - 16 - (-22) + 21 - 45$

**24.** $-10 + 7 - (-4) - (-11)$

When several additions and subtractions occur together, we can make them all additions.

**EXAMPLES** Simplify.

**10.** $8 - (-4) - 2 - (-4) + 2 = 8 + 4 + (-2) + 4 + 2$    Adding the
$$= 16$$    opposite

**11.** $3 - (-6) + 2 - (-4) = 3 + 6 + 2 + 4 = 15$

**12.** $9 - (-1) - 7 - 11 = 9 + 1 + (-7) + (-11) = -8$

Do Exercises 22–24.

## b Applications and Problem Solving

Let's now see how we can use addition and subtraction of integers to solve applied problems.

**EXAMPLE 13** *Surface Temperatures on Mars.* Surface temperatures on Mars vary from $-128°C$ during polar night to $27°C$ at the equator during midday at the closest point in orbit to the sun. Find the difference between the highest value and the lowest value in this temperature range.

**Source:** Mars Institute

We let $D$ = the difference in the temperatures. Then the problem translates to the following subtraction:

| Difference in temperatures | is | Highest temperature | minus | Lowest temperature |
|:---:|:---:|:---:|:---:|:---:|
| ↓ | ↓ | ↓ | ↓ | ↓ |
| $D$ | $=$ | $27$ | $-$ | $(-128)$. |

We then solve the equation: $D = 27 - (-128) = 27 + 128 = 155$.

The difference in the temperatures is $155°C$.

Do Exercise 25.

**25. Temperature Extremes.**
The highest temperature ever recorded in the United States was 134°F in Greenland Ranch, California, on July 10, 1913. The lowest temperature ever recorded was −80°F in Prospect Creek, Alaska, on January 23, 1971. How much higher was the temperature in Greenland Ranch than the temperature in Prospect Creek?

**Source:** National Oceanographic and Atmospheric Administration

*Answers*

**17.** −8   **18.** 7   **19.** −3   **20.** −17
**21.** 0   **22.** −9   **23.** 2   **24.** 12
**25.** 214°F

**2.3** Exercise Set

For Extra Help

MyMathLab

Math XL
PRACTICE

WATCH

DOWNLOAD

READ

REVIEW

**a** Subtract.

**1.** $3 - 7$

**2.** $5 - 10$

**3.** $0 - 7$

**4.** $0 - 8$

**5.** $-8 - (-2)$

**6.** $-6 - (-8)$

**7.** $-10 - (-10)$

**8.** $-8 - (-8)$

**9.** $12 - 16$

**10.** $14 - 19$

**11.** $20 - 27$

**12.** $26 - 7$

**13.** $-9 - (-3)$

**14.** $-6 - (-9)$

**15.** $-11 - (-11)$

**16.** $-14 - (-14)$

**17.** $8 - (-3)$

**18.** $-7 - 4$

**19.** $-6 - 8$

**20.** $6 - (-10)$

**21.** $-4 - (-9)$

**22.** $-14 - 2$

**23.** $2 - 9$

**24.** $2 - 8$

**25.** $0 - 5$

**26.** $0 - 10$

**27.** $-5 - (-2)$

**28.** $-3 - (-1)$

**29.** $2 - 25$

**30.** $18 - 63$

**31.** $-42 - 26$

**32.** $-18 - 63$

**33.** $-71 - 2$

**34.** $-49 - 3$

**35.** $24 - (-92)$

**36.** $48 - (-73)$

**37.** $-2 - 0$

**38.** $-6 - 1$

**39.** $3 - 8$

**40.** $5 - 9$

**41.** $2 - 3$

**42.** $4 - 12$

**43.** $-3 - 2$

**44.** $-8 - 5$

**45.** $13 - (-9)$

**46.** $15 - (-7)$

**47.** $6 - (-13)$

**48.** $2 - (-4)$

**49.** $-14 - 6$

**50.** $-7 - 9$

**51.** $1 - 9$

**52.** $1 - 7$

**53.** $11 - 21$

**54.** $30 - 51$

**55.** $7 - 10$

**56.** $8 - (-9)$

**57.** $16 - 23$

**58.** $-38 - (-12)$

**59.** $-47 - (-17)$

**60.** $-12 + 12$

**61.** $-9 - (-9)$

**62.** $8 - (-16)$

**63.** $122 - 123$

**64.** $20 - (-21)$

Simplify.

**65.** $18 - (-15) - 3 - (-5) + 2$

**66.** $22 - (-18) + 7 + (-42) - 27$

**67.** $-31 + (-28) - (-14) - 17$

**68.** $-43 - (-19) - (-21) + 25$

**69.** $-93 - (-84) - 41 - (-56)$

**70.** $84 + (-99) + 44 - (-18) - 43$

**71.** $-5 - (-30) + 30 + 40 - (-12)$

**72.** $14 - (-50) + 20 - (-32)$

**73.** $132 - (-21) + 45 - (-21)$

**74.** $81 - (-20) - 14 - (-50) + 53$

 **b** Solve.

**75.** *Difference in Elevation.* The lowest elevation in North America, Death Valley, California, is 282 ft below sea level. The highest elevation in North America, Mt. McKinley, Alaska, is 20,320 ft. Find the difference in elevation between the highest point and the lowest point.
**Source:** National Geographic Society

**76.** *Credit Card Bills.* On August 1, Lyle's credit card bill shows that he owes $470. During August, he sends a check to the credit card company for $45, charges another $160 in merchandise, and then pays off another $500 of his bill. What is the new balance of Lyle's account at the end of August (excluding interest for August)?

**77.** *Temperature Changes.* One day the temperature in Lawrence, Kansas, is 32° at 6:00 A.M. It rises 15° by noon, but falls 50° by midnight as a cold front moves in. What is the final temperature?

**78.** *Stock Price Changes.* On a recent day, the price of a stock opened at a value of $61. It rose $5, dropped $7, and then rose $3. Find the value of the stock at the end of the day.

**79.** *Lake Level.* In the course of one four-month period, the water level of Lake Champlain went down 2 ft, up 1 ft, down 5 ft, and up 3 ft. How much had the lake level changed at the end of the four months?

**80.** *Tallest Mountain.* The tallest mountain in the world, when measured from base to peak, is Mauna Kea (White Mountain) in Hawaii. From its base 19,684 ft below sea level in the Hawaiian Trough, it rises 33,480 ft. What is the elevation of the peak?
**Source:** The Guinness Book of Records

**81.** *Account Balance.* Leah has $460 in her checking account. She writes a check for $530, makes a deposit of $75, and then writes a check for $90. What is the balance in the account?

**82.** *Cell-Phone Bill.* Erika's cell-phone bill for July was $82. She made a payment of $50 and then made $37 worth of calls in August. How much did she then owe on her cell-phone bill?

**83.** *Temperature Records.* The greatest recorded temperature change in one 24-hr period occurred between January 23 and January 24, 1916, in Browning, Montana, where the temperature fell from 44°F to −56°F. Find the difference between these temperatures.
**Source:** *The Guinness Book of Records*

**84.** *Surface Temperature on Mercury.* Surface temperatures on Mercury vary from 840°F on the equator when the planet is closest to the sun to −290°F at night. Find the difference between these temperatures.
**Source:** Ian Ridpath, *Stars and Planets*, Princeton University Press, 1998

**85.** *Difference in Elevation.* At its highest point, the elevation of Denver, Colorado, is 5672 ft above sea level. At its lowest point, the elevation of New Orleans, Louisiana, is 4 ft below sea level. Find the difference between these elevations.
**Source:** *Information Please Almanac*

**86.** *Low Points on Continents.* The lowest point in Africa is Lake Assal, which is 512 ft below sea level. The lowest point in South America is the Valdes Peninsula, which is 131 ft below sea level. How much lower is Lake Assal than the Valdes Peninsula?
**Source:** National Geographic Society

## Skill Maintenance

Evaluate.  [1.9b]

**87.** $4^3$

**88.** $5^3$

Simplify.  [1.9c]

**89.** $5 \cdot 4 + 9$

**90.** $3 \cdot 16 - (7 - 1) \div 6 - (10 - 4)$

**91.** $2 + (5 + 3)^2$

**92.** $27 - 2^3 \cdot 3$

Solve.  [1.8a]

**93.** How many 12-oz cans of soda can be filled with 96 oz of soda?

**94.** A case of soda contains 24 bottles. If each bottle contains 12 oz, how many ounces of soda are in the case?

## Synthesis

Tell whether each statement is true or false for all integers $m$ and $n$. If false, give an example to show why.

**95.** $-n = 0 - n$

**96.** $n - 0 = 0 - n$

**97.** If $m \neq n$, then $m - n \neq 0$.

**98.** If $m = -n$, then $m + n = 0$.

**99.** If $m + n = 0$, then $m$ and $n$ are opposites.

**100.** If $m - n = 0$, then $m = -n$.

**101.** $m = -n$ if $m$ and $n$ are opposites.

**102.** If $m = -m$, then $m = 0$.

# Mid-Chapter Review

## Concept Reinforcement

Determine whether each statement is true or false.

_____ **1.** All of the natural numbers are integers.   [2.1a]

_____ **2.** If $a > b$, then $a$ lies to the left of $b$ on the number line.   [2.1c]

_____ **3.** The absolute value of a number is always nonnegative.   [2.1d]

## Guided Solutions

Fill in each blank with the number that creates a correct statement or solution.

**4.** Evaluate $-x$ and $-(-x)$ when $x = -4$.   [2.2b]

$$-x = -(\Box) = \Box;$$
$$-(-x) = -(-(\Box)) = -(\Box) = \Box$$

Subtract.   [2.3a]

**5.** $5 - 13 = 5 + (\Box) = \Box$

**6.** $-6 - (-7) = -6 + \Box = \Box$

## Mixed Review

Tell which integers correspond to each situation.   [2.1a]

**7.** Jerilyn deposited $450 in her checking account. Later that week she wrote a check for $79.

**8.** The temperature in Coatsville rose 20° in the morning. It fell 23° overnight.

Graph the number on the number line.   [2.1b]

**9.** $-3$

**10.** $0$

Use either $<$ or $>$ for $\Box$ to write a true sentence.   [2.1c]

**11.** $-6 \ \Box \ 6$ **12.** $-5 \ \Box \ -3$ **13.** $-9 \ \Box \ -10$ **14.** $5 \ \Box \ 0$

Find the absolute value.   [2.1d]

**15.** $|15|$ **16.** $|-18|$ **17.** $|0|$ **18.** $|-12|$

Find the opposite, or additive inverse, of the number. [2.2b]

**19.** $-5$          **20.** $7$          **21.** $0$          **22.** $-49$

**23.** Evaluate $-x$ when $x$ is $-19$. [2.2b]

**24.** Evaluate $-(-x)$ when $x$ is $2$. [2.2b]

Compute and simplify. [2.2a], [2.3a]

**25.** $7 + (-9)$     **26.** $-3 + 1$     **27.** $3 + (-3)$     **28.** $-8 + (-9)$

**29.** $2 + (-12)$    **30.** $-4 + (-3)$   **31.** $-14 + 5$     **32.** $19 + (-21)$

**33.** $-4 - 6$       **34.** $5 - (-11)$   **35.** $-1 - (-3)$   **36.** $12 - 24$

**37.** $-8 - (-4)$    **38.** $-1 - 5$      **39.** $12 - 14$     **40.** $6 - (-7)$

**41.** $16 - (-9) - 20 - (-4)$             **42.** $-4 + (-10) - (-3) - 12$

**43.** $17 - (-25) + 15 - (-18)$           **44.** $-9 + (-3) + 16 - (-10)$

Solve. [2.3b]

**45.** *Temperature Change.* In a chemistry lab, Ben works with a substance whose initial temperature is 25°C. During an experiment, the temperature falls to $-8$°C. Find the difference between the two temperatures.

**46.** *Stock Price Change.* The price of a stock opened at $56. During the day it dropped $3, then rose $1, and dropped $6. Find the value of the stock at the end of the day.

# Understanding Through Discussion and Writing

**47.** Find an example of a real-world situation that can be represented by a negative integer. [2.1a]

**48.** Explain why the absolute value of a number cannot be negative. [2.1d]

**49.** Explain in your own words why the sum of two negative numbers is always negative. [2.2a]

**50.** If a negative number is subtracted from a positive number, will the result always be positive? Why or why not? [2.3a]

# 2.4

## Multiplication of Integers

### OBJECTIVE

**a** Multiply integers.

### **a** Multiplication

Multiplication of integers is very much like multiplication of arithmetic numbers. The only difference is that we must determine whether the answer is positive or negative.

#### Multiplication of a Positive Number and a Negative Number

To see how to multiply a positive number and a negative number, consider the pattern of the following.

**1.** Complete, as in the example.

$$
\begin{aligned}
4 \cdot 10 &= 40 \\
3 \cdot 10 &= 30 \\
2 \cdot 10 &= \\
1 \cdot 10 &= \\
0 \cdot 10 &= \\
-1 \cdot 10 &= \\
-2 \cdot 10 &= \\
-3 \cdot 10 &=
\end{aligned}
$$

This number decreases by 1 each time.

$$
\begin{aligned}
4 \cdot 5 &= & 20 \\
3 \cdot 5 &= & 15 \\
2 \cdot 5 &= & 10 \\
1 \cdot 5 &= & 5 \\
0 \cdot 5 &= & 0 \\
-1 \cdot 5 &= & -5 \\
-2 \cdot 5 &= & -10 \\
-3 \cdot 5 &= & -15
\end{aligned}
$$

This number decreases by 5 each time.

Do Exercise 1.

According to this pattern, it looks as though the product of a negative number and a positive number is negative. That is the case, and we have the first part of the rule for multiplying integers.

Multiply.

**2.** $-3 \cdot 6$       **3.** $20 \cdot (-5)$

**4.** $4 \cdot (-20)$       **5.** $-7 \cdot 8$

**6.** $-4(7)$       **7.** $8(-4)$

> **THE PRODUCT OF A POSITIVE NUMBER AND A NEGATIVE NUMBER**
>
> To multiply a positive number and a negative number, multiply their absolute values. The answer is negative.

**EXAMPLES** Multiply.

**1.** $8(-5) = -40$       **2.** $-3 \cdot 7 = -21$       **3.** $(-8)6 = -48$

Do Exercises 2–7.

#### Multiplication of Two Negative Numbers

How do we multiply two negative numbers? Again we look for a pattern.

**8.** Complete, as in the example.

$$
\begin{aligned}
3 \cdot (-10) &= -30 \\
2 \cdot (-10) &= -20 \\
1 \cdot (-10) &= \\
0 \cdot (-10) &= \\
-1 \cdot (-10) &= \\
-2 \cdot (-10) &= \\
-3 \cdot (-10) &=
\end{aligned}
$$

This number decreases by 1 each time.

$$
\begin{aligned}
4 \cdot (-5) &= & -20 \\
3 \cdot (-5) &= & -15 \\
2 \cdot (-5) &= & -10 \\
1 \cdot (-5) &= & -5 \\
0 \cdot (-5) &= & 0 \\
-1 \cdot (-5) &= & 5 \\
-2 \cdot (-5) &= & 10 \\
-3 \cdot (-5) &= & 15
\end{aligned}
$$

This number increases by 5 each time.

Do Exercise 8.

*Answers*

**1.** 20; 10; 0; −10; −20; −30   **2.** −18
**3.** −100   **4.** −80   **5.** −56   **6.** −28
**7.** −32   **8.** −10; 0; 10; 20; 30

According to the pattern, it looks as though the product of two negative numbers is positive. That is actually so, and we have the second part of the rule for multiplying integers.

> ### THE PRODUCT OF TWO NEGATIVE NUMBERS
>
> To multiply two negative numbers, multiply their absolute values. The answer is positive.

Do Exercises 9–14.

> To multiply two nonzero numbers:
>
> a) Multiply the absolute values.
> b) If the signs are the same, the answer is positive.
> c) If the signs are different, the answer is negative.

**EXAMPLES** Multiply.

**4.** $(-3)(-4) = 12$      **5.** $-6(2) = -12$      **6.** $-5(-9) = 45$

Do Exercises 15–18.

## Multiplying More Than Two Numbers

When multiplying more than two integers, we can choose order and grouping as we please, using the commutative and associative laws.

**EXAMPLES** Multiply.

**7.** $-8 \cdot 2(-3) = -16(-3)$      Multiplying the first two numbers
$\qquad = 48$                     Multiplying the results

**8.** $-8 \cdot 2(-3) = 24 \cdot 2$      Multiplying the negative numbers. Every pair of
$\qquad = 48$                     negative numbers gives a positive product.

**9.** $-3(-2)(-5)(4) = 6(-5)(4)$      Multiplying the first two numbers
$\qquad = (-30)4 = -120$

**10.** $(-2)(2)(-3)(-4) = -4 \cdot 12$      Multiplying the first two numbers
$\qquad = -48$                     and the last two numbers

**11.** $-5 \cdot (-2) \cdot (-3) \cdot (-6) = 10 \cdot 18 = 180$

**12.** $(-3)(-5)(-2)(-3)(-6) = (-30)(18) = -540$

Considering that the product of a pair of negative numbers is positive, we can see the following pattern in the results of Examples 11 and 12.

> The product of an even number of negative numbers is positive.
> The product of an odd number of negative numbers is negative.

Do Exercises 19–24.

Multiply.

**9.** $-3 \cdot (-4)$

**10.** $-16 \cdot (-2)$

**11.** $-7 \cdot (-5)$

**12.** $-7(-9)$

**13.** $-3(-12)$

**14.** $-6(-4)$

Multiply.

**15.** $5(-6)$

**16.** $(-5)(-6)$

**17.** $(-3) \cdot 10$

**18.** $(-4)(15)$

Multiply.

**19.** $5 \cdot (-3) \cdot 2$

**20.** $-3 \times (-4) \times (-2)$

**21.** $-2 \cdot (-10) \cdot (-5)$

**22.** $-2 \cdot (-5) \cdot (-4) \cdot (-3)$

**23.** $(-4)(-5)(-2)(-3)(-1)$

**24.** $(-1)(-1)(-2)(-3)(-1)(-1)$

*Answers*

9. 12    10. 32    11. 35    12. 63    13. 36
14. 24    15. −30    16. 30    17. −30
18. −60    19. −30    20. −24    21. −100
22. 120    23. −120    24. 6

**a** Multiply.

**1.** $-8 \cdot 2$

**2.** $-3 \cdot 5$

**3.** $8 \cdot (-3)$

**4.** $-5 \cdot 2$

**5.** $-9 \cdot 8$

**6.** $-20 \cdot 3$

**7.** $-8 \cdot (-2)$

**8.** $-4 \cdot (-5)$

**9.** $-7 \cdot (-6)$

**10.** $-9 \cdot (-2)$

**11.** $15 \cdot (-8)$

**12.** $-11 \cdot (-10)$

**13.** $-14 \cdot 17$

**14.** $-13 \cdot (-15)$

**15.** $-25 \cdot (-48)$

**16.** $39 \cdot (-43)$

**17.** $-3 \cdot (-28)$

**18.** $97 \cdot (-2)$

**19.** $4 \cdot (-3)$

**20.** $3 \cdot (-22)$

**21.** $-6 \cdot (-4)$

**22.** $-5 \cdot (-6)$

**23.** $-7 \cdot (-3)$

**24.** $-4 \cdot (-32)$

**25.** $23 \cdot (-3)$

**26.** $7 \cdot (-9)$

**27.** $-3 \cdot (-9)$

**28.** $-25 \cdot (-8)$

**29.** $-6 \times 3$

**30.** $-8 \times 6$

**31.** $-9 \cdot 5$

**32.** $-8 \cdot 9$

**33.** $7 \cdot (-4) \cdot (-3) \cdot 5$

**34.** $9 \cdot (-2) \cdot (-6) \cdot 7$

**35.** $-3 \cdot 2 \cdot (-6)$

**36.** $-8 \cdot (-4) \cdot (-3)$

**37.** $-3 \cdot (-4) \cdot (-5)$

**38.** $-2 \cdot (-5) \cdot (-7)$

**39.** $-2 \cdot (-5) \cdot (-3) \cdot (-5)$

**40.** $-3 \cdot (-5) \cdot (-2) \cdot (-1)$

**41.** $5(-18)$

**42.** $-35(-2)$

**43.** $-7 \cdot (-21) \cdot 13$

**44.** $-14 \cdot 34 \cdot 12$

**45.** $-4 \cdot (-2) \cdot 7$

**46.** $-8 \cdot (-3) \cdot (-5)$

**47.** $-3(-2)(5)$

**48.** $-7(-5)(-2)$

**49.** $4 \cdot (-4) \cdot (-5) \cdot (-12)$

**50.** $-2 \cdot (-3) \cdot (-4) \cdot (-5)$

**51.** $7 \cdot (-7) \cdot 6 \cdot (-6)$

**52.** $8 \cdot (-8) \cdot (-9) \cdot (-9)$

**53.** $(-5)(8)(-3)(-2)$

**54.** $(4)(-2)(-3)(2)$

**55.** $(-14) \cdot (-27) \cdot (-2)$

**56.** $7 \cdot (-6) \cdot 5 \cdot (-4) \cdot 3 \cdot (-2) \cdot 1 \cdot (-1)$

**57.** $(-8)(-9)(-10)$

**58.** $(-7)(-8)(-9)(-10)$

**59.** $(-6)(-7)(-8)(-9)(-10)$

**60.** $(-5)(-6)(-7)(-8)(-9)(-10)$

## Skill Maintenance

In each of Exercises 61–68, fill in the blank with the correct term from the given list. Some of the choices may not be used.

**61.** The _____ of a set of numbers is the sum of the numbers divided by the number of addends. [1.9c]

**62.** $a$ _____ $b$ is true when $a$ is to the left of $b$ on the number line. [1.6c]

**63.** The statement $5 \cdot 4 = 4 \cdot 5$ illustrates the _____ law of multiplication. [1.4a]

**64.** In the sentence $60 \div 4 = 15$, the _____ is 4. [1.5a]

**65.** The _____ of a number is its distance from zero on the number line. [2.1d]

**66.** Two numbers whose sum is 0 are _____. [2.2b]

**67.** The _____ $a - b$ is the number $c$ for which $a = b + c$. [2.3a]

**68.** The _____ $a \div b$, where $b \neq 0$, is the unique number $c$ for which $a = b \cdot c$. [1.5a]

$<$
$>$
$=$
dividend
divisor
quotient
product
difference
sum
absolute value
associative
commutative
average
opposites
equivalent

## Synthesis

**69.** What must be true of $a$ and $b$ if $-ab$ is to be **(a)** positive? **(b)** zero? **(c)** negative?

# 2.5

## Division of Integers and Order of Operations

### OBJECTIVES

**a** Divide integers.

**b** Solve applied problems involving multiplication and division of integers.

**c** Simplify expressions using rules for order of operations.

**SKILL TO REVIEW**

Objective 1.9c: Simplify expressions using the rules for order of operations.

Simplify.

1. $100 \div 4 - 6$    2. $3 \cdot 8 - 2 \cdot 9$

We now consider division of integers. The definition of division results in rules for division that are the same as those for multiplication.

## a Division of Integers

> **DIVISION**
>
> The quotient $a \div b$, where $b \neq 0$, is that unique number $c$ for which $a = b \cdot c$.

**EXAMPLES**  Divide, if possible. Check your answer.

1. $14 \div (-7) = -2$    *Think*: What number multiplied by $-7$ gives 14? That number is $-2$. *Check*: $(-2)(-7) = 14$.

2. $-32 \div (-4) = 8$    *Think*: What number multiplied by $-4$ gives $-32$? That number is 8. *Check*: $8(-4) = -32$.

3. $-20 \div 5 = -4$    *Think*: What number multiplied by 5 gives $-20$? That number is $-4$. *Check*: $-4 \cdot 5 = -20$.

4. $-17 \div 0$ is **not defined**.    *Think*: What number multiplied by 0 gives $-17$? There is no such number because the product of 0 and *any* number is 0.

Do Margin Exercises 1–3.

The rules for division are the same as those for multiplication.

> To multiply or divide two numbers (where the divisor is nonzero):
>
> a) Multiply or divide the absolute values.
>
> b) If the signs are the same, the answer is positive.
>
> c) If the signs are different, the answer is negative.

Do Exercises 4–6.

Divide.

1. $6 \div (-3)$    *Think*: What number multiplied by $-3$ gives 6?

2. $\dfrac{-15}{-3}$    *Think*: What number multiplied by $-3$ gives $-15$?

3. $-24 \div 8$    *Think*: What number multiplied by 8 gives $-24$?

### Excluding Division by Zero

Example 4 shows why we cannot divide $-17$ by 0. We can use the same argument to show why we cannot divide any nonzero number $b$ by 0. Consider $b \div 0$. We look for a number that when multiplied by 0 gives $b$. There is no such number because the product of 0 and any number is 0. Thus we cannot divide a nonzero number $b$ by 0.

Now consider 0 divided by 0. We look for a number $c$ such that $0 \cdot c = 0$. But $0 \cdot c = 0$ for any number $c$. Thus it appears that $0 \div 0$ could be any number we choose. Getting any answer we want when we divide 0 by 0 would be very confusing. Thus we agree that division by zero is not defined.

Divide.

4. $\dfrac{-72}{-8}$    5. $\dfrac{30}{-5}$    6. $\dfrac{56}{-7}$

**Answers**

*Skill to Review:*
1. 19    2. 6

*Margin Exercises:*
1. $-2$    2. 5    3. $-3$    4. 9
5. $-6$    6. $-8$

### Dividing 0 by Other Numbers

Note that $0 \div 8 = 0$ because $0 = 0 \cdot 8$.

**DIVIDENDS OF 0**

Zero divided by any nonzero number is 0:

$0 \div a = 0, \quad a \neq 0.$

**EXAMPLES** Divide.

**5.** $0 \div (-6) = 0$    **6.** $0 \div 12 = 0$    **7.** $-3 \div 0$ is not defined.

Do Exercises 7 and 8.

Divide, if possible.

**7.** $-5 \div 0$      **8.** $0 \div (-3)$

## (b) Applications and Problem Solving

We can use multiplication and division of integers to solve applied problems.

**EXAMPLE 8** *Chemical Reaction.* During a chemical reaction, the temperature in a beaker decreased every minute by the same number of degrees. The temperature was 56°F at 10:10 A.M. By 10:42 A.M., the temperature had dropped to −8°F. By how many degrees did it change each minute?

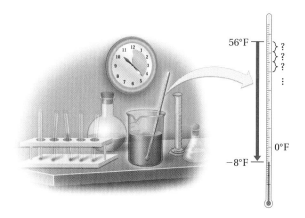

We first determine by how many degrees $d$ the temperature changed altogether. We subtract −8 from 56:

$d = 56 - (-8) = 56 + 8 = 64.$

*Answers*

**7.** Not defined    **8.** 0

Do Exercise 9.

**9. Chemical Reaction.** During a chemical reaction, the temperature in a beaker decreased every minute by the same number of degrees. The temperature was 71°F at 2:12 P.M. By 2:29 P.M., the temperature had dropped to −14°F. By how many degrees did it change each minute?

The temperature changed a total of 64°. We can express this as −64° since the temperature dropped.

The amount of time $t$ that passed was $42 - 10$, or 32 min. Thus the number of degrees $T$ that the temperature changed each minute is given by

$$T = d \div t = -64 \div 32 = -2.$$

The change was −2°F per minute.

## c Order of Operations

When several operations are to be done in a calculation or a problem, we apply the same rules that we did in Section 1.9. We repeat them here for review. If you did not study Section 1.9, you should do so before continuing.

> **RULES FOR ORDER OF OPERATIONS**
>
> 1. Do all calculations within grouping symbols before operations outside.
> 2. Evaluate all exponential expressions.
> 3. Do all multiplications and divisions in order from left to right.
> 4. Do all additions and subtractions in order from left to right.

**EXAMPLE 9**  Simplify: $-34 \cdot 56 - 17$.

There are no parentheses or exponents so we start with the third step.

$-34 \cdot 56 - 17 = -1904 - 17$  Carrying out all multiplications and divisions in order from left to right

$\qquad\qquad\qquad = -1921$  Carrying out all additions and subtractions in order from left to right

**EXAMPLE 10**  Simplify: $256 \div 2^4 \cdot (-5)$.

There are no calculations within grouping symbols so we start with the second step.

$256 \div 2^4 \cdot (-5)$

$= 256 \div 16 \cdot (-5)$  Evaluating the exponential expression

$= 16 \cdot (-5)$  Doing all multiplications and divisions in order from left to right

$= -80$

*Answer*

**9.** −5°F per minute

**EXAMPLE 11** Simplify: $2^4 + 51 \cdot 4 - (37 + 23 \cdot 2)$.

$2^4 + 51 \cdot 4 - (37 + 23 \cdot 2)$

$= 2^4 + 51 \cdot 4 - (37 + 46)$    Carrying out all operations inside parentheses first, multiplying 23 by 2, following the rules for order of operations within the parentheses

$= 2^4 + 51 \cdot 4 - 83$    Completing the addition inside parentheses

$= 16 + 51 \cdot 4 - 83$    Evaluating exponential expressions

$= 16 + 204 - 83$    Doing all multiplications

$= 220 - 83$    Doing all additions and subtractions in order from left to right

$= 137$

**EXAMPLE 12** Simplify: $[-64 \div (-16) \div (-2)] \div (2^3 - 3^2)$.

$[-64 \div (-16) \div (-2)] \div (2^3 - 3^2)$

$= [4 \div (-2)] \div (8 - 9)$

$= -2 \div (-1)$

$= 2$

Do Exercises 10–13.

Simplify.

**10.** $23 - 42 \cdot 30$

**11.** $-243 \div 3^3 \cdot 6$

**12.** $52 \cdot 5 + 5^3 - (4^2 - 48 \div 4)$

**13.** $(5 - 10 - 5 \cdot 23) \div (2^3 + 3^2 - 7)$

*Answers*

**10.** $-1237$    **11.** $-54$    **12.** $381$    **13.** $-12$

# Translating for Success

**1.** *Gas Mileage.* After driving his sports car for 759 mi on the first day of a recent trip, Nathan noticed that it took 23 gal of gasoline to fill the tank. How many miles per gallon did the sports car get?

**2.** *Apartment Rent.* Cecilia needs $4500 for tuition. This is two times as much as she needs for apartment rent. How much does she need for the rent?

**3.** *Purchase Price.* The price of an office copier is $4500 and the sales tax is $150. Find the total purchase price.

**4.** *Change in Elevation.* The lowest elevation in Australia, Lake Eyre, is 52 ft below sea level. The highest elevation in Australia, Mt. Kosciusko, is 7310 ft. Find the difference in elevation between the highest point and the lowest point.

**5.** *Cell-Phone Bill.* Jeff's cell-phone bill for September was $73. He made a payment of $52. How much did he then owe on his phone bill?

The goal of these matching questions is to practice step (2), *Translate*, of the five-step problem-solving process. Translate each problem to an equation and select a correct translation from equations A–O.

**A.** $7 \cdot x = 588$

**B.** $52 + x = 73$

**C.** $2 \cdot x = 4500$

**D.** $150 \cdot x = 4500$

**E.** $23(759) = x$

**F.** $4500 = 150 + x$

**G.** $4500 + 150 = x$

**H.** $x = 150 \cdot 4500$

**I.** $23 \cdot x = 759$

**J.** $7 \div 588 = x$

**K.** $x = 23 + 759$

**L.** $x = 7310 - (-52)$

**M.** $x = 588 \cdot 7$

**N.** $2(4500) = x$

**O.** $52 \cdot x = 7310$

*Answers on page A-3*

**6.** *Camp Sponsorships.* Donations of $52 per camper are needed for camp sponsorships at Lazy Day Summer Camp. Two weeks prior to camp, the sponsorship fund contained only $7310. How many campers can be enrolled?

**7.** *Drain Pipe.* A construction engineer has 588 ft of flexible drain pipe. How many 7-ft lengths can be cut from the total amount of pipe available?

**8.** *Savings-Account Balance.* After $150 interest has been added to Kelly's savings account, the new balance is $4500. Find the previous balance.

**9.** *Loan Payments.* Matt's student loan totals $4500. If he repays $150 each month, how many months will it take him to repay the loan?

**10.** *Drain Pipe.* An irrigation sub-contractor needs 588 pieces of pipe for a large development project. If each piece must be 7 ft long, how many feet need to be purchased?

**a**    Divide, if possible. Check each answer.

**1.** $36 \div (-6)$      **2.** $42 \div (-7)$      **3.** $26 \div (-2)$      **4.** $24 \div (-12)$

**5.** $-16 \div 8$      **6.** $-18 \div (-2)$      **7.** $-48 \div (-12)$      **8.** $-72 \div (-9)$

**9.** $-72 \div 9$      **10.** $-50 \div 25$      **11.** $-100 \div (-50)$      **12.** $-200 \div 8$

**13.** $-108 \div 9$      **14.** $-64 \div (-8)$      **15.** $200 \div (-25)$      **16.** $-390 \div (-13)$

**17.** $75 \div 0$      **18.** $0 \div (-5)$      **19.** $81 \div (-9)$      **20.** $-145 \div (-5)$

**b**    Solve.

**21.** *Chemical Reaction.* The temperature of a chemical compound was 0°C at 11:00 A.M. During a reaction, it dropped 3°C per minute until 11:18 A.M. What was the temperature at 11:18 A.M.?

**22.** *Chemical Reaction.* The temperature in a chemical compound was −5°C at 3:20 P.M. During a reaction, it increased 2°C per minute until 3:52 P.M. What was the temperature at 3:52 P.M.?

**23.** *Stock Price.* The price of ePDQ.com began the day at $32 per share and dropped $2 per hour for 3 hr. What was the price of the stock after 3 hr?

**24.** *Population Decrease.* The population of a rural town was 12,500. It decreased 380 each year for 4 yr. What was the population of the town after 4 yr?

**25.** *Diver's Position.* After diving 95 m below sea level, a diver rises at a rate of 7 meters per minute for 9 min. Where is the diver in relation to the surface?

**26.** *Debit Card Balance.* Karen had $234 in her checking account. After using her debit card to make seven purchases at $39 each, what was the balance in her checking account?

**C**   Simplify.

**27.** $8 - 2 \cdot 3 - 9$

**28.** $8 - (2 \cdot 3 - 9)$

**29.** $(8 - 2 \cdot 3) - 9$

**30.** $(8 - 2)(3 - 9)$

**31.** $16 \cdot (-24) + 50$

**32.** $10 \cdot 20 - 15 \cdot 24$

**33.** $2^4 + 2^3 - 10$

**34.** $40 - 3^2 - 2^3$

**35.** $5^3 + 26 \cdot 71 - (16 + 25 \cdot 3)$

**36.** $4^3 + 10 \cdot 20 + 8^2 - 23$

**37.** $4 \cdot 5 - 2 \cdot 6 + 4$

**38.** $4 \cdot (6 + 8) \div (4 + 3)$

**39.** $4^3 \div 8$

**40.** $5^3 - 7^2$

**41.** $8(-7) + 6(-5)$

**42.** $10(-5) + 1(-1)$

**43.** $19 - 5(-3) + 3$

**44.** $14 - 2(-6) + 7$

**45.** $9 \div (-3) + 16 \div 8$

**46.** $-32 - 8 \div 4 - (-2)$

**47.** $-4^2 + 6$

**48.** $-5^2 + 7$

**49.** $-8^2 - 3$

**50.** $-9^2 - 11$

**51.** $12 - 20^3$

**52.** $20 + 4^3 \div (-8)$

**53.** $2 \times 10^3 - 5000$

**54.** $-7(3^4) + 18$

**55.** $6[9 - (3 - 4)]$

**56.** $8[(6 - 13) - 11]$

**57.** $-1000 \div (-100) \div 10$

**58.** $256 \div (-32) \div (-4)$

**59.** $8 - (7 - 9)$

**60.** $(8 - 7) - 9$

**61.** $(10 - 6^2) \div (3^2 + 2^2)$

**62.** $(-3 - 5^3 - 4^3) \div (6^2 - 10^2)$

**63.** $[20(8 - 3) - 4(10 - 3)] \div [10(2 - 6) + 2(7 + 4)]$

**64.** $[(3 - 5)^2 - 4(5 - 13)] \div [(12 - 9)^2 + (11 - 14)^2]$

## Skill Maintenance

What does the digit 8 mean in each number? [1.1a]

**65.** 4,678,952

**66.** 8,473,901

**67.** 7148

**68.** 23,803

Subtract. [1.3a]

**69.** $9001 - 6798$

**70.** $2037 - 1189$

**71.** $67,113 - 29,874$

**72.** $12,327 - 476$

Solve. [1.8a]

**73.** A playing field is 78 ft long and 64 ft wide. What is its area? its perimeter?

**74.** A landscaper buys 13 small maple trees and 17 small oak trees for a project. A maple costs $23 and an oak costs $37. How much is spent altogether for the trees?

## Synthesis

Simplify.

**75.** ▦ $(19 - 17^2) \div (13^2 - 34)$

**76.** ▦ $[195 + (-15)^3] \div (195 - 7 \cdot 5^2)$

Determine the sign of each expression if $m$ is negative and $n$ is positive.

**77.** $-n \div m$

**78.** $-n \div (-m)$

**79.** $-(-n \div m)$

**80.** $-[n \div (-m)]$

**81.** $-[-n \div (-m)]$

## Key Terms and Properties

integers, p. 88
absolute value, p. 91

opposite, p. 96
additive inverse, p. 96

*Rules for addition of integers,*
p. 95

*Multiplication or division of two numbers*
*(when the divisor is nonzero),* p. 112

*Identity property of 0,* p. 95

*Rules for order of operations,* p. 114

## Concept Reinforcement

Determine whether each statement is true or false.

_____ **1.** The sum of two negative numbers is positive. [2.2a]

_____ **2.** The sum of a number and its opposite is 0. [2.2b]

_____ **3.** The product of an even number of negative numbers is negative. [2.4a]

_____ **4.** The opposite of the opposite of a number is the number itself. [2.2b]

## Important Concepts

**Objective 2.1b** Graph integers on the number line.

**Example** Graph −3 on the number line.

We locate −3 on the number line and mark its point with a dot.

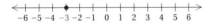

**Practice Exercise**

**1.** Graph 4 on the number line.

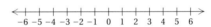

**Objective 2.1c** Determine which of two integers is greater and indicate which, using < or >.

**Example** Use < or > for ☐ to write a true sentence:

−6 ☐ −8.

Since −6 is to the right of −8 on the number line, we have

−6 > −8.

**Practice Exercise**

**2.** Use < or > for ☐ to write a true sentence:

−5 ☐ −4.

**Objective 2.1d** Find the absolute value of an integer.

**Example** Find each of the following: **(a)** |−9|; **(b)** |35|.

**(a)** The number is negative, so we make it positive.

|−9| = 9

**(b)** The number is positive, so the absolute value is the same as the number.

|35| = 35

**Practice Exercise**

**3.** Find each of the following: **(a)** |−17|; **(b)** |14|.

**Objective 2.2a**  Add integers without using the number line.

**Example**  Add without using the number line: $-15 + 9$.

We have a negative number and a positive number. The absolute values are 15 and 9. Their difference is 6. The negative number has the greater absolute value, so the answer is negative.

$$-15 + 9 = -6$$

**Example**  Add without using the number line: $-8 + (-9)$.

We have two negative numbers. We add the absolute values, 8 and 9. The answer is negative.

$$-8 + (-9) = -17$$

**Practice Exercises**

Add without using the number line.

**4.** $6 + (-9)$

**5.** $-5 + (-3)$

---

**Objective 2.3a**  Subtract integers.

**Example**  Subtract: $8 - 12$.

We add the opposite of the number being subtracted.

$$8 - 12 = 8 + (-12)$$
$$= -4$$

**Practice Exercise**

**6.** Subtract: $6 - (-8)$.

---

**Objective 2.4a**  Multiply integers.

**Example**  Multiply: $-6(-4)$.

The signs are the same, so the answer is positive.

$$-6(-4) = 24$$

**Example**  Multiply: $-5(7)$.

The signs are different, so the answer is negative.

$$-5(7) = -35$$

**Practice Exercises**

Multiply.

**7.** $-9(-8)$

**8.** $6(-15)$

---

**Objective 2.5a**  Divide integers.

**Example**  Divide: $-36 \div (-4)$.

The signs are the same, so the answer is positive.

$$-36 \div (-4) = 9$$

**Example**  Divide: $-21 \div 7$.

The signs are different, so the answer is negative.

$$-21 \div 7 = -3$$

**Practice Exercises**

Divide.

**9.** $-32 \div (-8)$

**10.** $48 \div (-12)$

## Objective 2.5c Simplify expressions using rules for order of operations.

**Example** Simplify: $3^2 - 24 \div 2 - (4 + 2 \cdot 8)$.

$$3^2 - 24 \div 2 - (4 + 2 \cdot 8)$$
$$= 3^2 - 24 \div 2 - (4 + 16)$$
$$= 3^2 - 24 \div 2 - 20$$
$$= 9 - 24 \div 2 - 20$$
$$= 9 - 12 - 20$$
$$= -3 - 20$$
$$= -23$$

**Practice Exercise**

**11.** Simplify: $4 - 8^2 \div (10 - 6)$.

## Review Exercises

**1.** State the integers that correspond to this situation: [2.1a]

David has a debt of $45 and Joe has $72 in his savings account.

Simplify. [2.1d]

**2.** $|-38|$

**3.** $|7|$

**4.** $|0|$

**5.** $-|-2|$

Use either $<$ or $>$ for ☐ to write a true sentence. [2.1c]

**6.** $-3$ ☐ $10$

**7.** $-1$ ☐ $-6$

**8.** $11$ ☐ $-12$

**9.** $-2$ ☐ $-1$

Graph the number on the number line. [2.1b]

**10.** 6

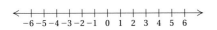

**11.** $-2$

Find the opposite, or additive inverse, of the number. [2.2b]

**12.** 8

**13.** $-14$

**14.** Evaluate $-x$ when $x$ is $-34$. [2.2b]

**15.** Evaluate $-(-x)$ when $x$ is 5. [2.2b]

Compute.

**16.** $4 + (-7)$ [2.2a]

**17.** $-8 + 1$ [2.2a]

**18.** $6 + (-9) + (-8) + 7$ [2.2a]

**19.** $-4 + 5 + (-12) + (-4) + 10$ [2.2a]

**20.** $-3 - (-7)$   [2.3a]

**21.** $-9 - 5$   [2.3a]

**22.** $-4 - 4$   [2.3a]

**23.** $-9 \cdot (-6)$   [2.4a]

**24.** $-3(13)$   [2.4a]

**25.** $7 \cdot (-8)$   [2.4a]

**26.** $3 \cdot (-7) \cdot (-2) \cdot (-5)$   [2.4a]

**27.** $35 \div (-5)$   [2.5a]

**28.** $-51 \div 17$   [2.5a]

**29.** $-42 \div (-7)$   [2.5a]

Simplify.   [2.5c]

**30.** $(-3 - 12) - 8(-7)$

**31.** $[-12(-3) - 2^3] - (-9)(-10)$

**32.** $625 \div (-25) \div 5$

**33.** $-16 \div 4 - 30 \div (-5)$

**34.** $9[(7 - 14) - 13]$

**35.** On the first, second, and third downs, a football team had these gains and losses: 5-yd gain, 12-yd loss, and 15-yd gain, respectively. Find the total gain (or loss).   [2.3b]

**36.** Kaleb's total assets are $170. He borrows $300. What are his total assets now?   [2.3b]

**37.** *Stock Price.*   The price of a stock opened at a value of $18 per share and dropped by $2 per hour for 4 hr. What was the price of the stock after 4 hr? [2.5b]

**38.** *Checking Account Balance.*   Yuri had $68 in his checking account. After writing checks to make seven purchases of DVDs at the same price for each, the balance in his account was −$65. What was the price of each DVD?   [2.5b]

**39.** Compute and simplify: $8 - (-5) - 7 - (-9)$. [2.3a]

   **A.** −13           **B.** −3
   **C.** 15            **D.** 29

**40.** Simplify: $-3 \cdot 4 - 12 \div 4$.   [2.5c]

   **A.** −16         **B.** −15
   **C.** 0            **D.** 6

## Synthesis

**41.** The following are examples of consecutive integers: 4, 5, 6, 7, 8; and −13, −12, −11, −10. [2.2a], [2.4a]

  **a)** Express the number 8 as the sum of 16 consecutive integers.
  **b)** Find the product of the 16 consecutive integers in part (a).

**42.** Insert one pair of parentheses to make the following a true statement:   [2.5c]

$$9 - 3 - 4 + 5 = 15.$$

Simplify.

**43.** $-|8 - (-4 \div 2) - 3 \cdot 5|$   [2.1d], [2.5c]

**44.** $(|-6 - 3| + 3^2 - |-3|) \div (-3)$   [2.1d], [2.5c]

## Understanding Through Discussion and Writing

**1.** Under what circumstances will the sum of a positive integer and a negative integer be negative?   [2.2a]

**2.** Write as many arguments as you can to convince a fellow classmate that $-(-a) = a$ for all numbers $a$. [2.2b]

**3.** Write a problem for a classmate to solve. Design the problem so that the solution is "The temperature dropped to −9°F."   [2.3b]

**4.** What rule have we developed that would tell you the sign of $(-7)^8$ and $(-7)^{11}$ without doing the computations? Explain.   [2.4a]

**5.** ▦ Jake keys $18 \div 2 \cdot 3$ into his calculator and expects the result to be 3. What mistake is he making?   [2.5c]

**6.** Explain how multiplication can be used to justify why the quotient of two negative integers is positive.   [2.5a]

# Test

For Extra Help

Step-by-step test solutions are found on the Chapter Test Prep Videos available via the Video Resources on DVD, in *MyMathLab*, and on YouTube (search "BittingerBasicMathEI" and click on "Channels").

Use either < or > for ☐ to write a true sentence.

**1.** $-4\ \square\ 0$

**2.** $-3\ \square\ -8$

**3.** $-7\ \square\ -8$

**4.** $-1\ \square\ 1$

Simplify.

**5.** $|-7|$

**6.** $|94|$

**7.** $-|-27|$

Find the opposite, or additive inverse.

**8.** $23$

**9.** $-14$

**10.** Evaluate $-x$ when $x$ is $-8$.

**11.** Graph $-6$ on the number line.

$$\xleftarrow{\hspace{0.2em}}\overset{\displaystyle +\ +\ +\ +\ +\ +\ +\ +\ +\ +\ +\ +\ +}{\underset{-6\ -5\ -4\ -3\ -2\ -1\ \ 0\ \ 1\ \ 2\ \ 3\ \ 4\ \ 5\ \ 6}{}}\xrightarrow{\hspace{0.2em}}$$

Compute.

**12.** $31 - (-47)$

**13.** $-8 + 4 + (-7) + 3$

**14.** $-13 + 15$

**15.** $2 - (-8)$

**16.** $32 - 57$

**17.** $18 + (-3)$

**18.** $4 \cdot (-12)$

**19.** $-8 \cdot (-3)$

**20.** $-45 \div 5$

**21.** $-63 \div (-7)$

**22.** $64 \div (-16)$

**23.** $-2(16) - [2(-8) - 5^3]$

**24.** *Antarctica Highs and Lows.* The continent of Antarctica, which lies in the southern hemisphere, experiences winter in July. The average high temperature is $-67°F$ and the average low temperature is $-81°F$. How much higher is the average high than the average low?

**Source:** National Climatic Data Center

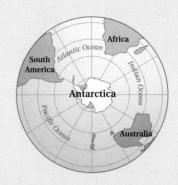

**25.** Maureen kept track of the changes in the stock market over a period of 5 weeks. By how many points had the market risen or fallen over this time?

| WEEK 1 | WEEK 2 | WEEK 3 | WEEK 4 | WEEK 5 |
|---|---|---|---|---|
| Down 13 pts | Down 16 pts | Up 36 pts | Down 11 pts | Up 19 pts |

**26.** *Population Decrease.* The population of Melville was 18,600. It decreased by 420 each year for 6 yr. What was the population of the city after 6 yr?

**27.** *Chemical Reaction.* During a chemical reaction, the temperature in the beaker decreased by the same number of degrees every minute. The temperature was $17°C$ at 11:08 A.M. By 11:25 A.M., the temperature had dropped to $-17°C$. By how many degrees did it drop each minute?

**28.** Evaluate $-(-x)$ when $x$ is 14.

**A.** $-14$

**B.** $-\dfrac{1}{14}$

**C.** $\dfrac{1}{14}$

**D.** $14$

## Synthesis

**29.** Simplify: $|-27 - 3(4)| - |-36| + |-12|$.

**30.** The deepest point in the Pacific Ocean is the Marianas Trench with a depth of 11,033 m. The deepest point in the Atlantic Ocean is the Puerto Rico Trench with a depth of 8648 m. How much higher is the Puerto Rico Trench than the Marianas Trench?

**Source:** Defense Mapping Agency, Hydrographic/Topographic Center

**31.** Find the next three numbers in each sequence.

**a)** 6, 5, 3, 0, _____ , _____ , _____

**b)** 14, 10, 6, 2, _____ , _____ , _____

**c)** $-4, -6, -9, -13,$ _____ , _____ , _____

**d)** $64, -32, 16, -8,$ _____ , _____ , _____

# Fraction Notation: Multiplication and Division

## Real-World Application

Of every 100 voters in the 2004 presidential election, 16 voters were in the 18–29 age group. In the 2008 presidential election, this number increased to 18.5 voters. What is the ratio of voters in the 18–29 age group to all voters in the 2004 presidential election? in the 2008 election?

*Sources:* The Center for Information & Research on Civic Learning & Engagement; www.civicyouth.org

***This problem appears as Exercise 41 in Section 3.3.***

# 3.1

# Factorizations

## OBJECTIVES

**a** Determine whether one number is a factor of another, and find the factors of a number.

**b** Find some multiples of a number, and determine whether a number is divisible by another.

**c** Given a number from 1 to 100, tell whether it is prime, composite, or neither.

**d** Find the prime factorization of a composite number.

---

**SKILL TO REVIEW**

Objective 1.5a: Divide whole numbers.

Divide.

**1.** $329 \div 8$

**2.** $2\,3\overline{)1\,0\,8\,1}$

---

In this chapter, we begin our work with fractions and fraction notation. Certain skills make such work easier. For example, in order to simplify $\frac{12}{32}$, it is important that we be able to *factor* 12 and 32, as follows:

$$\frac{12}{32} = \frac{4 \cdot 3}{4 \cdot 8}.$$

Then we "remove" a factor of 1:

$$\frac{4 \cdot 3}{4 \cdot 8} = \frac{4}{4} \cdot \frac{3}{8} = 1 \cdot \frac{3}{8} = \frac{3}{8}.$$

Thus factoring is an important skill in working with fractions.

## **a** Factors and Factorization

In Sections 3.1 and 3.2, we consider only the **natural numbers** 1, 2, 3, and so on.

Let's look at the product $3 \cdot 4 = 12$. We say that 3 and 4 are **factors** of 12. When we divide 12 by 3, we get a remainder of 0. We say that the divisor 3 is a **factor** of the dividend 12.

---

**FACTOR**

- In the product $a \cdot b$, $a$ and $b$ are called **factors**.
- If we divide $Q$ by $d$ and get a remainder of 0, then the divisor $d$ is a **factor** of the dividend $Q$.

---

**EXAMPLE 1** Determine by long division **(a)** whether 6 is a factor of 198 and **(b)** whether 15 is a factor of 198.

a) 
$$\begin{array}{r} 33 \\ 6\overline{)198} \\ \underline{18} \\ 18 \\ \underline{18} \\ 0 \end{array} \leftarrow \text{Remainder is 0.}$$

b)
$$\begin{array}{r} 13 \\ 15\overline{)198} \\ \underline{15} \\ 48 \\ \underline{45} \\ 3 \end{array} \leftarrow \text{Not 0}$$

The remainder is 0, so 6 is a factor of 198.

The remainder is not 0, so 15 is not a factor of 198.

Determine whether the second number is a factor of the first.

**1.** 72; 8          **2.** 2384; 28

Do Margin Exercises 1 and 2.

Consider $12 = 3 \cdot 4$. We say that $3 \cdot 4$ is a **factorization** of 12. Similarly, $6 \cdot 2$, $12 \cdot 1$, $2 \cdot 2 \cdot 3$, and $1 \cdot 3 \cdot 4$ are also factorizations of 12. Since $a = a \cdot 1$, every number has a factorization, and every number has factors. For some numbers, the factors consist of only the number itself and 1. For example, the only factorization of 17 is $17 \cdot 1$, so the only factors of 17 are 17 and 1.

*Answers*

*Skill to Review:*
**1.** 41 R 1     **2.** 47

*Margin Exercises:*
**1.** Yes     **2.** No

**EXAMPLE 2** Find all the factors of 70.

We find as many "two-factor" factorizations as we can. We check sequentially the numbers 1, 2, 3, and so on, to see if we can form any factorizations:

70

$1 \cdot 70$
$2 \cdot 35$
$5 \cdot 14$
$7 \cdot 10.$

Note that all of the factors of a natural number are *less than or equal to* the number.

Note that 3, 4, and 6 are not factors. If there are additional factors, they must be between 7 and 10. Since 8 and 9 are not factors, we are finished. The factors of 70 are 1, 2, 5, 7, 10, 14, 35, and 70.

Do Exercises 3–6.

Find all the factors of each number.
**3.** 10          **4.** 45

**5.** 62          **6.** 24

# b Multiples and Divisibility

A **multiple** of a natural number is a product of that number and some natural number. For example, some multiples of 2 are

2   (because $2 = 1 \cdot 2$);
4   (because $4 = 2 \cdot 2$);
6   (because $6 = 3 \cdot 2$);
8   (because $8 = 4 \cdot 2$).

Note that all but one of the multiples of a number are *larger* than the number.

We find multiples of 2 by counting by twos: 2, 4, 6, 8, and so on. We can find multiples of 3 by counting by threes: 3, 6, 9, 12, and so on.

**EXAMPLE 3** Show that each of the numbers 8, 12, 20, and 36 is a multiple of 4.

$$8 = 2 \cdot 4 \qquad 12 = 3 \cdot 4 \qquad 20 = 5 \cdot 4 \qquad 36 = 9 \cdot 4$$

Do Exercises 7 and 8.

**7.** Show that each of the numbers 5, 45, and 100 is a multiple of 5.

**8.** Show that each of the numbers 10, 60, and 110 is a multiple of 10.

**EXAMPLE 4** Multiply by 1, 2, 3, and so on, to find ten multiples of 7.

$1 \cdot 7 = 7 \qquad 3 \cdot 7 = 21 \qquad 5 \cdot 7 = 35 \qquad 7 \cdot 7 = 49 \qquad 9 \cdot 7 = 63$
$2 \cdot 7 = 14 \qquad 4 \cdot 7 = 28 \qquad 6 \cdot 7 = 42 \qquad 8 \cdot 7 = 56 \qquad 10 \cdot 7 = 70$

Do Exercise 9.

**9.** Multiply by 1, 2, 3, and so on, to find ten multiples of 5.

## DIVISIBILITY

The number $a$ is **divisible** by another number $b$ if there exists a number $c$ such that $a = b \cdot c$. The statements "$a$ is **divisible** by $b$," "$a$ is a **multiple** of $b$," and "$b$ is a **factor** of $a$" all have the same meaning.

Thus, 27 is *divisible* by 3 because $27 = 3 \cdot 9$. We can also say that 27 is a *multiple* of 3, and 3 is a *factor* of 27.

**EXAMPLE 5** Determine (a) whether 45 is divisible by 9 and (b) whether 45 is divisible by 4.

a) We divide 45 by 9 to determine whether 9 is a factor of 45.

$$\begin{array}{r} 5 \\ 9\overline{)45} \\ \underline{45} \\ 0 \end{array} \leftarrow \text{Remainder is 0.}$$

Because the remainder is 0, 45 is divisible by 9.

b) We divide 45 by 4 to determine whether 4 is a factor of 45.

$$\begin{array}{r} 11 \\ 4\overline{)45} \\ \underline{4} \\ 5 \\ \underline{4} \\ 1 \end{array} \leftarrow \text{Not 0}$$

Since the remainder is not 0, 45 is not divisible by 4.

**10.** Determine whether 16 is divisible by 2.

**11.** Determine whether 125 is divisible by 5.

**12.** Determine whether 125 is divisible by 6.

Do Exercises 10–12.

**Calculator Corner**

**Divisibility and Factors** We can use a calculator to determine whether one number is divisible by another number or whether one number is a factor of another number. For example, to determine whether 387 is divisible by 9, we press [3][8][7][÷] [9][=]. The display is [ 43 ]. Since 43 is a natural number, we know that 387 is a multiple of 9; that is, 387 = 43 · 9. Thus, 387 is divisible by 9, and 9 is a factor of 387.

**Exercises:** For each pair of numbers, determine whether the second number is a factor of the first number.

**1.** 502; 8

**2.** 651; 21

**3.** 3875; 25

**4.** 1047; 14

**13.** Tell whether each number is prime, composite, or neither.

1, 2, 6, 12, 13, 19, 41, 65, 73, 99

## c  Prime and Composite Numbers

### PRIME AND COMPOSITE NUMBERS

- A natural number that has exactly two *different* factors, only itself and 1, is called a **prime number**.
- The number 1 is *not* prime.
- A natural number, other than 1, that is not prime is **composite**.

**EXAMPLE 6** Determine whether the numbers 1, 2, 7, 8, 9, 11, 18, 27, 39, 43, 56, 59, and 77 are prime, composite, or neither.

The number 1 is not prime. It does not have *two* different factors.

The number 2 is prime. It has only the factors 2 and 1.

The numbers 7, 11, 43, and 59 are prime. Each has only two factors, itself and 1.

The number 8 is not prime. It has the factors 1, 2, 4, and 8 and is composite.

The numbers 9, 18, 27, 39, 56, and 77 are composite. Each has more than two factors.

Thus we have

Prime:        2, 7, 11, 43, 59;

Composite:   8, 9, 18, 27, 39, 56, 77;

Neither:      1.

The number 2 is the *only* even prime number. It is also the smallest prime number. The number 0 is neither prime nor composite, but 0 is *not* a natural number and thus is not considered here. We are considering only natural numbers.

Do Exercise 13.

**Answers**

**10.** Yes   **11.** Yes   **12.** No   **13.** 2, 13, 19, 41, 73 are prime; 6, 12, 65, 99 are composite; 1 is neither.

The table at right lists the prime numbers from 2 to 97. There are more extensive tables, but these prime numbers will be the most helpful to you in this text.

## (d) Prime Factorizations

When we factor a composite number into a product of primes, we find the **prime factorization** of the number. To do this, we consider the primes

2, 3, 5, 7, 11, 13, 17, 19, 23, and so on,

and determine whether a given number is divisible by the primes.

**EXAMPLE 7**   Find the prime factorization of 39.

**a)**  We divide by the first prime, 2.

$$
\begin{array}{r}
19 \\
2\overline{)39} \\
2 \\
\hline
19 \\
18 \\
\hline
1
\end{array}
$$

Because the remainder is not 0, 2 is not a factor of 39, and 39 is not divisible by 2.

**b)**  We divide by the next prime, 3.

$$
\begin{array}{r}
13 \quad R = 0 \\
3\overline{)39}
\end{array}
$$

The remainder is 0, so we know that $39 = 3 \cdot 13$. Because 13 is a prime, we are finished. The prime factorization is

$$39 = 3 \cdot 13.$$

**EXAMPLE 8**   Find the prime factorization of 220.

**a)**  We divide by the first prime, 2.

$$
\begin{array}{r}
110 \quad R = 0 \\
2\overline{)220}
\end{array}
\qquad 220 = 2 \cdot 110
$$

**b)**  Because 110 is composite, we continue to divide, starting with 2 again.

$$
\begin{array}{r}
55 \quad R = 0 \\
2\overline{)110}
\end{array}
\qquad 220 = 2 \cdot 2 \cdot 55
$$

**c)**  Since 55 is composite and is not divisible by 2 or 3, we divide by the next prime, 5.

$$
\begin{array}{r}
11 \quad R = 0 \\
5\overline{)55}
\end{array}
\qquad 220 = 2 \cdot 2 \cdot 5 \cdot 11
$$

Because 11 is prime, we are finished. The prime factorization is

$$220 = 2 \cdot 2 \cdot 5 \cdot 11.$$

We abbreviate our procedure as follows.

$$
\begin{array}{r}
11 \\
5\overline{)55} \\
2\overline{)110} \\
2\overline{)220}
\end{array}
$$

$$220 = 2 \cdot 2 \cdot 5 \cdot 11$$

Because multiplication is commutative, a factorization such as $2 \cdot 2 \cdot 5 \cdot 11$ could also be expressed as $5 \cdot 2 \cdot 2 \cdot 11$ or $2 \cdot 5 \cdot 11 \cdot 2$ (or, in exponential notation, as $2^2 \cdot 5 \cdot 11$ or $11 \cdot 2^2 \cdot 5$), but the prime factors are the same in each case. For this reason, we agree that any of these is "the" prime factorization of 220.

> Every number has just one (unique) prime factorization.

**EXAMPLE 9** Find the prime factorization of 72.

We can do divisions "up" as follows:

$$
\begin{array}{r}
3 \leftarrow \text{Prime quotient}\\
3\overline{)\,9} \\
2\overline{)18} \\
2\overline{)36} \\
2\overline{)72} \leftarrow \text{Begin here}
\end{array}
$$

$$72 = 2 \cdot 2 \cdot 2 \cdot 3 \cdot 3$$

Another way to find the prime factorization of 72 uses a **factor tree** as follows. Begin by determining any factorization you can, and then continue factoring.

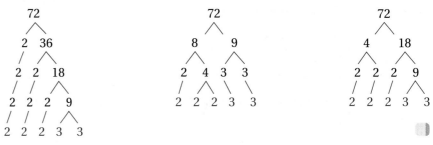

**EXAMPLE 10** Find the prime factorization of 189.

We can use a string of successive divisions or a factor tree. Since 189 is not divisible by 2, we begin with 3.

$$
\begin{array}{r}
7 \\
3\overline{)21} \\
3\overline{)63} \\
3\overline{)189}
\end{array}
$$

$$189 = 3 \cdot 3 \cdot 3 \cdot 7$$

```
     189
     / \
    3   63
    |  / \
    3 7   9
    | |  / \
    3 7 3   3
```

Do Exercises 14–21.

Find the prime factorization of each number.

**14.** 6          **15.** 12

**16.** 45         **17.** 98

**18.** 126        **19.** 144

**20.** 1960       **21.** 1925

*Answers*

**14.** $2 \cdot 3$   **15.** $2 \cdot 2 \cdot 3$   **16.** $3 \cdot 3 \cdot 5$
**17.** $2 \cdot 7 \cdot 7$   **18.** $2 \cdot 3 \cdot 3 \cdot 7$
**19.** $2 \cdot 2 \cdot 2 \cdot 2 \cdot 3 \cdot 3$   **20.** $2 \cdot 2 \cdot 2 \cdot 5 \cdot 7 \cdot 7$
**21.** $5 \cdot 5 \cdot 7 \cdot 11$

**a** Determine whether the second number is a factor of the first.

**1.** 52; 14

**2.** 52; 13

**3.** 625; 25

**4.** 680; 16

Find all the factors of each number.

**5.** 18

**6.** 16

**7.** 54

**8.** 48

**9.** 4

**10.** 9

**11.** 1

**12.** 13

**13.** 98

**14.** 100

**15.** 255

**16.** 120

**b** Multiply by 1, 2, 3, and so on, to find ten multiples of each number.

**17.** 4

**18.** 11

**19.** 20

**20.** 50

**21.** 3

**22.** 15

**23.** 12

**24.** 13

**25.** 10

**26.** 6

**27.** 9

**28.** 14

**29.** Determine whether 26 is divisible by 6.

**30.** Determine whether 48 is divisible by 8.

**31.** Determine whether 1880 is divisible by 8.

**32.** Determine whether 4227 is divisible by 3.

**33.** Determine whether 256 is divisible by 16.

**34.** Determine whether 102 is divisible by 4.

**35.** Determine whether 4227 is divisible by 9.

**36.** Determine whether 200 is divisible by 25.

**37.** Determine whether 8650 is divisible by 16.

**38.** Determine whether 4143 is divisible by 7.

**c** Determine whether each number is prime, composite, or neither.

**39.** 1

**40.** 2

**41.** 9

**42.** 19

**43.** 11

**44.** 27

**45.** 29

**46.** 49

**d**   Find the prime factorization of each number.

**47.** 8      **48.** 16      **49.** 14      **50.** 15      **51.** 42

**52.** 32      **53.** 25      **54.** 40      **55.** 50      **56.** 62

**57.** 169      **58.** 140      **59.** 100      **60.** 110      **61.** 35

**62.** 70      **63.** 72      **64.** 86      **65.** 77      **66.** 99

**67.** 2884      **68.** 484      **69.** 51      **70.** 91      **71.** 1200

**72.** 1800      **73.** 273      **74.** 675      **75.** 1122      **76.** 6435

## Skill Maintenance

Multiply. [1.4a]

**77.** $2 \cdot 13$      **78.** $8 \cdot 32$      **79.** $17 \cdot 25$      **80.** $25 \cdot 168$

Divide. [1.5a]

**81.** $0 \div 22$      **82.** $22 \div 1$      **83.** $22 \div 22$      **84.** $66 \div 22$

Solve. [1.8a]

**85.** *Morel Mushrooms.* During the spring, Kate's Country Market sold 43 pounds of fresh morel mushrooms. The mushrooms sold for $22 a pound. Find the total amount Kate took in from the sale of the mushrooms.

**86.** Sandy can type 62 words per minute. How long will it take her to type 12,462 words?

## Synthesis

**87.** *Factors and Sums.* The top number in each column of the table below can be factored as a product of two numbers whose sum is the bottom number in the column. For example, in the first column, 56 has been factored as $7 \cdot 8$, and $7 + 8 = 15$. Fill in the blank spaces in the table.

| PRODUCT | 56 | 63 | 36 | 72 | 140 | 96 |    | 168 | 110 |    |    |    |
|---------|----|----|----|----|-----|----|----|-----|-----|----|----|----|
| FACTOR  | 7  |    |    |    |     |    |    |     |     | 9  | 24 | 3  |
| FACTOR  | 8  |    |    |    |     |    | 8  | 8   |     | 10 | 18 |    |
| SUM     | 15 | 16 | 20 | 38 | 24  | 20 | 14 |     | 21  |    |    | 24 |

# 3.2

# Divisibility

Suppose you are asked to find the simplest fraction notation for

$$\frac{117}{225}.$$

Since the numbers are quite large, you might feel that the task is difficult. However, both the numerator and the denominator are divisible by 9. If you knew this, you could factor and simplify quickly as follows:

$$\frac{117}{225} = \frac{9 \cdot 13}{9 \cdot 25} = \frac{9}{9} \cdot \frac{13}{25} = 1 \cdot \frac{13}{25} = \frac{13}{25}.$$

How did we know that both numbers have 9 as a factor? There is a simple test for determining this.

In this section, you learn quick ways to determine whether a number is divisible by 2, 3, 4, 5, 6, 8, 9, or 10. This will make simplifying fraction notation much easier.

**OBJECTIVE**

**a** Determine whether a number is divisible by 2, 3, 4, 5, 6, 8, 9, or 10.

**SKILL TO REVIEW**
Objective 1.2a: Add whole numbers.

Add.

**1.** $9 + 8 + 2 + 1$
**2.** $7 + 3 + 0 + 2 + 6$

## a  Rules for Divisibility

### Divisibility by 2

You may already know the test for divisibility by 2.

---

**BY 2**

A number is **divisible by 2** (is *even*) if it has a ones digit of 0, 2, 4, 6, or 8 (that is, it has an even ones digit).

---

Let's see why this test works. Consider 354, which is

3 hundreds + 5 tens + 4.

Hundreds and tens are both multiples of 2. If the last digit is a multiple of 2, then the entire number is a multiple of 2.

**EXAMPLES**  Determine whether each number is divisible by 2.

**1.** 355   is *not* divisible by 2; 5 is *not* even.
**2.** 4786   is divisible by 2; 6 is even.
**3.** 8990   is divisible by 2; 0 is even.
**4.** 4261   is *not* divisible by 2; 1 is *not* even.

Do Margin Exercises 1–4.

Determine whether each number is divisible by 2.

**1.** 84          **2.** 59

**3.** 998        **4.** 2225

*Answers*

*Skill to Review:*
**1.** 20    **2.** 18

*Margin Exercises:*
**1.** Yes    **2.** No    **3.** Yes    **4.** No

## Divisibility by 3

> ### BY 3
>
> A number is **divisible by 3** if the sum of its digits is divisible by 3.

Let's illustrate why the test for divisibility by 3 works. Consider 852; since $852 = 3 \cdot 284$, 852 is divisible by 3.

$$852 = 8 \cdot 100 + 5 \cdot 10 + 2 \cdot 1$$
$$= 8(99 + 1) + 5(9 + 1) + 2(1)$$
$$= 8 \cdot 99 + 8 \cdot 1 + 5 \cdot 9 + 5 \cdot 1 + 2 \cdot 1 \qquad \text{Using the distributive law:}$$
$$a(b + c) = a \cdot b + a \cdot c$$

Since 99 and 9 are each a multiple of 3, we see that $8 \cdot 99$ and $5 \cdot 9$ are multiples of 3. This leaves $8 \cdot 1 + 5 \cdot 1 + 2 \cdot 1$, or $8 + 5 + 2$. If $8 + 5 + 2$, the sum of the digits, is divisible by 3, then 852 is divisible by 3.

**EXAMPLES** Determine whether each number is divisible by 3.

**5.** 18   $1 + 8 = 9$  ⎫
**6.** 93   $9 + 3 = 12$  ⎬ Each is divisible by 3 because the sum of its digits is divisible by 3.
**7.** 201   $2 + 0 + 1 = 3$  ⎭

**8.** 256   $2 + 5 + 6 = 13$   The sum of the digits, 13, is *not* divisible by 3, so 256 is *not* divisible by 3.

Do Exercises 5–8.

Determine whether each number is divisible by 3.

**5.** 111      **6.** 1111

**7.** 309      **8.** 17,216

## Divisibility by 6

A number divisible by 6 is a multiple of 6. But $6 = 2 \cdot 3$, so the number is also a multiple of 2 and 3. Thus we have the following.

> ### BY 6
>
> A number is **divisible by 6** if its ones digit is 0, 2, 4, 6, or 8 (is even) and the sum of its digits is divisible by 3.

**EXAMPLES** Determine whether each number is divisible by 6.

**9.** 720

Because 720 is even, it is divisible by 2. Also, $7 + 2 + 0 = 9$, so 720 is divisible by 3. Thus, 720 is divisible by 6.

$$\begin{array}{ccc} 720 & & 7 + 2 + 0 = 9 \\ \uparrow & & \uparrow \\ \text{Even} & & \text{Divisible by 3} \end{array}$$

**10.** 73

73 is *not* divisible by 6 because it is *not* even.

**11.** 256

Although 256 is even, it is *not* divisible by 6 because the sum of its digits, $2 + 5 + 6$, or 13, is *not* divisible by 3.

Do Exercises 9–12.

Determine whether each number is divisible by 6.

**9.** 420      **10.** 106

**11.** 321      **12.** 444

*Answers*

**5.** Yes   **6.** No   **7.** Yes   **8.** No   **9.** Yes
**10.** No   **11.** No   **12.** Yes

## Divisibility by 9

The test for divisibility by 9 is similar to the test for divisibility by 3.

> **BY 9**
>
> A number is **divisible by 9** if the sum of its digits is divisible by 9.

**EXAMPLES** Determine whether each number is divisible by 9.

**12.** 6984

Because $6 + 9 + 8 + 4 = 27$ and 27 is divisible by 9, 6984 is divisible by 9.

**13.** 322

Because $3 + 2 + 2 = 7$ and 7 is *not* divisible by 9, 322 is *not* divisible by 9.

Do Exercises 13–16.

Determine whether each number is divisible by 9.

**13.** 16            **14.** 117

**15.** 930          **16.** 29,223

## Divisibility by 10

> **BY 10**
>
> A number is **divisible by 10** if its ones digit is 0.

We know that this test works because the product of 10 and *any* number has a ones digit of 0.

**EXAMPLES** Determine whether each number is divisible by 10.

**14.** 3440   is divisible by 10 because the ones digit is 0.

**15.** 3447   is *not* divisible by 10 because the ones digit is not 0.

Do Exercises 17–20.

Determine whether each number is divisible by 10.

**17.** 305          **18.** 847

**19.** 300          **20.** 8760

## Divisibility by 5

> **BY 5**
>
> A number is **divisible by 5** if its ones digit is 0 or 5.

**EXAMPLES** Determine whether each number is divisible by 5.

**16.**   220   is divisible by 5 because the ones digit is 0.

**17.**   475   is divisible by 5 because the ones digit is 5.

**18.** 6514   is *not* divisible by 5 because the ones digit is neither 0 nor 5.

Do Exercises 21–24.

Determine whether each number is divisible by 5.

**21.** 5780        **22.** 3427

**23.** 34,678      **24.** 7775

Let's see why the test for 5 works. Consider 7830:

$$7830 = 10 \cdot 783 = 5 \cdot 2 \cdot 783.$$

Since 7830 is divisible by 10 and 5 is a factor of 10, 7830 is divisible by 5.

*Answers*

13. No   14. Yes   15. No   16. Yes
17. No   18. No   19. Yes   20. Yes
21. Yes  22. No   23. No    24. Yes

Consider 6734:

$$6734 = 673 \text{ tens} + 4.$$

Tens are multiples of 5, so the only number that must be checked is the ones digit. If the last digit is a multiple of 5, then the entire number is. In this case, 4 is *not* a multiple of 5, so 6734 is *not* divisible by 5.

## Divisibility by 4

The test for divisibility by 4 is similar to the test for divisibility by 2.

> ### BY 4
>
> A number is **divisible by 4** if the number named by its last *two* digits is divisible by 4.

**EXAMPLES**   Determine whether the number is divisible by 4.

**19.** 8212   is divisible by 4 because 12 is divisible by 4.

**20.** 5216   is divisible by 4 because 16 is divisible by 4.

**21.** 8211   is *not* divisible by 4 because 11 is *not* divisible by 4.

**22.** 7538   is *not* divisible by 4 because 38 is *not* divisible by 4.

Do Exercises 25–28.

To see why the test for divisibility by 4 works, consider 516:

$$516 = 5 \text{ hundreds} + 16.$$

Hundreds are multiples of 4. If the number named by the last two digits is a multiple of 4, then the entire number is a multiple of 4.

## Divisibility by 8

The test for divisibility by 8 is an extension of the tests for divisibility by 2 and 4.

> ### BY 8
>
> A number is **divisible by 8** if the number named by its last *three* digits is divisible by 8.

**EXAMPLES**   Determine whether the number is divisible by 8.

**23.**   5648   is divisible by 8 because 648 is divisible by 8.

**24.** 96,088   is divisible by 8 because 88 is divisible by 8.

**25.**   7324   is *not* divisible by 8 because 324 is *not* divisible by 8.

**26.** 13,420   is *not* divisible by 8 because 420 is *not* divisible by 8.

Do Exercises 29–32.

## A Note about Divisibility by 7

There are several tests for divisibility by 7, but all of them are more complicated than simply dividing by 7. So if you want to test for divisibility by 7, simply divide by 7, either by hand or using a calculator.

---

Determine whether each number is divisible by 4.

**25.** 216

**26.** 217

**27.** 5862

**28.** 23,524

---

Determine whether each number is divisible by 8.

**29.** 7564

**30.** 7864

**31.** 17,560

**32.** 25,716

---

*Answers*

**25.** Yes   **26.** No   **27.** No   **28.** Yes
**29.** No   **30.** Yes   **31.** Yes   **32.** No

**a**    To answer Exercises 1–8, consider the following numbers.

| | | | |
|---|---|---|---|
| 46 | 300 | 85 | 256 |
| 224 | 36 | 711 | 8064 |
| 19 | 45,270 | 13,251 | 1867 |
| 555 | 4444 | 254,765 | 21,568 |

**1.** Which of the above are divisible by 2?

**2.** Which of the above are divisible by 3?

**3.** Which of the above are divisible by 4?

**4.** Which of the above are divisible by 5?

**5.** Which of the above are divisible by 6?

**6.** Which of the above are divisible by 8?

**7.** Which of the above are divisible by 9?

**8.** Which of the above are divisible by 10?

To answer Exercises 9–16, consider the following numbers.

| | | | |
|---|---|---|---|
| 56 | 200 | 75 | 35 |
| 324 | 42 | 812 | 402 |
| 784 | 501 | 2345 | 111,111 |
| 55,555 | 3009 | 2001 | 1005 |

**9.** Which of the above are divisible by 3?

**10.** Which of the above are divisible by 2?

**11.** Which of the above are divisible by 5?

**12.** Which of the above are divisible by 4?

**13.** Which of the above are divisible by 8?

**14.** Which of the above are divisible by 6?

**15.** Which of the above are divisible by 10?

**16.** Which of the above are divisible by 9?

To answer Exercises 17–24, consider the following numbers.

| 305 | 313,332 | 876 | 64,000 |
|---|---|---|---|
| 1101 | 7624 | 1110 | 9990 |
| 13,205 | 111,126 | 5128 | 126,111 |

**17.** Which of the above are divisible by 2?

**18.** Which of the above are divisible by 3?

**19.** Which of the above are divisible by 6?

**20.** Which of the above are divisible by 5?

**21.** Which of the above are divisible by 9?

**22.** Which of the above are divisible by 8?

**23.** Which of the above are divisible by 10?

**24.** Which of the above are divisible by 4?

## Skill Maintenance

Solve.  [1.7b]

**25.** $56 + x = 194$

**26.** $y + 124 = 263$

**27.** $3008 = x + 2134$

**28.** $18 \cdot t = 1008$

**29.** $24 \cdot m = 624$

**30.** $338 = a \cdot 26$

Divide.  [1.5a]

**31.** $2106 \div 9$

**32.** $4\,5\,\overline{)\,1\,8\,0{,}1\,3\,5}$

Solve.  [1.8a]

**33.** Marty's automobile has a 5-speed transmission and gets 33 mpg in city driving. How many gallons of gas will it use to travel 1485 mi of city driving?

**34.** There are 60 min in 1 hr. How many minutes are there in 72 hr?

## Synthesis

Find the prime factorization of each number. Use divisibility tests where applicable.

**35.** 7800

**36.** 2520

**37.** 2772

**38.** 1998

**39.** ▦ Fill in the missing digits of the number

$$95,\ \square\ \square\ 8$$

so that it is divisible by 99.

**40.** A passenger in a taxicab asks for the driver's company number. The driver says abruptly, "Sure—you can have my number. Work it out: If you divide it by 2, 3, 4, 5, or 6, you will get a remainder of 1. If you divide it by 11, the remainder will be 0 and no driver has a company number that meets these requirements and is smaller than this one." Determine the number.

# 3.3 Fractions and Fraction Notation

The study of arithmetic begins with the set of whole numbers

0, 1, 2, 3, 4, 5, 6, 7, 8, 9, 10, 11, and so on.

But we also need to be able to use fractional parts of numbers such as halves, thirds, fourths, and so on. Here is an example.

The tread depth of an IRL Indy Car Series tire is $\frac{3}{32}$ in. Tires for a normal car have a tread depth of $\frac{10}{32}$ in. when new and are considered bald at $\frac{2}{32}$ in.

**Sources:** Indy500.com; *Consumer Reports*

## OBJECTIVES

**a** Identify the numerator and the denominator of a fraction and write fraction notation for part of an object or part of a set of objects and as a ratio.

**b** Simplify fraction notation like $n/n$ to 1, $0/n$ to 0, and $n/1$ to $n$.

## a Fractions and the Real World

Numbers like those above and the ones below are written in **fraction notation**. The top number is called the **numerator** and the bottom number is called the **denominator**.

$$\frac{1}{2}, \; \frac{7}{8}, \; \frac{7}{4}, \; \frac{11}{19}, \; \frac{3}{32}$$

**EXAMPLE 1** Identify the numerator and the denominator.

$$\frac{7}{8} \;\; \begin{array}{l} \leftarrow \text{Numerator} \\ \leftarrow \text{Denominator} \end{array}$$

> Do Exercises 1–3.

Let's look at various situations that involve fractions.

### Fractions as a Partition of an Object Divided into Equal Parts

Consider a candy bar divided into 5 equal sections. If you eat 2 sections, you have eaten $\frac{2}{5}$ of the candy bar.

The denominator 5 tells us the unit, $\frac{1}{5}$. The numerator 2 tells us the number of equal parts we are considering, 2.

For each fraction, identify the numerator and the denominator.

1. $\frac{3}{20}$ of people age 5 and older in Maryland speak a language other than English at home.
   **Source:** 2006 American Community Survey

2. In 2006, about $\frac{4}{25}$ of the motor vehicles produced in the world were produced in the United States.
   **Source:** Automotive News Data Center and R.L. Polk Marketing Systems GmbH

3. $\frac{19}{100}$ of the U.S. population in 2050 is projected to be foreign-born.
   **Source:** Pew Research Center

*Answers*

1. Numerator: 3; denominator: 20
2. Numerator: 4; denominator: 25
3. Numerator: 19; denominator: 100

**EXAMPLE 2** What part is shaded?

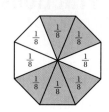

There are 8 equal parts. This tells us the unit, $\frac{1}{8}$. The *denominator* is 8. We have 5 of the units shaded. This tells us the *numerator*, 5. Thus,

$$\frac{5}{8} \begin{array}{l} \leftarrow \text{5 units are shaded.} \\ \leftarrow \text{The unit is } \frac{1}{8}. \end{array}$$

is shaded.

**EXAMPLE 3** What part is shaded?

There are 18 equal parts. Thus the unit is $\frac{1}{18}$. The denominator is 18. We have 7 units shaded. This tells us the numerator, 7. Thus, $\frac{7}{18}$ is shaded.

Do Exercises 4–7.

The markings on a ruler use fractions.

**EXAMPLE 4** What part of an inch is highlighted?

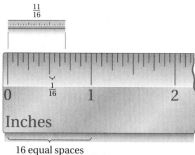

16 equal spaces

Each inch on the ruler shown above is divided into 16 equal parts. The highlighting extends to the 11th mark. Thus, $\frac{11}{16}$ of an inch is highlighted.

Do Exercise 8.

---

What part is shaded?

**4.**

**5.** 1 mile

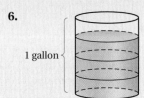

**6.**

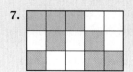

1 gallon

**7.**

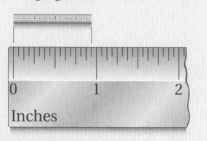

**8.** What part of an inch is highlighted?

Inches

---

*Answers*

4. $\frac{5}{6}$   5. $\frac{1}{3}$   6. $\frac{3}{4}$   7. $\frac{8}{15}$   8. $\frac{15}{16}$

Fractions greater than or equal to 1, such as $\frac{24}{24}$, $\frac{10}{3}$, and $\frac{5}{4}$, correspond to situations like the following.

**EXAMPLE 5** What part is shaded?

a)

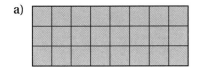

The rectangle is divided into 24 equal parts. Thus the unit is $\frac{1}{24}$. The denominator is 24. All 24 equal parts are shaded. This tells us that the numerator is 24. Thus, $\frac{24}{24}$ is shaded.

b)

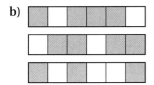

Each rectangle is divided into 6 parts. Thus the unit is $\frac{1}{6}$. The denominator is 6. We see that 11 of the equal units are shaded. This tells us that the numerator is 11. Thus, $\frac{11}{6}$ is shaded.

**EXAMPLE 6** *Ice-Cream Roll-up Cake.* What part of an ice-cream roll-up cake is shaded?

3 ice cream roll-up cakes

Each cake is divided into 6 equal slices. The unit is $\frac{1}{6}$. The *denominator* is 6. We see that 13 of the slices are shaded. This tells us that the *numerator* is 13. Thus, $\frac{13}{6}$ are shaded.

Do Exercises 9–11.

Negative fractions can also be used to represent real-world situations. If the water level in a reservoir drops five-eighths in., for example, the fraction $-\frac{5}{8}$ (read "negative five-eighths") represents this situation.

Do Exercise 12.

Fractions larger than or equal to 1, such as $\frac{13}{6}$ or $\frac{9}{9}$, are sometimes referred to as "improper" fractions. We will not use this terminology because notation such as $\frac{27}{8}$, $\frac{11}{3}$, and $\frac{4}{4}$ is quite "proper" and very common in algebra.

What part is shaded?

9.

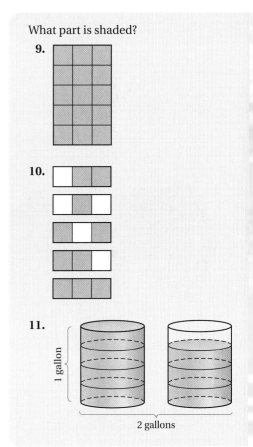

10.

11.

1 gallon

2 gallons

12. Write a fraction that represents the situation:

The blacktop on the playground sank one-fourth in.

*Answers*

9. $\frac{15}{15}$  10. $\frac{10}{3}$  11. $\frac{7}{4}$  12. $-\frac{1}{4}$

Fractions like $\frac{2}{3}$ are not integers. A number system called the **rational numbers** contains integers and fractions. The rational numbers consist of quotients of integers with nonzero divisors.

The following are rational numbers:

$$\frac{2}{3}, \quad -\frac{2}{3}, \quad \frac{7}{1}, \quad 4, \quad -3, \quad 0, \quad \frac{23}{-8}.$$

The number $-\frac{2}{3}$ (read "negative two-thirds") can also be named $\frac{-2}{3}$ or $\frac{2}{-3}$. That is,

$$-\frac{a}{b} = \frac{-a}{b} = \frac{a}{-b}.$$

---

### RATIONAL NUMBERS

The **rational numbers** consist of all numbers that can be named in the form $\frac{a}{b}$, where $a$ and $b$ are integers and $b$ is not 0.

---

### Fractions as Ratios

A **ratio** is a quotient of two quantities. We can express a ratio with fraction notation. (We will consider ratios in more detail in Chapter 6.)

**EXAMPLE 7**  *States East and West of the Mississippi River.*  What part of this group of states is east of the Mississippi River? west of the Mississippi River?

| | | | |
|---|---|---|---|
| Arkansas | Wisconsin | Illinois | Alabama |
| Nebraska | West Virginia | South Dakota | |

There are 7 states in the set. We know that 4 of them, Illinois, Alabama, Wisconsin, and West Virginia, are east of the Mississippi River. Thus, 4 of 7, or $\frac{4}{7}$, are east of the Mississippi River. The 3 remaining states are west of the Mississippi River. Thus, $\frac{3}{7}$ are west of the Mississippi River.

Do Exercise 13.

Mississippi River

**13.** What part of the set of states in Example 7 is north of Nashville, Tennessee? south of Branson, Missouri?

## b Some Fraction Notation for Integers

### Fraction Notation for 1

The number 1 corresponds to situations like those shown at left. If we divide an object into $n$ parts and take $n$ of them, we get all of the object (1 whole object). Similarly, when we divide a negative integer by itself, we get 1.

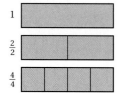

---

### THE NUMBER 1 IN FRACTION NOTATION

$\dfrac{n}{n} = 1,$   for any integer $n$ that is not 0.

---

*Answer*

13. $\frac{5}{7}; \frac{2}{7}$

**EXAMPLES** Simplify.

**8.** $\dfrac{5}{5} = 1$      **9.** $\dfrac{-9}{-9} = 1$      **10.** $\dfrac{23}{23} = 1$

Do Exercises 14–19.

## Fraction Notation for 0

Consider the fraction $\frac{0}{4}$. This corresponds to dividing an object into 4 parts and taking none of them. We get 0.

> **THE NUMBER 0 IN FRACTION NOTATION**
>
> $\dfrac{0}{n} = 0,$   for any integer $n$ that is not 0.

**EXAMPLES** Simplify.

**11.** $\dfrac{0}{1} = 0$      **12.** $\dfrac{0}{-9} = 0$      **13.** $\dfrac{0}{23} = 0$

Fraction notation with a denominator of 0, such as $n/0$, is meaningless because we cannot speak of an object being divided into *zero* parts. See also the discussion of excluding division by 0 in Section 1.5.

> **A DENOMINATOR OF 0**
>
> $\dfrac{n}{0}$   is not defined for any integer $n$.

Do Exercises 20–25.

## Other Integers

Consider the fraction $\frac{4}{1}$. This corresponds to taking 4 objects and dividing each into 1 part. (In other words, we do not divide them.) We have 4 objects. Similarly, when we divide a negative integer by 1, the result is also that integer.

> **ANY INTEGER IN FRACTION NOTATION**
>
> Any integer divided by 1 is the integer itself. That is,
>
> $\dfrac{n}{1} = n,$   for any integer $n$.

**EXAMPLES** Simplify.

**14.** $\dfrac{2}{1} = 2$      **15.** $\dfrac{9}{1} = 9$      **16.** $\dfrac{-34}{1} = -34$

Do Exercises 26–29.

---

Simplify.

**14.** $\dfrac{1}{1}$      **15.** $\dfrac{-4}{-4}$

**16.** $\dfrac{-34}{-34}$      **17.** $\dfrac{100}{100}$

**18.** $\dfrac{2347}{2347}$      **19.** $\dfrac{-103}{-103}$

Simplify, if possible.

**20.** $\dfrac{0}{1}$      **21.** $\dfrac{0}{-8}$

**22.** $\dfrac{0}{-107}$      **23.** $\dfrac{4-4}{567}$

**24.** $\dfrac{15}{0}$      **25.** $\dfrac{0}{3-3}$

Simplify.

**26.** $\dfrac{8}{1}$      **27.** $\dfrac{-10}{1}$

**28.** $\dfrac{-346}{1}$      **29.** $\dfrac{24-1}{23-22}$

*Answers*

14. 1   15. 1   16. 1   17. 1   18. 1
19. 1   20. 0   21. 0   22. 0   23. 0
24. Not defined   25. Not defined
26. 8   27. −10   28. −346   29. 23

**a** Identify the numerator and the denominator.

1. $\dfrac{3}{4}$    2. $\dfrac{9}{10}$    3. $\dfrac{11}{2}$    4. $\dfrac{18}{5}$    5. $\dfrac{0}{7}$    6. $\dfrac{1}{13}$

What part of each object or set of objects is shaded?

7.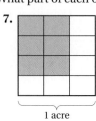

1 acre

8.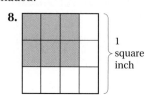

1 square inch

9.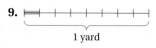

1 yard

10. 1 mile

11.

12.

13. 1 gold bar

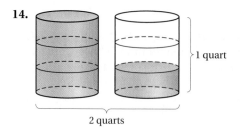

2 gold bars

14.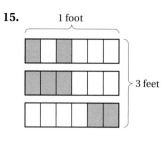

1 quart

2 quarts

15. 1 foot

3 feet

16. 1 mile

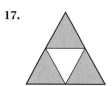

2 miles

17.

18.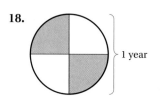

1 year

**19.**

**20.**

**21.**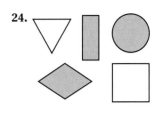
} 1 pie

**22.**

**23.**

**24.**

**25.**

**26.**

What part of an inch is highlighted?

**27.**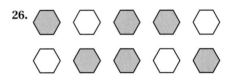

**28.**

**29.**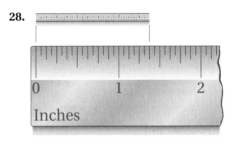

**30.**

For each of Exercises 31–34, give fraction notation for the amount of gas **(a)** in the tank and **(b)** used from a full tank.

**31.**

**32.**

**33.**

**34.**

**35.** For the following set of people, what is the ratio of:
  **a)** women to the total number of people?
  **b)** women to men?
  **c)** men to the total number of people?
  **d)** men to women?

**36.** For the following set of nuts and bolts, what is the ratio of:
  **a)** nuts to bolts?
  **b)** bolts to nuts?
  **c)** nuts to the total number of elements?
  **d)** total number of elements to nuts?

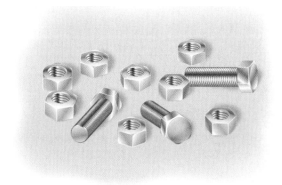

For Exercises 37 and 38, use the following bar graph, which shows the number of registered nurses per 100,000 residents in the District of Columbia and in several states.

**Registered Nurses per 100,000 Residents**

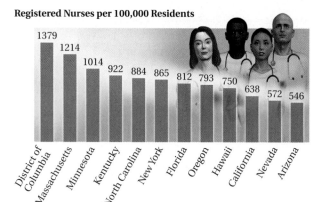

1379 District of Columbia
1214 Massachusetts
1014 Minnesota
922 Kentucky
884 North Carolina
865 New York
812 Florida
793 Oregon
750 Hawaii
638 California
572 Nevada
546 Arizona

**37.** What is the ratio of registered nurses to 100,000 residents in the given district or state?
  **a)** Minnesota      **b)** Hawaii
  **c)** Florida        **d)** Kentucky
  **e)** New York       **f)** District of Columbia

**38.** What is the ratio of registered nurses per 100,000 residents in the given state?
  **a)** North Carolina  **b)** California
  **c)** Massachusetts   **d)** Nevada
  **e)** Arizona         **f)** Oregon

**39.** Bryce delivers car parts to auto service centers. On Thursday he had 15 deliveries scheduled. By noon he had delivered only 4 orders. What is the ratio of:

**a)** orders delivered to total number of orders?
**b)** orders delivered to orders not delivered?
**c)** orders not delivered to total number of orders?

**40.** *Gas Mileage.* A Mazda MX-5 will travel 308 mi on 11 gal of gasoline in highway driving. What is the ratio of:

**a)** miles driven to gasoline used?
**b)** gasoline used to miles driven?

Source: *Road & Track*, August 2008, p. 52

**41.** *18–29 Voter Age Group.* Of every 100 voters in the 2004 presidential election, 16 voters were in the 18–29 age group. In the 2008 presidential election, this number increased to 18.5 voters. What is the ratio of voters in the 18–29 age group to all voters in the 2004 presidential election? in the 2008 election?

**Sources:** The Center for Information & Research on Civic Learning & Engagement; www.civicyouth.org

**42.** *Air-Conditioned Houses.* In 1971, of every 1000 new houses in the United States, 360 houses had air conditioning. In 2006, this ratio rose to 890 houses of every 1000 new houses. What is the ratio of new houses with air conditioning to all new houses in 1971? in 2006?

**Source:** U.S. Census Bureau

**b** Simplify.

**43.** $\dfrac{0}{8}$

**44.** $\dfrac{-8}{-8}$

**45.** $\dfrac{8-1}{9-8}$

**46.** $\dfrac{16}{1}$

**47.** $\dfrac{-20}{-20}$

**48.** $\dfrac{-20}{1}$

**49.** $\dfrac{45}{45}$

**50.** $\dfrac{11-1}{10-9}$

**51.** $\dfrac{0}{238}$

**52.** $\dfrac{238}{1}$

**53.** $\dfrac{238}{238}$

**54.** $\dfrac{0}{16}$

**55.** $\dfrac{3}{3}$

**56.** $\dfrac{56}{56}$

**57.** $\dfrac{0}{-87}$

**58.** $\dfrac{0}{98}$

**59.** $\dfrac{18}{18}$　　　　**60.** $\dfrac{0}{-18}$　　　　**61.** $\dfrac{-18}{1}$　　　　**62.** $\dfrac{8-8}{1247}$

**63.** $\dfrac{729}{0}$　　　　**64.** $\dfrac{1317}{0}$　　　　**65.** $\dfrac{5}{6-6}$　　　　**66.** $\dfrac{13}{10-10}$

## Skill Maintenance

Round 34,562 to the nearest: [1.6a]

**67.** Ten.　　　　**68.** Hundred.　　　　**69.** Thousand.　　　　**70.** Ten thousand.

Solve. [1.8a]

**71.** *New York Road Runners.* The New York Road Runners, a group that runs the New York City Marathon, had only 22,190 members in 1983. The membership increased to 46,655 in 2008. How many more members were there in 2008 than in 1983?

Source: New York Road Runners

**72.** *Gas Mileage.* The 2009 Chevrolet Corvette Z51 gets 26 mpg in highway driving. How many gallons will it use in 1846 mi of highway driving?

Source: *Road & Track*, August 2008, p. 52

Subtract. [1.3a]

**73.** $4035 - 2369$　　　　**74.** $8003 - 2642$

Add. [2.2a]

**75.** $7 + (-14)$　　　　**76.** $-6 + (-9)$

## Synthesis

What part of each object is shaded?

**77.** 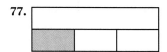　　**78.** 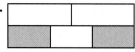　　**79.** 　　**80.**

Shade or mark each figure to show $\frac{3}{5}$.

**81.** 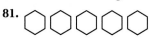　　**82.** 　　**83.** 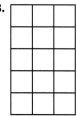　　**84.**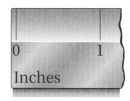

# 3.4 Multiplication and Applications

## a Multiplication Using Fraction Notation

Let's visualize the product of two fractions. We consider the multiplication

$$\frac{3}{5} \cdot \frac{3}{4}.$$

This is equivalent to finding $\frac{3}{5}$ of $\frac{3}{4}$. We first consider an object and take $\frac{3}{4}$ of it. We divide the object into 4 equal parts using vertical lines and take 3 of them. That is shown by the shading below.

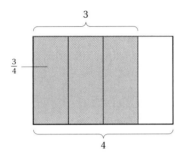

Next, we take $\frac{3}{5}$ of the shaded area. We divide it into 5 equal parts using horizontal lines and take 3 of them. That is shown by the darker shading below.

The entire object has now been divided into 20 parts, and we have shaded 9 of them twice. Thus we see that $\frac{3}{5}$ of $\frac{3}{4}$ is $\frac{9}{20}$, or

$$\frac{3}{5} \cdot \frac{3}{4} = \frac{9}{20}.$$

The figure above shows a rectangular array inside a rectangular array. The number of pieces in the entire array is $5 \cdot 4$ (the product of the denominators). The number of pieces shaded a second time is $3 \cdot 3$ (the product of the numerators). The product is represented by 9 pieces out of a set of 20, or $\frac{9}{20}$, which is the product of the numerators over the product of the denominators. This leads us to a statement of the procedure for multiplying a fraction by a fraction.

Do Margin Exercise 1.

## OBJECTIVES

**a** Multiply a fraction by a fraction, and multiply an integer and a fraction.

**b** Solve applied problems involving multiplication of fractions.

**SKILL TO REVIEW**
Objective 1.4a: Multiply whole numbers.

Multiply.

**1.** $24 \cdot 17$      **2.** $5(13)$

---

**1.** Draw a diagram like the one at left to show the multiplication

$$\frac{2}{3} \cdot \frac{4}{5}.$$

---

### Answers

*Skill to Review:*
**1.** 408    **2.** 65

*Margin Exercise:*
**1.**

We find a product such as $\frac{9}{7} \cdot \frac{3}{4}$ as follows.

> To multiply a fraction by a fraction,
>
> a) multiply the numerators to get the new numerator, and
>
> $$\frac{9}{7} \cdot \frac{3}{4} = \frac{9 \cdot 3}{7 \cdot 4} = \frac{27}{28}$$
>
> b) multiply the denominators to get the new denominator.

**EXAMPLES** Multiply.

1. $\dfrac{5}{6} \times \dfrac{7}{4} = \dfrac{5 \times 7}{6 \times 4} = \dfrac{35}{24}$

Skip writing this step whenever you can.

2. $\dfrac{3}{5} \cdot \dfrac{7}{8} = \dfrac{3 \cdot 7}{5 \cdot 8} = \dfrac{21}{40}$

3. $\dfrac{3}{5} \cdot \left(-\dfrac{4}{7}\right) = -\dfrac{3 \cdot 4}{5 \cdot 7} = -\dfrac{12}{35}$

4. $\dfrac{1}{4} \cdot \dfrac{1}{3} = \dfrac{1}{12}$

Do Exercises 2–5.

### Multiplication by an Integer

When multiplying a fraction by an integer, we first express the integer in fraction notation. We find a product such as $6 \cdot \frac{4}{5}$ as follows:

$$6 \cdot \frac{4}{5} = \frac{6}{1} \cdot \frac{4}{5} \qquad 6 = \frac{6}{1}$$

$$= \frac{6 \cdot 4}{1 \cdot 5} \qquad \text{Multiplying}$$

$$= \frac{24}{5}.$$

**EXAMPLES** Multiply.

5. $5 \times \dfrac{3}{8} = \dfrac{5 \times 3}{8} = \dfrac{15}{8}$

6. $\dfrac{2}{5} \cdot 13 = \dfrac{2 \cdot 13}{5} = \dfrac{26}{5}$

7. $10 \cdot \left(-\dfrac{1}{3}\right) = -\dfrac{10 \cdot 1}{3} = -\dfrac{10}{3}$

Do Exercises 6–8.

Multiply.

2. $\dfrac{3}{8} \cdot \dfrac{5}{7}$

3. $-\dfrac{4}{3} \times \dfrac{8}{5}$

4. $\dfrac{3}{10} \cdot \dfrac{1}{10}$

5. $-\dfrac{5}{2} \cdot \left(-\dfrac{9}{4}\right)$

Multiply.

6. $5 \times \dfrac{2}{3}$

7. $\dfrac{3}{8} \cdot 11$

8. $23 \cdot \left(-\dfrac{2}{5}\right)$

*Answers*

2. $\dfrac{15}{56}$   3. $-\dfrac{32}{15}$   4. $\dfrac{3}{100}$
5. $\dfrac{45}{8}$   6. $\dfrac{10}{3}$   7. $\dfrac{33}{8}$   8. $-\dfrac{46}{5}$

# b Applications and Problem Solving

Many problems that can be solved by multiplying fractions can be thought of in terms of rectangular arrays.

**EXAMPLE 8** A real estate developer owns a plot of land and plans to use $\frac{4}{5}$ of the plot for a small strip mall and parking lot. Of this, $\frac{2}{3}$ will be needed for the parking lot. What part of the plot will be used for parking?

1. **Familiarize.** We first make a drawing to help familiarize ourselves with the problem. The land may not be rectangular, but we can think of it as a rectangle. The strip mall, including the parking lot, uses $\frac{4}{5}$ of the plot. We shade $\frac{4}{5}$ as shown on the left below. The parking lot alone uses $\frac{2}{3}$ of the part we just shaded. We shade that as shown on the right below.

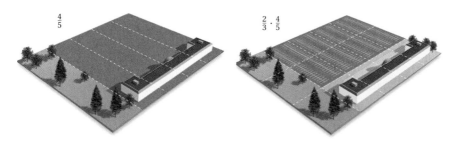

2. **Translate.** We let $n$ = the part of the plot that is used for parking. We are taking "two-thirds of four-fifths." Recall from Section 1.8 that the word "of" corresponds to multiplication. Thus the following multiplication sentence corresponds to the situation:

$$\frac{2}{3} \cdot \frac{4}{5} = n.$$

3. **Solve.** The number sentence tells us what to do. We multiply:

$$\frac{2}{3} \cdot \frac{4}{5} = \frac{2 \cdot 4}{3 \cdot 5} = \frac{8}{15}.$$

Thus, $\frac{8}{15} = n$.

4. **Check.** We can do a partial check by noting that the answer is a fraction less than 1, which we expect since the developer is using only part of the original plot of land. Thus, $\frac{8}{15}$ is a reasonable answer. We can also check this in the figure above, where we see that 8 of 15 parts represent the parking lot.

5. **State.** The parking lot takes up $\frac{8}{15}$ of the plot of land.

Do Exercise 9.

**STUDY TIPS**

**STUDY RESOURCES**

The new mathematical skills and concepts presented in class will be most easily learned if you begin your homework assignment as soon as possible after the lecture, while it is fresh in your mind. Then if you have difficulty with any of the exercises, you have time to access supplementary resources such as:

- *Student's Solutions Manual*
- Video Resources on DVD Featuring Chapter Test Prep Videos
- MyMathLab (and the Pearson Tutor Center)

9. A resort hotel uses $\frac{3}{4}$ of its extra land for recreational purposes. Of that, $\frac{1}{2}$ is used for swimming pools. What part of the land is used for swimming pools?

*Answer*

9. $\frac{3}{8}$

**EXAMPLE 9** *Area of a Cornfield.* The length of a rectangular cornfield is $\frac{15}{16}$ mi. The width is $\frac{5}{8}$ mi. What is the area of the cornfield?

1. **Familiarize.** Recall that area is length times width. We make a drawing and let $A$ = the area of the cornfield.

2. **Translate.** Then we translate:

$$\underbrace{\text{Area}}_{A} \quad \underbrace{\text{is}}_{=} \quad \underbrace{\text{Length}}_{\frac{15}{16}} \quad \underbrace{\text{times}}_{\times} \quad \underbrace{\text{Width}}_{\frac{5}{8}}.$$

3. **Solve.** The sentence tells us what to do. We multiply:

$$A = \frac{15}{16} \cdot \frac{5}{8} = \frac{15 \cdot 5}{16 \cdot 8} = \frac{75}{128}.$$

4. **Check.** We check by repeating the calculation. This is left to the student.

5. **State.** The area is $\frac{75}{128}$ square mile (mi²).

Do Exercise 10.

**10. Area of a Ceramic Tile.** The length of a rectangular ceramic tile on an inlaid ceramic counter is $\frac{4}{9}$ ft. The width is $\frac{2}{9}$ ft. What is the area of one tile?

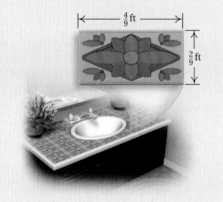

**EXAMPLE 10** A recipe for oatmeal chocolate chip cookies calls for $\frac{3}{4}$ cup of brown sugar. Monica is making $\frac{1}{2}$ of the recipe. How much brown sugar should she use?

1. **Familiarize.** We first make a drawing or at least visualize the situation. We let $n$ = the amount of brown sugar that Monica should use.

$\frac{3}{4}$ cup in the recipe        $\frac{1}{2} \cdot \frac{3}{4}$ cup in $\frac{1}{2}$ the recipe

2. **Translate.** We are finding $\frac{1}{2}$ of $\frac{3}{4}$, so the multiplication sentence $\frac{1}{2} \cdot \frac{3}{4} = n$ corresponds to the situation.

3. **Solve.** We carry out the multiplication:

$$\frac{1}{2} \cdot \frac{3}{4} = \frac{1 \cdot 3}{2 \cdot 4} = \frac{3}{8}.$$

Thus, $\frac{3}{8} = n$.

4. **Check.** We check by repeating the calculation. This is left to the student.

5. **State.** Monica should use $\frac{3}{8}$ cup of brown sugar.

Do Exercise 11.

**11.** Of the students at Overton Junior College, $\frac{1}{8}$ participate in sports and $\frac{3}{5}$ of these play football. What fractional part of the students play football?

*Answers*

10. $\frac{8}{81}$ ft²    11. $\frac{3}{40}$

**a** Multiply.

**1.** $\dfrac{2}{5} \cdot \dfrac{2}{3}$

**2.** $\dfrac{3}{4} \cdot \dfrac{3}{5}$

**3.** $10 \cdot \dfrac{7}{9}$

**4.** $9 \cdot \dfrac{5}{8}$

**5.** $-\dfrac{2}{3} \times \dfrac{1}{5}$

**6.** $-\dfrac{3}{5} \times \dfrac{1}{5}$

**7.** $-\dfrac{2}{11} \cdot 4$

**8.** $-\dfrac{2}{5} \cdot 3$

**9.** $\dfrac{7}{8} \cdot \dfrac{7}{8}$

**10.** $\dfrac{4}{5} \cdot \dfrac{4}{5}$

**11.** $5 \times \dfrac{1}{8}$

**12.** $4 \times \dfrac{1}{5}$

**13.** $-\dfrac{2}{3} \cdot \dfrac{7}{13}$

**14.** $-\dfrac{3}{11} \cdot \dfrac{4}{5}$

**15.** $\dfrac{1}{2} \cdot \dfrac{1}{3}$

**16.** $\dfrac{1}{6} \cdot \dfrac{1}{4}$

**17.** $\dfrac{2}{5} \cdot (-3)$

**18.** $\dfrac{3}{5} \cdot (-4)$

**19.** $3 \cdot \dfrac{1}{5}$

**20.** $2 \cdot \dfrac{1}{3}$

**21.** $7 \cdot \dfrac{3}{4}$

**22.** $7 \cdot \dfrac{2}{5}$

**23.** $\dfrac{3}{4} \cdot \dfrac{3}{4}$

**24.** $\dfrac{3}{7} \cdot \dfrac{4}{5}$

**25.** $\dfrac{1}{10} \cdot \left(-\dfrac{1}{100}\right)$

**26.** $\dfrac{3}{10} \cdot \left(-\dfrac{7}{100}\right)$

**27.** $-\dfrac{2}{5} \cdot (-1)$

**28.** $-\dfrac{3}{8} \cdot (-1)$

**29.** $\dfrac{6}{5} \cdot \dfrac{6}{5}$

**30.** $\dfrac{7}{4} \cdot \dfrac{3}{2}$

**31.** $\dfrac{11}{5} \cdot 6$

**32.** $\dfrac{5}{3} \cdot 2$

**33.** $\dfrac{1}{10} \cdot \left(-\dfrac{7}{10}\right)$

**34.** $\dfrac{3}{10} \cdot \left(-\dfrac{3}{10}\right)$

**b** Solve.

**35.** *Hair Bows.* It takes $\frac{5}{3}$ yd of ribbon to make a hair bow. How much ribbon is needed to make 8 bows?

**36.** A gasoline can holds $\frac{5}{2}$ gal. How much will the can hold when it is $\frac{1}{2}$ full?

**37.** *Basketball: High School to Pro.* One of 35 high school basketball players plays college basketball. One of 75 college players plays professional basketball. What fractional part of high school basketball players play professional basketball?

Source: National Basketball Association

**38.** *Football: High School to Pro.* One of 42 high school football players plays college football. One of 85 college players plays professional football. What fractional part of high school football players play professional football?

Source: National Football League

**39.** *Serving of Cheesecake.* At the Cheesecake Factory, a piece of cheesecake is $\frac{1}{12}$ of a cheesecake. How much of the cheesecake is $\frac{1}{2}$ piece?

Source: The Cheesecake Factory

**40.** *Tossed Salad.* The recipe for a tossed salad calls for $\frac{3}{8}$ cup of sliced almonds. How much is needed to make $\frac{1}{2}$ of the recipe?

**41.** *Floor Tiling.* The floor of a room is being covered with tile. An area $\frac{3}{5}$ of the length and $\frac{3}{4}$ of the width is covered. What fraction of the floor has been tiled?

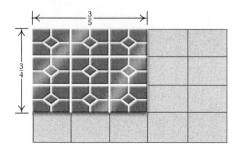

**42.** A rectangular table top measures $\frac{4}{5}$ m long by $\frac{3}{5}$ m wide. What is its area?

## Skill Maintenance

Divide.

**43.** $9 \overline{)27{,}009}$   [1.5a]

**44.** $35 \overline{)7148}$   [1.5a]

**45.** $7140 \div (-35)$   [2.5a]

**46.** $-32{,}200 \div 46$   [2.5a]

What does the digit 8 mean in each number?   [1.1a]

**47.** 4,678,952

**48.** 8,473,901

**49.** 7148

**50.** 23,803

Simplify.

**51.** $17 - 4^2$   [1.9c]

**52.** $(17 - 4)^2$   [1.9c]

**53.** $-144 \div 12 \div (-4)$   [2.5c]

**54.** $-144 \div [12 \div (-4)]$   [2.5c]

## Synthesis

Multiply. Write the answer using fraction notation.

**55.** 🖩 $\dfrac{341}{517} \cdot \dfrac{209}{349}$

**56.** 🖩 $\left(\dfrac{57}{61}\right)^3$

**57.** 🖩 $\left(\dfrac{2}{5}\right)^3 \left(-\dfrac{7}{9}\right)$

**58.** 🖩 $\left(-\dfrac{1}{2}\right)^5 \left(\dfrac{3}{5}\right)$

# 3.5 Simplifying

## a Multiplying by 1

Recall the following:

$$1 = \frac{1}{1} = \frac{2}{2} = \frac{3}{3} = \frac{4}{4} = \frac{10}{10} = \frac{45}{45} = \frac{100}{100} = \frac{n}{n}.$$

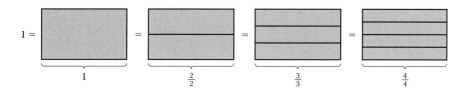

$$1 = \quad = \quad = \quad =$$
$$\underbrace{\qquad}_{1} \quad \underbrace{\qquad}_{\frac{2}{2}} \quad \underbrace{\qquad}_{\frac{3}{3}} \quad \underbrace{\qquad}_{\frac{4}{4}}$$

Any nonzero number divided by itself is 1. (See Section 1.5.)

Now recall the multiplicative identity from Section 1.4. For any whole number $a$, $1 \cdot a = a \cdot 1 = a$. This holds for fractions as well.

> ### MULTIPLICATIVE IDENTITY FOR FRACTIONS
>
> When we multiply a number by 1, we get the same number:
>
> $$\frac{3}{5} = \frac{3}{5} \cdot 1 = \frac{3}{5} \cdot \frac{4}{4} = \frac{12}{20}.$$

Since $\frac{3}{5} = \frac{12}{20}$, we know that $\frac{3}{5}$ and $\frac{12}{20}$ are two names for the same number. We also say that $\frac{3}{5}$ and $\frac{12}{20}$ are **equivalent fractions**.

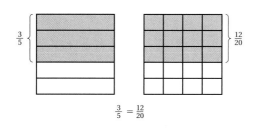

$$\frac{3}{5} = \frac{12}{20}$$

Do Margin Exercises 1–4.

Suppose we want to find a name for $\frac{2}{3}$, but one that has a denominator of 9. We can multiply by 1 to find equivalent fractions. Since $9 = 3 \cdot 3$, we choose $\frac{3}{3}$ for 1 in order to get a denominator of 9:

$$\frac{2}{3} = \frac{2}{3} \cdot 1 = \frac{2}{3} \cdot \frac{3}{3} = \frac{2 \cdot 3}{3 \cdot 3} = \frac{6}{9}.$$

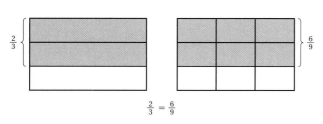

$$\frac{2}{3} = \frac{6}{9}$$

## OBJECTIVES

**a** Multiply a number by 1 to find fraction notation with a specified denominator.

**b** Simplify fraction notation.

**c** Use the test for equality to determine whether two fractions name the same number.

**SKILL TO REVIEW**
Objective 3.1d: Find the prime factorization of a composite number.

Find the prime factorization.
1. 84          2. 2250

Multiply.

1. $\frac{1}{2} \cdot \frac{8}{8}$          2. $\frac{3}{5} \cdot \frac{10}{10}$

3. $-\frac{13}{25} \cdot \frac{4}{4}$          4. $\frac{8}{3} \cdot \frac{25}{25}$

*Answers*

*Skill to Review:*
1. $2 \cdot 2 \cdot 3 \cdot 7$    2. $2 \cdot 3 \cdot 3 \cdot 5 \cdot 5 \cdot 5$

*Margin Exercises:*
1. $\frac{8}{16}$    2. $\frac{30}{50}$    3. $-\frac{52}{100}$    4. $\frac{200}{75}$

**EXAMPLE 1** Find a name for $\frac{1}{4}$ with a denominator of 24.

Since $4 \cdot 6 = 24$, we multiply by $\frac{6}{6}$:

$$\frac{1}{4} = \frac{1}{4} \cdot \frac{6}{6} = \frac{1 \cdot 6}{4 \cdot 6} = \frac{6}{24}.$$

We say that $\frac{1}{4}$ and $\frac{6}{24}$ represent the same number. They are equivalent.

**EXAMPLE 2** Find a name for $-\frac{2}{5}$ with a denominator of 35.

Since $5 \cdot 7 = 35$, we multiply by $\frac{7}{7}$:

$$-\frac{2}{5} = -\frac{2}{5} \cdot \frac{7}{7} = -\frac{2 \cdot 7}{5 \cdot 7} = -\frac{14}{35}.$$

The numbers $-\frac{2}{5}$ and $-\frac{14}{35}$ are equivalent.

Do Exercises 5–9.

Do Exercises 5–9.

## **b** Simplifying Fraction Notation

All of the following are names for three-fourths:

$$\frac{3}{4}, \frac{6}{8}, \frac{9}{12}, \frac{12}{16}, \frac{15}{20}.$$

We say that $\frac{3}{4}$ is **simplest** because it has the smallest numerator and the smallest denominator. That is, the numerator and the denominator have no common factor other than 1.

To simplify, we reverse the process of multiplying by 1:

$$\frac{12}{18} = \frac{2 \cdot 6}{3 \cdot 6} \quad \begin{matrix}\leftarrow \text{Factoring the numerator} \\ \leftarrow \text{Factoring the denominator}\end{matrix}$$

$$= \frac{2}{3} \cdot \frac{6}{6} \qquad \text{Factoring the fraction}$$

$$= \frac{2}{3} \cdot 1 \qquad \frac{6}{6} = 1$$

$$= \frac{2}{3}. \qquad \text{Removing a factor of 1: } \frac{2}{3} \cdot 1 = \frac{2}{3}$$

**EXAMPLES** Simplify.

**3.** $\dfrac{8}{20} = \dfrac{2 \cdot 4}{5 \cdot 4} = \dfrac{2}{5} \cdot \dfrac{4}{4} = \dfrac{2}{5}$

**4.** $\dfrac{2}{6} = \dfrac{1 \cdot 2}{3 \cdot 2} = \dfrac{1}{3} \cdot \dfrac{2}{2} = \dfrac{1}{3}$

> The number 1 allows for pairing of factors in the numerator and the denominator.

**5.** $\dfrac{30}{6} = \dfrac{5 \cdot 6}{1 \cdot 6} = \dfrac{5}{1} \cdot \dfrac{6}{6} = \dfrac{5}{1} = 5$

> We could also simplify $\frac{30}{6}$ by doing the division $30 \div 6$. That is, $\frac{30}{6} = 30 \div 6 = 5$.

Do Exercises 10–13.

Do Exercises 10–13.

---

Find another name for each number, but with the denominator indicated. Use multiplying by 1.

**5.** $\dfrac{4}{3} = \dfrac{?}{9}$

**6.** $\dfrac{3}{4} = \dfrac{?}{24}$

**7.** $-\dfrac{9}{10} = -\dfrac{?}{100}$

**8.** $\dfrac{3}{15} = \dfrac{?}{45}$

**9.** $-\dfrac{8}{7} = -\dfrac{?}{49}$

Simplify.

**10.** $\dfrac{2}{8}$

**11.** $-\dfrac{24}{18}$

**12.** $\dfrac{10}{12}$

**13.** $\dfrac{15}{80}$

*Answers*

**5.** $\dfrac{12}{9}$  **6.** $\dfrac{18}{24}$  **7.** $-\dfrac{90}{100}$  **8.** $\dfrac{9}{45}$  **9.** $-\dfrac{56}{49}$

**10.** $\dfrac{1}{4}$  **11.** $-\dfrac{4}{3}$  **12.** $\dfrac{5}{6}$  **13.** $\dfrac{3}{16}$

The use of prime factorizations can be helpful for simplifying when numerators and/or denominators are large numbers.

**EXAMPLE 6** Simplify: $\dfrac{90}{84}$.

$$\dfrac{90}{84} = \dfrac{2 \cdot 3 \cdot 3 \cdot 5}{2 \cdot 2 \cdot 3 \cdot 7} \qquad \text{Factoring the numerator and the denominator into primes}$$

$$= \dfrac{2 \cdot 3 \cdot 3 \cdot 5}{2 \cdot 3 \cdot 2 \cdot 7} \qquad \text{Changing the order so that like primes are above and below each other}$$

$$= \dfrac{2}{2} \cdot \dfrac{3}{3} \cdot \dfrac{3 \cdot 5}{2 \cdot 7} \qquad \text{Factoring the fraction}$$

$$= 1 \cdot 1 \cdot \dfrac{3 \cdot 5}{2 \cdot 7}$$

$$= \dfrac{3 \cdot 5}{2 \cdot 7} \qquad \text{Removing factors of 1}$$

$$= \dfrac{15}{14}$$

The tests for divisibility (Section 3.2) are very helpful in simplifying fraction notation. We could have shortened the preceding example had we noted that 6 is a factor of both the numerator and the denominator. Then we have

$$\dfrac{90}{84} = \dfrac{6 \cdot 15}{6 \cdot 14} = \dfrac{6}{6} \cdot \dfrac{15}{14} = \dfrac{15}{14}.$$

**EXAMPLE 7** Simplify: $\dfrac{603}{207}$.

At first glance this looks difficult. But note, using the test for divisibility by 9 (sum of digits is divisible by 9), that both the numerator and the denominator are divisible by 9. Thus we write both numbers with a factor of 9:

$$\dfrac{603}{207} = \dfrac{9 \cdot 67}{9 \cdot 23} = \dfrac{9}{9} \cdot \dfrac{67}{23} = \dfrac{67}{23}.$$

**EXAMPLE 8** Simplify: $-\dfrac{660}{1140}$.

Using the tests for divisibility, we have

$$-\dfrac{660}{1140} = -\dfrac{66 \cdot 10}{114 \cdot 10} = -\dfrac{66}{114} \cdot \dfrac{10}{10} = -\dfrac{66}{114} \qquad \text{Both 660 and 1140 are divisible by 10.}$$

$$= -\dfrac{11 \cdot 6}{19 \cdot 6} = -\dfrac{11}{19} \cdot \dfrac{6}{6} = -\dfrac{11}{19}. \qquad \text{Both 66 and 114 are divisible by 6.}$$

Do Exercises 14–20.

## Canceling

Canceling is a shortcut that you may have used for removing a factor of 1 when working with fraction notation. With *great* concern, we mention it as a possibility for speeding up your work. Canceling may be done only when removing common factors in numerators and denominators. Each such pair allows us to remove a factor of 1 in a fraction.

Simplify.

**14.** $\dfrac{35}{40}$

**15.** $\dfrac{801}{702}$

**16.** $-\dfrac{24}{21}$

**17.** $-\dfrac{75}{300}$

**18.** $\dfrac{280}{960}$

**19.** $\dfrac{1332}{2880}$

**20.** Simplify each fraction in this circle graph.

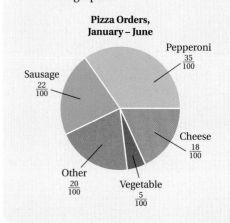

**Pizza Orders, January – June**

Pepperoni $\dfrac{35}{100}$

Sausage $\dfrac{22}{100}$

Cheese $\dfrac{18}{100}$

Other $\dfrac{20}{100}$

Vegetable $\dfrac{5}{100}$

**Simplifying Fraction Notation**   Fraction calculators are equipped with a key, often labeled $a^b/c$, that allows for simplification with fraction notation. To simplify

$$\frac{208}{256}$$

with such a fraction calculator, the following keystrokes can be used.

$$\boxed{2}\ \boxed{0}\ \boxed{8}\ \boxed{a^b/c}$$

$$\boxed{2}\ \boxed{5}\ \boxed{6}\ \boxed{=}.$$

The display that appears

$$\boxed{13 \lrcorner 16.}$$

represents simplified fraction notation $\frac{13}{16}$.

**Exercises:**   Use a fraction calculator to simplify each of the following.

1. $\dfrac{84}{90}$      2. $\dfrac{35}{40}$

3. $\dfrac{690}{835}$      4. $\dfrac{42}{150}$

Our concern is that canceling be done with care and understanding. In effect, slashes are used to indicate factors of 1 that have been removed. For instance, Example 6 might have been done faster as follows:

$$\frac{90}{84} = \frac{2 \cdot 3 \cdot 3 \cdot 5}{2 \cdot 2 \cdot 3 \cdot 7} \quad \text{Factoring the numerator and the denominator}$$

$$= \frac{2 \cdot \cancel{3} \cdot 3 \cdot 5}{2 \cdot 2 \cdot \cancel{3} \cdot 7} \quad \begin{array}{l}\text{When a factor of 1 is noted,}\\[4pt] \text{it is "canceled" as shown: } \dfrac{2}{2} \cdot \dfrac{3}{3} = 1.\end{array}$$

$$= \frac{3 \cdot 5}{2 \cdot 7} = \frac{15}{14}.$$

*— — — — — — — — — — — — — —* ***Caution!*** *— — — — — — — — — — — — — —*

The difficulty with canceling is that it is often applied incorrectly in situations like the following:

$$\frac{\cancel{2} + 3}{\cancel{2}} = 3; \qquad \frac{\cancel{4} + 1}{\cancel{4} + 2} = \frac{1}{2}; \qquad \frac{1\cancel{5}}{\cancel{5}4} = \frac{1}{4}.$$

$$\text{Wrong!} \qquad\qquad \text{Wrong!} \qquad\qquad \text{Wrong!}$$

The correct answers are

$$\frac{2 + 3}{2} = \frac{5}{2}; \qquad \frac{4 + 1}{4 + 2} = \frac{5}{6}; \qquad \frac{15}{54} = \frac{3 \cdot 5}{3 \cdot 18} = \frac{3}{3} \cdot \frac{5}{18} = \frac{5}{18}.$$

In each situation, the number canceled was not a factor of 1. Factors are parts of products. For example, in $2 \cdot 3$, 2 and 3 are factors, but in $2 + 3$, 2 and 3 are *not* factors. Canceling may not be done when sums or differences are in numerators or denominators, as shown here. **If you cannot factor, you cannot cancel! If in doubt, do not cancel!**

## c  A Test for Equality

When denominators are the same, we say that fractions have a **common denominator**. When fractions have a common denominator, we can compare them by comparing numerators. Suppose we want to compare $\frac{3}{6}$ and $\frac{2}{4}$. First, we find a common denominator. To do this, we multiply each fraction by 1, using the denominator of the other fraction to form the symbol for 1. We multiply $\frac{3}{6}$ by $\frac{4}{4}$ and $\frac{2}{4}$ by $\frac{6}{6}$:

$$\frac{3}{6} = \frac{3}{6} \cdot \frac{4}{4} = \frac{3 \cdot 4}{6 \cdot 4} = \frac{12}{24}; \qquad \text{Multiplying by } \frac{4}{4}$$

$$\frac{2}{4} = \frac{2}{4} \cdot \frac{6}{6} = \frac{2 \cdot 6}{4 \cdot 6} = \frac{12}{24}. \qquad \text{Multiplying by } \frac{6}{6}$$

Once we have a common denominator, 24, we compare the numerators. And since these numerators are both 12, the fractions are equal:

$$\frac{3}{6} = \frac{2}{4}.$$

Note in the preceding that if $\frac{3}{6} = \frac{2}{4}$, then $3 \cdot 4 = 6 \cdot 2$. This tells us that we need to check only the products $3 \cdot 4$ and $6 \cdot 2$ to compare the fractions.

## A TEST FOR EQUALITY

We multiply these two numbers: $3 \cdot 4$.

We multiply these two numbers: $6 \cdot 2$.

$$\frac{3}{6} \,\square\, \frac{2}{4}$$

We call $3 \cdot 4$ and $6 \cdot 2$ **cross products**. Since the cross products are the same—that is, $3 \cdot 4 = 6 \cdot 2$—we know that

$$\frac{3}{6} = \frac{2}{4}.$$

If a sentence $a = b$ is true, it means that $a$ and $b$ name the same number. If a sentence $a \neq b$ (read "$a$ is not equal to $b$") is true, it means that $a$ and $b$ do *not* name the same number.

**EXAMPLE 9**  Use $=$ or $\neq$ for $\square$ to write a true sentence:

$$\frac{6}{7} \,\square\, \frac{7}{8}.$$

We multiply these two numbers: $6 \cdot 8 = 48$.

We multiply these two numbers: $7 \cdot 7 = 49$.

$$\frac{6}{7} \,\square\, \frac{7}{8}$$

Because $48 \neq 49$, $\frac{6}{7}$ and $\frac{7}{8}$ do not name the same number. Thus,

$$\frac{6}{7} \neq \frac{7}{8}.$$

Recalling that $\dfrac{-a}{b} = \dfrac{a}{-b} = -\dfrac{a}{b}$ can be helpful when checking negative fractions for equality.

**EXAMPLE 10**  Use $=$ or $\neq$ for $\square$ to write a true sentence:

$$-\frac{6}{10} \,\square\, \frac{-3}{5}.$$

We rewrite $-\frac{6}{10}$ as $\frac{-6}{10}$ and then check cross products.

We multiply these two numbers: $-6 \cdot 5 = -30$.

We multiply these two numbers: $10 \cdot (-3) = -30$.

$$\frac{-6}{10} \,\square\, \frac{-3}{5}.$$

Because the cross products are the same, we have

$$\frac{-6}{10} = \frac{-3}{5}, \text{ or } -\frac{6}{10} = \frac{-3}{5}.$$

Do Exercises 21 and 22.

Use $=$ or $\neq$ for $\square$ to write a true sentence.

**21.** $\dfrac{2}{6} \,\square\, \dfrac{3}{9}$

**22.** $\dfrac{-2}{3} \,\square\, -\dfrac{14}{20}$

*Answers*

**21.** $=$    **22.** $\neq$

**a** Find another name for the given number, but with the denominator indicated. Use multiplying by 1.

**1.** $\dfrac{1}{2} = \dfrac{?}{10}$

**2.** $\dfrac{1}{6} = \dfrac{?}{18}$

**3.** $\dfrac{5}{8} = \dfrac{?}{32}$

**4.** $\dfrac{2}{9} = \dfrac{?}{18}$

**5.** $-\dfrac{9}{10} = -\dfrac{?}{30}$

**6.** $-\dfrac{5}{6} = -\dfrac{?}{48}$

**7.** $\dfrac{7}{8} = \dfrac{?}{32}$

**8.** $\dfrac{2}{5} = \dfrac{?}{25}$

**9.** $\dfrac{5}{12} = \dfrac{?}{48}$

**10.** $\dfrac{3}{8} = \dfrac{?}{56}$

**11.** $\dfrac{17}{18} = \dfrac{?}{-54}$

**12.** $\dfrac{11}{16} = \dfrac{?}{-256}$

**13.** $\dfrac{5}{3} = \dfrac{?}{45}$

**14.** $\dfrac{11}{5} = \dfrac{?}{30}$

**15.** $\dfrac{7}{22} = \dfrac{?}{132}$

**16.** $\dfrac{10}{21} = \dfrac{?}{126}$

**b** Simplify.

**17.** $\dfrac{2}{4}$

**18.** $\dfrac{4}{8}$

**19.** $-\dfrac{6}{8}$

**20.** $-\dfrac{8}{12}$

**21.** $\dfrac{3}{15}$

**22.** $\dfrac{8}{10}$

**23.** $\dfrac{-24}{8}$

**24.** $\dfrac{-36}{9}$

**25.** $\dfrac{18}{24}$

**26.** $\dfrac{42}{48}$

**27.** $\dfrac{14}{16}$

**28.** $\dfrac{15}{25}$

**29.** $\dfrac{12}{10}$

**30.** $\dfrac{16}{14}$

**31.** $\dfrac{16}{48}$

**32.** $\dfrac{100}{-20}$

**33.** $\dfrac{150}{-25}$

**34.** $\dfrac{19}{76}$

**35.** $-\dfrac{17}{51}$

**36.** $-\dfrac{425}{525}$

**37.** $\dfrac{390}{1410}$

**38.** $\dfrac{540}{810}$

**39.** $-\dfrac{1080}{2688}$

**40.** $-\dfrac{1000}{1080}$

**c** Use = or ≠ for ☐ to write a true sentence.

**41.** $\dfrac{3}{4}$ ☐ $\dfrac{9}{12}$

**42.** $\dfrac{4}{8}$ ☐ $\dfrac{3}{6}$

**43.** $\dfrac{1}{5}$ ☐ $\dfrac{2}{9}$

**44.** $\dfrac{1}{4}$ ☐ $\dfrac{2}{9}$

**45.** $\dfrac{3}{8}$ ☐ $\dfrac{6}{16}$

**46.** $\dfrac{2}{6}$ ☐ $\dfrac{6}{18}$

**47.** $\dfrac{-2}{5}$ ☐ $\dfrac{-3}{7}$

**48.** $\dfrac{-1}{3}$ ☐ $\dfrac{-1}{4}$

**49.** $\dfrac{12}{9}$ $\square$ $\dfrac{8}{6}$

**50.** $\dfrac{16}{14}$ $\square$ $\dfrac{8}{7}$

**51.** $-\dfrac{5}{2}$ $\square$ $\dfrac{-17}{7}$

**52.** $\dfrac{-3}{10}$ $\square$ $-\dfrac{7}{24}$

**53.** $\dfrac{3}{-10}$ $\square$ $-\dfrac{30}{100}$

**54.** $-\dfrac{700}{1000}$ $\square$ $\dfrac{70}{-100}$

**55.** $\dfrac{5}{10}$ $\square$ $\dfrac{520}{1000}$

**56.** $\dfrac{49}{100}$ $\square$ $\dfrac{50}{1000}$

What do college graduates do if they are not offered a full-time job in their first-choice field? The responses of 700 recent graduates are organized in the circle graph below. Use this graph in Exercises 57–60. Simplify the fraction representing each of the following.

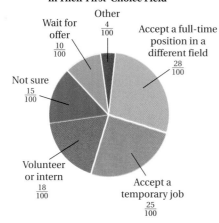

**What College Graduates Will Do If They Are Not Offered a Full-Time Job in Their First-Choice Field**

Other $\frac{4}{100}$

Wait for offer $\frac{10}{100}$

Accept a full-time position in a different field $\frac{28}{100}$

Not sure $\frac{15}{100}$

Volunteer or intern $\frac{18}{100}$

Accept a temporary job $\frac{25}{100}$

SOURCE: Yahoo Hotjobs survey

**57.** Accept a full-time position in a different field

**58.** Wait for offer

**59.** Accept a temporary job

**60.** Volunteer or intern

## Skill Maintenance

Solve.  [1.8a]

**61.** *Tiananmen Square.*  Tiananmen Square in Beijing, China, is the largest public square in the world. The length is 963 yd and the width is 547 yd. What is its area? its perimeter?

**62.** Rocco bought 3 shirts at $29 each and 2 sweaters at $38 each. How much did he spend altogether?

Subtract.

**63.** $803 - 617$  [1.3a]

**64.** $15 - 18$  [2.3a]

Solve.  [1.7b]

**65.** $5280 = 1760 + t$

**66.** $10{,}947 = 123 \cdot y$

## Synthesis

**67.** ▦ On a test of 82 questions, a student got 63 correct. On another test of 100 questions, she got 77 correct. Did she get the same portion of each test correct? Why or why not?

**68.** *Batting Averages.*  For the 2008 season, Joe Mauer, of the Minnesota Twins, won the American League batting title with 176 hits in 536 times at bat. Chipper Jones, of the Atlanta Braves, won the National League title with 160 hits in 439 times at bat. Did they have the same fraction of hits in times at bat (batting average)? Why or why not?

**Source:** Major League Baseball

# Mid-Chapter Review

## Concept Reinforcement

Determine whether each statement is true or false.

_____ **1.** A number $a$ is divisible by another number $b$ if $b$ is a factor of $a$.   [3.1b]

_____ **2.** If a number is not divisible by 6, then it is not divisible by 3.   [3.2a]

_____ **3.** The fraction $\dfrac{13}{7}$ is equal to the fraction $\dfrac{65}{34}$.   [3.5c]

_____ **4.** The number 1 is not prime.   [3.1c]

## Guided Solutions

Fill in each blank with the number that creates a correct statement or solution.

**5.** $\dfrac{25}{\square} = 1$   [3.3b]

**6.** $\dfrac{\square}{-9} = 0$   [3.3b]

**7.** $\dfrac{-8}{\square} = -8$   [3.3b]

**8.** $\dfrac{6}{13} = \dfrac{\square}{39}$   [3.5a]

**9.** Simplify: $\dfrac{70}{225}$.   [3.5b]

$$\frac{70}{225} = \frac{2 \cdot \square \cdot 7}{\square \cdot 3 \cdot 5 \cdot \square} \qquad \text{Factoring the numerator}$$
$$\text{Factoring the denominator}$$

$$= \frac{5}{5} \cdot \frac{\square \cdot 7}{3 \cdot \square \cdot 5} \qquad \text{Factoring the fraction}$$

$$= \square \cdot \frac{\square}{45} \qquad \frac{5}{5} = 1$$

$$= \frac{\square}{\square} \qquad \text{Removing a factor of 1}$$

## Mixed Review

To answer Exercises 10–14, consider the following numbers.   [3.2a]

| | | | |
|---|---|---|---|
| 84 | 132 | 594 | 350 |
| 300 | 500 | 120 | 14,850 |
| 17,576 | 180 | 1125 | 504 |
| 224 | 351 | 495 | 1632 |

**10.** Which of the above are divisible by 2 but not by 10?

**11.** Which of the above are divisible by 4 but not by 8?

**12.** Which of the above are divisible by 4 but not by 6?

**13.** Which of the above are divisible by 3 but not by 9?

**14.** Which of the above are divisible by 4, 5, and 6?

Determine whether each number is prime, composite, or neither.   [3.1c]

**15.** 61                **16.** 2                **17.** 91                **18.** 1

Find all the factors of each composite number. Then find the prime factorization of the number.   [3.1a], [3.1d]

**19.** 160                **20.** 222                **21.** 98                **22.** 315

What part of each object or set of objects is shaded?   [3.3a]

**23.**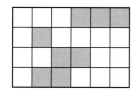

**24.**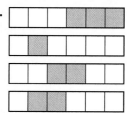

Multiply.   [3.4a]

**25.** $7 \cdot \dfrac{1}{9}$          **26.** $\dfrac{4}{15} \cdot \dfrac{2}{3}$          **27.** $\dfrac{5}{11} \cdot (-8)$          **28.** $-\dfrac{2}{9} \cdot \dfrac{4}{5}$

**29.** $-\dfrac{3}{4} \cdot \left(-\dfrac{7}{8}\right)$          **30.** $-5 \cdot \dfrac{2}{3}$          **31.** $\dfrac{1}{8} \cdot \dfrac{5}{2}$          **32.** $\dfrac{3}{7} \cdot \left(-\dfrac{9}{4}\right)$

Simplify.   [3.3b], [3.5b]

**33.** $\dfrac{24}{60}$                **34.** $\dfrac{220}{60}$                **35.** $\dfrac{17}{17}$

**36.** $\dfrac{0}{23}$                **37.** $\dfrac{315}{435}$                **38.** $\dfrac{14}{0}$

Use = or ≠ for ☐ to write a true sentence.   [3.5c]

**39.** $\dfrac{3}{7}$ ☐ $\dfrac{48}{112}$                **40.** $\dfrac{19}{3}$ ☐ $\dfrac{95}{18}$

**41.** *Veterinary Care.*   Of every 500 households that own pet birds, 60 obtain veterinary care for their birds. What is the ratio of households that own birds that seek veterinary care for their birds to all households that own birds?   [3.3a]
**Source:** American Veterinary Medical Association

**42.** *Area of an Ice-Skating Rink.*   The length of a rectangular ice-skating rink in the atrium of a shopping mall is $\dfrac{7}{100}$ mi. The width is $\dfrac{3}{100}$ mi. What is the area of the rink?   [3.4b]

# Understanding Through Discussion and Writing

**43.** Explain a method for finding a composite number that contains exactly two factors other than itself and 1. [3.1c]

**44.** Which of the years from 2000 to 2020, if any, also happen to be prime numbers? Explain at least two ways in which you might go about solving this problem.   [3.2a]

**45.** Explain in your own words when it *is* possible to "cancel" and when it *is not* possible to "cancel."   [3.5b]

**46.** Can fraction notation be simplified if the numerator and the denominator are two different prime numbers? Why or why not?   [3.5b]

# 3.6

## Multiplying, Simplifying, and Applications

### OBJECTIVES

**a** Multiply and simplify using fraction notation.

**b** Solve applied problems involving multiplication of fractions.

**SKILL TO REVIEW**

Objective 3.2a: Determine whether a number is divisible by 2, 3, 4, 5, 6, 8, 9, or 10.

Determine whether each number is divisible by 9.

1. 486          2. 129

### a Multiplying and Simplifying Using Fraction Notation

It is often possible to simplify after we multiply. To make such simplifying easier, it is usually best not to carry out the products in the numerator and the denominator immediately, but to factor and simplify first. Consider the product $\frac{3}{8} \cdot \frac{4}{9}$. We proceed as follows:

$$\frac{3}{8} \cdot \frac{4}{9} = \frac{3 \cdot 4}{8 \cdot 9}$$

We write the products in the numerator and the denominator, but we do not carry them out.

$$= \frac{3 \cdot 2 \cdot 2}{2 \cdot 2 \cdot 2 \cdot 3 \cdot 3}$$

Factoring the numerator and the denominator

$$= \frac{3 \cdot 2 \cdot 2 \cdot 1}{2 \cdot 2 \cdot 2 \cdot 3 \cdot 3}$$

Using the identity property of 1 to insert the number 1 as a factor

$$= \frac{3 \cdot 2 \cdot 2}{3 \cdot 2 \cdot 2} \cdot \frac{1}{2 \cdot 3}$$

Factoring the fraction

$$= 1 \cdot \frac{1}{2 \cdot 3}$$

$$= \frac{1}{2 \cdot 3}$$

Removing a factor of 1

$$= \frac{1}{6}.$$

The procedure could have been shortened had we noticed that 4 is a factor of the 8 in the denominator:

$$\frac{3}{8} \cdot \frac{4}{9} = \frac{3 \cdot 4}{8 \cdot 9} = \frac{3 \cdot 4}{4 \cdot 2 \cdot 3 \cdot 3} = \frac{3 \cdot 4}{3 \cdot 4} \cdot \frac{1}{2 \cdot 3} = 1 \cdot \frac{1}{2 \cdot 3} = \frac{1}{2 \cdot 3} = \frac{1}{6}.$$

> **To multiply and simplify:**
>
> a) Write the products in the numerator and the denominator, but do not carry them out.
>
> b) Factor the numerator and the denominator.
>
> c) Factor the fraction to remove a factor of 1, if possible.
>
> d) Carry out the remaining products.

### STUDY TIPS

**ASKING QUESTIONS**

Don't be afraid to ask questions in class. Most instructors welcome this and encourage students to ask them. Other students probably have the same questions you do.

**EXAMPLES**   Multiply and simplify.

**1.** $\frac{2}{3} \cdot \frac{9}{4} = \frac{2 \cdot 9}{3 \cdot 4} = \frac{2 \cdot 3 \cdot 3}{3 \cdot 2 \cdot 2} = \frac{2 \cdot 3}{2 \cdot 3} \cdot \frac{3}{2} = 1 \cdot \frac{3}{2} = \frac{3}{2}$

**2.** $\frac{6}{7} \cdot \frac{5}{3} = \frac{6 \cdot 5}{7 \cdot 3} = \frac{3 \cdot 2 \cdot 5}{7 \cdot 3} = \frac{3}{3} \cdot \frac{2 \cdot 5}{7} = 1 \cdot \frac{2 \cdot 5}{7} = \frac{2 \cdot 5}{7} = \frac{10}{7}$

**3.** $40 \cdot \left(-\frac{7}{8}\right) = -\frac{40 \cdot 7}{8} = -\frac{5 \cdot 8 \cdot 7}{8 \cdot 1} = -\frac{5 \cdot 7}{1} \cdot \frac{8}{8} = -\frac{5 \cdot 7}{1} \cdot 1$

$= -\frac{5 \cdot 7}{1} = -35$

*Answers*

*Skill to Review:*

1. Yes    2. No

--------------------------------- *Caution!* ---------------------------------

Canceling can be used as follows for these examples.

**1.** $\dfrac{2}{3} \cdot \dfrac{9}{4} = \dfrac{2 \cdot 9}{3 \cdot 4} = \dfrac{2 \cdot \cancel{3} \cdot 3}{\cancel{3} \cdot \cancel{2} \cdot 2} = \dfrac{3}{2}$    Removing a factor of 1:
$\dfrac{2 \cdot 3}{2 \cdot 3} = 1$

**2.** $\dfrac{6}{7} \cdot \dfrac{5}{3} = \dfrac{6 \cdot 5}{7 \cdot 3} = \dfrac{\cancel{3} \cdot 2 \cdot 5}{7 \cdot \cancel{3}} = \dfrac{2 \cdot 5}{7} = \dfrac{10}{7}$    Removing a factor of 1:
$\dfrac{3}{3} = 1$

**3.** $40 \cdot \left(-\dfrac{7}{8}\right) = -\dfrac{40 \cdot 7}{1 \cdot 8} = -\dfrac{5 \cdot \cancel{8} \cdot 7}{\cancel{8} \cdot 1} = -\dfrac{5 \cdot 7}{1}$    Removing a factor of 1:
$\dfrac{8}{8} = 1$

$= -35$

**Remember: if you can't factor, you can't cancel!**

--------------------------------------------------------

> Do Exercises 1–4.

Do Exercises 1–4.

> **Multiply and simplify.**
>
> **1.** $\dfrac{2}{3} \cdot \dfrac{7}{8}$
>
> **2.** $-\dfrac{4}{5} \cdot \dfrac{5}{12}$
>
> **3.** $16 \cdot \dfrac{3}{8}$
>
> **4.** $\left(-\dfrac{5}{8}\right) \cdot (-4)$

## ⓑ Applications and Problem Solving

**EXAMPLE 4** *Landscaping.* Celina's Landscaping uses $\frac{2}{3}$ lb of peat moss when planting a rosebush. How much will be needed to plant 21 rosebushes?

**1. Familiarize.** We first make a drawing or at least visualize the situation. We let $n =$ the number of pounds of peat moss needed.

21 rosebushes

$\frac{2}{3}$ pound of peat moss for each rosebush

**2. Translate.** The problem translates to the following equation:

$$n = 21 \cdot \frac{2}{3}.$$

**3. Solve.** To solve the equation, we carry out the multiplication:

$$n = 21 \cdot \frac{2}{3} = \frac{21}{1} \cdot \frac{2}{3} = \frac{21 \cdot 2}{1 \cdot 3}\qquad \text{Multiplying}$$

$$= \frac{3 \cdot 7 \cdot 2}{1 \cdot 3} = \frac{3}{3} \cdot \frac{7 \cdot 2}{1} = 14.$$

**4. Check.** We check by repeating the calculation. (This is left to the student.) We can also ask if the answer seems reasonable. We are putting less than a pound of peat moss on each bush, so the answer should be less than 21. Since 14 is less than 21, we have a partial check. The number 14 checks.

**5. State.** Celina's Landscaping will need 14 lb of peat moss to plant 21 rosebushes.

> Do Exercise 5.

Do Exercise 5.

> **5. Truffles.** Chocolate Delight sells $\frac{4}{5}$-lb boxes of truffles. How many pounds of truffles will be needed to fill 85 boxes?

> *Answers*
>
> **1.** $\dfrac{7}{12}$   **2.** $-\dfrac{1}{3}$   **3.** 6   **4.** $\dfrac{5}{2}$   **5.** 68 lb

# 3.6 Exercise Set

**For Extra Help**

MyMathLab

Math XL
PRACTICE

WATCH

DOWNLOAD

READ

REVIEW

**a** Multiply and simplify.  | Don't forget to simplify! |

1. $\dfrac{2}{3} \cdot \dfrac{1}{2}$

2. $\dfrac{3}{8} \cdot \dfrac{1}{3}$

3. $\dfrac{7}{8} \cdot \dfrac{1}{7}$

4. $\dfrac{4}{9} \cdot \dfrac{1}{4}$

5. $-\dfrac{1}{8} \cdot \dfrac{4}{5}$

6. $-\dfrac{2}{5} \cdot \dfrac{1}{6}$

7. $\dfrac{1}{4} \cdot \dfrac{2}{3}$

8. $\dfrac{4}{6} \cdot \dfrac{1}{6}$

9. $\dfrac{12}{5} \cdot \dfrac{9}{8}$

10. $\dfrac{16}{15} \cdot \dfrac{5}{4}$

11. $\dfrac{10}{9} \cdot \left(-\dfrac{7}{5}\right)$

12. $\dfrac{25}{12} \cdot \left(-\dfrac{4}{3}\right)$

13. $9 \cdot \dfrac{1}{9}$

14. $4 \cdot \dfrac{1}{4}$

15. $\dfrac{1}{3} \cdot 3$

16. $\dfrac{1}{6} \cdot 6$

17. $\left(-\dfrac{7}{10}\right) \cdot \left(-\dfrac{10}{7}\right)$

18. $\left(-\dfrac{8}{9}\right) \cdot \left(-\dfrac{9}{8}\right)$

19. $\dfrac{7}{5} \cdot \dfrac{5}{7}$

20. $\dfrac{2}{11} \cdot \dfrac{11}{2}$

21. $\dfrac{1}{4} \cdot 8$

22. $\dfrac{1}{3} \cdot 18$

23. $-24 \cdot \dfrac{1}{6}$

24. $-16 \cdot \dfrac{1}{2}$

**25.** $12 \cdot \dfrac{3}{4}$

**26.** $18 \cdot \dfrac{5}{6}$

**27.** $-\dfrac{3}{8} \cdot 24$

**28.** $-\dfrac{2}{9} \cdot 36$

**29.** $-13 \cdot \left(-\dfrac{2}{5}\right)$

**30.** $-15 \cdot \left(-\dfrac{1}{6}\right)$

**31.** $\dfrac{7}{10} \cdot 28$

**32.** $\dfrac{5}{8} \cdot 34$

**33.** $\dfrac{1}{6} \cdot 360$

**34.** $\dfrac{1}{3} \cdot 120$

**35.** $240 \cdot \left(-\dfrac{1}{8}\right)$

**36.** $150 \cdot \left(-\dfrac{1}{5}\right)$

**37.** $\dfrac{4}{10} \cdot \dfrac{5}{10}$

**38.** $\dfrac{7}{10} \cdot \dfrac{34}{150}$

**39.** $-\dfrac{8}{10} \cdot \dfrac{45}{100}$

**40.** $-\dfrac{3}{10} \cdot \dfrac{8}{10}$

**41.** $\dfrac{11}{24} \cdot \dfrac{3}{5}$

**42.** $\dfrac{15}{22} \cdot \dfrac{4}{7}$

**43.** $-\dfrac{10}{21} \cdot \left(-\dfrac{3}{4}\right)$

**44.** $-\dfrac{17}{18} \cdot \left(-\dfrac{3}{5}\right)$

 Solve.

The *pitch* of a screw is the distance between its threads. With each complete rotation, the screw goes in or out a distance equal to its pitch. Use this information to do Exercises 45 and 46.

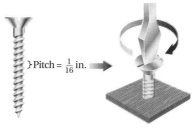

}Pitch = $\frac{1}{16}$ in.

Each rotation moves the screw in or out $\frac{1}{16}$ in.

**45.** The pitch of a screw is $\frac{1}{16}$ in. How far will it go into a piece of oak when it is turned 10 complete rotations clockwise?

**46.** The pitch of a screw is $\frac{3}{32}$ in. How far will it go out of a piece of plywood when it is turned 10 complete rotations counterclockwise?

**47.** *Level of Education and Median Income.* The median yearly income of someone with an associate's degree is approximately $\frac{2}{3}$ of the median income of someone with a bachelor's degree. If the median income for those with bachelor's degrees is $72,420, what is the median income of those with associate's degrees?

Source: U.S. Census Bureau

**48.** After Jack completes 60 hr of teacher training in college, he can earn $75 for working a full day as a substitute teacher. How much will he receive for working $\frac{3}{5}$ of a day?

**49.** *Mailing-List Addresses.* Researchers have determined that $\frac{1}{4}$ of the addresses on a mailing list will change in one year. A business has a mailing list of 2500 people. After one year, how many addresses on that list will be incorrect?

**50.** *Shy People.* Sociologists have determined that $\frac{2}{5}$ of the people in the world are shy. A sales manager is considering 650 people for an aggressive sales position. How many of these people might be shy?

**51.** A recipe for piecrust calls for $\frac{2}{3}$ cup of flour. A baker is making $\frac{1}{2}$ of the recipe. How much flour should the baker use?

**52.** Of the students in the freshman class, $\frac{4}{5}$ have cameras; $\frac{1}{4}$ of these students also join the college photography club. What fraction of the students in the freshman class join the photography club?

**53.** A house worth $154,000 is assessed for $\frac{3}{4}$ of its value. What is the assessed value of the house?

**54.** Roxanne's tuition was $4600. A loan was obtained for $\frac{3}{4}$ of the tuition. How much was the loan?

**55.** *Map Scaling.* On a map, 1 in. represents 240 mi. What distance does $\frac{2}{3}$ in. represent?

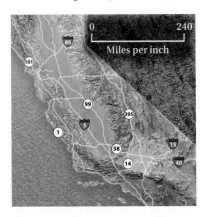

**56.** *Map Scaling.* On a map, 1 in. represents 120 mi. What distance does $\frac{3}{4}$ in. represent?

**57.** *Household Budgets.* A family has an annual income of $42,000. Of this, $\frac{1}{5}$ is spent for food, $\frac{1}{4}$ for housing, $\frac{1}{10}$ for clothing, $\frac{1}{14}$ for savings, $\frac{1}{5}$ for taxes, and the rest for other expenses. How much is spent for each?

**58.** *Household Budgets.* A family has an annual income of $28,140. Of this, $\frac{1}{5}$ is spent for food, $\frac{1}{4}$ for housing, $\frac{1}{10}$ for clothing, $\frac{1}{14}$ for savings, $\frac{1}{5}$ for taxes, and the rest for other expenses. How much is spent for each?

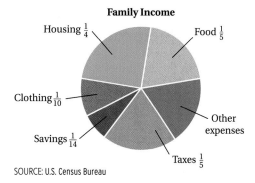

**Family Income**

Housing $\frac{1}{4}$    Food $\frac{1}{5}$

Clothing $\frac{1}{10}$

Savings $\frac{1}{14}$

Other expenses

Taxes $\frac{1}{5}$

SOURCE: U.S. Census Bureau

## Skill Maintenance

Solve.   [1.7b]

**59.** $48 \cdot t = 1680$

**60.** $74 \cdot x = 6290$

**61.** $3125 = 25 \cdot t$

**62.** $2880 = 24 \cdot y$

**63.** $t + 28 = 5017$

**64.** $456 + x = 9002$

**65.** $8797 = y + 2299$

**66.** $10{,}000 = 3593 + m$

Subtract.   [2.3a]

**67.** $-14 - 2$

**68.** $5 - (-6)$

**69.** $8 - 12$

**70.** $-4 - (-4)$

## Synthesis

Multiply and simplify. Use the list of prime numbers on p. 131 or a fraction calculator.

**71.** ▦ $\dfrac{201}{535} \cdot \dfrac{4601}{6499}$

**72.** ▦ $-\dfrac{5767}{3763} \cdot \dfrac{159}{395}$

**73.** *College Profile.* Of students entering a college, $\frac{7}{8}$ have completed high school and $\frac{2}{3}$ are older than 20. If $\frac{1}{7}$ of all students are left-handed, what fraction of students entering the college are left-handed high school graduates over the age of 20?

**74.** *College Profile.* Refer to the information in Exercise 73. If 480 students are entering the college, how many of them are left-handed high school graduates 20 yr old or younger?

**75.** *College Profile.* Refer to Exercise 73. What fraction of students entering the college did not graduate from high school, are 20 yr old or younger, and are left-handed?

# 3.7

## Division and Applications

**a** Find the reciprocal of a number.

**b** Divide and simplify using fraction notation.

**c** Solve equations of the type $a \cdot x = b$ and $x \cdot a = b$, where $a$ and $b$ may be fractions.

**d** Solve applied problems involving division of fractions.

### SKILL TO REVIEW

Objective 1.7b: Solve equations like $t + 28 = 54$, $28 \cdot x = 168$, and $98 \cdot 2 = y$.

Solve.

1. $88 = 8 \cdot y$

2. $16 \cdot x = 1152$

### a  Reciprocals

Look at these products:

$$8 \cdot \frac{1}{8} = \frac{8}{1} \cdot \frac{1}{8} = \frac{8 \cdot 1}{1 \cdot 8} = \frac{8}{8} = 1; \qquad \frac{2}{3} \cdot \frac{3}{2} = \frac{2 \cdot 3}{3 \cdot 2} = \frac{6}{6} = 1.$$

---

**RECIPROCALS**

If the product of two numbers is 1, we say that they are **reciprocals** of each other. To find the reciprocal of a fraction, interchange the numerator and the denominator.

$$\text{Number: } \frac{3}{4} \longrightarrow \text{Reciprocal: } \frac{4}{3}$$

---

**EXAMPLES**  Find the reciprocal.

1. The reciprocal of 24 is $\frac{1}{24}$.  Think of 24 as $\frac{24}{1}$: $\frac{24}{1} \cdot \frac{1}{24} = \frac{24}{24} = 1$.

2. The reciprocal of $\frac{1}{3}$ is 3.  $\frac{1}{3} \cdot 3 = \frac{1}{3} \cdot \frac{3}{1} = \frac{3}{3} = 1$

3. The reciprocal of $-\frac{3}{4}$ is $-\frac{4}{3}$.  The reciprocal of a negative number is negative: $-\frac{3}{4} \cdot \left(-\frac{4}{3}\right) = \frac{12}{12} = 1$.

Do Margin Exercises 1–4.

Does 0 have a reciprocal? If it did, it would have to be a number $x$ such that $0 \cdot x = 1$. But 0 times any number is 0. Thus we have the following.

---

**0 HAS NO RECIPROCAL**

The number 0, or $\frac{0}{n}$, has no reciprocal. $\left( \text{Recall that } \frac{n}{0} \text{ is not defined.} \right)$

---

Find the reciprocal.

1. 9

2. $\frac{10}{7}$

3. $-\frac{2}{5}$

4. $\frac{1}{5}$

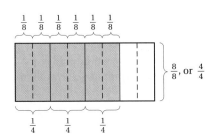

### b  Division

Consider the division $\frac{3}{4} \div \frac{1}{8}$. We are asking how many $\frac{1}{8}$'s are in $\frac{3}{4}$. We can answer this by looking at the figure at left. We see that there are six $\frac{1}{8}$'s in $\frac{3}{4}$. Thus,

$$\frac{3}{4} \div \frac{1}{8} = 6.$$

We can check this by multiplying:

$$6 \cdot \frac{1}{8} = \frac{6}{1} \cdot \frac{1}{8} = \frac{6}{8} = \frac{2 \cdot 3}{2 \cdot 4} = \frac{2}{2} \cdot \frac{3}{4} = \frac{3}{4}.$$

### Answers

*Skill to Review:*
1. 11   2. 72

*Margin Exercises:*
1. $\frac{1}{9}$   2. $\frac{7}{10}$   3. $-\frac{5}{2}$   4. 5

Here is a faster way to do this division:

$$\frac{3}{4} \div \frac{1}{8} = \frac{3}{4} \cdot \frac{8}{1} = \frac{3 \cdot 8}{4 \cdot 1} = \frac{24}{4} = 6.$$   Multiplying by the reciprocal of the divisor

---

To divide fractions, multiply the dividend by the reciprocal of the divisor:

$$\frac{2}{5} \div \frac{3}{4} = \frac{2}{5} \cdot \frac{4}{3} = \frac{2 \cdot 4}{5 \cdot 3} = \frac{8}{15}.$$

Multiply by the reciprocal of the divisor.

---

**EXAMPLES**   Divide and simplify.

**4.** $\dfrac{5}{6} \div \dfrac{2}{3} = \dfrac{5}{6} \cdot \dfrac{3}{2} = \dfrac{5 \cdot 3}{6 \cdot 2} = \dfrac{5 \cdot 3}{3 \cdot 2 \cdot 2} = \dfrac{3}{3} \cdot \dfrac{5}{2 \cdot 2} = \dfrac{5}{2 \cdot 2} = \dfrac{5}{4}$

**5.** $-\dfrac{7}{8} \div \dfrac{1}{16} = -\dfrac{7}{8} \cdot 16 = -\dfrac{7 \cdot 16}{8} = -\dfrac{7 \cdot 2 \cdot 8}{8 \cdot 1} = -\dfrac{7 \cdot 2}{1} \cdot \dfrac{8}{8}$

$= -\dfrac{7 \cdot 2}{1} = -14$

**6.** $\dfrac{2}{5} \div 6 = \dfrac{2}{5} \cdot \dfrac{1}{6} = \dfrac{2 \cdot 1}{5 \cdot 6} = \dfrac{2 \cdot 1}{5 \cdot 2 \cdot 3} = \dfrac{2}{2} \cdot \dfrac{1}{5 \cdot 3} = \dfrac{1}{5 \cdot 3} = \dfrac{1}{15}$

---

### Caution!

Canceling can be used as follows for Examples 4–6.

**4.** $\dfrac{5}{6} \div \dfrac{2}{3} = \dfrac{5}{6} \cdot \dfrac{3}{2} = \dfrac{5 \cdot 3}{6 \cdot 2} = \dfrac{5 \cdot \cancel{3}}{\cancel{3} \cdot 2 \cdot 2} = \dfrac{5}{2 \cdot 2} = \dfrac{5}{4}$   Removing a factor of 1: $\frac{3}{3} = 1$

**5.** $-\dfrac{7}{8} \div \dfrac{1}{16} = -\dfrac{7}{8} \cdot 16 = -\dfrac{7 \cdot 16}{8} = -\dfrac{7 \cdot \cancel{8} \cdot 2}{\cancel{8} \cdot 1}$   Removing a factor of 1: $\frac{8}{8} = 1$

$= -\dfrac{7 \cdot 2}{1} = -14$

**6.** $\dfrac{2}{5} \div 6 = \dfrac{2}{5} \cdot \dfrac{1}{6} = \dfrac{2 \cdot 1}{5 \cdot 6} = \dfrac{\cancel{2} \cdot 1}{5 \cdot \cancel{2} \cdot 3} = \dfrac{1}{5 \cdot 3} = \dfrac{1}{15}$   Removing a factor of 1: $\frac{2}{2} = 1$

**Remember: if you can't factor, you can't cancel!**

Do Exercises 5–8.

What is the explanation for multiplying by a reciprocal when dividing? Let's consider $\frac{2}{3} \div \frac{7}{5}$. We multiply by 1. The name for 1 that we will use is $(5/7)/(5/7)$; it comes from the reciprocal of $\frac{7}{5}$.

$$\frac{2}{3} \div \frac{7}{5} = \frac{\frac{2}{3}}{\frac{7}{5}} = \frac{\frac{2}{3}}{\frac{7}{5}} \cdot 1 = \frac{\frac{2}{3} \cdot \frac{5}{7}}{\frac{7}{5} \cdot \frac{5}{7}} = \frac{\frac{2}{3} \cdot \frac{5}{7}}{1} = \frac{2}{3} \cdot \frac{5}{7} = \frac{10}{21}$$

Thus,

$$\frac{2}{3} \div \frac{7}{5} = \frac{2}{3} \cdot \frac{5}{7} = \frac{10}{21}.$$

Do Exercise 9.

Divide and simplify.

**5.** $\dfrac{6}{7} \div \dfrac{3}{4}$   **6.** $-\dfrac{2}{3} \div \dfrac{1}{4}$

**7.** $\dfrac{4}{5} \div 8$   **8.** $-60 \div \left(-\dfrac{3}{5}\right)$

**9.** Divide by multiplying by 1:

$$\frac{\frac{4}{5}}{\frac{6}{7}}$$

*Answers*

**5.** $\dfrac{8}{7}$   **6.** $-\dfrac{8}{3}$   **7.** $\dfrac{1}{10}$   **8.** 100   **9.** $\dfrac{14}{15}$

## c  Solving Equations

Now let's solve equations $a \cdot x = b$ and $x \cdot a = b$, where $a$ and $b$ may be fractions. We proceed as we did with equations involving integers. We divide by $a$ on both sides.

**EXAMPLE 7**  Solve: $\frac{4}{3} \cdot x = \frac{6}{7}$.

We have

$$\frac{4}{3} \cdot x = \frac{6}{7}$$

$$\frac{\frac{4}{3} \cdot x}{\frac{4}{3}} = \frac{\frac{6}{7}}{\frac{4}{3}} \qquad \text{Dividing by } \tfrac{4}{3} \text{ on both sides}$$

$$x = \frac{6}{7} \cdot \frac{3}{4} \qquad \text{Multiplying by the reciprocal}$$

$$= \frac{6 \cdot 3}{7 \cdot 4} = \frac{2 \cdot 3 \cdot 3}{7 \cdot 2 \cdot 2} = \frac{2}{2} \cdot \frac{3 \cdot 3}{7 \cdot 2} = \frac{3 \cdot 3}{7 \cdot 2} = \frac{9}{14}.$$

The solution is $\frac{9}{14}$.

**EXAMPLE 8**  Solve: $t \cdot \frac{4}{5} = -80$.

Dividing by $\frac{4}{5}$ on both sides, we get

$$t = -80 \div \frac{4}{5} = -80 \cdot \frac{5}{4} = \frac{-80 \cdot 5}{4} = \frac{4 \cdot (-20) \cdot 5}{4 \cdot 1}$$

$$= \frac{4}{4} \cdot \frac{-20 \cdot 5}{1} = \frac{-20 \cdot 5}{1} = -100.$$

The solution is $-100$.

Do Exercises 10 and 11.

## d  Applications and Problem Solving

**EXAMPLE 9**  *Doses of an Antibiotic.*  How many doses, each containing $\frac{15}{4}$ milliliters (mL), can be obtained from a bottle of a children's antibiotic that contains 60 mL?

1. **Familiarize.**  We are asking the question "How many $\frac{15}{4}$'s are in 60?" Repeated addition will apply here. We make a drawing. We let $n =$ the number of doses in all.

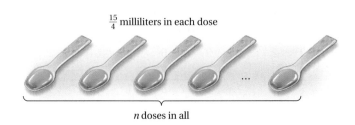

$\frac{15}{4}$ milliliters in each dose

$n$ doses in all

2. **Translate.**  The equation that corresponds to the situation is

$$n = 60 \div \frac{15}{4}.$$

Solve.

**10.** $\frac{5}{6} \cdot y = \frac{2}{3}$

**11.** $\frac{3}{4} \cdot n = -24$

*Answers*

10. $\frac{4}{5}$   11. $-32$

**3. Solve.** We solve the equation by carrying out the division:

$$n = 60 \div \frac{15}{4} = 60 \cdot \frac{4}{15} = \frac{60}{1} \cdot \frac{4}{15}$$

$$= \frac{60 \cdot 4}{1 \cdot 15} = \frac{4 \cdot 15 \cdot 4}{1 \cdot 15} = \frac{15}{15} \cdot \frac{4 \cdot 4}{1}$$

$$= 1 \cdot 16 = 16.$$

**4. Check.** We check by multiplying the number of doses by the size of the dose: $16 \cdot \frac{15}{4} = 60$. The answer checks.

**5. State.** There are 16 doses in a 60-mL bottle of the antibiotic.

> Do Exercise 12.

---

**EXAMPLE 10** *Brick Walkway.* Logan's Landscape specializes in brick and stone walkways. After they complete 63 ft of a brick walkway between two buildings on campus, $\frac{7}{8}$ of the walkway is complete. What is the total length of this walkway?

**1. Familiarize.** We ask: "63 ft is $\frac{7}{8}$ of what length?" We make a drawing or at least visualize the problem. We let $w =$ the length of the walkway.

$\frac{7}{8}$ of the walkway
63 ft

**2. Translate.** We translate to an equation:

| Fraction completed | of | Total length of walkway | is | Amount already completed |
|:---:|:---:|:---:|:---:|:---:|
| $\downarrow$ | $\downarrow$ | $\downarrow$ | $\downarrow$ | $\downarrow$ |
| $\frac{7}{8}$ | $\cdot$ | $w$ | $=$ | $63.$ |

**3. Solve.** We divide by $\frac{7}{8}$ on both sides and carry out the division:

$$w = 63 \div \frac{7}{8} = \frac{63}{1} \cdot \frac{8}{7} = \frac{63 \cdot 8}{1 \cdot 7} = \frac{7 \cdot 9 \cdot 8}{1 \cdot 7} = \frac{7}{7} \cdot \frac{9 \cdot 8}{1} = 72.$$

**4. Check.** We determine whether $\frac{7}{8}$ of 72 is 63:

$$\frac{7}{8} \cdot 72 = \frac{7 \cdot 72}{8 \cdot 1} = \frac{7 \cdot 8 \cdot 9}{8 \cdot 1} = \frac{8}{8} \cdot \frac{7 \cdot 9}{1} = 63.$$

The answer, 72, checks.

**5. State.** The total length of the walkway is 72 ft.

> Do Exercise 13.

---

**12.** Each loop in a spring uses $\frac{21}{8}$ in. of wire. How many loops can be made from 210 in. of wire?

**13. Sales Trip.** John sells soybean seeds to seed companies. After he had driven 210 mi, $\frac{5}{6}$ of his sales trip was completed. How long was the total trip?

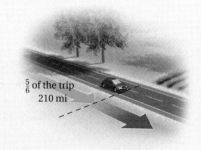

$\frac{5}{6}$ of the trip
210 mi

*Answers*

**12.** 80 loops   **13.** 252 mi

# Translating for Success

1. *Valentine Boxes.* Jane's Fudge Shop is preparing Valentine boxes. How many pounds of fudge will be needed to fill 80 boxes if each box contains $\frac{5}{16}$ lb?

2. *Gallons of Gasoline.* On the third day of a business trip, a sales representative used $\frac{4}{5}$ of a tank of gasoline. If the tank is a 20-gal tank, how many gallons were used on the third day?

3. *Purchasing a Shirt.* Tom received $36 for his birthday. If he spends $\frac{3}{4}$ of the gift on a new shirt, what is the cost of the shirt?

4. *Checkbook Balance.* The balance in Sam's checking account is $1456. He writes a check for $28 and makes a deposit of $52. What is the new balance?

5. *Valentine Boxes.* Jane's Fudge Shop prepared 80 lb of fudge for Valentine boxes. If each box contains $\frac{5}{16}$ lb, how many boxes can be filled?

The goal of these matching questions is to practice step (2), *Translate*, of the five-step problem-solving process. Translate each problem to an equation and select a correct translation from equations A–O.

**A.** $x = \frac{3}{4} \cdot 36$

**B.** $28 \cdot x = 52$

**C.** $x = 80 \cdot \frac{5}{16}$

**D.** $x = 1456 \div 28$

**E.** $\frac{5}{4} \cdot x = 20$

**F.** $20 = \frac{4}{5} \cdot x$

**G.** $x = 12 \cdot 28$

**H.** $x = \frac{4}{5} \cdot 20$

**I.** $\frac{3}{4} \cdot x = 36$

**J.** $x = 1456 - 52 - 28$

**K.** $x \div 28 = 1456$

**L.** $x = 52 - 28$

**M.** $x = 52 \cdot 28$

**N.** $x = 1456 - 28 + 52$

**O.** $\frac{5}{16} \cdot x = 80$

*Answers on page A-5*

6. *Gasoline Tank.* A gasoline tank contains 20 gal when it is $\frac{4}{5}$ full. How many gallons can it hold when full?

7. *Knitting a Scarf.* It takes Rachel 36 hr to knit a scarf. She can knit only $\frac{3}{4}$ hr per day because she is taking 16 hr of college classes. How many days will it take her to knit the scarf?

8. *Bicycle Trip.* On a recent 52-mi bicycle trip, David stopped to make a cell-phone call after completing 28 mi. How many more miles did he bicycle after the call?

9. *Crème de Menthe Thins.* Andes Candies L.P. makes Crème de Menthe Thins. How many 28-piece packages can be filled with 1456 pieces?

10. *Cereal Donations.* The Williams family donates 28 boxes of cereal weekly to the local Family in Crisis Center. How many boxes does this family donate in one year?

**a** Find the reciprocal.

**1.** $\dfrac{5}{6}$

**2.** $\dfrac{7}{8}$

**3.** 6

**4.** 4

**5.** $\dfrac{1}{6}$

**6.** $\dfrac{1}{4}$

**7.** $-\dfrac{10}{3}$

**8.** $-\dfrac{17}{4}$

**b** Divide and simplify.   Don't forget to simplify!

**9.** $\dfrac{3}{5} \div \dfrac{3}{4}$

**10.** $\dfrac{2}{3} \div \dfrac{3}{4}$

**11.** $-\dfrac{3}{5} \div \dfrac{9}{4}$

**12.** $-\dfrac{6}{7} \div \dfrac{3}{5}$

**13.** $\dfrac{4}{3} \div \dfrac{1}{3}$

**14.** $\dfrac{10}{9} \div \dfrac{1}{3}$

**15.** $-\dfrac{1}{3} \div \left(-\dfrac{1}{6}\right)$

**16.** $-\dfrac{1}{4} \div \left(-\dfrac{1}{5}\right)$

**17.** $\dfrac{3}{8} \div 3$

**18.** $\dfrac{5}{6} \div 5$

**19.** $\dfrac{12}{7} \div 4$

**20.** $\dfrac{18}{5} \div 2$

**21.** $12 \div \left(-\dfrac{3}{2}\right)$

**22.** $24 \div \left(-\dfrac{3}{8}\right)$

**23.** $28 \div \dfrac{4}{5}$

**24.** $40 \div \dfrac{2}{3}$

**25.** $-\dfrac{5}{8} \div \dfrac{5}{8}$

**26.** $-\dfrac{2}{5} \div \dfrac{2}{5}$

**27.** $\dfrac{8}{15} \div \left(-\dfrac{4}{5}\right)$

**28.** $\dfrac{6}{13} \div \left(-\dfrac{3}{26}\right)$

**29.** $-\dfrac{9}{5} \div \left(-\dfrac{4}{5}\right)$

**30.** $-\dfrac{5}{12} \div \left(-\dfrac{25}{36}\right)$

**31.** $120 \div \dfrac{5}{6}$

**32.** $360 \div \dfrac{8}{7}$

Solve.

**33.** $\dfrac{4}{5} \cdot x = 60$

**34.** $\dfrac{3}{2} \cdot t = 90$

**35.** $-\dfrac{5}{3} \cdot y = \dfrac{10}{3}$

**36.** $-\dfrac{4}{9} \cdot m = \dfrac{8}{3}$

**37.** $x \cdot \dfrac{25}{36} = \dfrac{5}{12}$

**38.** $p \cdot \dfrac{4}{5} = \dfrac{8}{15}$

**39.** $n \cdot \dfrac{8}{7} = -360$

**40.** $y \cdot \dfrac{5}{6} = -120$

  Solve.

**41.** *Extension Cords.* An electrical supplier sells rolls of SJO 14-3 cable to a company that makes extension cords. It takes $\frac{7}{3}$ ft of cable to make each cord. How many extension cords can be made with a roll of cable containing 2240 ft of cable?

**42.** Benny uses $\frac{2}{5}$ gram (g) of toothpaste each time he brushes his teeth. If Benny buys a 30-g tube, how many times will he be able to brush his teeth?

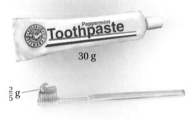

**43.** A pair of basketball shorts requires $\frac{3}{4}$ yd of nylon. How many pairs of shorts can be made from 24 yd of nylon?

**44.** A child's baseball shirt requires $\frac{5}{6}$ yd of fabric. How many shirts can be made from 25 yd of fabric?

**45.** How many $\frac{2}{3}$-cup sugar bowls can be filled from 16 cups of sugar?

**46.** How many $\frac{2}{3}$-cup cereal bowls can be filled from 10 cups of cornflakes?

**47.** A bucket had 12 L of water in it when it was $\frac{3}{4}$ full. How much could it hold when full?

**48.** A tank had 20 L of gasoline in it when it was $\frac{4}{5}$ full. How much could it hold when full?

**49.** Yoshi Teramoto sells hardware tools. After driving 180 kilometers (km), he has completed $\frac{5}{8}$ of a sales trip. How long is the total trip? How many kilometers are left to drive?

**50.** A piece of coaxial cable $\frac{4}{5}$ meter (m) long is to be cut into 8 pieces of the same length. What is the length of each piece?

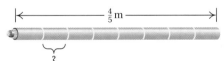

*Pitch of a Screw.* Refer to Exercises 45 and 46 in Exercise Set 3.6.

**51.** After a screw has been turned 8 complete rotations, it is extended $\frac{1}{2}$ in. into a piece of wallboard. What is the pitch of the screw?

**52.** The pitch of a screw is $\frac{3}{32}$ in. How many complete rotations are necessary to drive the screw $\frac{3}{4}$ in. into a piece of pine wood?

## Skill Maintenance

In each of Exercises 53–60, fill in the blank with the correct term from the given list. Some of the choices may not be used.

**53.** The equation $14 + (2 + 30) = (14 + 2) + 30$ illustrates the _____ law of addition.   [1.2a]

**54.** In the product $10 \cdot \frac{3}{4}$, 10 and $\frac{3}{4}$ are called _____.   [3.1a]

**55.** A natural number that has exactly two different factors, only itself and 1, is called a(n) _____ number.   [3.1c]

**56.** In the fraction $\frac{4}{17}$, we call 17 the _____.   [3.3a]

**57.** For any number $a$, $a + 0 = a$. The number 0 is the _____ identity.   [1.2a]

**58.** The product of 6 and $\frac{1}{6}$ is 1. We say that 6 and $\frac{1}{6}$ are _____.   [3.7a]

**59.** The set of _____ numbers is $0, 1, 2, 3, 4, \ldots$.   [1.1b]

**60.** A sentence with $=$ is called a(n) _____.   [1.7a]

associative

commutative

additive

multiplicative

reciprocals

factors

prime

composite

numerator

denominator

equation

variables

whole

natural

## Synthesis

Simplify. Use the list of prime numbers on p. 131.

**61.** ▦ $\dfrac{711}{1957} \div \dfrac{10{,}033}{13{,}081}$

**62.** ▦ $\dfrac{8633}{7387} \div \left( -\dfrac{485}{581} \right)$

**63.** $\left[ \dfrac{9}{10} \div \left( -\dfrac{2}{5} \right) \div \dfrac{3}{8} \right]^2$

**64.** $\dfrac{\left( \dfrac{3}{7} \right)^2 \div \dfrac{12}{5}}{\left( \dfrac{2}{9} \right)\left( \dfrac{9}{2} \right)}$

**65.** If $\frac{1}{3}$ of a number is $\frac{1}{4}$, what is $\frac{1}{2}$ of the number?

## Key Terms

factor (noun), p. 128
factor (verb), p. 128
factorization, p. 128
multiple, p. 129
divisible, p. 129

prime number, p. 130
composite number, p. 130
prime factorization, p. 131
even number, p. 135
fraction notation, p. 141

numerator, p. 141
denominator, p. 141
equivalent fractions, p. 157
cross products, p. 161

## Concept Reinforcement

Determine whether each statement is true or false.

_____ **1.** For any natural number $n$, $\dfrac{n}{n} > \dfrac{0}{n}$.   [3.3b]

_____ **2.** A number is divisible by 10 if its ones digit is 0 or 5.   [3.2a]

_____ **3.** If a number is divisible by 9, then it is also divisible by 3.   [3.2a]

_____ **4.** The fraction $\dfrac{6}{17}$ is equal to the fraction $\dfrac{66}{187}$.   [3.5c]

## Important Concepts

**Objective 3.1a**   Find the factors of a number.

**Example**   Find the factors of 84.

We find as many "two-factor" factorizations as we can.

$1 \cdot 84$    $4 \cdot 21$
$2 \cdot 42$    $6 \cdot 14$
$3 \cdot 28$    $7 \cdot 12$ ← Since 8, 9, 10, and 11 are not factors, we are finished.

The factors are 1, 2, 3, 4, 6, 7, 12, 14, 21, 28, 42, and 84.

**Practice Exercise**

**1.** Find the factors of 104.

**Objective 3.1d**   Find the prime factorization of a composite number.

**Example**   Find the prime factorization of 84.

To find the prime factorization, we can use either successive divisions or a factor tree.

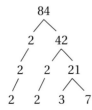

$$7$$
$$3\overline{)21}$$
$$2\overline{)42}$$
$$2\overline{)84}$$

Thus, $84 = 2 \cdot 2 \cdot 3 \cdot 7$.

**Practice Exercise**

**2.** Find the prime factorization of 104.

**Objective 3.3b**  Simplify fraction notation like $n/n$ to 1, $0/n$ to 0, and $n/1$ to $n$.

**Example**  Simplify $\dfrac{6}{6}, \dfrac{0}{6}$, and $\dfrac{-6}{1}$.

$$\dfrac{6}{6} = 1, \qquad \dfrac{0}{6} = 0, \qquad \dfrac{-6}{1} = -6$$

**Practice Exercise**

**3.** Simplify $\dfrac{0}{18}, \dfrac{-18}{-18}$, and $\dfrac{18}{1}$.

---

**Objective 3.5a**  Multiply a number by 1 to find fraction notation with a specified denominator.

**Example**  Find a name for $-\dfrac{2}{7}$ with a denominator of 63.

Since $7 \cdot 9 = 63$, we multiply by $\dfrac{9}{9}$:

$$-\dfrac{2}{7} = -\dfrac{2}{7} \cdot \dfrac{9}{9} = -\dfrac{2 \cdot 9}{7 \cdot 9} = -\dfrac{18}{63}.$$

**Practice Exercise**

**4.** Find a name for $\dfrac{7}{12}$ with a denominator of 96.

---

**Objective 3.5b**  Simplify fraction notation.

**Example**  Simplify: $\dfrac{315}{1650}$.

Using the test for divisibility by 5, we see that both the numerator and the denominator are divisible by 5:

$$\dfrac{315}{1650} = \dfrac{5 \cdot 63}{5 \cdot 330} = \dfrac{5}{5} \cdot \dfrac{63}{330} = 1 \cdot \dfrac{63}{330}$$

$$= \dfrac{63}{330} = \dfrac{3 \cdot 21}{3 \cdot 110} = \dfrac{3}{3} \cdot \dfrac{21}{110} = 1 \cdot \dfrac{21}{110} = \dfrac{21}{110}.$$

**Practice Exercise**

**5.** Simplify: $\dfrac{100}{280}$.

---

**Objective 3.5c**  Use the test for equality to determine whether two fractions name the same number.

**Example**  Use = or ≠ for ☐ to write a true sentence:

$$\dfrac{10}{54} \ \square \ \dfrac{15}{81}.$$

We find the cross products. We multiply 10 and 81:
$10 \cdot 81 = 810$. Then we multiply 54 and 15: $54 \cdot 15 = 810$.
Because the cross products are the same, we have

$$\dfrac{10}{54} = \dfrac{15}{81}.$$

If the cross products had been different, the fractions would not be equal.

**Practice Exercise**

**6.** Use = or ≠ for ☐ to write a true sentence:

$$\dfrac{8}{48} \ \square \ \dfrac{6}{44}.$$

---

**Objective 3.6a**  Multiply and simplify using fraction notation.

**Example**  Multiply and simplify: $-\dfrac{7}{16} \cdot \left(-\dfrac{40}{49}\right)$.

$$-\dfrac{7}{16} \cdot \left(-\dfrac{40}{49}\right) = \dfrac{7 \cdot 40}{16 \cdot 49} = \dfrac{7 \cdot 2 \cdot 2 \cdot 2 \cdot 5}{2 \cdot 2 \cdot 2 \cdot 2 \cdot 7 \cdot 7}$$

$$= \dfrac{2 \cdot 2 \cdot 2 \cdot 7}{2 \cdot 2 \cdot 2 \cdot 7} \cdot \dfrac{5}{2 \cdot 7} = 1 \cdot \dfrac{5}{14} = \dfrac{5}{14}$$

**Practice Exercise**

**7.** Multiply and simplify: $\dfrac{80}{3} \cdot \left(-\dfrac{21}{72}\right)$.

**Objective 3.7b** Divide and simplify using fraction notation.

**Example** Divide and simplify: $\dfrac{9}{20} \div \dfrac{18}{25}$.

$$\frac{9}{20} \div \frac{18}{25} = \frac{9}{20} \cdot \frac{25}{18} = \frac{9 \cdot 25}{20 \cdot 18} = \frac{3 \cdot 3 \cdot 5 \cdot 5}{2 \cdot 2 \cdot 5 \cdot 2 \cdot 3 \cdot 3}$$

$$= \frac{3 \cdot 3 \cdot 5}{3 \cdot 3 \cdot 5} \cdot \frac{5}{2 \cdot 2 \cdot 2} = 1 \cdot \frac{5}{8} = \frac{5}{8}$$

**Practice Exercise**

**8.** Divide and simplify: $\dfrac{9}{4} \div \dfrac{45}{14}$.

---

**Objective 3.7d** Solve applied problems involving division of fractions.

**Example** A rental car had 18 gal of gasoline when its gas tank was $\frac{6}{7}$ full. How much could the tank hold when full?

The equation that corresponds to the situation is

$$\frac{6}{7} \cdot g = 18.$$

We divide by $\frac{6}{7}$ on both sides and carry out the division:

$$g = 18 \div \frac{6}{7} = \frac{18}{1} \cdot \frac{7}{6} = \frac{18 \cdot 7}{1 \cdot 6}$$

$$= \frac{3 \cdot 6 \cdot 7}{1 \cdot 6} = \frac{6}{6} \cdot \frac{3 \cdot 7}{1} = 1 \cdot \frac{21}{1} = 21.$$

The rental car's tank can hold 21 gal of gasoline.

**Practice Exercise**

**9.** A flower vase has $\frac{7}{4}$ cups of water in it when it is $\frac{3}{4}$ full. How much can it hold when full?

---

# Review Exercises

Find all the factors of each number.   [3.1a]

**1.** 60                **2.** 176

**3.** Multiply by 1, 2, 3, and so on, to find ten multiples of 8.
[3.1b]

**4.** Determine whether 924 is divisible by 11.   [3.1b]

**5.** Determine whether 1800 is divisible by 16.   [3.1b]

Determine whether each number is prime, composite, or neither.   [3.1c]

**6.** 37        **7.** 1        **8.** 91

Find the prime factorization of each number.   [3.1d]

**9.** 70                **10.** 30

**11.** 45                **12.** 150

**13.** 648                **14.** 5250

To do Exercises 15–22, consider the following numbers:

| 140 | 716 | 93 | 2802 |
| 95 | 2432 | 330 | 711 |
| 182 | 4344 | 255,555 | |
| 475 | 600 | 780 | |

Which of the above are divisible by the given number?   [3.2a]

**15.** 3                **16.** 2

**17.** 4                **18.** 8

**19.** 5                **20.** 6

**21.** 9                **22.** 10

**23.** Identify the numerator and the denominator of $\frac{2}{7}$.
[3.3a]

What part of each object is shaded?   [3.3a]

**24.**                **25.**

**26.** What part of the set of objects is shaded? [3.3a]

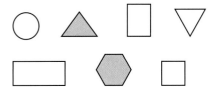

**27.** For a committee in the United States Senate that consists of 3 Democrats and 5 Republicans, what is the ratio of: [3.3a]
   **a)** Democrats to Republicans?
   **b)** Republicans to Democrats?
   **c)** Democrats to the total number of members of the committee?

Simplify. [3.3b], [3.5b]

**28.** $\dfrac{12}{30}$   **29.** $\dfrac{7}{28}$   **30.** $\dfrac{-23}{-23}$

**31.** $\dfrac{0}{-25}$   **32.** $\dfrac{1170}{1200}$   **33.** $\dfrac{18}{1}$

**34.** $-\dfrac{9}{27}$   **35.** $-\dfrac{88}{184}$   **36.** $\dfrac{18}{0}$

**37.** $\dfrac{48}{8}$   **38.** $\dfrac{140}{490}$   **39.** $-\dfrac{288}{2025}$

**40.** Simplify the fractions on this circle graph, if possible. [3.5b]

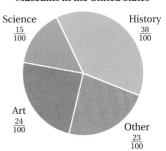

**Museums in the United States**

Science $\dfrac{15}{100}$

History $\dfrac{38}{100}$

Art $\dfrac{24}{100}$

Other $\dfrac{23}{100}$

Use = or ≠ for ☐ to write a true sentence. [3.5c]

**41.** $\dfrac{3}{5} \;\square\; \dfrac{4}{6}$   **42.** $\dfrac{-4}{7} \;\square\; -\dfrac{8}{14}$

**43.** $-\dfrac{4}{5} \;\square\; -\dfrac{5}{6}$   **44.** $\dfrac{4}{3} \;\square\; \dfrac{28}{21}$

Multiply and simplify. [3.6a]

**45.** $4 \cdot \dfrac{3}{8}$   **46.** $\dfrac{7}{3} \cdot 24$

**47.** $-9 \cdot \dfrac{5}{18}$   **48.** $\dfrac{6}{5} \cdot (-20)$

**49.** $\dfrac{3}{4} \cdot \dfrac{8}{9}$   **50.** $\dfrac{5}{7} \cdot \dfrac{1}{10}$

**51.** $-\dfrac{3}{7} \cdot \dfrac{14}{9}$   **52.** $\dfrac{1}{4} \cdot \dfrac{2}{11}$

**53.** $\dfrac{4}{25} \cdot \dfrac{15}{16}$   **54.** $-\dfrac{11}{3} \cdot \left(-\dfrac{30}{77}\right)$

Find the reciprocal. [3.7a]

**55.** $\dfrac{4}{5}$   **56.** $-3$

**57.** $\dfrac{1}{9}$   **58.** $-\dfrac{47}{36}$

Divide and simplify. [3.7b]

**59.** $6 \div \dfrac{4}{3}$   **60.** $-\dfrac{5}{9} \div \dfrac{5}{18}$

**61.** $\dfrac{1}{6} \div \dfrac{1}{11}$   **62.** $\dfrac{3}{14} \div \left(-\dfrac{6}{7}\right)$

**63.** $-\dfrac{1}{4} \div \left(-\dfrac{1}{9}\right)$   **64.** $180 \div \dfrac{3}{5}$

**65.** $\dfrac{23}{25} \div \dfrac{23}{25}$   **66.** $-\dfrac{2}{3} \div \left(-\dfrac{3}{2}\right)$

Solve. [3.7c]

**67.** $\dfrac{5}{4} \cdot t = \dfrac{3}{8}$   **68.** $x \cdot \dfrac{2}{3} = -160$

Solve. [3.6b], [3.7d]

69. A road crew repaves $\frac{1}{12}$ mi of road each day. How long will it take the crew to repave a $\frac{3}{4}$-mi stretch of road?

70. *Cotton Production.* In 2005–2006, the United States accounted for approximately $\frac{1}{5}$ of the world production of cotton. The total world cotton production was about 114,000,000 metric tons. How much cotton did the United States produce?

Sources: U.S. Department of Agriculture; Foreign Agricultural Service; *World Agricultural Production*, June 2007

71. After driving 600 km, the Buxton family has completed $\frac{3}{5}$ of their vacation. How long is the total trip?

72. Molly is making a pepper steak recipe that calls for $\frac{2}{3}$ cup of green bell peppers. How much would be needed to make $\frac{1}{2}$ recipe? 3 recipes?

73. Bernardo earns $105 for working a full day. How much does he receive for working $\frac{1}{7}$ of a day?

74. A book bag requires $\frac{4}{5}$ yd of fabric. How many bags can be made from 48 yd?

75. Solve: $\dfrac{2}{13} \cdot x = -\dfrac{1}{2}$. [3.7c]

**A.** $-\dfrac{1}{13}$     **B.** $-13$

**C.** $-\dfrac{4}{13}$     **D.** $-\dfrac{13}{4}$

76. Multiply and simplify: $\dfrac{15}{26} \cdot \dfrac{13}{90}$. [3.6a]

**A.** $\dfrac{195}{234}$     **B.** $\dfrac{1}{12}$

**C.** $\dfrac{3}{36}$     **D.** $\dfrac{13}{156}$

## Synthesis

77. ▦ In the division below, find $a$ and $b$. [3.7b]

$$\frac{19}{24} \div \frac{a}{b} = \frac{187,853}{268,224}$$

78. A prime number that remains a prime number when its digits are reversed is called a **palindrome prime**. For example, 17 is a palindrome prime because both 17 and 71 are primes. Which of the following numbers are palindrome primes? [3.1c]

13, 91, 16, 11, 15, 24, 29, 101, 201, 37

# Understanding Through Discussion and Writing

1. A student incorrectly insists that $\frac{2}{5} \div \frac{3}{4}$ is $\frac{15}{8}$. What mistake is he probably making? [3.7b]

2. Use the number 9432 to explain why the test for divisibility by 9 works. [3.2a]

3. A student claims that "taking $\frac{1}{2}$ of a number is the same as dividing by $\frac{1}{2}$." Explain the error in this reasoning. [3.7b]

4. On p. 151, we explained, using words and pictures, why $\frac{3}{5} \cdot \frac{3}{4}$ equals $\frac{9}{20}$. Present a similar explanation of why $\frac{2}{3} \cdot \frac{4}{7}$ equals $\frac{8}{21}$. [3.4a]

5. Without performing the division, explain why $5 \div \frac{1}{7}$ is a larger number than $5 \div \frac{2}{3}$. [3.5c], [3.7b]

6. If a fraction's numerator and denominator have no factors (other than 1) in common, can the fraction be simplified? Why or why not? [3.5b]

## CHAPTER
# 3

Test

**For Extra Help**
CHAPTER
Test Prep
VIDEOS

Step-by-step test solutions are found on the Chapter Test Prep Videos available via the Video Resources on DVD, in *MyMathLab*  , and on You Tube (search "BittingerBasicMathEI" and click on "Channels").

1. Find all the factors of 300.

Determine whether each number is prime, composite, or neither.

2. 41

3. 14

Find the prime factorization of the number.

4. 18

5. 60

6. Determine whether 1784 is divisible by 8.

7. Determine whether 784 is divisible by 9.

8. Determine whether 5552 is divisible by 5.

9. Determine whether 2322 is divisible by 6.

10. Identify the numerator and the denominator of $\frac{4}{5}$.

11. What part is shaded?

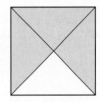

12. What part of the set is shaded?

13. *Pass Completion Ratio.* In the 2008 regular season, Peyton Manning, of the Indianapolis Colts, completed 371 of 555 passes to earn a third Associated Press NFL Most Valuable Player award.

a) What was the ratio of pass completions to attempts?
b) What was the ratio of incomplete passes to attempts?

Simplify.

14. $\frac{26}{1}$

15. $\frac{-12}{-12}$

16. $\frac{0}{16}$

17. $-\frac{12}{24}$

18. $\frac{42}{7}$

19. $\frac{9}{0}$

20. $\frac{7}{2-2}$

21. $-\frac{72}{108}$

Use = or ≠ for ☐ to write a true sentence.

22. $\frac{3}{4}$ ☐ $\frac{6}{8}$

23. $\frac{-5}{4}$ ☐ $\frac{-9}{7}$

Multiply and simplify.

24. $\frac{4}{3} \cdot 24$

25. $-5 \cdot \frac{3}{10}$

26. $\frac{2}{3} \cdot \frac{15}{4}$

27. $-\frac{22}{15} \cdot \left(-\frac{5}{33}\right)$

Find the reciprocal.

**28.** $\dfrac{5}{8}$

**29.** $-\dfrac{1}{4}$

**30.** 18

Divide and simplify.

**31.** $\dfrac{1}{5} \div \dfrac{1}{8}$

**32.** $12 \div \left(-\dfrac{2}{3}\right)$

**33.** $\dfrac{24}{5} \div \dfrac{28}{15}$

Solve.

**34.** $\dfrac{7}{8} \cdot x = -56$

**35.** $t \cdot \dfrac{2}{5} = \dfrac{7}{10}$

**36.** There are 7000 students at La Poloma College, and $\frac{5}{8}$ of them live in dorms. How many live in dorms?

**37.** A strip of taffy $\frac{9}{10}$ m long is cut into 12 equal pieces. What is the length of each piece?

**38.** A thermos of iced tea held 3 qt of tea when it was $\frac{3}{5}$ full. How much tea could it hold when full?

**39.** The pitch of a screw is $\frac{1}{8}$ in. How far will it go into a piece of walnut when it is turned 6 complete rotations?

**40.** In which figure does the shaded part represent $\frac{7}{6}$ of the object?

**A.**

**B.**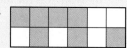

**C.**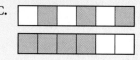

**D.**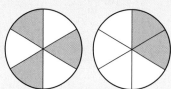

# Synthesis

**41.** Grandma Hammons left $\frac{2}{3}$ of her $\frac{7}{8}$-acre apple farm to Karl. Karl gave $\frac{1}{4}$ of his share to his oldest daughter, Eileen. How much land did Eileen receive?

**42.** Simplify: $\left(\dfrac{3}{8}\right)^2 \div \dfrac{6}{7} \cdot \dfrac{2}{9} \div (-5).$

1. Write a word name: 7,453,062.

2. What does the digit 4 mean in 23,746,591?

Simplify.

3. $\dfrac{5}{1}$

4. $\dfrac{-17}{-17}$

5. $-\dfrac{56}{42}$

6. $\dfrac{32}{0}$

Find the reciprocal.

7. 8

8. $-\dfrac{3}{2}$

Calculate and simplify.

9. $4658 + 729$

10. $-3 + (-9)$

11. $5042 - 3658$

12. $-9 - 7$

13. $457 \cdot 36$

14. $-8 \cdot 12$

15. $-\dfrac{5}{6} \cdot 15$

16. $\dfrac{3}{7} \cdot \dfrac{14}{9}$

17. $10{,}846 \div 24$

18. $-56 \div (-7)$

19. $-\dfrac{4}{7} \div 28$

20. $\dfrac{3}{4} \div \dfrac{9}{16}$

21. $8^2 \div 8 \cdot 2 - (2 + 2 \cdot 7)$

22. $108 \div 9 - [3(18 - 5 \cdot 3)]$

23. $-20 - 10 \div 5 + 2^3$

24. $(8 - 10^2) \div (5^2 - 2)$

25. Find the average of four test scores of 85, 91, 80, and 88.

26. Round 165,739 to the nearest thousand.

Evaluate.

27. $9^2$

28. $5^3$

29. $2^4$

Use either $<$ or $>$ for $\square$ to write a true sentence.

30. $-26 \ \square \ 2$

31. $19 \ \square \ 17$

Find the absolute value.

**32.** |33|    **33.** |−86|    **34.** |0|

Find the opposite, or additive inverse.

**35.** 29                    **36.** −144

**37.** Evaluate −(−x) when x is −7.

**38.** Graph 8 on the number line.

Determine whether each number is prime, composite, or neither.

**39.** 29         **40.** 1         **41.** 24

Find the prime factorization of each number.

**42.** 36         **43.** 135         **44.** 1350

**45.** Determine whether 6330 is divisible by 2; by 3; by 4; by 5; by 6; by 8; by 9; by 10.

Use = or ≠ for □ to write a true sentence.

**46.** $\frac{5}{8}$ □ $\frac{7}{11}$                    **47.** $-\frac{5}{4}$ □ $-\frac{20}{16}$

Solve

**48.** 17 + x = 61                    **49.** $\frac{3}{5} \cdot y = -\frac{9}{25}$

**50.** *Halley's Comet.*   Halley's Comet passes by Earth every 76 yr. It last appeared to people on Earth in 1986. When will it appear again?
**Source:** *The Cambridge Factfinder*

**51.** *Fundraiser.*   To support a school fundraiser, Mary Ann bought 4 rolls of wrapping paper at $5 each and 2 candles at $8 each. Find the total cost of Mary Ann's order.

**52.** *Chemical Reaction.*   The temperature of a chemical compound was 5°C at 9:00 A.M. During a reaction, it dropped 2°C per minute until 9:11 A.M. What was the temperature at 9:11 A.M.?

**53.** *Working Students.*   An instructor determines that $\frac{2}{5}$ of the 35 students in her accounting class at Beckwith Community College have full-time jobs. How many students have full-time jobs?

## Synthesis

**54.** Find a pair of factors whose product is 100 and:
   **a)** whose sum is 25.
   **b)** whose sum is −29.
   **c)** whose difference is 0.

# Fraction Notation and Mixed Numerals

## Real-World Application

Black bears typically have two cubs. In January 2007 in northern New Hampshire, a black bear sow gave birth to a litter of 5 cubs. This is so rare that Tom Sears, a wildlife photographer, spent 28 hr per week for six weeks watching for the perfect opportunity to photograph this family of six. At the time of this photo, an observer estimated that the cubs weighed $7\frac{1}{2}$ lb, 8 lb, $9\frac{1}{2}$ lb, $10\frac{5}{8}$ lb, and $11\frac{3}{4}$ lb. What was the average weight of the cubs?

*Source:* Andrew Timmins, New Hampshire Fish and Game Department, *Northcountry News,* Warren, NH; Tom Sears, photographer

***This problem appears as Exercise 27 in Section 4.7.***

# 4.1

# Least Common Multiples

## OBJECTIVE

**a** Find the least common multiple, or LCM, of two or more numbers.

**SKILL TO REVIEW**

Objective 3.1b: Find some multiples of a number.

Multiply by 1, 2, 3, and so on, to find six multiples of each number.

   **1.** 8            **2.** 25

In this chapter, we study addition and subtraction using fraction notation. Suppose we want to add $\frac{2}{3}$ and $\frac{1}{2}$. To do so, we rewrite the fractions with a common denominator. The number we choose for the common denominator is the least common multiple of the denominators, 6; $\frac{2}{3} + \frac{1}{2} = \frac{4}{6} + \frac{3}{6}$. Then we add the numerators and keep the common denominator. In order to do this, we must be able to find the **least common denominator (LCD)**, or **least common multiple (LCM)**, of the denominators.

## **a** Finding Least Common Multiples

> **LEAST COMMON MULTIPLE, LCM**
>
> The **least common multiple**, or LCM, of two natural numbers is the smallest number that is a multiple of both numbers.

**EXAMPLE 1** Find the LCM of 20 and 30.

First, we list some multiples of 20 by multiplying 20 by 1, 2, 3, and so on:

$20, 40, 60, 80, 100, 120, 140, 160, 180, 200, 220, 240, \ldots.$

Then we list some multiples of 30 by multiplying 30 by 1, 2, 3, and so on:

$30, 60, 90, 120, 150, 180, 210, 240, \ldots.$

Now we determine the smallest number *common* to both lists. The LCM of 20 and 30 is 60.

Do Margin Exercise 1.

**1.** Find the LCM of 9 and 15 by examining lists of multiples.

Next, we develop three methods that are more efficient for finding LCMs. You may choose to learn only one method. (Consult with your instructor.) If you are going to study algebra, you should definitely learn method 2.

### Method 1: Finding LCMs Using One List of Multiples

One method for finding LCMs uses *one* list of multiples. Let's consider finding the LCM of 9 and 12. The larger number, 12, is not a multiple of 9, so we check multiples of 12 until we find a number that is also a multiple of 9:

$1 \cdot 12 = 12, \quad$ not a multiple of 9;

$2 \cdot 12 = 24, \quad$ not a multiple of 9;

$3 \cdot 12 = 36, \quad$ a multiple of 9: $4 \cdot 9 = 36$.

The LCM of 9 and 12 is 36.

*Answers*

*Skill to Review:*

**1.** 8, 16, 24, 32, 40, 48
**2.** 25, 50, 75, 100, 125, 150

*Margin Exercise:*

**1.** 45

> *Method 1.* To find the LCM of a set of natural numbers using a list of multiples:
>
> a) Determine whether the largest number is a multiple of the others. If it is, it is the LCM. That is, if the largest number has the others as factors, the LCM is that number.
>
> b) If not, check multiples of the largest number until you get one that is a multiple of each of the others.

**EXAMPLE 2** Find the LCM of 12 and 15.

a) 15 is not a multiple of 12.

b) Check multiples of 15:

$1 \cdot 15 = 15,$    Not a multiple of 12. When we divide 15 by 12, we get a nonzero remainder.

$2 \cdot 15 = 30,$    Not a multiple of 12

$3 \cdot 15 = 45,$    Not a multiple of 12

$4 \cdot 15 = 60.$    A multiple of 12: $5 \cdot 12 = 60$

The LCM $= 60$.

Do Exercise 2.

**2.** Find the LCM of 8 and 10 by examining lists of multiples.

**EXAMPLE 3** Find the LCM of 8 and 32.

32 is a multiple of 8 $(4 \cdot 8 = 32)$, so the LCM $= 32$.

**EXAMPLE 4** Find the LCM of 10, 20, and 50.

a) 50 is a multiple of 10 but not a multiple of 20.

b) Check multiples of 50:

$1 \cdot 50 = 50,$

$2 \cdot 50 = 100.$    A multiple of 10 and of 20: $10 \cdot 10 = 100$ and $5 \cdot 20 = 100$

The LCM $= 100$.

Do Exercises 3–6.

Find the LCM.

**3.** 10, 15      **4.** 4, 10

**5.** 5, 10      **6.** 20, 45, 80

## Method 2: Finding LCMs Using Prime Factorizations

A second method for finding LCMs uses prime factorizations. Consider again 20 and 30. Their prime factorizations are $20 = 2 \cdot 2 \cdot 5$ and $30 = 2 \cdot 3 \cdot 5$. Let's look at these prime factorizations in order to find the LCM. Any multiple of 20 will have to have *two* 2's as factors and *one* 5 as a factor. Any multiple of 30 will need to have *one* 2, *one* 3, and *one* 5 as factors. The smallest number satisfying these conditions is

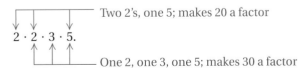

Two 2's, one 5; makes 20 a factor

$2 \cdot 2 \cdot 3 \cdot 5.$

One 2, one 3, one 5; makes 30 a factor

**TUNE OUT DISTRACTIONS**

Do you usually study in noisy places? If there is constant noise in your home, dorm, or other study area, consider finding a quiet place in the library—preferably a spot away from the main traffic areas so that distractions are kept to a minimum.

Thus the LCM of 20 and 30 is $2 \cdot 2 \cdot 3 \cdot 5$, or 60. It has all the factors of 20 and all the factors of 30, but the factors are not repeated when they are common to both numbers.

Note that the greatest number of times that 2 occurs as a factor of either 20 or 30 is two, and the LCM has 2 as a factor twice. The greatest number of times that 3 occurs as a factor of either 20 or 30 is one, and the LCM has 3 as a factor once. The greatest number of times that 5 occurs as a factor of either 20 or 30 is one, and the LCM has 5 as a factor once.

*Method 2.* To find the LCM of a set of numbers using prime factorizations:

a) Find the prime factorization of each number.

b) Create a product of factors, using each factor the greatest number of times that it occurs in any one factorization.

**EXAMPLE 5**　Find the LCM of 6 and 8.

a) Find the prime factorization of each number.

$$6 = 2 \cdot 3, \qquad 8 = 2 \cdot 2 \cdot 2$$

b) Create a product by writing factors that appear in the factorizations of 6 and 8, using each the greatest number of times that it occurs in any one factorization.

*Consider the factor 2.* The greatest number of times that 2 occurs in any one factorization is three. We write 2 as a factor three times.

$$2 \cdot 2 \cdot 2 \cdot ?$$

*Consider the factor 3.* The greatest number of times that 3 occurs in any one factorization is one. We write 3 as a factor one time.

$$2 \cdot 2 \cdot 2 \cdot 3 \cdot ?$$

Since there are no other prime factors in either factorization, the

LCM is $2 \cdot 2 \cdot 2 \cdot 3$, or 24.

**EXAMPLE 6**　Find the LCM of 24 and 36.

a) Find the prime factorization of each number.

$$24 = 2 \cdot 2 \cdot 2 \cdot 3, \qquad 36 = 2 \cdot 2 \cdot 3 \cdot 3$$

b) Create a product by writing factors, using each the greatest number of times that it occurs in any one factorization.

*Consider the factor 2.* The greatest number of times that 2 occurs in any one factorization is three. We write 2 as a factor three times:

$$2 \cdot 2 \cdot 2 \cdot ?$$

*Consider the factor 3.* The greatest number of times that 3 occurs in any one factorization is two. We write 3 as a factor two times:

$$2 \cdot 2 \cdot 2 \cdot 3 \cdot 3 \cdot ?$$

Since there are no other prime factors in either factorization, the

LCM is $2 \cdot 2 \cdot 2 \cdot 3 \cdot 3$, or 72.

Use prime factorizations to find each LCM.

**7.** 8, 10

**8.** 18, 40

**9.** 32, 54

*Answers*

**7.** 40　**8.** 360　**9.** 864

Do Exercises 7–9.

**EXAMPLE 7** Find the LCM of 81, 90, and 84.

a) Find the prime factorization of each number.

$$81 = 3 \cdot 3 \cdot 3 \cdot 3, \qquad 90 = 2 \cdot 3 \cdot 3 \cdot 5, \qquad 84 = 2 \cdot 2 \cdot 3 \cdot 7$$

b) Create a product by writing factors, using each the greatest number of times that it occurs in any one factorization.

*Consider the factor 2.* The greatest number of times that 2 occurs in any one factorization is two. We write 2 as a factor two times:

$$2 \cdot 2 \cdot ?$$

*Consider the factor 3.* The greatest number of times that 3 occurs in any one factorization is four. We write 3 as a factor four times:

$$2 \cdot 2 \cdot 3 \cdot 3 \cdot 3 \cdot 3 \cdot ?$$

*Consider the factor 5.* The greatest number of times that 5 occurs in any one factorization is one. We write 5 as a factor one time:

$$2 \cdot 2 \cdot 3 \cdot 3 \cdot 3 \cdot 3 \cdot 5 \cdot ?$$

*Consider the factor 7.* The greatest number of times that 7 occurs in any one factorization is one. We write 7 as a factor one time:

$$2 \cdot 2 \cdot 3 \cdot 3 \cdot 3 \cdot 3 \cdot 5 \cdot 7 \cdot ?$$

Since there are no other prime factors in any of the factorizations, the

LCM is $2 \cdot 2 \cdot 3 \cdot 3 \cdot 3 \cdot 3 \cdot 5 \cdot 7$, or 11,340.

Do Exercise 10.

**10.** Find the LCM of 24, 35, and 45.

Let's reconsider Example 7 using exponents. We want to find the LCM of 81, 90, and 84. We factor and then convert to exponential notation. The largest exponents indicate the greatest number of times that 2, 3, 5, and 7 occur as factors.

$81 = 3 \cdot 3 \cdot 3 \cdot 3 = 3^4;$    4 is the largest exponent of 3 in any of the factorizations.

$90 = 2 \cdot 3 \cdot 3 \cdot 5 = 2^1 \cdot 3^2 \cdot 5^1;$    1 is the largest exponent of 5 in any of the factorizations.

$84 = 2 \cdot 2 \cdot 3 \cdot 7 = 2^2 \cdot 3^1 \cdot 7^1.$    2 is the largest exponent of 2 and 1 is the largest exponent of 7 in any of the factorizations.

Thus the LCM is $2^2 \cdot 3^4 \cdot 5^1 \cdot 7^1$, or 11,340.

Do Exercises 11 and 12.

**11.** Use exponents to find the LCM of 24, 35, and 45.

**12.** Redo Margin Exercises 7–9 using exponents.

**EXAMPLE 8** Find the LCM of 8 and 9.

We find the prime factorization of each number.

$$8 = 2 \cdot 2 \cdot 2, \qquad 9 = 3 \cdot 3$$

Note that the two numbers, 8 and 9, have no common prime factor. When this is the case, the LCM is just the product of the two numbers. Thus the LCM is $2 \cdot 2 \cdot 2 \cdot 3 \cdot 3$, or $8 \cdot 9$, or 72.

Do Exercises 13 and 14.

Find the LCM.

**13.** 4, 9

**14.** 5, 6, 7

*Answers*

**10.** 2520    **11.** $2^3 \cdot 3^2 \cdot 5 \cdot 7$, or 2520
**12.** $2^3 \cdot 5$, or 40; $2^3 \cdot 3^2 \cdot 5$, or 360;
$2^5 \cdot 3^3$, or 864    **13.** 36    **14.** 210

**EXAMPLE 9** Find the LCM of 7 and 21.

We find the prime factorization of each number. Because 7 is prime, it has no prime factorization.

$$7 = 7, \qquad 21 = 3 \cdot 7$$

Note that 7 is a factor of 21. We stated earlier that if one number is a factor of another, the LCM is the larger of the numbers. Thus the LCM is $7 \cdot 3$, or 21.

Do Exercises 15 and 16.

Find the LCM.

**15.** 3, 18          **16.** 12, 24

### Method 3: Finding LCMs Using Division by Primes

Here is another method for finding LCMs that is especially useful for three or more numbers. For example, to find the LCM of 48, 72, and 80, we first look for any prime that divides any two of the numbers with no remainder. Then we divide as follows. We repeat the process, bringing down any numbers not divisible by the prime, until we can divide no more—that is, until there are no two numbers divisible by the same prime:

$$
\begin{array}{r|rrr}
2 & 48 & 72 & 80 \\
3 & 24 & 36 & 40 \\
2 & 8 & 12 & 40 \\
2 & 4 & 6 & 20 \\
2 & 2 & 3 & 10 \\
\hline
& 1 & 3 & 5
\end{array}
$$

40 is not divisible by 3.

3 is not divisible by 2.

No two of 1, 3, and 5 are divisible by the same prime. We stop here. The LCM is

$$2 \cdot 3 \cdot 2 \cdot 2 \cdot 2 \cdot 1 \cdot 3 \cdot 5, \quad \text{or} \quad 720.$$

---

*Method 3.* To find the LCM using division by primes:

a) First look for any prime that divides at least two of the numbers with no remainder. Then divide, bringing down any numbers not divisible by the prime. If you cannot find a prime that divides at least two of the numbers, then the LCM is the product of the numbers.

b) Repeat the process until you can divide no more—that is, until there are no two numbers divisible by the same prime.

---

**EXAMPLE 10** Find the LCM of 12, 18, 21, and 40.

$$
\begin{array}{r|rrrr}
3 & 12 & 18 & 21 & 40 \\
2 & 4 & 6 & 7 & 40 \\
2 & 2 & 3 & 7 & 20 \\
\hline
& 1 & 3 & 7 & 10
\end{array}
$$

No two of 1, 3, 7, and 10 are divisible by the same prime. We stop here.

The LCM is $3 \cdot 2 \cdot 2 \cdot 1 \cdot 3 \cdot 7 \cdot 10$, or 2520.

Find the LCM using division by primes.

**17.** 81, 90, 84

**18.** 24, 35, 45

**19.** 12, 75, 80, 120

Do Exercises 17–19.

*Answers*

**15.** 18   **16.** 24   **17.** 11,340   **18.** 2520
**19.** 1200

**a** Find the LCM of each set of numbers.

**1.** 2, 4

**2.** 3, 15

**3.** 10, 25

**4.** 10, 15

**5.** 20, 40

**6.** 8, 12

**7.** 18, 27

**8.** 9, 11

**9.** 30, 50

**10.** 24, 36

**11.** 30, 40

**12.** 21, 27

**13.** 18, 24

**14.** 12, 18

**15.** 60, 70

**16.** 35, 45

**17.** 16, 36

**18.** 18, 20

**19.** 32, 36

**20.** 36, 48

**21.** 2, 3, 5

**22.** 3, 5, 7

**23.** 5, 18, 3

**24.** 6, 12, 18

**25.** 24, 36, 12

**26.** 8, 16, 22

**27.** 5, 12, 15

**28.** 12, 18, 40

**29.** 9, 12, 6

**30.** 8, 16, 12

**31.** 180, 100, 450, 60

**32.** 18, 30, 50, 48

**33.** 8, 48

**34.** 16, 32

**35.** 5, 50

**36.** 12, 72

**37.** 11, 13

**38.** 13, 14

**39.** 12, 35

**40.** 23, 25

**41.** 54, 63

**42.** 56, 72

**43.** 81, 90

**44.** 75, 100

**45.** 36, 54, 80

**46.** 22, 42, 51

**47.** 39, 91, 108, 26

**48.** 625, 75, 500, 25

**49.** 2000, 3000

**50.** 300, 4000

*Applications of LCMs: Planet Orbits.*  Jupiter, Saturn, and Uranus all revolve around the sun. Jupiter takes 12 yr, Saturn 30 yr, and Uranus 84 yr to make a complete revolution. On a certain night, you look at Jupiter, Saturn, and Uranus and wonder how many years it will take before they have the same position again. (*Hint*: To find out, you find the LCM of 12, 30, and 84. It will be that number of years.)

Source: *The Handy Science Answer Book*

**51.** How often will Jupiter and Saturn appear in the same direction in the night sky as seen from the earth?

**52.** How often will Jupiter and Uranus appear in the same direction in the night sky as seen from the earth?

**53.** How often will Saturn and Uranus appear in the same direction in the night sky as seen from the earth?

**54.** How often will Jupiter, Saturn, and Uranus appear in the same direction in the night sky as seen from the earth?

## Skill Maintenance

**55.** *Tornadoes.*  The United States reports more tornadoes per year than any other country in the world. From January through May 2008, 1007 tornadoes were recorded. This was an increase of 348 tornadoes over the same time period in 2007. How many tornadoes occurred from January through May in 2007?  [1.8a]

Sources: NOAA's National Weather Service; www.tornadoarchive.com/History.aspx

**56.** *Population of Africa.*  The population of South America is approximately $\frac{2}{5}$ of the population of Africa. In 2007, the population of South America was about 376,000,000. What was the population of Africa in 2007?  [3.7d]

Sources: Population Division/International Programs Center; *The World Almanac* 2008

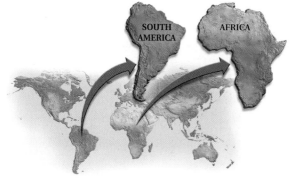

**57.** Divide and simplify: $-\dfrac{4}{5} \div \dfrac{7}{10}$.  [3.7b]

**58.** Add: 23,456 + 5677 + 4002.  [1.2a]

**59.** Multiply and simplify: $\dfrac{4}{5} \cdot \dfrac{10}{12}$.  [3.6a]

**60.** Subtract: 80,004 − 2305.  [1.3a]

## Synthesis

**61.** A pencil company uses two sizes of boxes, 5 in. by 6 in. and 5 in. by 8 in. These boxes are packed in bigger cartons for shipping. Find the width and the length of the smallest carton that will accommodate boxes of either size without any room left over. (Each carton can contain only one type of box and all boxes must point in the same direction.)

**62.** Consider 8 and 12. Determine whether each of the following is the LCM of 8 and 12. Tell why or why not.
a) $2 \cdot 2 \cdot 3 \cdot 3$
b) $2 \cdot 2 \cdot 3$
c) $2 \cdot 3 \cdot 3$
d) $2 \cdot 2 \cdot 2 \cdot 3$

# 4.2 Addition and Applications

## a) Addition Using Fraction Notation

### Like Denominators

Addition using fraction notation corresponds to combining or putting like things together, just as addition with whole numbers does. For example,

We combine two sets, each of which consists of fractional parts of one object that are the same size.

This is the resulting set.

$$\frac{2}{8} \quad + \quad \frac{3}{8} \quad = \quad \frac{5}{8}$$

2 eighths + 3 eighths = 5 eighths,

or $\quad 2 \cdot \frac{1}{8} + 3 \cdot \frac{1}{8} = 5 \cdot \frac{1}{8}$, $\quad$ or $\quad \frac{2}{8} + \frac{3}{8} = \frac{5}{8}$.

We see that to add when denominators are the same, we add the numerators and keep the denominator.

> Do Margin Exercise 1.

To add when denominators are the same,

**a)** add the numerators,

**b)** keep the denominator, and

$$\frac{2}{6} + \frac{5}{6} = \frac{2 + 5}{6} = \frac{7}{6}$$

**c)** simplify, if possible.

**EXAMPLES** Add and simplify.

**1.** $\dfrac{2}{4} + \dfrac{1}{4} = \dfrac{2 + 1}{4} = \dfrac{3}{4}$ $\qquad$ No simplifying is possible.

**2.** $\dfrac{11}{6} + \dfrac{3}{6} = \dfrac{11 + 3}{6} = \dfrac{14}{6} = \dfrac{2 \cdot 7}{2 \cdot 3} = \dfrac{2}{2} \cdot \dfrac{7}{3} = 1 \cdot \dfrac{7}{3} = \dfrac{7}{3}$ $\qquad$ Here we simplified.

**3.** $\dfrac{3}{12} + \dfrac{-5}{12} = \dfrac{3 + (-5)}{12} = \dfrac{-2}{12} = \dfrac{2 \cdot (-1)}{2 \cdot 6} = \dfrac{2}{2} \cdot \dfrac{-1}{6} = 1 \cdot \dfrac{-1}{6} = \dfrac{-1}{6}$, or $-\dfrac{1}{6}$

Recall that $\dfrac{-a}{b} = -\dfrac{a}{b}$.

> Do Exercises 2–5.

### Different Denominators

What do we do when denominators are different? We can find a common denominator by multiplying by 1. Consider adding $\frac{1}{6}$ and $\frac{3}{4}$. There are many common denominators that can be obtained. Let's look at two possibilities.

### OBJECTIVES

**a** Add using fraction notation.

**b** Solve applied problems involving addition with fraction notation.

**SKILL TO REVIEW**

Objective 3.1d: Find the prime factorization of a composite number.

Find the prime factorization of each number.

**1.** 96 $\qquad$ **2.** 1400

**1.** Find $\dfrac{1}{5} + \dfrac{3}{5}$.

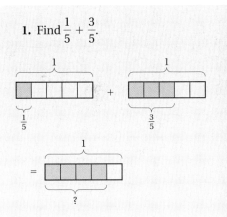

Add and simplify.

**2.** $\dfrac{5}{13} + \dfrac{9}{13}$ $\qquad$ **3.** $\dfrac{1}{3} + \dfrac{2}{3}$

**4.** $\dfrac{5}{12} + \dfrac{1}{12}$ $\qquad$ **5.** $\dfrac{9}{16} + \dfrac{-3}{16}$

*Answers*

*Skill to Review:*
**1.** $2 \cdot 2 \cdot 2 \cdot 2 \cdot 2 \cdot 3$
**2.** $2 \cdot 2 \cdot 2 \cdot 5 \cdot 5 \cdot 7$

*Margin Exercises:*
**1.** $\frac{4}{5}$ $\quad$ **2.** $\frac{14}{13}$ $\quad$ **3.** 1 $\quad$ **4.** $\frac{1}{2}$ $\quad$ **5.** $\frac{3}{8}$

**A.** $\dfrac{1}{6} + \dfrac{3}{4} = \dfrac{1}{6} \cdot 1 + \dfrac{3}{4} \cdot 1$

$= \dfrac{1}{6} \cdot \dfrac{4}{4} + \dfrac{3}{4} \cdot \dfrac{6}{6}$

$= \dfrac{4}{24} + \dfrac{18}{24}$

$= \dfrac{22}{24}$

$= \dfrac{11}{12}$   Simplifying

**B.** $\dfrac{1}{6} + \dfrac{3}{4} = \dfrac{1}{6} \cdot 1 + \dfrac{3}{4} \cdot 1$

$= \dfrac{1}{6} \cdot \dfrac{2}{2} + \dfrac{3}{4} \cdot \dfrac{3}{3}$

$= \dfrac{2}{12} + \dfrac{9}{12}$

$= \dfrac{11}{12}$

We had to simplify in (A) but not in (B). In (B), we used the least common multiple of the denominators, 12, as the common denominator. That number is called the **least common denominator**, or **LCD**. As we will see in Example 6, we may still need to simplify when using the LCD, but it is usually easier than when we use a larger common denominator.

> To add when denominators are different:
>
> **a)** Find the least common multiple of the denominators. That number is the least common denominator, LCD.
> **b)** Multiply by 1, using an appropriate notation, $n/n$, to express each number in terms of the LCD.
> **c)** Add the numerators, keeping the same denominator.
> **d)** Simplify, if possible.

**EXAMPLE 4**  Add: $\dfrac{3}{4} + \dfrac{1}{8}$.

The LCD is 8.   4 is a factor of 8, so the LCM of 4 and 8 is 8.

$\dfrac{3}{4} + \dfrac{1}{8} = \dfrac{3}{4} \cdot 1 + \dfrac{1}{8}$ ← This fraction already has the LCD as its denominator.

$= \dfrac{3}{4} \cdot \dfrac{2}{2} + \dfrac{1}{8}$   *Think*: $4 \times \square = 8$. The answer is 2, so we multiply by 1, using $\frac{2}{2}$.

$= \dfrac{6}{8} + \dfrac{1}{8} = \dfrac{7}{8}$

**EXAMPLE 5**  Add: $\dfrac{1}{9} + \dfrac{5}{6}$.

The LCD is 18.   $9 = 3 \cdot 3$ and $6 = 2 \cdot 3$, so the LCM of 9 and 6 is $2 \cdot 3 \cdot 3$, or 18.

$\dfrac{1}{9} + \dfrac{5}{6} = \dfrac{1}{9} \cdot 1 + \dfrac{5}{6} \cdot 1 = \dfrac{1}{9} \cdot \dfrac{2}{2} + \dfrac{5}{6} \cdot \dfrac{3}{3}$   *Think*: $6 \times \square = 18$. The answer is 3, so we multiply by 1 using $\frac{3}{3}$.

*Think*: $9 \times \square = 18$. The answer is 2, so we multiply by 1 using $\frac{2}{2}$.

$= \dfrac{2}{18} + \dfrac{15}{18} = \dfrac{17}{18}$

Do Exercises 6 and 7.

**Add.**

**6.** $\dfrac{2}{3} + \dfrac{1}{6}$

**7.** $\dfrac{3}{8} + \dfrac{5}{6}$

*Answers*

**6.** $\dfrac{5}{6}$   **7.** $\dfrac{29}{24}$

**EXAMPLE 6** Add: $\dfrac{5}{9} + \dfrac{11}{18}$.

The LCD is 18.  9 is a factor of 18, so the LCM is 18.

$$\dfrac{5}{9} + \dfrac{11}{18} = \dfrac{5}{9} \cdot \dfrac{2}{2} + \dfrac{11}{18} = \dfrac{10}{18} + \dfrac{11}{18}$$

$$\left. \begin{aligned} &= \dfrac{21}{18} = \dfrac{3 \cdot 7}{3 \cdot 6} = \dfrac{3}{3} \cdot \dfrac{7}{6} \\ &= \dfrac{7}{6} \end{aligned} \right\} \quad \text{Simplifying}$$

Do Exercise 8.

**8.** Add: $\dfrac{1}{6} + \dfrac{7}{18}$.

**EXAMPLE 7** Add: $\dfrac{1}{10} + \dfrac{3}{100} + \dfrac{-7}{1000}$.

Since 10 and 100 are factors of 1000, the LCD is 1000. Then

$$\dfrac{1}{10} + \dfrac{3}{100} + \dfrac{-7}{1000} = \dfrac{1}{10} \cdot \dfrac{100}{100} + \dfrac{3}{100} \cdot \dfrac{10}{10} + \dfrac{-7}{1000}$$

$$= \dfrac{100}{1000} + \dfrac{30}{1000} + \dfrac{-7}{1000} = \dfrac{100 + 30 + (-7)}{1000} = \dfrac{123}{1000}.$$

Do Exercise 9.

**9.** Add: $\dfrac{4}{10} + \dfrac{-1}{100} + \dfrac{3}{1000}$.

When denominators are large, we most often use the prime factorization of each denominator to find the LCD. This is shown in Example 8.

**EXAMPLE 8** Add: $\dfrac{13}{70} + \dfrac{11}{21} + \dfrac{8}{15}$.

We have

$$\dfrac{13}{70} + \dfrac{11}{21} + \dfrac{8}{15} = \dfrac{13}{2 \cdot 5 \cdot 7} + \dfrac{11}{3 \cdot 7} + \dfrac{8}{3 \cdot 5}. \quad \text{Factoring denominators}$$

The LCD is $2 \cdot 3 \cdot 5 \cdot 7$, or 210. Then

$$\dfrac{13}{70} + \dfrac{11}{21} + \dfrac{8}{15} = \dfrac{13}{2 \cdot 5 \cdot 7} \cdot \dfrac{3}{3} + \dfrac{11}{3 \cdot 7} \cdot \dfrac{2 \cdot 5}{2 \cdot 5} + \dfrac{8}{3 \cdot 5} \cdot \dfrac{7 \cdot 2}{7 \cdot 2}$$

The LCM of 70, 21, and 15 is $2 \cdot 3 \cdot 5 \cdot 7$. In each case, think of which factors are needed to get the LCD. Then multiply by 1 to obtain the LCD in each denominator.

$$= \dfrac{13 \cdot 3}{2 \cdot 5 \cdot 7 \cdot 3} + \dfrac{11 \cdot 2 \cdot 5}{3 \cdot 7 \cdot 2 \cdot 5} + \dfrac{8 \cdot 7 \cdot 2}{3 \cdot 5 \cdot 7 \cdot 2}$$

$$= \dfrac{39}{2 \cdot 3 \cdot 5 \cdot 7} + \dfrac{110}{2 \cdot 3 \cdot 5 \cdot 7} + \dfrac{112}{2 \cdot 3 \cdot 5 \cdot 7}$$

$$= \dfrac{261}{2 \cdot 3 \cdot 5 \cdot 7} = \dfrac{3 \cdot 3 \cdot 29}{2 \cdot 3 \cdot 5 \cdot 7} \quad \text{Factoring the numerator}$$

$$= \dfrac{3}{3} \cdot \dfrac{3 \cdot 29}{2 \cdot 5 \cdot 7} = \dfrac{3 \cdot 29}{2 \cdot 5 \cdot 7} = \dfrac{87}{70}.$$

Do Exercises 10 and 11.

Add.

**10.** $\dfrac{7}{10} + \dfrac{2}{21} + \dfrac{6}{7}$

**11.** $\dfrac{5}{18} + \dfrac{7}{24} + \dfrac{-11}{36}$

*Answers*

**8.** $\dfrac{5}{9}$   **9.** $\dfrac{393}{1000}$   **10.** $\dfrac{347}{210}$   **11.** $\dfrac{19}{72}$

## b Applications and Problem Solving

**EXAMPLE 9** *Subflooring.* A contractor requires his subcontractors to use two layers of subflooring under a ceramic tile floor. First, the subcontractors install a $\frac{3}{4}$-in. layer of oriented strand board (OSB). Then a $\frac{1}{2}$-in. sheet of cement board is mortared to the OSB. The mortar is $\frac{1}{8}$-in. thick. What is the total thickness of the two installed subfloors?

1. **Familiarize.** We first make a drawing. We let $T$ = the total thickness of the subfloors.

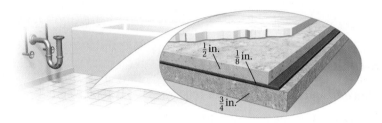

2. **Translate.** The problem can be translated to an equation as follows.

| OSB | plus | Mortar | plus | Cement board | is | Total thickness |
|-----|------|--------|------|--------------|-----|-----------------|
| ↓ | ↓ | ↓ | ↓ | ↓ | ↓ | ↓ |
| $\frac{3}{4}$ | + | $\frac{1}{8}$ | + | $\frac{1}{2}$ | = | $T$ |

3. **Solve.** To solve the equation, we carry out the addition. The LCM of the denominators is 8 because 2 and 4 are factors of 8. We multiply by 1 in order to obtain the LCD:

$$\frac{3}{4} + \frac{1}{8} + \frac{1}{2} = T$$

$$\frac{3}{4} \cdot \frac{2}{2} + \frac{1}{8} + \frac{1}{2} \cdot \frac{4}{4} = T$$

$$\frac{6}{8} + \frac{1}{8} + \frac{4}{8} = T$$

$$\frac{11}{8} = T.$$

4. **Check.** We check by repeating the calculation. We also note that the sum should be larger than any of the individual measurements, which it is. This tells us that the answer is reasonable.

5. **State.** The total thickness of the installed subfloors is $\frac{11}{8}$ in.

Do Exercise 12.

12. **Fruit Salad.** A caterer prepares a mixed berry salad with $\frac{7}{8}$ qt of strawberries, $\frac{3}{4}$ qt of raspberries, and $\frac{5}{16}$ qt of blueberries. What is the total amount of berries in the salad?

**Answer**

12. $\frac{31}{16}$ qt

**a** Add and simplify.

**1.** $\dfrac{7}{8} + \dfrac{1}{8}$

**2.** $\dfrac{2}{5} + \dfrac{3}{5}$

**3.** $\dfrac{1}{8} + \dfrac{5}{8}$

**4.** $\dfrac{3}{10} + \dfrac{3}{10}$

**5.** $\dfrac{2}{3} + \dfrac{-5}{6}$

**6.** $\dfrac{-5}{6} + \dfrac{1}{9}$

**7.** $\dfrac{1}{8} + \dfrac{1}{6}$

**8.** $\dfrac{1}{6} + \dfrac{3}{4}$

**9.** $\dfrac{-4}{5} + \dfrac{7}{10}$

**10.** $\dfrac{3}{4} + \dfrac{-1}{12}$

**11.** $\dfrac{5}{12} + \dfrac{3}{8}$

**12.** $\dfrac{7}{8} + \dfrac{1}{16}$

**13.** $\dfrac{3}{20} + \dfrac{3}{4}$

**14.** $\dfrac{2}{15} + \dfrac{2}{5}$

**15.** $\dfrac{5}{6} + \dfrac{-7}{9}$

**16.** $\dfrac{5}{8} + \dfrac{-5}{6}$

**17.** $\dfrac{3}{10} + \dfrac{1}{100}$

**18.** $\dfrac{9}{10} + \dfrac{3}{100}$

**19.** $\dfrac{5}{12} + \dfrac{4}{15}$

**20.** $\dfrac{3}{16} + \dfrac{1}{12}$

**21.** $\dfrac{-9}{10} + \dfrac{99}{100}$

**22.** $\dfrac{-3}{10} + \dfrac{27}{100}$

**23.** $\dfrac{7}{8} + \dfrac{0}{1}$

**24.** $\dfrac{0}{1} + \dfrac{5}{6}$

**25.** $\dfrac{3}{8} + \dfrac{1}{6}$

**26.** $\dfrac{5}{8} + \dfrac{1}{6}$

**27.** $\dfrac{5}{12} + \dfrac{7}{24}$

**28.** $\dfrac{1}{18} + \dfrac{7}{12}$

**29.** $\dfrac{3}{16} + \dfrac{5}{16} + \dfrac{4}{16}$

**30.** $\dfrac{3}{8} + \dfrac{1}{8} + \dfrac{2}{8}$

**31.** $\dfrac{8}{10} + \dfrac{7}{100} + \dfrac{4}{1000}$

**32.** $\dfrac{1}{10} + \dfrac{2}{100} + \dfrac{3}{1000}$

**33.** $\dfrac{3}{8} + \dfrac{-7}{12} + \dfrac{8}{15}$

**34.** $\dfrac{1}{2} + \dfrac{-3}{8} + \dfrac{1}{4}$

**35.** $\dfrac{15}{24} + \dfrac{7}{36} + \dfrac{91}{48}$

**36.** $\dfrac{5}{7} + \dfrac{25}{52} + \dfrac{7}{4}$

Solve.

**37.** *Riding a Segway®.*   Tate rode a Segway® Personal Transporter $\frac{5}{6}$ mi to the library, then $\frac{3}{4}$ mi to class, and then $\frac{3}{2}$ mi to his part-time job. How far did he ride his Segway®?

**38.** For a community project, an earth science class volunteered one hour per day for three days to join the state highway beautification project. The students collected trash along a $\frac{4}{5}$-mi stretch of highway the first day, $\frac{5}{8}$ mi the second day, and $\frac{1}{2}$ mi the third day. How many miles along the highway did they clean?

**39.** *Caffeine.*   To cut back on their caffeine intake, Michelle and Gerry mix caffeinated and decaffeinated coffee beans before grinding for a customized mix. They mix $\frac{3}{16}$ lb of decaffeinated beans with $\frac{5}{8}$ lb of caffeinated beans. What is the total amount of coffee beans in the mixture?

**40.** Alyse bought $\frac{1}{3}$ lb of orange pekoe tea and $\frac{1}{2}$ lb of English cinnamon tea. How many pounds of tea did she buy?

**41.** *Punch Recipe.*   A recipe for strawberry punch calls for $\frac{1}{5}$ qt of ginger ale and $\frac{3}{5}$ qt of strawberry soda. How much liquid is needed? If the recipe is doubled, how much liquid is needed? If the recipe is halved, how much liquid is needed?

**42.** *Concrete Mix.*   A cubic meter of concrete mix contains 420 kg of cement, 150 kg of stone, and 120 kg of sand. What is the total weight of a cubic meter of the mix? What part is cement? stone? sand? Add these fractional amounts. What is the result?

**43.** *Iced Brownies.*   The campus culinary arts department is preparing brownies for the international student reception. Students in the catering program iced the $\frac{11}{16}$-in.$\left(\frac{11}{16}''\right)$ butterscotch brownies with a $\frac{5}{32}$-in.$\left(\frac{5}{32}''\right)$ layer of icing. What is the thickness of the iced brownie?

**44.** *Carpentry.*   To cut expenses, a carpenter sometimes glues two kinds of plywood together. He glues a $\frac{1}{4}$-in.$\left(\frac{1}{4}''\right)$ piece of walnut plywood to a $\frac{3}{8}$-in.$\left(\frac{3}{8}''\right)$ piece of less expensive plywood. What is the total thickness of these pieces?

**45.** A tile $\frac{5}{8}$ in. thick is glued to a board $\frac{7}{8}$ in. thick. The glue is $\frac{3}{32}$ in. thick. How thick is the result?

**46.** A baker used $\frac{1}{2}$ lb of flour for rolls, $\frac{1}{4}$ lb for donuts, and $\frac{1}{3}$ lb for cookies. How much flour was used?

# Skill Maintenance

Multiply.   [1.4a], [2.4a]

**47.** $408 \cdot 516$

**48.** $1125 \cdot 3728$

**49.** $-8 \cdot 7$

**50.** $-9 \cdot (-12)$

*Presidential Elections.*   The results of the five closest presidential elections are listed in the following table. Use these data for Exercises 51–56.   [1.8a]

| DATE | PRESIDENT | ELECTORAL VOTES | POPULAR VOTES |
|------|-----------|-----------------|---------------|
| 1916 | Woodrow Wilson (D) | 277 | 9,126,300 |
|      | Charles E. Hughes (R) | 254 | 8,546,789 |
| 1960 | John F. Kennedy (D) | 303 | 34,226,731 |
|      | Richard M. Nixon (R) | 219 | 34,108,157 |
| 1968 | Richard M. Nixon (R) | 301 | 31,785,480 |
|      | Hubert H. Humphrey (D) | 191 | 31,275,166 |
|      | George C. Wallace (3rd Party) | 46 | 9,906,473 |
| 1976 | Jimmy Carter (D) | 297 | 40,830,763 |
|      | Gerald R. Ford (R) | 240 | 39,147,793 |
| 2000 | George W. Bush (R) | 271 | 50,459,211 |
|      | Albert A. Gore (D) | 266 | 51,003,894 |

SOURCE: *The World Almanac*, 2008, p. 544

**51.** How many more popular votes did Albert A. Gore have than George W. Bush in the 2000 presidential election?

**52.** How many more electoral votes did George W. Bush have than Albert A. Gore in the 2000 presidential election?

**53.** In the 1960 presidential election, how many more electoral votes did John F. Kennedy have than Richard M. Nixon?

**54.** In the 1968 presidential election, how many more popular votes did Richard M. Nixon have than Hubert H. Humphrey?

**55.** How much greater was the total of the popular vote counts in the 2000 presidential election than in the 1976 election?

**56.** How much greater was the total of the popular vote counts in the 2000 presidential election than in the 1960 election?

**57.** *Honey Production.*   In 2007, 154,907,000 lb of honey were produced in the United States. The two states with the greatest honey production were North Dakota and California. North Dakota produced 25,900,000 lb of honey, and California produced 19,760,000 lb. How many more pounds of honey were produced in North Dakota than in California?   [1.8a]

Source: www.cattlenetwork.com

**58.** The Bingham community garden is to be split into 16 equally sized plots. If the garden occupies $\frac{3}{4}$ acre of land, how large will each plot be?   [3.7d]

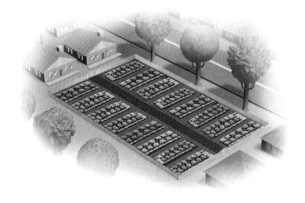

# Synthesis

**59.** A guitarist's band is booked for Friday and Saturday nights at a local club. The guitarist is part of a trio on Friday and part of a quintet on Saturday. Thus the guitarist is paid one-third of one-half the weekend's pay for Friday and one-fifth of one-half the weekend's pay for Saturday. What fractional part of the band's pay did the guitarist receive for the weekend's work? If the band was paid $1200, how much did the guitarist receive?

# 4.3

# Subtraction, Order, and Applications

## OBJECTIVES

**a** Subtract using fraction notation.

**b** Use < or > with fraction notation to write a true sentence.

**c** Solve equations of the type $x + a = b$ and $a + x = b$, where $a$ and $b$ may be fractions.

**d** Solve applied problems involving subtraction with fraction notation.

**SKILL TO REVIEW**

Objective 1.6c: Use < or > for ☐ to write a true sentence in a situation like 6 ☐ 10.

Use < or > for ☐ to write a true sentence.

**1.** 218 ☐ 128     **2.** 41 ☐ 95

Subtract and simplify.

**1.** $\dfrac{7}{8} - \dfrac{3}{8}$

**2.** $\dfrac{10}{16} - \dfrac{4}{16}$

**3.** $\dfrac{3}{10} - \dfrac{8}{10}$

## a Subtraction Using Fraction Notation

### Like Denominators

Let's consider the difference $\frac{4}{8} - \frac{3}{8}$.

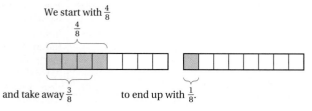

We start with $\frac{4}{8}$

and take away $\frac{3}{8}$   to end up with $\frac{1}{8}$.

We start with 4 eighths and take away 3 eighths:

4 eighths − 3 eighths = 1 eighth,

or   $4 \cdot \dfrac{1}{8} - 3 \cdot \dfrac{1}{8} = \dfrac{1}{8}$,   or   $\dfrac{4}{8} - \dfrac{3}{8} = \dfrac{1}{8}$.

To subtract when denominators are the same,

a) subtract the numerators,

b) keep the denominator, and

c) simplify, if possible.

$$\dfrac{7}{10} - \dfrac{4}{10} = \dfrac{7 - 4}{10} = \dfrac{3}{10}$$

**EXAMPLES** Subtract and simplify.

**1.** $\dfrac{32}{12} - \dfrac{25}{12} = \dfrac{32 - 25}{12} = \dfrac{7}{12}$

**2.** $\dfrac{7}{10} - \dfrac{3}{10} = \dfrac{7 - 3}{10} = \dfrac{4}{10} = \dfrac{2 \cdot 2}{5 \cdot 2} = \dfrac{2}{5} \cdot \dfrac{2}{2} = \dfrac{2}{5} \cdot 1 = \dfrac{2}{5}$

**3.** $\dfrac{25}{12} - \dfrac{32}{12} = \dfrac{25 - 32}{12} = \dfrac{-7}{12}$, or $-\dfrac{7}{12}$

Do Margin Exercises 1–3.

### Different Denominators

The procedure for subtraction with different denominators is similar to the procedure for addition.

To subtract when denominators are different:

a) Find the least common multiple of the denominators. That number is the least common denominator, LCD.

b) Multiply by 1, using an appropriate notation, $n/n$, to express each number in terms of the LCD.

c) Subtract the numerators, keeping the same denominator.

d) Simplify, if possible.

*Answers*

*Skill to Review:*
1. >     2. <

*Margin Exercises:*
1. $\dfrac{1}{2}$   2. $\dfrac{3}{8}$   3. $-\dfrac{1}{2}$

**EXAMPLE 4** Subtract: $\frac{2}{5} - \frac{3}{8}$.

The LCM of 5 and 8 is 40, so the LCD is 40.

$$\frac{2}{5} - \frac{3}{8} = \frac{2}{5} \cdot \frac{8}{8} - \frac{3}{8} \cdot \frac{5}{5} \longleftarrow$$

*Think*: $8 \times \square = 40$. The answer is 5, so we multiply by 1, using $\frac{5}{5}$.

*Think*: $5 \times \square = 40$. The answer is 8, so we multiply by 1, using $\frac{8}{8}$.

$$= \frac{16}{40} - \frac{15}{40} = \frac{16 - 15}{40} = \frac{1}{40}$$

Do Exercise 4.

**4.** Subtract: $\frac{3}{4} - \frac{2}{3}$.

**EXAMPLE 5** Subtract: $\frac{7}{12} - \frac{5}{6}$.

Since 12 is a multiple of 6, the LCM of 6 and 12 is 12. The LCD is 12.

$$\frac{7}{12} - \frac{5}{6} = \frac{7}{12} - \frac{5}{6} \cdot \frac{2}{2}$$

$$= \frac{7}{12} - \frac{10}{12} = \frac{7 - 10}{12} = \frac{-3}{12}$$

$$= \frac{3 \cdot (-1)}{3 \cdot 4} = \frac{3}{3} \cdot \frac{-1}{4} = \frac{-1}{4} \text{ or } -\frac{1}{4}$$

Do Exercises 5 and 6.

Subtract.

**5.** $\frac{5}{6} - \frac{1}{9}$    **6.** $\frac{3}{10} - \frac{4}{5}$

**EXAMPLE 6** Subtract: $\frac{17}{24} - \frac{4}{15}$.

We have

$$\frac{17}{24} - \frac{4}{15} = \frac{17}{2 \cdot 2 \cdot 2 \cdot 3} - \frac{4}{3 \cdot 5}.$$    Factoring denominators

The LCD is $2 \cdot 2 \cdot 2 \cdot 3 \cdot 5$, or 120. Then

$$\frac{17}{24} - \frac{4}{15} = \frac{17}{2 \cdot 2 \cdot 2 \cdot 3} \cdot \frac{5}{5} - \frac{4}{3 \cdot 5} \cdot \frac{2 \cdot 2 \cdot 2}{2 \cdot 2 \cdot 2}$$

The LCM of 24 and 15 is $2 \cdot 2 \cdot 2 \cdot 3 \cdot 5$. In each case, we multiply by 1 to obtain the LCD.

$$= \frac{17 \cdot 5}{2 \cdot 2 \cdot 2 \cdot 3 \cdot 5} - \frac{4 \cdot 2 \cdot 2 \cdot 2}{3 \cdot 5 \cdot 2 \cdot 2 \cdot 2}$$

$$= \frac{85}{120} - \frac{32}{120} = \frac{53}{120}.$$

Do Exercise 7.

**7.** Subtract: $\frac{11}{28} - \frac{5}{16}$.

## b Order

We see from this figure that $\frac{4}{5}$ is greater than $\frac{3}{5}$, and $\frac{3}{5}$ is less than $\frac{4}{5}$. That is, $\frac{4}{5} > \frac{3}{5}$, and $\frac{3}{5} < \frac{4}{5}$.

$\frac{4}{5}$ [figure]

$\frac{3}{5}$ [figure]

*Answers*

**4.** $\frac{1}{12}$    **5.** $\frac{13}{18}$    **6.** $-\frac{1}{2}$    **7.** $\frac{9}{112}$

**8.** Use $<$ or $>$ for $\square$ to write a true sentence:

$$\frac{3}{8} \,\square\, \frac{5}{8}.$$

**9.** Use $<$ or $>$ for $\square$ to write a true sentence:

$$\frac{7}{10} \,\square\, \frac{6}{10}.$$

To determine which of two numbers is greater when there is a common denominator, compare the numerators:

$$\frac{4}{5}, \ \frac{3}{5}; \qquad 4 > 3; \qquad \frac{4}{5} > \frac{3}{5}.$$

Do Exercises 8 and 9.

When denominators are different, we cannot compare numerators until we multiply by 1 to make the denominators the same.

**EXAMPLE 7** Use $<$ or $>$ for $\square$ to write a true sentence:

$$\frac{2}{5} \,\square\, \frac{3}{4}.$$

The LCD is 20. We have

$$\frac{2}{5} \cdot \frac{4}{4} = \frac{8}{20}; \qquad \text{We multiply by 1 using } \frac{4}{4} \text{ to get the LCD.}$$

$$\frac{3}{4} \cdot \frac{5}{5} = \frac{15}{20}. \qquad \text{We multiply by 1 using } \frac{5}{5} \text{ to get the LCD.}$$

Now that the denominators are the same, 20, we can compare the numerators. Since $8 < 15$, it follows that $\frac{8}{20} < \frac{15}{20}$, so

$$\frac{2}{5} < \frac{3}{4}.$$

**EXAMPLE 8** Use $<$ or $>$ for $\square$ to write a true sentence: $\dfrac{-89}{100} \,\square\, \dfrac{-9}{10}.$

The LCD is 100. We write $\frac{-9}{10}$ with a denominator of 100 to make the denominators the same.

$$\frac{-9}{10} \cdot \frac{10}{10} = \frac{-90}{100} \qquad \text{We multiply by } \frac{10}{10} \text{ to get the LCD.}$$

Since $-89 > -90$, it follows that $\frac{-89}{100} > \frac{-90}{100}$, so

$$\frac{-89}{100} > \frac{-9}{10}.$$

Do Exercises 10–12.

Use $<$ or $>$ for $\square$ to write a true sentence:

**10.** $\dfrac{2}{3} \,\square\, \dfrac{5}{8}$

**11.** $\dfrac{3}{4} \,\square\, \dfrac{8}{12}$

**12.** $\dfrac{-7}{8} \,\square\, \dfrac{-5}{6}$

## (c) Solving Equations

Now let's solve equations of the form $x + a = b$ or $a + x = b$, where $a$ and $b$ may be fractions. Proceeding as we have before, we subtract $a$ on both sides of the equation.

*Answers*

**8.** $<$    **9.** $>$    **10.** $>$    **11.** $>$    **12.** $<$

**EXAMPLE 9**  Solve: $x + \dfrac{1}{4} = -\dfrac{3}{5}$.

$$x + \dfrac{1}{4} - \dfrac{1}{4} = -\dfrac{3}{5} - \dfrac{1}{4} \qquad \text{Subtracting } \dfrac{1}{4} \text{ on both sides}$$

$$x + 0 = -\dfrac{3}{5} \cdot \dfrac{4}{4} - \dfrac{1}{4} \cdot \dfrac{5}{5} \qquad \begin{array}{l}\text{The LCD is 20. We multiply by 1}\\ \text{to get the LCD.}\end{array}$$

$$x = -\dfrac{12}{20} - \dfrac{5}{20}$$

$$x = -\dfrac{17}{20}$$

Do Exercises 13 and 14.

Solve.

**13.** $x + \dfrac{2}{3} = \dfrac{5}{6}$

**14.** $\dfrac{3}{5} + t = -\dfrac{7}{8}$

## d  Applications and Problem Solving

**EXAMPLE 10**  *Fraction of the Moon Illuminated.*  From anywhere on Earth, the moon appears to be a circular disk. At midnight on July 11, 2008 (Eastern Daylight Time), $\frac{3}{5}$ of the moon appeared illuminated. By July 22, 2008, the illuminated portion had increased to $\frac{17}{20}$. How much more of the moon appeared illuminated on July 22 than on July 11?

Source: Astronomical Applications Department, U.S. Naval Observatory, Washington DC 20392

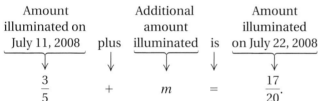

1. **Familiarize.**  We let $m =$ the additional part of the moon that appeared illuminated.

2. **Translate.**  We translate to an equation:

| Amount illuminated on July 11, 2008 | plus | Additional amount illuminated | is | Amount illuminated on July 22, 2008 |
|:---:|:---:|:---:|:---:|:---:|
| $\dfrac{3}{5}$ | $+$ | $m$ | $=$ | $\dfrac{17}{20}.$ |

3. **Solve.**  To solve the equation, we subtract $\frac{3}{5}$ on both sides:

$$\dfrac{3}{5} + m = \dfrac{17}{20}$$

$$\dfrac{3}{5} + m - \dfrac{3}{5} = \dfrac{17}{20} - \dfrac{3}{5} \qquad \text{Subtracting } \dfrac{3}{5} \text{ on both sides}$$

$$m + 0 = \dfrac{17}{20} - \dfrac{3}{5} \cdot \dfrac{4}{4} \qquad \begin{array}{l}\text{The LCD is 20. We multiply by 1}\\ \text{to obtain the LCD.}\end{array}$$

$$m = \dfrac{17}{20} - \dfrac{12}{20} = \dfrac{5}{20} = \dfrac{5 \cdot 1}{5 \cdot 4} = \dfrac{5}{5} \cdot \dfrac{1}{4} = \dfrac{1}{4}.$$

4. **Check.**  To check, we add:

$$\dfrac{3}{5} + \dfrac{1}{4} = \dfrac{3}{5} \cdot \dfrac{4}{4} + \dfrac{1}{4} \cdot \dfrac{5}{5} = \dfrac{12}{20} + \dfrac{5}{20} = \dfrac{17}{20}.$$

5. **State.**  On July 22, $\frac{1}{4}$ more of the moon was illuminated than on July 11.

Do Exercise 15.

**15. Fraction of the Moon Illuminated.**  At midnight on April 18, 2008 (Eastern Daylight Time), $\frac{19}{20}$ of the moon appeared illuminated. By April 27, 2008, the illuminated part had decreased to $\frac{16}{25}$. How much less of the moon appeared illuminated on April 27 than on April 18?

Source: Astronomical Applications Department, U.S. Naval Observatory, Washington DC 20392

*Answers*

13. $\dfrac{1}{6}$    14. $-\dfrac{59}{40}$    15. $\dfrac{31}{100}$

# Translating for Success

**1.** *Bubble Wrap.* One-Stop Postal Center orders bubble wrap in 64-yd rolls. On average, $\frac{3}{4}$ yd is used per small package. How many small packages can be prepared with 2 rolls of bubble wrap?

**2.** *Distance from College.* The post office is $\frac{7}{9}$ mi from the community college. The medical clinic is $\frac{2}{5}$ as far from the college as the post office is. How far is the clinic from the college?

**3.** *Swimming.* Andrew swims $\frac{7}{9}$ mi every day. One day he swims $\frac{2}{5}$ mi by 11:00 A.M. How much farther must Andrew swim to reach his daily goal?

**4.** *Tuition.* The average tuition at Waterside University is $12,000. If a loan is obtained for $\frac{1}{3}$ of the tuition, how much is the loan?

**5.** *Thermos Bottle Capacity.* A thermos bottle holds $\frac{11}{12}$ qt. How much is in the bottle when it is $\frac{4}{7}$ full?

The goal of these matching questions is to practice step (2), *Translate*, of the five-step problem-solving process. Translate each problem to an equation and select a correct translation from equations A–O.

**A.** $\frac{3}{4} \cdot 64 = x$

**B.** $\frac{1}{3} \cdot 12,000 = x$

**C.** $\frac{1}{3} + \frac{2}{5} = x$

**D.** $\frac{2}{5} + x = \frac{7}{9}$

**E.** $\frac{2}{5} \cdot \frac{7}{9} = x$

**F.** $\frac{3}{4} \cdot x = 64$

**G.** $\frac{4}{7} = x + \frac{11}{12}$

**H.** $\frac{2}{5} = x + \frac{7}{9}$

**I.** $\frac{4}{7} \cdot \frac{11}{12} = x$

**J.** $\frac{3}{4} \cdot x = 128$

**K.** $\frac{1}{3} \cdot x = 12,000$

**L.** $\frac{1}{3} + \frac{2}{5} + x = 1$

**M.** $\frac{4}{3} \cdot 64 = x$

**N.** $\frac{4}{7} + x = \frac{11}{12}$

**O.** $\frac{1}{3} + x = \frac{2}{5}$

*Answers on page A-6*

**6.** *Cutting Rope.* A piece of rope $\frac{11}{12}$ yd long is cut into two pieces. One piece is $\frac{4}{7}$ yd long. How long is the other piece?

**7.** *Planting Corn.* Each year, Prairie State Farm plants 64 acres of corn. With good weather, $\frac{3}{4}$ of the planting can be completed by April 20. How many acres can be planted with good weather?

**8.** *Painting Trim.* A painter used $\frac{1}{3}$ gal of white paint for the trim in the library and $\frac{2}{5}$ gal in the family room. How much paint was used for the trim in the two rooms?

**9.** *Lottery Winnings.* Sally won $12,000 in a state lottery and decided to give the net amount after taxes to three charities. One received $\frac{1}{3}$ of the money, and a second received $\frac{2}{5}$. What fractional part did the third charity receive?

**10.** *Reading Assignment.* When Lowell had read 64 pages of his political science assignment, he had completed $\frac{3}{4}$ of his required reading. How many total pages were assigned?

**a** Subtract and simplify.

1. $\dfrac{5}{6} - \dfrac{1}{6}$

2. $\dfrac{5}{8} - \dfrac{3}{8}$

3. $\dfrac{11}{12} - \dfrac{2}{12}$

4. $\dfrac{17}{18} - \dfrac{11}{18}$

5. $\dfrac{1}{8} - \dfrac{3}{4}$

6. $\dfrac{1}{9} - \dfrac{2}{3}$

7. $\dfrac{1}{8} - \dfrac{1}{12}$

8. $\dfrac{1}{6} - \dfrac{1}{8}$

9. $\dfrac{5}{6} - \dfrac{4}{3}$

10. $\dfrac{1}{16} - \dfrac{7}{8}$

11. $\dfrac{3}{4} - \dfrac{3}{28}$

12. $\dfrac{2}{5} - \dfrac{2}{15}$

13. $\dfrac{3}{4} - \dfrac{3}{20}$

14. $\dfrac{5}{6} - \dfrac{1}{2}$

15. $\dfrac{1}{20} - \dfrac{3}{4}$

16. $\dfrac{4}{16} - \dfrac{3}{4}$

17. $\dfrac{5}{12} - \dfrac{2}{15}$

18. $\dfrac{9}{10} - \dfrac{11}{16}$

19. $\dfrac{6}{10} - \dfrac{7}{100}$

20. $\dfrac{9}{10} - \dfrac{3}{100}$

21. $\dfrac{7}{15} - \dfrac{3}{25}$

22. $\dfrac{18}{25} - \dfrac{4}{35}$

23. $\dfrac{99}{100} - \dfrac{9}{10}$

24. $\dfrac{78}{100} - \dfrac{11}{20}$

25. $-\dfrac{2}{3} - \dfrac{1}{8}$

26. $-\dfrac{3}{4} - \dfrac{1}{2}$

27. $\dfrac{3}{5} - \dfrac{1}{2}$

28. $\dfrac{5}{6} - \dfrac{2}{3}$

29. $\dfrac{3}{8} - \dfrac{5}{12}$

30. $\dfrac{2}{9} - \dfrac{7}{12}$

31. $\dfrac{7}{8} - \dfrac{1}{16}$

32. $\dfrac{5}{12} - \dfrac{5}{16}$

33. $\dfrac{4}{15} - \dfrac{17}{25}$

34. $\dfrac{7}{24} - \dfrac{11}{18}$

35. $\dfrac{23}{25} - \dfrac{112}{150}$

36. $\dfrac{89}{90} - \dfrac{53}{120}$

**b** Use $<$ or $>$ for $\square$ to write a true sentence.

37. $\dfrac{5}{8} \ \square \ \dfrac{6}{8}$

38. $\dfrac{7}{9} \ \square \ \dfrac{5}{9}$

39. $\dfrac{1}{3} \ \square \ \dfrac{1}{4}$

40. $\dfrac{1}{8} \ \square \ \dfrac{1}{6}$

41. $\dfrac{-5}{7} \ \square \ \dfrac{-2}{3}$

42. $\dfrac{-4}{7} \ \square \ \dfrac{-3}{5}$

43. $\dfrac{4}{5} \ \square \ \dfrac{5}{6}$

44. $\dfrac{3}{2} \ \square \ \dfrac{7}{5}$

45. $\dfrac{-4}{5} \ \square \ \dfrac{-19}{20}$

46. $\dfrac{5}{6} \ \square \ \dfrac{13}{16}$

47. $\dfrac{19}{20} \ \square \ \dfrac{9}{10}$

48. $\dfrac{3}{4} \ \square \ \dfrac{11}{15}$

49. $\dfrac{-41}{13} \ \square \ \dfrac{-31}{21}$

50. $\dfrac{-132}{49} \ \square \ \dfrac{-12}{7}$

**c** Solve.

**51.** $x + \dfrac{1}{30} = \dfrac{1}{10}$

**52.** $y + \dfrac{9}{12} = \dfrac{11}{12}$

**53.** $\dfrac{2}{3} + t = -\dfrac{4}{5}$

**54.** $\dfrac{2}{3} + p = -\dfrac{7}{8}$

**55.** $x + \dfrac{1}{3} = \dfrac{5}{6}$

**56.** $m + \dfrac{5}{6} = \dfrac{9}{10}$

**d** Solve.

**57.** For a research paper, Kaitlyn spent $\frac{3}{4}$ hr searching the Internet on google.com and $\frac{1}{3}$ hr on chacha.com. How many more hours did she spend on google.com than on chacha.com?

**58.** As part of a fitness program, Deb swims $\frac{1}{2}$ mi every day. One day she had already swum $\frac{1}{5}$ mi. How much farther should Deb swim?

**59.** *Tire Tread.* A long-life tire has a tread depth of $\frac{3}{8}$ in. instead of a more typical $\frac{11}{32}$-in. depth. How much deeper is the long-life tread depth?

**Source:** *Popular Science*

$\frac{3}{8}$ in.

$\frac{11}{32}$ in.

**60.** A baker has a dispenser containing $\frac{15}{16}$ cup of icing and puts $\frac{1}{12}$ cup on a cinnamon roll. How much icing remains in the dispenser?

**61.** At a party, three friends, Ashley, Cole, and Lauren, shared a big tub of popcorn. Within 30 min, the tub was empty. Ashley ate $\frac{7}{12}$ of the tub while Lauren ate only $\frac{1}{6}$ of the tub. How much did Cole eat?

**62.** An estate was left to four children. One received $\frac{1}{4}$ of the estate, the second $\frac{1}{16}$, and the third $\frac{3}{8}$. How much did the fourth receive?

**63.** From a $\frac{4}{5}$-lb wheel of cheese, a $\frac{1}{4}$-lb piece was served. How much cheese remained on the wheel?

**64.** Jovan has an $\frac{11}{10}$-lb mixture of cashews and peanuts that includes $\frac{3}{5}$ lb of cashews. How many pounds of peanuts are in the mixture?

## Skill Maintenance

Divide, if possible. If not possible, write "not defined."   [1.5a], [3.3b]

**65.** $\dfrac{-38}{-38}$

**66.** $\dfrac{38}{0}$

**67.** $\dfrac{124}{0}$

**68.** $\dfrac{124}{31}$

Divide and simplify.   [3.7b]

**69.** $\dfrac{3}{7} \div \left(-\dfrac{9}{4}\right)$

**70.** $-\dfrac{9}{10} \div \left(-\dfrac{3}{5}\right)$

**71.** $7 \div \dfrac{1}{3}$

**72.** $\dfrac{1}{4} \div 8$

**73.** *Crayons.*   The Crayola 64 box of crayons with a built-in sharpener celebrated its 50th birthday in 2008. Since 1958, approximately 200 million Crayola 64 boxes have been sold. How many crayons in the 64 box have been sold altogether? [1.8a]
**Source:** Crayola LCC

**74.** A batch of fudge requires $\dfrac{3}{4}$ cup of sugar. How much sugar is needed to make 12 batches? [3.6b]

## Synthesis

Solve.

**75.** ▦ $x + \dfrac{16}{323} = \dfrac{10}{187}$

**76.** ▦ $x + \dfrac{7}{253} = \dfrac{12}{299}$

**77.** As part of a rehabilitation program, an athlete must swim and then walk a total of $\dfrac{9}{10}$ km each day. If one lap in the swimming pool is $\dfrac{3}{80}$ km, how far must the athlete walk after swimming 10 laps?

**78.** *Mountain Climbing.*   A mountain climber, beginning at sea level, climbs $\dfrac{3}{5}$ km, descends $\dfrac{1}{4}$ km, climbs $\dfrac{1}{3}$ km, and then descends $\dfrac{1}{7}$ km. At what elevation does the climber finish?

Simplify. Use the rules for order of operations given in Section 1.9.

**79.** $\dfrac{7}{8} - \dfrac{1}{10} \times \dfrac{5}{6}$

**80.** $\dfrac{2}{5} + \dfrac{1}{6} \div 3$

**81.** $\left(\dfrac{2}{3}\right)^2 - \left(\dfrac{3}{4}\right)^2$

**82.** $-5 \times \dfrac{3}{7} - \dfrac{1}{7} \times \dfrac{4}{5}$

Use $<$, $>$, or $=$ for □ to write a true sentence.

**83.** ▦ $\dfrac{37}{157} + \dfrac{19}{107}$ □ $\dfrac{6941}{16{,}799}$

**84.** ▦ $\dfrac{12}{97} + \dfrac{67}{139}$ □ $\dfrac{8167}{13{,}289}$

**85.** *Microsoft Interview.*   The following is a question taken from an employment interview with Microsoft. Try to answer it. "Given a gold bar that can be cut exactly twice and a contractor who must be paid one-seventh of a gold bar every day for seven days, how should the bar be cut?"
**Source:** *Fortune Magazine,* January 22, 2001

# Mixed Numerals

## OBJECTIVES

**a** Convert between mixed numerals and fraction notation.

**b** Divide whole numbers, writing the quotient as a mixed numeral.

**SKILL TO REVIEW**
Objective 1.5a: Divide whole numbers.

Divide.

**1.** $735 \div 16$

**2.** $23 \overline{)6023}$

Convert to a mixed numeral.

**1.** $1 + \dfrac{2}{3} = \Box\dfrac{\Box}{\Box}$

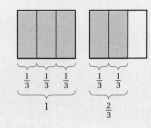

**2.** $2 + \dfrac{3}{4} = \Box\dfrac{\Box}{\Box}$

**3.** $12 + \dfrac{2}{7}$

## **a** Mixed Numerals

The following figure illustrates the use of a **mixed numeral**. The bolt shown is $2\frac{3}{8}$ in. long. The length is given as a whole-number part, 2, and a fractional part less than 1, $\frac{3}{8}$. We can also represent the measurement of the bolt with fraction notation as $\frac{19}{8}$, but the meaning or interpretation of such a symbol is less understandable or visual than that of mixed numeral notation.

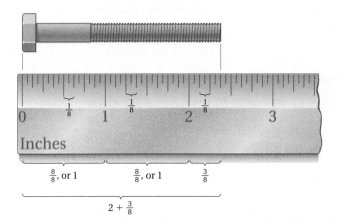

A mixed numeral $2\frac{3}{8}$ represents a sum:

$$2\frac{3}{8} \quad \text{means} \quad 2 + \frac{3}{8}.$$

This is a whole number.     This is a fraction less than 1.

**EXAMPLES** Convert to a mixed numeral.

**1.** $7 + \dfrac{2}{5} = 7\dfrac{2}{5}$        **2.** $4 + \dfrac{3}{10} = 4\dfrac{3}{10}$

> Do Margin Exercises 1–3.

The notation $2\frac{3}{4}$ has a plus sign left out. To aid in understanding, we sometimes write the missing plus sign. This is especially helpful when we convert a mixed numeral to fraction notation.

**EXAMPLES** Convert to fraction notation.

**3.** $2\dfrac{3}{4} = 2 + \dfrac{3}{4}$      Inserting the missing plus sign

$\qquad = \dfrac{2}{1} + \dfrac{3}{4}$      $2 = \dfrac{2}{1}$

$\qquad = \dfrac{2}{1} \cdot \dfrac{4}{4} + \dfrac{3}{4}$      Finding a common denominator

$\qquad = \dfrac{8}{4} + \dfrac{3}{4} = \dfrac{11}{4}$

*Answers*

*Skill to Review:*
**1.** 45 R 15    **2.** 261 R 20

*Margin Exercises:*
**1.** $1\frac{2}{3}$   **2.** $2\frac{3}{4}$   **3.** $12\frac{2}{7}$

**4.** $4\frac{3}{10} = 4 + \frac{3}{10} = \frac{4}{1} + \frac{3}{10} = \frac{4}{1} \cdot \frac{10}{10} + \frac{3}{10} = \frac{40}{10} + \frac{3}{10} = \frac{43}{10}$

Do Exercises 4 and 5.

Let's now consider a faster method for converting a mixed numeral to fraction notation.

> To convert from a mixed numeral to fraction notation:
> ⓐ Multiply the whole number by the denominator: $4 \cdot 10 = 40$.
> ⓑ Add the result to the numerator: $40 + 3 = 43$.
> ⓒ Keep the denominator.
>
> ⓑ ↗
> $4\frac{3}{10} = \frac{43}{10}$ ← ⓒ
> ⓐ ↙

**EXAMPLES** Convert to fraction notation.

**5.** $6\frac{2}{3} = \frac{20}{3}$     $6 \cdot 3 = 18, 18 + 2 = 20$

**6.** $8\frac{2}{9} = \frac{74}{9}$     $8 \cdot 9 = 72, 72 + 2 = 74$

**7.** $10\frac{7}{8} = \frac{87}{8}$     $10 \cdot 8 = 80, 80 + 7 = 87$

Do Exercises 6–9.

## Writing Mixed Numerals

We can find a mixed numeral for $\frac{5}{3}$ as follows:

$$\frac{5}{3} = \frac{3}{3} + \frac{2}{3} = 1 + \frac{2}{3} = 1\frac{2}{3}.$$

In terms of objects, we can think of $\frac{5}{3}$ as $\frac{3}{3}$, or 1, plus $\frac{2}{3}$, as shown below.

$\frac{5}{3} = \quad \frac{3}{3}$, or 1 $\quad + \quad \frac{2}{3}$

Fraction symbols like $\frac{5}{3}$ also indicate division; $\frac{5}{3}$ means $5 \div 3$. Let's divide the numerator by the denominator.

$$\begin{array}{r} 1 \\ 3\overline{)5} \\ \underline{3} \\ 2 \end{array} \leftarrow 2 \div 3 = \frac{2}{3}$$

Thus, $\frac{5}{3} = 1\frac{2}{3}$.

---

Convert to fraction notation.

**4.** $4\frac{2}{5}$     **5.** $6\frac{1}{10}$

Convert to fraction notation. Use the faster method.

**6.** $4\frac{5}{6}$     **7.** $9\frac{1}{4}$

**8.** $20\frac{2}{3}$     **9.** $1\frac{9}{13}$

*Answers*

**4.** $\frac{22}{5}$  **5.** $\frac{61}{10}$  **6.** $\frac{29}{6}$  **7.** $\frac{37}{4}$

**8.** $\frac{62}{3}$  **9.** $\frac{22}{13}$

To convert from fraction notation to a mixed numeral, divide.

$$\frac{13}{5} \qquad 5\overline{)13} \quad \frac{10}{3} \qquad 2\frac{3}{5}$$

The divisor — The quotient — The remainder

**EXAMPLES** Convert to a mixed numeral.

**8.** $\dfrac{69}{10}$ $\qquad 10\overline{)69}$ $\dfrac{60}{9}$ $\qquad \dfrac{69}{10} = 6\dfrac{9}{10}$

**9.** $\dfrac{122}{8}$ $\qquad 8\overline{)122}$ $\dfrac{8}{42}$ $\dfrac{40}{2}$ $\qquad \dfrac{122}{8} = 15\dfrac{2}{8} = 15\dfrac{1}{4}$

Do Exercises 10–13.

The same type of procedure also works with negative numbers. Of course, the result will be a negative mixed numeral.

**EXAMPLE 10** Convert $-\dfrac{17}{5}$ to a mixed numeral.

Since

$$5\overline{)17} \quad \dfrac{15}{2}$$

we have

$$\frac{17}{5} = 3\frac{2}{5}.$$

Then

$$-\frac{17}{5} = -3\frac{2}{5}.$$

Do Exercises 14 and 15.

A fraction larger than 1, such as $\frac{27}{8}$, is sometimes referred to as an "improper" fraction. We will not use this terminology because notation such as $\frac{27}{8}$, $\frac{11}{9}$, and $\frac{89}{10}$ is quite "proper" and very common in algebra.

Convert to a mixed numeral.

**10.** $\dfrac{7}{3}$ $\qquad$ **11.** $\dfrac{11}{10}$

**12.** $\dfrac{110}{6}$ $\qquad$ **13.** $\dfrac{229}{18}$

Convert to a mixed numeral.

**14.** $-\dfrac{25}{4}$ $\qquad$ **15.** $-\dfrac{33}{9}$

*Answers*

**10.** $2\dfrac{1}{3}$ **11.** $1\dfrac{1}{10}$ **12.** $18\dfrac{1}{3}$ **13.** $12\dfrac{13}{18}$

**14.** $-6\dfrac{1}{4}$ **15.** $-3\dfrac{2}{3}$

## b Writing Mixed Numerals for Quotients

It is quite common when dividing whole numbers to write the quotient using a mixed numeral. The remainder is the numerator of the fraction part of the mixed numeral.

**EXAMPLE 11**  Divide. Write a mixed numeral for the answer.

$$4\ 2\ )\overline{8\ 9\ 1\ 5}$$

We first divide as usual.

$$
\begin{array}{r}
2\ 1\ 2 \\
4\ 2\ )\overline{8\ 9\ 1\ 5} \\
8\ 4\phantom{\ \ \ } \\
\hline
5\ 1\phantom{\ \ } \\
4\ 2\phantom{\ \ } \\
\hline
9\ 5 \\
8\ 4 \\
\hline
1\ 1
\end{array}
$$

$$\frac{8915}{42} = 212\frac{11}{42}$$

The answer is $212\frac{11}{42}$.

Do Exercises 16 and 17.

Divide. Write a mixed numeral for the answer.

**16.** $6\ )\overline{4\ 8\ 4\ 6}$

**17.** $4\ 5\ )\overline{6\ 0\ 5\ 3}$

*Answers*

**16.** $807\frac{2}{3}$   **17.** $134\frac{23}{45}$

---

## 4.4  Exercise Set

### a  Solve.

**1.** *Greenhouse Dimensions.*  A community college horticulture department builds a greenhouse that measures $32\frac{1}{2}$ ft $\times$ $20\frac{5}{6}$ ft $\times$ $11\frac{3}{4}$ ft. Convert $32\frac{1}{2}$, $20\frac{5}{6}$, and $11\frac{3}{4}$ to fraction notation.

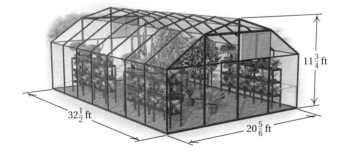

$11\frac{3}{4}$ ft

$32\frac{1}{2}$ ft

$20\frac{5}{6}$ ft

**2.** *Alarm Clock Dimensions.*  A world time clock/radio with NOAA (National Oceanic and Atmospheric Administration) Weather Band measures $14\frac{1}{4}$ in. $\times$ $6\frac{3}{4}$ in. $\times$ $2\frac{1}{4}$ in. Convert $14\frac{1}{4}$, $6\frac{3}{4}$, and $2\frac{1}{4}$ to fraction notation.

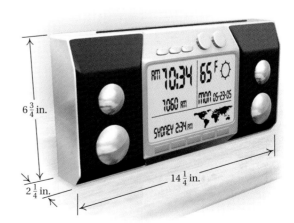

$6\frac{3}{4}$ in.

$2\frac{1}{4}$ in.

$14\frac{1}{4}$ in.

**3.** A quilt design requires three different fabrics. The quilter determines that she needs $\frac{17}{4}$ yd of a dominant fabric, $\frac{10}{3}$ yd of a contrasting fabric, and $\frac{9}{8}$ yd of a border fabric. Convert $\frac{17}{4}$, $\frac{10}{3}$, and $\frac{9}{8}$ to mixed numerals.

**4.** *Bake Sale.* The Valley Township Fire Department organized a bake sale for a fund-raiser for the local library. Each pie was cut into 8 pieces and each cake into 12 pieces. They sold 73 pieces of pie, or $\frac{73}{8}$ pies, and 55 pieces of cake, or $\frac{55}{12}$ cakes. Convert $\frac{73}{8}$ and $\frac{55}{12}$ to mixed numerals.

Convert to fraction notation.

**5.** $5\frac{2}{3}$

**6.** $3\frac{4}{5}$

**7.** $3\frac{1}{4}$

**8.** $6\frac{1}{2}$

**9.** $-10\frac{1}{8}$

**10.** $-20\frac{1}{5}$

**11.** $5\frac{1}{10}$

**12.** $9\frac{1}{10}$

**13.** $20\frac{3}{5}$

**14.** $30\frac{4}{5}$

**15.** $-9\frac{5}{6}$

**16.** $-8\frac{7}{8}$

**17.** $7\frac{3}{10}$

**18.** $6\frac{9}{10}$

**19.** $1\frac{5}{8}$

**20.** $1\frac{3}{5}$

**21.** $-12\frac{3}{4}$

**22.** $-15\frac{2}{3}$

**23.** $-4\frac{3}{10}$

**24.** $-5\frac{7}{10}$

**25.** $2\frac{3}{100}$

**26.** $5\frac{7}{100}$

**27.** $66\frac{2}{3}$

**28.** $33\frac{1}{3}$

**29.** $-5\frac{29}{50}$

**30.** $-84\frac{3}{8}$

**31.** $101\frac{5}{16}$

**32.** $205\frac{3}{14}$

Convert to a mixed numeral.

**33.** $\frac{18}{5}$

**34.** $\frac{17}{4}$

**35.** $\frac{14}{3}$

**36.** $\frac{39}{8}$

**37.** $-\frac{27}{6}$

**38.** $-\dfrac{30}{9}$  **39.** $\dfrac{57}{10}$  **40.** $\dfrac{89}{10}$  **41.** $-\dfrac{53}{7}$  **42.** $-\dfrac{59}{8}$

**43.** $\dfrac{45}{6}$  **44.** $\dfrac{50}{8}$  **45.** $\dfrac{46}{4}$  **46.** $\dfrac{39}{9}$  **47.** $-\dfrac{12}{8}$

**48.** $-\dfrac{28}{6}$  **49.** $\dfrac{757}{100}$  **50.** $\dfrac{467}{100}$  **51.** $-\dfrac{345}{8}$  **52.** $-\dfrac{223}{4}$

 Divide. Write a mixed numeral for the answer.

**53.** $8\,)\overline{8\ 6\ 9}$  **54.** $3\,)\overline{2\ 1\ 2\ 6}$  **55.** $5\,)\overline{3\ 0\ 9\ 1}$  **56.** $9\,)\overline{9\ 1\ 1\ 0}$

**57.** $2\ 1\,)\overline{8\ 5\ 2}$  **58.** $8\ 5\,)\overline{7\ 6\ 7\ 0}$  **59.** $1\ 0\ 2\,)\overline{5\ 6\ 1\ 2}$  **60.** $4\ 6\,)\overline{1\ 0\ 8\ 1}$

**61.** $3\ 5\,)\overline{8\ 0\ ,\ 2\ 4\ 3}$  **62.** $1\ 5\ 2\,)\overline{2\ 6\ ,\ 1\ 0\ 7}$

## Skill Maintenance

**63.** Round to the nearest hundred: 45,765.  [1.6a]

**64.** Round to the nearest ten: 45,765.  [1.6a]

Simplify.  [3.5b]

**65.** $\dfrac{200}{375}$  **66.** $\dfrac{63}{75}$  **67.** $-\dfrac{160}{270}$  **68.** $-\dfrac{6996}{8028}$

Multiply and simplify.  [3.4a], [3.6a]

**69.** $\dfrac{6}{5}\cdot 15$  **70.** $\dfrac{5}{12}\cdot(-6)$  **71.** $-\dfrac{7}{10}\cdot\left(-\dfrac{5}{14}\right)$  **72.** $\dfrac{1}{10}\cdot\dfrac{20}{5}$

Divide and simplify.  [3.7b]

**73.** $-\dfrac{2}{3}\div\dfrac{1}{36}$  **74.** $28\div\dfrac{4}{7}$  **75.** $200\div\dfrac{15}{64}$  **76.** $\dfrac{3}{4}\div\left(-\dfrac{9}{16}\right)$

## Synthesis

Write a mixed numeral.

**77.** 📱 $\dfrac{128,236}{541}$  **78.** 📱 $\dfrac{103,676}{349}$  **79.** $\dfrac{56}{7}+\dfrac{2}{3}$  **80.** $-\dfrac{72}{12}-\dfrac{5}{6}$

**81.** There are $\frac{366}{7}$ weeks in a leap year.

**82.** There are $\frac{365}{7}$ weeks in a year.

# Mid-Chapter Review

## Concept Reinforcement

Determine whether each statement is true or false.

_____ **1.** If $\dfrac{a}{b} > \dfrac{c}{b}$, $b \neq 0$, then $a > c$.  [4.3b]

_____ **2.** All positive mixed numerals represent numbers larger than 1.  [4.4a]

_____ **3.** The least common multiple of two natural numbers is the smallest number that is a factor of both.  [4.1a]

_____ **4.** To add fractions when denominators are the same, we keep the numerator and add the denominators.  [4.2a]

## Guided Solutions

Fill in each blank with the number that creates a correct solution.

**5.** Subtract: $\dfrac{11}{42} - \dfrac{3}{35}$.  [4.3a]

$$\dfrac{11}{42} - \dfrac{3}{35} = \dfrac{11}{2 \cdot \square \cdot 7} - \dfrac{3}{\square \cdot 7} \qquad \text{Factoring the denominators}$$

$$= \dfrac{11}{2 \cdot \square \cdot 7} \cdot \left(\dfrac{\square}{\square}\right) - \dfrac{3}{\square \cdot 7} \cdot \left(\dfrac{\square \cdot \square}{\square \cdot \square}\right) \qquad \text{Multiplying by 1 to get the LCD}$$

$$= \dfrac{11 \cdot \square}{2 \cdot 3 \cdot 7 \cdot \square} - \dfrac{3 \cdot \square \cdot \square}{5 \cdot 7 \cdot \square \cdot \square} \qquad \text{Multiplying}$$

$$= \dfrac{\square}{2 \cdot 3 \cdot 5 \cdot 7} - \dfrac{\square}{2 \cdot 3 \cdot 5 \cdot 7} \qquad \text{Simplifying}$$

$$= \dfrac{\square - \square}{2 \cdot 3 \cdot 5 \cdot 7} = \dfrac{\square}{\square} \qquad \text{Subtracting and simplifying}$$

**6.** Solve: $x + \dfrac{1}{8} = \dfrac{2}{3}$.  [4.3c]

$$x + \dfrac{1}{8} = \dfrac{2}{3}$$

$$x + \dfrac{1}{8} - \square = \dfrac{2}{3} - \square \qquad \text{Subtracting on both sides}$$

$$x + \square = \dfrac{2}{3} \cdot \dfrac{\square}{\square} - \dfrac{1}{8} \cdot \dfrac{\square}{\square} \qquad \text{Multiplying by 1 to get the LCD}$$

$$x = \dfrac{\square}{\square} - \dfrac{\square}{\square} \qquad \text{Simplifying and multiplying}$$

$$x = \dfrac{\square}{\square} \qquad \text{Subtracting}$$

# Mixed Review

**7.** Match each set of numbers in the first column with its least common multiple in the second column by drawing connecting lines.   [4.1a]

45 and 50

50 and 80   120

30 and 24   720

18, 24, and 80   400

30, 45, and 50   450

Calculate and simplify.   [4.2a], [4.3a]

**8.** $\frac{1}{5} + \frac{7}{45}$

**9.** $\frac{5}{6} + \frac{2}{3} + \frac{7}{12}$

**10.** $\frac{1}{6} - \frac{2}{9}$

**11.** $\frac{-5}{18} + \frac{1}{15}$

**12.** $\frac{19}{48} - \frac{11}{30}$

**13.** $\frac{3}{7} + \frac{15}{17}$

**14.** $\frac{229}{720} - \frac{5}{24}$

**15.** $\frac{2}{35} - \frac{8}{65}$

Solve.

**16.** Miguel jogs for $\frac{4}{5}$ mi, rests, and then jogs for another $\frac{2}{3}$ mi. How far does he jog in all?   [4.2b]

**17.** One weekend, Kirby spent $\frac{39}{5}$ hr playing two iPod games—Brain Challenge: Cerebral Burn and Scrabble: Go for a Triple Word Score. She spent $\frac{11}{4}$ hr playing Scrabble. How many hours did she spend playing Brain Challenge?   [4.3d]

**18.** Arrange in order from smallest to largest: $\frac{4}{9}, \frac{3}{10}, \frac{2}{7}$, and $\frac{1}{5}$.

[4.3b]

**19.** Solve: $\frac{2}{5} + x = \frac{9}{16}$.   [4.3c]

**20.** Divide:  $15\overline{)263}$. Write a mixed numeral for the answer.   [4.4b]

**21.** Fraction notation for $9\frac{3}{8}$ is which of the following?   [4.4a]

**A.** $\frac{27}{8}$   **B.** $\frac{93}{8}$   **C.** $\frac{75}{8}$   **D.** $\frac{80}{3}$

**22.** Mixed numeral notation for $\frac{39}{4}$ is which of the following?   [4.4a]

**A.** $35\frac{1}{4}$   **B.** $\frac{4}{39}$   **C.** $9\frac{3}{4}$   **D.** $36\frac{3}{4}$

# Understanding Through Discussion and Writing

**23.** Is the LCM of two numbers always larger than either number? Why or why not?   [4.1a]

**24.** Explain the role of multiplication when adding using fraction notation with different denominators.   [4.2a]

**25.** A student made the following error:

$$\frac{8}{5} - \frac{8}{2} = \frac{8}{3}.$$

Find at least two ways to convince him of the mistake.
[4.3a]

**26.** Are the numbers $2\frac{1}{3}$ and $2 \cdot \frac{1}{3}$ equal? Why or why not?
[3.4a], [4.4a]

# Addition and Subtraction Using Mixed Numerals; Applications

## OBJECTIVES

**a** Add using mixed numerals.

**b** Subtract using mixed numerals.

**c** Solve applied problems involving addition and subtraction with mixed numerals.

**SKILL TO REVIEW**

Objective 3.5b: Simplify fraction notation.

Simplify.

1. $\dfrac{18}{32}$    2. $\dfrac{78}{117}$

## a Addition Using Mixed Numerals

To find the sum $1\frac{5}{8} + 3\frac{1}{8}$, we first add the fractions. Then we add the whole numbers.

$$
\begin{array}{rcl}
1\dfrac{5}{8} & = & 1\dfrac{5}{8} \\[2mm]
+\,3\dfrac{1}{8} & = & +\,3\dfrac{1}{8} \\[2mm]
\hline
\dfrac{6}{8} & & 4\dfrac{6}{8} = 4\dfrac{3}{4}
\end{array}
$$

Simplifying: $\dfrac{6}{8} = \dfrac{3}{4}$

Add the fractions.    Add the whole numbers.

Sometimes we must write the fractional parts with a common denominator before we can add.

**EXAMPLE 1** Add: $5\frac{2}{3} + 3\frac{5}{6}$. Write a mixed numeral for the answer.

The LCD is 6.

$$
\begin{array}{rcl}
5\dfrac{2}{3} \cdot \dfrac{2}{2} & = & 5\dfrac{4}{6} \\[2mm]
+\,3\dfrac{5}{6} & = & +\,3\dfrac{5}{6} \\[2mm]
\hline
& & 8\dfrac{9}{6} = 8 + \dfrac{9}{6} \\[3mm]
& & = 8 + 1\dfrac{1}{2} \quad \text{Writing } \tfrac{9}{6} \text{ as a mixed numeral, } 1\tfrac{1}{2} \\[3mm]
& & = 9\dfrac{1}{2}
\end{array}
$$

Do Margin Exercises 1 and 2.

**EXAMPLE 2** Add: $10\frac{5}{6} + 7\frac{3}{8}$.

The LCD is 24.

$$
\begin{array}{rcl}
10\dfrac{5}{6} \cdot \dfrac{4}{4} & = & 10\dfrac{20}{24} \\[2mm]
+\,7\dfrac{3}{8} \cdot \dfrac{3}{3} & = & +\,7\dfrac{9}{24} \\[2mm]
\hline
& & 17\dfrac{29}{24} = 17 + \dfrac{29}{24} \\[3mm]
& & = 17 + 1\dfrac{5}{24} \quad \text{Writing } \tfrac{29}{24} \text{ as a mixed numeral, } 1\tfrac{5}{24} \\[3mm]
& & = 18\dfrac{5}{24}
\end{array}
$$

Do Exercise 3.

Add.

1. $\begin{array}{r} 2\frac{3}{10} \\ +\,5\frac{1}{10} \\ \hline \end{array}$    2. $\begin{array}{r} 8\frac{2}{5} \\ +\,3\frac{7}{10} \\ \hline \end{array}$

3. Add.

$\begin{array}{r} 9\frac{3}{4} \\ +\,3\frac{5}{6} \\ \hline \end{array}$

*Answers*

*Skill to Review:*

1. $\dfrac{9}{16}$    2. $\dfrac{2}{3}$

*Margin Exercises:*

1. $7\frac{2}{5}$    2. $12\frac{1}{10}$    3. $13\frac{7}{12}$

## b Subtraction Using Mixed Numerals

Subtraction is a lot like addition; we subtract the fractions and then the whole numbers.

**EXAMPLE 3** Subtract: $7\frac{3}{4} - 2\frac{1}{4}$.

$$
\begin{array}{r}
7\ \dfrac{3}{4} \\[4pt]
-\ 2\ \dfrac{1}{4} \\[2pt]
\hline
\dfrac{2}{4}
\end{array}
\quad = \quad
\begin{array}{r}
7\ \dfrac{3}{4} \\[4pt]
-\ 2\ \dfrac{1}{4} \\[2pt]
\hline
5\ \dfrac{2}{4} = 5\dfrac{1}{2}
\end{array}
$$

Simplifying: $\dfrac{2}{4} = \dfrac{1}{2}$

Subtract the fractions.  Subtract the whole numbers.

**EXAMPLE 4** Subtract: $9\frac{4}{5} - 3\frac{1}{2}$.

The LCD is 10.

$$
\begin{array}{r}
9\ \dfrac{4}{5}\cdot\dfrac{2}{2} = \quad 9\dfrac{8}{10} \\[6pt]
-\ 3\ \dfrac{1}{2}\cdot\dfrac{5}{5} = -\ 3\dfrac{5}{10} \\[2pt]
\hline
6\dfrac{3}{10}
\end{array}
$$

Do Exercises 4 and 5.

Subtract.

**4.** $\begin{array}{r} 10\dfrac{7}{8} \\[4pt] -\ 9\dfrac{3}{8} \\ \hline \end{array}$

**5.** $\begin{array}{r} 8\dfrac{2}{3} \\[4pt] -\ 5\dfrac{1}{2} \\ \hline \end{array}$

**EXAMPLE 5** Subtract: $7\frac{1}{6} - 2\frac{1}{4}$.

The LCD is 12.

$$
\left.
\begin{array}{r}
7\ \dfrac{1}{6}\cdot\dfrac{2}{2} = \quad 7\dfrac{2}{12} \\[8pt]
-\ 2\ \dfrac{1}{4}\cdot\dfrac{3}{3} = -\ 2\dfrac{3}{12}
\end{array}
\right\}
$$

We cannot subtract $\frac{3}{12}$ from $\frac{2}{12}$.
← We borrow 1, or $\frac{12}{12}$, from 7:
$7\frac{2}{12} = 6 + 1 + \frac{2}{12} = 6 + \frac{12}{12} + \frac{2}{12} = 6\frac{14}{12}$.

We can write this as

$$
\begin{array}{r}
7\ \dfrac{2}{12} = \quad 6\dfrac{14}{12} \\[8pt]
-\ 2\ \dfrac{3}{12} = -\ 2\dfrac{3}{12} \\[2pt]
\hline
4\dfrac{11}{12}.
\end{array}
$$

Do Exercise 6.

**6.** Subtract.

$$
\begin{array}{r}
8\dfrac{1}{9} \\[6pt]
-\ 4\dfrac{5}{6} \\ \hline
\end{array}
$$

*Answers*

**4.** $1\dfrac{1}{2}$  **5.** $3\dfrac{1}{6}$  **6.** $3\dfrac{5}{18}$

**EXAMPLE 6**  Subtract: $12 - 9\frac{3}{8}$.

$$
\begin{array}{rl}
12 & = \quad 11\frac{8}{8} \qquad 12 = 11 + 1 = 11 + \frac{8}{8} = 11\frac{8}{8} \\
-\ 9\frac{3}{8} & = -\ 9\frac{3}{8} \\
\hline
& \qquad 2\frac{5}{8}
\end{array}
$$

**7.** Subtract.

$$
\begin{array}{r}
5 \\
-\ 1\frac{1}{3} \\
\hline
\end{array}
$$

Do Exercise 7.

## c  Applications and Problem Solving

**EXAMPLE 7**  *Widening a Driveway.*  Sherry and Woody are widening their existing $17\frac{1}{4}$-ft driveway by adding $5\frac{9}{10}$ ft on one side. What is the width of the new driveway?

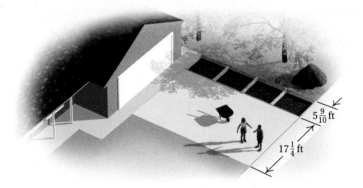

1. **Familiarize.**  We let $w$ = the width of the new driveway.
2. **Translate.**  We translate as follows:

$$
\underbrace{\text{Width of existing driveway}}_{\displaystyle 17\frac{1}{4}} \; + \; \underbrace{\text{Width of addition}}_{\displaystyle 5\frac{9}{10}} \; = \; \underbrace{\text{Width of new driveway}}_{\displaystyle w.}
$$

3. **Solve.**  The translation tells us what to do. We add. The LCD is 20.

$$
\begin{array}{rcl}
17\frac{1}{4} & = \quad 17\ \dfrac{1}{4}\cdot\dfrac{5}{5} = & 17\frac{5}{20} \\[2mm]
+\ 5\frac{9}{10} & = +\ 5\ \dfrac{9}{10}\cdot\dfrac{2}{2} = & +\ 5\frac{18}{20} \\
\hline
& & 22\frac{23}{20} = 23\frac{3}{20}
\end{array}
$$

**8. Travel Distance.**  On a two-day business trip, Paul drove $213\frac{7}{10}$ mi the first day and $107\frac{5}{8}$ mi the second day. What was the total distance that Paul drove?

Thus, $w = 23\frac{3}{20}$.

4. **Check.**  We check by repeating the calculation. We also note that the answer is larger than either of the widths, which means that the answer is reasonable.
5. **State.**  The width of the new driveway is $23\frac{3}{20}$ ft.

Do Exercise 8.

*Answers*

**7.** $3\frac{2}{3}$  **8.** $321\frac{13}{40}$ mi

**EXAMPLE 8** *Communication Cable.* Perfection Cable has two crews who install Internet communication cable. Crew A can install $38\frac{1}{8}$ ft per hour. Crew B can install $31\frac{2}{3}$ ft per hour. How many fewer feet can Crew B install per hour than Crew A?

1. **Familiarize.** The phrase "how many fewer" indicates subtraction. We let $c =$ the difference in the numbers of feet per hour.

2. **Translate.** We translate as follows:

$$
\underbrace{\begin{array}{c}\text{Crew A:}\\ \text{ft/hr}\end{array}}_{} \;\; - \;\; \underbrace{\begin{array}{c}\text{Crew B:}\\ \text{ft/hr}\end{array}}_{} \;\; = \;\; \underbrace{\begin{array}{c}\text{Difference in}\\ \text{ft/hr}\end{array}}_{}
$$

$$
38\frac{1}{8} \;\; - \;\; 31\frac{2}{3} \;\; = \;\; c.
$$

3. **Solve.** To solve the equation, we carry out the subtraction. The LCD = 24.

$$
\begin{aligned}
38\frac{1}{8} &= 38\frac{1}{8}\cdot\frac{3}{3} = 38\frac{3}{24} = 37\frac{27}{24}\\
-\;31\frac{2}{3} &= -31\frac{2}{3}\cdot\frac{8}{8} = -31\frac{16}{24} = -31\frac{16}{24}\\
\hline
& \hspace{8.5cm} 6\frac{11}{24}
\end{aligned}
$$

Thus, $c = 6\frac{11}{24}$.

4. **Check.** To check, we add the difference, $6\frac{11}{24}$, to Crew B's rate:

$$
31\frac{2}{3} + 6\frac{11}{24} = 31\frac{16}{24} + 6\frac{11}{24}
$$

$$
= 37\frac{27}{24} = 38\frac{3}{24} = 38\frac{1}{8}.
$$

This checks.

5. **State.** Crew B installs $6\frac{11}{24}$ fewer feet of cable per hour than Crew A.

Do Exercise 9.

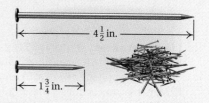

9. **Nail Length.** A 30d nail is $4\frac{1}{2}$ in. long. A 5d nail is $1\frac{3}{4}$ in. long. How much longer is the 30d nail than the 5d nail? (The "d" stands for "penny," which was used years ago in England to specify the number of pennies needed to buy 100 nails. Today, "penny" is used only to indicate the length of the nail.)

Source: *Pocket Ref,* 2nd ed., by Thomas J. Glover, p. 280, Sequoia Publishing, Inc., Littleton, CO

## Multistep Problems

**EXAMPLE 9** *Fly Fishing.* Henry is putting together a fly fishing line and uses $58\frac{5}{8}$ ft of slow-sinking fly line and $8\frac{3}{4}$ ft of leader line. The leader line provides a low visibility link between the heavier fly line and the fly. He uses $\frac{3}{8}$ ft of the slow-sinking fly line to connect the two lines. The knot used to connect the fly to the leader line uses $\frac{1}{6}$ ft of the leader line. How long is the finished fly fishing line?

*Answer*

9. $2\frac{3}{4}$ in.

1. **Familiarize.** Let's make a drawing to organize the given information. We let $l$ = the length of the finished fly fishing line, in feet.

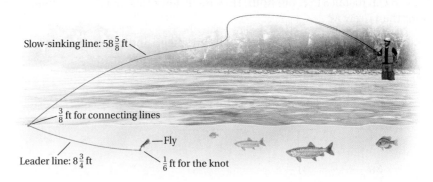

Slow-sinking line: $58\frac{5}{8}$ ft

$\frac{3}{8}$ ft for connecting lines

Leader line: $8\frac{3}{4}$ ft

—Fly

$\frac{1}{6}$ ft for the knot

2. **Translate.** From the drawing, we see that the length $l$ is the sum of the lengths of the two lines, $58\frac{5}{8}$ ft and $8\frac{3}{4}$ ft, minus the sum of the lengths, $\frac{3}{8}$ ft and $\frac{1}{6}$ ft, needed for connecting the lines and attaching the fly with a knot. Thus we have

$$l = \left(58\frac{5}{8} + 8\frac{3}{4}\right) - \left(\frac{3}{8} + \frac{1}{6}\right).$$

3. **Solve.** This is a three-step problem.

   a) We first add the two lengths $58\frac{5}{8}$ and $8\frac{3}{4}$.

   $$58\frac{5}{8} = \quad 58\frac{5}{8}$$
   $$+ \; 8\frac{3}{4} = + \; 8\frac{6}{8}$$
   $$\overline{\qquad\qquad\qquad} \quad \overline{66\frac{11}{8} = 67\frac{3}{8}}$$

   b) Next, we add the two lengths $\frac{3}{8}$ and $\frac{1}{6}$.

   $$\frac{3}{8} + \frac{1}{6} = \frac{3}{8} \cdot \frac{3}{3} + \frac{1}{6} \cdot \frac{4}{4} = \frac{9}{24} + \frac{4}{24} = \frac{13}{24}$$

   c) Finally, we subtract $\frac{13}{24}$ from $67\frac{3}{8}$.

   $$67\frac{3}{8} = \quad 67\frac{9}{24} = \quad 66\frac{33}{24}$$
   $$- \; \frac{13}{24} = - \quad \frac{13}{24} = - \quad \frac{13}{24}$$
   $$\overline{\qquad\qquad} \quad \overline{\qquad\qquad} \quad \overline{66\frac{20}{24} = 66\frac{5}{6}}$$

   Thus, $l = 66\frac{5}{6}$.

4. **Check.** We can check by adding the finished length to the lengths needed for the connection and the knot:

   $$66\frac{5}{6} + \frac{3}{8} + \frac{1}{6} = \left(66\frac{5}{6} + \frac{1}{6}\right) + \frac{3}{8} = 67 + \frac{3}{8} = 67\frac{3}{8}.$$

   This is the sum of the given lengths.

5. **State.** The finished length of the fly line is $66\frac{5}{6}$ ft.

Do Exercise 10.

---

10. **Liquid Fertilizer.** There is $283\frac{5}{8}$ gal of liquid fertilizer in a fertilizer application tank. After applying $178\frac{2}{3}$ gal to a soybean field, the farmer requests that Braden's Farm Supply deliver an additional 250 gal. How many gallons of fertilizer are in the tank after the delivery?

*Answer*

10. $354\frac{23}{24}$ gal

**a** Add. Write a mixed numeral for the answer.

1. $\begin{array}{r} 20 \\ + 8\frac{3}{4} \\ \hline \end{array}$

2. $\begin{array}{r} 37 \\ + 18\frac{2}{3} \\ \hline \end{array}$

3. $\begin{array}{r} 129\frac{7}{8} \\ + 56 \\ \hline \end{array}$

4. $\begin{array}{r} 2003\frac{4}{11} \\ + 59 \\ \hline \end{array}$

5. $\begin{array}{r} 2\frac{7}{8} \\ + 3\frac{5}{8} \\ \hline \end{array}$

6. $\begin{array}{r} 4\frac{5}{6} \\ + 3\frac{5}{6} \\ \hline \end{array}$

7. $1\frac{1}{4} + 1\frac{2}{3}$

8. $4\frac{1}{3} + 5\frac{2}{9}$

9. $\begin{array}{r} 8\frac{3}{4} \\ + 5\frac{5}{6} \\ \hline \end{array}$

10. $\begin{array}{r} 4\frac{3}{8} \\ + 6\frac{5}{12} \\ \hline \end{array}$

11. $\begin{array}{r} 3\frac{2}{5} \\ + 8\frac{7}{10} \\ \hline \end{array}$

12. $\begin{array}{r} 5\frac{1}{2} \\ + 3\frac{7}{10} \\ \hline \end{array}$

13. $\begin{array}{r} 5\frac{3}{8} \\ + 10\frac{5}{6} \\ \hline \end{array}$

14. $\begin{array}{r} \frac{5}{8} \\ + 1\frac{5}{6} \\ \hline \end{array}$

15. $\begin{array}{r} 12\frac{4}{5} \\ + 8\frac{7}{10} \\ \hline \end{array}$

16. $\begin{array}{r} 15\frac{5}{8} \\ + 11\frac{3}{4} \\ \hline \end{array}$

17. $\begin{array}{r} 14\frac{5}{8} \\ + 13\frac{1}{4} \\ \hline \end{array}$

18. $\begin{array}{r} 16\frac{1}{4} \\ + 15\frac{7}{8} \\ \hline \end{array}$

19. $\begin{array}{r} 7\frac{1}{8} \\ 9\frac{2}{3} \\ + 10\frac{3}{4} \\ \hline \end{array}$

20. $\begin{array}{r} 45\frac{2}{3} \\ 31\frac{3}{5} \\ + 12\frac{1}{4} \\ \hline \end{array}$

**b** Subtract. Write a mixed numeral for the answer.

21. $\begin{array}{r} 4\frac{1}{5} \\ - 2\frac{3}{5} \\ \hline \end{array}$

22. $\begin{array}{r} 5\frac{1}{8} \\ - 2\frac{3}{8} \\ \hline \end{array}$

23. $6\frac{3}{5} - 2\frac{1}{2}$

24. $7\frac{2}{3} - 6\frac{1}{2}$

25. $\begin{array}{r} 34\frac{1}{3} \\ - 12\frac{5}{8} \\ \hline \end{array}$

26. $\begin{array}{r} 23\frac{5}{16} \\ - 16\frac{3}{4} \\ \hline \end{array}$

27. $\begin{array}{r} 21 \\ - 8\frac{3}{4} \\ \hline \end{array}$

28. $\begin{array}{r} 42 \\ - 3\frac{7}{8} \\ \hline \end{array}$

**29.**  $\begin{array}{r} 34 \\ -\ 18\dfrac{5}{8} \\ \hline \end{array}$

**30.**  $\begin{array}{r} 23 \\ -\ 19\dfrac{3}{4} \\ \hline \end{array}$

**31.**  $\begin{array}{r} 21\dfrac{1}{6} \\ -\ 13\dfrac{3}{4} \\ \hline \end{array}$

**32.**  $\begin{array}{r} 42\dfrac{1}{10} \\ -\ 23\dfrac{7}{12} \\ \hline \end{array}$

**33.**  $\begin{array}{r} 14\dfrac{1}{8} \\ -\ \ \dfrac{3}{4} \\ \hline \end{array}$

**34.**  $\begin{array}{r} 28\dfrac{1}{6} \\ -\ \ 5 \\ \hline \end{array}$

**35.**  $\begin{array}{r} 25\dfrac{1}{9} \\ -\ 13\dfrac{5}{6} \\ \hline \end{array}$

**36.**  $\begin{array}{r} 23\dfrac{5}{16} \\ -\ 14\dfrac{7}{12} \\ \hline \end{array}$

 Solve.

**37.** *Planting Flowers.* A landscaper planted $4\frac{1}{2}$ flats of impatiens, $6\frac{2}{3}$ flats of snapdragons, and $3\frac{3}{8}$ flats of phlox. How many flats did she plant altogether?

**38.** A plumber uses two pipes, each of length $51\frac{5}{16}$ in., and one pipe of length $34\frac{3}{4}$ in. in the installation of a shower. How much pipe was used in all?

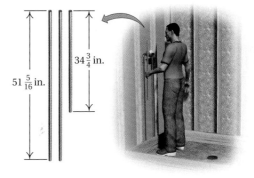

**39.** For a family barbecue, Cayla bought packages of hamburger weighing $1\frac{2}{3}$ lb and $5\frac{3}{4}$ lb. What was the total weight of the meat?

**40.** Marsha's Butcher Shop sold packages of sliced turkey breast weighing $1\frac{1}{3}$ lb and $4\frac{3}{5}$ lb. What was the total weight of the meat?

**41.** Tara is 66 in. tall and her son, Tom, is $59\frac{7}{12}$ in. tall. How much taller is Tara?

**42.** Nicholas is $73\frac{2}{3}$ in. tall and his daughter, Kendra, is $71\frac{5}{16}$ in. tall. How much shorter is Kendra?

**43.** *Upholstery Fabric.* Executive Car Care sells 45-in. upholstery fabric for car restoration. Art buys $9\frac{1}{4}$ yd and $10\frac{5}{6}$ yd for two car projects. How many yards did Art buy?

**44.** *Cutco Cutlery.* The Essentials 5-piece set sold by Cutco contains three knives: $7\frac{5}{8}"$ Petite Chef, $6\frac{3}{4}"$ Petite Carver, and $2\frac{3}{4}"$ Paring Knife. How much larger is the blade of the Petite Chef than that of the Petite Carver? than that of the Paring Knife?

**Source:** Cutco Cutlery Corporation, Fall 2008

**45.** *Stone Bench.* Baytown Village Stone Creations is making a custom stone bench as shown below. The recommended height for the bench is 18 in. The slab that forms the seat is $3\frac{3}{8}$ in. thick. Each of the two supporting legs is made up of three stacked stones. Two of the stones measure $3\frac{1}{2}$ in. and $5\frac{1}{4}$ in. How much must the third stone measure?

$3\frac{3}{8}$ in.
$3\frac{1}{2}$ in.
$5\frac{1}{4}$ in.
?
18 in.

**46.** *Winterizing a Swimming Pool.* To winterize their swimming pool, the Jablonskis are draining the water into a nearby field. The distance to the field is $103\frac{1}{2}$ ft. Because their only hose measures $62\frac{3}{4}$ ft, they need to buy an additional hose. How long must the new hose be?

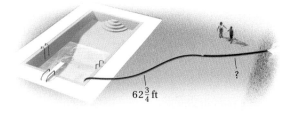

$62\frac{3}{4}$ ft
?

**47.** Kim Park is a computer technician. One day, she drove $180\frac{7}{10}$ mi away from Los Angeles for a service call. The next day, she drove $85\frac{1}{2}$ mi back toward Los Angeles for another service call. How far was she then from Los Angeles?

**48.** Jose is $4\frac{1}{2}$ in. taller than his daughter, Teresa. Teresa is $66\frac{2}{3}$ in. tall. How tall is Jose?

**49.** Creative Glass sells a framed beveled mirror as shown below. Its dimensions are $30\frac{1}{2}"$ wide by $36\frac{5}{8}"$ high. What is the perimeter of (total distance around) the framed mirror?

$36\frac{5}{8}$ in.
$30\frac{1}{2}$ in.

**50.** *Book Size.* One standard book size is $8\frac{1}{2}$ in. by $9\frac{3}{4}$ in. What is the perimeter of (total distance around) the front cover of such a book?

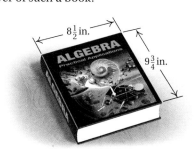

$8\frac{1}{2}$ in.
$9\frac{3}{4}$ in.

**51. Carpentry.** When cutting wood with a saw, a carpenter must take into account the thickness of the saw blade. Suppose that from a piece of wood 36 in. long, a carpenter cuts a $15\frac{3}{4}$-in. length with a saw blade that is $\frac{1}{8}$ in. thick. How long is the piece that remains?

**52. Painting.** A painter used $1\frac{3}{4}$ gal of paint for the Garcias' living room and $1\frac{1}{3}$ gal for their family room. How much paint was used in all?

**53.** Rene is $5\frac{1}{4}$ in. taller than his son, who is $72\frac{5}{6}$ in. tall. How tall is Rene?

**54.** A Boeing 767 flew 640 mi on a nonstop flight. On the return flight, it landed after having flown $320\frac{3}{10}$ mi. How far was the plane from its original point of departure?

**55. Interior Design.** Sue worked $10\frac{1}{2}$ hr over a three-day period on an interior design project. If she worked $2\frac{1}{2}$ hr on the first day and $4\frac{1}{5}$ hr on the second, how many hours did Sue work on the third day?

**56. Painting.** Geri had $3\frac{1}{2}$ gal of paint. It took $2\frac{3}{4}$ gal to paint the family room. She estimated that it would take $2\frac{1}{4}$ gal to paint the living room. How much more paint did Geri need?

Find the perimeter of (distance around) each figure.

**57.**

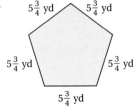

$5\frac{3}{4}$ yd  $5\frac{3}{4}$ yd

$5\frac{3}{4}$ yd  $5\frac{3}{4}$ yd

$5\frac{3}{4}$ yd

**58.**

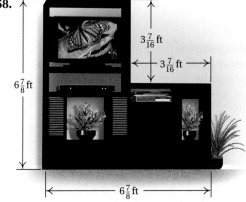

$3\frac{7}{16}$ ft

$3\frac{7}{16}$ ft

$6\frac{7}{8}$ ft

$6\frac{7}{8}$ ft

Find the length $d$ in each figure.

**59.**

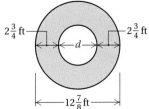

$2\frac{3}{4}$ ft  $2\frac{3}{4}$ ft

$d$

$12\frac{7}{8}$ ft

**60.**

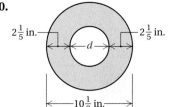

$2\frac{1}{5}$ in.  $2\frac{1}{5}$ in.

$d$

$10\frac{1}{2}$ in.

**61.** *Sewing from a Pattern.* Suppose you want to make a formal dress with jacket in size 8. Using 45-in. fabric, you need $1\frac{3}{8}$ yd for the dress, $\frac{5}{8}$ yd of contrasting fabric for the band at the bottom, and $3\frac{3}{8}$ yd for the jacket. How many yards of 45-in. fabric are needed to make the outfit?

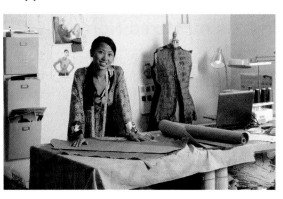

**62.** *Sewing from a Pattern.* Suppose you want to make an outfit in size 12. Using 45-in. fabric, you need $2\frac{3}{4}$ yd for the dress and $3\frac{1}{2}$ yd for the jacket. How many yards of 45-in. fabric are needed to make the outfit?

**63.** *Window Dimensions.* The Sanchez family is replacing a window in their lake house. The original window measures $4\frac{5}{6}$ ft $\times$ $8\frac{1}{4}$ ft. The new window is $2\frac{1}{3}$ ft wider. What are the dimensions of the new window?

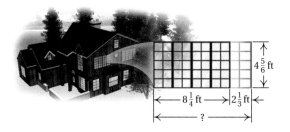

**64.** Find the smallest length of a bolt that will pass through a piece of tubing with an outside diameter of $\frac{1}{2}$ in., a washer $\frac{1}{16}$ in. thick, a piece of tubing with a $\frac{3}{4}$-in. outside diameter, another washer, and a nut $\frac{3}{16}$ in. thick.

## Skill Maintenance

Solve.

**65.** Rick's Market sells Swiss cheese in $\frac{3}{4}$-lb packages. How many packages can be made from a 12-lb slab of cheese? [3.7d]

**66.** Holstein's Dairy produced 4578 oz of milk one morning. How many 16-oz cartons could be filled? How much milk would be left over? [1.8a]

Determine whether the first number is divisible by the second. [3.2a]

**67.** 9993 by 3

**68.** 9993 by 9

**69.** 2345 by 9

**70.** 2345 by 5

**71.** 2335 by 10

**72.** 7764 by 6

**73.** 18,888 by 8

**74.** 18,888 by 4

**75.** Multiply and simplify: $\dfrac{15}{9} \cdot \dfrac{18}{39}$. [3.6a]

**76.** Divide and simplify: $\dfrac{12}{25} \div \left( -\dfrac{24}{5} \right)$. [3.7b]

## Synthesis

Perform the indicated operation. Write a mixed numeral for the answer.

**77.** $-5\dfrac{1}{4} + \left( -3\dfrac{3}{8} \right)$

**78.** $-8\dfrac{1}{3} + \left( -2\dfrac{5}{6} \right)$

**79.** $-4\dfrac{5}{12} - 6\dfrac{5}{8}$

**80.** $-9\dfrac{1}{2} - 7\dfrac{3}{5}$

# 4.6

# Multiplication and Division Using Mixed Numerals; Applications

## OBJECTIVES

**a** Multiply using mixed numerals.

**b** Divide using mixed numerals.

**c** Solve applied problems involving multiplication and division with mixed numerals.

## **a** Multiplication Using Mixed Numerals

Carrying out addition and subtraction with mixed numerals is usually easier if the numbers are left as mixed numerals. With multiplication and division, however, it is easier to convert the numbers to fraction notation first.

> **MULTIPLICATION USING MIXED NUMERALS**
>
> To multiply using mixed numerals, first convert to fraction notation and multiply. Then convert the answer to a mixed numeral, if appropriate.

### SKILL TO REVIEW
Objective 3.7b: Divide and simplify using fraction notation.

Divide and simplify.

**1.** $85 \div \dfrac{17}{5}$      **2.** $\dfrac{7}{65} \div \left(-\dfrac{21}{25}\right)$

**EXAMPLE 1** Multiply: $6 \cdot 2\dfrac{1}{2}$.

$$6 \cdot 2\frac{1}{2} = \frac{6}{1} \cdot \frac{5}{2} = \frac{6 \cdot 5}{1 \cdot 2} = \frac{2 \cdot 3 \cdot 5}{2 \cdot 1} = \frac{2}{2} \cdot \frac{3 \cdot 5}{1} = 1 \cdot \frac{3 \cdot 5}{1} = 15$$

> Converting the numbers to fraction notation first makes it easier to carry out the multiplication.

**EXAMPLE 2** Multiply: $3\dfrac{1}{2} \cdot \dfrac{3}{4}$.

$$3\frac{1}{2} \cdot \frac{3}{4} = \frac{7}{2} \cdot \frac{3}{4} = \frac{21}{8} = 2\frac{5}{8}$$

Here we write fraction notation.

**Multiply.**

**1.** $6 \cdot 3\dfrac{1}{3}$      **2.** $2\dfrac{1}{2} \cdot \dfrac{3}{4}$

Do Margin Exercises 1 and 2.

**EXAMPLE 3** Multiply: $-8 \cdot 4\dfrac{2}{3}$.

$$-8 \cdot 4\frac{2}{3} = -\frac{8}{1} \cdot \frac{14}{3} = -\frac{112}{3} = -37\frac{1}{3}$$

**Multiply.**

**3.** $-2 \cdot 6\dfrac{2}{5}$      **4.** $3\dfrac{1}{3} \cdot 2\dfrac{1}{2}$

**EXAMPLE 4** Multiply: $2\dfrac{1}{4} \cdot 3\dfrac{2}{5}$.

$$2\frac{1}{4} \cdot 3\frac{2}{5} = \frac{9}{4} \cdot \frac{17}{5} = \frac{153}{20} = 7\frac{13}{20}$$

---------- *Caution!* ----------

$2\dfrac{1}{4} \cdot 3\dfrac{2}{5} \neq 6\dfrac{2}{20}$. A common error is to multiply the whole numbers and then the fractions. This does not give the correct answer, $7\dfrac{13}{20}$, which is found by converting to fraction notation first.

Do Exercises 3 and 4.

### Answers

*Skill to Review:*

**1.** $-25$    **2.** $-\dfrac{5}{39}$

*Margin Exercises:*

**1.** $20$    **2.** $1\dfrac{7}{8}$    **3.** $-12\dfrac{4}{5}$    **4.** $8\dfrac{1}{3}$

## b  Division Using Mixed Numerals

The division $1\frac{1}{2} \div \frac{1}{6}$ is shown here. *Think*: "How many $\frac{1}{6}$'s are in $1\frac{1}{2}$?" We see that the answer is 9.

When we divide using mixed numerals, we convert to fraction notation first.

$$1\frac{1}{2} \div \frac{1}{6} = \frac{3}{2} \div \frac{1}{6} = \frac{3}{2} \cdot \frac{6}{1}$$

$$= \frac{3 \cdot 6}{2 \cdot 1} = \frac{3 \cdot 3 \cdot 2}{2 \cdot 1} = \frac{3 \cdot 3}{1} \cdot \frac{2}{2} = \frac{3 \cdot 3}{1} \cdot 1 = 9$$

> **DIVISION USING MIXED NUMERALS**
>
> To divide using mixed numerals, first write fraction notation and divide. Then convert the answer to a mixed numeral, if appropriate.

**EXAMPLE 5**  Divide: $32 \div 3\frac{1}{5}$.

$$32 \div 3\frac{1}{5} = \frac{32}{1} \div \frac{16}{5} \quad \text{Writing fraction notation}$$

$$= \frac{32}{1} \cdot \frac{5}{16} = \frac{32 \cdot 5}{1 \cdot 16} = \frac{2 \cdot 16 \cdot 5}{1 \cdot 16} = \frac{16}{16} \cdot \frac{2 \cdot 5}{1} = 1 \cdot \frac{2 \cdot 5}{1} = 10$$

$\uparrow$
Remember to multiply by the reciprocal.

Do Exercises 5 and 6.

**5.** Divide: $84 \div 5\frac{1}{4}$.

**6.** Divide: $26 \div 3\frac{1}{2}$.

**EXAMPLE 6**  Divide: $2\frac{1}{3} \div 1\frac{3}{4}$.

$$2\frac{1}{3} \div 1\frac{3}{4} = \frac{7}{3} \div \frac{7}{4} = \frac{7}{3} \cdot \frac{4}{7} = \frac{7 \cdot 4}{3 \cdot 7} = \frac{7}{7} \cdot \frac{4}{3} = 1 \cdot \frac{4}{3} = \frac{4}{3} = 1\frac{1}{3}$$

**EXAMPLE 7**  Divide: $-1\frac{3}{5} \div \left(-3\frac{1}{3}\right)$.

$$-1\frac{3}{5} \div \left(-3\frac{1}{3}\right) = -\frac{8}{5} \div \left(-\frac{10}{3}\right) = -\frac{8}{5} \cdot \left(-\frac{3}{10}\right)$$

$$= \frac{8 \cdot 3}{5 \cdot 10} = \frac{2 \cdot 4 \cdot 3}{5 \cdot 2 \cdot 5} = \frac{2}{2} \cdot \frac{4 \cdot 3}{5 \cdot 5} = \frac{12}{25}$$

Do Exercises 7 and 8.

Divide.

**7.** $2\frac{1}{4} \div 1\frac{1}{5}$          **8.** $1\frac{3}{4} \div \left(-2\frac{1}{2}\right)$

*Answers*

**5.** 16    **6.** $7\frac{3}{7}$    **7.** $1\frac{7}{8}$    **8.** $-\frac{7}{10}$

## (c) Applications and Problem Solving

**EXAMPLE 8** *Average Speed in Indianapolis 500.* Arie Luyendyk won the Indianapolis 500 in 1990 with a record high average speed of about 186 mph. This speed is about $2\frac{12}{25}$ times the average speed of the first winner, Ray Harroun, in 1911. What was the average speed of the winner in the first Indianapolis 500?

Source: Indianapolis Motor Speedway

1. **Familiarize.** We ask the question "186 is $2\frac{12}{25}$ times what number?" We let $s$ = the average speed in 1911. Then the average speed in 1990 was $2\frac{12}{25} \cdot s$.

2. **Translate.** The problem can be translated to an equation as follows:

$$\underbrace{\begin{array}{c}\text{Average speed}\\\text{of winner}\\\text{in 1990}\end{array}}\quad \text{is}\quad 2\frac{12}{25} \quad \text{times} \quad \underbrace{\begin{array}{c}\text{Average speed}\\\text{of winner}\\\text{in 1911}\end{array}}$$

$$\downarrow \qquad\qquad \downarrow \quad \downarrow \quad \downarrow \qquad\qquad \downarrow$$

$$186 \qquad\quad = \quad 2\frac{12}{25} \quad \cdot \qquad\quad s.$$

3. **Solve.** To solve the equation, we divide on both sides.

$$186 = \frac{62}{25} \cdot s \qquad \text{Converting } 2\frac{12}{25} \text{ to fraction notation}$$

$$\frac{186}{\dfrac{62}{25}} = \frac{\dfrac{62}{25} \cdot s}{\dfrac{62}{25}} \qquad \text{Dividing by } \frac{62}{25} \text{ on both sides}$$

$$\frac{186}{\dfrac{62}{25}} = s \qquad \begin{array}{l}\text{Factoring and removing a factor of 1:}\\(62/25)/(62/25) = 1\end{array}$$

$$186 \cdot \frac{25}{62} = s \qquad \text{Multiplying by the reciprocal}$$

$$75 = s \qquad \text{Simplifying: } 186 \cdot \frac{25}{62} = \frac{3 \cdot 62 \cdot 25}{1 \cdot 62} = 75$$

4. **Check.** If the average winning speed in 1911 was about 75 mph, we find the average winning speed in 1990 by multiplying 75 by $2\frac{12}{25}$:

$$2\frac{12}{25} \cdot 75 = \frac{62}{25} \cdot 75 = \frac{62 \cdot 75}{25 \cdot 1} = \frac{62 \cdot 25 \cdot 3}{25 \cdot 1} = 62 \cdot 3 = 186.$$

The answer checks.

5. **State.** The average speed of the winner in the first Indianapolis 500 was about 75 mph.

Do Exercises 9 and 10.

Solve.

9. Kyle's pickup truck travels on an interstate highway at 65 mph for $3\frac{1}{2}$ hr. How far does it travel?

10. Holly's minivan travels 302 mi on $15\frac{1}{10}$ gal of gas. How many miles per gallon did it get?

*Answers*

**9.** $227\frac{1}{2}$ mi  **10.** 20 mpg

**EXAMPLE 9** *Koi Pond.* Colleen designed a koi fish pond for her back-yard. Using the dimensions shown in the diagram below, determine the area of Colleen's backyard before the pond was constructed and the area of the yard remaining after the pond was completed.

**Sources:** en.wikipedia.org; pondliner.com

The brightly colored koi originated in Japan in the 18th century. The hobby of keeping koi increased greatly after the plastic bag was available for safe shipping.

1. **Familiarize.** From the diagram, we know that the dimensions of the backyard are $40\frac{2}{3}$ yd by $27\frac{1}{3}$ yd and the dimensions of the pond are $15\frac{3}{4}$ yd by $11\frac{1}{2}$ yd. We let $B =$ the area of the yard before the construction, $P =$ the area of the pond, and $R =$ the area of the yard that remains after the pond is complete.

2. **Translate.** This is a multistep problem. To find $R$, the area of the yard remaining, we first find $B$, the area of the yard before construction, and $P$, the area of the pond. Then $R$ is the area of the yard $B$ minus the area of the pond $P$:

$$R = B - P.$$

We find each area using the formula for the area of a rectangle, $A = l \times w$.

3. **Solve.** We carry out the calculations as follows:

$$B = \text{length} \times \text{width} \qquad\qquad P = \text{length} \times \text{width}$$
$$= 40\frac{2}{3} \cdot 27\frac{1}{3} \qquad\qquad\qquad = 15\frac{3}{4} \cdot 11\frac{1}{2}$$
$$= \frac{122}{3} \cdot \frac{82}{3} \qquad\qquad\qquad\quad = \frac{63}{4} \cdot \frac{23}{2}$$
$$= \frac{10{,}004}{9} = 1111\frac{5}{9} \text{ sq yd}; \qquad = \frac{1449}{8} = 181\frac{1}{8} \text{ sq yd}.$$

Then

$$R = B - P$$
$$= 1111\frac{5}{9} - 181\frac{1}{8}$$
$$= 1111\frac{40}{72} - 181\frac{9}{72}$$
$$= 930\frac{31}{72} \text{ sq yd}.$$

**11.** A room measures $22\frac{1}{2}$ ft by $15\frac{1}{2}$ ft. A 9-ft by 12-ft rug is placed in the center of the room. How much floor area is not covered by the rug?

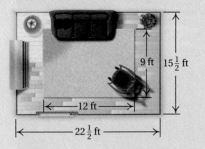

9 ft    $15\frac{1}{2}$ ft

12 ft

$22\frac{1}{2}$ ft

**4. Check.** We can perform a check by repeating the calculations or we can round the measurements to the nearest yard and estimate a solution.

$$B \approx 41 \times 27 = 1107;$$
$$P \approx 16 \times 12 = 192;$$
$$R \approx 1107 - 192 = 915 \text{ sq yd}$$

Since 915 sq yd is close to $930\frac{31}{72}$ sq yd, the answer seems reasonable.

**5. State.** The area of the yard before construction of the pond was $1111\frac{5}{9}$ sq yd. The area of the yard remaining after the pond was completed was $930\frac{31}{72}$ sq yd.

Do Exercise 11.

## Calculator Corner

**Operations on Fractions and Mixed Numerals**   Fraction calculators can add, subtract, multiply, and divide fractions and mixed numerals. The $\boxed{a^b/c}$ key is used to enter fractions and mixed numerals. To find $\frac{3}{4} + \frac{1}{2}$, for example, we press $\boxed{3}\;\boxed{a^b/c}\;\boxed{4}\;\boxed{+}\;\boxed{1}\;\boxed{a^b/c}\;\boxed{2}\;\boxed{=}$. Note that 3/4 and 1/2 appear on the display as $\boxed{\quad 3\;\lrcorner 4 \quad}$ and $\boxed{\quad 1\;\lrcorner 2 \quad}$, respectively. The result is given as the mixed numeral $1\frac{1}{4}$ and is displayed as $\boxed{\quad 1\;\lrcorner 1\;\lrcorner 4 \quad}$. Fraction results that are greater than 1 are always displayed as mixed numerals. To express this result as a fraction, we press $\boxed{\text{SHIFT}}\;\boxed{d/c}$. We get $\boxed{\quad 5\;\lrcorner 4 \quad}$, or 5/4.

To find $3\frac{2}{3} \cdot 4\frac{1}{5}$, we press $\boxed{3}\;\boxed{a^b/c}\;\boxed{2}\;\boxed{a^b/c}\;\boxed{3}\;\boxed{\times}\;\boxed{4}\;\boxed{a^b/c}\;\boxed{1}\;\boxed{a^b/c}\;\boxed{5}\;\boxed{=}$. The calculator displays $\boxed{\quad 15\;\lrcorner 2\;\lrcorner 5 \quad}$, so the product is $15\frac{2}{5}$.

Some calculators are capable of displaying mixed numerals in the way in which we write them, as shown below.

**Exercises:**   Perform each calculation. Give the answer in fraction notation.

**1.** $\frac{1}{3} + \frac{1}{4}$

**2.** $\frac{7}{5} - \frac{3}{10}$

**3.** $\frac{15}{4} \cdot \frac{7}{12}$

**4.** $\frac{4}{5} \div \frac{8}{3}$

Perform each calculation. Give the answer as a mixed numeral.

**5.** $4\frac{1}{3} + 5\frac{4}{5}$

**6.** $9\frac{2}{7} - 8\frac{1}{4}$

**7.** $2\frac{1}{3} \cdot 4\frac{3}{5}$

**8.** $10\frac{7}{10} \div 3\frac{5}{6}$

*Answer*

**11.** $240\frac{3}{4}$ sq ft

# Translating
# for Success

1. *Raffle Tickets.* At the Happy Hollow Camp Fall Festival, Rico and Becca, together, spent $270 on raffle tickets that sell for $$\frac{9}{20}$ each. How many tickets did they buy?

2. *Irrigation Pipe.* Jed uses two pipes, one of which measures $5\frac{1}{3}$ ft, to repair the irrigation system in the Buxtons' lawn. The total length of the two pipes is $8\frac{7}{12}$ ft. How long is the other pipe?

3. *Vacation Days.* Together, Helmut and Claire have 36 vacation days a year. Helmut has 22 vacation days per year. How many does Claire have?

4. *Enrollment in Japanese Classes.* Last year at the Lakeside Community College, 225 students enrolled in basic mathematics. This number is $4\frac{1}{2}$ times as many as the number who enrolled in Japanese. How many enrolled in Japanese?

5. *Bicycling.* Cade rode his bicycle $5\frac{1}{3}$ mi on Saturday and $8\frac{7}{12}$ mi on Sunday. How far did he ride on the weekend?

The goal of these matching questions is to practice step (2), *Translate,* of the five-step problem-solving process. Translate each problem to an equation and select a correct translation from equations A–O.

**A.** $13\frac{11}{12} = x + 5\frac{1}{3}$

**B.** $\frac{3}{4} \cdot x = 1\frac{2}{3}$

**C.** $\frac{20}{9} \cdot 270 = x$

**D.** $225 = 4\frac{1}{2} \cdot x$

**E.** $98 \div 2\frac{1}{3} = x$

**F.** $22 + x = 36$

**G.** $x = 4\frac{1}{2} \cdot 225$

**H.** $x = 5\frac{1}{3} + 8\frac{7}{12}$

**I.** $22 \cdot x = 36$

**J.** $x = \frac{3}{4} \cdot 1\frac{2}{3}$

**K.** $5\frac{1}{3} + x = 8\frac{7}{12}$

**L.** $\frac{9}{20} \cdot 270 = x$

**M.** $1\frac{2}{3} + \frac{3}{4} = x$

**N.** $98 - 2\frac{1}{3} = x$

**O.** $\frac{9}{20} \cdot x = 270$

*Answers on page A-7*

6. *Deli Order.* For a promotional open house for contractors last year, the Bayside Builders Association ordered 225 turkey sandwiches. Due to increased registrations this year, $4\frac{1}{2}$ times as many sandwiches are needed. How many sandwiches are ordered?

7. *Dog Ownership.* In Sam's community, $\frac{9}{20}$ of the households own at least one dog. There are 270 households. How many own dogs?

8. *Magic Tricks.* Samantha has 98 ft of rope and needs to cut it into $2\frac{1}{3}$-ft pieces to be used in a magic trick. How many pieces can be cut from the rope?

9. *Painting.* Laura needs $1\frac{2}{3}$ gal of paint to paint the ceiling of the exercise room and $\frac{3}{4}$ gal of the same paint for the bathroom. How much paint does Laura need?

10. *Chocolate Fudge Bars.* A recipe for chocolate fudge bars that serves 16 includes $1\frac{2}{3}$ cups of sugar. How much sugar is needed for $\frac{3}{4}$ of this recipe?

**a** Multiply. Write a mixed numeral for the answer.

**1.** $8 \cdot 2\dfrac{5}{6}$

**2.** $5 \cdot 3\dfrac{3}{4}$

**3.** $3\dfrac{5}{8} \cdot \left(-\dfrac{2}{3}\right)$

**4.** $6\dfrac{2}{3} \cdot \left(-\dfrac{1}{4}\right)$

**5.** $3\dfrac{1}{2} \cdot 2\dfrac{1}{3}$

**6.** $4\dfrac{1}{5} \cdot 5\dfrac{1}{4}$

**7.** $3\dfrac{2}{5} \cdot 2\dfrac{7}{8}$

**8.** $2\dfrac{3}{10} \cdot 4\dfrac{2}{5}$

**9.** $-4\dfrac{7}{10} \cdot 5\dfrac{3}{10}$

**10.** $-6\dfrac{3}{10} \cdot 5\dfrac{7}{10}$

**11.** $-20\dfrac{1}{2} \cdot \left(-4\dfrac{2}{3}\right)$

**12.** $-21\dfrac{1}{3} \cdot \left(-3\dfrac{5}{8}\right)$

**b** Divide. Write a mixed numeral for the answer.

**13.** $20 \div 3\dfrac{1}{5}$

**14.** $18 \div 2\dfrac{1}{4}$

**15.** $8\dfrac{2}{5} \div 7$

**16.** $3\dfrac{3}{8} \div 3$

**17.** $-4\dfrac{3}{4} \div 1\dfrac{1}{3}$

**18.** $-5\dfrac{4}{5} \div 2\dfrac{1}{2}$

**19.** $1\dfrac{7}{8} \div 1\dfrac{2}{3}$

**20.** $4\dfrac{3}{8} \div 2\dfrac{5}{6}$

**21.** $5\dfrac{1}{10} \div 4\dfrac{3}{10}$

**22.** $4\dfrac{1}{10} \div 2\dfrac{1}{10}$

**23.** $-20\dfrac{1}{4} \div (-90)$

**24.** $-12\dfrac{1}{2} \div (-50)$

**C** Solve.

**25.** *Beagles.* There are about 155,000 Labrador retrievers registered with The American Kennel Club. This is $3\frac{4}{9}$ times the number of beagles registered. How many beagles are registered?

Source: The American Kennel Club

**26.** *Spreading Grass Seed.* Emily seeds lawns for Sam's Superior Lawn Care. When she walks at a rapid pace, the wheel on the broadcast spreader completes $150\frac{2}{3}$ revolutions per minute. How many revolutions does the wheel complete in 15 min?

**27.** *Population.* The population of Alabama is $6\frac{4}{5}$ times that of Alaska. The population of Alabama is approximately 4,700,000. What is the population of Alaska?

Source: U.S. Census Bureau

**28.** *Population.* The population of Iowa is $2\frac{1}{3}$ times the population of Hawaii. The population of Hawaii is approximately 1,300,000. What is the population of Iowa?

Source: U.S. Census Bureau

**29.** *Mural.* Cecilia hired an artist to paint a mural on the wall in her twin sons' bedroom. The dimensions of the mural are $6\frac{2}{3}$ ft by $9\frac{3}{8}$ ft. What is the area of the mural?

**30.** *Sidewalk.* A sidewalk alongside a garden at the conservatory is to be $14\frac{2}{5}$ yd long. Rectangular stone tiles that are each $1\frac{1}{8}$ yd long are used to form the sidewalk. How many tiles are used?

**31.** *Sodium Consumption.* The average American woman consumes $1\frac{1}{3}$ tsp of sodium each day. How much sodium do 10 average American women consume in one day?

Source: Nutrition Action Health Letter, March 1994

**32.** *Aeronautics.* Most space shuttles orbit the earth once every $1\frac{1}{2}$ hr. How many orbits are made every 24 hr?

**33.** *Weight of Water.* The weight of water is $62\frac{1}{2}$ lb per cubic foot. What is the weight of $5\frac{1}{2}$ cubic feet of water?

**34.** *Weight of Water.* The weight of water is $62\frac{1}{2}$ lb per cubic foot. What is the weight of $2\frac{1}{4}$ cubic feet of water?

**35.** *Temperature.*  Fahrenheit temperature can be obtained from Celsius (Centigrade) temperature by multiplying by $1\frac{4}{5}$ and adding 32°. What Fahrenheit temperature corresponds to a Celsius temperature of 20°?

**36.** *Temperature.*  Fahrenheit temperature can be obtained from Celsius (Centigrade) temperature by multiplying by $1\frac{4}{5}$ and adding 32°. What Fahrenheit temperature corresponds to the Celsius temperature of boiling water, 100°?

**37.** *Newspaper Circulation.*  In 2002, daily circulation of the *Wall Street Journal* (New York) was about $5\frac{1}{4}$ times the daily circulation of the *Star Tribune* (Minneapolis). The circulation of the *Star Tribune* was about 343,000. What was the daily circulation of the *Wall Street Journal*?

Source: *Editor and Publisher International Year Book* 2003

**38.** *Language Enrollments.*  In institutions of higher education in the fall of 2002, the number of students who enrolled in Spanish was $14\frac{3}{10}$ times the number who enrolled in Japanese. About 746,270 students enrolled in Spanish. About how many enrolled in Japanese?

Source: Association of Departments of Foreign Languages at the Modern Language Association

**39.** *Population.*  The population of India is about $3\frac{3}{4}$ times the population of the United States. In 2007, the population of India was approximately 1,130,100,000. What was the population of the United States in 2007?

Source: Population Division/International Programs Center, U.S. Census Bureau, U.S. Dept. of Commerce

**40.** *Population.*  The population of Cleveland is about $1\frac{1}{3}$ times the population of Cincinnati. In 2007, the population of Cincinnati was approximately 332,400. What was the population of Cleveland in 2007?

Source: U.S. Census Bureau

**41.** *Doubling a Recipe.*  The chef of a 5-star hotel is doubling a recipe for chocolate cake. The cake requires $2\frac{3}{4}$ cups of flour and $1\frac{1}{3}$ cups of sugar. How much flour and how much sugar will she need?

**42.** *Half of a Recipe.*  A caterer is following a salad dressing recipe that calls for $1\frac{7}{8}$ cups of mayonnaise and $1\frac{1}{6}$ cups of sugar. How much mayonnaise and how much sugar will he need if he prepares $\frac{1}{2}$ of the recipe?

**43.** A car traveled 213 mi on $14\frac{2}{10}$ gal of gas. How many miles per gallon did it get?

**44.** A car traveled 385 mi on $15\frac{4}{10}$ gal of gas. How many miles per gallon did it get?

**45.** *Weight of Water.* The weight of water is $62\frac{1}{2}$ lb per cubic foot. How many cubic feet would be occupied by 25,000 lb of water?

**46.** *Weight of Water.* The weight of water is $8\frac{1}{3}$ lb per gallon. Harry rolls his lawn with an 800-lb capacity roller. Express the water capacity of the roller in gallons.

**47.** *Servings of Salmon.* A serving of filleted fish is generally considered to be about $\frac{1}{3}$ lb. How many servings can be prepared from $5\frac{1}{2}$ lb of salmon fillet?

**48.** *Servings of Tuna.* A serving of fish steak (cross section) is generally $\frac{1}{2}$ lb. How many servings can be prepared from a cleaned $18\frac{3}{4}$-lb tuna?

Find the area of each shaded region.

**49.**

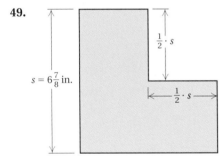

**50.**

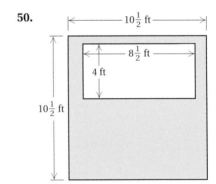

**51.** *Construction.* A rectangular lot has dimensions of $302\frac{1}{2}$ ft by $205\frac{1}{4}$ ft. A building with dimensions of 100 ft by $25\frac{1}{2}$ ft is built on the lot. How much area is left over?

**52.** *Word Processing.* Kelly wants to create a table using Microsoft® Word. She needs to have two columns each $1\frac{1}{2}$ in. wide and five columns each $\frac{3}{4}$ in. wide. Will this table fit on a piece of standard paper that is $8\frac{1}{2}$ in. wide? If so, how wide will each margin be if the margins on each side are to be of equal width?

## Skill Maintenance

In each of Exercises 53–60, fill in the blank with the correct term from the given list. Some of the choices may not be used and some may be used more than once.

53. In the equation $420 \div 60 = 7$, 60 is the _____, 7 the _____ , and 420 the _____ .  [1.5a]

54. When denominators are the same, we say that fractions have a _____ denominator.   [3.5c]

55. The numbers 38, 54, and 91 are examples of _____ numbers.   [3.1c]

56. The number 22,223,133 is _____ by 9 because the sum of its digits is _____ by 9.   [3.2a]

57. When simplifying $24 \div 4 + 4 \times 12 - 6 \div 2$, do all _____ and _____ in order from left to right before doing all _____ and _____ in order from left to right.   [1.9c]

58. In the equation $2 + 3 = 5$, 2 and 3 are called _____.   [1.2a]

59. In the expression $\dfrac{c}{d}$, we call $c$ the _____.   [3.3a]

60. The number 0 has no _____.   [3.7a]

identity

reciprocal

divisor

dividend

quotient

additions

subtractions

multiplications

divisions

numerator

denominator

equal

common

prime

composite

product

divisible

digits

factors

addends

## Synthesis

Multiply. Write the answer as a mixed numeral whenever possible.

61. ▦  $15\dfrac{2}{11} \cdot 23\dfrac{31}{43}$

62. ▦  $17\dfrac{23}{31} \cdot 19\dfrac{13}{15}$

Simplify. Write a mixed numeral for the answer where appropriate.

63. $8 \div \dfrac{1}{2} + \dfrac{3}{4} - \left(5 - \dfrac{5}{8}\right)^2$

64. $\left(\dfrac{5}{9} - \dfrac{1}{4}\right) \times 12 + \left(4 - \dfrac{3}{4}\right)^2$

65. $\dfrac{1}{3} \div \left(\dfrac{1}{2} - \dfrac{1}{5}\right) \times \dfrac{1}{4} + \dfrac{1}{6}$

66. $\dfrac{7}{8} - 1\dfrac{1}{8} \times \dfrac{2}{3} - \dfrac{9}{10} \div \dfrac{3}{5}$

67. $4\dfrac{1}{2} \div 2\dfrac{1}{2} + 8 - 4 \div \dfrac{1}{2}$

68. $6 - 2\dfrac{1}{3} \times \dfrac{3}{4} + \dfrac{5}{8} \div \dfrac{2}{3}$

# 4.7 Order of Operations; Estimation

## a Order of Operations; Fraction Notation and Mixed Numerals

The rules for order of operations that we use with whole numbers (see Section 1.9) apply when we are simplifying expressions involving fraction notation and mixed numerals. For review, these rules are listed below.

> **RULES FOR ORDER OF OPERATIONS**
> 1. Do all calculations within parentheses before operations outside.
> 2. Evaluate all exponential expressions.
> 3. Do all multiplications and divisions in order from left to right.
> 4. Do all additions and subtractions in order from left to right.

**EXAMPLE 1** Simplify: $\frac{1}{6} + \frac{2}{3} \div \frac{1}{2} \cdot \frac{5}{8}$.

$$\frac{1}{6} + \frac{2}{3} \div \frac{1}{2} \cdot \frac{5}{8} = \frac{1}{6} + \frac{2}{3} \cdot \frac{2}{1} \cdot \frac{5}{8}$$    Doing the division first by multiplying by the reciprocal of $\frac{1}{2}$

$$= \frac{1}{6} + \frac{2 \cdot 2 \cdot 5}{3 \cdot 1 \cdot 8}$$    Doing the multiplications in order from left to right

$$= \frac{1}{6} + \frac{2 \cdot 2 \cdot 5}{3 \cdot 1 \cdot 2 \cdot 2 \cdot 2}$$    Factoring in order to simplify

$$= \frac{1}{6} + \frac{5}{6}$$    Removing a factor of 1: $\frac{2 \cdot 2}{2 \cdot 2} = 1$; simplifying

$$= \frac{6}{6}, \text{ or } 1$$    Doing the addition

Do Margin Exercises 1 and 2.

**EXAMPLE 2** Simplify: $\frac{2}{3} \cdot 24 - 11\frac{1}{2}$.

$$\frac{2}{3} \cdot 24 - 11\frac{1}{2} = \frac{2 \cdot 24}{3 \cdot 1} - 11\frac{1}{2}$$    Doing the multiplication first

$$= \frac{2 \cdot 3 \cdot 8}{3 \cdot 1} - 11\frac{1}{2}$$    Factoring the fraction

$$= 2 \cdot 8 - 11\frac{1}{2}$$    Removing a factor of 1: $\frac{3}{3} = 1$

$$= 16 - 11\frac{1}{2}$$    Completing the multiplication

$$= 4\frac{1}{2}, \text{ or } \frac{9}{2}$$    Doing the subtraction

Do Exercise 3.

### OBJECTIVES

**a** Simplify expressions using the rules for order of operations.

**b** Estimate with fraction notation and mixed numerals.

**SKILL TO REVIEW**
Objective 1.9c: Simplify expressions using the rules for order of operations.

Simplify.
1. $22 - 3 \cdot 4$
2. $6(4 + 1)^2 - 3^4 \div 3$

Simplify.
1. $\frac{2}{5} \cdot \frac{5}{8} + \frac{1}{4}$

2. $\frac{1}{3} \cdot \frac{3}{4} \div \frac{5}{8} - \frac{1}{10}$

3. Simplify: $\frac{3}{4} \cdot 16 + 8\frac{2}{3}$.

*Answers*

*Skill to Review:*
1. 10    2. 123

*Margin Exercises:*
1. $\frac{1}{2}$    2. $\frac{3}{10}$    3. $20\frac{2}{3}$, or $\frac{62}{3}$

**EXAMPLE 3** *Harvesting Walnut Trees.* A woodland owner decided to harvest five walnut trees in order to improve the growing conditions of the remaining trees. The logs she sold measured $7\frac{5}{8}$ ft, $8\frac{1}{4}$ ft, $8\frac{3}{4}$ ft, $9\frac{1}{8}$ ft, and $10\frac{1}{2}$ ft. What is the average length of the logs?

Recall that to compute an average, we add the numbers and then divide the sum by the number of addends. (See Section 1.9.) We have

$$\frac{7\frac{5}{8}+8\frac{1}{4}+8\frac{3}{4}+9\frac{1}{8}+10\frac{1}{2}}{5}=\frac{7\frac{5}{8}+8\frac{2}{8}+8\frac{6}{8}+9\frac{1}{8}+10\frac{4}{8}}{5}$$

$$=\frac{42\frac{18}{8}}{5}=\frac{42\frac{9}{4}}{5}=\frac{\frac{177}{4}}{5}\qquad\begin{array}{l}\text{Adding,}\\ \text{simplifying, and}\\ \text{converting}\\ \text{to fraction notation}\end{array}$$

$$=\frac{177}{4}\cdot\frac{1}{5}\qquad\begin{array}{l}\text{Multiplying}\\ \text{by the reciprocal}\end{array}$$

$$=\frac{177}{20}=8\frac{17}{20}.\qquad\begin{array}{l}\text{Converting}\\ \text{to a mixed numeral}\end{array}$$

The average length of the logs is $8\frac{17}{20}$ ft.

Do Exercises 4–6.

**4.** Rachel has triplets. Their birth weights are $3\frac{1}{2}$ lb, $2\frac{3}{4}$ lb, and $3\frac{1}{8}$ lb. What is the average weight of her babies?

**5.** Find the average of
$$\frac{1}{2},\ \frac{1}{3},\ \text{and}\ \frac{5}{6}.$$

**6.** Find the average of $\frac{3}{4}$ and $\frac{4}{5}$.

**EXAMPLE 4** Simplify: $\left(\dfrac{7}{8}-\dfrac{1}{3}\right)\times 48+\left(13+\dfrac{4}{5}\right)^2.$

$$\left(\frac{7}{8}-\frac{1}{3}\right)\times 48+\left(13+\frac{4}{5}\right)^2$$

$$=\left(\frac{7}{8}\cdot\frac{3}{3}-\frac{1}{3}\cdot\frac{8}{8}\right)\times 48+\left(13\cdot\frac{5}{5}+\frac{4}{5}\right)^2\qquad\begin{array}{l}\text{Carrying out operations}\\ \text{inside parentheses first.}\\ \text{To do so, we first multiply}\\ \text{by 1 to obtain the LCD.}\end{array}$$

$$=\left(\frac{21}{24}-\frac{8}{24}\right)\times 48+\left(\frac{65}{5}+\frac{4}{5}\right)^2$$

$$=\frac{13}{24}\times 48+\left(\frac{69}{5}\right)^2\qquad\text{Completing the operations within parentheses}$$

$$=\frac{13}{24}\times 48+\frac{4761}{25}\qquad\text{Evaluating the exponential expression next}$$

$$=26+\frac{4761}{25}\qquad\text{Doing the multiplication}$$

$$=26+190\frac{11}{25}\qquad\text{Converting to a mixed numeral}$$

$$=216\frac{11}{25},\ \text{ or }\ \frac{5411}{25}\qquad\text{Adding}$$

Answers can be given using either fraction notation or mixed numerals.

Do Exercise 7.

**7.** Simplify:
$$\left(\frac{2}{3}+\frac{3}{4}\right)\div 2\frac{1}{3}-\left(\frac{1}{2}\right)^3.$$

**Answers**

**4.** $3\frac{1}{8}$ lb  **5.** $\frac{5}{9}$  **6.** $\frac{31}{40}$  **7.** $\frac{27}{56}$

## (b) Estimation with Fraction Notation and Mixed Numerals

We now estimate with fraction notation and mixed numerals.

**EXAMPLES** Estimate each of the following as $0$, $\frac{1}{2}$, or $1$.

**5.** $\dfrac{2}{17}$

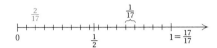

A fraction is close to 0 when the numerator is small in comparison to the denominator. Thus, 0 is an estimate for $\frac{2}{17}$ because 2 is small in comparison to 17. Thus, $\frac{2}{17} \approx 0$.

**6.** $\dfrac{11}{23}$

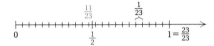

A fraction is close to $\frac{1}{2}$ when the denominator is about twice the numerator. Thus, $\frac{1}{2}$ is an estimate for $\frac{11}{23}$ because $2 \cdot 11 = 22$ and 22 is close to 23. Thus, $\frac{11}{23} \approx \frac{1}{2}$.

**7.** $\dfrac{37}{38}$

A fraction is close to 1 when the numerator is nearly equal to the denominator. Thus, 1 is an estimate for $\frac{37}{38}$ because 37 is nearly equal to 38. Thus, $\frac{37}{38} \approx 1$.

**8.** $\dfrac{43}{41}$

As in the preceding example, the numerator 43 is nearly equal to the denominator 41. Thus, $\frac{43}{41} \approx 1$.

> Do Exercises 8–11.

**EXAMPLE 9** Estimate $16\frac{8}{9} + 11\frac{2}{13} - 4\frac{22}{43}$ by first estimating each mixed numeral as a whole number or as a mixed numeral where the fractional part is $\frac{1}{2}$.

We estimate each fraction as $0$, $\frac{1}{2}$, or $1$. Then we calculate:

$$16\frac{8}{9} + 11\frac{2}{13} - 4\frac{22}{43} \approx 17 + 11 - 4\frac{1}{2}$$

$$= 28 - 4\frac{1}{2}$$

$$= 23\frac{1}{2}.$$

> Do Exercises 12–14.

---

Estimate each of the following as $0$, $\frac{1}{2}$, or $1$.

**8.** $\dfrac{3}{59}$     **9.** $\dfrac{61}{59}$

**10.** $\dfrac{29}{59}$     **11.** $\dfrac{57}{59}$

Estimate each of the following by first estimating each mixed numeral as a whole number or as a mixed numeral where the fractional part is $\frac{1}{2}$ and each fraction as $0$, $\frac{1}{2}$, or $1$.

**12.** $5\dfrac{9}{10} + 26\dfrac{1}{2} - 10\dfrac{3}{29}$

**13.** $10\dfrac{7}{8} \cdot \left(25\dfrac{11}{13} - 14\dfrac{1}{9}\right)$

**14.** $\left(10\dfrac{4}{5} + 7\dfrac{5}{9}\right) \div \dfrac{17}{30}$

*Answers*

**8.** 0   **9.** 1   **10.** $\dfrac{1}{2}$   **11.** 1

**12.** $22\dfrac{1}{2}$   **13.** 132   **14.** 37

**a**   Simplify.

**1.** $\dfrac{1}{2} \cdot \dfrac{1}{3} \cdot \dfrac{1}{4}$

**2.** $\dfrac{1}{3} \cdot \dfrac{1}{4} \cdot \dfrac{1}{5}$

**3.** $6 \div 3 \div 5$

**4.** $12 \div 4 \div 8$

**5.** $\dfrac{2}{3} \div \left(-\dfrac{4}{3}\right) \div \dfrac{7}{8}$

**6.** $\dfrac{5}{6} \div \left(-\dfrac{3}{4}\right) \div \dfrac{5}{2}$

**7.** $\dfrac{5}{8} \div \dfrac{1}{4} - \dfrac{2}{3} \cdot \dfrac{4}{5}$

**8.** $\dfrac{4}{7} \cdot \dfrac{7}{15} + \dfrac{2}{3} \div 8$

**9.** $\dfrac{3}{4} - \dfrac{2}{3} \cdot \left(\dfrac{1}{2} + \dfrac{2}{5}\right)$

**10.** $\dfrac{3}{4} \div \dfrac{1}{2} \cdot \left(\dfrac{8}{9} - \dfrac{2}{3}\right)$

**11.** $28\dfrac{1}{8} - 5\dfrac{1}{4} + 3\dfrac{1}{2}$

**12.** $10\dfrac{3}{5} - 4\dfrac{1}{10} - 1\dfrac{1}{2}$

**13.** $\dfrac{7}{8} \div \dfrac{1}{2} \cdot \dfrac{1}{4}$

**14.** $\dfrac{7}{10} \cdot \dfrac{4}{5} \div \dfrac{2}{3}$

**15.** $\left(\dfrac{2}{3}\right)^2 - \dfrac{1}{3} \cdot 1\dfrac{1}{4}$

**16.** $\left(\dfrac{3}{4}\right)^2 + 3\dfrac{1}{2} \div 1\dfrac{1}{4}$

**17.** $-\dfrac{1}{2} - \left(\dfrac{1}{2}\right)^2 + \left(\dfrac{1}{2}\right)^3$

**18.** $-1 + \dfrac{1}{4} + \left(\dfrac{1}{4}\right)^2 - \left(\dfrac{1}{4}\right)^3$

**19.** Find the average of $\dfrac{2}{3}$ and $\dfrac{7}{8}$.

**20.** Find the average of $\dfrac{1}{4}$ and $\dfrac{1}{5}$.

**21.** Find the average of $\dfrac{1}{6}, \dfrac{1}{8}$, and $\dfrac{3}{4}$.

**22.** Find the average of $\dfrac{4}{5}, \dfrac{1}{2}$, and $\dfrac{1}{10}$.

**23.** Find the average of $3\dfrac{1}{2}$ and $9\dfrac{3}{8}$.

**24.** Find the average of $10\dfrac{2}{3}$ and $24\dfrac{5}{6}$.

**25.** *Hiking the Appalachian Trail.* Ellen camped and hiked for three consecutive days along a section of the Appalachian Trail. The distances she hiked on the three days were $15\frac{5}{32}$ mi, $20\frac{3}{16}$ mi, and $12\frac{7}{8}$ mi. Find the average of these distances.

**26.** *Vertical Leaps.* Eight-year-old Zachary registered vertical leaps of $12\frac{3}{4}$ in., $13\frac{3}{4}$ in., $13\frac{1}{2}$ in., and 14 in. Find his average vertical leap.

**27.** *Black Bear Cubs.* Black bears typically have two cubs. In January 2007 in northern New Hampshire, a black bear sow gave birth to a litter of 5 cubs. This is so rare that Tom Sears, a wildlife photographer, spent 28 hr per week for six weeks watching for the perfect opportunity to photograph this family of six. At the time of this photo, an observer estimated that the cubs weighed $7\frac{1}{2}$ lb, 8 lb, $9\frac{1}{2}$ lb, $10\frac{5}{8}$ lb, and $11\frac{3}{4}$ lb. What was the average weight of the cubs?

**Source:** Andrew Timmins, New Hampshire Fish and Game Department, *Northcountry News*, Warren, NH; Tom Sears, photographer

**28.** *Acceleration.* The results of a *Road & Track* road acceleration test for five cars are given in the graph below. The test measures the time in seconds required to go from 0 mph to 60 mph. What was the average time?

**Acceleration: 0 mph to 60 mph**

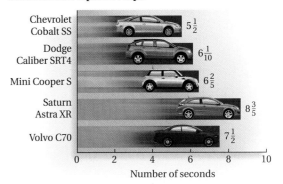

SOURCE: *Road & Track*, October 2008, pp.156–157

**b** Estimate each of the following as $0$, $\frac{1}{2}$, or $1$.

**29.** $\dfrac{2}{47}$   **30.** $\dfrac{5}{12}$   **31.** $\dfrac{7}{8}$   **32.** $\dfrac{1}{13}$   **33.** $\dfrac{6}{11}$   **34.** $\dfrac{11}{13}$

**35.** $\dfrac{7}{15}$   **36.** $\dfrac{1}{16}$   **37.** $\dfrac{7}{100}$   **38.** $\dfrac{5}{9}$   **39.** $\dfrac{20}{19}$   **40.** $\dfrac{5}{4}$

Estimate each of the following by estimating each mixed numeral as a whole number or as a mixed numeral where the fractional part is $\frac{1}{2}$ and by estimating each fraction as $0$, $\frac{1}{2}$, or $1$. Then carry out any computations that are indicated.

**41.** $2\dfrac{7}{8}$

**42.** $1\dfrac{1}{3}$

**43.** $12\dfrac{5}{6}$

**44.** $26\dfrac{6}{13}$

**45.** $\dfrac{4}{5} + \dfrac{7}{8}$

**46.** $\dfrac{1}{12} \cdot \dfrac{7}{15}$

**47.** $\dfrac{3}{8} + \dfrac{7}{13} + \dfrac{5}{9}$

**48.** $\dfrac{8}{9} + \dfrac{4}{5} + \dfrac{11}{12}$

**49.** $\dfrac{43}{100} + \dfrac{1}{10} - \dfrac{11}{1000}$

**50.** $\dfrac{23}{24} + \dfrac{37}{39} + \dfrac{51}{50}$

**51.** $7\dfrac{29}{60} + 10\dfrac{12}{13} \cdot 24\dfrac{2}{17}$

**52.** $5\dfrac{13}{14} - 1\dfrac{5}{8} + 1\dfrac{23}{28} \cdot 6\dfrac{35}{74}$

**53.** $24 \div 7\dfrac{8}{9}$

**54.** $43\dfrac{16}{17} \div 11\dfrac{2}{13}$

**55.** $76\dfrac{3}{14} + 23\dfrac{19}{20}$

**56.** $76\dfrac{13}{14} \cdot 23\dfrac{17}{20}$

**57.** $16\dfrac{1}{5} \div 2\dfrac{1}{11} + 25\dfrac{9}{10} - 4\dfrac{11}{23}$

**58.** $96\dfrac{2}{13} \div 5\dfrac{19}{20} + 3\dfrac{1}{7} \cdot 5\dfrac{18}{21}$

## Skill Maintenance

Divide and simplify.  [3.7b]

**59.** $-\dfrac{4}{5} \div \left( -\dfrac{3}{10} \right)$

**60.** $\dfrac{3}{10} \div \dfrac{4}{5}$

**61.** Classify the given numbers as prime, composite, or neither.

$\quad$ 1, 5, 7, 9, 14, 23, 43  [3.1c]

**62.** Divide: $7865 \div 132$.  [1.5a]

Solve.

**63.** *Luncheon Servings.*  Ian purchased 6 lb of cold cuts for a luncheon. If he has allowed $\frac{3}{8}$ lb per person, how many people did he invite to the luncheon?  [3.7d]

**64.** *Cholesterol.*  A 3-oz serving of crabmeat contains 85 milligrams (mg) of cholesterol. A 3-oz serving of shrimp contains 128 mg of cholesterol. How much more cholesterol is in the shrimp?  [1.8a]

## Synthesis

**65.** ▦ In the sum below, $a$ and $b$ are digits. Find $a$ and $b$.

$$\dfrac{a}{17} + \dfrac{1b}{23} = \dfrac{35a}{391}$$

**66.** ▦ Consider only the numbers 3, 4, 5, and 6. Assume each can be placed in a blank in the following.

$$\square + \dfrac{\square}{\square} \cdot \square = ?$$

What placement of the numbers in the blanks yields the largest sum?

**67.** ▦ Consider only the numbers 2, 3, 4, and 5. Assume each is placed in a blank in the following.

$$\dfrac{\square}{\square} + \dfrac{\square}{\square} = ?$$

What placement of the numbers in the blanks yields the largest sum?

**68.** ▦ Use a calculator to arrange the following in order from smallest to largest.

$$\dfrac{3}{4}, \; \dfrac{17}{21}, \; \dfrac{13}{15}, \; \dfrac{7}{9}, \; \dfrac{15}{17}, \; \dfrac{13}{12}, \; \dfrac{19}{22}$$

# Summary and Review

## Key Terms

least common multiple, p. 190
least common denominator, pp. 190, 198
mixed numeral, p. 212

## Concept Reinforcement

Determine whether each statement is true or false.

_____ **1.** The mixed numeral $5\frac{2}{3}$ can be represented by the sum $5 \cdot \frac{3}{3} + \frac{2}{3}$.  [4.4a]

_____ **2.** The least common multiple of two numbers is always larger than or equal to the larger number.  [4.1a]

_____ **3.** If $\frac{a}{b} < 1$, then $a > b$.  [4.7b]

_____ **4.** The product of any two positive mixed numerals is greater than 1.  [4.6a]

## Important Concepts

**Objective 4.1a**  Find the least common multiple, or LCM, of two or more numbers.

**Example**  Find the LCM of 105 and 90.

$$105 = 3 \cdot 5 \cdot 7,$$
$$90 = 2 \cdot 3 \cdot 3 \cdot 5;$$
$$\text{LCM} = 2 \cdot 3 \cdot 3 \cdot 5 \cdot 7 = 630$$

**Practice Exercise**

**1.** Find the LCM of 52 and 78.

**Objective 4.2a**  Add using fraction notation.

**Example**  Add: $\dfrac{5}{24} + \dfrac{7}{45}$.

$$\frac{5}{24} + \frac{7}{45} = \frac{5}{2 \cdot 2 \cdot 2 \cdot 3} + \frac{7}{3 \cdot 3 \cdot 5}$$

$$= \frac{5}{2 \cdot 2 \cdot 2 \cdot 3} \cdot \frac{3 \cdot 5}{3 \cdot 5} + \frac{7}{3 \cdot 3 \cdot 5} \cdot \frac{2 \cdot 2 \cdot 2}{2 \cdot 2 \cdot 2}$$

$$= \frac{75}{360} + \frac{56}{360} = \frac{131}{360}$$

**Practice Exercise**

**2.** Add: $\dfrac{19}{60} + \dfrac{11}{36}$.

**Objective 4.3a**  Subtract using fraction notation.

**Example**  Subtract: $\dfrac{7}{12} - \dfrac{11}{60}$.

$$\frac{7}{12} - \frac{11}{60} = \frac{7}{12} \cdot \frac{5}{5} - \frac{11}{60} = \frac{35}{60} - \frac{11}{60}$$

$$= \frac{35 - 11}{60} = \frac{24}{60} = \frac{2 \cdot \cancel{12}}{5 \cdot \cancel{12}} = \frac{2}{5}$$

**Practice Exercise**

**3.** Subtract: $\dfrac{29}{35} - \dfrac{5}{7}$.

**Objective 4.3b**   Use $<$ or $>$ with fraction notation to write a true sentence.

**Example**   Use $<$ or $>$ for $\square$ to write a true sentence:

$$\frac{5}{12} \,\square\, \frac{9}{16}.$$

The LCD $= 48$. Thus we have

$$\frac{5}{12} \cdot \frac{4}{4} \,\square\, \frac{9}{16} \cdot \frac{3}{3}$$

$$\frac{20}{48} \,\square\, \frac{27}{48}.$$

Since $20 < 27$, $\dfrac{20}{48} < \dfrac{27}{48}$ and thus $\dfrac{5}{12} < \dfrac{9}{16}$.

**Practice Exercise**

4. Use $<$ or $>$ for $\square$ to write a true sentence:

$$\frac{3}{13} \,\square\, \frac{5}{12}.$$

---

**Objective 4.3c**   Solve equations of the type $x + a = b$ and $a + x = b$, where $a$ and $b$ may be fractions.

**Example**   Solve: $x - \dfrac{1}{6} = -\dfrac{5}{8}$.

$$x - \frac{1}{6} = -\frac{5}{8}$$

$$x - \frac{1}{6} + \frac{1}{6} = -\frac{5}{8} + \frac{1}{6}$$

$$x = -\frac{5}{8} \cdot \frac{3}{3} + \frac{1}{6} \cdot \frac{4}{4}$$

$$x = -\frac{15}{24} + \frac{4}{24} = -\frac{11}{24}$$

**Practice Exercise**

5. Solve: $\dfrac{2}{9} + x = -\dfrac{9}{11}$.

---

**Objective 4.4a**   Convert between mixed numerals and fraction notation.

**Example**   Convert $2\dfrac{5}{13}$ to fraction notation: $2\dfrac{5}{13} = \dfrac{31}{13}$.

**Example**   Convert $-\dfrac{40}{9}$ to a mixed numeral: $-\dfrac{40}{9} = -4\dfrac{4}{9}$.

**Practice Exercises**

6. Convert $8\dfrac{2}{3}$ to fraction notation.

7. Convert $-\dfrac{47}{6}$ to a mixed numeral.

---

**Objective 4.5b**   Subtract using mixed numerals.

**Example**   Subtract: $3\dfrac{3}{8} - 1\dfrac{4}{5}$.

$$3\frac{3}{8} = 3\frac{15}{40} = 2\frac{55}{40}$$

$$-1\frac{4}{5} = -1\frac{32}{40} = -1\frac{32}{40}$$

$$\rule{3cm}{0.4pt}$$

$$1\frac{23}{40}$$

**Practice Exercise**

8. Subtract: $10\dfrac{5}{7} - 2\dfrac{3}{4}$.

**Objective 4.6a**  Multiply using mixed numerals.

**Example**  Multiply: $7\frac{1}{4} \cdot \left(-5\frac{3}{10}\right)$. Write a mixed numeral for the answer.

$$7\frac{1}{4} \cdot \left(-5\frac{3}{10}\right) = \frac{29}{4} \cdot \left(-\frac{53}{10}\right)$$

$$= -\frac{1537}{40} = -38\frac{17}{40}$$

**Practice Exercise**

**9.** Multiply: $-4\frac{1}{5} \cdot 3\frac{7}{15}$.

---

**Objective 4.6c**  Solve applied problems involving multiplication and division with mixed numerals.

**Example**  The population of New York is $3\frac{1}{3}$ times that of Missouri. The population of New York is approximately 19,000,000. What is the population of Missouri?

Translate:

$$\underbrace{\text{Population of New York}}_{} \quad \underset{\downarrow}{\text{is}} \quad \underset{\downarrow}{3\frac{1}{3}} \quad \underset{\downarrow}{\text{times}} \quad \underbrace{\text{Population of Missouri}}_{}$$

$$19,000,000 \quad = \quad 3\frac{1}{3} \quad \cdot \quad x.$$

Solve:

$$19,000,000 = \frac{10}{3} \cdot x$$

$$\frac{19,000,000}{\frac{10}{3}} = \frac{\frac{10}{3} \cdot x}{\frac{10}{3}}$$

$$19,000,000 \cdot \frac{3}{10} = x$$

$$5,700,000 = x.$$

The population of Missouri is about 5,700,000.

**Practice Exercise**

**10.** The population of Louisiana is $2\frac{1}{2}$ times the population of West Virginia. The population of West Virginia is approximately 1,800,000. What is the population of Louisiana?

---

**Objective 4.7a**  Simplify expressions using the rules for order of operations.

**Example**  Simplify: $\left(\frac{4}{5}\right)^2 - \frac{1}{5} \cdot 2\frac{1}{8}$.

$$\left(\frac{4}{5}\right)^2 - \frac{1}{5} \cdot 2\frac{1}{8} = \frac{16}{25} - \frac{1}{5} \cdot 2\frac{1}{8}$$

$$= \frac{16}{25} - \frac{1}{5} \cdot \frac{17}{8} = \frac{16}{25} - \frac{17}{40}$$

$$= \frac{16}{25} \cdot \frac{8}{8} - \frac{17}{40} \cdot \frac{5}{5}$$

$$= \frac{128}{200} - \frac{85}{200}$$

$$= \frac{43}{200}$$

**Practice Exercise**

**11.** Simplify: $\frac{3}{2} \cdot 1\frac{1}{3} \div \left(\frac{2}{3}\right)^2$.

**Objective 4.7b**   Estimate with fraction notation and mixed numerals.

**Example**   Estimate
$$3\tfrac{2}{5} + \tfrac{7}{10} + 6\tfrac{5}{9}$$
by first estimating each mixed numeral as a whole number or as a mixed numeral where the fractional part is $\tfrac{1}{2}$ and by estimating each fraction as $0$, $\tfrac{1}{2}$, or 1.

$$3\tfrac{2}{5} + \tfrac{7}{10} + 6\tfrac{5}{9} \approx 3\tfrac{1}{2} + \tfrac{1}{2} + 6\tfrac{1}{2} = 10\tfrac{1}{2}$$

**Practice Exercise**

**12.** Estimate
$$1\tfrac{19}{20} + 3\tfrac{1}{8} - \tfrac{8}{17}$$
by first estimating each mixed numeral as a whole number or as a mixed numeral where the fractional part is $\tfrac{1}{2}$ and by estimating each fraction as $0$, $\tfrac{1}{2}$, or 1.

# Review Exercises

Find the LCM.   [4.1a]

**1.** 12 and 18      **2.** 18 and 45

**3.** 3, 6, and 30      **4.** 26, 36, and 54

Add and simplify.   [4.2a]

**5.** $\dfrac{6}{5} + \dfrac{3}{8}$      **6.** $\dfrac{-5}{16} + \dfrac{1}{12}$

**7.** $\dfrac{6}{5} + \dfrac{11}{15} + \dfrac{3}{20}$      **8.** $\dfrac{1}{1000} + \dfrac{19}{100} + \dfrac{7}{10}$

Subtract and simplify.   [4.3a]

**9.** $\dfrac{5}{9} - \dfrac{2}{9}$      **10.** $\dfrac{7}{8} - \dfrac{3}{4}$

**11.** $\dfrac{2}{9} - \dfrac{11}{27}$      **12.** $-\dfrac{5}{6} - \dfrac{2}{9}$

Use $<$ or $>$ for $\square$ to write a true sentence.   [4.3b]

**13.** $\dfrac{4}{7} \ \square \ \dfrac{5}{9}$      **14.** $\dfrac{-11}{13} \ \square \ \dfrac{-8}{9}$

Solve. [4.3c]

**15.** $x + \dfrac{2}{5} = \dfrac{7}{8}$      **16.** $\dfrac{1}{2} + y = -\dfrac{9}{10}$

Convert to fraction notation.   [4.4a]

**17.** $7\dfrac{1}{2}$      **18.** $8\dfrac{3}{8}$

**19.** $4\dfrac{1}{3}$      **20.** $-10\dfrac{5}{7}$

Convert to a mixed numeral.   [4.4a]

**21.** $\dfrac{7}{3}$      **22.** $\dfrac{27}{4}$

**23.** $-\dfrac{63}{5}$      **24.** $\dfrac{7}{2}$

Divide. Write a mixed numeral for the answer.   [4.4b]

**25.** $9 \overline{)\ 7\ 8\ 9\ 6}$      **26.** $2\ 3 \overline{)\ 1\ 0{,}4\ 9\ 3}$

Add. Write a mixed numeral for the answer.   [4.5a]

**27.** $\begin{aligned}5\tfrac{3}{5}\\[-2pt]+\,4\tfrac{4}{5}\\ \hline\end{aligned}$      **28.** $\begin{aligned}8\tfrac{1}{3}\\[-2pt]+\,3\tfrac{2}{5}\\ \hline\end{aligned}$

**29.** $\begin{aligned}5\tfrac{5}{6}\\[-2pt]+\,4\tfrac{5}{6}\\ \hline\end{aligned}$      **30.** $\begin{aligned}2\tfrac{3}{4}\\[-2pt]+\,5\tfrac{1}{2}\\ \hline\end{aligned}$

Subtract. Write a mixed numeral for the answer where appropriate.   [4.5b]

**31.** $\begin{aligned}12\\[-2pt]-\,4\tfrac{2}{9}\\ \hline\end{aligned}$      **32.** $\begin{aligned}9\tfrac{3}{5}\\[-2pt]-\,4\tfrac{13}{15}\\ \hline\end{aligned}$

**33.** $\begin{aligned}10\tfrac{1}{4}\\[-2pt]-\,6\tfrac{1}{10}\\ \hline\end{aligned}$      **34.** $\begin{aligned}20\tfrac{7}{24}\\[-2pt]-\,6\tfrac{11}{12}\\ \hline\end{aligned}$

Multiply. Write a mixed numeral for the answer where appropriate.   [4.6a]

**35.** $6 \cdot 2\dfrac{2}{3}$      **36.** $5\dfrac{1}{4} \cdot \left(-\dfrac{2}{3}\right)$

**37.** $2\dfrac{1}{5} \cdot 1\dfrac{1}{10}$      **38.** $-2\dfrac{2}{5} \cdot 2\dfrac{1}{2}$

Divide. Write a mixed numeral for the answer where appropriate.   [4.6b]

**39.** $-27 \div 2\frac{1}{4}$

**40.** $2\frac{2}{5} \div 1\frac{7}{10}$

**41.** $3\frac{1}{4} \div 26$

**42.** $\left(-4\frac{1}{5}\right) \div \left(-4\frac{2}{3}\right)$

Solve.   [4.2b], [4.5c], [4.6c]

**43.** *Sewing.*   Gloria wants to make a dress and a jacket. She needs $1\frac{5}{8}$ yd of 60-in. fabric for the dress and $2\frac{5}{8}$ yd for the jacket. How many yards in all does Gloria need to make the outfit?

**44.** What is the sum of the areas in the figure below?

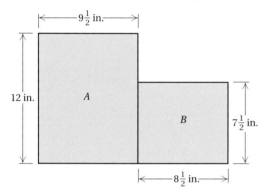

**45.** In the figure above, how much larger is the area of rectangle *A* than the area of rectangle *B*?

**46.** *Carpentry.*   A board $\frac{9}{10}$ in. thick is glued to a board $\frac{4}{5}$ in. thick. The glue is $\frac{3}{100}$ in. thick. How thick is the result?

**47.** *Turkey Servings.*   Turkey contains $1\frac{1}{3}$ servings per pound. How many pounds are needed for 32 servings?

**48.** *Painting a Border.*   Katie hired an artist to paint a decorative border around the top of her son's bedroom. The artist charges $20 per foot. The room measures $11\frac{3}{4}$ ft $\times$ $9\frac{1}{2}$ ft. What is Katie's cost for the project?

**49.** *Cake Recipe.*   A wedding-cake recipe requires 12 cups of shortening. Being calorie-conscious, the wedding couple decides to reduce the shortening by $3\frac{5}{8}$ cups and replace it with prune purée. How many cups of shortening are used in their new recipe?

**50.** *Building a Ziggurat.*   The dimensions of all of the square bricks that King Nebuchadnezzar used over 2500 yr ago to build ziggurats were $13\frac{1}{4}$ in. $\times$ $13\frac{1}{4}$ in. $\times$ $3\frac{1}{4}$ in. What is the perimeter and the area of the $13\frac{1}{4}$ in. $\times$ $13\frac{1}{4}$ in. side? of the $13\frac{1}{4}$ in. $\times$ $3\frac{1}{4}$ in. side?
**Source:** www.eartharchitecture.org

**51.** *Humane Society Pie Sale.* Green River's Humane Society recently conducted its annual pie sale. Each of the 83 pies donated was cut into 6 pieces. At the end of the evening, 382 pieces of pie had been sold. How many pies were sold? How many were left over? Express your answers in mixed numerals.

Simplify each expression using the rules for order of operations. [4.7a]

**52.** $\dfrac{1}{8} \div \dfrac{1}{4} + \dfrac{1}{2}$

**53.** $\dfrac{4}{5} - \dfrac{1}{2} \cdot \left(1 + \dfrac{1}{4}\right)$

**54.** $20\dfrac{3}{4} - 1\dfrac{1}{2} \times 12 + \left(\dfrac{1}{2}\right)^2$

**55.** Find the average of $\dfrac{1}{2}, \dfrac{1}{4}, \dfrac{1}{3}$, and $\dfrac{1}{5}$. [4.7a]

Estimate each of the following as $0, \frac{1}{2}$, or 1. [4.7b]

**56.** $\dfrac{29}{59}$    **57.** $\dfrac{2}{59}$    **58.** $\dfrac{61}{59}$

Estimate by estimating each mixed numeral as a whole number or as a mixed numeral where the fractional part is $\frac{1}{2}$ and by estimating each fraction as $0, \frac{1}{2}$, or 1. Then carry out any computations indicated. [4.7b]

**59.** $6\dfrac{7}{8}$    **60.** $10\dfrac{2}{17}$

**61.** $\dfrac{11}{12} \cdot 5\dfrac{6}{13}$    **62.** $\dfrac{1}{15} \cdot \dfrac{2}{3}$

**63.** $\dfrac{6}{11} + \dfrac{5}{6} + \dfrac{31}{29}$

**64.** $32\dfrac{14}{15} + 27\dfrac{6}{7} - 4\dfrac{25}{28} \cdot 6\dfrac{37}{76}$

**65.** Simplify: $\dfrac{1}{4} + \dfrac{2}{5} \div 5^2$. [4.7a]

**A.** $\dfrac{133}{500}$    **B.** $\dfrac{3}{500}$
**C.** $\dfrac{117}{500}$    **D.** $\dfrac{5}{2}$

**66.** Solve: $x + \dfrac{2}{3} = -5$. [4.3c]

**A.** $\dfrac{17}{3}$    **B.** $\dfrac{13}{3}$
**C.** $-\dfrac{13}{3}$    **D.** $-\dfrac{17}{3}$

## Synthesis

**67.** *Orangutan Circus Act.* Yuri and Olga are orangutans that perform in a circus by riding bicycles around a circular track. It takes Yuri 6 min and Olga 4 min to make one trip around the track. Suppose they start at the same point and then complete their act when they again reach the same point. How long is their act? [4.1a]

**68.** Place the numbers 3, 4, 5, and 6 in the boxes in order to make a true equation: [4.5a]

$$\dfrac{\square}{\square} + \dfrac{\square}{\square} = 3\dfrac{1}{4}.$$

# Understanding Through Discussion and Writing

**1.** Is the sum of two mixed numerals always a mixed numeral? Why or why not? [4.5a]

**2.** Write a problem for a classmate to solve. Design the problem so that its solution is found by performing the multiplication $4\dfrac{1}{2} \cdot 33\dfrac{1}{3}$. [4.6c]

**3.** A student insists that $3\dfrac{2}{5} \cdot 1\dfrac{3}{7} = 3\dfrac{6}{35}$. What mistake is he making and how should he have proceeded? [4.6a]

**4.** Discuss the role of least common multiples in adding and subtracting with fraction notation. [4.2a], [4.3a]

**5.** Describe a real-world situation that fits this equation:
$$2 \cdot 15\dfrac{3}{4} + 2 \cdot 28\dfrac{5}{8} = 88\dfrac{3}{4}.$$ [4.5c], [4.6c]

**6.** A student insists that $5 \cdot 3\dfrac{2}{7} = (5 \cdot 3) \cdot \left(5 \cdot \dfrac{2}{7}\right)$. What mistake is she making and how should she have proceeded? [4.6a]

Test

For Extra Help

Step-by-step test solutions are found on the Chapter Test Prep Videos available via the Video Resources on DVD, in **MyMathLab**, and on  (search "BittingerBasicMathEI" and click on "Channels").

Find the LCM.

**1.** 16 and 12

**2.** 15, 40, and 50

Add and simplify.

**3.** $\dfrac{1}{2} + \dfrac{5}{2}$

**4.** $\dfrac{-7}{8} + \dfrac{2}{3}$

**5.** $\dfrac{7}{10} + \dfrac{19}{100} + \dfrac{31}{1000}$

Subtract and simplify.

**6.** $\dfrac{5}{6} - \dfrac{3}{6}$

**7.** $\dfrac{5}{6} - \dfrac{3}{4}$

**8.** $-\dfrac{17}{24} - \dfrac{1}{15}$

Solve.

**9.** $\dfrac{1}{4} + y = 4$

**10.** $x + \dfrac{2}{3} = \dfrac{11}{12}$

**11.** Use < or > for ☐ to write a true sentence:

$\dfrac{6}{7}$ ☐ $\dfrac{21}{25}$.

Convert to fraction notation.

**12.** $3\dfrac{1}{2}$

**13.** $-9\dfrac{7}{8}$

Convert to a mixed numeral.

**14.** $\dfrac{9}{2}$

**15.** $-\dfrac{74}{9}$

Divide. Write a mixed numeral for the answer.

**16.** $1\,1\,\overline{)1\,7\,8\,9}$

Add. Write a mixed numeral for the answer.

**17.**    $6\dfrac{2}{5}$

   $+\;7\dfrac{4}{5}$
   _____

**18.**    $9\dfrac{1}{4}$

   $+\;5\dfrac{1}{6}$
   _____

Subtract. Write a mixed numeral for the answer.

**19.**    $10\dfrac{1}{6}$

   $-\;5\dfrac{7}{8}$
   _____

**20.**    $14$

   $-\;7\dfrac{5}{6}$
   _____

Multiply. Write a mixed numeral for the answer.

**21.** $9 \cdot 4\dfrac{1}{3}$

**22.** $-6\dfrac{3}{4} \cdot \dfrac{2}{3}$

Divide. Write a mixed numeral for the answer.

**23.** $2\dfrac{1}{3} \div 1\dfrac{1}{6}$

**24.** $2\dfrac{1}{12} \div (-75)$

**25.** *Weightlifting.*   In 2002, Hossein Rezazadeh of Iran did a clean and jerk of 263 kg. This amount was about $2\dfrac{1}{2}$ times his body weight. How much did Rezazadeh weigh? Round to the nearest kilogram.
   **Source:** *The Guinness Book of Records,* 2005

**26.** *Book Order.*   An order of books for a math course weighs 220 lb. Each book weighs $2\dfrac{3}{4}$ lb. How many books are in the order?

**27. Carpentry.** The following diagram shows a middle drawer support guide for a cabinet drawer. Find each of the following.

   **a)** The length $a$ across the top
   **b)** The length $b$ across the bottom

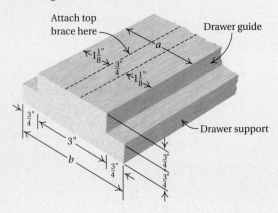

Attach top brace here

Drawer guide

$1\frac{1}{8}"$

$\frac{3}{4}"$

$1\frac{1}{8}"$

$a$

$\frac{3}{4}"$

$3"$

$b$

$\frac{3}{4}"$

Drawer support

$\frac{3}{4}"$

$\frac{3}{4}"$

**28. Carpentry.** In carpentry, some pieces of plywood that are called "$\frac{3}{4}$-inch" plywood are actually $\frac{11}{16}$ in. thick. How much thinner is such a piece than its name indicates?

**29. Women's Dunks.** The first three women in the history of college basketball able to dunk a basketball are listed below, along with their heights and universities:

Michelle Snow, $6\frac{5}{12}$ ft, Tennessee;

Charlotte Smith, $5\frac{11}{12}$ ft, North Carolina;

Georgeann Wells, $6\frac{7}{12}$ ft, West Virginia.

Find the average height of these women.

**Source:** *USA Today,* 11/30/00, p. 3C

Simplify.

**30.** $\frac{2}{3} + 1\frac{1}{3} \cdot 2\frac{1}{8}$

**31.** $1\frac{1}{2} - \frac{1}{2}\left(\frac{1}{2} \div \frac{1}{4}\right) + \left(\frac{1}{2}\right)^2$

Estimate each of the following as $0$, $\frac{1}{2}$, or $1$.

**32.** $\frac{3}{82}$

**33.** $\frac{93}{91}$

Estimate each of the following by first estimating each mixed numeral as a whole number or as a mixed numeral where the fractional part is $\frac{1}{2}$ and by estimating each fraction as $0$, $\frac{1}{2}$, or $1$.

**34.** $256 \div 15\frac{19}{21}$

**35.** $43\frac{15}{31} \cdot 27\frac{5}{6} - 9\frac{15}{28} + 6\frac{5}{76}$

**36.** Find the LCM of 12, 36, and 60.

   **A.** 6            **B.** 12
   **C.** 60           **D.** 180

# Synthesis

**37.** The students in a math class can be organized into study groups of 8 each so that no students are left out. The same class of students can also be organized into groups of 6 so that no students are left out.

   **a)** Find some class sizes for which this will work.
   **b)** Find the smallest such class size.

**38.** Rebecca walks 17 laps at her health club. Trent walks 17 laps at his health club. If the track at Rebecca's health club is $\frac{1}{7}$ mi long, and the track at Trent's is $\frac{1}{8}$ mi long, who walks farther? How much farther?

# Cumulative Review

**CHAPTERS 1-4**

Solve.

**1.** *Cross-Country Skiing.* During a three-day holiday weekend trip, David and Sally Jean cross-country skied $3\frac{2}{3}$ mi on Friday, $6\frac{1}{8}$ mi on Saturday, and $4\frac{3}{4}$ mi on Sunday.

   **a)** Find the total number of miles that they skied.

   **b)** Find the average number of miles that they skied per day. Express your answer as a mixed numeral.

**2.** How many people can receive equal $16 shares from a total of $496?

**3.** A recipe calls for $\frac{4}{5}$ tsp of salt. How much salt should be used for $\frac{1}{2}$ recipe? for 5 recipes?

**4.** How many pieces, each $2\frac{3}{8}$ ft long, can be cut from a piece of wire 38 ft long?

**5.** An emergency food pantry fund contains $423. From this fund, $148 and $167 are withdrawn for expenses. How much is left in the fund?

**6.** In a walkathon, Jermaine walked $\frac{9}{10}$ mi and Oleta walked $\frac{3}{4}$ mi. What was the total distance that they walked?

What part is shaded?

**7.**

**8.**

Calculate and simplify.

**9.**
$$\begin{array}{r} 3\ 7\ 0\ 4 \\ +\ 5\ 2\ 7\ 8 \\ \hline \end{array}$$

**10.**
$$\begin{array}{r} 7\ 6\ 0\ 5 \\ -\ 3\ 0\ 8\ 7 \\ \hline \end{array}$$

**11.** $-27 + 12$

**12.** $\dfrac{3}{8} + \dfrac{1}{24}$

**13.** $-20 - (-6)$

**14.** $-\dfrac{3}{4} - \dfrac{1}{3}$

**15.**
$$\begin{array}{r} 2\dfrac{3}{4} \\ +\ 5\dfrac{1}{2} \\ \hline \end{array}$$

**16.**
$$\begin{array}{r} 2\dfrac{1}{3} \\ -\ 1\dfrac{1}{6} \\ \hline \end{array}$$

**17.** $15 \cdot (-5)$

**18.** $\dfrac{9}{10} \cdot \dfrac{5}{3}$

**19.** $-18 \cdot \left(-\dfrac{5}{6}\right)$

**20.** $2\dfrac{1}{3} \cdot 3\dfrac{1}{7}$

Divide. Write the answer in the form 34 R 7.

**21.** $6\overline{)4290}$ **22.** $45\overline{)2531}$

**23.** Write a mixed numeral for the answer in Exercise 22.

**24.** In the number 2753, what digit names tens?

**25.** *Room Carpeting.* The Chandlers are carpeting an L-shaped family room consisting of one rectangle that is $8\frac{1}{2}$ ft by 11 ft and another rectangle that is $6\frac{1}{2}$ ft by $7\frac{1}{2}$ ft.

 **a)** Find the area of the carpet.
 **b)** Find the perimeter of the carpet.

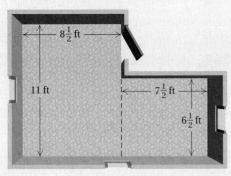

**26.** Round 38,478 to the nearest hundred.

**27.** Find the LCM of 18 and 24.

**28.** Simplify:

$$\left(\frac{1}{2}+\frac{2}{5}\right)^2 \div 3 + 6 \times \left(2+\frac{1}{4}\right).$$

Use $<$, $>$, or $=$ for ☐ to write a true sentence.

**29.** $\frac{4}{5}\ \square\ \frac{4}{6}$ **30.** $\frac{3}{13}\ \square\ \frac{9}{39}$ **31.** $\frac{-3}{7}\ \square\ \frac{-5}{12}$

Estimate each of the following as $0, \frac{1}{2}$, or 1.

**32.** $\frac{29}{30}$ **33.** $\frac{15}{29}$ **34.** $\frac{2}{43}$

Simplify.

**35.** $\frac{36}{45}$ **36.** $\frac{0}{-27}$ **37.** $\frac{320}{10}$

**38.** Convert to fraction notation: $4\frac{5}{8}$.

**39.** Convert to a mixed numeral: $\frac{17}{3}$.

Solve.

**40.** $x + 24 = 117$ **41.** $x + \frac{7}{9} = \frac{4}{3}$

**42.** $\frac{7}{9} \cdot t = -\frac{4}{3}$ **43.** $y = 32,580 \div 36$

**44.** *Matching.* Match each item in the first column with the appropriate item in the second column by drawing connecting lines. There can be more than one correct correspondence for an item.

Factors of 68            12, 54, 72, 300
Factorization of 68      2, 3, 17, 19, 23, 31, 47, 101
Prime factorization of 68   2 · 2 · 17
Numbers divisible by 6   2 · 34
Numbers divisible by 8   8, 16, 24, 32, 40, 48, 64, 864
Numbers divisible by 5   1, 2, 4, 17, 34, 68
Prime numbers            70, 95, 215

## Synthesis

**45.** Find the smallest prime number that is larger than 2000.

# Decimal Notation

## Real-World Application

Camping in national parks has declined steadily in recent years. The numbers of overnight camping stays in Park Service campgrounds in the National Park System from 2002 to 2006 were 5.8 million, 5.7 million, 5.4 million, 5.2 million, and 5.0 million, respectively. Find the average number of stays per year during this period.

*Source:* U.S. National Park Service

***This problem appears as Exercise 77 in Section 5.4.***

# 5.1

# Decimal Notation, Order, and Rounding

## OBJECTIVES

**a** Given decimal notation, write a word name.

**b** Convert between decimal notation and fraction notation.

**c** Given a pair of numbers in decimal notation, tell which is larger.

**d** Round decimal notation to the nearest thousandth, hundredth, tenth, one, ten, hundred, or thousand.

**SKILL TO REVIEW**
Objective 1.6a: Round to the nearest ten, hundred, or thousand.

Round 4735 to the nearest:

1. Ten.          2. Hundred.

Recall that the set of **rational numbers** consists of the integers, . . . , $-3$, $-2$, $-1$, $0$, $1$, $2$, $3$, . . . , and fractions like $\frac{1}{2}$, $-\frac{5}{3}$, $\frac{9}{8}$, $-\frac{17}{10}$, and so on. (See Section 3.3.) In this chapter, we use **decimal notation** to represent the set of rational numbers. For example, $\frac{1}{2}$ can be written as 0.5 in decimal notation and $3\frac{1}{10}$ can be written as 3.1.

We picture the rational numbers on the number line, as follows.

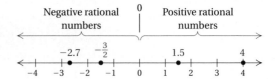

The word *decimal* comes from the Latin word *decima*, meaning a tenth part. Since our counting system is based on tens, decimal notation is a natural extension of a system with which we are already familiar.

## a Decimal Notation and Word Names

One model of the Magellan GPS navigation system sells for $249.98. The dot in $249.98 is called a **decimal point**. Since $0.98, or 98¢, is $\frac{98}{100}$ of a dollar, it follows that

$$\$249.98 = 249 + \frac{98}{100} \text{ dollars.}$$

Also, since $0.98, or 98¢, has the same value as

9 dimes + 8 cents

and 1 dime is $\frac{1}{10}$ of a dollar and 1 cent is $\frac{1}{100}$ of a dollar, we can write

$$249.98 = 2 \cdot 100 + 4 \cdot 10 + 9 \cdot 1 + 9 \cdot \frac{1}{10} + 8 \cdot \frac{1}{100}.$$

This is an extension of the expanded notation for whole numbers that we used in Chapter 1. The place values are 100, 10, 1, $\frac{1}{10}$, $\frac{1}{100}$, and so on. We can see this on a **place-value chart**. The value of each place is $\frac{1}{10}$ as large as that of the one to its left.

Let's consider decimal notation using a place-value chart to represent 26.2922 min, the men's 10,000-meter run record held by Kenenisa Bekele from Ethiopia.

| PLACE-VALUE CHART | | | | | | | |
|---|---|---|---|---|---|---|---|
| Hundreds | Tens | Ones | Ten*ths* | Hundred*ths* | Thousand*ths* | Ten-Thousand*ths* | Hundred-Thousand*ths* |
| 100 | 10 | 1 | $\frac{1}{10}$ | $\frac{1}{100}$ | $\frac{1}{1000}$ | $\frac{1}{10,000}$ | $\frac{1}{100,000}$ |
| 2 | 6 | . 2 | 9 | 2 | 2 | | |

*Answers*
*Skill to Review:*
1. 4740    2. 4700

The decimal notation 26.2922 means

$$20 + 6 + \frac{2}{10} + \frac{9}{100} + \frac{2}{1000} + \frac{2}{10{,}000}, \quad \text{or} \quad 26\frac{2922}{10{,}000}.$$

We read both 26.2922 and $26\frac{2922}{10{,}000}$ as

"Twenty-six *and* two thousand, nine hundred twenty-two ten-thousandths."

We can also read 26.2922 as

"Two six *point* two nine two two," or "twenty-six *point* two nine two two."

To write a word name from decimal notation,

**a)** write a word name for the integer (the number named to the left of the decimal point),

397.685
└──→ Three hundred ninety-seven

**b)** write the word "and" for the decimal point, and

397.685
Three hundred ninety-seven and ↑

**c)** write a word name for the number named to the right of the decimal point, followed by the place value of the last digit.

397.685
Three hundred ninety-seven and six hundred eighty-five *thousandths*

**EXAMPLE 1** *Birth Rate.* There were 14.3 births per 1000 Americans in a recent year. Write a word name for 14.3.

**Source:** U.S. Centers for Disease Control and Prevention

Fourteen and three tenths

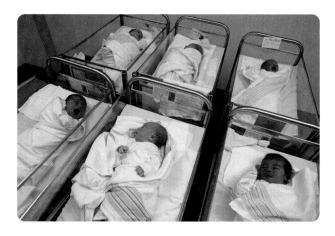

**EXAMPLE 2** Write a word name for 410.87.

Four hundred ten and eighty-seven hundredths

1. **Life Expectancy.** The life expectancy at birth in the United States is projected to be 79.2 yr in 2015. Write a word name for 79.2.
   Source: U.S. Census Bureau

2. **Olympic Swimming.** American swimmer Michael Phelps won eight gold medals in the 2008 Summer Olympics, including one for the 200-m freestyle, which he won in the world record time of 1.716 min. Write a word name for 1.716.
   Source: beijing2008.cn

Write a word name for each number.

3. 245.89

4. −34.0064

5. 31,079.756

6. **Temperature Extremes.** The lowest average annual temperature in the United States is −12.6° C. It occurs in Barrow, Alaska.
   Source: Time Almanac 2005

*Answers*

1. Seventy-nine and two tenths
2. One and seven hundred sixteen thousandths
3. Two hundred forty-five and eighty-nine hundredths
4. Negative thirty-four and sixty-four ten-thousandths
5. Thirty-one thousand, seventy-nine and seven hundred fifty-six thousandths
6. Negative twelve and six tenths

**EXAMPLE 3** *Indianapolis 500.* In the 2009 Indianapolis 500-mile race, winner Helio Castroneves had a 1.9819-second margin of victory over second-place finisher Dan Wheldon. Write a word name for 1.9819.
Source: Indianapolis Motor Speedway, historian Donald Davidson

One and nine thousand, eight hundred nineteen ten-thousandths

**EXAMPLE 4** Write a word name for −1788.405.

Negative one thousand, seven hundred eighty-eight and four hundred five thousandths

Do Exercises 1–6.

## b Converting Between Decimal Notation and Fraction Notation

Given decimal notation, we can convert to fraction notation as follows:

$$9.875 = 9 + \frac{8}{10} + \frac{7}{100} + \frac{5}{1000}$$

$$= 9 \cdot \frac{1000}{1000} + \frac{8}{10} \cdot \frac{100}{100} + \frac{7}{100} \cdot \frac{10}{10} + \frac{5}{1000}$$

$$= \frac{9000}{1000} + \frac{800}{1000} + \frac{70}{1000} + \frac{5}{1000} = \frac{9875}{1000}.$$

Decimal notation ⌐ Fraction notation

$$9.\underline{875} \qquad \frac{9875}{1000}$$

3 decimal places    3 zeros

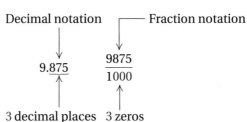

To convert from decimal notation to fraction notation,

a) count the number of decimal places,  
   $4.98$  
   2 places

b) move the decimal point that many places to the right, and  
   $4.98.$  Move 2 places.

c) write the result over a denominator with a 1 followed by that number of zeros.  
   $\frac{498}{100}$  2 zeros

**EXAMPLE 5** Write fraction notation for 0.876. Do not simplify.

$$0.876 \qquad 0.876. \qquad 0.876 = \frac{876}{1000}$$

3 places          3 zeros

For a number like 0.876, we generally write a 0 before the decimal point to draw attention to the presence of the decimal point.

**EXAMPLE 6** Write fraction notation for 56.23. Do not simplify.

56.23     56.23.     $56.23 = \dfrac{5623}{100}$

           2 places        2 zeros

**EXAMPLE 7** Write fraction notation for −1.5018. Do not simplify.

−1.5018     −1.5018.     $-1.5018 = -\dfrac{15{,}018}{10{,}000}$

           4 places        4 zeros

> Do Exercises 7–10.

> Write fraction notation.
> **7.** 0.896    **8.** 23.78
> **9.** −5.6789    **10.** 1.9

If a number in fraction notation has a denominator that is a power of ten, such as 10, 100, 1000, and so on, we reverse the procedure we used before.

> To convert from fraction notation to decimal notation when the denominator is 10, 100, 1000, and so on,
>
> **a)** count the number of zeros, and
> $$\dfrac{8679}{1000}$$
> 3 zeros
>
> **b)** move the decimal point that number of places to the left. Leave off the denominator.
> 8.679.     Move 3 places.
> $$\dfrac{8679}{1000} = 8.679$$

**EXAMPLE 8** Write decimal notation for $\dfrac{47}{10}$.

$\dfrac{47}{10}$      4.7.      $\dfrac{47}{10} = 4.7$

1 zero    1 place

**EXAMPLE 9** Write decimal notation for $\dfrac{13}{1000}$.

$\dfrac{13}{1000}$      0.013.      $\dfrac{13}{1000} = 0.013$

3 zeros    3 places

> Write decimal notation.
> **11.** $\dfrac{743}{100}$    **12.** $\dfrac{406}{1000}$
> **13.** $\dfrac{67{,}089}{10{,}000}$    **14.** $-\dfrac{9}{10}$
> **15.** $\dfrac{57}{1000}$    **16.** $-\dfrac{830}{10{,}000}$

**EXAMPLE 10** Write decimal notation for $-\dfrac{570}{100{,}000}$.

$-\dfrac{570}{100{,}000}$      −0.00570.      $-\dfrac{570}{100{,}000} = -0.0057$

5 zeros    5 places

> Do Exercises 11–16.

*Answers*

**7.** $\dfrac{896}{1000}$   **8.** $\dfrac{2378}{100}$   **9.** $-\dfrac{56{,}789}{10{,}000}$   **10.** $\dfrac{19}{10}$
**11.** 7.43   **12.** 0.406   **13.** 6.7089
**14.** −0.9   **15.** 0.057   **16.** −0.083

When denominators are numbers other than 10, 100, and so on, we will use another method for conversion. It will be considered in Section 5.5.

If a positive mixed numeral has a fractional part with a denominator that is a power of ten, such as 10, 100, or 1000, and so on, we first write the mixed numeral as a sum of a whole number and a fraction. Then we convert to decimal notation.

**EXAMPLE 11** Write decimal notation for $23\frac{59}{100}$.

$$23\frac{59}{100} = 23 + \frac{59}{100} = 23 \text{ and } \frac{59}{100} = 23.59$$

**EXAMPLE 12** Write decimal notation for $-772\frac{129}{10,000}$.

First we consider $772\frac{129}{10,000}$.

$$772\frac{129}{10,000} = 772 + \frac{129}{10,000} = 772 \text{ and } \frac{129}{10,000} = 772.0129$$

Since $772\frac{129}{10,000} = 772.0129$, we have $-772\frac{129}{10,000} = -772.0129$.

Do Exercises 17–19.

Write decimal notation.

**17.** $4\frac{3}{10}$

**18.** $283\frac{71}{100}$

**19.** $-456\frac{13}{1000}$

## (c) Order

To understand how to compare numbers in decimal notation, consider 0.85 and 0.9. First note that $0.9 = 0.90$ because $\frac{9}{10} = \frac{90}{100}$. Then $0.85 = \frac{85}{100}$ and $0.9 = \frac{90}{100}$. Since $\frac{85}{100} < \frac{90}{100}$, it follows that $0.85 < 0.9$. This leads us to a quick way to compare two positive numbers in decimal notation.

> **COMPARING POSITIVE DECIMALS**
>
> To compare two positive numbers in decimal notation, start at the left and compare corresponding digits moving from left to right. If two digits differ, the number with the larger digit is the larger of the two numbers. To ease the comparison, extra zeros can be written to the right of the last decimal place.

**EXAMPLE 13** Which of 2.109 and 2.1 is larger?

*Think.*

2.109 $\longrightarrow$ 2.109

2.1 $\longrightarrow$ 2.100

Same — Different; 9 > 0

Thus, 2.109 is larger than 2.1. That is, 2.109 > 2.1.

**EXAMPLE 14**   Which of 0.09 and 0.108 is larger?

*Think.*

0.09 → 0.090
0.108    0.108

Same └── Different; 1 > 0

Thus, 0.108 is larger than 0.09. That is, 0.108 > 0.09.

> Do Exercises 20–23.

How do we compare negative decimals? Using the number line, we see that −1 > −2. Similarly, −1.4 > −1.5.

$$-1.5 \rightarrow \quad \leftarrow -1.4$$

```
      |     |   |||  |     |
     -3    -2   -1   0     1
```

This leads to the following procedure.

> **COMPARING NEGATIVE DECIMALS**
>
> To compare two negative numbers in decimal notation, start at the left and compare corresponding digits. When two digits differ, the number with the smaller digit is the larger of the two numbers.

**EXAMPLE 15**   Which of −5.6 and −5.63 is larger?

*Think.*

−5.6 → −5.60
−5.63    −5.63

Same └── Different; 0 < 3

Thus, −5.6 is larger than −5.63. That is, −5.6 > −5.63.

> Do Exercises 24–27.

## (d) Rounding

Rounding is done as for whole numbers. To understand, we first consider an example using the number line. It might help to review Section 1.6.

**EXAMPLE 16**   Round 0.37 to the nearest tenth.

Here is part of the number line.

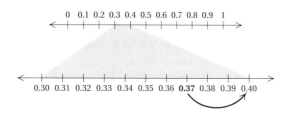

We see that 0.37 is closer to 0.40 than to 0.30. Thus, 0.37 rounded to the nearest tenth is 0.4.

---

**Which number is larger?**

**20.** 2.04, 2.039

**21.** 0.06, 0.008

**22.** 0.5, 0.58

**23.** 1, 0.9999

**Which number is larger?**

**24.** −7.56, −6.18

**25.** −21.006, −21.05

**26.** −0.44, −0.46

**27.** −13.94, −13.79

*Answers*

**20.** 2.04   **21.** 0.06   **22.** 0.58   **23.** 1
**24.** −6.18   **25.** −21.006   **26.** −0.44
**27.** −13.79

> **ROUNDING DECIMAL NOTATION**
>
> To round to a certain place:
>
> a) Locate the digit in that place.
>
> b) Consider the next digit to the right.
>
> c) If the digit to the right is 5 or higher, round the digit to its left up;
> If the digit to the right is 4 or lower, round down.

After rounding up or down, we change all digits to the right of the rounding location to zeros just as we do when we round whole numbers.

**EXAMPLE 17** Round 3872.2459 to the nearest tenth.

a) Locate the digit in the tenths place, 2.

3 8 7 2.2 4 5 9
      ↑

b) Consider the next digit to right, 4.

3 8 7 2.2 4 5 9
       ↑

------------- *Caution!* -------------

3872.3 is not a correct answer to Example 17. It is *incorrect* to round from the ten-thousandths digit over to the tenths digit, as follows:

3872.246 → 3872.25 → 3872.3.

c) Since that digit, 4, is less than 5, round down.

3 8 7 2.2 ← This is the answer.

**EXAMPLE 18** Round 1.008 to the nearest tenth.

a) Locate the digit in the tenths place, 0.

−1 . 0 0 8
     ↑

b) Consider the next digit to the right, 0.

−1 . 0 0 8
      ↑

c) Since that digit, 0, is less than 5, round down.

The answer is 1.0.

**EXAMPLE 19** Round −14.8973 to the nearest hundredth.

a) Locate the digit in the hundredths place, 9.

−1 4.8 9 7 3
        ↑

b) Consider the next digit to the right, 7.

−1 4.8 9 7 3
         ↑

c) Since that digit, 7, is 5 or higher, round the digit 9 up. When we make the hundredths digit a 10, we carry 1 to the tenths place.

The answer is −14.90. The 0 in −14.90 indicates that the answer is correct to the nearest hundredth. Note that when we rounded the digit 9 up, we actually rounded −14.8973 *down*. That is, −14.90 < −14.8973.

**EXAMPLE 20** Round 3872.2459 to the nearest thousandth, hundredth, tenth, one, ten, hundred, and thousand.

| | | | |
|---|---|---|---|
| Thousandth: | 3872.246 | Ten: | 3870 |
| Hundredth: | 3872.25 | Hundred: | 3900 |
| Tenth: | 3872.2 | Thousand: | 4000 |
| One: | 3872 | | |

Do Exercises 28–46.

---

Round to the nearest tenth.

**28.** 2.76      **29.** 13.85

**30.** 234.448      **31.** −7.009

Round to the nearest hundredth.

**32.** 0.636      **33.** 7.834

**34.** −34.675      **35.** 0.025

Round to the nearest thousandth.

**36.** 0.9434      **37.** −8.0038

**38.** 43.1119      **39.** 37.4005

Round 7459.3548 to the nearest:

**40.** Thousandth.

**41.** Hundredth.

**42.** Tenth.      **43.** One.

**44.** Ten. (*Caution:* "Tens" are not "tenths.")

**45.** Hundred.      **46.** Thousand.

*Answers*

**28.** 2.8   **29.** 13.9   **30.** 234.4   **31.** −7.0
**32.** 0.64   **33.** 7.83   **34.** −34.68   **35.** 0.03
**36.** 0.943   **37.** −8.004   **38.** 43.112
**39.** 37.401   **40.** 7459.355   **41.** 7459.35
**42.** 7459.4   **43.** 7459   **44.** 7460
**45.** 7500   **46.** 7000

**a** Write a word name for the number in each sentence.

**1.** *Stock Price.* Google's stock was priced at $486.34 per share.
**Source:** New York Stock Exchange

**2.** *Soft-Drink Consumption.* The average American consumed 51.5 gal of carbonated soft drinks in a recent year.
**Source:** U.S. Department of Agriculture

**3.** *Currency Conversion.* One Chinese yuan was worth about $0.146 in U.S. currency recently.
**Source:** Interactive Data

**4.** *Currency Conversion.* One U.S. dollar was worth 0.6714 euros recently.
**Source:** Interactive Data

**5.** *Game System.* The Nintendo Wii game system sold for $249.89 recently.
**Source:** Best Buy

**6.** *Cell-Phone Plan.* The Sprint Talk/Message cell-phone plan with unlimited anytime minutes was priced at $89.99 per month recently.
**Source:** Sprint

**7.** One gallon of paint is equal to 3.785 L of paint.

**8.** *Water Weight.* One gallon of water weighs 8.35 lb.

Write a word name.

**9.** −34.891

**10.** −27.1245

**b** Write fraction notation. Do not simplify.

**11.** 8.3

**12.** 203.6

**13.** 3.56

**14.** 0.17

**15.** 46.03

**16.** 1.509

**17.** −0.00013

**18.** −0.0109

**19.** −1.0008

**20.** −2.0114

**21.** 20.003

**22.** 4567.2

Write decimal notation.

**23.** $\dfrac{8}{10}$

**24.** $\dfrac{51}{10}$

**25.** $\dfrac{889}{100}$

**26.** $\dfrac{92}{100}$

**27.** $-\dfrac{3798}{1000}$

**28.** $-\dfrac{780}{1000}$

**29.** $\dfrac{78}{10,000}$

**30.** $\dfrac{56,788}{100,000}$

**31.** $\dfrac{19}{100,000}$

**32.** $\dfrac{2173}{100}$

**33.** $-\dfrac{376,193}{1,000,000}$

**34.** $-\dfrac{8,953,074}{1,000,000}$

**35.** $99\dfrac{44}{100}$

**36.** $4\dfrac{909}{1000}$

**37.** $3\dfrac{798}{1000}$

**38.** $67\dfrac{83}{100}$

**39.** $-2\dfrac{1739}{10,000}$

**40.** $-9243\dfrac{1}{10}$

**41.** $8\dfrac{953,073}{1,000,000}$

**42.** $2256\dfrac{3059}{10,000}$

**c** Which number is larger?

**43.** 0.06, 0.58

**44.** 0.008, 0.8

**45.** 0.905, 0.91

**46.** 42.06, 42.1

**47.** −0.0009, −0.001

**48.** −7.067, −7.054

**49.** 234.07, 235.07

**50.** 0.99999, 1

**51.** 0.004, $\dfrac{4}{100}$

**52.** $\dfrac{73}{10}$, 0.73

**53.** −0.432, −0.4325

**54.** −0.8437, −0.84384

**d** Round to the nearest tenth.

**55.** 0.11

**56.** 0.85

**57.** −0.49

**58.** −0.5794

**59.** 2.7449

**60.** 4.78

**61.** −123.65

**62.** −36.049

Round to the nearest hundredth.

**63.** 0.893

**64.** 0.675

**65.** 0.6666

**66.** 6.529

**67.** −0.995

**68.** −207.9976

**69.** −0.094

**70.** −11.4246

Round to the nearest thousandth.

**71.** 0.3246

**72.** 0.6666

**73.** −17.0015

**74.** −123.4562

**75.** 10.1011

**76.** 0.1161

**77.** −9.9989

**78.** −67.100602

Round 809.4732 to the nearest:

**79.** Hundred.

**80.** Tenth.

**81.** Thousandth.

**82.** Hundredth.

**83.** One.

**84.** Ten.

Round 34.54389 to the nearest:

**85.** Ten-thousandth.

**86.** Thousandth.

**87.** Hundredth.

**88.** Tenth.

**89.** One.

**90.** Ten.

## Skill Maintenance

Round 6172 to the nearest:  [1.6a]

**91.** Ten.

**92.** Hundred.

**93.** Thousand.

Find the prime factorization.  [3.1d]

**94.** 2000

**95.** 1530

**96.** 2002

**97.** 4312

## Synthesis

**98.** Arrange the following numbers in order from smallest to largest.

−0.99, −0.099, −1, −0.9999, −0.89999, −1.00009, −0.909, −0.9889

**99.** Arrange the following numbers in order from smallest to largest.

−2.1, −2.109, −2.108, −2.018, −2.0119, −2.0302, −2.000001

*Truncating.*  There are other methods of rounding decimal notation. To round using **truncating**, we drop off all decimal places past the rounding place, which is the same as changing all digits to the right to zeros. For example, rounding 6.78093456285102 to the ninth decimal place, using truncating, gives us 6.780934562. Use truncating to round each of the following to the fifth decimal place, that is, the hundred-thousandth.

**100.** 6.78346623

**101.** 6.783465902

**102.** 99.999999999

**103.** 0.030303030303

# 5.2

# Addition and Subtraction

## OBJECTIVES

**a** Add using decimal notation for positive numbers.

**b** Subtract using decimal notation for positive numbers.

**c** Add and subtract negative decimals.

**d** Solve equations of the type $x + a = b$ and $a + x = b$, where $a$ and $b$ may be in decimal notation.

**e** Balance a checkbook.

### SKILLS TO REVIEW

Objective 1.2a: Add whole numbers.

Objective 1.3a: Subtract whole numbers.

Add or subtract.

1.  ```
      3 8 7
    + 2 5 4
    ```

2.  ```
      4 0 2 3
    − 1 6 6 7
    ```

## a  Adding Positive Decimals

Adding positive decimals is similar to adding whole numbers. First, we line up the decimal points so that we can add corresponding place-value digits. Then we add digits from the right. For example, we add the thousandths, then the hundredths, and so on, carrying if necessary. If desired, we can write extra zeros to the right of the decimal point so that the number of places is the same in all of the addends.

**EXAMPLE 1**   Add: 56.314 + 17.78.

```
  5 6 . 3 1 4       Lining up the decimal points in order to add
+ 1 7 . 7 8 0       Writing an extra zero to the right
                    of the decimal point
```

```
  5 6 . 3 1 4
+ 1 7 . 7 8 0
            4       Adding thousandths
```

```
  5 6 . 3 1 4
+ 1 7 . 7 8 0
          9 4       Adding hundredths
```

```
        1
  5 6 . 3 1 4       Adding tenths
+ 1 7 . 7 8 0       We get 10 tenths = 1 one + 0 tenths,
      . 0 9 4       so we carry the 1 to the ones column.
                    Writing a decimal point in the answer
```

```
  1   1
  5 6 . 3 1 4       Adding ones
+ 1 7 . 7 8 0       We get 14 ones = 1 ten + 4 ones,
      4 . 0 9 4     so we carry the 1 to the tens column.
```

```
  1   1
  5 6 . 3 1 4
+ 1 7 . 7 8 0
  7 4 . 0 9 4       Adding tens
```

Do Margin Exercises 1 and 2.

**EXAMPLE 2**   Add: 3.42 + 0.237 + 14.1.

```
    3 . 4 2 0       Lining up the decimal points
    0 . 2 3 7       and writing extra zeros
+ 1 4 . 1 0 0
  1 7 . 7 5 7       Adding
```

Do Exercises 3–5.

Add.

1.  ```
      0 . 8 4 7
    + 1 0 . 0 7
    ```

2.  ```
        2 . 1
        0 . 7 3
    + 3 1 . 3 6 8
    ```

Add.

3. 0.02 + 4.3 + 0.649

4. 0.12 + 3.006 + 0.4357

5. 0.4591 + 0.2374 + 8.70894

### Answers

*Skills to Review:*

1. 641   2. 2356

*Margin Exercises:*

1. 10.917   2. 34.198   3. 4.969

4. 3.5617   5. 9.40544

Now we consider the addition $3456 + 19.347$. Keep in mind that any whole number has an "unwritten" decimal point at the right that can be followed by zeros. For example, 3456 can also be written 3456.000. When adding, we can always write in the decimal point and extra zeros if desired.

**EXAMPLE 3**  Add: $3456 + 19.347$.

$$
\begin{array}{r}
\overset{\scriptstyle1}{\phantom{0}}\\
3\,4\,\overset{\scriptstyle1}{5}\,6.0\,0\,0\\
+\phantom{000}1\,9.3\,4\,7\\
\hline
3\,4\,7\,5.3\,4\,7
\end{array}
$$

Writing in the decimal point and extra zeros
Lining up the decimal points
Adding

Do Exercises 6 and 7.

Add.

**6.** $789 + 123.67$

**7.** $45.78 + 2467 + 1.993$

## b) Subtracting Positive Decimals

Subtracting positive decimals is similar to subtracting whole numbers. First, we line up the decimal points so that we can subtract corresponding place-value digits. Then we subtract digits from the right. For example, we subtract the thousandths, then the hundredths, the tenths, and so on, borrowing if necessary.

**EXAMPLE 4**  Subtract: $56.314 - 17.78$.

$$
\begin{array}{r}
5\,6.3\,1\,4\\
-\,1\,7.7\,8\,0\\
\hline
\end{array}
$$

Lining up the decimal points in order to subtract
Writing an extra 0

$$
\begin{array}{r}
5\,6.3\,1\,4\\
-\,1\,7.7\,8\,0\\
\hline
4
\end{array}
$$

Subtracting thousandths

$$
\begin{array}{r}
\phantom{5\,6.}{}^{2}\phantom{.}{}^{11}\\
5\,6.\cancel{3}\,\cancel{1}\,4\\
-\,1\,7.7\,8\,0\\
\hline
3\,4
\end{array}
$$

Borrowing tenths to subtract hundredths
Subtracting hundredths

$$
\begin{array}{r}
{}^{12}\\
5\,\,{}^{2}\,{}^{11}\\
5\,\cancel{6}.\cancel{3}\,\cancel{1}\,4\\
-\,1\,7.7\,8\,0\\
\hline
.5\,3\,4
\end{array}
$$

Borrowing ones to subtract tenths
Subtracting tenths; writing a decimal point

$$
\begin{array}{r}
{}^{15}\,{}^{12}\\
4\,\,\cancel{5}\,\,{}^{2}\,{}^{11}\\
\cancel{5}\,\cancel{6}.\cancel{3}\,\cancel{1}\,4\\
-\,1\,7.7\,8\,0\\
\hline
8.5\,3\,4
\end{array}
$$

Borrowing tens to subtract ones
Subtracting ones

$$
\begin{array}{r}
{}^{15}\,{}^{12}\\
4\,\,\cancel{5}\,\,{}^{2}\,{}^{11}\\
\cancel{5}\,\cancel{6}.\cancel{3}\,\cancel{1}\,4\\
-\,1\,7.7\,8\,0\\
\hline
3\,8.5\,3\,4
\end{array}
$$

Subtracting tens

Check by adding:
$$
\begin{array}{r}
{}^{1}\,{}^{1}\,{}^{1}\phantom{.5}\\
3\,8.5\,3\,4\\
+\,1\,7.7\,8\,0\\
\hline
5\,6.3\,1\,4
\end{array}
$$

The answer checks because this is the top number in the subtraction.

Do Exercises 8 and 9.

Subtract.

**8.** $37.428 - 26.674$

**9.**
$$
\begin{array}{r}
0.3\,4\,7\\
-\,0.0\,0\,8\\
\hline
\end{array}
$$

*Answers*

**6.** 912.67  **7.** 2514.773  **8.** 10.754
**9.** 0.339

**EXAMPLE 5**  Subtract: 13.07 − 9.205.

$$
\begin{array}{r}
\overset{12}{\cancel{2}}\ ^{10}\ ^{6}\ ^{10} \\
\cancel{1}\ 3.\cancel{0}\ 7\ \cancel{0} \\
-\ \ \ 9.2\ 0\ 5 \\
\hline
3.8\ 6\ 5
\end{array}
$$

Writing an extra zero

Subtracting

**EXAMPLE 6**  Subtract: 23.08 − 5.0053.

$$
\begin{array}{r}
^{1}\ ^{13}\ \ \ \ ^{7}\ ^{9}\ ^{10} \\
2\ \cancel{3}.0\ \cancel{8}\ \cancel{0}\ \cancel{0} \\
-\ \ \ 5.0\ 0\ 5\ 3 \\
\hline
1\ 8.0\ 7\ 4\ 7
\end{array}
$$

Writing two extra zeros

Subtracting

Check by adding:

$$
\begin{array}{r}
^{1}\ \ \ \ \ ^{1}\ ^{1} \\
1\ 8.0\ 7\ 4\ 7 \\
+\ \ \ 5.0\ 0\ 5\ 3 \\
\hline
2\ 3.0\ 8\ 0\ 0
\end{array}
$$

Do Exercises 10–12.

When subtraction involves a whole number, again keep in mind that there is an "unwritten" decimal point that can be written in if desired. Extra zeros can also be written in to the right of the decimal point.

**EXAMPLE 7**  Subtract: 456 − 2.467.

$$
\begin{array}{r}
^{5}\ ^{9}\ ^{9}\ ^{10} \\
4\ 5\ \cancel{6}.\cancel{0}\ \cancel{0}\ \cancel{0} \\
-\ \ \ \ \ 2.4\ 6\ 7 \\
\hline
4\ 5\ 3.5\ 3\ 3
\end{array}
$$

Writing in the decimal point and extra zeros

Subtracting

Do Exercises 13 and 14.

Subtract.

**10.** 1.2345 − 0.7

**11.** 0.9564 − 0.4392

**12.** 7.37 − 0.00008

Subtract.

**13.** 1277 − 82.78

**14.** 5 − 0.0089

---

**Calculator Corner**

**Addition and Subtraction with Decimal Notation**  To use a calculator to add and subtract with decimal notation, we use the $\boxed{\cdot}$, $\boxed{+}$, $\boxed{-}$, and $\boxed{=}$ keys. To find 47.046 − 28.193, for example, we press $\boxed{4}$ $\boxed{7}$ $\boxed{\cdot}$ $\boxed{0}$ $\boxed{4}$ $\boxed{6}$ $\boxed{-}$ $\boxed{2}$ $\boxed{8}$ $\boxed{\cdot}$ $\boxed{1}$ $\boxed{9}$ $\boxed{3}$ $\boxed{=}$. The display reads $\boxed{18.853}$, so 47.046 − 28.193 = 18.853.

**Exercises:**

Use a calculator to add.

**1.**  274.159
    + 43.486

**2.**  19.805
    + 486.748

**3.** 1.7 + 14.56 + 0.89

**4.** 3.4 + 45 + 0.68

Use a calculator to subtract.

**5.**  9.2
    − 4.8

**6.**  52.34
    − 18.51

**7.** 489 − 34.26

**8.** 6.09 − 5.1

---

**Answers**

**10.** 0.5345   **11.** 0.5172   **12.** 7.36992
**13.** 1194.22   **14.** 4.9911

## (c) Adding and Subtracting Negative Decimals

Negative decimals are added or subtracted just like negative integers.

> ### ADDITION INVOLVING NEGATIVE NUMBERS
>
> 1. *Negative numbers*: Add absolute values. The answer is negative.
> 2. *A positive and a negative number*:
>    - If the numbers have the same absolute value, the answer is 0.
>    - If the numbers have different absolute values, subtract the smaller absolute value from the larger. Then:
>      a) If the positive number has the greater absolute value, the answer is positive.
>      b) If the negative number has the greater absolute value, the answer is negative.
> 3. *One number is zero*: The sum is the other number.

**EXAMPLE 8** Add: $-5.103 + (-6.275)$.

$$\begin{array}{r} 5.1\ 0\ 3 \\ +\ 6.2\ 7\ 5 \\ \hline 1\ 1.3\ 7\ 8 \end{array}$$

The absolute values are 5.103 and 6.275.

Adding the absolute values

The answer is negative: $-5.103 + (-6.275) = -11.378$.

**EXAMPLE 9** Add: $-9.48 + 2.37$.

$$\begin{array}{r} 9.4\ 8 \\ -\ 2.3\ 7 \\ \hline 7.1\ 1 \end{array}$$

Finding the difference of the absolute values

The negative number has the greater absolute value, so the answer is negative: $-9.48 + 2.37 = -7.11$.

> Do Exercises 15–18.

To subtract, we add the opposite of the number being subtracted.

**EXAMPLE 10** Subtract: $-3.1 - 4.8$.

$$-3.1 - 4.8 = -3.1 + (-4.8)$$    Adding the opposite of 4.8
$$= -7.9$$    The sum of two negative numbers is negative.

**EXAMPLE 11** Subtract: $-7.9 - (-8.5)$.

$$-7.9 - (-8.5) = -7.9 + 8.5$$    Adding the opposite of $-8.5$
$$= 0.6$$    Finding the difference of the absolute values. The answer is positive since 8.5 has the greater absolute value.

> Do Exercises 19–22.

Add.

**15.** $12.064 + (-7.591)$

**16.** $-2.81 + (-8.432)$

**17.** $7.42 + (-9.38)$

**18.** $-4.201 + 7.36$

Subtract.

**19.** $9.25 - 13.41$

**20.** $-5.72 - 4.19$

**21.** $9.8 - (-2.6)$

**22.** $-5.9 - (-3.2)$

*Answers*

**15.** 4.473  **16.** $-11.242$  **17.** $-1.96$
**18.** 3.159  **19.** $-4.16$  **20.** $-9.91$
**21.** 12.4  **22.** $-2.7$

## (d) Solving Equations

Now let's solve equations $x + a = b$ and $a + x = b$, where $a$ and $b$ may be in decimal notation. Proceeding as we have before, we subtract $a$ on both sides.

**EXAMPLE 12**  Solve: $x + 28.89 = 74.567$.

We have

$$x + 28.89 - 28.89 = 74.567 - 28.89 \qquad \text{Subtracting 28.89 on both sides}$$
$$x = 45.677.$$

$$\begin{array}{r} {\scriptstyle 13\ 14} \\ {\scriptstyle 6\ \not{3}\ \not{4}\ 16} \\ 7\ 4.\not{5}\ \not{6}\ 7 \\ -\ 2\ 8.8\ 9\ 0 \\ \hline 4\ 5.6\ 7\ 7 \end{array}$$

The solution is 45.677.

**EXAMPLE 13**  Solve: $0.8879 + y = 9.0026$.

We have

$$0.8879 + y - 0.8879 = 9.0026 - 0.8879 \qquad \text{Subtracting 0.8879 on both sides}$$
$$y = 8.1147.$$

$$\begin{array}{r} {\scriptstyle 11} \\ {\scriptstyle 8\ 9\ 9\ \not{1}\ 16} \\ 9.\not{0}\ \not{0}\ 2\ \not{6} \\ -\ 0.8\ 8\ 7\ 9 \\ \hline 8.1\ 1\ 4\ 7 \end{array}$$

The solution is 8.1147.

Do Exercises 23 and 24.

**EXAMPLE 14**  Solve: $x + 7.4 = 5.6$.

We have

$$x + 7.4 - 7.4 = 5.6 - 7.4 \qquad \text{Subtracting 7.4 on both sides}$$
$$x = -1.8. \qquad 5.6 - 7.4 = 5.6 + (-7.4) = -1.8$$

The solution is $-1.8$.

Do Exercise 25.

## (e) Balancing a Checkbook

Let's use addition and subtraction with decimals to balance a checkbook.

**EXAMPLE 15**  Find the errors, if any, in the balances in this checkbook.

| 20___ | | RECORD ALL CHARGES OR CREDITS THAT AFFECT YOUR ACCOUNT | | | | | | |
|---|---|---|---|---|---|---|---|---|
| DATE | CHECK NUMBER | TRANSACTION DESCRIPTION | √ T | (−) PAYMENT/ DEBIT | (+ OR −) OTHER | (+) DEPOSIT/ CREDIT | BALANCE FORWARD | |
| | | | | | | | 8767 | 73 |
| 8/16 | 432 | Burch Laundry | | 23 56 | | | 8744 | 16 |
| 8/19 | 433 | Rogers Bakery | | 20 49 | | | 8764 | 65 |
| 8/20 | | Deposit | | | | 85 00 | 8848 | 65 |
| 8/21 | 434 | Galaxy Video | | 48 60 | | | 8801 | 05 |
| 8/22 | 435 | Benson's Garage | | 267 95 | | | 8533 | 09 |

*Solve.*

**23.** $x + 17.78 = 56.314$

**24.** $8.906 + t = 23.07$

**25.** Solve: $10.6 + y = 2.8$.

*Answers*

**23.** 38.534  **24.** 14.164  **25.** $-7.8$

There are two ways to determine whether there are errors. We assume that the amount $8767.73 in the "Balance forward" column is correct. If we can determine that the ending balance is correct, we have some assurance that the checkbook is correct. But two errors could offset each other to give us that balance.

**METHOD 1**

a) We add the debits: $23.56 + 20.49 + 48.60 + 267.95 = 360.60$.

b) We add the deposits/credits. In this case, there is only one deposit, 85.00.

c) We add the total of the deposits to the balance brought forward:

$$8767.73 + 85.00 = 8852.73.$$

d) We subtract the total of the debits: $8852.73 - 360.60 = 8492.13$.

The result should be the ending balance, 8533.09. We see that $8492.13 \neq 8533.09$. Since the numbers are not equal, we proceed to method 2.

**METHOD 2** We successively add or subtract deposit/credits and debits, and check the result in the "Balance forward" column.

$$8767.73 - 23.56 = 8744.17.$$

We have found our first error. The subtraction was incorrect. We correct it and continue, using 8744.17 as the corrected balance forward:

$$8744.17 - 20.49 = 8723.68.$$

It looks as though 20.49 was added instead of subtracted. Actually, we would have to correct this line even if it had been subtracted, because the error of 1¢ in the first step has been carried through successive calculations. We correct that balance line and continue, using 8723.68 as the balance and adding the deposit 85.00:

$$8723.68 + 85.00 = 8808.68.$$

We make the correction and continue subtracting the last two debits:

$$8808.68 - 48.60 = 8760.08.$$

Then

$$8760.08 - 267.95 = 8492.13.$$

The corrected checkbook is below.

| DATE | CHECK NUMBER | TRANSACTION DESCRIPTION | √ T | (−) PAYMENT/ DEBIT | (+ OR −) OTHER | (+) DEPOSIT/ CREDIT | BALANCE FORWARD 8767 73 | |
|---|---|---|---|---|---|---|---|---|
| 8/16 | 432 | Burch Laundry | | 23 56 | | | 8744 16 | → 8744.17 |
| 8/19 | 433 | Rogers Bakery | | 20 49 | | | 8764 65 | → 8723.68 |
| 8/20 | | Deposit | | | | 85 00 | 8848 65 | → 8808.68 |
| 8/21 | 434 | Galaxy Video | | 48 60 | | | 8801 05 | → 8760.08 |
| 8/22 | 435 | Benson's Garage | | 267 95 | | | 8533 09 | → 8492.13 |

Do Exercise 26.

There are other ways in which errors can be made in checkbooks, such as forgetting to record a transaction or writing the amounts incorrectly, but we will not consider those here.

**26.** Find the errors, if any, in this checkbook.

| DATE | CHECK NUMBER | TRANSACTION DESCRIPTION | √ T | (−) PAYMENT/ DEBIT | (+ OR −) OTHER | (+) DEPOSIT/ CREDIT | BALANCE FORWARD 3078 92 |
|---|---|---|---|---|---|---|---|
| 12/1 | 888 | H.H. Gregg Appliances | | 340 69 | | | 2738 23 |
| 12/3 | 889 | Marie Callendar's Pies | | 78 56 | | | 2659 66 |
| 12/5 | | Deposit <Paycheck> | | | | 230 80 | 2890 46 |
| 12/6 | 890 | Chili's Restaurant | | 13 14 | | | 2877 32 |
| 12/8 | 891 | Stonecreek Golf Course | | 48 00 | | | 2829 32 |
| 12/8 | | Deposit <Molly> | | | | 39 58 | 2848 90 |
| 12/10 | 892 | Frontier Trading Post | | 102 87 | | | 2766 83 |
| 12/14 | 893 | Goody's Music | | 48 59 | | | 2697 45 |
| 12/15 | 894 | Salvation Army | | 100 00 | | | 2497 45 |

*Answer*

**26.** The "Balance forward" column should read:
$3078.92
2738.23
2659.67
2890.47
2877.33
2829.33
2868.91
2766.04
2697.45
2597.45

**a** Add.

**1.**    3 1 6.2 5
      +   1 8.1 2

**2.**    6 4 1.8 0 3
      +   1 4.9 3 5

**3.**    6 5 9.4 0 3
      + 9 1 6.8 1 2

**4.**    4 2 0 3.2 8
      +        3.3 9

**5.**        9.1 0 4
      + 1 2 3.4 5 6

**6.**    8 1.0 0 8
      +   3.4 0 9

**7.** 20.0124 + 30.0124

**8.** 0.687 + 0.9

**9.** 39 + 1.007

**10.** 2.3 + 0.729 + 23

**11.**    4 7.8
       2 1 9.8 5 2
         4 3.5 9
      + 6 6 6.7 1 3

**12.**    1 3.7 2
             9.1 1 2
       6 5 4 2.7 9 0 8
      +      2 3.9 0 1

**13.** 0.34 + 3.5 + 0.127 + 768

**14.** 17 + 3.24 + 0.256 + 0.3689

**15.** 99.6001 + 7285.18 + 500.042 + 870

**16.** 65.987 + 9.4703 + 6744.02 + 1.0003 + 200.895

**b** Subtract.

**17.**    5 1.3 1
       −     2.2 9

**18.**    4 4.3 4 5
       −     3.1 0 5

**19.**    9 2.3 4 1
       −      6.4 2

**20.**    9 7.0 1
       −     3.1 5

**21.**    2.5
       − 0.0 0 2 5

**22.**    3 9.0
       −    0.2 8

**23.**    3.4
       − 0.0 0 3

**24.**    2.8
       − 2.0 8

**25.** 28.2 − 19.35

**26.** $100.16 - 0.118$

**27.** $34.07 - 30.7$

**28.** $36.2 - 16.28$

**29.** $8.45 - 7.405$

**30.** $3.801 - 2.81$

**31.** $6.003 - 2.3$

**32.** $9.087 - 8.807$

**33.** $1 - 0.0098$

**34.** $2 - 1.0908$

**35.** $100 - 0.34$

**36.** $624 - 18.79$

**37.** $7.48 - 2.6$

**38.** $18.4 - 5.92$

**39.** $3 - 2.006$

**40.** $263.7 - 102.08$

**41.** $19 - 1.198$

**42.** $2548.98 - 2.007$

**43.** $65 - 13.87$

**44.** $45 - 0.999$

**45.**
$$\begin{array}{r} 3\,2.7\,9\,7\,8 \\ -\quad 0.0\,5\,9\,2 \\ \hline \end{array}$$

**46.**
$$\begin{array}{r} 0.4\,9\,6\,3\,4 \\ -\,0.1\,2\,6\,7\,8 \\ \hline \end{array}$$

**47.**
$$\begin{array}{r} 6.0\,7 \\ -\,2.0\,0\,7\,8 \\ \hline \end{array}$$

**48.**
$$\begin{array}{r} 1.0 \\ -\,0.9\,9\,9\,9 \\ \hline \end{array}$$

**c** Add or subtract, as indicated.

**49.** $-5.02 + 1.73$

**50.** $-4.31 + 7.66$

**51.** $12.9 - 15.4$

**52.** $27.2 - 31.9$

**53.** $-2.9 + (-4.3)$

**54.** $-7.49 - 1.82$

**55.** $-4.301 + 7.68$

**56.** $-5.952 + 7.98$

**57.** $-12.9 - 3.7$

**58.** $-8.7 - 12.4$

**59.** $-2.1 - (-4.6)$

**60.** $-4.3 - (-2.5)$

**61.** $14.301 + (-17.82)$

**62.** $13.45 + (-18.701)$

**63.** $7.201 - (-2.4)$

**64.** $2.901 - (-5.7)$

**65.** $96.9 + (-21.4)$

**66.** $43.2 + (-10.9)$

**67.** $-8.9 - (-12.7)$

**68.** $-4.5 - (-7.3)$

**69.** $-4.9 - 5.392$

**70.** $89.3 - 92.1$

**71.** $14.7 - 23.5$

**72.** $-7.201 - 1.9$

**d** Solve.

**73.** $x + 17.5 = 29.15$

**74.** $t + 50.7 = 54.07$

**75.** $17.95 + p = 402.63$

**76.** $w + 1.3004 = 47.8$

**77.** $13{,}083.3 = x + 12{,}500.33$

**78.** $100.23 = 67.8 + z$

**79.** $x + 2349 = -17{,}684.3$

**80.** $1830.4 + t = -23{,}067$

**e** Find the errors, if any, in each checkbook.

**81.**

| 20___ | | RECORD ALL CHARGES OR CREDITS THAT AFFECT YOUR ACCOUNT | | | | | | |
|---|---|---|---|---|---|---|---|---|
| DATE | CHECK NUMBER | TRANSACTION DESCRIPTION | √ T | (-) PAYMENT/ DEBIT | (+ OR -) OTHER | (+) DEPOSIT/ CREDIT | BALANCE FORWARD |  |
|  |  |  |  |  |  |  | 9704 | 56 |
| 8/8 | 342 | Bill Rydman |  | 27 44 |  |  | 9677 | 12 |
| 8/9 |  | Deposit |  |  |  | 1000 00 | 10,677 | 12 |
| 8/12 | 343 | Chuck Taylor |  | 123 95 |  |  | 10,553 | 17 |
| 8/14 | 344 | Jennifer Crum |  | 124 02 |  |  | 10,677 | 19 |
| 8/22 | 345 | Neon Johnny's Pizza |  | 12 43 |  |  | 10,664 | 76 |
| 8/24 |  | Deposit |  |  |  | 2500 00 | 13,164 | 76 |
| 8/29 | 346 | Border's Bookstore |  | 137 78 |  |  | 13,302 | 54 |
| 9/2 |  | Deposit |  |  |  | 18 88 | 13,283 | 66 |
| 9/3 | 347 | Fireman's Fund |  | 2800 00 |  |  | 10,483 | 66 |

**82.**

| 20___ | | RECORD ALL CHARGES OR CREDITS THAT AFFECT YOUR ACCOUNT | | | | | | |
|---|---|---|---|---|---|---|---|---|
| DATE | CHECK NUMBER | TRANSACTION DESCRIPTION | √ T | (-) PAYMENT/ DEBIT | (+ OR -) OTHER | (+) DEPOSIT/ CREDIT | BALANCE FORWARD |  |
|  |  |  |  |  |  |  | 1876 | 43 |
| 4/1 | 500 | Bart Kaufman |  | 500 12 |  |  | 1376 | 31 |
| 4/3 | 501 | Jim Lawler |  | 28 56 |  |  | 1347 | 75 |
| 4/3 |  | Deposit |  |  |  | 10,000 00 | 11,347 | 75 |
| 4/3 | 502 | Victoria Montoya |  | 464 00 |  |  | 10,883 | 75 |
| 4/3 |  | Deposit |  |  |  | 2500 00 | 8383 | 75 |
| 4/4 | 503 | Art's Garage |  | 1600 00 |  |  | 6783 | 75 |
| 4/8 | 504 | Golf Galaxy |  | 1349 98 |  |  | 5433 | 77 |
| 4/12 | 505 | Don Mitchell Pro Shops |  | 658 97 |  |  | 4774 | 80 |
| 4/13 |  | Deposit |  |  |  | 900 00 | 5674 | 80 |
| 4/15 | 506 | Kroger |  | 185 58 |  |  | 5479 | 22 |

## Skill Maintenance

**83.** Round 34,567 to the nearest thousand.  [1.6a]

**84.** Round 34,496 to the nearest thousand.  [1.6a]

Subtract.

**85.** $\dfrac{13}{24} - \dfrac{3}{8}$  [4.3a]

**86.** $\dfrac{8}{9} - \dfrac{2}{15}$  [4.3a]

**87.** $\dfrac{1}{5} - \left(-\dfrac{1}{3}\right)$  [4.3a]

**88.** $-\dfrac{7}{4} - \dfrac{5}{6}$  [4.3a]

Solve.

**89.** A serving of filleted fish is generally considered to be about $\frac{1}{3}$ lb. How many servings can be prepared from $5\frac{1}{2}$ lb of flounder fillet?  [4.6c]

**90.** A photocopier technician drove $125\frac{7}{10}$ mi away from Scottsdale for a repair call. The next day he drove $65\frac{1}{2}$ mi back toward Scottsdale for another service call. How far was the technician from Scottsdale?  [4.5c]

## Synthesis

**91.** A student presses the wrong button when using a calculator and adds 235.7 instead of subtracting it. The incorrect answer is 817.2. What is the correct answer?

# 5.3

## Multiplication

### OBJECTIVES

**a** Multiply using decimal notation.

**b** Convert from notation like 45.7 million to standard notation, and convert between dollars and cents.

**SKILL TO REVIEW**

Objective 1.4a: Multiply whole numbers.

Multiply.

1. $\begin{array}{r} 4\,2 \\ \times\ 6\,3 \\ \hline \end{array}$

2. $\begin{array}{r} 7\,1\,6 \\ \times\ \ \ 5\,8 \\ \hline \end{array}$

### STUDY TIPS

#### HAVE A POSITIVE ATTITUDE

You can choose to improve your attitude and raise the academic goals that you have set for yourself. Projecting a positive attitude toward your study of mathematics and expecting a positive outcome can make it easier for you to learn and to perform well in this course.

### a  Multiplication

Let's find the product

$$2.3 \times 1.12.$$

To understand how we find such a product, we first convert each factor to fraction notation. Next, we multiply the numerators and then divide by the product of the denominators.

$$2.3 \times 1.12 = \frac{23}{10} \times \frac{112}{100} = \frac{23 \times 112}{10 \times 100} = \frac{2576}{1000} = 2.576$$

Note the number of decimal places.

$$\begin{array}{r} 1.1\,2 \\ \times\ \ \ 2.3 \\ \hline 2.5\,7\,6 \end{array}$$ 
(2 decimal places)
(1 decimal place)
(3 decimal places)

Now consider

$$0.011 \times 15.0002 = \frac{11}{1000} \times \frac{150{,}002}{10{,}000} = \frac{1{,}650{,}022}{10{,}000{,}000} = 0.1650022.$$

Note the number of decimal places.

$$\begin{array}{r} 1\,5.0\,0\,0\,2 \\ \times\ \ \ \ \ \ 0.0\,1\,1 \\ \hline 0.1\,6\,5\,0\,0\,2\,2 \end{array}$$ 
(4 decimal places)
(3 decimal places)
(7 decimal places)

---

To multiply using decimals:

a) Ignore the decimal points and multiply as though both factors were integers.

b) Then place the decimal point in the result. The number of decimal places in the product is the sum of the numbers of places in the factors. (Count places from the right.)

$$0.8 \times 0.43$$

$$\begin{array}{r} \overset{2}{0.4}\,3 \\ \times\ \ \ \ 0.8 \\ \hline 3\,4\,4 \end{array}$$ Ignore the decimal points for now.

$$\begin{array}{r} 0.4\,3 \\ \times\ \ \ 0.8 \\ \hline 0.3\,4\,4 \end{array}$$ 
(2 decimal places)
(1 decimal place)
(3 decimal places)

---

**EXAMPLE 1** Multiply: $8.3 \times 74.6$.

a) We ignore the decimal points and multiply as though factors were integers.

$$\begin{array}{r} 7\,4.6 \\ \times\ \ \ \ 8.3 \\ \hline 2\,2\,3\,8 \\ 5\,9\,6\,8\,0 \\ \hline 6\,1\,9\,1\,8 \end{array}$$

*Answers*

*Skill to Review:*

**1.** 2646  **2.** 41,528

**b)** We place the decimal point in the result. The number of decimal places in the product is the sum of the numbers of places in the factors, $1 + 1$, or 2.

$$
\begin{array}{r}
7\ 4\ .6 \quad \text{(1 decimal place)} \\
\times \qquad 8\ .3 \quad \text{(1 decimal place)} \\
\hline
2\ 2\ 3\ 8 \\
5\ 9\ 6\ 8\ 0 \\
\hline
6\ 1\ 9\ .1\ 8 \quad \text{(2 decimal places)}
\end{array}
$$

Do Exercise 1.

**EXAMPLE 2** Multiply: $0.0032 \times 2148$.

$$
\begin{array}{r}
2\ 1\ 4\ 8 \quad \text{(0 decimal places)} \\
\times\ 0.0\ 0\ 3\ 2 \quad \text{(4 decimal places)} \\
\hline
4\ 2\ 9\ 6 \\
6\ 4\ 4\ 4\ 0 \\
\hline
6.8\ 7\ 3\ 6 \quad \text{(4 decimal places)}
\end{array}
$$

**EXAMPLE 3** Multiply: $-0.14 \times 0.867$.

First we multiply the absolute values.

$$
\begin{array}{r}
0.8\ 6\ 7 \quad \text{(3 decimal places)} \\
\times \qquad 0.1\ 4 \quad \text{(2 decimal places)} \\
\hline
3\ 4\ 6\ 8 \\
8\ 6\ 7\ 0 \\
\hline
0.1\ 2\ 1\ 3\ 8 \quad \text{(5 decimal places)}
\end{array}
$$

Since the product of a negative number and a positive number is negative, the answer is $-0.12138$.

Do Exercises 2 and 3.

## Multiplying by 0.1, 0.01, 0.001, and So On

Now let's consider some special kinds of products. The first involves multiplying by a tenth, hundredth, thousandth, ten-thousandth, and so on. Let's look at those products.

$$0.1 \times 38 = \frac{1}{10} \times 38 = \frac{38}{10} = 3.8$$

$$0.01 \times 38 = \frac{1}{100} \times 38 = \frac{38}{100} = 0.38$$

$$0.001 \times 38 = \frac{1}{1000} \times 38 = \frac{38}{1000} = 0.038$$

$$0.0001 \times 38 = \frac{1}{10,000} \times 38 = \frac{38}{10,000} = 0.0038$$

Note in each case that the product is *smaller* than 38. That is, the decimal point in each product is farther to the left than the unwritten decimal point in 38.

**1.** Multiply.

$$
\begin{array}{r}
8\ 5\ .4 \\
\times \qquad 6\ .2 \\
\hline
\end{array}
$$

Multiply.

**2.**
$$
\begin{array}{r}
1\ 2\ 3\ 4 \\
\times\ 0.0\ 0\ 4\ 1 \\
\hline
\end{array}
$$

**3.** $-0.804 \times 42.65$

### Calculator Corner

**Multiplication with Decimal Notation** To use a calculator to multiply with decimal notation, we use the $\boxed{\cdot}$, $\boxed{\times}$, and $\boxed{=}$ keys. To find $4.78 \times 0.34$, for example, we press $\boxed{4}\ \boxed{\cdot}\ \boxed{7}\ \boxed{8}\ \boxed{\times}\ \boxed{\cdot}\ \boxed{3}$ $\boxed{4}\ \boxed{=}$. The display reads $\boxed{1.6252}$, so $4.78 \times 0.34 = 1.6252$.

**Exercises:** Use a calculator to multiply.

**1.**
$$
\begin{array}{r}
5.4 \\
\times \qquad 9 \\
\hline
\end{array}
$$

**2.**
$$
\begin{array}{r}
415 \\
\times\ 16.7 \\
\hline
\end{array}
$$

**3.**
$$
\begin{array}{r}
17.63 \\
\times \qquad 8.1 \\
\hline
\end{array}
$$

**4.** $0.04 \times 12.69$

**5.** $586.4 \times 13.5$

**6.** $4.003 \times 5.1$

To multiply any number by 0.1, 0.01, 0.001, and so on,

a) count the number of decimal places in the tenth, hundredth, or thousandth, and so on, and

$0.001 \times 34.45678$

→ 3 places

b) move the decimal point in the other number that many places to the left.

$0.001 \times 34.45678 = 0.034.45678$

Move 3 places to the left.

$0.001 \times 34.45678 = 0.03445678$

**EXAMPLES** Multiply.

**4.** $0.1 \times 14.605 = 1.4605$    1.4.605

**5.** $0.01 \times 39.051 = 0.39051$

**6.** $0.001 \times (-86.432) = -0.086432$

└ We write an extra zero.

**7.** $0.0001 \times 23.4 = 0.00234$

└ We write two extra zeros.

Do Exercises 4–7.

Do Exercises 4–7.

Multiply.

**4.** $0.1 \times 3.48$

**5.** $0.01 \times 12.49$

**6.** $0.001 \times (-7.6)$

**7.** $0.0001 \times 176.02$

## Multiplying by 10, 100, 1000, and So On

Next, let's consider multiplying by 10, 100, 1000, and so on. Let's look at those products.

$$10 \times 97.34 = 973.4$$
$$100 \times 97.34 = 9734$$
$$1000 \times 97.34 = 97,340$$

Note in each case that the product is *larger* than 97.34. That is, the decimal point in each product is farther to the right than the decimal point in 97.34.

To multiply any number by 10, 100, 1000, and so on,

a) count the number of zeros, and

$1000 \times 34.45678$

→ 3 zeros

b) move the decimal point in the other number that many places to the right.

$1000 \times 34.45678 = 34.456.78$

Move 3 places to the right.

$1000 \times 34.45678 = 34,456.78$

Multiply.

**8.** $10 \times 3.48$

**9.** $100 \times 12.49$

**10.** $1000 \times (-7.6)$

**11.** $10,000 \times 176.02$

**EXAMPLES** Multiply.

**8.** $10 \times 14.605 = 146.05$    14.6.05

**9.** $100 \times 39.051 = 3905.1$

**10.** $1000 \times (-86.432) = -86,432$

**11.** $10,000 \times 23.4 = 234,000$    23.4000. We write three extra zeros.

Do Exercises 8–11.

**Answers**

**4.** 0.348  **5.** 0.1249  **6.** −0.0076
**7.** 0.017602  **8.** 34.8  **9.** 1249
**10.** −7600  **11.** 1,760,200

## b Naming Large Numbers; Money Conversion

### Naming Large Numbers

We often see notation like the following in newspapers and magazines and on television and the Internet.

- U.S. venture-capital firms invested $2.6 billion in alternative-energy startup enterprises during the first three quarters of 2007.
  Source: National Venture Capital Association

- Each day about 39.7 billion person-to-person e-mails, 17.1 billion automated e-mails, and 40.5 billion pieces of spam are sent worldwide.
  Source: IDC

- The largest building in the world is the Pentagon, which has 3.7 million square feet of floor space.

To understand such notation, consider the information in the following table.

---

**NAMING LARGE NUMBERS**

1 hundred $= 100 = 10 \cdot 10 = 10^2$
$\phantom{1 hundred = 100 = 10}$ → 2 zeros

1 thousand $= 1000 = 10 \cdot 10 \cdot 10 = 10^3$
$\phantom{1 thousand = 1000 = 10}$ → 3 zeros

1 million $= 1{,}000{,}000 = 10 \cdot 10 \cdot 10 \cdot 10 \cdot 10 \cdot 10 = 10^6$
$\phantom{1 million = 1,000}$ → 6 zeros

1 billion $= 1{,}000{,}000{,}000 = 10^9$
$\phantom{1 billion = 1,000}$ → 9 zeros

1 trillion $= 1{,}000{,}000{,}000{,}000 = 10^{12}$
$\phantom{1 trillion = 1,000}$ → 12 zeros

---

To convert a large number to standard notation, we proceed as follows.

**EXAMPLE 12**  Losses due to identity fraud totaled $49.3 billion in 2007. Convert 49.3 billion to standard notation.

Source: Javelin Strategy & Research

$$49.3 \text{ billion} = 49.3 \times 1 \text{ billion}$$
$$= 49.3 \times 1,\underbrace{000,000,000}_{\text{9 zeros}}$$
$$= 49,300,000,000 \qquad \text{Moving the decimal point 9 places to the right}$$

Do Exercises 12 and 13.

## Money Conversion

Converting from dollars to cents is like multiplying by 100. To see why, consider $19.43.

| | |
|---|---|
| $19.43 = 19.43 \times \$1 | We think of $19.43 as 19.43 × 1 dollar, or 19.43 × $1. |
| $= 19.43 \times 100¢ | Substituting 100¢ for $1: $1 = 100¢ |
| $= 1943¢ | Multiplying |

> **DOLLARS TO CENTS**
>
> To convert from dollars to cents, move the decimal point two places to the right and change the $ sign in front to a ¢ sign at the end.

**EXAMPLES**  Convert from dollars to cents.

**13.** $189.64 = 18,964¢

**14.** $0.75 = 75¢

Do Exercises 14 and 15.

Converting from cents to dollars is like multiplying by 0.01. To see why, consider 65¢.

| | |
|---|---|
| 65¢ = 65 × 1¢ | We think of 65¢ as 65 × 1 cent, or 65 × 1¢. |
| = 65 × $0.01 | Substituting $0.01 for 1¢: 1¢ = $0.01 |
| = $0.65 | Multiplying |

> **CENTS TO DOLLARS**
>
> To convert from cents to dollars, move the decimal point two places to the left and change the ¢ sign at the end to a $ sign in front.

**EXAMPLES**  Convert from cents to dollars.

**15.** 395¢ = $3.95

**16.** 8503¢ = $85.03

Do Exercises 16 and 17.

---

Convert the number in each sentence to standard notation.

**12.** The largest building in the world is the Pentagon, which has 3.7 million square feet of floor space.

**13.** Each day about 39.7 billion personal e-mails are sent.

Convert from dollars to cents.

**14.** $15.69          **15.** $0.17

Convert from cents to dollars.

**16.** 35¢          **17.** 577¢

*Answers*

**12.** 3,700,000   **13.** 39,700,000,000
**14.** 1569¢   **15.** 17¢   **16.** $0.35   **17.** $5.77

**a** Multiply.

**1.**  8.6
× 7

**2.**  5.7
× 0.8

**3.**  0.8 4
× 8

**4.**  9.4
× 0.6

**5.**  6.3
× 0.0 4

**6.**  9.8
× 0.0 8

**7.**  8 7
× 0.0 0 6

**8.**  1 8.4
× 0.0 7

**9.** $10 \times 23.76$

**10.** $100 \times 3.8798$

**11.** $-1000 \times 583.686852$

**12.** $-0.34 \times 1000$

**13.** $-7.8 \times 100$

**14.** $0.00238 \times (-10)$

**15.** $0.1 \times 89.23$

**16.** $0.01 \times 789.235$

**17.** $0.001 \times 97.68$

**18.** $8976.23 \times 0.001$

**19.** $-78.2 \times 0.01$

**20.** $-0.0235 \times 0.1$

**21.**  3 2.6
× 1 6

**22.**  9.2 8
× 8.6

**23.**  0.9 8 4
× 3.3

**24.**  8.4 8 9
× 7.4

**25.** $(374)(-2.4)$

**26.** $(865)(-1.08)$

**27.** $-749(-0.43)$

**28.** $-978(-20.5)$

**29.**  0.8 7
× 6 4

**30.**  7.2 5
× 6 0

**31.**  4 6.5 0
× 7 5

**32.**  8.2 4
× 7 0 3

**33.**
$$\begin{array}{r} 8\ 1.7 \\ \times\ 0.6\ 1\ 2 \\ \hline \end{array}$$

**34.**
$$\begin{array}{r} 3\ 1.8\ 2 \\ \times\ \ \ 7.1\ 5 \\ \hline \end{array}$$

**35.**
$$\begin{array}{r} 1\ 0.1\ 0\ 5 \\ \times\ 1\ 1.3\ 2\ 4 \\ \hline \end{array}$$

**36.**
$$\begin{array}{r} 1\ 5\ 1.2 \\ \times\ 4.5\ 5\ 5 \\ \hline \end{array}$$

**37.**
$$\begin{array}{r} 1\ 2.3 \\ \times\ 1.0\ 8 \\ \hline \end{array}$$

**38.**
$$\begin{array}{r} 7.8\ 2 \\ \times\ 0.0\ 2\ 4 \\ \hline \end{array}$$

**39.**
$$\begin{array}{r} 3\ 2.4 \\ \times\ \ \ 2.8 \\ \hline \end{array}$$

**40.**
$$\begin{array}{r} 8.0\ 9 \\ \times\ 0.0\ 0\ 7\ 5 \\ \hline \end{array}$$

**41.**
$$\begin{array}{r} 0.0\ 0\ 3\ 4\ 2 \\ \times\ \ \ \ \ \ 0.8\ 4 \\ \hline \end{array}$$

**42.**
$$\begin{array}{r} 2.0\ 0\ 5\ 6 \\ \times\ \ \ \ \ 3.8 \\ \hline \end{array}$$

**43.**
$$\begin{array}{r} 0.3\ 4\ 7 \\ \times\ \ \ 2.0\ 9 \\ \hline \end{array}$$

**44.**
$$\begin{array}{r} 2.5\ 3\ 2 \\ \times\ 1.0\ 6\ 7 \\ \hline \end{array}$$

**45.** $3.005 \times (-0.623)$

**46.** $16.34 \times (-0.000512)$

**47.** $(-6.4)(-15.6)$

**48.** $(-12.7)(-8.9)$

**49.** $1000 \times 45.678$

**50.** $0.001 \times 45.678$

 Convert from dollars to cents.

**51.** $28.88

**52.** $67.43

**53.** $0.66

**54.** $1.78

Convert from cents to dollars.

**55.** 34¢

**56.** 95¢

**57.** 3445¢

**58.** 933¢

**59.** *Back-to-School Spending.* It is estimated that the amount spent on back-to-school supplies by college students and their families was $47.3 billion in 2007. Convert 47.3 billion to standard notation.

**Source:** National Retail Federation

**60.** *Cell-Phone Ads.* Global spending on advertisements appearing on cell phones is projected to be $16.2 billion in 2011. Convert 16.2 billion to standard notation.

**Source:** eMarketer

**61.** *Identity Fraud.* About 9.3 million adults were victims of identity fraud in the United States in 2007. Convert 9.3 million to standard notation.

**Source:** Javelin Strategy & Research

**62.** *Low-Cost Laptops.* It is estimated that 9.2 million low-cost laptop computers will be shipped worldwide in 2012. Convert 9.2 million to standard notation.

**Source:** IDC

**63.** *DVD Sales.* Spending on DVDs totaled $23.4 billion in 2007. Convert 23.4 billion to standard notation.

**Source:** The Digital Entertainment Group

**64.** *Text Messages.* In June 2007, 28.9 billion text messages were sent in the United States. Convert 28.9 billion to standard notation.

**Source:** CTIA

## Skill Maintenance

Calculate.

**65.** $2\frac{1}{3} \cdot 4\frac{4}{5}$  [4.6a]

**66.** $-2\frac{1}{3} \div 4\frac{4}{5}$  [4.6b]

**67.** $4\frac{4}{5} - 2\frac{1}{3}$  [4.5b]

**68.** $4\frac{4}{5} + 2\frac{1}{3}$  [4.5a]

Divide.  [1.5a], [2.5a]

**69.** $2\,4\,\overline{)\,8\,2\,0\,8}$

**70.** $4\,\overline{)\,3\,4\,8}$

**71.** $7\,\overline{)\,3\,1{,}9\,6\,2}$

**72.** $1\,8\,\overline{)\,2\,2{,}6\,2\,6}$

**73.** $49 \div (-7)$

**74.** $-56 \div 4$

## Synthesis

Consider the following names for large numbers in addition to those already discussed in this section:

$$1 \text{ quadrillion} = 1{,}000{,}000{,}000{,}000{,}000 = 10^{15};$$
$$1 \text{ quintillion} = 1{,}000{,}000{,}000{,}000{,}000{,}000 = 10^{18};$$
$$1 \text{ sextillion} = 1{,}000{,}000{,}000{,}000{,}000{,}000{,}000 = 10^{21};$$
$$1 \text{ septillion} = 1{,}000{,}000{,}000{,}000{,}000{,}000{,}000{,}000 = 10^{24}.$$

Find each of the following. Express the answer with a name that is a power of 10.

**75.** (1 trillion) · (1 billion)

**76.** (1 million) · (1 billion)

**77.** (1 trillion) · (1 trillion)

**78.** Is a billion millions the same as a million billions? Explain.

# 5.4

# Division

## OBJECTIVES

**a** Divide using decimal notation.

**b** Solve equations of the type $a \cdot x = b$, where $a$ and $b$ may be in decimal notation.

**c** Simplify expressions using the rules for order of operations.

## a Division

### Whole-Number Divisors

We use the following method when we divide a decimal quantity by a whole number.

To divide by a whole number,

a) place the decimal point directly above the decimal point in the dividend, and

b) divide as though dividing whole numbers.

$$
\begin{array}{r}
0.8\,4 \leftarrow \text{Quotient} \\
\text{Divisor} \rightarrow 7\,\overline{)\,5.8\,8\,} \leftarrow \text{Dividend} \\
5\,6 \\
\overline{\phantom{0}2\,8} \\
2\,8 \\
\overline{\phantom{0}0} \leftarrow \text{Remainder}
\end{array}
$$

**SKILL TO REVIEW**

Objective 1.5a: Divide whole numbers.

Divide.

1. $5\,\overline{)\,2\,4\,5\,}$

2. $2\,3\,\overline{)\,1\,9\,7\,8\,}$

**EXAMPLE 1** Divide: $379.2 \div 8$.

Place the decimal point.

$$
\begin{array}{r}
4\,7.4 \\
8\,\overline{)\,3\,7\,9.2\,} \\
3\,2 \\
\overline{\phantom{0}5\,9} \\
5\,6 \\
\overline{\phantom{00}3\,2} \\
3\,2 \\
\overline{\phantom{00}0}
\end{array}
$$

Divide as though dividing whole numbers.

**EXAMPLE 2** Divide: $82.08 \div (-24)$.

First we consider $82.08 \div 24$.

Place the decimal point.

$$
\begin{array}{r}
3.4\,2 \\
2\,4\,\overline{)\,8\,2.0\,8\,} \\
7\,2 \\
\overline{\phantom{0}1\,0\,0} \\
9\,6 \\
\overline{\phantom{00}4\,8} \\
4\,8 \\
\overline{\phantom{00}0}
\end{array}
$$

Divide as though dividing whole numbers.

Divide.

1. $9\,\overline{)\,5.4\,}$

2. $1\,5\,\overline{)\,2\,2.5\,}$

3. $38.54 \div (-82)$

Since a positive number divided by a negative number is negative, the answer is $-3.42$.

Do Margin Exercises 1–3.

*Answers*

*Skill to Review:*
1. 49    2. 86

*Margin Exercises:*
1. 0.6    2. 1.5    3. −0.47

We can think of a whole-number dividend as having a decimal point at the end with as many zeros as we wish after the decimal point. For example, $12 = 12.0 = 12.00 = 12.000$, and so on. We can also add zeros after the last digit in the decimal portion of a number: $3.6 = 3.60 = 3.600$, and so on.

**EXAMPLE 3** Divide: $30 \div 8$.

$$
\begin{array}{r}
3. \\
8 \overline{)\,3\ 0.} \\
2\ 4 \\
\hline
6
\end{array}
$$
Place the decimal point and divide to find how many ones.

$$
\begin{array}{r}
3. \\
8 \overline{)\,3\ 0.0} \\
2\ 4\ \downarrow \\
\hline
6\ 0
\end{array}
$$
Write an extra zero.

$$
\begin{array}{r}
3.7 \\
8 \overline{)\,3\ 0.0} \\
2\ 4 \\
\hline
6\ 0 \\
5\ 6 \\
\hline
4
\end{array}
$$
Divide to find how many tenths.

$$
\begin{array}{r}
3.7 \\
8 \overline{)\,3\ 0.0\ 0} \\
2\ 4 \\
\hline
6\ 0 \\
5\ 6\ \downarrow \\
\hline
4\ 0
\end{array}
$$
Write another zero.

$$
\begin{array}{r}
3.7\ 5 \\
8 \overline{)\,3\ 0.0\ 0} \\
2\ 4 \\
\hline
6\ 0 \\
5\ 6 \\
\hline
4\ 0 \\
4\ 0 \\
\hline
0
\end{array}
$$
Divide to find how many hundredths.

Check:
$$
\begin{array}{r}
\overset{6\ \ 4}{3.7\ 5} \\
\times \qquad 8 \\
\hline
3\ 0.0\ 0
\end{array}
$$

**EXAMPLE 4** Divide: $-4 \div 25$.

First we consider $4 \div 25$.

$$
\begin{array}{r}
0.1\ 6 \\
2\ 5 \overline{)\,4.0\ 0} \\
2\ 5 \\
\hline
1\ 5\ 0 \\
1\ 5\ 0 \\
\hline
0
\end{array}
$$

Check:
$$
\begin{array}{r}
\overset{1}{\underset{3}{0.1\ 6}} \\
\times \qquad 2\ 5 \\
\hline
8\ 0 \\
3\ 2\ 0 \\
\hline
4.0\ 0
\end{array}
$$

Since a negative number divided by a positive number is negative, the answer is $-0.16$.

Do Exercises 4–6.

**Calculator Corner**

**Division with Decimal Notation**  To use a calculator to divide with decimal notation, we use the $\boxed{\cdot}$, $\boxed{\div}$, and $\boxed{=}$ keys. To find $237.12 \div 5.2$, for example, we press $\boxed{2}\ \boxed{3}\ \boxed{7}\ \boxed{\cdot}\ \boxed{1}\ \boxed{2}\ \boxed{\div}\ \boxed{5}\ \boxed{\cdot}$ $\boxed{2}\ \boxed{=}$. The display reads $\boxed{\quad 45.6 \quad}$, so $237.12 \div 5.2 = 45.6$.

**Exercises:**  Use a calculator to divide.
1. $12.4 \overline{)177.32}$
2. $49 \overline{)125.44}$
3. $1.6 \div 25$
4. $518.472 \div 6.84$

Divide.
4. $25 \overline{)\,8}$

5. $-15 \div 4$

6. $8\ 6 \overline{)\,2\ 1.5}$

*Answers*
4. 0.32   5. $-3.75$   6. 0.25

## Divisors That Are Not Whole Numbers

Consider the division

$$0.2\,4\,\overline{)\,8.2\,0\,8}$$

We write the division as $\dfrac{8.208}{0.24}$. Then we multiply by 1 to change to a whole-number divisor:

The division $0.24\overline{)8.208}$ is the same as $24\overline{)820.8}$.

$$\frac{8.208}{0.24} = \frac{8.208}{0.24} \times \frac{100}{100} = \frac{820.8}{24}.$$

The divisor is now a whole number.

---

To divide when the divisor is not a whole number,

a) move the decimal point (multiply by 10, 100, and so on) to make the divisor a whole number;

$$0.2\,4\,\overline{)\,8.2\,0\,8}$$

Move 2 places to the right.

b) move the decimal point in the dividend the same number of places (multiply the same way); and

$$0.2\,4\,\overline{)\,8.2\,0\,8}$$

Move 2 places to the right.

c) place the decimal point in the quotient directly above the new decimal point in the dividend and divide as though dividing whole numbers.

$$
\begin{array}{r}
3\,4.2 \\
0.2\,4\,\overline{)\,8.2\;0_\wedge 8} \\
7\,2 \\
\hline
1\,0\,0 \\
9\,6 \\
\hline
4\,8 \\
4\,8 \\
\hline
0
\end{array}
$$

(The new decimal point in the dividend is indicated by a caret.)

---

**EXAMPLE 5** Divide: $5.848 \div 8.6$.

$$8.6\,\overline{)\,5.8\,4\,8}$$

Multiply the divisor by 10. (Move the decimal point 1 place.) Multiply the same way in the dividend. (Move 1 place.)

$$
\begin{array}{r}
0.6\,8 \\
8.6\,\overline{)\,5.8_\wedge 4\;8} \\
5\,1\,6 \\
\hline
6\,8\,8 \\
6\,8\,8 \\
\hline
0
\end{array}
$$

Place a decimal point above the new decimal point and then divide.

*Note:* $\dfrac{5.848}{8.6} = \dfrac{5.848}{8.6} \cdot \dfrac{10}{10} = \dfrac{58.48}{86}.$

Do Exercises 7–9.

---

**7. a)** Complete.

$$\frac{3.75}{0.25} = \frac{3.75}{0.25} \times \frac{100}{100}$$

$$= \frac{(\quad)}{25}$$

**b)** Divide.

$$0.2\,5\,\overline{)\,3.7\,5}$$

Divide.

**8.** $0.8\,3\,\overline{)\,4.0\,6\,7}$

**9.** $3.5\,\overline{)\,4\,4.8}$

---

*Answers*

**7. (a)** 375; **(b)** 15  **8.** 4.9  **9.** 12.8

**EXAMPLE 6** Divide: $-12 \div 0.64$.

First we consider $12 \div 0.64$.

$$0.6\,4\,\overline{)\,1\,2.}$$

Place a decimal point at the end of the whole number.

$$0.6\,4\,\overline{)\,1\,2.0\,0}$$

Multiply the divisor by 100 (move the decimal point 2 places). Multiply the same way in the dividend (move 2 places).

$$
\begin{array}{r}
1\ 8.7\ 5 \\
0.6\,4\,\overline{)\,1\,2.0\,0\,0\,0} \\
6\ 4\phantom{000000} \\
\hline
5\ 6\ 0\phantom{0000} \\
5\ 1\ 2\phantom{0000} \\
\hline
4\ 8\ 0\phantom{00} \\
4\ 4\ 8\phantom{00} \\
\hline
3\ 2\ 0 \\
3\ 2\ 0 \\
\hline
0
\end{array}
$$

Place a decimal point above the new decimal point and then divide.

Since $12 \div 0.64 = 18.75$, we have $-12 \div 0.64 = -18.75$.

Do Exercise 10.

**10.** Divide. $-25 \div 1.6$.

### Dividing by 10, 100, 1000, and So On

It is often helpful to be able to divide quickly by a ten, hundred, or thousand, or by a tenth, hundredth, or thousandth. Each procedure we use is based on multiplying by 1. Consider the following example:

$$\frac{23.789}{1000} = \frac{23.789}{1000} \cdot \frac{1000}{1000} = \frac{23,789}{1,000,000} = 0.023789.$$

We are dividing by a number greater than 1: The result is *smaller* than 23.789.

> To divide by 10, 100, 1000, and so on,
>
> **a)** count the number of zeros in the divisor, and
>
> $$\frac{713.49}{100}$$
>
> 2 zeros
>
> **b)** write the quotient by moving the decimal point in the dividend that number of places to the left.
>
> $$\frac{713.49}{100}, \qquad 7.13.49, \qquad \frac{713.49}{100} = 7.1349$$
>
> 2 places to the left

**EXAMPLE 7** Divide: $\dfrac{0.0104}{10}$.

$$\frac{0.0104}{10}, \qquad 0.0.0104, \qquad \frac{0.0104}{10} = 0.00104$$

1 zero       1 place to the left

*Answer*

**10.** $-15.625$

### Dividing by 0.1, 0.01, 0.001, and So On

Now consider the following example:

$$\frac{23.789}{0.01} = \frac{23.789}{0.01} \cdot \frac{100}{100} = \frac{2378.9}{1} = 2378.9.$$

We are dividing by a number less than 1: The result is *larger* than 23.789. We use the following procedure.

---

To divide by 0.1, 0.01, 0.001, and so on,

a) count the number of decimal places in the divisor, and

$$\frac{713.49}{0.001}$$

→ 3 places

b) write the quotient by moving the decimal point in the dividend that number of places to the right.

$$\frac{713.49}{0.001},$$   713.490.   $$\frac{713.49}{0.001} = 713,490$$

3 places to the right to change 0.001 to 1

---

**EXAMPLE 8**   Divide: $\dfrac{23.738}{0.001}$.

$$\frac{23.738}{0.001},$$   23.738.   $$\frac{23.738}{0.001} = 23,738$$

3 places   3 places to the right

Do Exercises 11–14.

**Divide.**

11. $\dfrac{0.1278}{0.01}$

12. $\dfrac{0.1278}{100}$

13. $\dfrac{98.47}{1000}$

14. $\dfrac{-6.7832}{0.1}$

---

### (b) Solving Equations

Now let's solve equations of the type $a \cdot x = b$, where $a$ and $b$ may be in decimal notation. Proceeding as before, we divide by $a$ on both sides.

**EXAMPLE 9**   Solve: $8 \cdot x = 27.2$.

We have

$$\frac{8 \cdot x}{8} = \frac{27.2}{8} \qquad \text{Dividing by 8 on both sides}$$

$$x = 3.4.$$

$$
\begin{array}{r}
3.4 \\
8 \overline{)\ 2\ 7.2} \\
2\ 4 \phantom{.2}\\
\hline
3\ 2 \\
3\ 2 \\
\hline
0
\end{array}
$$

The solution is 3.4.

**EXAMPLE 10**  Solve: $2.9 \cdot t = -0.14616$.

We have

$$\frac{2.9 \cdot t}{2.9} = \frac{-0.14616}{2.9} \qquad \text{Dividing by 2.9 on both sides}$$

$$t = -0.0504.$$

```
        0. 0 5 0 4
2.9 ) 0.1ʌ4 6 1 6
        1 4 5
        ———
          1 1 6
          1 1 6
          ———
              0
```

Since $0.14616 \div 2.9 = 0.0504$, then $-0.14616 \div 2.9 = -0.0504$. Thus the solution is $-0.0504$.

Do Exercises 15 and 16.

Solve.
**15.** $100 \cdot x = 78.314$

**16.** $0.25 \cdot y = -276.4$

## (c) Order of Operations: Decimal Notation

The same rules for order of operations used with integers and fraction notation apply when simplifying expressions with decimal notation.

---

**RULES FOR ORDER OF OPERATIONS**

1. Do all calculations within grouping symbols before operations outside.
2. Evaluate all exponential expressions.
3. Do all multiplications and divisions in order from left to right.
4. Do all additions and subtractions in order from left to right.

---

**EXAMPLE 11**  Simplify: $2.56 \times 25.6 \div 25{,}600 \times 256$.

There are no exponents or parentheses, so we multiply and divide from left to right:

$$2.56 \times 25.6 \div 25{,}600 \times 256 = 65.536 \div 25{,}600 \times 256$$

Doing all multiplications and divisions in order from left to right

$$= 0.00256 \times 256$$

$$= 0.65536.$$

**EXAMPLE 12**  Simplify: $(5 - 0.06) \div (-2) + 3.42 \times 0.1$.

$$(5 - 0.06) \div (-2) + 3.42 \times 0.1 = 4.94 \div (-2) + 3.42 \times 0.1$$

Carrying out the operation inside parentheses

$$= -2.47 + 0.342$$

Doing all multiplications and divisions in order from left to right

$$= -2.128$$

Adding

*Answers*
**15.** 0.78314   **16.** −1105.6

**Simplify.**

**17.** $625 \div 62.5 \times 25 \div 6250$

**18.** $0.25 \cdot (-1 + 0.08) - 0.0274$

**19.** $20^2 - 3.4^2 +$
$\{2.5[20(9.2 - 5.6)] + 5(10 - 5)\}$

**20. Mountains in Peru.** Refer to the figure in Example 14. Find the average height of the mountains, in meters.

**EXAMPLE 13**   Simplify: $10^2 \times \{[(3 - 0.24) \div 2.4] - (0.21 - 0.092)\}$.

$10^2 \times \{[(3 - 0.24) \div 2.4] - (0.21 - 0.092)\}$

$= 10^2 \times \{[2.76 \div 2.4] - 0.118\}$    Doing the calculations in the innermost parentheses first

$= 10^2 \times \{1.15 - 0.118\}$    Again, doing the calculation in the innermost grouping symbols

$= 10^2 \times 1.032$    Subtracting inside the grouping symbols

$= 100 \times 1.032$    Evaluating the exponential expression

$= 103.2$    Multiplying

Do Exercises 17–19.

**EXAMPLE 14**   *Mountains in Peru.*   The figure below shows a range of very high mountains in Peru, together with their altitudes, given both in feet and in meters. Find the average height of these mountains, in feet.
Source: *National Geographic*, July 1968, p. 130

Nev. Sara Sara, 18,060 ft 5,505 m
Nev. Coropuna, 21,079 ft 6,425 m
Nevado Ampato, 20,700 ft 6,309 m
Nev. Chachani, 19,931 ft 6,075 m
Volcan Misti, 19,101 ft 5,822 m
Nev. Pichu Pichu, 18,600 ft 5,669 m
Arequipa
*Pacific Ocean*
Peru
*South America*
Area enlarged
Scale varies in this perspective.
SOURCE: WOOD RONASVILLE HARLIN INC/NGS Image Collection

The **average** of a set of numbers is the sum of the numbers divided by the number of addends. (See Section 1.9.) We find the sum of the heights divided by the number of addends, 6:

$$\frac{18{,}060 + 21{,}079 + 20{,}700 + 19{,}931 + 19{,}101 + 18{,}600}{6} = \frac{117{,}471}{6} = 19{,}578.5.$$

Thus the average height of these mountains is 19,578.5 ft.

Do Exercise 20.

*Answers*

**17.** 0.04   **18.** −0.2574   **19.** 593.44
**20.** 5967.5 m

**a** Divide.

1. $2 \overline{)\, 5.9\, 8}$

2. $5 \overline{)\, 1\, 8}$

3. $4 \overline{)\, 9\, 5.1\, 2}$

4. $8 \overline{)\, 2\, 5.9\, 2}$

5. $1\, 2 \overline{)\, 8\, 9.7\, 6}$

6. $2\, 3 \overline{)\, 2\, 5.0\, 7}$

7. $3\, 3 \overline{)\, 2\, 3\, 7.6}$

8. $-12.4 \div 4$

9. $-9.144 \div 8$

10. $4.5 \div 9$

11. $12.123 \div (-3)$

12. $5.6 \div (-7)$

13. $5 \overline{)\, 0.3\, 5}$

14. $0.0\, 4 \overline{)\, 1.6\, 8}$

15. $0.1\, 2 \overline{)\, 8.4}$

16. $0.3\, 6 \overline{)\, 2.8\, 8}$

17. $3.4 \overline{)\, 6\, 8}$

18. $0.2\, 5 \overline{)\, 5}$

19. $-6 \div (-15)$

20. $-1.8 \div (-12)$

21. $3\, 6 \overline{)\, 1\, 4.7\, 6}$

22. $5\, 2 \overline{)\, 1\, 1\, 9.6}$

23. $3.2 \overline{)\, 2\, 7.2}$

24. $8.5 \overline{)\, 2\, 7.2}$

25. $4.2 \overline{)\, 3\, 9.0\, 6}$

26. $4.8 \overline{)\, 0.1\, 1\, 0\, 4}$

27. $-5 \div 8$

28. $-3 \div 8$

29. $0.4\, 7 \overline{)\, 0.1\, 2\, 2\, 2}$

30. $1.0\, 8 \overline{)\, 0.5\, 4}$

31. $4.8 \overline{)\, 7\, 5}$

**32.** $0.2\ 8\ \overline{)\ 6\ 3}$

**33.** $0.0\ 3\ 2\ \overline{)\ 0.0\ 7\ 4\ 8\ 8}$

**34.** $0.0\ 1\ 7\ \overline{)\ 1.5\ 8\ 1}$

**35.** $8\ 2\ \overline{)\ 3\ 8.5\ 4}$

**36.** $3\ 4\ \overline{)\ 0.1\ 4\ 6\ 2}$

**37.** $\dfrac{213.4567}{1000}$

**38.** $\dfrac{769.3265}{1000}$

**39.** $\dfrac{-23.59}{10}$

**40.** $\dfrac{-83.57}{10}$

**41.** $\dfrac{426.487}{100}$

**42.** $\dfrac{591.348}{100}$

**43.** $\dfrac{-16.94}{0.1}$

**44.** $\dfrac{-100.7604}{0.1}$

**45.** $\dfrac{1.0237}{0.001}$

**46.** $\dfrac{3.4029}{0.001}$

**47.** $\dfrac{-42.561}{0.01}$

**48.** $\dfrac{-98.473}{0.01}$

 Solve.

**49.** $4.2 \cdot x = 39.06$

**50.** $36 \cdot y = 14.76$

**51.** $1000 \cdot y = -9.0678$

**52.** $-789.23 = 0.25 \cdot q$

**53.** $1048.8 = 23 \cdot t$

**54.** $28.2 \cdot x = 423$

 Simplify.

**55.** $14 \times (82.6 + 67.9)$

**56.** $(26.2 - 14.8) \times 12$

**57.** $0.003 - 3.03 \div 0.01$

**58.** $9.94 + 118.8 \div (-6.2 - 4.6) - 0.9$

**59.** $42 \times (10.6 + 0.024)$

**60.** $(18.6 - 4.9) \times 13$

**61.** $4.2 \times 5.7 + 0.7 \div 3.5$

**62.** $123.3 - 4.24 \times 1.01$

**63.** $-9.0072 + 0.04 \div 0.1^2$

**64.** $-12 \div 0.03 + 12 \times 0.03^2$

**65.** $(8 - 0.04)^2 \div 4 + 8.7 \times 0.4$

**66.** $(5 - 2.5)^2 \div 100 + 0.1 \times 6.5$

**67.** $86.7 + 4.22 \times (9.6 - 0.03)^2$

**68.** $2.48 \div (1 - 0.504) + 24.3 - 11 \times 2$

**69.** $4 \div (-0.4) + 0.1 \times 5 - 0.1^2$

**70.** $6 \times (-0.9) + 0.1 \div 4 - 0.2^3$

**71.** $5.5^2 \times [(6 - 4.2) \div 0.06 + 0.12]$

**72.** $12^2 \div (12 + 2.4) - [(2 - 1.6) \div 0.8]$

**73.** $200 \times \{[(4 - 0.25) \div 2.5] - (4.5 - 4.025)\}$

**74.** $0.03 \times \{10 \times 50.2 - [(8 - 7.5) \div 0.05]\}$

**75.** Find the average of $1276.59, $1350.49, $1123.78, and $1402.58.

**76.** Find the average weight of two wrestlers who weigh 308 lb and 296.4 lb.

**77.** *Camping in National Parks.* Camping in national parks has declined steadily in recent years. The graph below shows the numbers of overnight camping stays in Park Service campgrounds in the National Park System from 2002 to 2006. Find the average number of stays per year during this period.

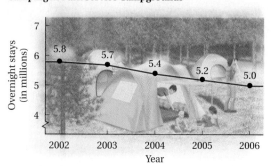

**Camping in Park Service Campgrounds**

SOURCE: U.S. National Park Service

**78.** *Unclaimed Prizes.* Each year millions of lottery prize dollars go unclaimed. The table below shows the five states with the most unclaimed prize money for a recent year. Find the average amount that went unclaimed in these states.

| STATE | UNCLAIMED PRIZES (in millions) |
|---|---|
| New York | $73.6 |
| Texas | 58.9 |
| Florida | 48.4 |
| Ohio | 33.4 |
| California | 29.4 |

SOURCE: Individual state lotteries

## Skill Maintenance

Simplify.  [3.5b]

**79.** $\dfrac{36}{42}$

**80.** $\dfrac{56}{64}$

**81.** $-\dfrac{38}{146}$

**82.** $-\dfrac{92}{124}$

Find the prime factorization.  [3.1d]

**83.** 684

**84.** 162

**85.** 2007

**86.** 2005

**87.** Add: $10\frac{1}{2} + 4\frac{5}{8}$.  [4.5a]

**88.** Subtract: $10\frac{1}{2} - 4\frac{5}{8}$.  [4.5b]

## Synthesis

Simplify.

**89.** ▦ $9.0534 - 2.041^2 \times 0.731 \div 1.043^2$

**90.** ▦ $23.042(7 - 4.037 \times 1.46 - 0.932^2)$

In Exercises 91–94, find the missing value.

**91.** $439.57 \times 0.01 \div 1000 \times \square = 4.3957$

**92.** $5.2738 \div 0.01 \times 1000 \div \square = 52.738$

**93.** $0.0329 \div 0.001 \times 10^4 \div \square = 3290$

**94.** $0.0047 \times 0.01 \div 10^4 \times \square = 4.7$

# Mid-Chapter Review

## Concept Reinforcement

Determine whether each statement is true or false.

_____ **1.** In the number 308.00567, the digit 6 names the tens place.   [5.1a]

_____ **2.** When writing a word name for decimal notation, we write the word "and" for the decimal point.   [5.1a]

_____ **3.** To multiply any number by 10, 100, 1000, and so on, count the number of zeros and move the decimal point that many places to the right.   [5.3a]

## Guided Solutions

Fill in each blank with the number that creates a correct statement or solution.

**4.** Solve: $y + 12.8 = 23.35$.   [5.2d]

$$y + 12.8 = 23.35$$
$$y + 12.8 - \square = 23.35 - \square \qquad \text{Subtracting 12.8 on both sides}$$
$$y + \square = \square \qquad \text{Carrying out the subtractions}$$
$$y = \square$$

**5.** Simplify: $5.6 + 4.3 \times (6.5 - 0.25)^2$.   [5.4c]

$$5.6 + 4.3 \times (6.5 - 0.25)^2 = 5.6 + 4.3 \times (\square)^2 \qquad \text{Carrying out the operation inside parentheses}$$
$$= 5.6 + 4.3 \times \square \qquad \text{Evaluating the exponential expression}$$
$$= 5.6 + \square \qquad \text{Multiplying}$$
$$= \square \qquad \text{Adding}$$

## Mixed Review

**6.** Usain Bolt of Jamaica set a world record of 9.69 sec in the men's 100-m race at the 2008 Summer Olympics. Write a word name for 9.69.   [5.1a]

**Source:** beijing2008.cn

**7.** After removing coins from their pockets at security checkpoints in U.S. airports, travelers forgot to reclaim coins totaling about $1.05 million from 2005 to 2007. Convert 1.05 million to standard notation.   [5.3b]

**Source:** Transportation Security Administration

Write fraction notation.   [5.1b]

**8.** 4.53

**9.** −0.287

Which number is larger?   [5.1c]

**10.** 0.07, 0.13

**11.** −5.2, −5.09

Write decimal notation.   [5.1b]

**12.** $\dfrac{7}{10}$

**13.** $\dfrac{-639}{100}$

**14.** $35\dfrac{67}{100}$

**15.** $8\dfrac{2}{1000}$

Round 28.4615 to the nearest:   [5.1d]

**16.** Thousandth.

**17.** Hundredth.

**18.** Tenth.

**19.** One

Add. [5.2a], [5.2c]

**20.**
```
  4 7.6 3 8
+    2.4 5 7
```

**21.**
```
    1 5.6
  2 3 4.7 2 9
      3.0 8
+ 9 6 1.4 5 3
```

**22.** $-4.5 + 0.728$

**23.** $16 + 0.34 + 1.9$

Subtract. [5.2b], [5.2c]

**24.**
```
  3 2 1.5 7
-    4 9.3 8
```

**25.**
```
  5.6
- 0.0 0 7
```

**26.** $34.3 - 18.75$

**27.** $-49.07 - 9.7$

Multiply. [5.3a]

**28.**
```
    4.6
×   0.9
```

**29.**
```
    1 5.3
×   6.0 7
```

**30.** $-100 \times 81.236$

**31.** $0.1 \times 29.37$

Divide. [5.4a]

**32.** $20.24 \div 4$

**33.** $-21.76 \div 6.8$

**34.** $76.34 \div 0.1$

**35.** $914.036 \div 1000$

**36.** Convert $20.45 to cents. [5.3b]

**37.** Convert 147¢ to dollars. [5.3b]

**38.** Solve: $46.3 + x = -59$. [5.2d]

**39.** Solve: $42.84 = 5.1 \cdot y$. [5.4b]

Simplify. [5.4c]

**40.** $6.594 + 0.5318 \div 0.01$

**41.** $7.3 \times 4.6 - 0.8 \div (-3.2)$

# Understanding Through Discussion and Writing

**42.** A fellow student rounds 236.448 to the nearest one and gets 237. Explain the possible error. [5.1d]

**43.** Explain the error in the following: [5.2b]
Subtract.
$$73.089 - 5.0061 = 2.3028$$

**44.** Explain why $10 \div 0.2 = 100 \div 2$. [5.4a]

**45.** Kayla made these two computational mistakes:
$$0.247 \div 0.1 = 0.0247; \quad 0.247 \div 10 = 2.47.$$
In each case, how could you convince her that a mistake has been made? [5.4a]

# 5.5

## Converting from Fraction Notation to Decimal Notation

### a Fraction Notation to Decimal Notation

When a denominator has no prime factors other than 2's and 5's, we can find decimal notation by multiplying by 1. We multiply to get a denominator that is a power of ten, like 10, 100, or 1000.

**EXAMPLE 1** Find decimal notation for $\frac{3}{5}$.

$$\frac{3}{5} = \frac{3}{5} \cdot \frac{2}{2} = \frac{6}{10} = 0.6$$    $5 \cdot 2 = 10$, so we use $\frac{2}{2}$ for 1 to get a denominator of 10.

**EXAMPLE 2** Find decimal notation for $\frac{7}{20}$.

$$\frac{7}{20} = \frac{7}{20} \cdot \frac{5}{5} = \frac{35}{100} = 0.35$$    $20 \cdot 5 = 100$, so we use $\frac{5}{5}$ for 1 to get a denominator of 100.

**EXAMPLE 3** Find decimal notation for $-\frac{87}{25}$.

$$-\frac{87}{25} = -\frac{87}{25} \cdot \frac{4}{4} = -\frac{348}{100} = -3.48$$    We use $\frac{4}{4}$ for 1 to get a denominator of 100.

**EXAMPLE 4** Find decimal notation for $\frac{9}{40}$.

$$\frac{9}{40} = \frac{9}{40} \cdot \frac{25}{25} = \frac{225}{1000} = 0.225$$    $40 \cdot 25 = 1000$, so we use $\frac{25}{25}$ for 1 to get a denominator of 1000.

> Do Margin Exercises 1–4.

We can always divide to find decimal notation.

**EXAMPLE 5** Find decimal notation for $\frac{3}{5}$.

$$\frac{3}{5} = 3 \div 5 \qquad \begin{array}{r} 0.6 \\ 5 \overline{)\ 3.0} \\ \underline{3\ 0} \\ 0 \end{array} \qquad \frac{3}{5} = 0.6$$

**EXAMPLE 6** Find decimal notation for $-\frac{7}{8}$.

First we consider $\frac{7}{8}$.

$$\frac{7}{8} = 7 \div 8 \qquad \begin{array}{r} 0.8\ 7\ 5 \\ 8 \overline{)\ 7.0\ 0\ 0} \\ \underline{6\ 4} \\ 6\ 0 \\ \underline{5\ 6} \\ 4\ 0 \\ \underline{4\ 0} \\ 0 \end{array} \qquad \frac{7}{8} = 0.875$$

Since $\frac{7}{8} = 0.875$, we have $-\frac{7}{8} = -0.875$.

> Do Exercises 5 and 6.

## OBJECTIVES

**a** Convert from fraction notation to decimal notation.

**b** Round numbers named by repeating decimals in problem solving.

**c** Calculate using fraction notation and decimal notation together.

**SKILL TO REVIEW**
Objective 5.4a: Divide using decimal notation.

Divide.
  1. $3 \div 4$        2. $25 \div 8$

Find decimal notation. Use multiplying by 1.

  1. $\frac{4}{5}$        2. $\frac{9}{20}$

  3. $\frac{11}{40}$        4. $-\frac{33}{25}$

Find decimal notation.

  5. $-\frac{2}{5}$        6. $\frac{3}{8}$

*Answers*

*Skill to Review:*
**1.** 0.75   **2.** 3.125

*Margin Exercises:*
**1.** 0.8   **2.** 0.45   **3.** 0.275
**4.** −1.32   **5.** −0.4   **6.** 0.375

STUDY TIPS

**MAKING POSITIVE CHOICES**

Making these choices will contribute to your success in this course.

- Choose to make a strong commitment to learning.
- Choose to place the primary responsibility for learning on yourself.
- Choose to allocate the proper amount of time to learn.

In Examples 5 and 6, the division *terminated,* meaning that eventually we got a remainder of 0. A **terminating decimal** occurs when the denominator has only 2's or 5's, or both, as factors, as in $\frac{17}{25}$, $\frac{5}{8}$, or $\frac{83}{100}$. This assumes that the fraction notation has been simplified.

If division does *not* lead to a remainder of 0, but instead leads to a repeating pattern of nonzero remainders, we have a **repeating decimal**.

**EXAMPLE 7**  Find decimal notation for $\frac{5}{6}$.

$$\frac{5}{6} = 5 \div 6 \qquad 6\overline{)\begin{array}{l} 0.8\ 3\ 3 \\ 5.0\ 0\ 0 \end{array}}$$

$$\begin{array}{r} 4\ 8 \\ \hline 2\ 0 \\ 1\ 8 \\ \hline 2\ 0 \\ 1\ 8 \\ \hline 2 \end{array}$$

Since 2 keeps reappearing as a remainder, the digit 3 repeats in the quotient and will continue to do so; therefore,

$$\frac{5}{6} = 0.83333\dots.$$

The red dots indicate an endless sequence of digits in the quotient. When there is a repeating pattern, the dots are often replaced by a bar to indicate the repeating part—in this case, only the 3:

$$\frac{5}{6} = 0.8\overline{3}.$$

Do Exercises 7 and 8.

**EXAMPLE 8**  Find decimal notation for $-\frac{4}{11}$.

First we consider $\frac{4}{11}$.

$$\frac{4}{11} = 4 \div 11 \qquad 11\overline{)\begin{array}{l} 0.3\ 6\ 3\ 6 \\ 4.0\ 0\ 0\ 0 \end{array}}$$

$$\begin{array}{r} 3\ 3 \\ \hline 7\ 0 \\ 6\ 6 \\ \hline 4\ 0 \\ 3\ 3 \\ \hline 7\ 0 \\ 6\ 6 \\ \hline 4 \end{array}$$

Since 7 and 4 keep repeating as remainders, the sequence of digits "36" repeats in the quotient, and

$$\frac{4}{11} = 0.363636\dots, \text{ so we have } -\frac{4}{11} = -0.363636\dots, \text{ or } -0.\overline{36}.$$

Do Exercises 9 and 10.

Find decimal notation.

7. $\dfrac{1}{6}$          8. $\dfrac{2}{3}$

Find decimal notation.

9. $\dfrac{5}{11}$          10. $-\dfrac{12}{11}$

*Answers*

7. $0.1\overline{6}$   8. $0.\overline{6}$   9. $0.\overline{45}$   10. $-1.\overline{09}$

**EXAMPLE 9** Find decimal notation for $\frac{5}{7}$.

$$
\begin{array}{r}
0.7\ 1\ 4\ 2\ 8\ 5 \\
7\ )\overline{5.0\ 0\ 0\ 0\ 0\ 0} \\
\underline{4\ 9} \\
1\ 0 \\
\underline{7} \\
3\ 0 \\
\underline{2\ 8} \\
2\ 0 \\
\underline{1\ 4} \\
6\ 0 \\
\underline{5\ 6} \\
4\ 0 \\
\underline{3\ 5} \\
5
\end{array}
$$

Since 5 appears again as a remainder, the sequence of digits "714285" repeats in the quotient, and

$$\frac{5}{7} = 0.714285714285\ldots, \text{ or } 0.\overline{714285}.$$

The length of a repeating part can be very long—too long to find on a calculator. An example is $\frac{5}{97}$, which has a repeating part of 96 digits.

> Do Exercise 11.

**11.** Find decimal notation for $\frac{3}{7}$.

## b  Rounding in Problem Solving

In applied problems, repeating decimals are rounded to get approximate answers. To round a repeating decimal, we can extend the decimal notation at least one place past the rounding digit, and then round as before.

**EXAMPLES** Round each of the following to the nearest tenth, hundredth, and thousandth.

|  | Nearest tenth | Nearest hundredth | Nearest thousandth |
|---|---|---|---|
| **10.** $0.8\overline{3} = 0.83333\ldots$ | 0.8 | 0.83 | 0.833 |
| **11.** $0.\overline{09} = 0.090909\ldots$ | 0.1 | 0.09 | 0.091 |
| **12.** $-0.\overline{714285} = -0.714285714285\ldots$ | −0.7 | −0.71 | −0.714 |

> Do Exercises 12-14.

Round each to the nearest tenth, hundredth, and thousandth.

**12.** $0.\overline{6}$

**13.** $0.\overline{80}$

**14.** $-6.\overline{245}$

### Converting Ratios to Decimal Notation

When solving applied problems, we often convert ratios to decimal notation.

**EXAMPLE 13** *Gas Mileage.* A car travels 457 mi on 16.4 gal of gasoline. The ratio of number of miles driven to amount of gasoline used is *gas mileage.* Find the gas mileage and convert the ratio to decimal notation rounded to the nearest tenth.

$$\frac{\text{Miles driven}}{\text{Gasoline used}} = \frac{457}{16.4} \approx 27.86 \qquad \text{Dividing to 2 decimal places}$$

$$\approx 27.9 \qquad \text{Rounding to 1 decimal place}$$

The gas mileage is about 27.9 miles to the gallon.

*Answers*

**11.** $0.\overline{428571}$   **12.** $0.7; 0.67; 0.667$
**13.** $0.8; 0.81; 0.808$   **14.** $-6.2; -6.25; -6.245$

## EXAMPLE 14  *Wildfires.*

A severe thunderstorm system that moved through northern and central California on June 20, 2008, produced over 6000 lightning strikes. This sparked 2096 wildfires that burned about 1,200,000 acres. Find the ratio of the number of acres burned to the number of fires and convert it to decimal notation rounded to the nearest thousandth.

**Source:** California Department of Forestry and Fire Prevention

We have

$$\frac{\text{Acres burned}}{\text{Number of fires}} = \frac{1,200,000 \text{ acres}}{2096 \text{ fires}}$$
$$\approx 572.51908.$$

About 572.519 acres burned per fire.

Do Exercises 15 and 16.

15. **Coin Tossing.** A coin is tossed 51 times. It lands heads 26 times. Find the ratio of heads to tosses and convert it to decimal notation rounded to the nearest thousandth. (This is also the experimental probability of getting heads.)

16. **Gas Mileage.** A car travels 380 mi on 15.7 gal of gasoline. Find the gasoline mileage and convert the ratio to decimal notation rounded to the nearest tenth.

### Averages

When finding an average, we may at times need to round an answer.

## EXAMPLE 15  *Peer-to-Peer Loans.*  Peer-to-peer lending, in which individuals make loans to other individuals via the Internet, is beginning to soar. The graph below shows the amounts of peer-to-peer loans made from 2005 to 2010. Find the average amount loaned per year during this period. Round to the nearest thousandth.

**Peer-to-Peer Loans**

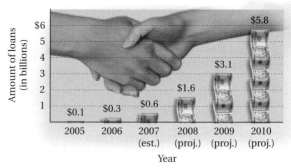

SOURCE: Celent

We add the amounts of the loans and divide by the number of addends, 6. The average is

$$\frac{0.1 + 0.3 + 0.6 + 1.6 + 3.1 + 5.8}{6} = \frac{11.5}{6} = 1.91\overline{6} \approx 1.917.$$

Peer-to-peer loans made from 2005 to 2010 averaged a total of about $1.917 billion per year.

Do Exercise 17.

17. **Peer-to-Peer Loans.** Refer to the data on peer-to-peer loans in Example 15. Find the average amount loaned per year from 2007 to 2009. Round to the nearest thousandth.

*Answers*

**15.** 0.510  **16.** 24.2 miles per gallon
**17.** About $1.767 billion

## (c) Calculations with Fraction Notation and Decimal Notation Together

In certain kinds of calculations, fraction notation and decimal notation might occur together. In such cases, there are at least three ways in which we might proceed.

**EXAMPLE 16** Calculate: $\frac{2}{3} \times 0.576$.

METHOD 1: One way to do this calculation is to convert the fraction notation to decimal notation so that both numbers are in decimal notation. Since $\frac{2}{3}$ converts to repeating decimal notation, it is first rounded to some chosen decimal place. We choose three decimal places because 0.576 has three decimal places. Then, using decimal notation, we multiply.

$$\frac{2}{3} \times 0.576 = 0.\overline{6} \times 0.576 \approx 0.667 \times 0.576 = 0.384192$$

METHOD 2: A second way to do this calculation is to convert the decimal notation to fraction notation so that both numbers are in fraction notation. The answer can be left in fraction notation and simplified, or we can convert to decimal notation and round, if appropriate.

$$\frac{2}{3} \times 0.576 = \frac{2}{3} \cdot \frac{576}{1000} = \frac{2 \cdot 576}{3 \cdot 1000}$$

$$= \frac{2 \cdot 2 \cdot 2 \cdot 2 \cdot 2 \cdot 2 \cdot 2 \cdot 3 \cdot 3}{2 \cdot 2 \cdot 2 \cdot 3 \cdot 5 \cdot 5 \cdot 5}$$

$$= \frac{2 \cdot 2 \cdot 2 \cdot 3}{2 \cdot 2 \cdot 2 \cdot 3} \cdot \frac{2 \cdot 2 \cdot 2 \cdot 2 \cdot 3}{5 \cdot 5 \cdot 5}$$

$$= 1 \cdot \frac{2 \cdot 2 \cdot 2 \cdot 3}{5 \cdot 5 \cdot 5}$$

$$= \frac{2 \cdot 2 \cdot 2 \cdot 2 \cdot 3}{5 \cdot 5 \cdot 5} = \frac{48}{125}, \text{ or } 0.384$$

METHOD 3: A third way to do this calculation is to treat 0.576 as $\frac{0.576}{1}$. Then we multiply 0.576 by 2, and divide the result by 3.

$$\frac{2}{3} \times 0.576 = \frac{2}{3} \times \frac{0.576}{1} = \frac{2 \times 0.576}{3} = \frac{1.152}{3} = 0.384$$

Note that we get an exact answer with methods 2 and 3, but method 1 gives an approximation since we rounded decimal notation for $\frac{2}{3}$.

Do Exercise 18.

**EXAMPLE 17** Calculate: $\frac{2}{3} \times 0.576 - 3.287 \div \frac{4}{5}$.

We use the rules for order of operations, first doing the multiplication and then the division. Then we subtract.

$$\frac{2}{3} \times 0.576 - 3.287 \div \frac{4}{5} = 0.384 - 3.287 \cdot \frac{5}{4}$$

$$= 0.384 - 4.10875$$

$$= -3.72475$$

Method 3:
$\frac{2}{3} \times \frac{0.576}{1} = 0.384$;
$\frac{3.287}{1} \times \frac{5}{4} = 4.10875$

Do Exercises 19 and 20.

**18.** Calculate: $\frac{5}{6} \times 0.864$.

Calculate.

**19.** $\frac{1}{3} \times 0.384 + \frac{5}{8} \times 0.6784$

**20.** $\frac{5}{6} \times 0.864 - 14.3 \div \frac{8}{5}$

*Answers*

**18.** 0.72    **19.** 0.552    **20.** −8.2175

For Extra Help

MyMathLab

PRACTICE    WATCH    DOWNLOAD    READ    REVIEW

**a**    Find decimal notation.

**1.** $\dfrac{23}{100}$        **2.** $\dfrac{9}{100}$        **3.** $\dfrac{3}{5}$        **4.** $\dfrac{19}{20}$        **5.** $-\dfrac{13}{40}$        **6.** $-\dfrac{3}{16}$

**7.** $\dfrac{1}{5}$        **8.** $\dfrac{4}{5}$        **9.** $-\dfrac{17}{20}$        **10.** $-\dfrac{11}{20}$        **11.** $\dfrac{3}{8}$        **12.** $\dfrac{7}{8}$

**13.** $-\dfrac{39}{40}$        **14.** $-\dfrac{31}{40}$        **15.** $\dfrac{13}{25}$        **16.** $\dfrac{61}{125}$        **17.** $-\dfrac{2502}{125}$        **18.** $-\dfrac{181}{200}$

**19.** $\dfrac{1}{4}$        **20.** $\dfrac{1}{2}$        **21.** $-\dfrac{29}{25}$        **22.** $-\dfrac{37}{25}$        **23.** $\dfrac{19}{16}$        **24.** $\dfrac{5}{8}$

**25.** $\dfrac{4}{15}$        **26.** $\dfrac{7}{9}$        **27.** $\dfrac{1}{3}$        **28.** $\dfrac{1}{9}$        **29.** $-\dfrac{4}{3}$        **30.** $-\dfrac{8}{9}$

**31.** $-\dfrac{7}{6}$        **32.** $-\dfrac{7}{11}$        **33.** $\dfrac{4}{7}$        **34.** $\dfrac{14}{11}$        **35.** $-\dfrac{11}{12}$        **36.** $-\dfrac{5}{12}$

**b**

**37.–47.** *Odds.*    Round each answer of the odd-numbered Exercises 25–35 to the nearest tenth, hundredth, and thousandth.

**38.–48.** *Evens.*    Round each answer of the even-numbered Exercises 26–36 to the nearest tenth, hundredth, and thousandth.

Round each to the nearest tenth, hundredth, and thousandth.

**49.** $0.1\overline{8}$        **50.** $0.8\overline{3}$        **51.** $-0.2\overline{7}$        **52.** $-3.5\overline{4}$

**53.** For this set of people, what is the ratio, in decimal notation rounded to the nearest thousandth, where appropriate, of:

  **a)** women to the total number of people?
  **b)** women to men?
  **c)** men to the total number of people?
  **d)** men to women?

**54.** For this set of pennies and quarters, what is the ratio, in decimal notation rounded to the nearest thousandth, where appropriate, of:

  **a)** pennies to quarters?
  **b)** quarters to pennies?
  **c)** pennies to total number of coins?
  **d)** total number of coins to pennies?

*Gas Mileage.* In each of Exercises 55–58, find the gas mileage rounded to the nearest tenth.

**55.** 285 mi; 18 gal

**56.** 396 mi; 17 gal

**57.** 324.8 mi; 18.2 gal

**58.** 264.8 mi; 12.7 gal

**59.** *Windy Locations.* Although nicknamed the Windy City, Chicago is not the windiest location in the United States. Listed in the table below are the six windiest locations and their average wind speeds. Find the average of these wind speeds and round your answer to the nearest tenth.
**Source:** *The Handy Geography Answer Book*

| LOCATION | AVERAGE WIND SPEED (in miles per hour) |
|---|---|
| Mt. Washington, NH | 35.3 |
| Boston, MA | 12.5 |
| Honolulu, HI | 11.3 |
| Dallas, TX | 10.7 |
| Kansas City, MO | 10.7 |
| Chicago, IL | 10.4 |

**60.** *Areas of the New England States.* The table below lists the areas of the New England states. Find the average area and round your answer to the nearest tenth.
**Source:** *The New York Times Almanac*

| STATE | TOTAL AREA (in square miles) |
|---|---|
| Maine | 33,265 |
| New Hampshire | 9,279 |
| Vermont | 9,614 |
| Massachusetts | 8,284 |
| Connecticut | 5,018 |
| Rhode Island | 1,211 |

*Stock Prices.* At one time, stock prices were given using mixed numerals involving halves, fourths, eighths, and so on. The Securities and Exchange Commission now mandates the use of decimal notation. Thus a price of $23\frac{13}{16}$ is expressed in decimal notation, rounded to the nearest hundredth, as $23.81. Complete the following table.

| | STOCK | PRICE PER SHARE | DECIMAL NOTATION | ROUNDED TO NEAREST HUNDREDTH |
|---|---|---|---|---|
| **61.** | General Electric | $29\frac{9}{16}$ | | |
| **62.** | Intel | $24\frac{7}{16}$ | | |
| **63.** | Microsoft | $27\frac{7}{8}$ | | |
| **64.** | Home Depot | $27\frac{1}{8}$ | | |
| **65.** | Verisign | $31\frac{31}{64}$ | | |
| **66.** | Hewlett-Packard | $45\frac{53}{64}$ | | |

SOURCE: New York Stock Exchange

 Calculate.

**67.** $\frac{7}{8} \times 12.64$

**68.** $\frac{4}{5} \times 384.8$

**69.** $2\frac{3}{4} + 5.65$

**70.** $4\frac{4}{5} + 3.25$

**71.** $\frac{47}{9} \times (-79.95)$

**72.** $\frac{7}{11} \times (-2.7873)$

**73.** $\frac{1}{2} - 0.5$

**74.** $3\frac{1}{8} - 2.75$

**75.** $4.875 - 2\frac{1}{16}$

**76.** $55\frac{3}{5} - 12.22$

**77.** $\frac{5}{6} \times 0.0765 - \frac{5}{4} \times 0.1124$

**78.** $\frac{2}{5} \times 114.8 - \frac{3}{8} \times 156.56$

**79.** $\frac{4}{5} \times 384.8 + 24.8 \div \frac{8}{3}$

**80.** $102.4 \div \frac{2}{5} - 12 \times \frac{5}{6}$

**81.** $\frac{7}{8} \times 0.86 - 0.76 \times \frac{3}{4}$

**82.** $17.95 \div \frac{5}{8} + \frac{3}{4} \times 16.2$

**83.** $3.375 \times 5\frac{1}{3}$

**84.** $2.5 \times 3\frac{5}{8}$

**85.** $6.84 \div \left(-2\frac{1}{2}\right)$

**86.** $-8\frac{1}{2} \div 2.125$

## Skill Maintenance

Multiply. [4.6a]

**87.** $9 \cdot 2\frac{1}{3}$

**88.** $\left(-10\frac{1}{2}\right) \cdot \left(-22\frac{3}{4}\right)$

Divide. [4.6b]

**89.** $-84 \div 8\frac{2}{5}$

**90.** $8\frac{3}{5} \div 10\frac{2}{5}$

Add. [4.5a]

**91.** $17\frac{5}{6} + 32\frac{3}{8}$

**92.** $14\frac{3}{5} + 16\frac{1}{10}$

Subtract. [4.5b]

**93.** $16\frac{1}{10} - 14\frac{3}{5}$

**94.** $32\frac{3}{8} - 17\frac{5}{6}$

Solve. [2.3b], [4.4a]

**95.** A recipe for bread calls for $\frac{2}{3}$ cup of water, $\frac{1}{4}$ cup of milk, and $\frac{1}{8}$ cup of oil. How many cups of liquid ingredients does the recipe call for?

**96.** A board $\frac{7}{10}$ in. thick is glued to a board $\frac{3}{5}$ in. thick. The glue is $\frac{3}{100}$ in. thick. How thick is the result?

## Synthesis

⊞ Find decimal notation.

**97.** $\frac{1}{7}$

**98.** $\frac{2}{7}$

**99.** $\frac{3}{7}$

**100.** $\frac{4}{7}$

**101.** $\frac{5}{7}$

**102.** ⊞ From the pattern of Exercises 97–101, guess the decimal notation for $\frac{6}{7}$. Check on your calculator.

⊞ Find decimal notation.

**103.** $\frac{1}{9}$

**104.** $\frac{1}{99}$

**105.** $\frac{1}{999}$

**106.** ⊞ From the pattern of Exercises 103–105, guess the decimal notation for $\frac{1}{9999}$. Check on your calculator.

# 5.6

## Estimating

**a** Estimate sums, differences, products, and quotients.

### SKILL TO REVIEW

Objective 1.6b: Estimate sums, differences, products, and quotients by rounding.

Estimate by first rounding to the nearest ten.

1.  
$$\begin{array}{r} 4\ 6\ 7 \\ -\ 2\ 8\ 4 \end{array}$$

2.  
$$\begin{array}{r} 5\ 4 \\ \times\ 2\ 9 \end{array}$$

### (a) Estimating Sums, Differences, Products, and Quotients

Estimating has many uses. It can be done before a problem is even attempted in order to get an idea of the answer. It can be done afterward as a check, even when we are using a calculator. In many situations, an estimate is all we need. We usually estimate by rounding the numbers so that there are one or two nonzero digits, depending on how accurate we want our estimate. Consider the following prices for Examples 1–3.

$289.95

8.2 Megapixel Digital Camera

19" Flat-Panel HDTV

Four Burner Gas Grill

$139.97

$449.99

**EXAMPLE 1** Estimate by rounding to the nearest ten the total cost of one grill and one TV.

We are estimating the sum

$289.95 + $449.99 = Total cost.

The estimate found by rounding the addends to the nearest ten is

$290 + $450 = $740.   (Estimated total cost)

Do Margin Exercise 1.

**EXAMPLE 2** About how much more does the TV cost than the camera? Estimate by rounding to the nearest ten.

We are estimating the difference

$449.99 − $139.97 = Price difference.

The estimate, found by rounding the minuend and the subtrahend to the nearest ten, is

$450 − $140 = $310.   (Estimated price difference)

Do Exercise 2.

1. Estimate by rounding to the nearest ten the total cost of one grill and one camera. Which of the following is an appropriate estimate?

a) $43    b) $400
c) $410    d) $430

2. About how much more does the TV cost than the grill? Estimate by rounding to the nearest ten. Which of the following is an appropriate estimate?

a) $100    b) $150
c) $160    d) $300

*Answers*

*Skill to Review:*
1. 190    2. 1500

*Margin Exercises:*
1. (d)    2. (c)

**EXAMPLE 3** Estimate the total cost of 4 cameras.

We are estimating the product

$4 \times \$139.97 = $ Total cost.

The estimate is found by rounding 139.97 to the nearest ten:

$4 \times \$140 = \$560.$

Do Exercise 3.

**3.** Estimate the total cost of 6 TVs. Which of the following is an appropriate estimate?
a) $450        b) $2700
c) $4500       d) $27,000

**EXAMPLE 4** About how many Blu-ray discs at $29.99 each can be purchased for $154?

We estimate the quotient

$\$154 \div \$29.99.$

Since we want a whole-number estimate, we choose our rounding appropriately. Rounding $29.99 to the nearest one, we get $30. Since $154 is close to $150, which is a multiple of 30, we estimate

$\$150 \div \$30,$

so the answer is 5.

Do Exercise 4.

**4.** About how many Blu-ray discs can be purchased for $485? Choose an appropriate estimate from the following.
a) 16        b) 21
c) 25        d) 30

**EXAMPLE 5** Estimate: $4.8 \times 52$. Do not find the actual product. Which of the following is an appropriate estimate?

a) 25          b) 250          c) 2500          d) 360

We round 4.8 to the nearest one and 52 to the nearest ten:

$5 \times 50 = 250.$     (Estimated product)

Thus an appropriate estimate is (b).

Other estimates that we might have used in Example 5 are

$5 \times 52 = 260$     or     $4.8 \times 50 = 240.$

The estimate in Example 5, $5 \times 50 = 250$, is the easiest to do because the factors have the fewest nonzero digits. You could probably do it mentally. In general, we try to round so that a computation has as few nonzero digits as possible while still keeping the estimated value close to the original value.

Do Exercises 5–10.

Estimate each product. Do not find the actual product. Which of the following is an appropriate estimate?

**5.** $2.4 \times 8$
a) 16        b) 34
c) 125       d) 5

**6.** $24 \times 0.6$
a) 200       b) 5
c) 110       d) 20

**7.** $0.86 \times 0.432$
a) 0.04      b) 0.4
c) 1.1       d) 4

**8.** $0.82 \times 0.1$
a) 800       b) 8
c) 0.08      d) 80

**9.** $0.12 \times 18.248$
a) 180       b) 1.8
c) 0.018     d) 18

**10.** $24.234 \times 5.2$
a) 200       b) 120
c) 12.5      d) 234

*Answers*

3. (b)    4. (a)    5. (a)    6. (d)    7. (b)
8. (c)    9. (b)    10. (b)

Estimate each quotient. Which of the following is an appropriate estimate?

**11.** 59.78 ÷ 29.1

    **a)** 200        **b)** 20

    **c)** 2          **d)** 0.2

**12.** 82.08 ÷ 2.4

    **a)** 40        **b)** 4.0

    **c)** 400      **d)** 0.4

**13.** 0.1768 ÷ 0.08

    **a)** 8         **b)** 10

    **c)** 2         **d)** 20

**14.** Estimate: 0.0069 ÷ 0.15. Which of the following is an appropriate estimate?

    **a)** 0.5      **b)** 50

    **c)** 0.05   **d)** 0.004

**EXAMPLE 6** Estimate: 82.08 ÷ 24. Which of the following is an appropriate estimate?

    **a)** 400        **b)** 16        **c)** 40        **d)** 4

This is about 80 ÷ 20, so the answer is about 4. Thus an appropriate estimate is (d).

**EXAMPLE 7** Estimate: 94.18 ÷ 3.2. Which of the following is an appropriate estimate?

    **a)** 30        **b)** 300      **c)** 3        **d)** 60

This is about 90 ÷ 3, so the answer is about 30. Thus an appropriate estimate is (a).

**EXAMPLE 8** Estimate: 0.0156 ÷ 1.3. Which of the following is an appropriate estimate?

    **a)** 0.2      **b)** 0.002     **c)** 0.02     **d)** 20

This is about 0.02 ÷ 1, so the answer is about 0.02. Thus an appropriate estimate is (c).

Do Exercises 11–13.

In some cases, it is easier to estimate a quotient directly rather than by rounding the divisor and the dividend.

**EXAMPLE 9** Estimate: 0.0074 ÷ 0.23. Which of the following is an appropriate estimate?

    **a)** 0.3      **b)** 0.03     **c)** 300     **d)** 3

We estimate 3 for a quotient. We check by multiplying.

$$0.23 \times 3 = 0.69$$

We make the estimate smaller. We estimate 0.3 and check by multiplying.

$$0.23 \times 0.3 = 0.069$$

We make the estimate smaller. We estimate 0.03 and check by multiplying.

$$0.23 \times 0.03 = 0.0069$$

This is about 0.0074, so the quotient is about 0.03. Thus an appropriate estimate is (b).

Do Exercise 14.

*Answers*

**11.** (c)    **12.** (a)    **13.** (c)    **14.** (c)

**a**   Consider the following prices for Exercises 1–8. Estimate the sums, differences, products, or quotients involved in these problems. Indicate which of the choices is an appropriate estimate.

1. Estimate the total cost of one printer and one satellite radio.

   **a)** $43     **b)** $4300     **c)** $360     **d)** $430

2. Estimate the total cost of one satellite radio and one ice cream maker.

   **a)** $230     **b)** $23     **c)** $2300     **d)** $400

3. About how much more does the printer cost than the satellite radio?

   **a)** $1300     **b)** $200     **c)** $130     **d)** $13

4. About how much more does the satellite radio cost than the ice cream maker?

   **a)** $7000     **b)** $70     **c)** $130     **d)** $700

5. Estimate the total cost of 6 ice cream makers.

   **a)** $480     **b)** $48     **c)** $240     **d)** $4800

6. Estimate the total cost of 4 printers.

   **a)** $1200     **b)** $1120     **c)** $11,200     **d)** $600

7. About how many ice cream makers can be purchased for $830?

   **a)** 120     **b)** 100     **c)** 10     **d)** 1000

8. About how many printers can be purchased for $5627?

   **a)** 200     **b)** 20     **c)** 1800     **d)** 2000

Estimate by rounding as directed.

9. $0.02 + 1.31 + 0.34$;
   nearest tenth

10. $0.88 + 2.07 + 1.54$;
    nearest one

11. $6.03 + 0.007 + 0.214$;
    nearest one

12. $1.11 + 8.888 + 99.94$;
    nearest one

13. $52.367 + 1.307 + 7.324$;
    nearest one

14. $12.9882 + 1.2115$;
    nearest tenth

**15.** $2.678 - 0.445$;
nearest tenth

**16.** $12.9882 - 1.0115$;
nearest one

**17.** $198.67432 - 24.5007$;
nearest ten

Estimate. Choose a rounding digit that gives one or two nonzero digits. Indicate which of the choices is an appropriate estimate.

**18.** $234.12321 - 200.3223$
   **a)** 600     **b)** 60
   **c)** 300     **d)** 30

**19.** $49 \times 7.89$
   **a)** 400     **b)** 40
   **c)** 4       **d)** 0.4

**20.** $7.4 \times 8.9$
   **a)** 95     **b)** 63
   **c)** 124    **d)** 6

**21.** $98.4 \times 0.083$
   **a)** 80     **b)** 12
   **c)** 8      **d)** 0.8

**22.** $78 \times 5.3$
   **a)** 400     **b)** 800
   **c)** 40      **d)** 8

**23.** $3.6 \div 4$
   **a)** 10     **b)** 1
   **c)** 0.1    **d)** 0.01

**24.** $0.0713 \div 1.94$
   **a)** 3.5     **b)** 0.35
   **c)** 0.035   **d)** 35

**25.** $74.68 \div 24.7$
   **a)** 9     **b)** 3
   **c)** 12    **d)** 120

**26.** $914 \div 0.921$
   **a)** 10      **b)** 100
   **c)** 1000   **d)** 1

**27.** *Fence Posts.* A zoo plans to construct a fence around its proposed animals of the Great Plains exhibit. The perimeter of the area to be fenced is 1760 ft. Estimate the number of wooden fence posts needed if the posts are placed 8.625 ft apart.

**28.** *Ticketmaster.* Recently, Ticketmaster stock sold for $22.25 per share. Estimate how many shares can be purchased for $4400.

**29.** *Day-Care Supplies.*   Helen wants to buy 12 boxes of crayons at $1.89 per box for the day-care center that she runs. Estimate the total cost of the crayons.

**30.** *Batteries.*   Oscar buys 6 packages of AAA batteries at $5.29 per package. Estimate the total cost of the purchase.

## Skill Maintenance

In each of Exercises 31–38, fill in the blank with the correct term from the given list. Some of the choices may not be used and some may be used more than once.

**31.** The decimal $0.57\overline{3}$ is an example of a(n) _____ decimal.   [5.5a]

**32.** The least common _____ of two natural numbers is the smallest number that is a multiple of both.   [4.1a]

**33.** The sentence $5(3 + 8) = 5 \cdot 3 + 5 \cdot 8$ illustrates the _____ law.   [1.4a]

**34.** A _____ of an equation is a replacement for the variable that makes the equation true.   [1.7a]

**35.** The number 1 is the _____ identity.   [1.4a]

**36.** The sentence $13 + 7 = 7 + 13$ illustrates the _____ law of addition.   [1.2a]

**37.** The least common _____ of two or more fractions is the least common _____ of their denominators.   [4.2a]

**38.** The number 3728 is _____ by 4 if the number named by the last two digits is _____ by 4.   [3.2a]

additive

multiplicative

numerator

denominator

commutative

associative

distributive

solution

divisible

terminating

repeating

multiple

factor

## Synthesis

The following were done on a calculator. Estimate to determine whether the decimal point was placed correctly.

**39.** $178.9462 \times 61.78 = 11,055.29624$

**40.** $14,973.35 \div 298.75 = 501.2$

**41.** $19.7236 - 1.4738 \times 4.1097 = 1.366672414$

**42.** $28.46901 \div 4.9187 - 2.5081 = 3.279813473$

**43.** ▦ Use one of $+, -, \times$, and $\div$ in each blank to make a true sentence.

   **a)** $(0.37 \boxed{\phantom{x}} 18.78) \boxed{\phantom{x}} 2^{13} = 156,876.8$

   **b)** $2.56 \boxed{\phantom{x}} 6.4 \boxed{\phantom{x}} 51.2 \boxed{\phantom{x}} 17.4 = 312.84$

# 5.7

## Applications and Problem Solving

### **a** Solving Applied Problems

Solving applied problems with decimals is like solving applied problems with integers. We translate first to an equation that corresponds to the situation. Then we solve the equation.

**EXAMPLE 1** *Canals.* The Panama Canal in Panama is 50.7 mi long. The Suez Canal in Egypt is 119.9 mi long. How much longer is the Suez Canal?

Panama Canal    Suez Canal

**1. Debit-Card Transactions.**
Debit-card transactions in 2000 totaled 9.8 billion. The number of transactions projected to be made in 2010 is 42.5 billion. How many more debit-card transactions are expected in 2010 than in 2000?

**Source:** *The Nilson Report*

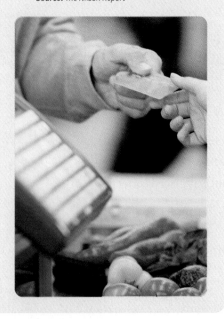

1. **Familiarize.** We let $l =$ the distance in miles that the length of the longer canal differs from the length of the shorter canal.

2. **Translate.** We translate as follows, using the given information:

| Length of Panama Canal, the shorter canal | plus | Additional length | is | Length of Suez Canal, the longer canal |
|:---:|:---:|:---:|:---:|:---:|
| ↓ | ↓ | ↓ | ↓ | ↓ |
| 50.7 mi | + | $l$ | = | 119.9 mi. |

3. **Solve.** We solve the equation by subtracting 50.7 mi on both sides:

$$50.7 + l = 119.9$$
$$50.7 + l - 50.7 = 119.9 - 50.7$$
$$l = 69.2.$$

4. **Check.** We can check by adding.

```
  5 0.7
+ 6 9.2
-------
1 1 9.9
```

The answer checks.

5. **State.** The Suez Canal is 69.2 mi longer than the Panama Canal.

Do Exercise 1.

*Answer*
1. 32.7 billion transactions

**EXAMPLE 2** *Meat and Seafood Consumption.* In a recent year, the average American consumed about 62.8 lb of beef, 47.3 lb of pork and lamb, 73.6 lb of poultry, and 16.1 lb of seafood. Find the total amount of meat and seafood consumed by the average American that year.

**Meat and Seafood Consumption**

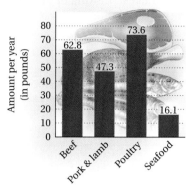

SOURCE: U.S. Department of Agriculture

1. **Familiarize.** The figure above helps us visualize the situation. We let $t$ = the total amount of meat and seafood consumed by the average American during the year.

2. **Translate.** We are combining amounts. We translate to an equation that does this.

| Beef | plus | Pork and lamb | plus | Poultry | plus | Seafood | is | Total |
|------|------|------|------|------|------|------|------|------|
| ↓ | ↓ | ↓ | ↓ | ↓ | ↓ | ↓ | ↓ | ↓ |
| 62.8 | + | 47.3 | + | 73.6 | + | 16.1 | = | $t$ |

3. **Solve.** To solve, we carry out the addition.

$$\begin{array}{r} {}^{1}\ {}^{1}\phantom{0}\\ 6\ 2.8 \\ 4\ 7.3 \\ 7\ 3.6 \\ +\ \ 1\ 6.1 \\ \hline 1\ 9\ 9.8 \end{array}$$

Thus, $t = 199.8$.

4. **Check.** We can check by repeating our addition. We can also see whether our answer is reasonable by first noting that it is indeed larger than any of the numbers being added. We can also do a check by rounding and estimating:

$$62.8 + 47.3 + 73.6 + 16.1 \approx 60 + 50 + 70 + 20$$
$$= 200 \approx 199.8.$$

If we had gotten an answer like 19.98 or 1998, then the estimate, 200, would have told us that we had made a mistake, such as not lining up the decimal points.

5. **State.** The average American consumed about 199.8 lb of meat and seafood in a recent year.

Do Exercise 2.

**2. Amount of Medication.** Over a 24-hr period, a patient received injections of 2.8 mL, 1.35 mL, 2.0 mL, and 1.88 mL of a medication. What was the total amount of medication received?

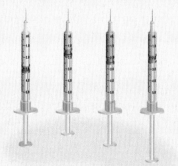

2.8 mL   1.35 mL   2.0 mL   1.88 mL

*Answer*

**2.** 8.03 mL

**EXAMPLE 3** *IRS Driving Allowance.* The Internal Revenue Service allowed a tax deduction of 58.5¢ per mile driven for business purposes in the last six months of 2008. What deduction, in dollars, would be allowed for driving 9143 mi during this time period?

Source: Internal Revenue Service

1. **Familiarize.** We first make a drawing or at least visualize the situation. Repeated addition fits this situation. We let $d$ = the deduction, in dollars, allowed for driving 9143 mi.

9143 mi

2. **Translate.** We translate as follows, writing 58.5¢ as $0.585:

| Deduction for each mile | times | Number of miles driven | is | Total deduction |
|---|---|---|---|---|
| $0.585 | × | 9143 | = | $d$. |

3. **Solve.** To solve the equation, we carry out the multiplication.

$$\begin{array}{r} 9\ 1\ 4\ 3 \\ \times\ 0.5\ 8\ 5 \\ \hline 4\ 5\ 7\ 1\ 5 \\ 7\ 3\ 1\ 4\ 4\ 0 \\ 4\ 5\ 7\ 1\ 5\ 0\ 0 \\ \hline 5\ 3\ 4\ 8.6\ 5\ 5 \end{array}$$

Thus, $d$ = 5348.655 ≈ 5348.66.

4. **Check.** We can obtain a partial check by rounding and estimating:

$$9143 \times 0.585 \approx 9000 \times 0.6 = 5400 \approx 5348.66.$$

5. **State.** The total allowable deduction would be $5348.66.

**3. Printing Costs.** At Kwik Copy, the cost of copying is 11 cents per page. How much, in dollars, would it cost to make 466 copies?

Do Exercise 3.

**EXAMPLE 4** *Loan Payments.* A car loan of $7382.52 is to be paid off in 36 equal monthly payments. How much is each payment?

1. **Familiarize.** We first make a drawing. We let $n$ = the amount of each payment.

There may be some fractional part of $1.

$7382.52

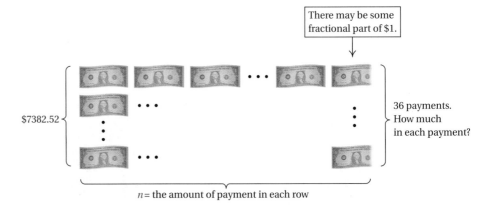

36 payments. How much in each payment?

$n$ = the amount of payment in each row

*Answer*

3. $51.26

**2. Translate.** The problem can be translated to the following equation, thinking that

(Total loan) ÷ (Number of payments) = Amount of each payment

$$7382.52 \div 36 = n.$$

**3. Solve.** To solve the equation, we carry out the division.

```
        2 0 5.0 7
3 6 ) 7 3 8 2.5 2
      7 2
        1 8 2
        1 8 0
            2 5 2
            2 5 2
                0
```

Thus, $n = 205.07$.

**4. Check.** A partial check can be obtained by estimating the quotient: $7382.52 \div 36 \approx 8000 \div 40 = 200 \approx 205.07$. The estimate checks.

**5. State.** Each payment is $205.07.

Do Exercise 4.

**4. Loan Payments.** A loan of $4425 is to be paid off in 12 equal monthly payments. How much is each payment?

**EXAMPLE 5** *Jackie Robinson Poster.* A limited-edition poster by sports artist Leroy Neiman commemorates Jackie Robinson's entrance into Major League Baseball in 1947. Robinson was the first African-American player in the major leagues. The dimensions of the poster are 19.3 in. by 27.4 in. Find the area.

Source: Barton L. Kaufman, private collection

**1. Familiarize.** We first make a drawing. We let $A =$ the area.

**2. Translate.** We use the formula $A = l \cdot w$ and substitute:

$A = l \cdot w$

$A = 27.4 \times 19.3.$

**5. Index Cards.** A standard-size index card measures 12.7 cm by 7.6 cm. What is its area?

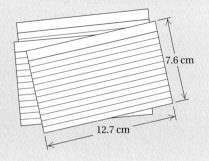

7.6 cm

12.7 cm

**3. Solve.** We solve by carrying out the multiplication.

$$
\begin{array}{r}
2\ 7.4 \\
\times\ 1\ 9.3 \\
\hline
8\ 2\ 2 \\
2\ 4\ 6\ 6\ 0 \\
2\ 7\ 4\ 0\ 0 \\
\hline
5\ 2\ 8.8\ 2
\end{array}
$$

Thus, $A = 528.82$.

**4. Check.** We obtain a partial check by estimating the product:

$$A = 27.4 \times 19.3 \approx 25 \times 20 = 500.$$

This estimate is close to 528.82, so the answer is probably correct.

**5. State.** The area of the Jackie Robinson poster is 528.82 in$^2$.

> Do Exercise 5.

**EXAMPLE 6** *Digital Camera Purchase.* Dawson Real Estate spent $10,399.74 on 26 Nikon Coolpix 10.1 megapixel digital cameras, so that its agents could place photos of properties on the firm's Web site. How much did each camera cost?

**1. Familiarize.** We let $c$ = the cost of each camera.

**2. Translate.** We translate as follows:

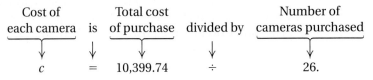

| Cost of each camera | is | Total cost of purchase | divided by | Number of cameras purchased |
|:---:|:---:|:---:|:---:|:---:|
| $c$ | $=$ | 10,399.74 | $\div$ | 26. |

**3. Solve.** To solve, we carry out the division.

$$
\begin{array}{r}
3\ 9\ 9.9\ 9 \\
2\ 6\ )\overline{\ 1\ 0,3\ 9\ 9.7\ 4} \\
\underline{7\ 8}\phantom{00000} \\
2\ 5\ 9\phantom{0000} \\
\underline{2\ 3\ 4}\phantom{000} \\
2\ 5\ 9\phantom{000} \\
\underline{2\ 3\ 4}\phantom{00} \\
2\ 5\ 7\phantom{0} \\
\underline{2\ 3\ 4}\phantom{0} \\
2\ 3\ 4 \\
\underline{2\ 3\ 4} \\
0
\end{array}
$$

**4. Check.** We check by estimating

$$10,399.74 \div 26 \approx 10,000 \div 25 = 400.$$

Since 400 is close to 399.99, the answer is probably correct.

**5. State.** The cost of each camera was $399.99.

> Do Exercise 6.

**6.** One pound of lean boneless ham contains 4.5 servings. It costs $7.99 per pound. What is the cost per serving? Round to the nearest cent.

*Answers*

**5.** 96.52 cm$^2$  **6.** $1.78

## Multistep Problems

**EXAMPLE 7**  *Gas Mileage.*  Ava filled her gas tank and noted that the odometer read 67,507.8. After the next filling, the odometer read 68,006.1. It took 16.5 gal to fill the tank. How many miles per gallon did Ava get?

1. **Familiarize.**   We first make a drawing.

This is a two-step problem. First, we find the number of miles that have been driven between fillups. We let $n =$ the number of miles driven.

**2., 3. Translate** and **Solve.**   We translate and solve as follows:

$$\underbrace{\text{First odometer reading}}_{67{,}507.8} \quad \overset{\text{plus}}{+} \quad \underbrace{\text{Number of miles driven}}_{n} \quad \overset{\text{is}}{=} \quad \underbrace{\text{Second odometer reading}}_{68{,}006.1.}$$

To solve the equation, we subtract 67,507.8 on both sides:

$n = 68{,}006.1 - 67{,}507.8$

$n = 498.3.$

$$\begin{array}{r} 6\ 8{,}0\ 0\ 6.1 \\ -\ 6\ 7{,}5\ 0\ 7.8 \\ \hline 4\ 9\ 8.3 \end{array}$$

Next, we divide the total number of miles driven by the number of gallons. This gives us $m$, the number of miles per gallon—that is, the mileage. The division that corresponds to the situation is

$498.3 \div 16.5 = m.$

To find the number $m$, we divide.

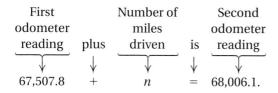

Thus, $m = 30.2.$

4. **Check.**   To check, we first multiply the number of miles per gallon times the number of gallons to find the number of miles driven:

$16.5 \times 30.2 = 498.3.$

Then we add 498.3 to 67,507.8 to find the new odometer reading:

$67{,}507.8 + 498.3 = 68{,}006.1.$

The mileage 30.2 checks.

5. **State.**   Ava got 30.2 miles per gallon.

Do Exercise 7.

**7. Gas Mileage.**   John filled his gas tank and noted that the odometer read 38,320.8. After the next filling, the odometer read 38,735.5. It took 14.5 gal to fill the tank. How many miles per gallon did John get?

*Answer*

**7.**  28.6 mpg

FOR SALE

**EXAMPLE 8** *Home-Cost Comparison.* Suppose you own a home like the one shown here and it is valued at $225,000 in Tulsa, Oklahoma. What would it cost to buy a similar home in Boston, Massachusetts? To find out, we can use an index table prepared by Coldwell Banker Real Estate Corporation. (For a complete index table, contact your local representative.) We use the following formula:

$$\left(\begin{array}{c}\text{Cost of your}\\\text{home in new city}\end{array}\right) = \frac{\left(\begin{array}{c}\text{Value of}\\\text{your home}\end{array}\right) \times \left(\begin{array}{c}\text{Index of}\\\text{new city}\end{array}\right)}{\left(\begin{array}{c}\text{Index of}\\\text{your city}\end{array}\right)}.$$

Find the cost of your Tulsa home in Boston.

| STATE | CITY | INDEX | STATE | CITY | INDEX |
|---|---|---|---|---|---|
| Massachusetts | Boston | 327 | Texas | Dallas | 72 |
| Illinois | Chicago | 173 | Oklahoma | Tulsa | 36 |
| California | Palo Alto | 397 | Florida | Miami | 151 |
|  | Sacramento | 90 |  | Orlando | 96 |
|  | San Diego | 147 |  | Key West | 220 |
| Alaska | Juneau | 110 | Colorado | Boulder | 146 |
| Kentucky | Louisville | 56 | North Carolina | Charlotte | 61 |

SOURCE: Coldwell Banker Real Estate Corporation

1. **Familiarize.** We let $C$ = the cost of the home in Boston. We use the table and look up the indexes of Tulsa and the new city, Boston. We see that Tulsa's index is 36 and Boston's index is 327.

2. **Translate.** Using the formula, we translate to the following equation:

$$C = \frac{\$225{,}000 \times 327}{36}.$$

3. **Solve.** To solve, we carry out the computations using the rules for order of operations. (See Section 5.4.)

$$C = \frac{\$225{,}000 \times 327}{36}$$
$$= \frac{\$73{,}575{,}000}{36} \qquad \text{Carrying out the multiplication first}$$
$$= \$2{,}043{,}750 \qquad \text{Carrying out the division}$$

On a calculator, the computation could be done in one step.

4. **Check.** We can repeat our computations. The answer checks.

5. **State.** A home that sells for $225,000 in Tulsa would cost $2,043,750 in Boston.

Do Exercises 8 and 9.

Refer to the table in Example 8 to answer Exercises 8 and 9. Round to the nearest dollar.

8. **Home-Cost Comparison.** Suppose your home in Charlotte, North Carolina, is valued at $180,000. What would it cost to buy a similar home in Chicago, IL?

9. **Home-Cost Comparison.** Suppose your home in Boulder, Colorado, is valued at $315,000. What would it cost to buy a similar home in Louisville, KY?

*Answers*

**8.** $510,492   **9.** $120,822

# Translating for Success

1. *Gas Mileage.* Art filled his SUV's gas tank and noted that the odometer read 38,271.8. At the next filling, the odometer read 38,677.92. It took 28.4 gal to fill the tank. How many miles per gallon did the SUV get?

2. *Dimensions of a Parking Lot.* Seals' parking lot is a rectangle that measures 85.2 ft by 52.3 ft. What is the area of the parking lot?

3. *Game Snacks.* Three students pay $18.40 for snacks at a football game. What is each person's share?

4. *Electrical Wiring.* An electrician needs 1314 ft of wire cut into $2\frac{1}{2}$-ft pieces. How many pieces will she have?

5. *College Tuition.* Wayne needs $4638 for the fall semester's tuition. On the day of registration, he has only $3092. How much does he need to borrow?

The goal of these matching questions is to practice step (2), *Translate*, of the five-step problem-solving process. Translate each problem to an equation and select a correct translation from equations A–O.

**A.** $2\frac{1}{2} \cdot n = 1314$

**B.** $18.4 \times 3.87 = n$

**C.** $n = 85.2 \times 52.3$

**D.** $1314.28 - 437 = n$

**E.** $3 \times 18.40 = n$

**F.** $2\frac{1}{2} \cdot 1314 = n$

**G.** $3092 + n = 4638$

**H.** $18.4 \cdot n = 3.87$

**I.** $\dfrac{406.12}{28.4} = n$

**J.** $52.3 \cdot n = 85.2$

**K.** $n = 1314.28 + 437$

**L.** $52.3 + n = 85.2$

**M.** $3092 + 4638 = n$

**N.** $3 \cdot n = 18.40$

**O.** $85.2 + 52.3 = n$

*Answers on page A-8*

6. *Cost of Gasoline.* What is the cost, in dollars, of 18.4 gal of gasoline at $3.87 per gallon?

7. *Savings Account Balance.* Margaret has $1314.28 in her savings account. Before using her debit card to buy an office chair, she transferred $437 to her checking account. How much was left in her savings account?

8. *Acres Planted.* This season Sam planted 85.2 acres of corn and 52.3 acres of soybeans. Find the total number of acres that he planted.

9. *Amount Inherited.* Tara inherited $2\frac{1}{2}$ times as much as her cousin. Her cousin received $1314. How much did Tara receive?

10. *Travel Funds.* The athletic department needs travel funds of $4638 for the tennis team and $3092 for the golf team. What is the total amount needed for travel?

## 5.7 Exercise Set

For Extra Help

MyMathLab

Math XL
PRACTICE

WATCH

DOWNLOAD

READ

REVIEW

**a** Solve.

*Hurricane Damage.* The amount of damage caused by the five most costly Atlantic hurricanes in the United States is shown in the table below. Use this table to do Exercises 1 and 2.

**Most Costly Hurricanes**

| RANK | HURRICANE | YEAR | COST IN 2007 DOLLARS (in billions) |
|------|-----------|------|-----------------------------------|
| 1 | Katrina | 2005 | $81.2 |
| 2 | Andrew | 1992 | 38.1 |
| 3 | Wilma | 2005 | 30.4 |
| 4 | Ivan | 2004 | 18.1 |
| 5 | Charley | 2004 | 16.2 |

SOURCE: National Hurricane Center

**1.** How much more costly was Hurricane Katrina than Hurricane Andrew?

**2.** What was the total cost of the two hurricanes that occurred in 2005?

*Top American Sky Routes.* The figure below shows the numbers of airline passengers carried on the top five sky routes in the United States from December 2006 to November 2007. Note that these routes go in both directions. Use this figure to do Exercises 3 and 4.

**Top American Sky Routes**

| Route | Number of passengers (in millions) |
|-------|-----------------------------------|
| New York City–Chicago | 3.47 |
| Washington–Chicago | 2.82 |
| Atlanta–Orlando | 2.77 |
| Los Angeles–Chicago | 2.71 |
| Atlanta–New York City | 2.69 |

SOURCE: U.S. Bureau of Transportation Statistics

**3.** What is the total number of passengers carried on the top two routes?

**4.** How many more passengers were carried on the New York City–Chicago route than on the Los Angeles–Chicago route?

**5.** *Record Movie Openings.* The movie *The Dark Knight* took in $155.34 million on its first weekend. This topped the previous high opening weekend revenue set by *Spider-Man 3* by $4.24 million. How much did *Spider-Man 3* take in on its opening weekend?

Source: Associated Press

**6.** *Counterfeit Money.* The amount of counterfeit money passed in the United States is on the rise. About $62.0 million was passed in 2006. This amount was $22.8 million more than entered circulation in 1999. How much counterfeit money was passed in 1999?

Source: U.S. Secret Service

**7.** *Cost of Bottled Water.* The cost of a year's supply of a popular brand of bottled water, based on the recommended consumption of 64 oz per day, at the supermarket price of $3.99 for a six-pack of half-liter bottles is $918.82. This is $918.31 more than the cost of drinking the same amount of tap water for a year. What is the cost of drinking tap water for a year?

**Source:** American Water Works Association

**8.** *Bottled Water Consumption.* The annual consumption of bottled water in the United States was 27.6 gal per person in 2006. This was an increase of 26 gal per person over the amount consumed in 1976. What was the annual consumption of bottled water in 1976?

**Source:** Beverage Marketing Corporation

**9.** *Body Temperature.* Normal body temperature is 98.6°F. During an illness, a patient's temperature rose 4.2°. What was the new temperature?

**10.** *Gasoline Cost.* What is the cost, in dollars, of 12.6 gal of gasoline at $3.79 per gallon? Round the answer to the nearest cent.

**11.** *Lottery Winnings.* The largest lotto jackpot ever won in California totaled $193,000,000 and was shared equally by 3 winners. How much was each winner's share? Round to the nearest cent.

**Source:** California State Lottery

**12.** *Lunch Costs.* A group of 4 students pays $47.84 for lunch and splits the cost equally. What is each person's share?

**13.** *Stamp.* Find the area and the perimeter of the stamp shown here.

2.5 cm

3.25 cm

**14.** *Pole Vault Pit.* Find the area and the perimeter of the landing area of the pole vault pit shown here.

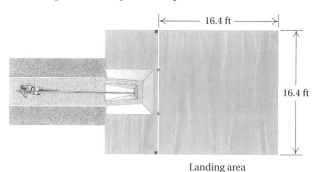

16.4 ft

16.4 ft

Landing area

**15.** *Odometer Reading.* The Binford family's odometer reads 22,456.8 at the beginning of a trip. The family's online driving directions tell them that they will be driving 234.7 mi. What will the odometer read at the end of the trip?

**16.** *Miles Driven.* Petra bought gasoline when the odometer read 14,296.3. At the next gasoline purchase, the odometer read 14,515.8. How many miles had been driven?

**17.** *Gas Mileage.* Peggy filled her van's gas tank and noted that the odometer read 26,342.8. After the next filling, the odometer read 26,736.7. It took 19.5 gal to fill the tank. How many miles per gallon did the van get?

**18.** *Gas Mileage.* Henry filled his Honda's gas tank and noted that the odometer read 18,943.2. After the next filling, the odometer read 19,306.2. It took 13.2 gal to fill the tank. How many miles per gallon did the car get?

**19.** *Chemistry.* The water in a filled tank weighs 748.45 lb. One cubic foot of water weighs 62.5 lb. How many cubic feet of water does the tank hold?

**20.** *Highway Routes.* You can drive from home to work using either of two routes:

> *Route A*: Via interstate highway, 7.6 mi, with a speed limit of 65 mph.

> *Route B*: Via a country road, 5.6 mi, with a speed limit of 50 mph.

Assuming you drive at the posted speed limit, which route takes less time? (Use the formula *Distance = Speed × Time.*)

Find the perimeter of each figure.

**21.**

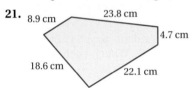

8.9 cm    23.8 cm    4.7 cm    18.6 cm    22.1 cm

**22.**

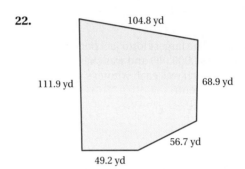

104.8 yd    111.9 yd    68.9 yd    56.7 yd    49.2 yd

**23.**

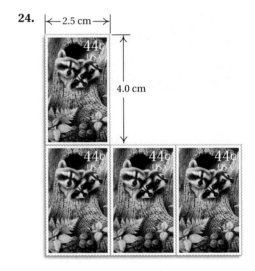

2.5 cm

2.25 cm

**24.**

2.5 cm

4.0 cm

**25.** Andrew bought a DVD of the movie *Horton Hears a Who* for his nephew for $23.99 plus $1.68 sales tax. He paid for it with a $50 bill. How much change did he receive?

**26.** Claire bought a copy of the book *Make Way for Ducklings* for her daughter for $16.95 plus $0.85 sales tax. She paid for it with a $20 bill. How much change did she receive?

Find the length *d* in each figure.

**27.**

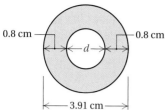

0.8 cm — | — 0.8 cm

← *d* →

← 3.91 cm →

**28.**

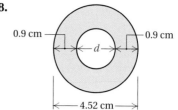

0.9 cm — | — 0.9 cm

← *d* →

← 4.52 cm →

**29.** *Calories Burned Mowing.* A person weighing 150 lb burns 7.3 calories per minute while mowing a lawn with a power lawnmower. How many calories would be burned in 2 hr of mowing?

**Source:** *The Handy Science Answer Book*

**30.** Lot A measures 250.1 ft by 302.7 ft. Lot B measures 389.4 ft by 566.2 ft. What is the total area of the two lots?

**31.** Holly had $1123.56 in her checking account. She used her debit card to make purchases of $23.82, $507.88, and $98.32. She then deposited a bonus check of $678.20. How much was in her account after these changes?

**32.** Natalie had $185.00 to spend for fall clothes: $44.95 was spent on shoes, $71.95 for a jacket, and $55.35 for pants. How much was left?

**33.** A rectangular yard is 20 ft by 15 ft. The yard is covered with grass except for an 8.5-ft-square flower garden. How much grass is in the yard?

**34.** Rita earns a gross paycheck (before deductions) of $495.72. Her deductions are $59.60 for federal income tax, $29.00 for FICA, and $29.00 for medical insurance. What is her take-home paycheck?

**35.** *Batting Average.* Chipper Jones of the Atlanta Braves won the 2008 National League batting title with 160 hits in 439 times at bat. What part of his at-bats were hits? Give decimal notation rounded to the nearest thousandth. (This is a *batting average.*)

**Source:** Major League Baseball

**36.** *Batting Average.* Joe Mauer of the Minnesota Twins won the 2008 American League batting title with 176 hits in 536 times at bat. What part of his at-bats were hits? Give decimal notation rounded to the nearest thousandth.

**Source:** Major League Baseball

**37.** *Field Dimensions.* The dimensions of a World Cup soccer field are 114.9 yd by 74.4 yd. The dimensions of a standard football field are 120 yd by 53.3 yd. How much greater is the area of a World Cup soccer field?

**38.** *Loan Payment.* In order to make money on loans, financial institutions are paid back more money than they loan. Suppose you borrow $120,000 to buy a house and agree to make monthly payments of $880.52 for 30 yr. How much do you pay back altogether? How much more do you pay back than the amount of the loan?

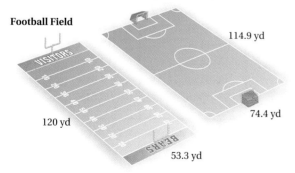

**World Cup Soccer Field**

**Football Field**

114.9 yd

120 yd

74.4 yd

53.3 yd

**39.** *Egg Costs.*   A restaurant owner bought 20 dozen eggs for $25.80. Find the cost of each egg to the nearest tenth of a cent (thousandth of a dollar).

**40.** *Weight Loss.*   A person weighing 170 lb burns 8.6 calories per minute while mowing a lawn. One must burn about 3500 calories in order to lose 1 lb. How many pounds would be lost by mowing for 2 hr? Round to the nearest tenth.

**41.** *Construction Pay.*   A construction worker is paid $18.50 per hour for the first 40 hr of work, and time and a half, or $27.75 per hour, for any overtime exceeding 40 hr per week. One week she works 46 hr. How much is her pay?

**42.** *Summer Work.*   Zachary worked 53 hr during a week one summer. He earned $7.50 per hour for the first 40 hr and $11.25 per hour for overtime (hours exceeding 40). How much did Zachary earn during the week?

**43.** *Projected World Population.*   Using the information in the bar graph below, determine the average population of the world for the years 1950 through 2050. Round to the nearest thousandth of a billion.

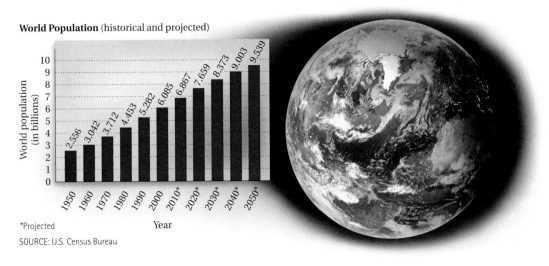

World Population (historical and projected)

*Projected

SOURCE: U.S. Census Bureau

**44.** *Number of Volunteers.*   The table at right lists the number of Americans in various age groups who do volunteer work. What is the average number of volunteers in all the age groups shown?

| AGE | NUMBER OF VOLUNTEERS (in millions) |
|---|---|
| 16–24 yr | 8.044 |
| 25–34 yr | 9.096 |
| 35–44 yr | 13.308 |
| 45–54 yr | 13.415 |
| 55–64 yr | 8.819 |
| 65 yr and older | 8.518 |

SOURCE: U.S. Census Bureau

**45.** *Body Temperature.* Normal body temperature is 98.6°F. A baby's bath water should be 100°F. How many degrees above normal body temperature is this?

**46.** *Body Temperature.* Normal body temperature is 98.6°F. The lowest temperature at which a patient has survived is 69°F. How many degrees below normal is this?

*Home-Cost Comparison.* Use the table and the formula in Example 8 to do Exercises 47–54. In each of the following cases, find the value of the house in the new location. Round to the nearest dollar.

Source: Coldwell Banker Real Estate Corporation

|  | VALUE | PRESENT LOCATION | NEW LOCATION | NEW VALUE |
|---|---|---|---|---|
| **47.** | $125,000 | Dallas | Miami | |
| **48.** | $180,000 | San Diego | Key West | |
| **49.** | $96,000 | Juneau | Orlando | |
| **50.** | $300,000 | Palo Alto | Dallas | |
| **51.** | $240,000 | Louisville | Boston | |
| **52.** | $160,000 | Charlotte | Sacramento | |
| **53.** | $140,000 | Chicago | Tulsa | |
| **54.** | $99,000 | Boulder | Orlando | |

**55.** *Cell-Phone Plan.* In 2008, at&t offered an individual cell-phone plan with 450 anytime minutes for a monthly access fee of $39.99. Minutes in excess of 450 were charged at the rate of $0.45 per minute. In June, Leila used her cell phone for 479 min. What was she charged?

Source: at&t

**56.** *Cell-Phone Plan.* In 2008, at&t offered an individual cell-phone plan with 900 anytime minutes per month for a monthly access fee of $59.99. Minutes in excess of 900 were charged at the rate of $0.40 per minute. One month Jeff used his cell phone for 946 min. What was the charge?

Source: at&t

*IRS Driving Allowance.* The Internal Revenue Service allowed a tax deduction of 50.5¢ per mile driven for business purposes from January 1, 2008, through June 30, 2008. In response to increasing gas prices, this deduction was increased to 58.5¢ per mile driven from July 1, 2008, through December 31, 2008. Use this information for Exercises 57 and 58.

Source: Internal Revenue Service

**57.** Sheila drove 3156 mi for business purposes between January 1, 2008, and June 30, 2008. She drove an additional 3678 mi between July 1, 2008, and December 31, 2008. What deduction, in dollars, was she allowed to take?

**58.** Manolo drove 1250 mi for business purposes between January 1, 2008, and June 30, 2008. He drove an additional 1872 mi between July 1, 2008, and December 31, 2008. What deduction, in dollars, was he allowed to take?

**59.** *Property Taxes.* The Brunners own a house with an assessed value of $165,500. For every $1000 of assessed value, they pay $8.50 in property taxes each year. How much do they pay in property taxes each year?

**60.** *Property Taxes.* The Colavitos own a house with an assessed value of $184,500. For every $1000 of assessed value, they pay $7.68 in property taxes each year. How much do they pay in property taxes each year?

## Skill Maintenance

Add.

**61.** $4569 + 1766$  [1.2a]

**62.** $\dfrac{2}{3} + \dfrac{5}{8}$  [4.2a]

**63.** $4\dfrac{1}{3} + 2\dfrac{1}{2}$  [4.5a]

**64.** $-\dfrac{5}{6} + \dfrac{7}{10}$  [4.2a]

Subtract.

**65.** $\dfrac{2}{3} - \dfrac{5}{8}$  [4.3a]

**66.** $4569 - 1766$  [1.3a]

**67.** $-\dfrac{5}{6} - \dfrac{7}{10}$  [4.3a]

**68.** $4\dfrac{1}{3} - 2\dfrac{1}{2}$  [4.5b]

Simplify.  [3.5b]

**69.** $\dfrac{3225}{6275}$

**70.** $\dfrac{125}{400}$

**71.** $-\dfrac{325}{625}$

**72.** $-\dfrac{625}{475}$

Solve.

**73.** If a water wheel made 469 revolutions at a rate of $16\dfrac{3}{4}$ revolutions per minute, for how long did it rotate? [4.6c]

**74.** If a bicycle wheel made 480 revolutions at a rate of $66\dfrac{2}{3}$ revolutions per minute, for how long did it rotate? [4.6c]

**75.** *Calories in Pie.* A piece of pecan pie $\left(\dfrac{1}{8}\text{ of a 9-in. pie}\right)$ has 502 calories. A piece of pumpkin pie has 316 calories. How many more calories does a piece of pecan pie have than a piece of pumpkin pie? [1.8a]

**76.** *Calories in Turkey.* Dark meat turkey contains 187 calories per 3.5-oz serving. White meat turkey contains 157 calories per 3.5-oz serving. How many more calories per 3.5-oz serving does the dark meat turkey contain than the white meat turkey? [1.8a]

## Synthesis

**77.** Reggie buys a half-dozen packs of basketball cards with a dozen cards in each pack. The cost is twelve dozen cents for each half-dozen cards. How much does Reggie pay for the cards?

# Summary and Review

## Key Terms

rational numbers, p. 258          terminating decimal, p. 300          repeating decimal, p. 300

## Concept Reinforcement

Determine whether each statement is true or false.

_____ **1.** One thousand billion is one trillion.   [5.3b]

_____ **2.** The number of decimal places in the product of two numbers is the product of the numbers of places in the factors.   [5.3a]

_____ **3.** When we divide a positive number by 0.1, 0.01, 0.001, and so on, the quotient is larger than the divisor.   [5.4a]

_____ **4.** For a fraction with a prime factor other than 2 or 5 in its denominator, decimal notation terminates.   [5.5a]

_____ **5.** An estimate found by rounding to the nearest ten is usually more accurate than one found by rounding to the nearest hundred.   [5.6a]

## Important Concepts

**Objective 5.1b**   Convert between decimal notation and fraction notation.

**Example**   Write fraction notation for 5.347.

$$5.347 \qquad\qquad 5.347. \qquad \frac{5347}{1000}$$

$\uparrow$ 3 decimal places     Move 3 places to the right.     3 zeros

$$5.347 = \frac{5347}{1000}$$

**Example**   Write decimal notation for $-\dfrac{29}{1000}$.

$$-\frac{29}{1000} \qquad -0.029.$$

$\uparrow$ 3 zeros     Move 3 places to the left.

$$-\frac{29}{1000} = -0.029$$

**Example**   Write decimal notation for $4\dfrac{63}{100}$.

$$4\frac{63}{100} = 4 + \frac{63}{100} = 4 \text{ and } \frac{63}{100} = 4.63$$

**Practice Exercise**

**1.** Write fraction notation for 50.93.

**Practice Exercise**

**2.** Write decimal notation for $-\dfrac{817}{10}$.

**Practice Exercise**

**3.** Write decimal notation for $42\dfrac{159}{1000}$.

**Objective 5.1d**  Round decimal notation to the nearest thousandth, hundredth, tenth, one, ten, hundred, or thousand.

**Example**  Round 19.7625 to the nearest hundredth.

Locate the digit in the hundredths place, 6. Consider the next digit to the right, 2. Since that digit, 2, is 4 or lower, round down.

$$19.7625$$
$$\downarrow$$
$$19.76$$

**Practice Exercise**

**4.** Round 153.346 to the nearest hundredth.

---

**Objective 5.2a**  Add using decimal notation for positive numbers.

**Example**  Add: 14.26 + 63.589.

$$\begin{array}{r} \overset{1}{1}\,4.2\,6\,0 \\ +\ 6\,3.5\,8\,9 \\ \hline 7\,7.8\,4\,9 \end{array}$$  Writing an extra zero

**Practice Exercise**

**5.** Add: 5.54 + 33.071.

---

**Objective 5.2b**  Subtract using decimal notation for positive numbers.

**Example**  Subtract: 67.345 − 24.28.

$$\begin{array}{r} 6\,7.\overset{2}{3}\,\overset{14}{4}\,5 \\ -\ 2\,4.2\,8\,0 \\ \hline 4\,3.0\,6\,5 \end{array}$$  Writing an extra zero

**Practice Exercise**

**6.** Subtract: 221.04 − 13.192.

---

**Objective 5.3a**  Multiply using decimal notation.

**Example**  Multiply: 1.8 × 0.04.

$$\begin{array}{r} 1.8 \\ \times\ 0.0\,4 \\ \hline 0.0\,7\,2 \end{array}$$
(1 decimal place)
(2 decimal places)
(3 decimal places)

**Example**  Multiply: 0.001 × 87.1.

$$0.001 \times 87.1 \qquad 0.087.1$$
$$\uparrow \qquad \qquad \curvearrowleft$$

3 decimal places     Move 3 places to the left.
We write an extra zero.

$$0.001 \times 87.1 = 0.0871$$

**Example**  Multiply: −63.4 × 100.

$$-63.4 \times 100 \qquad -63.40.$$
$$\uparrow \qquad \qquad \curvearrowright$$

2 zeros     Move 2 places to the right.
We write an extra zero.

$$-63.4 \times 100 = -6340$$

**Practice Exercise**

**7.** Multiply: 5.46 × 3.5.

**Practice Exercise**

**8.** Multiply: −17.6 × 0.01.

**Practice Exercise**

**9.** Multiply: 1000 × 60.437.

**Objective 5.4a** Divide using decimal notation.

**Example**  Divide: $21.35 \div 6.1$.

$$
\begin{array}{r}
3.5 \\
6.1\,\overline{\smash{)}\,2\,1.3_{\wedge}5} \\
\underline{1\,8\,3\phantom{0}} \\
3\,0\,5 \\
\underline{3\,0\,5} \\
0
\end{array}
$$

**Example**  Divide: $\dfrac{16.7}{1000}$.

$\dfrac{16.7}{1000}$      $0.016.7$

$\uparrow$          $\curvearrowleft$

3 zeros     Move 3 places to the left.

$\dfrac{16.7}{1000} = 0.0167$

**Example**  Divide: $\dfrac{-42.93}{0.001}$.

$\dfrac{-42.93}{0.001}$      $-42.930.$

$\uparrow$          $\curvearrowright$

3 decimal places     Move 3 places to the right.

$\dfrac{-42.93}{0.001} = -42{,}930$

**Practice Exercise**

**10.** Divide: $26.64 \div 3.6$.

**Practice Exercise**

**11.** Divide: $\dfrac{4.7}{100}$.

**Practice Exercise**

**12.** Divide: $\dfrac{-156.9}{0.01}$.

# Review Exercises

Convert the number in each sentence to standard notation. [5.3b]

**1.** Russia has the largest total area of any country in the world, at 6.59 million square miles.

**2.** Americans eat more than 3.1 billion lb of chocolate each year.
**Source:** Chocolate Manufacturers' Association

Write a word name. [5.1a]

**3.** 3.47

**4.** 0.031

**5.** $-13.4$

**6.** 0.0007

Write fraction notation. [5.1b]

**7.** 0.09

**8.** 4.561

**9.** $-0.089$

**10.** $-3.0227$

Write decimal notation. [5.1b]

**11.** $\dfrac{34}{1000}$

**12.** $-\dfrac{42{,}603}{10{,}000}$

**13.** $27\dfrac{91}{100}$

**14.** $-867\dfrac{6}{1000}$

Which number is larger? [5.1c]

**15.** 0.034, 0.0185

**16.** 0.91, 0.19

**17.** 0.741, 0.6943

**18.** $-1.038$, $-1.041$

Round 17.4287 to the nearest: [5.1d]

**19.** Tenth.

**20.** Hundredth.

**21.** Thousandth.

**22.** One.

**Add.**   [5.2a, c]

**23.**
```
      2.0 4 8
    6 5.3 7 1
  + 5 0 7.1
```

**24.**
```
     0.6
     0.0 0 4
     0.0 7
  + 0.0 0 9 8
```

**25.** $-219.3 + 2.8 + 7$

**26.** $0.41 + 4.1 + 41 + 0.041$

**Subtract.**   [5.2b, c]

**27.**
```
    3 0.0
  −   0.7 9 0 8
```

**28.**
```
    8 4 5.0 8
  −   5 4.7 9
```

**29.** $37.645 - (-8.497)$

**30.** $-70.8 - 0.0109$

**Multiply.**   [5.3a]

**31.**
```
      4 8
  ×  0.2 7
```

**32.** $-0.174 \cdot (-0.83)$

**33.** $100 \times 0.043$

**34.** $0.001 \times -24.68$

**Divide.**   [5.4a]

**35.** $5\,2\,)\overline{\,2\,3.4\,}$

**36.** $2.6\,)\overline{\,1\,1\,7.5\,2\,}$

**37.** $2.1\,4\,)\overline{\,2.1\,8\,7\,0\,8\,}$

**38.** $-60 \div 8$

**39.** $\dfrac{276.3}{1000}$

**40.** $\dfrac{-13.892}{0.01}$

**Solve.**   [5.2d], [5.4b]

**41.** $x + 51.748 = 548.0275$

**42.** $3 \cdot x = -20.85$

**43.** $10 \cdot y = 425.4$

**44.** $0.0089 + y = 5$

**Solve.**   [5.7a]

**45.** Stacia earned $620.80 working as a coronary intensive-care nurse during a 40-hr week. What is her hourly wage?

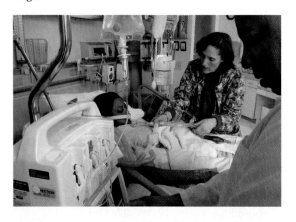

**46.** *Nutrition.*   The average person eats 688.6 lb of fruits and vegetables in a year. What is the average consumption in one day? (Use 1 year = 365 days.) Round to the nearest tenth of a pound.
**Source:** U.S. Department of Agriculture

**47.** Derek had $1034.46 in his checking account. He used his debit card to buy a Wii game system for $249.99. How much was left in his account?

**48.** *Cell-Phone Plan.*   In 2008, at&t offered its FamilyTalk cell-phone plan with 2 lines and 700 anytime minutes for a monthly access fee of $69.99. Minutes in excess of 700 were charged at the rate of $0.45 per minute. One month, Mr. and Mrs. Wu used their cell phones for 925 min. What was the charge?
**Source:** at&t

**49.** *Gas Mileage.*   Ellie wants to estimate the gas mileage of her car. At 36,057.1 mi, she fills the tank with 10.7 gal. At 36,217.6 mi, she fills the tank with 11.1 gal. Find the mileage per gallon. Round to the nearest tenth.

**50.** *Seafood Consumption.*   The graph below shows the annual consumption, in pounds, of seafood per person in the United States in some recent years.

**a)** Find the total consumption per person for the seven years.
**b)** Find the average annual consumption per person for the years shown, rounded to the nearest tenth.

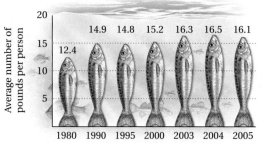

**Seafood Consumption**

SOURCE: U.S. Department of Agriculture

Estimate each of the following.   [5.6a]

**51.** The product $7.82 \times 34.487$ by rounding to the nearest one

**52.** The difference $219.875 - 4.478$ by rounding to the nearest one

**53.** The sum $\$45.78 + \$78.99$ by rounding to the nearest one

Find decimal notation. Use multiplying by 1.   [5.5a]

**54.** $\dfrac{13}{5}$       **55.** $-\dfrac{32}{25}$       **56.** $\dfrac{11}{4}$

Find decimal notation. Use division.   [5.5a]

**57.** $-\dfrac{13}{4}$       **58.** $\dfrac{7}{6}$       **59.** $\dfrac{17}{11}$

Round the answer to Exercise 59 to the nearest:   [5.5b]

**60.** Tenth.       **61.** Hundredth.       **62.** Thousandth.

Convert from cents to dollars.   [5.3b]

**63.** 8273¢                   **64.** 487¢

Convert from dollars to cents.   [5.3b]

**65.** $24.93                  **66.** $9.86

Calculate.   [5.4c], [5.5c]

**67.** $(8 - 1.23) \div (-4) + 5.6 \times 0.02$

**68.** $(1 + 0.07)^2 + 10^3 \div 10^2$
$+ [4(10.1 - 5.6) + 8(11.3 - 7.8)]$

**69.** $\dfrac{3}{4} \times (-20.85)$

**70.** Divide: $\dfrac{346.295}{0.001}$.   [5.4a]

**A.** 0.346295              **B.** 3.46295
**C.** 34,629.5             **D.** 346,295

**71.** Estimate the quotient $82.304 \div 17.287$ by rounding to the nearest ten.   [5.6a]

**A.** 0.4                  **B.** 4
**C.** 40                   **D.** 400

## Synthesis

**72.** 🖩 In each of the following, use one of $+$, $-$, $\times$, and $\div$ in each blank to make a true sentence.   [5.4c]
**a)** $2.56 \square 6.4 \square 51.2 \square 17.4 \square 89.7 = 72.62$
**b)** $(11.12 \square 0.29) \square 3^4 = 877.23$

**73.** Use the fact that $\frac{1}{3} = 0.\overline{3}$ to find repeating decimal notation for 1. Explain how you got your answer.   [5.5a]

# Understanding Through Discussion and Writing

**1.** Describe in your own words a procedure for converting from decimal notation to fraction notation.   [5.1b]

**2.** A student insists that $346.708 \times 0.1 = 3467.08$. How could you convince him that a mistake had been made without checking on a calculator?   [5.3a]

**3.** When is long division *not* the fastest way to convert from fraction notation to decimal notation?   [5.1b], [5.5a]

**4.** Consider finding decimal notation for $\frac{44}{125}$. Discuss as many ways as you can for finding such notation and give the answer.   [5.5a]

CHAPTER

**5**

**Test**

For Extra Help

CHAPTER
**Test Prep**
VIDEOS

Step-by-step test solutions are found on the Chapter Test Prep Videos available via the Video Resources on DVD, in **MyMathLab** , and on **You Tube** (search "BittingerBasicMathEI" and click on "Channels").

Convert the number in the sentence to standard notation.

**1.** The annual sales of antibiotics in the United States is $8.9 billion.
**Source:** IMS Health

**2.** There are 3.756 million people enrolled in bowling organizations in the United States.
**Source:** *Bowler's Journal International*, December 2000

Write a word name.

**3.** 2.34

**4.** 105.0005

Write fraction notation.

**5.** 0.91

**6.** −2.769

Write decimal notation.

**7.** $\dfrac{74}{1000}$

**8.** $-\dfrac{37{,}047}{10{,}000}$

**9.** $756\dfrac{9}{100}$

**10.** $-91\dfrac{703}{1000}$

Which number is larger?

**11.** 0.07, 0.162

**12.** −0.078, −0.06

**13.** 0.09, 0.9

Round 5.6783 to the nearest:

**14.** One.

**15.** Hundredth.

**16.** Thousandth.

**17.** Tenth.

Calculate.

**18.**
$$
\begin{array}{r}
0.7 \\
0.0\ 8 \\
0.0\ 0\ 9 \\
+\ 0.0\ 0\ 1\ 2 \\
\hline
\end{array}
$$

**19.** −102.4 + 6.1 + 78

**20.** 0.93 + 9.3 + 93 + 930

**21.**
$$
\begin{array}{r}
5\ 2.6\ 7\ 8 \\
-\ \ \ \ 4.3\ 2\ 1 \\
\hline
\end{array}
$$

**22.**
$$
\begin{array}{r}
2\ 0.0 \\
-\ \ \ 0.9\ 0\ 9\ 9 \\
\hline
\end{array}
$$

**23.** −234.6788 − 81.7854

**24.**
$$
\begin{array}{r}
0.1\ 2\ 5 \\
\times\ \ \ 0.2\ 4 \\
\hline
\end{array}
$$

**25.** 0.001 × (−213.45)

**26.** 1000 × 73.962

**27.** −19 ÷ 4

**28.** $3.3 \overline{)\ 1\ 0\ 0.3\ 2}$

**29.** $8\ 2 \overline{)\ 1\ 5.5\ 8}$

**30.** $\dfrac{-346.89}{1000}$

**31.** $\dfrac{346.89}{0.01}$

Solve.

**32.** $-4.8 \cdot y = 404.448$

**33.** $x + 0.018 = 9$

**34.** *Cell-Phone Plan.* In 2008, at&t offered its FamilyTalk cell-phone plan with 2 lines and 1400 anytime minutes for a monthly access fee of $89.99. Minutes in excess of 1400 were charged at the rate of $0.40 per minute. One month, Mr. and Mrs. Tews used their cell phones for 1510 min. What was the charge?

Source: at&t

**35.** *Gas Mileage.* Tina wants to estimate the gas mileage in her economy car. At 76,843 mi, she fills the tank with 14.3 gal of gasoline. At 77,310 mi, she fills the tank with 16.5 gal of gasoline. Find the mileage per gallon. Round to the nearest tenth.

**36.** *Checking Account Balance.* Nicholas has a balance of $820 in his checking account before making purchases of $123.89, $56.68, and $46.98 with his debit card. What was the balance after the purchases had been made?

**37.** The office manager for the Drake, Smith, and Hartner law firm buys 7 cases of copy paper at $41.99 per case. What is the total cost?

**38.** *Busiest Airports.* The graph below shows the numbers of passengers in 2007 who traveled through the country's busiest airports. Find the average number of passengers through these airports.

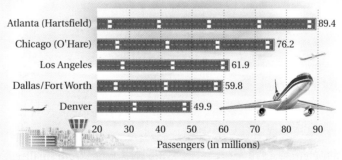

**Busiest Airports in the United States**

Atlanta (Hartsfield) — 89.4
Chicago (O'Hare) — 76.2
Los Angeles — 61.9
Dallas/Fort Worth — 59.8
Denver — 49.9

Passengers (in millions)

SOURCE: Airports Council International

Estimate each of the following.

**39.** The product $8.91 \times 22.457$ by rounding to the nearest one

**40.** The quotient $78.2209 \div 16.09$ by rounding to the nearest ten

Find decimal notation. Use multiplying by 1.

**41.** $\dfrac{8}{5}$

**42.** $\dfrac{22}{25}$

**43.** $-\dfrac{21}{4}$

Find decimal notation. Use division.

**44.** $\dfrac{3}{4}$

**45.** $-\dfrac{11}{9}$

**46.** $\dfrac{15}{7}$

Round the answer to Exercise 46 to the nearest:

**47.** Tenth.

**48.** Hundredth.

**49.** Thousandth.

Calculate.

**50.** $256 \div 3.2 \div 2 - 1.56 + 78.325 \times 0.02$

**51.** $(1 - 0.08)^2 + 6[5(12.1 - 8.7) + 10(14.3 - 9.6)]$

**52.** $-\dfrac{7}{8} \times (-345.6)$

**53.** $\dfrac{2}{3} \times 79.95 - \dfrac{7}{9} \times 1.235$

**54.** Convert from cents to dollars: 949¢.

**A.** 0.949¢

**B.** $9.49

**C.** $94.90

**D.** $949

## Synthesis

**55.** The Silver's Health Club generally charges a $79 membership fee and $42.50 a month. Allise has a coupon that will allow her to join the club for $299 for six months. How much will Allise save if she uses the coupon?

**56.** ▦ Arrange from smallest to largest.

$$-\dfrac{2}{3}, \ -\dfrac{15}{19}, \ -\dfrac{11}{13}, \ -\dfrac{5}{7}, \ -\dfrac{13}{15}, \ -\dfrac{17}{20}$$

# Cumulative Review

Convert to fraction notation.

**1.** $2\dfrac{2}{9}$

**2.** $-3.051$

Find decimal notation.

**3.** $-\dfrac{7}{5}$

**4.** $\dfrac{6}{11}$

**5.** Determine whether 43 is prime, composite, or neither.

**6.** Determine whether 2,053,752 is divisible by 4.

Calculate.

**7.** $48 + 12 \div 4 - 10 \times 2 + 6892 \div 4$

**8.** $0.2 - \{0.1[1.2(3.95 - 1.65) + 1.5 \div 2.5]\}$

Round to the nearest hundredth.

**9.** 584.973

**10.** $218.\overline{5}$

**11.** Estimate the product $16.392 \times 9.715$ by rounding to the nearest one.

**12.** Estimate by rounding to the nearest tenth:
$2.714 + 4.562 - 3.31 - 0.0023$.

**13.** Estimate the product $6418 \times 1984$ by rounding to the nearest hundred.

**14.** Estimate the quotient $717.832 \div 124.998$ by rounding to the nearest ten.

Add and simplify.

**15.**
$$\begin{array}{r} 2\dfrac{1}{4} \\[2mm] + \; 3\dfrac{4}{5} \\ \hline \end{array}$$

**16.**
$$\begin{array}{r} 3\,4{,}9\,2\,1 \\ 9\,3{,}0\,9\,2 \\ + \; 1\,1{,}1\,0\,3 \\ \hline \end{array}$$

**17.** $\dfrac{1}{6} + \dfrac{2}{3} + \dfrac{8}{9}$

**18.** $-143.9 + 2.053$

Subtract and simplify.

**19.** $723{,}041 - 12{,}904$

**20.** $19 - 5.903$

**21.** $5\dfrac{1}{7} - 4\dfrac{3}{7}$

**22.** $\dfrac{9}{10} - \dfrac{10}{11}$

Multiply and simplify.

**23.** $\dfrac{3}{8} \cdot \left(-\dfrac{4}{9}\right)$

**24.**
$$\begin{array}{r} 2\,5\,3\,2 \\ \times \; 2\,1\,0\,0 \\ \hline \end{array}$$

**25.**
$$\begin{array}{r} 2\,3.9 \\ \times \quad 0.2 \\ \hline \end{array}$$

**26.**
$$\begin{array}{r} 2\,7.9\,4\,3\,1 \\ \times \qquad 0.0\,0\,1 \\ \hline \end{array}$$

Divide and simplify.

**27.** $1\,6.5\,\overline{)\,3\,5.0\,1\,3}$

**28.** $2\,6\,\overline{)\,4\,7{,}9\,1\,8}$

**29.** $13.8621 \div 0.001$

**30.** $-\dfrac{4}{9} \div \left(-\dfrac{8}{15}\right)$

Solve.

**31.** $8.32 + x = 9.1$

**32.** $75 \cdot x = 2100$

**33.** $y \cdot 9.47 = -81.6314$

**34.** $1062 + y = 368,313$

**35.** $t + \dfrac{5}{6} = \dfrac{8}{9}$

**36.** $-\dfrac{7}{8} \cdot t = \dfrac{7}{16}$

**37.** *Organ Transplants.* There were 16,646 kidney transplants, 6136 liver transplants, and 2147 heart transplants in the United States in a recent year. How many transplants of these three organs were performed that year?

Source: U.S. Department of Health and Human Services

**38.** *Hospital Revenue.* Hospitals received $265 billion in revenue from private health insurance payments in a recent year. This was $88 billion more than was received from Medicare payments. How much revenue was received from Medicare payments?

Source: U.S. Census Bureau

**39.** After Lauren made a $450 down payment on a sofa, $\dfrac{3}{10}$ of the total cost was paid. How much did the sofa cost?

**40.** Joshua's tuition was $3600. He obtained a loan for $\dfrac{2}{3}$ of the tuition. How much was the loan?

**41.** The balance in Elliott's checking account is $314.79. After a check is written for $56.02, what is the balance in the account?

**42.** A clerk in Leah's Delicatessen sold $1\frac{1}{2}$ lb of ham, $2\frac{3}{4}$ lb of turkey, and $2\frac{1}{4}$ lb of roast beef. How many pounds of meat were sold altogether?

**43.** A baker used $\frac{1}{2}$ lb of sugar for cookies, $\frac{2}{3}$ lb of sugar for pie, and $\frac{5}{6}$ lb of sugar for cake. How much sugar was used in all?

**44.** The Currys' rectangular family room measures 19.8 ft by 23.6 ft. What is its area?

Simplify.

**45.** $\left(\dfrac{3}{4}\right)^2 - \dfrac{1}{8} \cdot \left(3 - 1\dfrac{1}{2}\right)^2$

**46.** $-1.2 \times 12.2 \div 0.1 \times 3.6$

## Synthesis

**47.** Using a manufacturer's coupon, Lucy bought 2 cartons of orange juice and received a third carton free. The price of each carton was $3.59. What was the cost per carton with the coupon? Round to the nearest cent.

**48.** A carton of gelatin mix packages weighs $15\frac{3}{4}$ lb. Each package weighs $1\frac{3}{4}$ oz. How many packages are in the carton? ($1$ lb $= 16$ oz)

# Ratio and Proportion

## Real-World Application

Of the 71 shark attacks recorded worldwide in 2007, 50 occurred in U.S. waters. What is the ratio of the number of shark attacks in U.S. waters to the number of shark attacks worldwide?

*Source:* University of Florida

***This problem appears as Example 7 in Section 6.1.***

# 6.1

# Introduction to Ratios

## OBJECTIVES

**a** Find fraction notation for ratios.

**b** Simplify ratios.

**SKILL TO REVIEW**

Objective 3.5b: Simplify fraction notation.

Simplify.

1. $\dfrac{16}{64}$     2. $-\dfrac{40}{24}$

## a  Ratios

> **RATIO**
>
> A **ratio** is the quotient of two quantities.

In the 2007–2008 regular basketball season, the Boston Celtics scored a total of 8245 points and allowed their opponents a total of 7404 points. The *ratio* of points scored to points allowed is given by the fraction notation

$$\dfrac{8245}{7404} \quad \begin{array}{l}\leftarrow \text{Points scored} \\ \leftarrow \text{Points allowed}\end{array}$$

or by the colon notation

$$8245 : 7404.$$

Points scored ———⤴  ⤴——— Points allowed

We read both forms of notation as "the ratio of 8245 to 7404."

> **RATIO NOTATION**
>
> The **ratio** of $a$ to $b$ is given by the fraction notation $\frac{a}{b}$, where $a$ is the numerator and $b$ is the denominator, or by the colon notation $a : b$.

**EXAMPLE 1**  Find the ratio of 7 to 8.

The ratio is $\dfrac{7}{8}$,  or  $7 : 8$.

**EXAMPLE 2**  Find the ratio of 31.4 to 100.

The ratio is $\dfrac{31.4}{100}$,  or  $31.4 : 100$.

**EXAMPLE 3**  Find the ratio of $4\frac{2}{3}$ to $5\frac{7}{8}$. You need not simplify.

The ratio is $\dfrac{4\frac{2}{3}}{5\frac{7}{8}}$,  or  $4\frac{2}{3} : 5\frac{7}{8}$.

1. Find the ratio of 5 to 11.

2. Find the ratio of 57.3 to 86.1.

3. Find the ratio of $6\dfrac{3}{4}$ to $7\dfrac{2}{5}$.

Do Margin Exercises 1–3.

In most of our work, we will use fraction notation for ratios.

*Answers*

*Skill to Review:*

1. $\dfrac{1}{4}$   2. $-\dfrac{5}{3}$

*Margin Exercises:*

1. $\dfrac{5}{11}$, or $5 : 11$   2. $\dfrac{57.3}{86.1}$, or $57.3 : 86.1$

3. $\dfrac{6\frac{3}{4}}{7\frac{2}{5}}$, or $6\dfrac{3}{4} : 7\dfrac{2}{5}$

**EXAMPLE 4** *Wind Speeds.* The average wind speed in Chicago is 10.4 mph. The average wind speed in Boston is 12.5 mph. Find the ratio of the wind speed in Chicago to the wind speed in Boston.

Source: *The Handy Geography Answer Book*

The ratio is $\dfrac{10.4}{12.5}$.

**EXAMPLE 5** *Record Rainfall.* The greatest rainfall ever recorded in the United States during a 12-month period was 739 in. in Kukui, Maui, Hawaii, from December 1981 to December 1982. What is the ratio of amount of rainfall, in inches, to time, in months? of time, in months, to amount of rainfall, in inches?

Source: *Time Almanac*

The ratio of amount of rainfall, in inches, to time, in months, is

$\dfrac{739}{12}$. $\begin{array}{l}\leftarrow \text{Rainfall} \\ \leftarrow \text{Time}\end{array}$

The ratio of time, in months, to amount of rainfall, in inches, is

$\dfrac{12}{739}$. $\begin{array}{l}\leftarrow \text{Time} \\ \leftarrow \text{Rainfall}\end{array}$

**EXAMPLE 6** Refer to the triangle below.

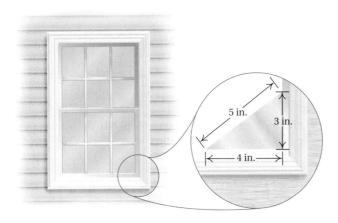

a) What is the ratio of the length of the longest side to the length of the shortest side?

$\dfrac{5}{3}$ $\begin{array}{l}\leftarrow \text{Longest side} \\ \leftarrow \text{Shortest side}\end{array}$

b) What is the ratio of the length of the shortest side to the length of the longest side?

$\dfrac{3}{5}$ $\begin{array}{l}\leftarrow \text{Shortest side} \\ \leftarrow \text{Longest side}\end{array}$

Do Exercises 4–6; Exercise 6 is on the next page.

**4. Record Snowfall.** The greatest snowfall recorded in North America during a 24-hr period was 76 in. in Silver Lake, Colorado, on April 14–15, 1921. What is the ratio of amount of snowfall, in inches, to time, in hours?

Source: U.S. Army Corps of Engineers

**5. Frozen Fruit Drinks.** A Berries & Kreme Chiller from Krispy Kreme contains 960 calories, while Smoothie King's MangoFest drink contains 258 calories. What is the ratio of the number of calories in the Krispy Kreme drink to the number of calories in the Smoothie King drink? of the number of calories in the Smoothie King drink to the number of calories in the Krispy Kreme drink?

Source: Physicians Committee for Responsible Medicine

*Answers*

4. $\dfrac{76}{24}$  5. $\dfrac{960}{258}$; $\dfrac{258}{960}$

**6.** In the triangle below, what is the ratio of the length of the shortest side to the length of the longest side?

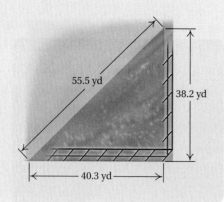

55.5 yd

38.2 yd

40.3 yd

**EXAMPLE 7** *Shark Attacks.* Of the 71 shark attacks recorded worldwide in 2007, 50 occurred in U.S. waters. The bar graph below shows the breakdown by state.

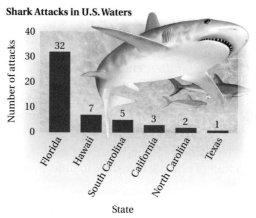

**Shark Attacks in U.S. Waters**

Number of attacks

40
32
30
20
10  7  5  3  2  1
0
Florida  Hawaii  South Carolina  California  North Carolina  Texas

State

SOURCE: University of Florida

a) What is the ratio of the number of shark attacks in U.S. waters to the number of shark attacks worldwide?

b) What is the ratio of the number of shark attacks in North Carolina to the number of shark attacks in South Carolina?

c) What is the ratio of the number of shark attacks in Florida to the total number of shark attacks in the other five states?

a) The ratio of the number of shark attacks in U.S. waters to the number of shark attacks worldwide is

$$\frac{50}{71}. \quad \begin{array}{l}\leftarrow \text{Attacks in U.S. waters} \\ \leftarrow \text{Attacks worldwide}\end{array}$$

b) The ratio of the number of shark attacks in North Carolina to the number of shark attacks in South Carolina is

$$\frac{2}{5}. \quad \begin{array}{l}\leftarrow \text{Attacks in North Carolina} \\ \leftarrow \text{Attacks in South Carolina}\end{array}$$

c) Of the 50 shark attacks in U.S. waters, 32 occurred in Florida. We subtract to determine how many shark attacks took place in the other five states. We have

$$50 - 32 = 18.$$

Thus the ratio of the number of shark attacks in Florida to the number of shark attacks in the other five states is

$$\frac{32}{18}. \quad \begin{array}{l}\leftarrow \text{Attacks in Florida} \\ \leftarrow \text{Attacks in other five states}\end{array}$$

Do Exercise 7.

**7. Soap Box Derby.** Of the 538 participants in the 2008 All-American Soap Box Derby, 296 were boys and 242 were girls. What was the ratio of girls to boys? of boys to girls? of boys to total number of participants?

**Source:** All-American Soap Box Derby

## b Simplifying Notation for Ratios

Sometimes a ratio can be simplified. This provides a means of finding other numbers with the same ratio.

**EXAMPLE 8** Find the ratio of 6 to 8. Then simplify to find two other numbers in the same ratio.

*Answers*

6. $\dfrac{38.2}{55.5}$  7. $\dfrac{242}{296}, \dfrac{296}{242}, \dfrac{296}{538}$

We write the ratio in fraction notation and then simplify:

$$\frac{6}{8} = \frac{2 \cdot 3}{2 \cdot 4} = \frac{2}{2} \cdot \frac{3}{4} = 1 \cdot \frac{3}{4} = \frac{3}{4}.$$

Thus, 3 and 4 have the same ratio as 6 and 8. We can express this by saying "6 is to 8 as 3 is to 4."

Do Exercise 8.

**8.** Find the ratio of 18 to 27. Then simplify to find two other numbers in the same ratio.

**EXAMPLE 9** Find the ratio of 2.4 to 10. Then simplify to find two other numbers in the same ratio.

We first write the ratio in fraction notation. Next, we multiply by 1 to clear the decimal from the numerator. Then we simplify.

$$\frac{2.4}{10} = \frac{2.4}{10} \cdot \frac{10}{10} = \frac{24}{100} = \frac{4 \cdot 6}{4 \cdot 25} = \frac{4}{4} \cdot \frac{6}{25} = \frac{6}{25}$$

Thus, 2.4 is to 10 as 6 is to 25.

Do Exercises 9 and 10.

**9.** Find the ratio of 3.6 to 12. Then simplify to find two other numbers in the same ratio.

**10.** Find the ratio of 1.2 to 1.5. Then simplify to find two other numbers in the same ratio.

**EXAMPLE 10** An HDTV screen that measures 46 in. diagonally has a width of 40 in. and a height of $22\frac{1}{2}$ in. Find the ratio of width to height and simplify.

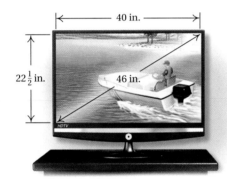

The ratio is $\dfrac{40}{22\frac{1}{2}} = \dfrac{40}{22.5} = \dfrac{40}{22.5} \cdot \dfrac{10}{10} = \dfrac{400}{225}$

$$= \frac{25 \cdot 16}{25 \cdot 9} = \frac{25}{25} \cdot \frac{16}{9}$$

$$= \frac{16}{9}.$$

Thus we can say that the ratio of width to height is 16 to 9, which can also be expressed as 16 : 9.

Do Exercise 11.

**11.** An HDTV screen that measures 44 in. diagonally has a width of 38.4 in. and a height of 21.6 in. Find the ratio of height to width and simplify.

*Answers*

**8.** 18 is to 27 as 2 is to 3.   **9.** 3.6 is to 12 as 3 is to 10.   **10.** 1.2 is to 1.5 as 4 is to 5.   **11.** $\dfrac{9}{16}$

**6.1** **Exercise Set**

For Extra Help

*MyMathLab*

Math XL
PRACTICE

WATCH

DOWNLOAD

READ

REVIEW

**a** Find fraction notation for each ratio. You need not simplify.

**1.** 4 to 5

**2.** 3 to 2

**3.** 178 to 572

**4.** 329 to 967

**5.** 0.4 to 12

**6.** 2.3 to 22

**7.** 3.8 to 7.4

**8.** 0.6 to 0.7

**9.** 56.78 to 98.35

**10.** 456.2 to 333.1

**11.** $8\frac{3}{4}$ to $9\frac{5}{6}$

**12.** $10\frac{1}{2}$ to $43\frac{1}{4}$

**13.** *Space Plane.* It is estimated that it will take 4 hr to fly from London to Sydney on the beyond-the-atmosphere space plane being developed in Europe. This 10,600-mi trip currently takes about 21 hr. What is the ratio of the time of the current trip to the time of the trip on the space plane? of the time of the trip on the space plane to the time of the current trip?

**Source:** EADS Atrium

**14.** *Silicon in the Earth's Crust.* Every 100 tons of the earth's crust contains about 28 tons of silicon. What is the ratio of the weight of silicon to the weight of crust? of the weight of crust to the weight of silicon?

**Source:** *The Handy Science Answer Book*

**15.** *Health Insurance Coverage.* Of every 100 people in the United States, 84.7 are covered by health insurance. What is the ratio of all people to those covered by health insurance? of those covered by health insurance to all people?

**Source:** U.S. Census Bureau

**16.** *Price of a Book.* The list price of the book *The Last Lecture* by Randy Pausch is $21.95, but the book recently sold for $12.07 on amazon.com. What is the ratio of the list price to the Amazon price? of the Amazon price to the list price?

**Source:** amazon.com

**17.** *Tax Freedom Day.* Of the 366 days in 2008 (a leap year), the average American worked 113 days to pay his or her federal, state, and local taxes. Find the ratio of the number of days worked to pay taxes in 2008 to the number of days in the year.

**Source:** Tax Foundation

**18.** *Spending Categories.* The average American worked 60 days in 2008 to pay housing and household operating expenses and 35 days to pay for food. Find the ratio of the number of days worked to pay for food to the number of days worked to pay housing and household operating expenses.

**Source:** Tax Foundation

**19.** *Population Estimates.* It is estimated that, of every 1000 people in the United States in 2050, 126 will be from 20 to 29 years old. What is the ratio of all people to those in the 20- to 29-year age group? of those in the 20- to 29-year age group to all people?

**Source:** U.S. Census Bureau

**20.** *Population Estimates.* It is estimated that, of every 1000 people in the United States in 2050, 118 will be age 75 and older. What is the ratio of all people to those age 75 and older? of those age 75 and older to all people?

**Source:** U.S. Census Bureau

**21.** *Field Hockey.* A diagram of the playing area for field hockey is shown below. What is the ratio of width to length? of length to width?

**Source:** *Sports: The Complete Visual Reference*

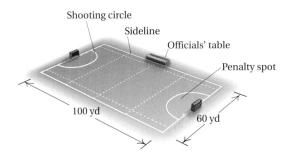

Shooting circle
Sideline
Officials' table
Penalty spot
100 yd
60 yd

**22.** *The Leaning Tower of Pisa.* The Leaning Tower of Pisa was reopened to the public in 2001 following a 10-yr stabilization project. The 184.5-ft tower now leans about 13 ft from its base. What is the ratio of the distance that it leans to its height? of its height to the distance that it leans?

**Source:** CNN

184.5 ft

13 ft

**b** Find the ratio of the first number to the second and simplify.

**23.** 4 to 6

**24.** 6 to 10

**25.** 18 to 24

**26.** 28 to 36

**27.** 4.8 to 10

**28.** 5.6 to 10

**29.** 2.8 to 3.6

**30.** 4.8 to 6.4

**31.** 20 to 30

**32.** 40 to 60

**33.** 56 to 100

**34.** 42 to 100

**35.** 128 to 256

**36.** 232 to 116

**37.** 0.48 to 0.64

**38.** 0.32 to 0.96

**39.** In this rectangle, find the ratios of length to width and of width to length.

478 ft

213 ft

**40.** In this right triangle, find the ratios of shortest length to longest length and of longest length to shortest length.

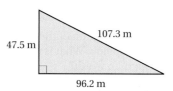

107.3 m

47.5 m

96.2 m

# Skill Maintenance

Use = or ≠ for ☐ to write a true sentence.   [3.5c]

**41.** $\frac{12}{8}$ ☐ $\frac{6}{4}$

**42.** $\frac{4}{7}$ ☐ $\frac{5}{9}$

**43.** $\frac{-7}{2}$ ☐ $\frac{-31}{9}$

**44.** $\frac{17}{25}$ ☐ $\frac{68}{100}$

Divide. Write decimal notation for the answer.   [5.4a]

**45.** $200 \div 0.4$

**46.** $95 \div 10$

**47.** $232 \div (-16)$

**48.** $-342 \div 2.25$

Solve.   [4.5c]

**49.** Rocky is $187\frac{1}{10}$ cm tall and his daughter is $180\frac{3}{4}$ cm tall. How much taller is Rocky?

**50.** Aunt Louise is $168\frac{1}{4}$ cm tall and her son is $150\frac{7}{10}$ cm tall. How much taller is Aunt Louise?

# Synthesis

**51.** Find the ratio of $3\frac{3}{4}$ to $5\frac{7}{8}$ and simplify.

*Fertilizer.*   Exercises 52 and 53 refer to a common lawn fertilizer known as "5, 10, 15." This mixture contains 5 parts of potassium for every 10 parts of phosphorus and 15 parts of nitrogen. (This is often denoted $5:10:15$.)

**52.** Find and simplify the ratio of potassium to nitrogen and of nitrogen to phosphorus.

**53.** Simplify the ratio $5:10:15$.

# 6.2 Rates and Unit Prices

## a Rates

A 2008 Kia Sportage EX can travel 414 mi on 18 gal of gasoline. Let's consider the ratio of miles to gallons:

**Source:** Kia Motors America, Inc.

$$\frac{414 \text{ mi}}{18 \text{ gal}} = \frac{414}{18} \frac{\text{miles}}{\text{gallon}} = \frac{23}{1} \frac{\text{miles}}{\text{gallon}}$$

$$= 23 \text{ miles per gallon} = 23 \text{ mpg}.$$

"per" means "division," or "for each."

The ratio

$$\frac{414 \text{ mi}}{18 \text{ gal}}, \quad \text{or} \quad \frac{414}{18} \frac{\text{mi}}{\text{gal}}, \quad \text{or } 23 \text{ mpg},$$

is called a **rate**.

### RATE

When a ratio is used to compare two different kinds of measure, we call it a **rate**.

Suppose David says his car travels 392.4 mi on 16.8 gal of gasoline. Is the mpg (mileage) of his car better than that of the Kia Sportage above? To determine this, it helps to convert the ratio to decimal notation and perhaps round. Then we have

$$\frac{392.4 \text{ miles}}{16.8 \text{ gallons}} = \frac{392.4}{16.8} \text{ mpg} \approx 23.357 \text{ mpg}.$$

Since $23.357 > 23$, David's car gets better mileage than the Kia Sportage does.

**EXAMPLE 1**   It takes 60 oz of grass seed to seed 3000 sq ft of lawn. What is the rate in ounces per square foot?

$$\frac{60 \text{ oz}}{3000 \text{ sq ft}} = \frac{1}{50} \frac{\text{oz}}{\text{sq ft}}, \quad \text{or} \quad 0.02 \frac{\text{oz}}{\text{sq ft}}$$

**EXAMPLE 2**   Martina bought 5 lb of organic russet potatoes for $4.99. What was the rate in cents per pound?

$$\frac{\$4.99}{5 \text{ lb}} = \frac{499 \text{ cents}}{5 \text{ lb}} = 99.8\text{¢/lb}$$

**EXAMPLE 3** A pharmacy student employed as a pharmacist's assistant earned $3930 for working 3 months one summer. What was the rate of pay per month?

The rate of pay is the ratio of money earned to length of time worked, or

$$\frac{\$3930}{3 \text{ mo}} = 1310 \frac{\text{dollars}}{\text{month}}, \quad \text{or} \quad \$1310 \text{ per month.}$$

**EXAMPLE 4** *Ratio of Strike-outs to Home Runs.* In the 2008 baseball season, Dustin Pedroia of the Boston Red Sox had 52 strikeouts and 17 home runs. What was his strikeout to home-run rate?

Source: Major League Baseball

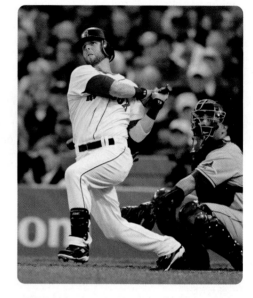

$$\frac{52 \text{ strikeouts}}{17 \text{ home runs}} = \frac{52 \text{ strikeouts}}{17 \text{ home runs}} = \frac{52}{17} \text{ strikeouts per home run}$$

$$\approx 3.06 \text{ strikeouts per home run}$$

Do Exercises 1–5.

A ratio of distance traveled to time is called *speed*. What is the rate, or speed, in miles per hour?

**1.** 45 mi, 9 hr

**2.** 120 mi, 10 hr

What is the rate, or speed, in feet per second?

**3.** 2200 ft, 2 sec

**4.** 52 ft, 13 sec

**5. Henry Aaron.** In his baseball career, Henry "Hank" Aaron had 1383 strikeouts and 755 home runs. What was his home-run to strikeout rate?

Source: Major League Baseball

## b Unit Pricing

**UNIT PRICE**

A **unit price**, or **unit rate**, is the ratio of price to the number of units.

**EXAMPLE 5** *Unit Price of Pears.* Ruth bought a 15.2-oz can of pears for $1.30. What is the unit price in cents per ounce?

$$\text{Unit price} = \frac{\text{Price}}{\text{Number of units}}$$

$$= \frac{\$1.30}{15.2 \text{ oz}} = \frac{130 \text{ cents}}{15.2 \text{ oz}} = \frac{130}{15.2} \frac{\text{cents}}{\text{oz}}$$

$$\approx 8.553 \text{ cents per ounce}$$

Do Exercise 6.

**6. Unit Price of Pasta Sauce.** Gregory bought a 26-oz jar of pasta sauce for $2.79. What is the unit price in cents per ounce?

*Answers*

**1.** 5 mi/hr, or 5 mph  **2.** 12 mi/hr, or 12 mph
**3.** 1100 ft/sec  **4.** 4 ft/sec  **5.** $\frac{755}{1383}$ home run per strikeout ≈ 0.546 home run per strikeout  **6.** 10.731 cents per ounce

Unit prices enable us to do comparison shopping and determine the best buy of a product on the basis of price. It is often helpful to change all prices to cents so that we can compare unit prices more easily.

**EXAMPLE 6** *Unit Price of Salad Dressing.* At the request of his customers, Angelo started bottling and selling the popular basil and lemon salad dressing that he serves in his neighborhood café. The dressing is sold in four sizes of containers as listed in the table below. Compute the unit price of each size of container and determine which size is the best buy on the basis of unit price alone.

| SIZE | PRICE | UNIT PRICE |
|------|-------|------------|
| 10 oz | $2.49 | 24.900¢/oz |
| 16 oz | $3.59 | 22.438¢/oz |
| 20 oz | $4.09 | 20.450¢/oz<br>Lowest unit price |
| 32 oz | $6.79 | 21.219¢/oz |

We compute the unit price of each size and fill in the chart:

10 oz: $\dfrac{\$2.49}{10\,\text{oz}} = \dfrac{249\,\text{cents}}{10\,\text{oz}} = \dfrac{249}{10}\dfrac{\text{cents}}{\text{oz}} = 24.900¢/\text{oz};$

16 oz: $\dfrac{\$3.59}{16\,\text{oz}} = \dfrac{359\,\text{cents}}{16\,\text{oz}} = \dfrac{359}{16}\dfrac{\text{cents}}{\text{oz}} \approx 22.438¢/\text{oz};$

20 oz: $\dfrac{\$4.09}{20\,\text{oz}} = \dfrac{409\,\text{cents}}{20\,\text{oz}} = \dfrac{409}{20}\dfrac{\text{cents}}{\text{oz}} = 20.450¢/\text{oz};$

32 oz: $\dfrac{\$6.79}{32\,\text{oz}} = \dfrac{679\,\text{cents}}{32\,\text{oz}} = \dfrac{679}{32}\dfrac{\text{cents}}{\text{oz}} \approx 21.219¢/\text{oz}.$

On the basis of unit price alone, we see that the 20-oz container is the best buy.

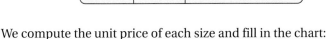

Do Exercise 7.

Although we often think that "bigger is cheaper," this is not always the case, as we see in Example 6. In addition, even when a larger package has a lower unit price than a smaller package, it still might not be the best buy for you. For example, some of the food in a large package could spoil before it is used, or you might not have room to store a large package.

**7. Cost of Mayonnaise.** Complete the following table for Hellmann's mayonnaise sold on an online shopping site. Which size has the lowest unit price?

Source: Peapod

| SIZE | PRICE | UNIT PRICE |
|------|-------|------------|
| 8 oz | $2.79 | |
| 10 oz | $3.69 | |
| 30 oz | $5.39 | |

**a** In Exercises 1–4, find each rate, or speed, as a ratio of distance to time. Round to the nearest hundredth where appropriate.

**1.** 120 km, 3 hr

**2.** 18 mi, 9 hr

**3.** 217 mi, 29 sec

**4.** 443 m, 48 sec

**5.** *Chevrolet Cobalt LS—City Driving.* A 2008 Chevrolet Cobalt LS will travel 312.5 mi on 12.5 gal of gasoline in city driving. What is the rate in miles per gallon?
**Source:** Chevrolet

**6.** *Mazda A3—City Driving.* A 2008 Mazda A3 will travel 348 mi on 14.5 gal of gasoline in city driving. What is the rate in miles per gallon?
**Source:** Ford Motor Company

**7.** *Mazda A3—Highway Driving.* A 2008 Mazda A3 will travel 624 mi on 19.5 gal of gasoline in highway driving. What is the rate in miles per gallon?
**Source:** Ford Motor Company

**8.** *Chevrolet Cobalt LS—Highway Driving.* A 2008 Chevrolet Cobalt LS will travel 486 mi on 13.5 gal of gasoline in highway driving. What is the rate in miles per gallon?
**Source:** Chevrolet

**9.** *Population Density of Monaco.* Monaco is a tiny country on the Mediterranean coast of France. It has an area of 0.75 square mile and a population of 32,796 people. What is the rate of number of people per square mile? The rate per square mile is called the *population density.* Monaco has the highest population density of any country in the world.
**Source:** The World Factbook

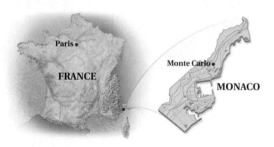

**10.** *Population Density of Australia.* The continent of Australia, with the island state of Tasmania, has an area of 2,967,893 sq mi and a population of 21,007,310 people. What is the rate of number of people per square mile? The rate per square mile is called the *population density.* Australia has one of the lowest population densities in the world.
**Source:** The World Factbook

**11.** A car is driven 500 mi in 20 hr. What is the rate in miles per hour? in hours per mile?

**12.** A student eats 3 hamburgers in 15 min. What is the rate in hamburgers per minute? in minutes per hamburger?

**13.** *Points per Game.* Dwayne Wade of the Miami Heat scored 1254 points in 51 games during the 2007–2008 basketball season. What was the rate in points per game?

**Source:** National Basketball Association

**14.** *Rebounds per Game.* Dwight Howard of the Orlando Magic got 1161 rebounds in 82 games during the 2007–2008 basketball season. What was the rate in rebounds per game?

**Source:** National Basketball Association

**15.** *Lawn Watering.* Watering a lawn adequately requires 623 gal of water for every 1000 ft$^2$. What is the rate in gallons per square foot?

**16.** A car is driven 200 km on 40 L of gasoline. What is the rate in kilometers per liter?

**17.** *Speed of Light.* Light travels 186,000 mi in 1 sec. What is its rate, or speed, in miles per second?

**Source:** *The Handy Science Answer Book*

**18.** *Speed of Sound.* Sound travels 1100 ft in 1 sec. What is its rate, or speed, in feet per second?

**Source:** *The Handy Science Answer Book*

**19.** Impulses in nerve fibers travel 310 km in 2.5 hr. What is the rate, or speed, in kilometers per hour?

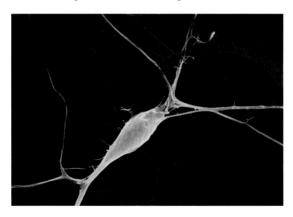

**20.** A black racer snake can travel 4.6 km in 2 hr. What is its rate, or speed, in kilometers per hour?

**21.** *Elephant Heart Rate.*   The heart of an elephant, at rest, will beat an average of 1500 beats in 60 min. What is the rate in beats per minute?

**Source:** *The Handy Science Answer Book*

**22.** *Human Heart Rate.*   The heart of a human, at rest, will beat an average of 4200 beats in 60 min. What is the rate in beats per minute?

**Source:** *The Handy Science Answer Book*

**b**   Find each unit price in Exercises 23–30. Then, in each exercise, determine which size is the better buy based on unit price alone.

**23.** *Hidden Valley Ranch Dressing.*

| SIZE | PRICE | UNIT PRICE |
|------|-------|-----------|
| 16 oz | $4.19 | |
| 20 oz | $5.29 | |

**24.** *Miracle Whip.*

| SIZE | PRICE | UNIT PRICE |
|------|-------|-----------|
| 32 oz | $5.29 | |
| 48 oz | $8.29 | |

**25.** *Cascade Powder Detergent.*

| SIZE | PRICE | UNIT PRICE |
|------|-------|-----------|
| 45 oz | $3.99 | |
| 75 oz | $5.49 | |

**26.** *Bush's Homestyle Baked Beans.*

| SIZE | PRICE | UNIT PRICE |
|------|-------|-----------|
| 16 oz | $1.59 | |
| 28 oz | $2.09 | |

**27.** *Jif Creamy Peanut Butter.*

| SIZE | PRICE | UNIT PRICE |
|------|-------|-----------|
| 18 oz | $2.50 | |
| 28 oz | $4.89 | |

**28.** *Hills Brothers Coffee.*

| SIZE | PRICE | UNIT PRICE |
|------|-------|-----------|
| 26 oz | $8.99 | |
| 39 oz | $12.99 | |

**29.** *Campbell's Condensed Tomato Soup.*

| SIZE | PRICE | UNIT PRICE |
|------|-------|-----------|
| 10.7 oz | $1.39 | |
| 26 oz | $2.69 | |

**30.** *Nabisco Saltines.*

| SIZE | PRICE | UNIT PRICE |
|------|-------|-----------|
| 16 oz | $3.59 | |
| 32 oz | $5.19 | |

Find the unit price of each brand in Exercises 31–34. Then, in each exercise, determine which brand is the best buy based on unit price alone.

**31.** *Vanilla Ice Cream.*

| BRAND | SIZE | PRICE |
| --- | --- | --- |
| B | 32 oz | $5.99 |
| E | 48 oz | $6.99 |

**32.** *Orange Juice.*

| BRAND | SIZE | PRICE |
| --- | --- | --- |
| M | 54 oz | $4.79 |
| T | 59 oz | $5.99 |

**33.** *Tomato Ketchup.*

| BRAND | SIZE | PRICE |
| --- | --- | --- |
| A | 24 oz | $2.49 |
| B | 36 oz | $3.29 |
| H | 46 oz | $3.69 |

**34.** *Yellow Mustard.*

| BRAND | SIZE | PRICE |
| --- | --- | --- |
| F | 14 oz | $1.29 |
| G | 19 oz | $1.99 |
| P | 20 oz | $2.49 |

## Skill Maintenance

Solve.

**35.** *Drive-in Movie Theaters.* The number of outdoor movie theaters has declined steadily since the 1950s. In 2007, there were 405 drive-in theaters. This is 3658 fewer than in 1958. How many drive-in movie theaters were there in 1958? [1.8a]

Source: Drive-Ins.com

**36.** A serving of fish steak (cross section) is generally $\frac{1}{2}$ lb. How many servings can be prepared from a cleaned $18\frac{3}{4}$-lb tuna? [4.6c]

**37.** *Gift-Card Sales.* Gift-card sales rose from $17.2 billion in 2003 to $26.3 billion in 2007. By how much did gift-card sales rise from 2003 to 2007? [5.7a]

Source: National Retail Federation

**38.** *Shifting Music Sales.* Sales of individual digital music tracks outpaced sales of albums by 176.9 million in the first quarter of 2008. During that quarter, 104.5 million albums were sold. How many individual tracks were sold during the same time period? [5.7a]

Source: Nielsen SoundScan

Multiply. [5.3a]

**39.**
$$\begin{array}{r} 4\,5.6\,7 \\ \times\quad 2.4 \\ \hline \end{array}$$

**40.**
$$\begin{array}{r} 6\,7\,8.1\,9 \\ \times\quad 1\,0\,0 \\ \hline \end{array}$$

**41.** $84.3 \times (-69.2)$

**42.** $(-1002.56) \times (-465)$

## Synthesis

**43.** Recently, some manufacturers have been changing the size of their containers in such a way that the consumer thinks the price of a product has been lowered when, in reality, a higher unit price is being charged.

Some aluminum juice cans are now concave (curved in) on the bottom. Suppose the volume of the can in the figure has been reduced from a fluid capacity of 6 oz to 5.5 oz, and the price of each can has been reduced from 65¢ to 60¢. Find the unit price of each container in cents per ounce.

$\frac{5}{16}$ in.

$1\frac{13}{16}$ in.

$2\frac{1}{16}$ in.

# 6.3

# Proportions

## OBJECTIVES

**a** Determine whether two pairs of numbers are proportional.

**b** Solve proportions.

**SKILL TO REVIEW**

Objective 3.5c: Use the test for equality to determine whether two fractions name the same number.

Use $=$ or $\neq$ for $\square$ to write a true sentence.

**1.** $\dfrac{18}{12} \square \dfrac{9}{6}$  **2.** $\dfrac{-5}{8} \square \dfrac{-4}{7}$

Determine whether the two pairs of numbers are proportional.

**1.** $3, 4$ and $6, 8$

**2.** $1, 4$ and $10, 39$

**3.** $1, 2$ and $20, 39$

Determine whether the two pairs of numbers are proportional.

**4.** $6.4, 12.8$ and $5.3, 10.6$

**5.** $6.8, 7.4$ and $3.4, 4.2$

## a Proportions

When two pairs of numbers, such as 3, 2 and 6, 4, have the same ratio, we say that they are **proportional**. The equation

$$\frac{3}{2} = \frac{6}{4}$$

states that the pairs 3, 2 and 6, 4 are proportional. Such an equation is called a **proportion**. We sometimes read $\frac{3}{2} = \frac{6}{4}$ as "3 is to 2 as 6 is to 4."

Since ratios can be written using fraction notation, we can use the test for equality of fractions discussed in Section 3.5 to determine whether two ratios are the same.

**EXAMPLE 1** Determine whether 1, 2 and 3, 6 are proportional.

We can use cross products:

$$1 \cdot 6 = 6 \qquad \frac{1}{2} \overset{?}{=} \frac{3}{6} \qquad 2 \cdot 3 = 6.$$

Since the cross products are the same, $6 = 6$, we know that $\frac{1}{2} = \frac{3}{6}$, so the numbers are proportional.

**EXAMPLE 2** Determine whether 2, 5 and 4, 7 are proportional.

We can use cross products:

$$2 \cdot 7 = 14 \qquad \frac{2}{5} \overset{?}{=} \frac{4}{7} \qquad 5 \cdot 4 = 20.$$

Since the cross products are not the same, $14 \neq 20$, we know that $\frac{2}{5} \neq \frac{4}{7}$, so the numbers are not proportional.

Do Margin Exercises 1–3.

**EXAMPLE 3** Determine whether 3.2, 4.8 and 0.16, 0.24 are proportional.

We can use cross products:

$$3.2 \times 0.24 = 0.768 \qquad \frac{3.2}{4.8} \overset{?}{=} \frac{0.16}{0.24} \qquad 4.8 \times 0.16 = 0.768.$$

Since the cross products are the same, $0.768 = 0.768$, we know that $\frac{3.2}{4.8} = \frac{0.16}{0.24}$, so the numbers are proportional.

Do Exercises 4 and 5.

**Answers**

*Skill to Review:*
**1.** $=$  **2.** $\neq$

*Margin Exercises:*
**1.** Yes  **2.** No  **3.** No  **4.** Yes  **5.** No

**EXAMPLE 4**  Determine whether $4\frac{2}{3}, 5\frac{1}{2}$ and $8\frac{7}{8}, 16\frac{1}{3}$ are proportional.

We can use cross products:

$$4\frac{2}{3} \cdot 16\frac{1}{3} = \frac{14}{3} \cdot \frac{49}{3} \qquad \frac{4\frac{2}{3}}{5\frac{1}{2}} \overset{?}{=} \frac{8\frac{7}{8}}{16\frac{1}{3}} \qquad 5\frac{1}{2} \cdot 8\frac{7}{8} = \frac{11}{2} \cdot \frac{71}{8}$$

$$= \frac{686}{9} \qquad\qquad\qquad\qquad\qquad = \frac{781}{16}$$

$$= 76\frac{2}{9}; \qquad\qquad\qquad\qquad\qquad = 48\frac{13}{16}.$$

Since the cross products are not the same, $76\frac{2}{9} \neq 48\frac{13}{16}$, we know that the numbers are not proportional.

> Do Exercise 6.

**6.** Determine whether $4\frac{2}{3}, 5\frac{1}{2}$ and $14, 16\frac{1}{2}$ are proportional.

# b  Solving Proportions

Let's now look at solving proportions. Consider the proportion

$$\frac{x}{3} = \frac{4}{6}.$$

One way to solve a proportion is to use cross products. Then we can divide on both sides to get the variable alone:

$$\frac{x}{3} = \frac{4}{6}$$

$$x \cdot 6 = 3 \cdot 4 \qquad \text{Equating cross products (finding cross products and setting them equal)}$$

$$\frac{x \cdot 6}{6} = \frac{3 \cdot 4}{6} \qquad \text{Dividing by 6 on both sides}$$

$$x = \frac{3 \cdot 4}{6} = \frac{12}{6} = 2.$$

We can check that 2 is the solution by replacing $x$ with 2 and finding cross products:

$$2 \cdot 6 = 12 \qquad \frac{2}{3} \overset{?}{=} \frac{4}{6} \qquad 3 \cdot 4 = 12.$$

Since the cross products are the same, it follows that $\frac{2}{3} = \frac{4}{6}$. Thus the pairs of numbers 2, 3 and 4, 6 are proportional, and 2 is the solution of the proportion.

> **SOLVING PROPORTIONS**
>
> To solve $\frac{x}{a} = \frac{c}{d}$ for $x$, equate *cross products* and divide on both sides to get $x$ alone.

> Do Exercise 7.

**7.** Solve: $\frac{x}{63} = \frac{2}{9}$.

*Answers*

**6.** Yes    **7.** 14

**EXAMPLE 5** Solve: $\dfrac{x}{7} = \dfrac{5}{3}$. Write a mixed numeral for the answer.

We have

$$\frac{x}{7} = \frac{5}{3}$$

$$x \cdot 3 = 7 \cdot 5 \qquad \text{Equating cross products}$$

$$\frac{x \cdot 3}{3} = \frac{7 \cdot 5}{3} \qquad \text{Dividing by 3}$$

$$x = \frac{7 \cdot 5}{3} = \frac{35}{3}, \text{ or } 11\frac{2}{3}.$$

The solution is $11\frac{2}{3}$.

Do Exercise 8.

**8.** Solve: $\dfrac{x}{9} = \dfrac{5}{4}$. Write a mixed numeral for the answer.

**EXAMPLE 6** Solve: $\dfrac{7.7}{15.4} = \dfrac{y}{2.2}$.

We have

$$\frac{7.7}{15.4} = \frac{y}{2.2}$$

$$7.7 \times 2.2 = 15.4 \times y \qquad \text{Equating cross products}$$

$$\frac{7.7 \times 2.2}{15.4} = \frac{15.4 \times y}{15.4} \qquad \text{Dividing by 15.4}$$

$$\frac{7.7 \times 2.2}{15.4} = y$$

$$\frac{16.94}{15.4} = y \qquad \text{Multiplying}$$

$$1.1 = y. \qquad \text{Dividing:}$$

$$
\begin{array}{r}
1.1 \\
15.4 \overline{\smash{)}\,16.9{\scriptstyle\wedge}4} \\
\underline{15\ 4\phantom{.}} \\
1\ 5\ 4 \\
\underline{1\ 5\ 4} \\
0
\end{array}
$$

The solution is 1.1.

**EXAMPLE 7** Solve: $\dfrac{8}{x} = \dfrac{5}{3}$. Write decimal notation for the answer.

We have

$$\frac{8}{x} = \frac{5}{3}$$

$$8 \cdot 3 = x \cdot 5 \qquad \text{Equating cross products}$$

$$\frac{8 \cdot 3}{5} = x \qquad \text{Dividing by 5}$$

$$\frac{24}{5} = x \qquad \text{Multiplying}$$

$$4.8 = x. \qquad \text{Simplifying}$$

The solution is 4.8.

**9.** Solve: $\dfrac{21}{5} = \dfrac{n}{2.5}$.

**10.** Solve: $\dfrac{6}{x} = \dfrac{25}{11}$. Write decimal notation for the answer.

Do Exercises 9 and 10.

*Answers*

**8.** $11\frac{1}{4}$   **9.** 10.5   **10.** 2.64

**EXAMPLE 8** Solve: $\dfrac{3.4}{4.93} = \dfrac{10}{n}$.

We have

$$\frac{3.4}{4.93} = \frac{10}{n}$$

$$3.4 \times n = 4.93 \times 10 \qquad \text{Equating cross products}$$

$$\frac{3.4 \times n}{3.4} = \frac{4.93 \times 10}{3.4} \qquad \text{Dividing by 3.4}$$

$$n = \frac{4.93 \times 10}{3.4}$$

$$n = \frac{49.3}{3.4} \qquad \text{Multiplying}$$

$$n = 14.5. \qquad \text{Dividing}$$

The solution is 14.5.

Do Exercise 11.

**11.** Solve: $\dfrac{0.4}{0.9} = \dfrac{4.8}{t}$.

**EXAMPLE 9** Solve: $\dfrac{4\frac{2}{3}}{5\frac{1}{2}} = \dfrac{14}{x}$.

We have

$$\frac{4\frac{2}{3}}{5\frac{1}{2}} = \frac{14}{x}$$

$$4\frac{2}{3} \cdot x = 5\frac{1}{2} \cdot 14 \qquad \text{Equating cross products}$$

$$\frac{14}{3} \cdot x = \frac{11}{2} \cdot 14 \qquad \text{Converting to fraction notation}$$

$$\frac{\dfrac{14}{3} \cdot x}{\dfrac{14}{3}} = \frac{\dfrac{11}{2} \cdot 14}{\dfrac{14}{3}} \qquad \text{Dividing by } \frac{14}{3}$$

$$x = \frac{11}{2} \cdot 14 \div \frac{14}{3}$$

$$x = \frac{11}{2} \cdot 14 \cdot \frac{3}{14} \qquad \text{Multiplying by the reciprocal of the divisor}$$

$$x = \frac{11 \cdot 3}{2} \qquad \text{Simplifying by removing a factor of 1: } \frac{14}{14} = 1$$

$$x = \frac{33}{2}, \text{ or } 16\frac{1}{2}.$$

The solution is $\frac{33}{2}$, or $16\frac{1}{2}$.

Do Exercise 12.

**12.** Solve:

$$\frac{8\frac{1}{3}}{x} = \frac{10\frac{1}{2}}{3\frac{3}{4}}.$$

# 6.3

## Exercise Set

**For Extra Help**

*MyMathLab*

Math XL
PRACTICE   WATCH   DOWNLOAD   READ   REVIEW

**a** Determine whether the two pairs of numbers are proportional.

**1.** 5, 6 and 7, 9

**2.** 7, 5 and 6, 4

**3.** 1, 2 and 10, 20

**4.** 7, 3 and 21, 9

**5.** 2.4, 3.6 and 1.8, 2.7

**6.** 4.5, 3.8 and 6.7, 5.2

**7.** $5\frac{1}{3}, 8\frac{1}{4}$ and $2\frac{1}{5}, 9\frac{1}{2}$

**8.** $2\frac{1}{3}, 3\frac{1}{2}$ and 14, 21

**b** Solve.

**9.** $\dfrac{18}{4} = \dfrac{x}{10}$

**10.** $\dfrac{x}{45} = \dfrac{20}{25}$

**11.** $\dfrac{x}{8} = \dfrac{9}{6}$

**12.** $\dfrac{8}{10} = \dfrac{n}{5}$

**13.** $\dfrac{t}{12} = \dfrac{5}{6}$

**14.** $\dfrac{12}{4} = \dfrac{x}{3}$

**15.** $\dfrac{2}{5} = \dfrac{8}{n}$

**16.** $\dfrac{10}{6} = \dfrac{5}{x}$

**17.** $\dfrac{n}{15} = \dfrac{10}{30}$

**18.** $\dfrac{2}{24} = \dfrac{x}{36}$

**19.** $\dfrac{16}{12} = \dfrac{24}{x}$

**20.** $\dfrac{8}{12} = \dfrac{20}{x}$

**21.** $\dfrac{6}{11} = \dfrac{12}{x}$

**22.** $\dfrac{8}{9} = \dfrac{32}{n}$

**23.** $\dfrac{20}{7} = \dfrac{80}{x}$

**24.** $\dfrac{36}{x} = \dfrac{9}{5}$

**25.** $\dfrac{12}{9} = \dfrac{x}{7}$

**26.** $\dfrac{x}{20} = \dfrac{16}{15}$

**27.** $\dfrac{x}{13} = \dfrac{2}{9}$

**28.** $\dfrac{8}{11} = \dfrac{x}{5}$

**29.** $\dfrac{100}{25} = \dfrac{20}{n}$

**30.** $\dfrac{35}{125} = \dfrac{7}{m}$

**31.** $\dfrac{6}{y} = \dfrac{18}{15}$

**32.** $\dfrac{15}{y} = \dfrac{3}{4}$

**33.** $\dfrac{x}{3} = \dfrac{0}{9}$

**34.** $\dfrac{x}{6} = \dfrac{1}{6}$

**35.** $\dfrac{1}{2} = \dfrac{7}{x}$

**36.** $\dfrac{2}{5} = \dfrac{12}{x}$

**37.** $\dfrac{1.2}{4} = \dfrac{x}{9}$

**38.** $\dfrac{x}{11} = \dfrac{7.1}{2}$

**39.** $\dfrac{8}{2.4} = \dfrac{6}{y}$

**40.** $\dfrac{3}{y} = \dfrac{5}{4.5}$

**41.** $\dfrac{t}{0.16} = \dfrac{0.15}{0.40}$

**42.** $\dfrac{0.12}{0.04} = \dfrac{t}{0.32}$

**43.** $\dfrac{0.5}{n} = \dfrac{2.5}{3.5}$

**44.** $\dfrac{6.3}{0.9} = \dfrac{0.7}{n}$

**45.** $\dfrac{1.28}{3.76} = \dfrac{4.28}{y}$

**46.** $\dfrac{10.4}{12.4} = \dfrac{6.76}{t}$

**47.** $\dfrac{7}{\frac{1}{4}} = \dfrac{28}{x}$

**48.** $\dfrac{5}{\frac{1}{3}} = \dfrac{3}{x}$

**49.** $\dfrac{\frac{1}{5}}{\frac{1}{10}} = \dfrac{\frac{1}{10}}{x}$

**50.** $\dfrac{\frac{1}{4}}{\frac{1}{2}} = \dfrac{\frac{1}{2}}{x}$

**51.** $\dfrac{y}{\frac{3}{5}} = \dfrac{\frac{7}{12}}{\frac{14}{15}}$

**52.** $\dfrac{\frac{5}{8}}{\frac{5}{4}} = \dfrac{y}{\frac{3}{2}}$

**53.** $\dfrac{x}{1\frac{3}{5}} = \dfrac{2}{15}$

**54.** $\dfrac{1}{7} = \dfrac{x}{4\frac{1}{2}}$

**55.** $\dfrac{2\frac{1}{2}}{3\frac{1}{3}} = \dfrac{x}{4\frac{1}{4}}$

**56.** $\dfrac{3\frac{1}{2}}{y} = \dfrac{6\frac{1}{2}}{4\frac{2}{3}}$

**57.** $\dfrac{5\frac{1}{5}}{6\frac{1}{6}} = \dfrac{y}{3\frac{1}{2}}$

**58.** $\dfrac{10\frac{3}{8}}{12\frac{2}{3}} = \dfrac{5\frac{3}{4}}{y}$

## Skill Maintenance

In each of Exercises 59–66, fill in the blank with the correct term from the given list. Some of the choices may not be used.

**59.** A ratio is the _____ of two quantities.   [6.1a]

**60.** A number is divisible by 9 if the _____ of the digits is divisible by 9.   [3.2a]

**61.** To compute a(n) _____ of a set of numbers, add the numbers and then divide by the number of addends.   [1.9c], [4.7a]

**62.** To convert from _____ to _____, move the decimal point two places to the right and change the $ sign in front to the ¢ sign at the end.   [5.3b]

**63.** In the equation $103 - 13 = 90$, the _____ is 13.   [1.3a]

**64.** The number 0.125 is an example of a(n) _____ decimal.   [5.5a]

**65.** The sentence $\dfrac{2}{5} \cdot \dfrac{4}{9} = \dfrac{4}{9} \cdot \dfrac{2}{5}$ illustrates the _____ law of multiplication.   [1.4a]

**66.** To solve $\dfrac{x}{a} = \dfrac{c}{d}$ for $x$, equate the _____ and divide on both sides to get $x$ alone.   [6.3b]

cross products
cents
dollars
terminating
repeating
sum
difference
minuend
subtrahend
associative
commutative
average
product
quotient

## Synthesis

Solve.

**67.** $\dfrac{1728}{5643} = \dfrac{836.4}{x}$

**68.** $\dfrac{328.56}{627.48} = \dfrac{y}{127.66}$

**69.** *Strikeouts per Home Run.*   Baseball Hall-of-Famer Babe Ruth had 1330 strikeouts and 714 home runs in his career. Hall-of-Famer Mike Schmidt had 1883 strikeouts and 548 home runs in his career. Find the rate of strikeouts per home run for each player. (These rates were considered among the highest in the history of the game and yet each made the Hall of Fame.)

# Mid-Chapter Review

## Concept Reinforcement

Determine whether each statement is true or false.

_____ **1.** A ratio is the quotient of two quantities. [6.1a]

_____ **2.** A rate is a ratio. [6.2a]

_____ **3.** The largest size package of an item always has the lowest unit price. [6.2b]

_____ **4.** If $\dfrac{x}{t} = \dfrac{y}{s}$, then $xy = ts$. [6.3b]

## Guided Solutions

**5.** What is the rate, or speed, in miles per hour? [6.2a]

120 mi, 2 hr

$$\frac{120\ \square}{\square\ \text{hr}} = \frac{120}{\square}\ \frac{\square}{\text{hr}} = \square\, \text{mi/hr}$$

**6.** Solve: $\dfrac{x}{4} = \dfrac{3}{6}$. [6.3b]

$$\frac{x}{4} = \frac{3}{6}$$

$x \cdot \square = \square \cdot 3$     Equating cross products

$\dfrac{x \cdot 6}{\square} = \dfrac{4 \cdot 3}{\square}$     Dividing on both sides

$x = \square$     Simplifying

## Mixed Review

Find fraction notation for each ratio. [6.1a]

**7.** 4 to 7

**8.** 313 to 199

**9.** 35 to 17

**10.** 59 to 101

Find the ratio of the first number to the second and simplify. [6.1b]

**11.** 8 to 12

**12.** 25 to 75

**13.** 32 to 28

**14.** 100 to 76

**15.** 112 to 56

**16.** 15 to 3

**17.** 2.4 to 8.4

**18.** 0.27 to 0.45

Find each rate, or speed, as a ratio of distance to time. Round to the nearest hundredth where appropriate.   [6.2a]

**19.** 243 mi, 4 hr          **20.** 146 km, 3 hr          **21.** 65 m, 5 sec          **22.** 97 ft, 6 sec

**23.** *Record Snowfall.*   The greatest recorded snowfall in a single storm occurred in the Mt. Shasta Ski Bowl in California, when 189 in. fell during a seven-day storm in 1959. What is the rate in inches per day?   [6.2a]
Source: U.S. Army Corps of Engineers

**24.** *Free Throws.*   During the 2007–2008 basketball season, Peja Stojakovic of the New Orleans Hornets attempted 140 free throws and made 130 of them. What is the rate in number of free throws made to number of free throws attempted? Round to the nearest thousandth.   [6.2a]
Source: National Basketball Association

**25.** Jerome bought an 18-oz jar of grape jelly for $2.09. What is the unit price in cents per ounce?   [6.2b]

**26.** Martha bought 12 oz of deli honey ham for $5.99. What is the unit price in cents per ounce?   [6.2b]

Determine whether the two pairs of numbers are proportional.   [6.3a]

**27.** 3, 7  and  15, 35          **28.** 9, 7  and  7, 5          **29.** 2.4, 1.5  and  3.2, 2.1          **30.** $1\frac{3}{4}, 1\frac{1}{3}$  and  $8\frac{3}{4}, 6\frac{2}{3}$

Solve.   [6.3b]

**31.** $\dfrac{9}{15} = \dfrac{x}{20}$

**32.** $\dfrac{x}{24} = \dfrac{30}{18}$

**33.** $\dfrac{12}{y} = \dfrac{20}{15}$

**34.** $\dfrac{2}{7} = \dfrac{10}{y}$

**35.** $\dfrac{y}{1.2} = \dfrac{1.1}{0.6}$

**36.** $\dfrac{0.24}{0.02} = \dfrac{y}{0.36}$

**37.** $\dfrac{\frac{1}{4}}{x} = \dfrac{\frac{1}{8}}{\frac{1}{4}}$

**38.** $\dfrac{1\frac{1}{2}}{3\frac{1}{4}} = \dfrac{7\frac{1}{2}}{x}$

# Understanding Through Discussion and Writing

**39.** Can every ratio be written as the ratio of some number to 1? Why or why not?   [6.1a]

**40.** What can be concluded about a rectangle's width if the ratio of length to perimeter is 1 to 3? Make some sketches and explain your reasoning.   [6.1a, b]

**41.** Instead of equating cross products, a student solves $\frac{x}{7} = \frac{5}{3}$ by multiplying on both sides by the least common denominator, 21. Is his approach a good one? Why or why not?   [6.3b]

**42.** An instructor predicts that a student's test grade will be proportional to the amount of time the student spends studying. What is meant by this? Write an example of a proportion that involves the grades of two students and their study times.   [6.3b]

# 6.4 Applications of Proportions

## a Applications and Problem Solving

Proportions have applications in such diverse fields as business, chemistry, health sciences, and home economics, as well as in many areas of daily life. Proportions are useful in making predictions.

**EXAMPLE 1** *Predicting Total Distance.* Donna drives her delivery van 800 mi in 3 days. At this rate, how far will she drive in 15 days?

1. **Familiarize.** We let $d$ = the distance traveled in 15 days.

2. **Translate.** We translate to a proportion. We make each side the ratio of distance to time, with distance in the numerator and time in the denominator.

$$\text{Distance in 15 days} \rightarrow \frac{d}{15} = \frac{800}{3} \leftarrow \text{Distance in 3 days} \\ \text{Time} \rightarrow \phantom{\frac{d}{15}} \phantom{=} \phantom{\frac{800}{3}} \leftarrow \text{Time}$$

It may help to verbalize the proportion above as "the unknown distance $d$ is to 15 days as the known distance 800 mi is to 3 days."

3. **Solve.** Next, we solve the proportion:

$$\frac{d}{15} = \frac{800}{3}$$

$$d \cdot 3 = 15 \cdot 800 \qquad \text{Equating cross products}$$

$$\frac{d \cdot 3}{3} = \frac{15 \cdot 800}{3} \qquad \text{Dividing by 3 on both sides}$$

$$d = \frac{15 \cdot 800}{3}$$

$$d = 4000. \qquad \text{Multiplying and dividing}$$

4. **Check.** We substitute into the proportion and check cross products:

$$\frac{4000}{15} = \frac{800}{3};$$

$$4000 \cdot 3 = 12{,}000; \qquad 15 \cdot 800 = 12{,}000.$$

The cross products are the same.

5. **State.** Donna will drive 4000 mi in 15 days.

Do Exercise 1.

Problems involving proportions can be translated in more than one way. For Example 1, any one of the following is also a correct translation:

$$\frac{15}{d} = \frac{3}{800}, \qquad \frac{15}{3} = \frac{d}{800}, \qquad \frac{800}{d} = \frac{3}{15}.$$

Equating the cross products in each proportion gives us the equation $d \cdot 3 = 15 \cdot 800$, which is the equation we obtained in Example 1.

### OBJECTIVE

a Solve applied problems involving proportions.

1. **Burning Calories.** The readout on Mary's treadmill indicates that she burns 108 calories when she walks for 24 min. How many calories will she burn if she walks at the same rate for 30 min?

*Answer*

1. 135 calories

**EXAMPLE 2** *Recommended Dosage.* To control a fever, a doctor suggests that a child who weighs 28 kg be given 320 mg of a liquid pain reliever. If the dosage is proportional to the child's weight, how much of the medication is recommended for a child who weighs 35 kg?

1. **Familiarize.** We let $t$ = the number of milligrams of the liquid pain reliever recommended for a child who weighs 35 kg.

2. **Translate.** We translate to a proportion, keeping the amount of medication in the numerators.

$$\text{Medication suggested} \rightarrow \frac{320}{28} = \frac{t}{35} \leftarrow \text{Medication suggested}$$
$$\text{Child's weight} \rightarrow \qquad\qquad \leftarrow \text{Child's weight}$$

3. **Solve.** Next, we solve the proportion:

$$320 \cdot 35 = 28 \cdot t \qquad \text{Equating cross products}$$

$$\frac{320 \cdot 35}{28} = \frac{28 \cdot t}{28} \qquad \text{Dividing by 28 on both sides}$$

$$\frac{320 \cdot 35}{28} = t$$

$$400 = t. \qquad \text{Multiplying and dividing}$$

4. **Check.** We substitute into the proportion and check cross products:

$$\frac{320}{28} = \frac{400}{35};$$

$$320 \cdot 35 = 11{,}200; \qquad 28 \cdot 400 = 11{,}200.$$

The cross products are the same.

5. **State.** The dosage for a child who weighs 35 kg is 400 mg.

　Do Exercise 2.

**2. Determining Paint Needs.** Lowell and Chris run a summer painting company to pay for their college expenses. They can paint 1600 ft² of clapboard with 4 gal of paint. How much paint would be needed for a building with 6000 ft² of clapboard?

**EXAMPLE 3** *Purchasing Tickets.* Carey bought 8 tickets to an international food festival for $52. How many tickets could she purchase with $90?

1. **Familiarize.** We let $n$ = the number of tickets that can be purchased with $90.

2. **Translate.** We translate to a proportion, keeping the number of tickets in the numerators.

$$\text{Tickets} \rightarrow \frac{8}{52} = \frac{n}{90} \leftarrow \text{Tickets}$$
$$\text{Cost} \rightarrow \qquad\qquad \leftarrow \text{Cost}$$

*Answer*

2. 15 gal

**3. Solve.** Next, we solve the proportion:

$$\frac{8}{52} = \frac{n}{90}$$

$8 \cdot 90 = 52 \cdot n$     Equating cross products

$$\frac{8 \cdot 90}{52} = \frac{52 \cdot n}{52}$$     Dividing by 52 on both sides

$$\frac{8 \cdot 90}{52} = n$$

$13.8 \approx n.$     Multiplying and dividing

Because it is impossible to buy a fractional part of a ticket, we must round our answer *down* to 13.

**4. Check.** As a check, we use a different approach: We find the cost per ticket and then divide $90 by that price. Since $52 \div 8 = 6.50$ and $90 \div 6.50 \approx 13.8$, we have a check.

**5. State.** Carey could purchase 13 tickets with $90.

Do Exercise 3.

**3. Purchasing Shirts.** If 2 shirts can be bought for $47, how many shirts can be bought with $200?

**EXAMPLE 4** *Waist-to-Hip Ratio.* To reduce the risk of heart disease, it is recommended that a man's waist-to-hip ratio be 0.9 or lower. Mac's hip measurement is 40 in. To meet the recommendation, what should his waist measurement be?

**Source:** Mayo Clinic

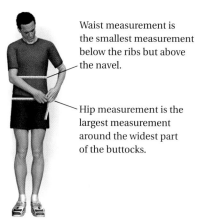

Waist measurement is the smallest measurement below the ribs but above the navel.

Hip measurement is the largest measurement around the widest part of the buttocks.

**1. Familiarize.** Note that $0.9 = \frac{9}{10}$. We let $w$ = Mac's waist measurement.

**2. Translate.** We translate to a proportion as follows:

Waist measurement $\rightarrow \dfrac{w}{40} = \dfrac{9}{10} \begin{matrix} \nwarrow \text{Recommended} \\ \swarrow \text{waist-to-hip ratio} \end{matrix}$
Hip measurement $\rightarrow$

**3. Solve.** Next, we solve the proportion:

$$\frac{w}{40} = \frac{9}{10}$$

$w \cdot 10 = 40 \cdot 9$     Equating cross products

$$\frac{w \cdot 10}{10} = \frac{40 \cdot 9}{10}$$     Dividing by 10 on both sides

$$w = \frac{40 \cdot 9}{10}$$

$w = 36.$     Multiplying and dividing

*Answer*

**3.** 8 shirts

**4. Waist-to-Hip Ratio.** It is recommended that a woman's waist-to-hip ratio be 0.85 or lower. Martina's hip measurement is 40 in. To meet the recommendation, what should her waist measurement be?
**Source:** Mayo Clinic

**4. Check.** As a check, we divide 36 by 40: $36 \div 40 = 0.9$. This is the desired ratio.

**5. State.** Mac's recommended waist measurement is 36 in. or less.

Do Exercise 4.

**EXAMPLE 5** *Construction Plans.* Architects make blueprints of projects to be constructed. These are scale drawings in which lengths are in proportion to actual sizes. The Hennesseys are adding a rectangular deck to their house. The architectural blueprints are rendered such that $\frac{3}{4}$ in. on the drawing is actually 2.25 ft on the deck. The width of the deck on the drawing is 4.3 in. How wide is the deck in reality?

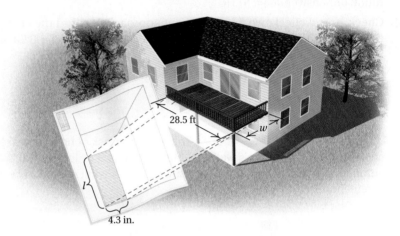

1. **Familiarize.** We let $w = $ the width of the deck.

2. **Translate.** Then we translate to a proportion, using 0.75 for $\frac{3}{4}$ in.

$$\text{Measure on drawing} \rightarrow \frac{0.75}{2.25} = \frac{4.3}{w} \leftarrow \text{Width on drawing} \\ \text{Measure on deck} \rightarrow \quad\quad\quad\quad \leftarrow \text{Width on deck}$$

3. **Solve.** Next, we solve the proportion:

$$0.75 \times w = 2.25 \times 4.3 \quad \text{Equating cross products}$$

$$\frac{0.75 \times w}{0.75} = \frac{2.25 \times 4.3}{0.75} \quad \text{Dividing by 0.75 on both sides}$$

$$w = \frac{2.25 \times 4.3}{0.75}$$

$$w = 12.9. \quad \text{Multiplying and dividing}$$

4. **Check.** We substitute into the proportion and check cross products:

$$\frac{0.75}{2.25} = \frac{4.3}{12.9};$$

$$0.75 \times 12.9 = 9.675; \quad 2.25 \times 4.3 = 9.675.$$

The cross products are the same.

5. **State.** The width of the deck is 12.9 ft.

Do Exercise 5.

**5. Construction Plans.** In Example 5, the length of the actual deck is 28.5 ft. What is the length of the deck on the blueprints?

*Answers*
**4.** 34 in. or less    **5.** 9.5 in.

**EXAMPLE 6** *Estimating a Wildlife Population.* To determine the number of fish in a lake, a conservationist catches 225 fish, tags them, and throws them back into the lake. Later, 108 fish are caught, and it is found that 15 of them are tagged. Estimate how many fish are in the lake.

1. **Familiarize.** We let $F$ = the number of fish in the lake. We assume that the ratio of the number of tagged fish to the total number of fish in the lake is the same as the ratio of the number of tagged fish caught later to the total number of fish caught later.

2. **Translate.** We translate to a proportion as follows:

   Fish tagged originally → $\dfrac{225}{F} = \dfrac{15}{108}$ ← Tagged fish caught later
   Fish in lake → $\phantom{\dfrac{225}{F}}$ ← Fish caught later

3. **Solve.** Next, we solve the proportion:

   $225 \cdot 108 = F \cdot 15$   Equating cross products

   $\dfrac{225 \cdot 108}{15} = \dfrac{F \cdot 15}{15}$   Dividing by 15 on both sides

   $\dfrac{225 \cdot 108}{15} = F$

   $1620 = F.$   Multiplying and dividing

4. **Check.** We substitute into the proportion and check cross products:

   $\dfrac{225}{1620} = \dfrac{15}{108};$

   $225 \cdot 108 = 24{,}300;$   $1620 \cdot 15 = 24{,}300.$

   The cross products are the same.

5. **State.** We estimate that there are 1620 fish in the lake.

Do Exercise 6.

6. **Estimating a Deer Population.** To determine the number of deer in a forest, a conservationist catches 153 deer, tags them, and releases them. Later, 62 deer are caught, and it is found that 18 of them are tagged. Estimate how many deer are in the forest.

*Answer*
**6.** 527 deer

# Translating
# for Success

1. *Calories in Cereal.* There are 140 calories in a $1\frac{1}{2}$-cup serving of Brand A cereal. How many calories are there in 6 cups of the cereal?

2. *Calories in Cereal.* There are 140 calories in 6 cups of Brand B cereal. How many calories are there in a $1\frac{1}{2}$-cup serving of the cereal?

3. *Gallons of Gasoline.* Jared's SUV traveled 310 mi on 15.5 gal of gasoline. At this rate, how many gallons would be needed to travel 465 mi?

4. *Gallons of Gasoline.* Elizabeth's new fuel-efficient car traveled 465 mi on 15.5 gal of gasoline. At this rate, how many gallons will be needed to travel 310 mi?

5. *Perimeter.* What is the perimeter of a rectangular field that measures 83.7 m by 62.4 m?

The goal of these matching questions is to practice step (2), *Translate*, of the five-step problem-solving process. Translate each word problem to an equation and select a correct translation from equations A–O.

**A.** $\dfrac{310}{15.5} = \dfrac{465}{x}$

**B.** $180 = 1\frac{1}{2} \cdot x$

**C.** $x = 71\frac{1}{8} - 76\frac{1}{2}$

**D.** $71\frac{1}{8} \cdot x = 74$

**E.** $74 \cdot 71\frac{1}{8} = x$

**F.** $x = 83.7 + 62.4$

**G.** $71\frac{1}{8} + x = 76\frac{1}{2}$

**H.** $x = 1\frac{2}{3} \cdot 180$

**I.** $\dfrac{140}{6} = \dfrac{x}{1\frac{1}{2}}$

**J.** $x = 2(83.7 + 62.4)$

**K.** $\dfrac{465}{15.5} = \dfrac{310}{x}$

**L.** $x = 83.7 \cdot 62.4$

**M.** $x = 180 \div 1\frac{1}{2}$

**N.** $\dfrac{140}{1\frac{1}{2}} = \dfrac{x}{6}$

**O.** $x = 1\frac{2}{3} \div 180$

*Answers on page A-10*

6. *Electric Bill.* Last month Todd's electric bills for his two rentals were $83.70 and $62.40. What was the total electric bill for the two properties?

7. *Package Tape.* A postal service center uses rolls of package tape that each contain 180 ft of tape. If it takes an average of $1\frac{1}{2}$ ft of tape per package, how many packages can be taped with one roll?

8. *Online Price.* Jane spent $180 for an area rug in a department store. Later, she saw the same rug for sale online and realized she had paid $1\frac{1}{2}$ times the online price. What was the online price?

9. *Heights of Sons.* Henry's three sons play basketball on three different college teams. Jeff's, Jason's, and Jared's heights are 74 in., $71\frac{1}{8}$ in., and $76\frac{1}{2}$ in., respectively. How much taller is Jared than Jason?

10. *Total Investment.* An investor bought 74 shares of stock at $71\frac{1}{8}$ per share. What was the total investment?

**a**   Solve.

**1.** *Study Time and Test Grades.*   An English instructor asserted that students' test grades are directly proportional to the amount of time spent studying. Lisa studies for 9 hr for a particular test and gets a score of 75. At this rate, for how many hours would she have had to study to get a score of 92?

**2.** *Study Time and Test Grades.*   A mathematics instructor asserted that students' test grades are directly proportional to the amount of time spent studying. Brent studies for 15 hr for a final exam and gets a score of 75. At this rate, what score would he have received if he had studied for 18 hr?

**3.** *Lefties.*   In a class of 40 students, on average, 6 will be left-handed. If a class includes 9 "lefties," how many students would you estimate are in the class?

**4.** *Sugaring.*   When 20 gal of maple sap are boiled down, the result is $\frac{1}{2}$ gal of maple syrup. How much sap is needed to produce 9 gal of syrup?

Source: University of Maine

**5.** *Gas Mileage.*   A 2008 Ford Mustang GT Convertible will travel 341 mi on 15.5 gal of gasoline in highway driving.

**a)** How many gallons of gasoline will it take to drive 2690 mi from Boston to Phoenix?

**b)** How far can the car be driven on 140 gal of gasoline?

Source: Ford Motor Company

**6.** *Gas Mileage.*   A 2008 Volkswagen Beetle will travel 462 mi on 16.5 gal of gasoline in highway driving.

**a)** How many gallons of gasoline will it take to drive 1650 mi from Pittsburgh to Albuquerque?

**b)** How far can the car be driven on 130 gal of gasoline?

Source: Volkswagen of America, Inc.

**7.** *Overweight Americans.*   A recent study determined that of every 100 Americans, 66 are overweight or obese. It is estimated that the U.S. population will be about 322 million in 2015. At the given rate, how many Americans would be considered overweight or obese in 2015?

Source: U.S. Centers for Disease Control and Prevention

**8.** *Prevalence of Diabetes.*   A recent study determined that of every 1000 Americans in the 65- to 74-year age group, 185 have been diagnosed with diabetes. It is estimated that there will be about 26.6 million Americans in this age group in 2015. At the given rate, how many in this age group will be diagnosed with diabetes in 2015?

Source: U.S. Centers for Disease Control and Prevention

**9.** *Quality Control.* A quality-control inspector examined 100 lightbulbs and found 7 of them to be defective. At this rate, how many defective bulbs will there be in a lot of 2500?

**10.** *Grading.* A high school math teacher must grade 32 reports on famous mathematicians. She can grade 4 reports in 90 min. At this rate, how long will it take her to grade all 32 reports?

**11.** *Painting.* Fred uses 3 gal of paint to cover 1275 ft$^2$ of siding. How much siding can Fred paint with 7 gal of paint?

**12.** *Waterproofing.* Bonnie can waterproof 450 ft$^2$ of decking with 2 gal of sealant. How many gallons should Bonnie buy for a 1200-ft$^2$ deck?

**13.** *Publishing.* Every 6 pages of an author's manuscript corresponds to 5 published pages. How many published pages will a 540-page manuscript become?

**14.** *Turkey Servings.* An 8-lb turkey breast contains 36 servings of meat. How many pounds of turkey breast would be needed for 54 servings?

**15.** *Currency Exchange.* On 12 September 2008, 1 U.S. dollar was worth about 0.5613 British pound.
  **a)** How much would 50 U.S. dollars be worth in British pounds?
  **b)** How much would a car that costs 8640 British pounds cost in U.S. dollars?

**16.** *Currency Exchange.* On 12 September 2008, 1 U.S. dollar was worth about 1.133 Swiss francs.
  **a)** How much would 360 U.S. dollars be worth in Swiss francs?
  **b)** How much would a pair of jeans that costs 80 Swiss francs cost in U.S. dollars?

**17.** *Currency Exchange.* On 12 September 2008, 1 U.S. dollar was worth about 107.197 Japanese yen.
  **a)** How much would 125 U.S. dollars be worth in Japanese yen?
  **b)** While traveling in Japan, Dana bought a vase for 5120 Japanese yen. How much did it cost in U.S. dollars?

**18.** *Currency Exchange.* On 12 September 2008, 1 U.S. dollar was worth about 10.5859 Mexican pesos.
  **a)** How much would 150 U.S. dollars be worth in Mexican pesos?
  **b)** While traveling in Mexico, Jake bought a watch that cost 3600 Mexican pesos. How much did it cost in U.S. dollars?

**19.** *Mileage.* Jean bought a new car. In the first 8 months, it was driven 9000 mi. At this rate, how many miles will the car be driven in 1 yr?

**20.** *Coffee Production.* Coffee beans from 18 trees are required to produce enough coffee each year for a person who drinks 2 cups of coffee per day. Jared brews 15 cups of coffee each day for himself and his coworkers. How many coffee trees are required for this?

**21.** *Cap'n Crunch's Peanut Butter Crunch® Cereal.* The nutritional chart on the side of a box of Quaker Cap'n Crunch's Peanut Butter Crunch® Cereal states that there are 110 calories in a $\frac{3}{4}$-cup serving. How many calories are there in 6 cups of the cereal?

**22.** *Rice Krispies® Cereal.* The nutritional chart on the side of a box of Kellogg's Rice Krispies® Cereal states that there are 130 calories in a $1\frac{1}{4}$-cup serving. How many calories are there in 5 cups of the cereal?

**Nutrition Facts**
Serving Size 3/4 Cup (27g)

| Amount Per Serving | Cereal Alone | with 1/2 Cup Vitamin A&D Fortified Skim Milk |
|---|---|---|
| **Calories** | 110 | 150 |
| Calories from Fat | 25 | 25 |
| | % Daily Value | |
| **Total Fat** 2.5g | 4% | 4% |
| Saturated Fat 1g | 5% | 6% |
| Trans Fat 0g | | |
| Polyunsaturated Fat 0.5g | | |
| Monounsaturated Fat 1g | | |
| **Cholesterol** 0mg | 0% | 1% |
| **Sodium** 200mg | 8% | 10% |
| **Potassium** 65mg | 2% | 7% |
| **Total Carbohydrate** 21g | 7% | 9% |
| Dietary Fiber 1g | 3% | 3% |
| Sugars 9g | | |
| Other Carbohydrate 11g | | |
| **Protein** 2g | | |

**Nutrition Facts**
Serving Size $1\frac{1}{4}$ Cups (33g/1.2oz )

| Amount Per Serving | Cereal | Cereal with 1/2 Cup Vitamins A&D Fat Free Milk |
|---|---|---|
| **Calories** | 130 | 170 |
| Calories from Fat | 0 | 0 |
| | % Daily Value | |
| **Total Fat** 0g | 0% | 0% |
| Saturated Fat 1g | 0% | 0% |
| Trans Fat 0g | | |
| **Cholesterol** 0mg | 0% | 0% |
| **Sodium** 220mg | 9% | 12% |
| **Potassium** 30mg | 1% | 7% |
| **Total Carbohydrate** 29g | 10% | 11% |
| Dietary Fiber 0g | 0% | 0% |
| Sugars 4g | | |
| Other Carbohydrate 25g | | |
| **Protein** 2g | | |

**23.** *Metallurgy.* In a metal alloy, the ratio of zinc to copper is 3 to 13. If there are 520 lb of copper, how many pounds of zinc are there?

**24.** *Class Size.* A college advertises that its student-to-faculty ratio is 14 to 1. If 56 students register for Introductory Spanish, how many sections of the course would you expect to see offered?

**25.** *Painting.* Helen can paint 950 ft$^2$ with 2 gal of paint. How many 1-gal cans does she need in order to paint a 30,000-ft$^2$ wall?

**26.** *Snow to Water.* Under typical conditions, $1\frac{1}{2}$ ft of snow will melt to 2 in. of water. To how many inches of water will $5\frac{1}{2}$ ft of snow melt?

**27.** *Grass-Seed Coverage.*   It takes 60 oz of grass seed to seed 3000 ft² of lawn. At this rate, how much would be needed for 5000 ft² of lawn?

**28.** *Grass-Seed Coverage.*   In Exercise 27, how much seed would be needed for 7000 ft² of lawn?

**29.** *Estimating a Deer Population.*   To determine the number of deer in a game preserve, a forest ranger catches 318 deer, tags them, and releases them. Later, 168 deer are caught, and it is found that 56 of them are tagged. Estimate how many deer are in the game preserve.

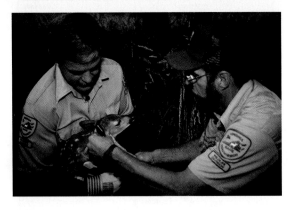

**30.** *Estimating a Trout Population.*   To determine the number of trout in a lake, a conservationist catches 112 trout, tags them, and throws them back into the lake. Later, 82 trout are caught, and it is found that 32 of them are tagged. Estimate how many trout there are in the lake.

**31.** *Map Scaling.*   On a road atlas map, 1 in. represents 16.6 mi. If two cities are 3.5 in. apart on the map, how far apart are they in reality?

**32.** *Map Scaling.*   On a map, $\frac{1}{4}$ in. represents 50 mi. If two cities are $3\frac{1}{4}$ in. apart on the map, how far apart are they in reality?

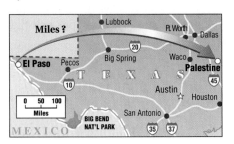

**33.** *Gasoline Mileage.* Nancy's van traveled 84 mi on 6.5 gal of gasoline. At this rate, how many gallons would be needed to travel 126 mi?

**34.** *Bicycling.* Roy bicycled 234 mi in 14 days. At this rate, how far would Roy bicycle in 42 days?

**35.** *Home Runs.* After playing 147 games in the 2008 Major League Baseball season, Ryan Howard of the Philadelphia Phillies had hit 43 home runs.
   **a)** At this rate, how many games would it take him to hit 50 home runs?
   **b)** At this rate, how many home runs would Howard hit in the 162-game baseball season?

Source: Major League Baseball

**36.** *Hits.* After playing 145 games in the 2008 Major League Baseball season, Ichiro Suzuki of the Seattle Mariners had 191 hits.
   **a)** At this rate, how many games would it take him to get 200 hits?
   **b)** At this rate, how many hits would Suzuki get in the 162-game baseball season?

Source: Major League Baseball

## Skill Maintenance

Determine whether each number is prime, composite, or neither.   [3.1c]

**37.** 1　　　　　　**38.** 83　　　　　　**39.** 28　　　　　　**40.** 93　　　　　　**41.** 47

Find the prime factorization of each number.   [3.1d]

**42.** 808　　　　**43.** 56　　　　**44.** 866　　　　**45.** 93　　　　**46.** 2020

## Synthesis

**47.** ▦ Carney College is expanding from 850 to 1050 students. To avoid any rise in the student-to-faculty ratio, the faculty of 69 professors must also increase. How many new faculty positions should be created?

**48.** ▦ In recognition of her outstanding work, Sheri's salary has been increased from $26,000 to $29,380. Tim is earning $23,000 and is requesting a proportional raise. How much more should he ask for?

**49.** *Baseball Statistics.* Cy Young, one of the greatest baseball pitchers of all time, gave up an average of 2.63 earned runs every 9 innings. Young pitched 7356 innings, more than anyone else in the history of baseball. How many earned runs did he give up?

**50.** ▦ *Real-Estate Values.* According to Coldwell Banker Real Estate Corporation, a home selling for $189,000 in Austin, Texas, would sell for $437,850 in Denver, Colorado. How much would a $350,000 home in Denver sell for in Austin? Round to the nearest $1000.

Source: Coldwell Banker Real Estate Corporation

**51.** ▦ The ratio 1:3:2 is used to estimate the relative costs of a CD player, receiver, and speakers when shopping for a sound system. That is, the receiver should cost three times the amount spent on the CD player and the speakers should cost twice the amount spent on the CD player. If you had $900 to spend, how would you allocate the money, using this ratio?

# 6.5

# Geometric Applications

## OBJECTIVES

**a** Find lengths of sides of similar triangles using proportions.

**b** Use proportions to find lengths in pairs of figures that differ only in size.

## **a** Proportions and Similar Triangles

Look at the pair of triangles below. Note that they appear to have the same shape, but their sizes are different. These are examples of **similar triangles**. You can imagine using a magnifying glass to enlarge the smaller triangle and get the larger. This process works because the corresponding sides of each triangle have the same ratio. That is, the following proportion is true.

$$\frac{a}{d} = \frac{b}{e} = \frac{c}{f}$$

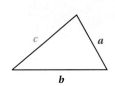 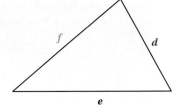

> **SIMILAR TRIANGLES**
>
> **Similar triangles** have the same shape. The lengths of their corresponding sides have the same ratio—that is, they are proportional.

**EXAMPLE 1** The triangles below are similar triangles. Find the missing length $x$.

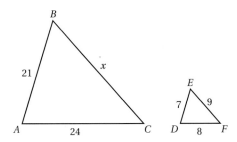

The ratio of $x$ to 9 is the same as the ratio of 24 to 8 or 21 to 7. We get the proportions

$$\frac{x}{9} = \frac{24}{8} \quad \text{and} \quad \frac{x}{9} = \frac{21}{7}.$$

We can solve either one of these proportions. We use the first one:

$$\frac{x}{9} = \frac{24}{8}$$

$$x \cdot 8 = 9 \cdot 24 \qquad \text{Equating cross products}$$

$$\frac{x \cdot 8}{8} = \frac{9 \cdot 24}{8} \qquad \text{Dividing by 8 on both sides}$$

$$x = 27. \qquad \text{Simplifying}$$

The missing length $x$ is 27. Other proportions could also be used.

Do Exercise 1.

**1.** This pair of triangles is similar. Find the missing length $x$.

*Answer*

1. 15

Similar triangles and proportions can often be used to find lengths that would ordinarily be difficult to measure. For example, we could find the height of a flagpole without climbing it or the distance across a river without crossing it.

**EXAMPLE 2** How high is a flagpole that casts a 56-ft shadow at the same time that a 6-ft man casts a 5-ft shadow?

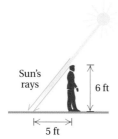
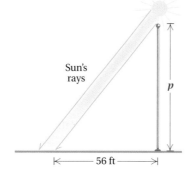

If we use the sun's rays to represent the third side of the triangle in our drawing of the situation, we see that we have similar triangles. Let $p$ = the height of the flagpole. The ratio of 6 to $p$ is the same as the ratio of 5 to 56. Thus we have the proportion

Height of man $\longrightarrow \dfrac{6}{p} = \dfrac{5}{56} \longleftarrow$ Length of shadow of man
Height of pole $\longrightarrow \dfrac{6}{p} = \dfrac{5}{56} \longleftarrow$ Length of shadow of pole

*Solve:* $\quad 6 \cdot 56 = p \cdot 5 \qquad$ Equating cross products

$$\frac{6 \cdot 56}{5} = \frac{p \cdot 5}{5} \qquad \text{Dividing by 5 on both sides}$$

$$\frac{6 \cdot 56}{5} = p \qquad \text{Simplifying}$$

$$67.2 = p.$$

The height of the flagpole is 67.2 ft.

Do Exercise 2.

**EXAMPLE 3** *Rafters of a House.* Carpenters use similar triangles to determine the lengths of rafters for a house. They first choose the pitch of the roof, or the ratio of the rise over the run. Then using a triangle with that ratio, they calculate the length of the rafter needed for the house. Loren is constructing rafters for a roof with a 6/12 pitch on a house that is 30 ft wide. Using a rafter guide, Loren knows that the rafter length corresponding to the 6/12 pitch is 13.4. Find the length $x$ of the rafter of the house to the nearest tenth of a foot.

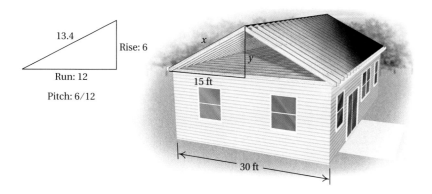

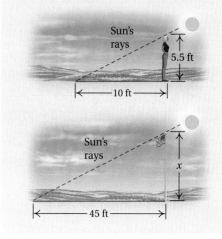

**2.** How high is a flagpole that casts a 45-ft shadow at the same time that a 5.5-ft woman casts a 10-ft shadow?

*Answer*

**2.** 24.75 ft

We have the proportion

Length of rafter
in 6/12 triangle $\longrightarrow$  $\dfrac{13.4}{x} = \dfrac{12}{15}$.  $\leftarrow$ Run in 6/12 triangle
Length of rafter $\longrightarrow$ $\leftarrow$ Run in similar
on the house $\qquad\qquad$ triangle on the house

*Solve:* $13.4 \cdot 15 = x \cdot 12$ $\qquad$ Equating cross products

$\dfrac{13.4 \cdot 15}{12} = \dfrac{x \cdot 12}{12}$ $\qquad$ Dividing by 12 on both sides

$\dfrac{13.4 \cdot 15}{12} = x$

$16.8 \text{ ft} \approx x.$ $\qquad$ Rounding to the nearest tenth of a foot

The length $x$ of the rafter of the house is about 16.8 ft.

---

**3. Rafters of a House.** Referring to Example 3, find the length $y$ of the rise of the rafter of the house to the nearest tenth of a foot.

Do Exercise 3.

## b Proportions and Other Geometric Shapes

When one geometric figure is a magnification of another, the figures are similar. Thus the corresponding lengths are proportional.

**EXAMPLE 4** The sides in the photographs below are proportional. Find the width of the larger photograph.

2.5 cm
$\longleftarrow$ 3.5 cm $\longrightarrow$

$x$
$\longleftarrow$ 10.5 cm $\longrightarrow$

We let $x = $ the width of the photograph. Then we translate to a proportion.

Larger width $\longrightarrow$ $\dfrac{x}{2.5} = \dfrac{10.5}{3.5}$ $\leftarrow$ Larger length
Smaller width $\longrightarrow$ $\leftarrow$ Smaller length

*Solve:* $x \times 3.5 = 2.5 \times 10.5$ $\qquad$ Equating cross products

$\dfrac{x \times 3.5}{3.5} = \dfrac{2.5 \times 10.5}{3.5}$ $\qquad$ Dividing by 3.5 on both sides

$x = \dfrac{2.5 \times 10.5}{3.5}$ $\qquad$ Simplifying

$x = 7.5.$

Thus the width of the larger photograph is 7.5 cm.

Do Exercise 4.

---

**4.** The sides in the photographs below are proportional. Find the width of the larger photograph.

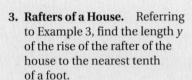
6 cm
$\longleftarrow$ 10 cm $\longrightarrow$

$x$
$\longleftarrow$ 35 cm $\longrightarrow$

---

*Answers*

**3.** 7.5 ft $\quad$ **4.** 21 cm

**EXAMPLE 5** A scale model of an addition to an athletic facility is 12 cm wide at the base and rises to a height of 15 cm. If the actual base is to be 52 ft, what will the actual height of the addition be?

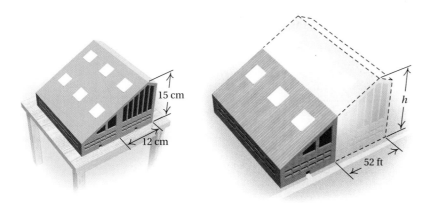

We let $h$ = the height of the addition. Then we translate to a proportion.

$$\begin{array}{c} \text{Width in model} \rightarrow \\ \text{Actual width} \rightarrow \end{array} \frac{12}{52} = \frac{15}{h} \begin{array}{l} \leftarrow \text{Height in model} \\ \leftarrow \text{Actual height} \end{array}$$

*Solve:* $12 \cdot h = 52 \cdot 15$      Equating cross products

$$\frac{12 \cdot h}{12} = \frac{52 \cdot 15}{12} \qquad \text{Dividing by 12 on both sides}$$

$$h = \frac{52 \cdot 15}{12} = 65.$$

Thus the height of the addition will be 65 ft.

Do Exercise 5.

**5.** Refer to the figures in Example 5. If a skylight on the model is 3 cm wide, how wide will the actual skylight be?

*Answer*

**5.** 13 ft

**a**   The triangles in each exercise are similar. Find the missing lengths.

**1.**

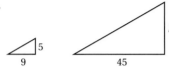

**2.**

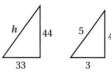

**3.**

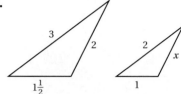

**4.**

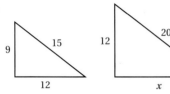

**5.**

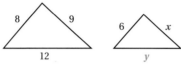

**6.**

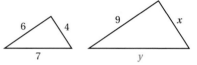

**7.**

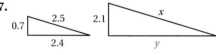

**8.**

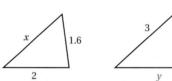

**9.** When a tree 8 m high casts a shadow 5 m long, how long a shadow is cast by a person 2 m tall?

**10.** How high is a flagpole that casts a 42-ft shadow at the same time that a $5\frac{1}{2}$-ft woman casts a 7-ft shadow?

**11.** How high is a tree that casts a 27-ft shadow at the same time that a 4-ft fence post casts a 3-ft shadow?

**12.** How high is a tree that casts a 32-ft shadow at the same time that an 8-ft light pole casts a 9-ft shadow?

**13.** Find the height $h$ of the wall.

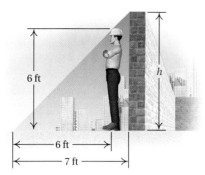

**14.** Find the length $L$ of the lake. Assume that the ratio of $L$ to 120 yd is the same as the ratio of 720 yd to 30 yd.

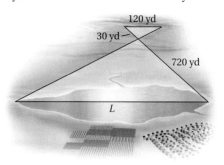

**15.** Find the distance across the river. Assume that the ratio of $d$ to 25 ft is the same as the ratio of 40 ft to 10 ft.

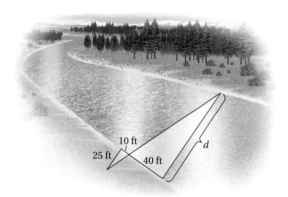

**16.** To measure the height of a hill, a string is stretched from level ground to the top of the hill. A 3-ft stick is placed under the string, touching it at point $P$, a distance of 5 ft from point $G$, where the string touches the ground. The string is then detached and found to be 120 ft long. How high is the hill?

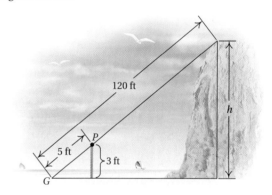

**b** In each of Exercises 17–26, the sides in each pair of figures are proportional. Find the missing lengths.

**17.**

**18.**

**19.**

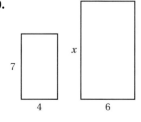

**20.**

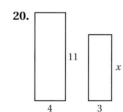

**21.**

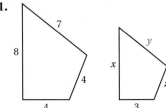

**22.**

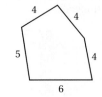

**23.**

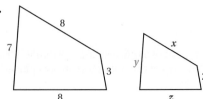

**24.**

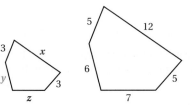

**25.**

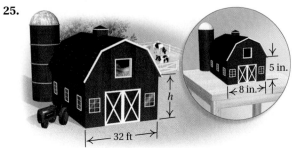

**26.**

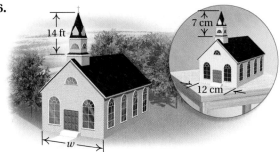

**27.** A scale model of an addition to an athletic facility is 15 cm wide at the base and rises to a height of 19 cm. If the actual base is to be 120 ft, what will the actual height of the addition be?

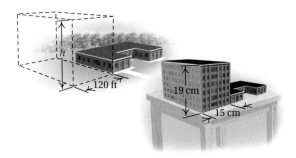

**28.** Refer to the figures in Exercise 27. If a skylight on the model is 3 cm wide, how wide will the actual skylight be?

## Skill Maintenance

**29.** *Expense Needs.* A student has $34.97 to spend for a book at $49.95, a CD at $14.88, and a sweatshirt at $29.95. How much more money does the student need to make these purchases? [5.7a]

**30.** Divide: $80.892 \div 8.4$. [5.4a]

Multiply. [5.3a]

**31.** $8.4 \times 80.892$

**32.** $0.01 \times (-274.568)$

**33.** $100 \times 274.568$

**34.** $-0.002 \times 274.568$

Find decimal notation and round to the nearest thousandth, if appropriate. [5.5a, b]

**35.** $\dfrac{17}{20}$

**36.** $\dfrac{73}{40}$

**37.** $\dfrac{10}{11}$

**38.** $-\dfrac{43}{51}$

## Synthesis

*Hockey Goals.* An official hockey goal is 6 ft wide. To make scoring more difficult, goalies often locate themselves far in front of the goal to "cut down the angle." In Exercises 39 and 40, suppose that a slapshot from point $A$ is attempted and that the goalie is 2.7 ft wide. Determine how far from the goal the goalie should be located if point $A$ is the given distance from the goal. (*Hint*: First find how far the goalie should be from point $A$.)

**39.** ▦ 25 ft

**40.** ▦ 35 ft

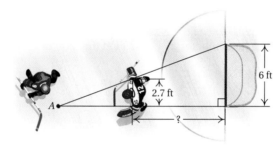

**41.** A miniature basketball hoop is built for the model referred to in Exercise 27. An actual hoop is 10 ft high. How high should the model hoop be?

▦ Solve. Round the answer to the nearest thousandth.

**42.** $\dfrac{8664.3}{10{,}344.8} = \dfrac{x}{9776.2}$

**43.** $\dfrac{12.0078}{56.0115} = \dfrac{789.23}{y}$

▦ The triangles in each exercise are similar triangles. Find the lengths not given.

**44.**

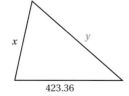

**45.**

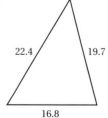

## Key Terms

ratio, p. 340
rate, p. 347

unit price, p. 348
unit rate, p. 348

proportion, p. 354
similar triangles, p. 374

## Concept Reinforcement

Determine whether each statement is true or false.

_____ **1.** When we simplify a ratio like $\frac{8}{12}$, we find two other numbers in the same ratio.   [6.1b]

_____ **2.** The proportion $\frac{a}{b} = \frac{c}{d}$ can also be written as $\frac{c}{a} = \frac{d}{b}$.   [6.3a]

_____ **3.** Similar triangles must be the same size.   [6.5a]

_____ **4.** Lengths of corresponding sides of similar figures have the same ratio.   [6.5b]

## Important Concepts

**Objective 6.1a**   Find fraction notation for ratios.

**Example**   Find the ratio of 7 to 18.

    Write a fraction with a numerator of 7 and a denominator of 18: $\frac{7}{18}$.

**Practice Exercise**

**1.** Find the ratio of 17 to 3.

**Objective 6.1b**   Simplify ratios.

**Example**   Simplify the ratio 8 to 2.5.

    We first write the ratio in fraction notation. Next, we multiply by 1 to clear the decimal from the denominator. Then we simplify.

$$\frac{8}{2.5} = \frac{8}{2.5} \cdot \frac{10}{10} = \frac{80}{25} = \frac{5 \cdot 16}{5 \cdot 5}$$

$$= \frac{5}{5} \cdot \frac{16}{5} = \frac{16}{5}$$

**Practice Exercise**

**2.** Simplify the ratio 3.2 to 2.8.

**Objective 6.2a**   Give the ratio of two different measures as a rate.

**Example**   A driver travels 156 mi on 6.5 gal of gas. What is the rate in miles per gallon?

$$\frac{156 \text{ mi}}{6.5 \text{ gal}} = \frac{156}{6.5} \frac{\text{mi}}{\text{gal}} = 24 \frac{\text{mi}}{\text{gal}}, \text{ or 24 mpg}$$

**Practice Exercise**

**3.** A student earned $120 for working 16 hr. What was the rate of pay per hour?

**Objective 6.2b** Find unit prices and use them to compare purchases.

**Example** A 16-oz can of Garcia diced tomatoes costs $1.00. A 20-oz can of Del's diced tomatoes costs $1.23. Which has the lower unit price?

We find each unit price:

For Garcia: $\dfrac{\$1.00}{16\,oz} = \dfrac{100\ cents}{16\,oz} = 6.25\,\dfrac{cents}{oz}$;

For Del's: $\dfrac{\$1.23}{20\,oz} = \dfrac{123\ cents}{20\,oz} = 6.15\,\dfrac{cents}{oz}$.

Thus Del's has the lower unit price.

**Practice Exercise**

4. A 28-oz jar of Brand A spaghetti sauce costs $2.79. A 32-oz jar of Brand B spaghetti sauce costs $3.29. Find the unit price of each brand and determine which is a better buy based on unit price alone.

---

**Objective 6.3a** Determine whether two pairs of numbers are proportional.

**Example** Determine whether 3, 4 and 7, 9 are proportional.

We have

$$3 \cdot 9 = 27 \qquad \dfrac{3}{4} \overset{?}{=} \dfrac{7}{9} \qquad 4 \cdot 7 = 28.$$

Since the cross products are not the same ($27 \neq 28$), $\dfrac{3}{4} \neq \dfrac{7}{9}$ and the numbers are not proportional.

**Practice Exercise**

5. Determine whether 7, 9 and 21, 27 are proportional.

---

**Objective 6.3b** Solve proportions.

**Example** Solve: $\dfrac{3}{4} = \dfrac{y}{7}$.

$$\dfrac{3}{4} = \dfrac{y}{7}$$

$3 \cdot 7 = 4 \cdot y$     Equating cross products

$\dfrac{3 \cdot 7}{4} = \dfrac{4 \cdot y}{4}$     Dividing by 4 on both sides

$\dfrac{21}{4} = y$

The solution is $\dfrac{21}{4}$.

**Practice Exercise**

6. Solve: $\dfrac{9}{x} = \dfrac{8}{3}$.

---

**Objective 6.4a** Solve applied problems involving proportions.

**Example** Martina bought 3 tickets to a campus theater production for $16.50. How much would 8 tickets cost?

We translate to a proportion.

$$\text{Tickets} \longrightarrow \dfrac{3}{16.50} = \dfrac{8}{c} \longleftarrow \text{Tickets}$$
$$\text{Cost} \longrightarrow \qquad\qquad \longleftarrow \text{Cost}$$

$3 \cdot c = 16.50 \cdot 8$     Equating cross products

$c = \dfrac{16.50 \cdot 8}{3}$

$c = 44$

Eight tickets would cost $44.

**Practice Exercise**

7. On a map, $\frac{1}{2}$ in. represents 50 mi. If two cities are $1\frac{3}{4}$ in. apart on the map, how far apart are they in reality?

**Objective 6.5a** Find lengths of sides of similar triangles using proportions.

**Example** The triangles below are similar. Find the missing length $x$.

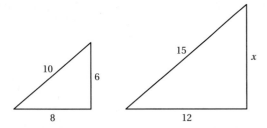

The ratio of 6 to $x$ is the same as the ratio of 8 to 12 (and also as the ratio of 10 to 15). We write and solve a proportion:

$$\frac{6}{x} = \frac{8}{12}$$

$$6 \cdot 12 = x \cdot 8$$

$$\frac{6 \cdot 12}{8} = \frac{x \cdot 8}{8}$$

$$9 = x.$$

The missing length is 9. (We could also have used other proportions, including $\frac{6}{x} = \frac{10}{15}$, to find $x$.)

**Practice Exercise**

**8.** The triangles below are similar. Find the missing length $y$.

# Review Exercises

Write fraction notation for each ratio. Do not simplify. [6.1a]

**1.** 47 to 84

**2.** 46 to 1.27

**3.** 83 to 100

**4.** 0.72 to 197

**5.** At Preston Seafood Market, 12,480 lb of tuna and 16,640 lb of salmon were sold one year. [6.1a, b]

   **a)** Write fraction notation for the ratio of tuna sold to salmon sold.

   **b)** Write fraction notation for the ratio of salmon sold to the total number of pounds of both kinds of fish sold.

Find the ratio of the first number to the second number and simplify. [6.1b]

**6.** 9 to 12

**7.** 3.6 to 6.4

**8.** *Gas Mileage.* The Chrysler PT Cruiser will travel 377 mi on 14.5 gal of gasoline in highway driving. What is the rate in miles per gallon? [6.2a]
Source: Chrysler Motor Corporation

**9.** *Flywheel Revolutions.* A certain flywheel makes 472,500 revolutions in 75 min. What is the rate of spin in revolutions per minute? [6.2a]

**10.** A lawn requires 319 gal of water for every 500 ft². What is the rate in gallons per square foot? [6.2a]

**11.** *Calcium Supplement.* The price for a particular calcium supplement is $18.99 for 300 tablets. Find the unit price in cents per tablet. [6.2b]

**12.** Nola bought a 24-oz loaf of 12-grain bread for $2.69. Find the unit price in cents per ounce. [6.2b]

**13.** *Vegetable Oil.* Find the unit price. Then determine which size has the lowest unit price. [6.2b]

| PACKAGE | PRICE | UNIT PRICE |
|---------|-------|------------|
| 32 oz | $4.79 | |
| 48 oz | $5.99 | |
| 64 oz | $9.99 | |

Determine whether the two pairs of numbers are proportional. [6.3a]

**14.** 9, 15 and 36, 60      **15.** 24, 37 and 40, 46.25

Solve. [6.3b]

**16.** $\dfrac{8}{9} = \dfrac{x}{36}$          **17.** $\dfrac{6}{x} = \dfrac{48}{56}$

**18.** $\dfrac{120}{\frac{3}{7}} = \dfrac{7}{x}$       **19.** $\dfrac{4.5}{120} = \dfrac{0.9}{x}$

Solve. [6.4a]

**20.** *Quality Control.* A factory manufacturing computer circuits found 3 defective circuits in a lot of 65 circuits. At this rate, how many defective circuits can be expected in a lot of 585 circuits?

**21.** *Exchanging Money.* On 16 September 2008, 1 U.S. dollar was worth about 1.068 Canadian dollars.
   **a)** How much would 250 U.S. dollars be worth in Canada?
   **b)** While traveling in Canada, Jamal saw a sweatshirt that cost 50 Canadian dollars. How much would it cost in U.S. dollars?

**22.** A train travels 448 mi in 7 hr. At this rate, how far will it travel in 13 hr?

**23.** Fifteen acres of land is required to produce 54 bushels of tomatoes. At this rate, how many acres are required to produce 97.2 bushels of tomatoes?

**24.** *Trash Production.* A study shows that 5 people generate 23 lb of trash each day. The population of Austin, Texas, is 743,074. How many pounds of trash are produced in Austin in one day?
**Sources:** U.S. Environmental Protection Agency; U.S. Census Bureau

**25.** *Snow to Water.* Under typical conditions, $1\frac{1}{2}$ ft of snow will melt to 2 in. of water. To how many inches of water will $4\frac{1}{2}$ ft of snow melt?

**26.** *Lawyers in Chicago.* In Illinois, there are about 4.8 lawyers for every 1000 people. The population of Chicago is 2,842,518. How many lawyers would you expect there to be in Chicago?
**Sources:** American Bar Association; U.S. Census Bureau

Each pair of triangles in Exercises 27 and 28 is similar. Find the missing length(s). [6.5a]

**27.**

**28.**

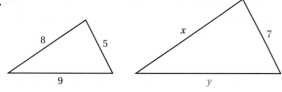

**29.** How high is a billboard that casts a 25-ft shadow at the same time that an 8-ft sapling casts a 5-ft shadow?  [6.5a]

**30.** The lengths in the figures below are proportional. Find the missing lengths.  [6.5b]

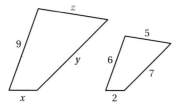

**31.** *Turkey Servings.*  A 25-lb turkey serves 18 people. Find the rate in servings per pound.  [6.2a]

   **A.** 0.36 serving/lb      **B.** 0.72 serving/lb
   **C.** 0.98 serving/lb      **D.** 1.39 servings/lb

**32.** If 3 dozen eggs cost $5.04, how much will 5 dozen eggs cost?  [6.4a]

   **A.** $6.72           **B.** $6.96
   **C.** $8.40           **D.** $10.08

## Synthesis

**33.** *Paper Towels.*  Find the unit price and determine which item would be the best buy based on unit price alone.  [6.2b]

| PACKAGE | PRICE | UNIT PRICE PER SHEET |
|---|---|---|
| 8 rolls, 60 (2 ply) sheets per roll | $6.38 | |
| 15 rolls, 60 (2 ply) sheets per roll | $13.99 | |
| 6 big rolls, 165 (2 ply) sheets per roll | $10.99 | |

**34.** ⊞ The following triangles are similar. Find the missing lengths.  [6.5a]

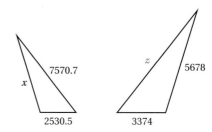

**35.** Shine-and-Glo Painters uses 2 gal of finishing paint for every 3 gal of primer. Each gallon of finishing paint covers 450 ft$^2$. If a surface of 4950 ft$^2$ needs both primer and finishing paint, how many gallons of each should be purchased?  [6.4a]

# Understanding Through Discussion and Writing

**1.** If you were a college president, which would you prefer: a low or high faculty-to-student ratio? Why?  [6.1a]

**2.** Can unit prices be used to solve proportions that involve money? Explain why or why not.  [6.2b], [6.4a]

**3.** Write a proportion problem for a classmate to solve. Design the problem so that the solution is "Leslie would need 16 gal of gasoline in order to travel 368 mi."  [6.4a]

**4.** Is it possible for two triangles to have two pairs of sides that are proportional without the triangles being similar? Why or why not?  [6.5a]

Test    For Extra Help

CHAPTER
Test Prep
VIDEOS

Step-by-step test solutions are found on the Chapter Test Prep Videos available via the Video Resources on DVD, in *MyMathLab*, and on You Tube (search "BittingerBasicMathEI" and click on "Channels").

Write fraction notation for each ratio. Do not simplify.

**1.** 85 to 97

**2.** 0.34 to 124

Find the ratio of the first number to the second number and simplify.

**3.** 18 to 20

**4.** 0.75 to 0.96

**5.** What is the rate in feet per second?

10 feet,  16 seconds

**6.** *Ham Servings.*    A 12-lb shankless ham contains 16 servings. What is the rate in servings per pound?

**7.** *Gas Mileage.*    The 2008 Chevrolet Malibu LTZ will travel 464 mi on 14.5 gal of gasoline in highway driving. What is the rate in miles per gallon?
Source: General Motors Corporation

**8.** *Bagged Salad Greens.*    Ron bought a 16-oz bag of salad greens for $2.49. Find the unit price in cents per ounce.

**9.** The table below lists prices for concentrated liquid laundry detergent. Find the unit price of each size in cents per ounce. Then determine which has the lowest unit price.

| SIZE | PRICE | UNIT PRICE |
|------|-------|-----------|
| 40 oz | $6.59 | |
| 50 oz | $6.99 | |
| 100 oz | $11.49 | |
| 150 oz | $24.99 | |

Determine whether the two pairs of numbers are proportional.

**10.** 7, 8  and  63, 72

**11.** 1.3, 3.4  and  5.6, 15.2

Solve.

**12.** $\dfrac{9}{4} = \dfrac{27}{x}$

**13.** $\dfrac{150}{2.5} = \dfrac{x}{6}$

**14.** $\dfrac{x}{100} = \dfrac{27}{64}$

**15.** $\dfrac{68}{y} = \dfrac{17}{25}$

Solve.

**16.** *Distance Traveled.*    An ocean liner traveled 432 km in 12 hr. At this rate, how far would the boat travel in 42 hr?

**17.** *Time Loss.*    A watch loses 2 min in 10 hr. At this rate, how much will it lose in 24 hr?

**18.** *Map Scaling.* On a map, 3 in. represents 225 mi. If two cities are 7 in. apart on the map, how far are they apart in reality?

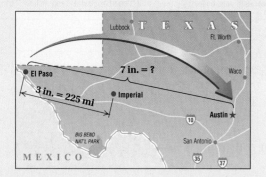

**19.** *Tower Height.* A birdhouse built on a pole that is 3 m high casts a shadow 5 m long. At the same time, the shadow of a tower is 110 m long. How high is the tower?

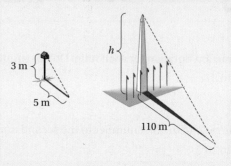

**20.** *Exchanging Money.* On 18 September 2008, 1 U.S. dollar was worth about 7.781 Hong Kong dollars.

a) How much would 450 U.S. dollars be worth in Hong Kong dollars?

b) While traveling in Hong Kong, Mitchell saw a DVD player that cost 795 Hong Kong dollars. How much would it cost in U.S. dollars?

**21.** *Thanksgiving Dinner.* A traditional turkey dinner for 8 people cost about $33.81 in a recent year. How much would it cost to serve a turkey dinner for 14 people?

**Source:** American Farm Bureau Federation

The sides in each pair of figures are proportional. Find the missing lengths.

**22.**

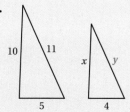

**23.**

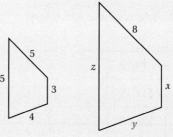

**24.** Lucita walks $4\frac{1}{2}$ mi in $1\frac{1}{2}$ hr. What is her rate in miles per hour?

**A.** $\frac{1}{3}$ mph        **B.** $1\frac{1}{2}$ mph

**C.** 3 mph        **D.** $4\frac{1}{2}$ mph

# Synthesis

**25.** Nancy wants to win a gift card from the campus bookstore by guessing the number of marbles in an 8-gal jar. She knows that there are 128 oz in a gallon. She goes home and fills an 8-oz jar with 46 marbles. How many marbles should she guess are in the 8-gal jar?

## CHAPTERS 1-6

# Cumulative Review

Calculate and simplify.

**1.**
$$\begin{array}{r} 2\,7.6\,8 \\ 3.0\,1\,9 \\ +\,4\,8\,3.2\,9\,7 \\ \hline \end{array}$$

**2.**
$$\begin{array}{r} 2\frac{1}{3} \\ +\,4\frac{5}{12} \\ \hline \end{array}$$

**3.** $-\dfrac{6}{35} + \left(-\dfrac{5}{28}\right)$

**4.**
$$\begin{array}{r} 4\,0.2 \\ -\quad 9.7\,0\,9 \\ \hline \end{array}$$

**5.** $73.82 - 0.908$

**6.** $-\dfrac{4}{15} - \dfrac{3}{20}$

**7.**
$$\begin{array}{r} 3\,7.6\,4 \\ \times\quad\ 5.9 \\ \hline \end{array}$$

**8.** $5.678 \times 100$

**9.** $2\dfrac{1}{3} \cdot \left(-1\dfrac{2}{7}\right)$

**10.** $2.3\,)\overline{\,9\,8.9\,}$

**11.** $5\,4\,)\overline{\,4\,8{,}5\,4\,6\,}$

**12.** $-\dfrac{7}{11} \div \dfrac{14}{33}$

**13.** Write expanded notation: 30,074.

**14.** Write a word name for 120.07.

Which number is larger?

**15.** 0.7, 0.698

**16.** $-0.8,\ -0.799$

**17.** Find the prime factorization of 144.

**18.** Find the LCM of 18 and 30.

**19.** What part is shaded?

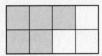

**20.** Simplify: $\dfrac{90}{144}$.

Calculate.

**21.** $\dfrac{3}{5} \times 9.53$

**22.** $\dfrac{1}{3} \times 0.645 - \dfrac{3}{4} \times 0.048$

**23.** Write fraction notation for the ratio 0.3 to 15. Do not simplify.

**24.** Determine whether the pairs 3, 9 and 25, 75 are proportional.

**25.** What is the rate in meters per second?

660 meters,  12 seconds

**26.** *Unit Prices.* An 8-oz can of pineapple chunks costs $0.99. A 24.5-oz jar of pineapple chunks costs $3.29. Which has the lower unit price?

Solve.

**27.** $\dfrac{14}{25} = \dfrac{x}{54}$

**28.** $423 = 16 \cdot t$

**29.** $\dfrac{2}{3} \cdot y = \dfrac{16}{27}$

**30.** $\dfrac{7}{16} = \dfrac{56}{x}$

**31.** $34.56 + n = -67.9$

**32.** $t + \dfrac{7}{25} = \dfrac{5}{7}$

**33.** Ramona's recipe for fettuccini Alfredo has 520 calories in 1 cup. How many calories are there in $\frac{3}{4}$ cup?

**34.** A machine can stamp out 925 washers in 5 min. An order is placed for 1295 washers. How long will it take to stamp them out?

**35.** A 46-oz juice can contains $5\frac{3}{4}$ cups of juice. A recipe calls for $3\frac{1}{2}$ cups of juice. How many cups are left over?

**36.** It takes a carpenter $\frac{2}{3}$ hr to hang a door. How many doors can the carpenter hang in 8 hr?

**37.** *Car Travel.* A car travels 337.62 mi in 8 hr. How far does it travel in 1 hr?

**38.** *Shuttle Orbits.* A space shuttle made 16 orbits a day during an 8.25-day mission. How many orbits were made during the entire mission?

**39.** How many even prime numbers are there?
  **A.** 5      **B.** 3      **C.** 2
  **D.** 1      **E.** None

**40.** The gas mileage of a car is 28.16 mpg. How many gallons per mile is this?
  **A.** $\dfrac{704}{25}$      **B.** $\dfrac{25}{704}$      **C.** $\dfrac{2816}{100}$
  **D.** $\dfrac{250}{704}$      **E.** None

## Synthesis

**41.** A soccer goalie wishing to block an opponent's shot moves toward the shooter to reduce the shooter's view of the goal. If the goalie can only defend a region 10 ft wide, how far in front of the goal should the goalie be? (See the figure at right.)

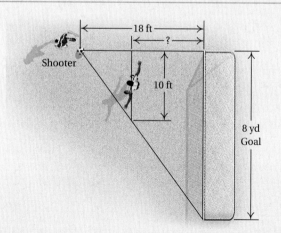

# Percent Notation

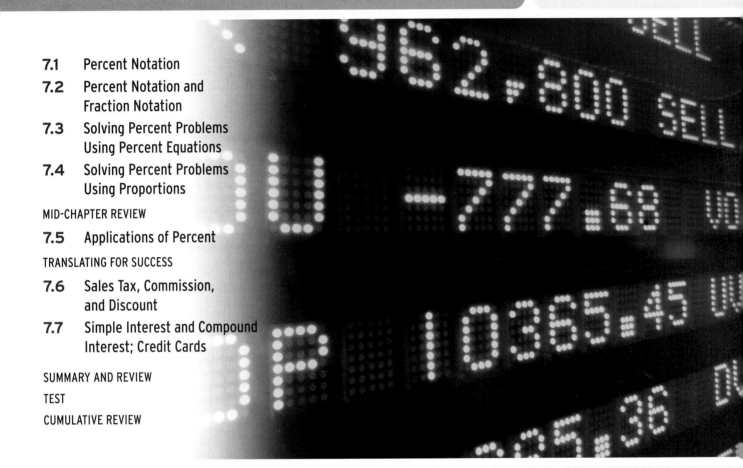

## Real-World Application

The Dow Jones Industrial Average (DJIA) plunged from 11,143 to 10,365 on September 29, 2008. This was the largest one-day drop in its history. What was the percent of decrease?

*Sources: Nightly Business Reports, September 29, 2008; DJIA*

***This problem appears as Example 3 in Section 7.5.***

# Percent Notation

## (a) Understanding Percent Notation

Of all the surface area of the earth, 70% is covered by water. What does this mean? It means that of every 100 square miles of the earth's surface area, 70 square miles are covered by water. Thus, 70% is a ratio of 70 to 100, or $\frac{70}{100}$.

**Source:** *The Handy Geography Answer Book*

70 of 100 squares are shaded.

70% or $\frac{70}{100}$ or 0.70 of the large square is shaded.

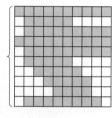

Percent notation is used extensively in our everyday lives. Here are some examples.

25.2% of the adult population in Washington, D.C., has a graduate degree.

Almost 22% of the glass that is produced is recycled.

A blood alcohol level of 0.08% is the standard used by most states as the legal limit for drunk driving.

71% of people in the United States use the Internet; 17% of people in China use the Internet.

In California, 42.5% of people age 5 and older speak a language other than English at home. In Kansas, the percentage is 10.3%.

Percent notation is often represented using a circle graph, or pie chart, to show how the parts of a quantity are related. For example, the circle graph at left illustrates the percentage of people in the United States with each of the four blood types.

**Blood Types in the United States**

Type AB 4%
Type B 10%
Type O 44%
Type A 42%

SOURCE: bloodcenter.stanford.edu/about_blood/blood_types.html

> **PERCENT NOTATION**
>
> The notation *n%* means "*n* per hundred."

This definition leads us to the following equivalent ways of defining percent notation.

---

**NOTATION FOR *n*%**

**Percent notation, *n*%,** can be expressed using:

ratio $\longrightarrow n\% = $ the ratio of $n$ to $100 = \dfrac{n}{100}$,

fraction notation $\longrightarrow n\% = n \times \dfrac{1}{100}$, or

decimal notation $\longrightarrow n\% = n \times 0.01$.

---

At the airport in Oslo, Norway, 64% of all passengers arrive and leave using trains and buses. At the airport in San Francisco, the percentage is 23%.

SOURCE: Transportation Research Board

**EXAMPLE 1** Write three kinds of notation for 35%.

Using ratio: $\qquad\qquad 35\% = \dfrac{35}{100}$ $\qquad$ A ratio of 35 to 100

Using fraction notation: $\quad 35\% = 35 \times \dfrac{1}{100}$ $\qquad$ Replacing % with $\times \dfrac{1}{100}$

Using decimal notation: $\quad 35\% = 35 \times 0.01$ $\qquad$ Replacing % with $\times 0.01$

**EXAMPLE 2** Write three kinds of notation for 67.8%.

Using ratio: $\qquad\qquad 67.8\% = \dfrac{67.8}{100}$ $\qquad$ A ratio of 67.8 to 100

Using fraction notation: $\quad 67.8\% = 67.8 \times \dfrac{1}{100}$ $\qquad$ Replacing % with $\times \dfrac{1}{100}$

Using decimal notation: $\quad 67.8\% = 67.8 \times 0.01$ $\qquad$ Replacing % with $\times 0.01$

Do Exercises 1–4.

---

Write three kinds of notation as in Examples 1 and 2.

**1.** 70% $\qquad\qquad$ **2.** 23.4%

**3.** 100% $\qquad\qquad$ **4.** 0.6%

---

## (b) Converting Between Percent Notation and Decimal Notation

Consider 78%. To convert to decimal notation, we can think of percent notation as a ratio and write

$78\% = \dfrac{78}{100}$ $\qquad$ Using the definition of percent as a ratio

$\phantom{78\%}= 0.78.$ $\qquad$ Dividing

Similarly,

$4.9\% = \dfrac{4.9}{100}$ $\qquad$ Using the definition of percent as a ratio

$\phantom{4.9\%}= 0.049.$ $\qquad$ Dividing

We could also convert 78% to decimal notation by replacing "%" with "$\times 0.01$" and write

$78\% = 78 \times 0.01$ $\qquad$ Replacing % with $\times 0.01$

$\phantom{78\%}= 0.78.$ $\qquad$ Multiplying

---

*Answers*

**1.** $\dfrac{70}{100}; 70 \times \dfrac{1}{100}; 70 \times 0.01$

**2.** $\dfrac{23.4}{100}; 23.4 \times \dfrac{1}{100}; 23.4 \times 0.01$

**3.** $\dfrac{100}{100}; 100 \times \dfrac{1}{100}; 100 \times 0.01$

**4.** $\dfrac{0.6}{100}; 0.6 \times \dfrac{1}{100}; 0.6 \times 0.01$

---

Find decimal notation.

**5.** 34%          **6.** 78.9%

Find decimal notation for the percent notation in each sentence.

**7. Working Online.** Of all American adults who use the Internet, 42% have gone online to work from home.

Sources: U.S. Bureau of Labor Statistics; Nielsen Home Technology Report

**8. Blood Alcohol Level.** A blood alcohol level of 0.08% is the standard used by most states as the legal limit for drunk driving.

*Answers*

**5.** 0.34    **6.** 0.789
**7.** 0.42    **8.** 0.0008

Similarly,

$$4.9\% = 4.9 \times 0.01 \qquad \text{Replacing \% with } \times 0.01$$
$$= 0.049. \qquad \text{Multiplying}$$

Dividing by 100 amounts to moving the decimal point two places to the left, which is the same as multiplying by 0.01. This leads us to a quick way to convert from percent notation to decimal notation: We drop the percent symbol and move the decimal point two places to the left.

| To convert from percent notation to decimal notation, | 36.5% |
|---|---|
| a) replace the percent symbol % with × 0.01, and | 36.5 × 0.01 |
| b) multiply by 0.01, which means move the decimal point two places to the left. | 0.36.5  Move 2 places to the left. |
|  | 36.5% = 0.365 |

**EXAMPLE 3** Find decimal notation for 99.44%.

a) Replace the percent symbol with × 0.01.          99.44 × 0.01

b) Move the decimal point two places to the left.          0.99.44

Thus, 99.44% = 0.9944.

**EXAMPLE 4** The interest rate on a $2\frac{1}{2}$-year certificate of deposit is $6\frac{3}{8}\%$. Find decimal notation for $6\frac{3}{8}\%$.

a) Convert $6\frac{3}{8}$ to decimal notation and replace the percent symbol with × 0.01.          $6\frac{3}{8}\%$  6.375 × 0.01

b) Move the decimal point two places to the left.          0.06.375

Thus, $6\frac{3}{8}\% = 0.06375$.

⟦Do Exercises 5–8.⟧

To convert 0.38 to percent notation, we can first write fraction notation, as follows:

$$0.38 = \frac{38}{100} \qquad \text{Converting to fraction notation}$$
$$= 38\%. \qquad \text{Using the definition of percent as a ratio}$$

Note that 100% = 100 × 0.01 = 1. Thus to convert 0.38 to percent notation, we can multiply by 1, using 100% as a symbol for 1.

$$0.38 = 0.38 \times 1$$
$$= 0.38 \times 100\%$$
$$= 0.38 \times 100 \times 0.01 \qquad \text{Replacing 100\% with } 100 \times 0.01$$
$$= (0.38 \times 100) \times 0.01 \qquad \text{Using the associative law of multiplication}$$
$$= 38 \times 0.01$$
$$= 38\% \qquad \text{Replacing } \times 0.01 \text{ with \%}$$

Even more quickly, since $0.38 = 0.38 \times 100\%$, we can simply multiply 0.38 by 100 and write the % symbol.

To convert from decimal notation to percent notation, we multiply by 100%. That is, we move the decimal point two places to the right and write a percent symbol.

> To convert from decimal notation to percent notation, multiply by 100%. That is,
>
> a) move the decimal point two places to the right, and
>
> b) write a % symbol.
>
> $0.675 = 0.675 \times 100\%$
>
> 0.67.5    Move 2 places to the right.
>
> 67.5%
>
> $0.675 = 67.5\%$

It is thought that the Roman emperor Augustus began percent notation by taxing goods sold at a rate of $\frac{1}{100}$. In time, the symbol "%" evolved by interchanging the parts of the symbol "100" to "0/0" and then to "%."

**EXAMPLE 5**  Find percent notation for 1.27.

a) Move the decimal point two places to the right.

1.27.

b) Write a % symbol.

127%

Thus, $1.27 = 127\%$.

**EXAMPLE 6**  Of the time that people declare as sick leave, 0.21 is actually used for family issues. Find percent notation for 0.21.
Source: CCH Inc.

a) Move the decimal point two places to the right.

0.21.

b) Write a % symbol.

21%

Thus, $0.21 = 21\%$.

**EXAMPLE 7**  Find percent notation for 5.6.

a) Move the decimal point two places to the right, adding an extra zero.

5.60.

b) Write a % symbol.

560%

Thus, $5.6 = 560\%$.

**EXAMPLE 8**  Of those who play golf, 0.149 play 8–24 rounds per year. Find percent notation for 0.149.
Source: U.S. Golf Association

a) Move the decimal point two places to the right.

0.14.9

b) Write a % symbol.

14.9%

Thus, $0.149 = 14.9\%$.

Do Exercises 9–14.

Find percent notation.

**9.** 0.24          **10.** 3.47

**11.** 1          **12.** 0.05

Find percent notation for the decimal notation in each sentence.

**13. Women in Congress.**  In 2008, 0.16 of the members of the United States Congress were women.
Source: Center for American Women and Politics at Rutgers University

**14. Soccer.**  For Americans in the 18–24 age group, 0.311 have played soccer; for the 12–17 age group, 0.396 have played.
Source: ESPN Sports Poll, a service of TNS Sport

*Answers*

**9.** 24%    **10.** 347%    **11.** 100%    **12.** 5%
**13.** 16%    **14.** 31.1%; 39.6%

**7.1** **Exercise Set**

For Extra Help

 *MyMathLab*

 Math XL
PRACTICE

 WATCH

 DOWNLOAD

 READ

 REVIEW

**a** Write three kinds of notation as in Examples 1 and 2 on p. 393.

**1.** 90%　　　　**2.** 58.7%　　　　**3.** 12.5%　　　　**4.** 130%

**b** Find decimal notation.

**5.** 67%　　　　**6.** 17%　　　　**7.** 45.6%　　　　**8.** 76.3%

**9.** 59.01%　　　**10.** 30.02%　　　**11.** 10%　　　　**12.** 80%

**13.** 1%　　　　**14.** 100%　　　**15.** 200%　　　**16.** 300%

**17.** 0.1%　　　**18.** 0.4%　　　**19.** 0.09%　　　**20.** 0.12%

**21.** 0.18%　　　**22.** 5.5%　　　**23.** 23.19%　　　**24.** 87.99%

**25.** $14\frac{7}{8}\%$　　　**26.** $93\frac{1}{8}\%$　　　**27.** $56\frac{1}{2}\%$　　　**28.** $61\frac{3}{4}\%$

Find decimal notation for the percent notation(s) in each sentence.

**29.** *Video Games.* According to a recent survey, 97% of the 12–17 age group play video games.
　　Sources: Pew Survey; *Time*, September 29, 2008

**30.** *Female Astronauts.* Of the 466 astronauts who have flown in space, 10.52% are female.
　　Source: *Encyclopedia Astronautica*

**31.** *Fuel Efficiency.* Speeding up by only 5 mph on the highway cuts fuel efficiency by approximately 7% to 8%.
**Source:** *Wall Street Journal*, "Pain Relief," by A. J. Miranda, September 15, 2008

**32.** *Foreign-Born Population.* In 2008, the U.S. foreign-born population was 12.6%, the highest since 1920.
**Source:** U.S. Census Bureau

**33.** *High School Sports.* During the 2007–2008 academic year, 54.8% of all high school students were involved in high school sports.
**Source:** National Federation of State High School Associations

**34.** *Eating Out.* On a given day, 58% of all Americans eat meals and snacks away from home.
**Source:** U.S. Department of Agriculture

Find percent notation.

**35.** 0.47　　　**36.** 0.87　　　**37.** 0.03　　　**38.** 0.01　　　**39.** 8.7

**40.** 4　　　**41.** 0.334　　　**42.** 0.889　　　**43.** 0.75　　　**44.** 0.99

**45.** 0.4　　　**46.** 0.5　　　**47.** 0.006　　　**48.** 0.008　　　**49.** 0.017

**50.** 0.024　　　**51.** 0.2718　　　**52.** 0.8911　　　**53.** 0.0239　　　**54.** 0.00073

Find percent notation for the decimal notation(s) in each sentence.

**55.** *Wasting Food.* Americans waste an estimated 0.27 of the food available for consumption. The waste occurs in restaurants, supermarkets, cafeterias, and household kitchens.
**Source:** *New York Times*, "One Country's Table Scraps, Another Country's Meal," by Andrew Martin, May 18, 2008

**56.** *Recycling Newspapers.* Over 0.73 of all newspapers are recycled.
**Source:** Newspaper Association of America

**57.** *Age 65 and Older.* In Alaska, 0.057 of the residents are age 65 and older. In Florida, 0.176 are age 65 and older.
**Source:** U.S. Census Bureau

**58.** *Dining Together.* In 2008, 0.2 of families dined together every evening. This rate declined from 0.59 in 1987.
**Source:** Online polls at USATODAY.com

**59.** *Cancer Survival.* In 2005, the estimated 10-yr survival rate was 0.906 for children diagnosed with non-Hodgkin's lymphoma (NHL) and 0.88 for those diagnosed with acute lymphoblastic leukemia (ALL).

Source: *Journal of the National Cancer Institute*, news release, September 8, 2008

**60.** *Postsecondary Degrees.* In 2007, 0.296 of those 25 to 29 years old in the United States had attained a bachelor's degree or higher.

Source: National Center for Education Statistics, U.S. Department of Commerce, U.S. Census Bureau, Current Population Survey, March Supplement 1971–2007

Find decimal notation for each percent notation in the graph.

**61.**

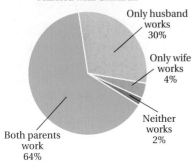

**U.S. Households: Married with Children**

Only husband works 30%

Only wife works 4%

Neither works 2%

Both parents work 64%

SOURCE: U.S. Census Bureau; U.S. Bureau of Labor Statistics

**62.**

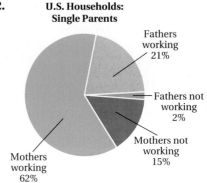

**U.S. Households: Single Parents**

Fathers working 21%

Fathers not working 2%

Mothers not working 15%

Mothers working 62%

SOURCE: U.S. Census Bureau; U.S. Bureau of Labor Statistics

## Skill Maintenance

Convert to a mixed numeral.   [4.4a]

**63.** $\dfrac{100}{3}$

**64.** $\dfrac{75}{2}$

**65.** $\dfrac{75}{8}$

**66.** $\dfrac{297}{16}$

**67.** $-\dfrac{567}{98}$

**68.** $-\dfrac{2345}{21}$

Convert to decimal notation.   [5.5a]

**69.** $\dfrac{2}{3}$

**70.** $\dfrac{1}{3}$

**71.** $-\dfrac{5}{6}$

**72.** $\dfrac{17}{12}$

**73.** $\dfrac{8}{3}$

**74.** $-\dfrac{15}{16}$

## Synthesis

Find percent notation. (*Hint*: Multiply by a form of 1 and obtain a denominator of 100.)

**75.** $\dfrac{1}{2}$

**76.** $\dfrac{3}{4}$

**77.** $\dfrac{7}{10}$

**78.** $\dfrac{2}{5}$

Find percent notation for each shaded area.

**79.**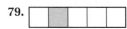

**80.**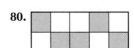

# 7.2 Percent Notation and Fraction Notation

## a Converting from Fraction Notation to Percent Notation

Consider the fraction notation $\frac{7}{8}$. To convert to percent notation, we use two skills that we already have. We first find decimal notation by dividing: $7 \div 8$.

$$\begin{array}{r} 0.8\,7\,5 \\ 8\,\overline{)\,7.0\,0\,0} \\ \underline{6\,4} \\ 6\,0 \\ \underline{5\,6} \\ 4\,0 \\ \underline{4\,0} \\ 0 \end{array} \qquad \frac{7}{8} = 0.875$$

Then we convert the decimal notation to percent notation. We move the decimal point two places to the right

$$0.8\,7.5$$

and write a % symbol:

$$\frac{7}{8} = 87.5\%, \text{ or } 87\tfrac{1}{2}\%. \qquad \left(0.5 = \tfrac{1}{2}\right)$$

To convert from fraction notation to percent notation,

a) find decimal notation by division, and

b) convert the decimal notation to percent notation.

$$\frac{3}{5} \qquad \text{Fraction notation}$$

$$\begin{array}{r} 0.6 \\ 5\,\overline{)\,3.0} \\ \underline{3\,0} \\ 0 \end{array}$$

$$0.6 = 0.60 = 60\% \qquad \text{Percent notation}$$

$$\frac{3}{5} = 60\%$$

**EXAMPLE 1** Find percent notation for $\frac{9}{16}$.

a) We first find decimal notation by division.

$$\begin{array}{r} 0.5\,6\,2\,5 \\ 1\,6\,\overline{)\,9.0\,0\,0\,0} \\ \underline{8\,0} \\ 1\,0\,0 \\ \underline{9\,6} \\ 4\,0 \\ \underline{3\,2} \\ 8\,0 \\ \underline{8\,0} \\ 0 \end{array} \qquad \frac{9}{16} = 0.5625$$

## OBJECTIVES

a Convert from fraction notation to percent notation.

b Convert from percent notation to fraction notation.

**SKILL TO REVIEW**
Objective 5.5a: Convert from fraction notation to decimal notation.

Find decimal notation.

1. $\frac{11}{16}$      2. $\frac{5}{9}$

### Calculator Corner

**Converting from Fraction Notation to Percent Notation** A calculator can be used to convert from fraction notation to percent notation. We simply perform the division on the calculator and then use the percent key. To convert $\frac{17}{40}$ to percent notation, for example, we press

$\boxed{1}\,\boxed{7}\,\boxed{\div}\,\boxed{4}\,\boxed{0}\,\boxed{\text{2nd}}\,\boxed{\%}$ , or

$\boxed{1}\,\boxed{7}\,\boxed{\div}\,\boxed{4}\,\boxed{0}\,\boxed{\text{SHIFT}}\,\boxed{\%}$ .

The display reads $\boxed{\phantom{xx}42.5\phantom{xx}}$ , so $\frac{17}{40} = 42.5\%$.

**Exercises:** Use a calculator to find percent notation. Round to the nearest hundredth of a percent.

1. $\frac{13}{25}$      2. $\frac{5}{13}$

3. $\frac{43}{39}$      4. $\frac{12}{7}$

5. $\frac{217}{364}$      6. $\frac{2378}{8401}$

*Answers*

*Skill to Review:*
1. 0.6875    2. $0.\overline{5}$

**b)** Next, we convert the decimal notation to percent notation. We move the decimal point two places to the right and write a % symbol.

$$0.56.25$$

$$\frac{9}{16} = 56.25\%, \text{ or } 56\tfrac{1}{4}\% \qquad \left(0.25 = \tfrac{1}{4}\right)$$

Don't forget the % symbol.

Do Exercises 1 and 2.

Find percent notation.

**1.** $\dfrac{1}{4}$      **2.** $\dfrac{5}{8}$

Fractions named by repeating decimals also can be converted to percent notation.

**EXAMPLE 2** *Without Health Insurance.* Approximately $\frac{1}{6}$ of all people in the United States are without health insurance. Find percent notation for $\frac{1}{6}$.
**Source:** U.S. Census Bureau, *Current Population Survey,* March 2003

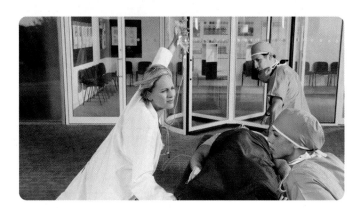

**a)** Find decimal notation by division.

$$
\begin{array}{r}
0.1\ 6\ 6 \\
6\ )\overline{\ 1.0\ 0\ 0} \\
\underline{6}\phantom{000} \\
4\ 0\phantom{0} \\
\underline{3\ 6}\phantom{0} \\
4\ 0 \\
\underline{3\ 6} \\
4
\end{array}
$$

We get a repeating decimal: $0.16\overline{6}$.

**b)** Convert the answer to percent notation.

$$0.16.\overline{6}$$

$$\frac{1}{6} = 16.\overline{6}\%, \text{ or } 16\tfrac{2}{3}\% \qquad \left(0.\overline{6} = \tfrac{2}{3}\right)$$

Do Exercises 3 and 4.

**3.** Water is the single most abundant chemical in the body. The human body is about $\frac{2}{3}$ water. Find percent notation for $\frac{2}{3}$.

**4.** Find percent notation: $\dfrac{5}{6}$.

**Answers**

**1.** 25%    **2.** 62.5%, or $62\dfrac{1}{2}\%$

**3.** $66.\overline{6}\%$, or $66\dfrac{2}{3}\%$

**4.** $83.\overline{3}\%$, or $83\dfrac{1}{3}\%$

In some cases, division is not the fastest way to convert. The following are some optional ways in which conversion might be done.

**EXAMPLE 3** Find percent notation for $\frac{69}{100}$.

We use the definition of percent as a ratio.

$$\frac{69}{100} = 69\%$$

**EXAMPLE 4** Find percent notation for $\frac{17}{20}$.

We can multiply by 1 to get 100 in the denominator. We think of what we must multiply 20 by in order to get 100. That number is 5, so we multiply by 1 using $\frac{5}{5}$.

$$\frac{17}{20} \cdot \frac{5}{5} = \frac{85}{100} = 85\%$$

Note that this shortcut works only when the denominator is a factor of 100.

**EXAMPLE 5** Find percent notation for $\frac{18}{25}$.

$$\frac{18}{25} = \frac{18}{25} \cdot \frac{4}{4} = \frac{72}{100} = 72\%$$

Do Exercises 5–8.

Find percent notation.

**5.** $\frac{57}{100}$

**6.** $\frac{19}{25}$

**7.** $\frac{7}{10}$

**8.** $\frac{1}{4}$

## b) Converting from Percent Notation to Fraction Notation

To convert from percent notation to fraction notation,

a) use the definition of percent as a ratio, and

b) simplify, if possible.

30%  Percent notation

$\frac{30}{100}$

$\frac{3}{10}$  Fraction notation

**EXAMPLE 6** Find fraction notation for 75%.

$$75\% = \frac{75}{100} \qquad \text{Using the definition of percent}$$

$$= \frac{3 \cdot 25}{4 \cdot 25} = \frac{3}{4} \cdot \frac{25}{25} \Bigg\} \quad \text{Simplifying}$$

$$= \frac{3}{4}$$

**EXAMPLE 7**  Find fraction notation for 62.5%.

$$62.5\% = \frac{62.5}{100}$$  Using the definition of percent

$$= \frac{62.5}{100} \times \frac{10}{10}$$  Multiplying by 1 to eliminate the decimal point in the numerator

$$= \frac{625}{1000}$$

$$= \frac{5 \cdot 125}{8 \cdot 125} = \frac{5}{8} \cdot \frac{125}{125} \Bigg\}$$  Simplifying

$$= \frac{5}{8}$$

**EXAMPLE 8**  Find fraction notation for $16\frac{2}{3}\%$.

$$16\frac{2}{3}\% = \frac{50}{3}\%$$  Converting from the mixed numeral to fraction notation

$$= \frac{50}{3} \times \frac{1}{100}$$  Using the definition of percent

$$= \frac{50 \cdot 1}{3 \cdot 50 \cdot 2} = \frac{1}{3 \cdot 2} \cdot \frac{50}{50} \Bigg\}$$  Simplifying

$$= \frac{1}{6}$$

Do Exercises 9–12.

**Find fraction notation.**

**9.** 60%          **10.** 3.25%

**11.** $66\frac{2}{3}\%$      **12.** $12\frac{1}{2}\%$

The table below lists fraction, decimal, and percent equivalents used so often that it would speed up your work if you memorized them. For example, $\frac{1}{3} = 0.\overline{3}$, so we say that the **decimal equivalent** of $\frac{1}{3}$ is $0.\overline{3}$, or that $0.\overline{3}$ has the **fraction equivalent** $\frac{1}{3}$. This table also appears on the inside back cover.

## FRACTION, DECIMAL, AND PERCENT EQUIVALENTS

| FRACTION NOTATION | $\frac{1}{10}$ | $\frac{1}{8}$ | $\frac{1}{6}$ | $\frac{1}{5}$ | $\frac{1}{4}$ | $\frac{3}{10}$ | $\frac{1}{3}$ | $\frac{3}{8}$ | $\frac{2}{5}$ | $\frac{1}{2}$ | $\frac{3}{5}$ | $\frac{5}{8}$ | $\frac{2}{3}$ | $\frac{7}{10}$ | $\frac{3}{4}$ | $\frac{4}{5}$ | $\frac{5}{6}$ | $\frac{7}{8}$ | $\frac{9}{10}$ | $\frac{1}{1}$ |
|---|---|---|---|---|---|---|---|---|---|---|---|---|---|---|---|---|---|---|---|---|
| DECIMAL NOTATION | 0.1 | 0.125 | $0.16\overline{6}$ | 0.2 | 0.25 | 0.3 | $0.33\overline{3}$ | 0.375 | 0.4 | 0.5 | 0.6 | 0.625 | $0.66\overline{6}$ | 0.7 | 0.75 | 0.8 | $0.83\overline{3}$ | 0.875 | 0.9 | 1 |
| PERCENT NOTATION | 10% | 12.5%, or $12\frac{1}{2}\%$ | $16.\overline{6}\%$, or $16\frac{2}{3}\%$ | 20% | 25% | 30% | $33.\overline{3}\%$, or $33\frac{1}{3}\%$ | 37.5%, or $37\frac{1}{2}\%$ | 40% | 50% | 60% | 62.5%, or $62\frac{1}{2}\%$ | $66.\overline{6}\%$, or $66\frac{2}{3}\%$ | 70% | 75% | 80% | $83.\overline{3}\%$, or $83\frac{1}{3}\%$ | 87.5%, or $87\frac{1}{2}\%$ | 90% | 100% |

**EXAMPLE 9**  Find fraction notation for $16.\overline{6}\%$.

We can use the table above or recall that $16.\overline{6}\% = 16\frac{2}{3}\% = \frac{1}{6}$. We can also recall from our work with repeating decimals in Chapter 5 that $0.\overline{6} = \frac{2}{3}$. Then we have $16.\overline{6}\% = 16\frac{2}{3}\%$ and can proceed as in Example 8.

Do Exercises 13 and 14.

**Find fraction notation.**

**13.** $33.\overline{3}\%$          **14.** $83.\overline{3}\%$

*Answers*

9. $\frac{3}{5}$   10. $\frac{13}{400}$   11. $\frac{2}{3}$

12. $\frac{1}{8}$   13. $\frac{1}{3}$   14. $\frac{5}{6}$

**a** Find percent notation.

1. $\dfrac{41}{100}$   2. $\dfrac{36}{100}$   3. $\dfrac{5}{100}$   4. $\dfrac{1}{100}$   5. $\dfrac{2}{10}$   6. $\dfrac{7}{10}$

7. $\dfrac{3}{10}$   8. $\dfrac{9}{10}$   9. $\dfrac{1}{2}$   10. $\dfrac{3}{4}$   11. $\dfrac{7}{8}$   12. $\dfrac{1}{8}$

13. $\dfrac{4}{5}$   14. $\dfrac{2}{5}$   15. $\dfrac{2}{3}$   16. $\dfrac{1}{3}$   17. $\dfrac{1}{6}$   18. $\dfrac{5}{6}$

19. $\dfrac{3}{16}$   20. $\dfrac{11}{16}$   21. $\dfrac{13}{16}$   22. $\dfrac{7}{16}$   23. $\dfrac{4}{25}$   24. $\dfrac{17}{25}$

25. $\dfrac{1}{20}$   26. $\dfrac{31}{50}$   27. $\dfrac{17}{50}$   28. $\dfrac{3}{20}$

Find percent notation for the fraction notation in each sentence.

29. *Heart Transplants.*   In the United States in 2006, $\frac{2}{25}$ of the organ transplants were heart transplants and $\frac{59}{100}$ were kidney transplants.
    **Source:** 2007 OPTN/SRTR Annual Report, Table 1.7

30. *Car Colors.*   The four most popular colors for 2006 compact/sports cars were silver, gray, black, and red. Of all cars in this category, $\frac{9}{50}$ were silver, $\frac{3}{20}$ gray, $\frac{3}{20}$ black, and $\frac{3}{20}$ red.
    **Sources:** Ward's Automotive Group; DuPont Automotive Products

In Exercises 31–36, write percent notation for the fractions in the pie chart below. (The sum of the fractions is greater than 1 because of rounding.)

**How Food Dollars Are Spent**

Dairy $\frac{3}{25}$

Beverages (nonalcoholic) $\frac{3}{25}$

Sugar $\frac{1}{25}$

Cereal and baked goods $\frac{13}{100}$

Fats and oils $\frac{3}{100}$

Fruits and vegetables $\frac{3}{20}$

Other $\frac{9}{50}$

Meat, poultry, and fish $\frac{11}{50}$

Eggs $\frac{1}{50}$

SOURCES: U.S. Bureau of Labor Statistics; Consumer Price Index; *The Hoosier Farmer*, Summer 2008

31. $\dfrac{11}{50}$   32. $\dfrac{9}{50}$

33. $\dfrac{3}{25}$   34. $\dfrac{1}{25}$

35. $\dfrac{3}{20}$   36. $\dfrac{13}{100}$

**b** Find fraction notation. Simplify.

**37.** 85%

**38.** 55%

**39.** 62.5%

**40.** 12.5%

**41.** $33\frac{1}{3}$%

**42.** $83\frac{1}{3}$%

**43.** $16.\overline{6}$%

**44.** $66.\overline{6}$%

**45.** 7.25%

**46.** 4.85%

**47.** 0.8%

**48.** 0.2%

**49.** $25\frac{3}{8}$%

**50.** $48\frac{7}{8}$%

**51.** $78\frac{2}{9}$%

**52.** $16\frac{5}{9}$%

**53.** $64\frac{7}{11}$%

**54.** $73\frac{3}{11}$%

**55.** 150%

**56.** 110%

**57.** 0.0325%

**58.** 0.419%

**59.** $33.\overline{3}$%

**60.** $83.\overline{3}$%

In Exercises 61–66, find fraction notation for the percent notations in the table below.

**U.S. POPULATION BY SELECTED AGE CATEGORIES (Data have been rounded to the nearest percent.)**

| AGE CATEGORY | PERCENT OF POPULATION |
|---|---|
| 5–17 years | 18% |
| 18–24 years | 10 |
| 15–44 years | 42 |
| 18 years and older | 75 |
| 65 years and older | 12 |
| 75 years and older | 6 |

SOURCES: U.S. Census Bureau; 2006 American Community Survey

**61.** 6%

**62.** 18%

**63.** 12%

**64.** 42%

**65.** 75%

**66.** 10%

Find fraction notation for the percent notation in each sentence.

**67.** A $\frac{3}{4}$-cup serving of Post Selects Great Grains cereal with $\frac{1}{2}$ cup fat-free milk satisfies 15% of the minimum daily requirement for calcium.
Source: Kraft Foods Global, Inc.

**68.** A 1.8-oz serving of Frosted Mini-Wheats®, Blueberry Muffin, with $\frac{1}{2}$ cup fat-free milk satisfies 35% of the minimum daily requirement for Vitamin B$_{12}$.
Source: Kellogg, Inc.

**69.** In 2006, 20.9% of Americans age 18 and older smoked cigarettes.
Sources: *Washington Post*, March 9, 2006; U.S. Centers for Disease Control and Prevention

**70.** In 2006, 14.9% of the adults age 18 and older in California smoked cigarettes.
Sources: U.S. Centers for Disease Control and Prevention; *AARP Magazine*, March 2008

Complete each table.

**71.**

| FRACTION NOTATION | DECIMAL NOTATION | PERCENT NOTATION |
|---|---|---|
| $\frac{1}{8}$ | | 12.5%, or $12\frac{1}{2}$% |
| $\frac{1}{6}$ | | |
| | | 20% |
| | 0.25 | |
| | | 33.$\overline{3}$%, or $33\frac{1}{3}$% |
| | | 37.5%, or $37\frac{1}{2}$% |
| | | 40% |
| $\frac{1}{2}$ | | |

**72.**

| FRACTION NOTATION | DECIMAL NOTATION | PERCENT NOTATION |
|---|---|---|
| $\frac{3}{5}$ | | |
| | 0.625 | |
| $\frac{2}{3}$ | | |
| | 0.75 | 75% |
| $\frac{4}{5}$ | | |
| $\frac{5}{6}$ | | 83.$\overline{3}$%, or $83\frac{1}{3}$% |
| $\frac{7}{8}$ | | 87.5%, or $87\frac{1}{2}$% |
| | | 100% |

**73.**

| FRACTION NOTATION | DECIMAL NOTATION | PERCENT NOTATION |
|---|---|---|
| | 0.5 | |
| $\frac{1}{3}$ | | |
| | | 25% |
| | | 16.$\overline{6}$%, or $16\frac{2}{3}$% |
| | 0.125 | |
| $\frac{3}{4}$ | | |
| | 0.8$\overline{3}$ | |
| $\frac{3}{8}$ | | |

**74.**

| FRACTION NOTATION | DECIMAL NOTATION | PERCENT NOTATION |
|---|---|---|
| | | 40% |
| | | 62.5%, or $62\frac{1}{2}$% |
| | 0.875 | |
| $\frac{1}{1}$ | | |
| | 0.6 | |
| | 0.$\overline{6}$ | |
| $\frac{1}{5}$ | | |

## Skill Maintenance

Solve.

**75.** $13 \cdot x = 910$   [1.7b]

**76.** $15 \cdot y = 75$   [1.7b]

**77.** $0.05 \times b = -20$   [5.4b]

**78.** $3 = 0.16 \times b$   [5.4b]

**79.** $\dfrac{24}{37} = \dfrac{15}{x}$   [6.3b]

**80.** $\dfrac{17}{18} = \dfrac{x}{27}$   [6.3b]

**81.** $\dfrac{9}{10} = \dfrac{x}{5}$   [6.3b]

**82.** $\dfrac{7}{x} = \dfrac{4}{5}$   [6.3b]

Convert to a mixed numeral.   [4.4a]

**83.** $\dfrac{100}{3}$

**84.** $\dfrac{75}{2}$

**85.** $\dfrac{250}{3}$

**86.** $\dfrac{123}{6}$

**87.** $-\dfrac{345}{8}$

**88.** $-\dfrac{373}{6}$

**89.** $\dfrac{75}{4}$

**90.** $\dfrac{67}{9}$

Convert from a mixed numeral to fraction notation.   [4.4a]

**91.** $1\dfrac{1}{17}$

**92.** $20\dfrac{9}{10}$

**93.** $-101\dfrac{1}{2}$

**94.** $-32\dfrac{3}{8}$

## Synthesis

Write percent notation.

**95.** $\dfrac{41}{369}$

**96.** $\dfrac{54}{999}$

**97.** $2.5\overline{74631}$

**98.** $3.2\overline{93847}$

Write decimal notation.

**99.** $\dfrac{14}{9}\%$

**100.** $\dfrac{19}{12}\%$

**101.** $\dfrac{729}{7}\%$

**102.** $\dfrac{637}{6}\%$

**103.** Arrange the following numbers from smallest to largest.

$$16\dfrac{1}{6}\%, \; 1.6, \; \dfrac{1}{6}\%, \; \dfrac{1}{2}, \; 0.2, \; 1.6\%, \; 1\dfrac{1}{6}\%, \; 0.5\%, \; \dfrac{2}{7}\%, \; 0.\overline{54}$$

# 7.3 Solving Percent Problems Using Percent Equations

## a   Translating to Equations

To solve a problem involving percents, it is helpful to translate first to an equation. To distinguish the method discussed in this section from that of Section 7.4, we will call these *percent equations*.

**EXAMPLES**   Translate each of the following.

**1.** 23%   of   5   is   what?
   23%   ·   5   =   $a$     This is a *percent equation*.

**2.** What   is   11%   of   49?
   $a$   =   11%   ·   49     Any letter can be used.

*Do Margin Exercises 1 and 2.*

**EXAMPLES**   Translate each of the following.

**3.** 3   is   10%   of   what?
   3   =   10%   ·   $b$

**4.** 45%   of   what   is   23?
   45%   ×   $b$   =   23

*Do Exercises 3 and 4.*

**EXAMPLES**   Translate each of the following.

**5.** 10   is   what percent   of   20?
   10   =   $p$   ×   20

**6.** What percent   of   50   is   7?
   $p$   ·   50   =   7

*Do Exercises 5 and 6.*

Translate to an equation. Do not solve.

**1.** 12% of 50 is what?

**2.** What is 40% of 60?

Translate to an equation. Do not solve.

**3.** 45 is 20% of what?

**4.** 120% of what is 60?

Translate to an equation. Do not solve.

**5.** 16 is what percent of 40?

**6.** What percent of 84 is 10.5?

*Answers*

*Skill to Review:*
**1.** 16,600    **2.** 5.081

*Margin Exercises:*
**1.** 12% × 50 = $a$    **2.** $a$ = 40% × 60
**3.** 45 = 20% × $b$    **4.** 120% × $b$ = 60
**5.** 16 = $p$ × 40    **6.** $p$ × 84 = 10.5

## b Solving Percent Problems

In solving percent problems, we use the *Translate* and *Solve* steps in the problem-solving strategy used throughout this text.

Percent problems are actually of three different types. Although the method we present does *not* require that you be able to identify which type you are solving, it is helpful to know them. Each of the three types of percent problems depends on which of the three pieces of information is missing.

1. **Finding the *amount* (the result of taking the percent)**

   *Example:*     **What is 25% of 60?**
   
   *Translation:*     $a = 25\% \cdot 60$

2. **Finding the *base* (the number you are taking the percent of)**

   *Example:*     15 is 25% of what?
   
   *Translation:*     $15 = 25\% \cdot b$

3. **Finding the *percent number* (the percent itself)**

   *Example:*     15 is what percent of 60?
   
   *Translation:*     $15 = p \cdot 60$

### Finding the Amount

**EXAMPLE 7**   What is 23.9% of $40,000,000,000?

*Translate:* $a = 23.9\% \times 40{,}000{,}000{,}000.$

*Solve:* The letter is by itself. To solve the equation, we just convert 23.9% to decimal notation and multiply:

$$a = 23.9\% \times 40{,}000{,}000{,}000$$
$$a = 0.239 \times 40{,}000{,}000{,}000 = 9{,}560{,}000{,}000.$$

Thus, $9,560,000,000 is 23.9% of $40,000,000,000. The answer is $9,560,000,000.

Each year, Americans spend about $40 billion on pets; approximately 23.9% of that amount is spent on veterinary care. What is spent per year on veterinary care? (See Example 7.)

SOURCE: American Pet Products Manufacturers Association

Do Exercise 7.

**7.** Solve:

What is 12% of $50?

**EXAMPLE 8**   120% of 42 is what?

*Translate:* $120\% \times 42 = a.$

*Solve:* The letter is by itself. To solve the equation, we carry out the calculation:

$$a = 120\% \times 42$$
$$a = 1.2 \times 42 \qquad 120\% = 1.2$$
$$a = 50.4.$$

Thus, 120% of 42 is 50.4. The answer is 50.4.

Do Exercise 8.

**8.** Solve:

64% of 55 is what?

*Answers*

**7.** $6    **8.** 35.2

## Finding the Base

**EXAMPLE 9** 5% of what is 20?

*Translate:* $5\% \times b = 20$.

*Solve:* This time the letter is *not* by itself. To solve the equation, we divide by 5% on both sides:

$$\frac{5\% \times b}{5\%} = \frac{20}{5\%} \qquad \text{Dividing by 5\% on both sides}$$

$$b = \frac{20}{0.05} \qquad 5\% = 0.05$$

$$b = 400.$$

Thus, 5% of 400 is 20. The answer is 400.

In a survey of a group of people, it was found that 5%, or 20 people, chose strawberry as their favorite ice cream flavor. How many people were surveyed? (See Example 9.)
SOURCE: International Ice Cream Association

**EXAMPLE 10** $3 is 16% of what?

*Translate:*

| $3 | is | 16% | of | what? |
| :-: | :-: | :-: | :-: | :-: |
| ↓ | ↓ | ↓ | ↓ | ↓ |
| 3 | = | 16% | × | $b$ |

*Solve:* To solve the equation, we divide by 16% on both sides:

$$\frac{3}{16\%} = \frac{16\% \times b}{16\%} \qquad \text{Dividing by 16\% on both sides}$$

$$\frac{3}{0.16} = b \qquad 16\% = 0.16$$

$$18.75 = b.$$

Thus, $3 is 16% of $18.75. The answer is $18.75.

Solve.
**9.** 20% of what is 45?

**10.** $60 is 120% of what?

Do Exercises 9 and 10.

## Finding the Percent Number

In solving these problems, you *must* remember to convert to percent notation after you have solved the equation.

**EXAMPLE 11** 2100 is what percent of 30,000?

*Translate:*

| 2100 | is | what percent | of | 30,000? |
| :-: | :-: | :-: | :-: | :-: |
| ↓ | ↓ | ↓ | ↓ | ↓ |
| 2100 | = | $p$ | × | 30,000 |

*Solve:* To solve the equation, we divide by 30,000 on both sides and convert the result to percent notation:

$$p \times 30,000 = 2100$$

$$\frac{p \times 30,000}{30,000} = \frac{2100}{30,000} \qquad \text{Dividing by 30,000 on both sides}$$

$$p = 0.07 \qquad \text{Dividing}$$

$$p = 7\%. \qquad \text{Converting to percent notation}$$

Thus, 2100 is 7% of 30,000. The answer is 7%.

In 2007, there were about 30,000 earthquakes worldwide. Of this number, 2100 earthquakes had magnitudes greater than 5.0. What percent of the 30,000 earthquakes had magnitudes greater than 5.0? (See Example 11.)
SOURCE: National Earthquake Information Center, U.S. Geological Survey

*Answers*
**9.** 225 **10.** $50

**EXAMPLE 12** What percent of $50 is $16?

*Translate:* $\underbrace{\text{What percent}}_{p} \quad \underset{\times}{\text{of}} \quad \underset{50}{\$50} \quad \underset{=}{\text{is}} \quad \underset{16}{\$16?}$

*Solve:* To solve the equation, we divide by 50 on both sides and convert the answer to percent notation:

$$\frac{p \times 50}{50} = \frac{16}{50} \qquad \text{Dividing by 50 on both sides}$$

$$p = \frac{16}{50}$$

$$p = 0.32$$

$$p = 32\%. \qquad \text{Converting to percent notation}$$

Thus, 32% of $50 is $16. The answer is 32%.

Do Exercises 11 and 12.

Solve.

**11.** 16 is what percent of 40?

**12.** What percent of $84 is $10.50?

------ *Caution!* ------

When a question asks "what percent?", be sure to give the answer in percent notation.

---

## Calculator Corner

**Using Percents in Computations** Many calculators have a `%` key that can be used in computations. (See the Calculator Corner on page 394.) For example, to find 11% of 49, we press `1` `1` `2nd` `%` `×` `4` `9` `=`, or `4` `9` `×` `1` `1` `SHIFT` `%`. The display reads [ 5.39 ], so 11% of 49 is 5.39.

In Example 9, we perform the computation 20/5%. To use the `%` key in this computation, we press `2` `0` `÷` `5` `2nd` `%` `=`, or `2` `0` `÷` `5` `SHIFT` `%`. The result is 400.

We can also use the `%` key to find the percent number in a problem. In Example 11, for instance, we answer the question "2100 is what percent of 30,000?" On a calculator, we press `2` `1` `0` `0` `÷` `3` `0` `0` `0` `0` `2nd` `%` `=`, or `2` `1` `0` `0` `÷` `3` `0` `0` `0` `0` `SHIFT` `%`. The result is 7, so 2100 is 7% of 30,000.

**Exercises:** Use a calculator to find each of the following.

**1.** What is 12.6% of $40?

**2.** 0.04% of 28 is what?

**3.** 8% of what is 36?

**4.** $45 is 4.5% of what?

**5.** 23 is what percent of 920?

**6.** What percent of $442 is $53.04?

---

*Answers*

11. 40%  12. 12.5%

**a** Translate to an equation. Do not solve.

**1.** What is 32% of 78?

**2.** 98% of 57 is what?

**3.** 89 is what percent of 99?

**4.** What percent of 25 is 8?

**5.** 13 is 25% of what?

**6.** 21.4% of what is 20?

**b** Translate to an equation and solve.

**7.** What is 85% of 276?

**8.** What is 74% of 53?

**9.** 150% of 30 is what?

**10.** 100% of 13 is what?

**11.** What is 6% of $300?

**12.** What is 4% of $45?

**13.** 3.8% of 50 is what?

**14.** $33\frac{1}{3}$% of 480 is what?
$\left(Hint: 33\frac{1}{3}\% = \frac{1}{3}.\right)$

**15.** $39 is what percent of $50?

**16.** $16 is what percent of $90?

**17.** 20 is what percent of 10?

**18.** 60 is what percent of 20?

**19.** What percent of $300 is $150?

**20.** What percent of $50 is $40?

**21.** What percent of 80 is 100?

**22.** What percent of 60 is 15?

**23.** 20 is 50% of what?

**24.** 57 is 20% of what?

**25.** 40% of what is $16?

**26.** 100% of what is $74?

**27.** 56.32 is 64% of what?

**28.** 71.04 is 96% of what?

**29.** 70% of what is 14?

**30.** 70% of what is 35?

**31.** What is $62\frac{1}{2}$% of 10?

**32.** What is $35\frac{1}{4}$% of 1200?

**33.** What is 8.3% of $10,200?

**34.** What is 9.2% of $5600?

**35.** 2.5% of what is 30.4?

**36.** 8.2% of what is 328?

## Skill Maintenance

Write fraction notation. [5.1b]

**37.** 0.09

**38.** 1.79

**39.** 0.875

**40.** 0.125

**41.** −0.9375

**42.** −0.6875

Write decimal notation. [5.1b]

**43.** $\dfrac{89}{100}$

**44.** $\dfrac{7}{100}$

**45.** $-\dfrac{3}{10}$

**46.** $-\dfrac{17}{1000}$

## Synthesis

Solve.

**47.** What is 7.75% of $10,880?

Estimate _____

Calculate _____

**48.** 50,951.775 is what percent of 78,995?

Estimate _____

Calculate _____

**49.** $2496 is 24% of what amount?

Estimate _____

Calculate _____

**50.** What is 38.2% of $52,345.79?

Estimate _____

Calculate _____

**51.** 40% of $18\frac{3}{4}$% of $25,000 is what?

# 7.4 Solving Percent Problems Using Proportions*

## a) Translating to Proportions

A percent is a ratio of some number to 100. For example, 46% is the ratio $\frac{46}{100}$. The numbers 67,620,000 and 147,000,000 have the same ratio as 46 and 100.

$$\frac{46}{100} = \frac{67,620,000}{147,000,000}$$

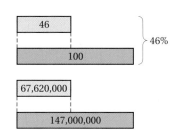

To solve a percent problem using a proportion, we translate as follows:

$$\text{Number} \rightarrow \frac{N}{100} \underset{\leftarrow 100}{} = \frac{a}{b} \begin{matrix} \leftarrow \text{Amount} \\ \leftarrow \text{Base} \end{matrix}$$

You might find it helpful to read this as "part is to whole as part is to whole."

For example, 60% of 25 is 15 translates to

$$\frac{60}{100} = \frac{15}{25} \begin{matrix} \leftarrow \text{Amount} \\ \leftarrow \text{Base} \end{matrix}$$

A clue in translating is that the base, $b$, corresponds to 100 and usually follows the wording "percent of." Also, $N\%$ always translates to $N/100$. Another aid in translating is to make a comparison drawing. To do this, we start with the percent side and list 0% at the top and 100% near the bottom. Then we estimate where the specified percent—in this case, 60%—is located. The corresponding quantities are then filled in. The base—in this case, 25—always corresponds to 100%, and the amount—in this case, 15—corresponds to the specified percent.

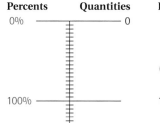 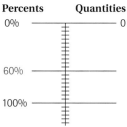

The proportion can then be read easily from the drawing: $\frac{60}{100} = \frac{15}{25}$.

In the United States, 46% of the labor force is women. In 2007, there were approximately 147,000,000 people in the labor force. This means that about 67,620,000 were women.

SOURCES: U.S. Department of Labor; U.S. Bureau of Labor Statistics

## OBJECTIVES

**a** Translate percent problems to proportions.

**b** Solve basic percent problems.

**SKILL TO REVIEW**
Objective 6.3b: Solve proportions.

Solve.

**1.** $\frac{3}{100} = \frac{27}{b}$

**2.** $\frac{4.3}{20} = \frac{N}{100}$

---

*Note: This section presents an alternative method for solving basic percent problems. You can use either equations or proportions to solve percent problems, but you might prefer one method over the other, or your instructor may direct you to use one method over the other.

*Answers*
*Skill to Review:*
**1.** 900   **2.** 21.5

**EXAMPLE 1** Translate to a proportion.

23% of 5 is what?

$$\frac{23}{100} = \frac{a}{5}$$

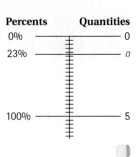

**EXAMPLE 2** Translate to a proportion.

What is 124% of 49?

$$\frac{124}{100} = \frac{a}{49}$$

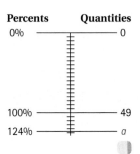

Translate to a proportion. Do not solve.

**1.** 12% of 50 is what?

**2.** What is 40% of 60?

**3.** 130% of 72 is what?

Do Exercises 1–3.

**EXAMPLE 3** Translate to a proportion.

3 is 10% of what?

$$\frac{10}{100} = \frac{3}{b}$$

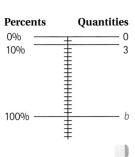

**EXAMPLE 4** Translate to a proportion.

45% of what is 23?

$$\frac{45}{100} = \frac{23}{b}$$

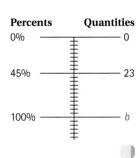

Translate to a proportion. Do not solve.

**4.** 45 is 20% of what?

**5.** 120% of what is 60?

Do Exercises 4 and 5.

**EXAMPLE 5** Translate to a proportion.

10 is what percent of 20?

$$\frac{N}{100} = \frac{10}{20}$$

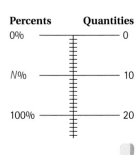

*Answers*

**1.** $\frac{12}{100} = \frac{a}{50}$   **2.** $\frac{40}{100} = \frac{a}{60}$   **3.** $\frac{130}{100} = \frac{a}{72}$

**4.** $\frac{20}{100} = \frac{45}{b}$   **5.** $\frac{120}{100} = \frac{60}{b}$

**EXAMPLE 6** Translate to a proportion.

What percent of 50 is 7?

$$\frac{N}{100} = \frac{7}{50}$$

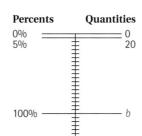

Percents    Quantities

0%  —————— 0
N% ┬ —————— 7
│
│
100% ┼ ————— 50
│

Do Exercises 6 and 7.

Translate to a proportion. Do not solve.

**6.** 16 is what percent of 40?

**7.** What percent of 84 is 10.5?

## **b** Solving Percent Problems

After a percent problem has been translated to a proportion, we solve as in Section 6.3.

**EXAMPLE 7** 5% of what is $20?

*Translate:* $\dfrac{5}{100} = \dfrac{20}{b}$

*Solve:* $5 \cdot b = 100 \cdot 20$    Equating cross products

$\dfrac{5 \cdot b}{5} = \dfrac{100 \cdot 20}{5}$    Dividing by 5

$b = \dfrac{2000}{5}$

$b = 400$    Simplifying

Thus, 5% of $400 is $20. The answer is $400.

Percents    Quantities

0%  —————— 0
5% ┬ —————— 20
│
│
100% ┼ ————— b
│

Do Exercise 8.

**8.** Solve:

20% of what is $45?

**EXAMPLE 8** 120% of 42 is what?

*Translate:* $\dfrac{120}{100} = \dfrac{a}{42}$

*Solve:* $120 \cdot 42 = 100 \cdot a$    Equating cross products

$\dfrac{120 \cdot 42}{100} = \dfrac{100 \cdot a}{100}$    Dividing by 100

$\dfrac{5040}{100} = a$

$50.4 = a$    Simplifying

Thus, 120% of 42 is 50.4. The answer is 50.4.

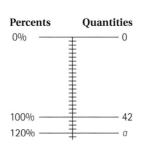

Percents    Quantities

0%  —————— 0
│
│
100% ┼ ————— 42
120% ┼ ——— a

Do Exercises 9 and 10.

Solve.

**9.** 64% of 55 is what?

**10.** What is 12% of 50?

*Answers*

**6.** $\dfrac{N}{100} = \dfrac{16}{40}$   **7.** $\dfrac{N}{100} = \dfrac{10.5}{84}$   **8.** $225

**9.** 35.2   **10.** 6

**EXAMPLE 9**   210 is $10\frac{1}{2}\%$ of what?

*Translate:* $\dfrac{210}{b} = \dfrac{10.5}{100}$ $\qquad 10\frac{1}{2}\% = 10.5\%$

*Solve:* $210 \cdot 100 = b \cdot 10.5$ $\qquad$ Equating cross products

$\dfrac{210 \cdot 100}{10.5} = \dfrac{b \cdot 10.5}{10.5}$ $\qquad$ Dividing by 10.5

$\dfrac{21{,}000}{10.5} = b$ $\qquad$ Multiplying and simplifying

$2000 = b$ $\qquad$ Dividing

Thus, 210 is $10\frac{1}{2}\%$ of 2000. The answer is 2000.

Do Exercise 11.

**11.** Solve:

60 is 120% of what?

**EXAMPLE 10**   $10 is what percent of $20?

*Translate:* $\dfrac{10}{20} = \dfrac{N}{100}$

*Solve:* $10 \cdot 100 = 20 \cdot N$ $\qquad$ Equating cross products

$\dfrac{10 \cdot 100}{20} = \dfrac{20 \cdot N}{20}$ $\qquad$ Dividing by 20

$\dfrac{1000}{20} = N$ $\qquad$ Multiplying and simplifying

$50 = N$ $\qquad$ Dividing

Thus, $10 is 50% of $20. The answer is 50%.

> Note when solving percent problems using proportions that $N$ is a percent and need not be converted.

**12.** Solve:

$12 is what percent of $40?

Do Exercise 12.

**EXAMPLE 11**   What percent of 50 is 16?

*Translate:* $\dfrac{N}{100} = \dfrac{16}{50}$

*Solve:* $N \cdot 50 = 100 \cdot 16$ $\qquad$ Equating cross products

$\dfrac{N \cdot 50}{50} = \dfrac{100 \cdot 16}{50}$ $\qquad$ Dividing by 50

$N = \dfrac{1600}{50}$ $\qquad$ Multiplying and simplifying

$N = 32$ $\qquad$ Dividing

Thus, 32% of 50 is 16. The answer is 32%.

**13.** Solve:

What percent of 84 is 10.5?

Do Exercise 13.

*Answers*

**11.** 50   **12.** 30%   **13.** 12.5%

**a**    Translate to a proportion. Do not solve.

**1.** What is 37% of 74?

**2.** 66% of 74 is what?

**3.** 4.3 is what percent of 5.9?

**4.** What percent of 6.8 is 5.3?

**5.** 14 is 25% of what?

**6.** 133% of what is 40?

**b**    Translate to a proportion and solve.

**7.** What is 76% of 90?

**8.** What is 32% of 70?

**9.** 70% of 660 is what?

**10.** 80% of 920 is what?

**11.** What is 4% of 1000?

**12.** What is 6% of 2000?

**13.** 4.8% of 60 is what?

**14.** 63.1% of 80 is what?

**15.** $24 is what percent of $96?

**16.** $14 is what percent of $70?

**17.** 102 is what percent of 100?

**18.** 103 is what percent of 100?

**19.** What percent of $480 is $120?

**20.** What percent of $80 is $60?

**21.** What percent of 160 is 150?

**22.** What percent of 33 is 11?

**23.** $18 is 25% of what?

**24.** $75 is 20% of what?

**25.** 60% of what is 54?

**26.** 80% of what is 96?

**27.** 65.12 is 74% of what?

**28.** 63.7 is 65% of what?

**29.** 80% of what is 16?

**30.** 80% of what is 10?

**31.** What is $62\frac{1}{2}$% of 40?

**32.** What is $43\frac{1}{4}$% of 2600?

**33.** What is 9.4% of $8300?

**34.** What is 8.7% of $76,000?

**35.** 80.8 is $40\frac{2}{5}$% of what?

**36.** 66.3 is $10\frac{1}{5}$% of what?

## Skill Maintenance

Solve.　[6.3b]

**37.** $\dfrac{x}{188} = \dfrac{2}{47}$

**38.** $\dfrac{15}{x} = \dfrac{3}{800}$

**39.** $\dfrac{4}{7} = \dfrac{x}{14}$

**40.** $\dfrac{612}{t} = \dfrac{72}{244}$

**41.** $\dfrac{5000}{t} = \dfrac{3000}{60}$

**42.** $\dfrac{75}{100} = \dfrac{n}{20}$

**43.** $\dfrac{x}{1.2} = \dfrac{36.2}{5.4}$

**44.** $\dfrac{y}{1\frac{1}{2}} = \dfrac{2\frac{3}{4}}{22}$

Solve.

**45.** A recipe for muffins calls for $\frac{1}{2}$ qt of buttermilk, $\frac{1}{3}$ qt of skim milk, and $\frac{1}{16}$ qt of oil. How many quarts of liquid ingredients does the recipe call for?　[4.2b]

**46.** The Ferristown School District purchased $\frac{3}{4}$ ton (T) of clay. If the clay is to be shared equally among the district's 6 art departments, how much will each art department receive?　[3.7d]

## Synthesis

Solve.

**47.** ▦ What is 8.85% of $12,640?

Estimate _____

Calculate _____

**48.** ▦ 78.8% of what is 9809.024?

Estimate _____

Calculate _____

# Mid-Chapter Review

## Concept Reinforcement

Determine whether each statement is true or false.

_____ **1.** When converting decimal notation to percent notation, move the decimal point two places to the right and write a percent symbol.   [7.1b]

_____ **2.** The symbol % is equivalent to $\times\ 0.10$.   [7.1a]

_____ **3.** Of the numbers $\frac{1}{10}$, 1%, 0.1%, 10%, and $\frac{1}{100}$, the smallest number is 0.1%.   [7.1b], [7.2a, b]

## Guided Solutions

Fill in each blank with the number that creates a correct statement or solution.    [7.1b], [7.2a, b]

**4.** $\frac{1}{2}\% = \frac{1}{2} \cdot \frac{1}{\square} = \frac{1}{\square}$

**5.** $\frac{80}{1000} = \frac{\square}{100} = \square\ \%$

**6.** $5.5\% = \frac{\square}{100} = \frac{\square}{1000} = \frac{11}{\square}$

**7.** $0.375 = \frac{\square}{1000} = \frac{\square}{100} = \square\ \%$

**8.** Solve:  15 is what percent of 80?   [7.3b]

$15 = p \times \square$     Translating

$\dfrac{15}{\square} = \dfrac{p \times \square}{\square}$     Dividing on both sides

$\dfrac{15}{\square} = p$     Simplifying

$\square = p$     Dividing

$\square\% = p$     Converting to percent notation

## Mixed Review

Find decimal notation.   [7.1b]

**9.** 28%

**10.** 0.15%

**11.** $5\frac{3}{8}\%$

**12.** 240%

Find percent notation.   [7.1b], [7.2a]

**13.** 0.71

**14.** $\dfrac{9}{100}$

**15.** 0.3891

**16.** $\dfrac{3}{16}$

**17.** 0.005

**18.** $\dfrac{37}{50}$

**19.** 6

**20.** $\dfrac{5}{6}$

Find fraction notation. Simplify. [7.2b]

**21.** 85%

**22.** 0.048%

**23.** $22\frac{3}{4}$%

**24.** $16.\overline{6}$%

Write percent notation for the shaded area. [7.2a]

**25.**

**26.**

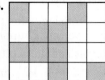

Solve. [7.3b], [7.4b]

**27.** 25% of what is 14.5?

**28.** 220 is what percent of 1320?

**29.** What is 3.2% of 80,000?

**30.** $17.50 is 35% of what?

**31.** What percent of $800 is $160?

**32.** 130% of $350 is what?

**33.** Arrange the following numbers from smallest to largest. [7.1b], [7.2a, b]

$$\frac{1}{2}\%, \ 5\%, \ 0.275, \ \frac{13}{100}, \ 1\%, \ 0.1\%, \ 0.05\%, \ \frac{3}{10}, \ \frac{7}{20}, \ 10\%$$

**34.** Solve: 8.5 is $2\frac{1}{2}$% of what? [7.3b], [7.4b]

**A.** 3.4

**B.** 21.25

**C.** 0.2125

**D.** 340

**35.** Solve: $102,000 is what percent of $3.6 million? [7.3b], [7.4b]

**A.** $2.8\overline{3} million

**B.** $2\frac{5}{6}$%

**C.** $0.028\overline{3}$%

**D.** $28.\overline{3}$

# Understanding Through Discussion and Writing

**36.** Is it always best to convert from fraction notation to percent notation by first finding decimal notation? Why or why not? [7.2a]

**37.** Suppose we know that 40% of 92 is 36.8. What is a quick way to find 4% of 92? 400% of 92? Explain. [7.3b]

**38.** In solving Example 10 in Section 7.4 a student simplifies $\frac{10}{20}$ before solving. Is this a good idea? Why or why not? [7.4b]

**39.** What do the following have in common? Explain. [7.1b], [7.2a, b]

$$\frac{23}{16}, \ 1\frac{875}{2000}, \ 1.4375, \ \frac{207}{144}, \ 1\frac{7}{16}, \ 143.75\%, \ 1\frac{4375}{10,000}$$

# 7.5

# Applications of Percent

## (a) Applied Problems Involving Percent

Applied problems involving percent are not always stated in a manner easily translated to an equation. In such cases, it is helpful to rephrase the problem before translating. Sometimes it also helps to make a drawing.

**EXAMPLE 1** *Extinction of Mammals.* According to a study conducted for the International Union for the Conservation of Nature (IUCN), the world's mammals are in danger of an extinction crisis. Of the 5487 species of mammals on Earth, 1141 are on the IUCN Red List of Threatened Species. Six of those mammals are shown below. What percent of all mammals are threatened with extinction?

**Sources:** Environment News Service, October 6, 2008; IUCN

*Top row, left to right:* Tasmanian devils, Père David's deer, African elephant. *Bottom row, left to right:* Iberian lynx, black-footed ferret, giant panda.

1. **Familiarize.** The question asks for a percent of the world's mammals that are in danger of extinction. We note that 5487 is approximately 5500 and 1141 is approximately 1100. Since 1100 is $\frac{1100}{5500}$, or $\frac{1}{5}$, or 20% of 5500, our answer is close to 20%. We let $p =$ the percent of mammals that are in danger of extinction.

2. **Translate.** There are two ways in which we can translate this problem.

*Percent equation (see Section 7.3):*

$$\underbrace{1141}_{\downarrow} \quad \underbrace{\text{is}}_{\downarrow} \quad \underbrace{\text{what percent}}_{\downarrow} \quad \underbrace{\text{of}}_{\downarrow} \quad \underbrace{5487?}_{\downarrow}$$

$$1141 \quad = \quad p \quad \cdot \quad 5487$$

*Proportion (see Section 7.4):*

$$\frac{N}{100} = \frac{1141}{5487}$$

For proportions, $N\% = p$.

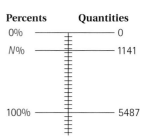

| Percents | Quantities |
|----------|-----------|
| 0% | 0 |
| N% | 1141 |
| 100% | 5487 |

*Answers*

*Skill to Review:*
1. 6   2. 11.3

3. **Solve.** We now have two ways in which to solve this problem.

*Percent equation (see Section 7.3):*

$$1141 = p \cdot 5487$$

$$\frac{1141}{5487} = \frac{p \cdot 5487}{5487} \qquad \text{Dividing by 5487 on both sides}$$

$$\frac{1141}{5487} = p$$

$$0.208 \approx p \qquad \text{Finding decimal notation and rounding to the nearest thousandth}$$

$$20.8\% \approx p \qquad \text{Remember to find percent notation.}$$

Note here that the solution, $p$, includes the % symbol.

*Proportion (see Section 7.4):*

$$\frac{N}{100} = \frac{1141}{5487}$$

$$N \cdot 5487 = 100 \cdot 1141 \qquad \text{Equating cross products}$$

$$\frac{N \cdot 5487}{5487} = \frac{114{,}100}{5487} \qquad \text{Dividing by 5487 on both sides}$$

$$N = \frac{114{,}100}{5487}$$

$$N \approx 20.8 \qquad \text{Dividing and rounding to the nearest tenth}$$

We use the solution of the proportion to express the answer to the problem as 20.8%. Note that in the proportion method, $N\% = p$.

4. **Check.** To check, we note that the answer 20.8% is close to 20%, as estimated in the *Familiarize* step. We can also find 20.8% of 5487: $20.8\% \times 5487 = 0.208 \times 5487 \approx 1141$. This is the number of mammals threatened with extinction, so the answer checks.

5. **State.** About 20.8% of the world's mammals are threatened with extinction.

Do Exercise 1.

**EXAMPLE 2** *Transportation to Work.* In the United States, there are about 147,000,000 workers 16 years and older. Approximately 77% drive to work alone. How many workers drive to work alone?

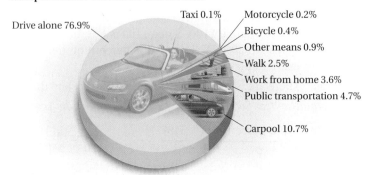

**Transportation to Work in the United States**

Drive alone 76.9%
Taxi 0.1%
Motorcycle 0.2%
Bicycle 0.4%
Other means 0.9%
Walk 2.5%
Work from home 3.6%
Public transportation 4.7%
Carpool 10.7%

SOURCES: U.S. Census Bureau; American Community Survey

---

1. **Presidential Assassinations in Office.** Of the 43 different U.S. presidents, 4 have been assassinated in office. These were James A. Garfield, William McKinley, Abraham Lincoln, and John F. Kennedy. What percent have been assassinated in office?

*Answer*

1. About 9.3%

1. **Familiarize.** We can simplify the pie chart shown on the preceding page to help familiarize ourselves with the problem. We let $a =$ the total number of workers who drive to work alone.

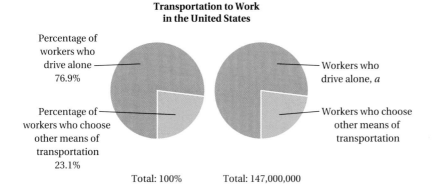

**Transportation to Work
in the United States**

Percentage of workers who drive alone 76.9%

Percentage of workers who choose other means of transportation 23.1%

Workers who drive alone, $a$

Workers who choose other means of transportation

Total: 100%      Total: 147,000,000

2. **Translate.** There are two ways in which we can translate this problem.

*Percent equation:*

What number is 76.9% of 147,000,000?

$$a = 76.9\% \cdot 147,000,000$$

*Proportion:*

$$\frac{76.9}{100} = \frac{a}{147,000,000}$$

3. **Solve.** We now have two ways in which to solve this problem.

*Percent equation:*

$$a = 76.9\% \cdot 147,000,000$$

We convert 76.9% to decimal notation and multiply:

$$a = 0.769 \times 147,000,000 = 113,043,000.$$

*Proportion:*

$$\frac{76.9}{100} = \frac{a}{147,000,000}$$

$$76.9 \times 147,000,000 = 100 \cdot a \qquad \text{Equating cross products}$$

$$\frac{76.9 \cdot 147,000,000}{100} = \frac{100 \cdot a}{100} \qquad \text{Dividing by 100}$$

$$\frac{11,304,300,000}{100} = a$$

$$113,043,000 = a \qquad \text{Simplifying}$$

4. **Check.** To check, we can repeat the calculations. We can also do a partial check by estimating. Since 76.9% is about 80%, or $\frac{4}{5}$, and $\frac{4}{5}$ of 147,000,000 is 117,600,000 and 117,600,000 is close to 113,043,000, our answer is reasonable.

5. **State.** The number of workers who drive to work alone is 113,043,000.

Do Exercise 2.

2. **Transportation to Work.** There are about 147,000,000 workers 16 years or older in the United States. Approximately 10.7% carpool to work. How many workers carpool to work?
Sources: U.S. Census Bureau; American Community Survey

*Answer*

2. 15,729,000 workers

New price: $21,682.30

Increase ↑

Former price: $20,455

## b Percent of Increase or Decrease

Percent is often used to state increase or decrease. Let's consider an example of each, using the price of a car as the original number.

### Percent of Increase

One year a car sold for $20,455. The manufacturer decides to raise the price of the following year's model by 6%. The increase is $0.06 \times \$20,455$, or $1227.30. The new price is $20,455 + $1227.30, or $21,682.30. Note that the new price is 106% of the *former* price.

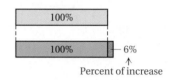

New price: $21,682.30    Increase    Percent of increase

The increase, $1227.30, is 6% of the *former* price, $20,455. The *percent of increase* is 6%.

### Percent of Decrease

Original price: $20,455

Decrease ↓

New value: $15,341.25

Abigail buys the car listed above for $20,455. After one year, the car depreciates in value by 25%. The decrease is $0.25 \times \$20,455$, or $5113.75. This lowers the value of the car to $20,455 − $5113.75, or $15,341.25. Note that the new value is 75% of the original price. If Abigail decides to sell the car after one year, $15,341.25 might be the most she could expect to get for it.

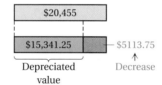

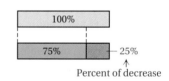

Depreciated value    Decrease    Percent of decrease

The decrease, $5113.75, is 25% of the *original* price, $20,455. The *percent of decrease* is 25%.

Do Exercises 3 and 4.

When a quantity is decreased by a certain percent, we say that we have **percent of decrease**.

**EXAMPLE 3**    *Dow Jones Industrial Average.*    The Dow Jones Industrial Average (DJIA) plunged from 11,143 to 10,365 on September 29, 2008. This was the largest one-day drop in its history. What was the percent of decrease?

Sources: *Nightly Business Reports,* September 29, 2008; DJIA

---

**3. Percent of Increase.**    The price of a car is $36,875. The price of the following year's model is increased by 4%.

a) How much is the increase?

b) What is the new price?

**4. Percent of Decrease.**    The value of a car is $36,875. The car depreciates in value by 25% after one year.

a) How much is the decrease?

b) What is the depreciated value of the car?

*Answers*

**3.** (a) $1475; (b) $38,350
**4.** (a) $9218.75; (b) $27,656.25

**1. Familiarize.** We first determine the amount of decrease and then make a drawing.

$$
\begin{array}{rl}
11{,}143 & \text{Opening average} \\
-10{,}365 & \text{Closing average} \\
\hline
778 & \text{Decrease}
\end{array}
$$

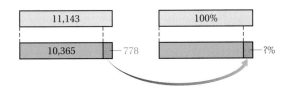

We are asking this question: The decrease is what percent of the opening average? We let $p$ = the percent of decrease.

**2. Translate.** There are two ways in which we can translate this problem.

*Percent equation:*

$$
\underbrace{778}_{} \;\; \underbrace{\text{is}}_{} \;\; \underbrace{\text{what percent}}_{} \;\; \underbrace{\text{of}}_{} \;\; \underbrace{11{,}143?}_{}
$$

$$
778 \;\; = \;\; p \;\; \times \;\; 11{,}143
$$

*Proportion:*

$$
\frac{N}{100} = \frac{778}{11{,}143}
$$

For proportions, $N\% = p$.

**3. Solve.** We have two ways in which to solve this problem.

*Percent equation:*

$$
778 = p \times 11{,}143
$$

$$
\frac{778}{11{,}143} = \frac{p \times 11{,}143}{11{,}143} \qquad \text{Dividing by 11,143 on both sides}
$$

$$
\frac{778}{11{,}143} = p
$$

$$
0.07 \approx p
$$

$$
7\% \approx p \qquad \text{Converting to percent notation}
$$

*Proportion:*

$$
\frac{N}{100} = \frac{778}{11{,}143}
$$

$$
N \times 11{,}143 = 100 \times 778 \qquad \text{Equating cross products}
$$

$$
\frac{N \times 11{,}143}{11{,}143} = \frac{100 \times 778}{11{,}143} \qquad \text{Dividing by 11,143 on both sides}
$$

$$
N = \frac{77{,}800}{11{,}143}
$$

$$
N \approx 7
$$

We use the solution of the proportion to express the answer to the problem as 7%.

**5. Volume of Mail.** The volume of U.S. mail decreased from about 213,138 million pieces of mail in 2006 to 212,234 million pieces in 2007. What was the percent of decrease?

**Source:** U.S. Postal Service

**4. Check.** To check, we note that, with a 7% decrease, the closing Dow average should be 93% of the opening average. Since

$$93\% \times 11{,}143 = 0.93 \times 11{,}143 \approx 10{,}363,$$

and 10,363 is close to 10,365, our answer checks. (Remember that we rounded to get 7%.)

**5. State.** The percent of decrease in the Dow Jones Industrial Average was approximately 7%.

> Do Exercise 5.

When a quantity is increased by a certain percent, we say we have **percent of increase**.

**EXAMPLE 4** *Costs for Moviegoers.* The average cost of movie tickets for a family of four was $16.56 in 1993. The cost rose to $28.32 in 2008. What was the percent of increase in the cost for a family of four to attend a movie?

**Source:** Motion Picture Association of America

**1. Familiarize.** We first determine the increase in the cost and then make a drawing.

$$\begin{array}{r r l} & \$28.32 & \text{Cost in 2008} \\ - & 16.56 & \text{Cost in 1993} \\ \hline & \$11.76 & \text{Increase} \end{array}$$

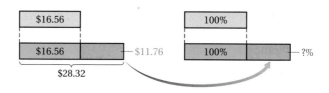

We are asking this question: The increase is what percent of the *original* cost? We let $p =$ the percent of increase.

**2. Translate.** There are two ways in which we can translate this problem.

*Percent equation:*

$$\underbrace{11.76}_{\displaystyle 11.76} \;\; \underbrace{\text{is}}_{\displaystyle =} \;\; \underbrace{\text{what percent}}_{\displaystyle p} \;\; \underbrace{\text{of}}_{\displaystyle \cdot} \;\; \underbrace{16.56?}_{\displaystyle 16.56}$$

*Proportion:*

$$\frac{N}{100} = \frac{11.76}{16.56}$$

For proportions, $N\% = p$.

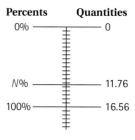

**3. Solve.**   We have two ways in which to solve this problem.

*Percent equation:*

$$11.76 = p \times 16.56$$

$$\frac{11.76}{16.56} = \frac{p \times 16.56}{16.56} \qquad \text{Dividing by 16.56 on both sides}$$

$$\frac{11.76}{16.56} = p$$

$$0.71 \approx p$$

$$71\% \approx p \qquad \text{Converting to percent notation}$$

*Proportion:*

$$\frac{N}{100} = \frac{11.76}{16.56}$$

$$N \times 16.56 = 100 \times 11.76 \qquad \text{Equating cross products}$$

$$\frac{N \times 16.56}{16.56} = \frac{100 \times 11.76}{16.56} \qquad \text{Dividing by 16.56 on both sides}$$

$$N = \frac{1176}{16.56}$$

$$N \approx 71$$

We use the solution of the proportion to express the answer to the problem as 71%.

**4. Check.**   To check, we take 71% of 16.56:

$$71\% \times 16.56 = 0.71 \times 16.56 = 11.7576.$$

The approximation 11.7576 is close to 11.76, so the answer checks. (Remember that we rounded to get 71%.)

**5. State.**   Between 1993 and 2008, the percent of increase in the cost for a family of four to attend a movie was 71%.

Do Exercise 6.

**6. Shipping Costs.**   The total cost of shipping a 40-ft container from Shanghai to New York rose from $5100 in 2005 to $8350 in 2008. Find the percent of increase.

**Source:** CIBC World Markets, IMF

# Translating for Success

1. *Distance Walked.* After knee replacement, Alex walked $\frac{1}{8}$ mi each morning and $\frac{1}{5}$ mi each afternoon. How much farther did he walk in the afternoon?

2. *Stock Prices.* A stock sold for $5 per share on Monday and only $2.125 per share on Friday. What was the percent of decrease from Monday to Friday?

3. *SAT Score.* After attending a class titled *Improving Your SAT Scores*, Jacob raised his total score from 884 to 1040. What was the percent of increase?

4. *Change in Population.* The population of a small farming community decreased from 1040 to 884. What was the percent of decrease?

5. *Lawn Mowing.* During the summer, brothers Steve and Rob earned money for college by mowing lawns. The largest lawn that they mowed was $2\frac{1}{8}$ acres. Steve can mow $\frac{1}{5}$ acre per hour, and Rob can mow only $\frac{1}{8}$ acre per hour. Working together, how many acres did they mow per hour?

The goal of these matching questions is to practice step (2), *Translate*, of the five-step problem-solving process. Translate each problem to an equation and select a correct translation from equations A–O.

**A.** $x + \dfrac{1}{5} = \dfrac{1}{8}$

**B.** $250 = x \cdot 1040$

**C.** $884 = x \cdot 1040$

**D.** $\dfrac{250}{16.25} = \dfrac{1000}{x}$

**E.** $156 = x \cdot 1040$

**F.** $16.25 = 250 \cdot x$

**G.** $\dfrac{1}{5} + \dfrac{1}{8} = x$

**H.** $2\dfrac{1}{8} = x \cdot 5$

**I.** $5 = 2.875 \cdot x$

**J.** $\dfrac{1}{8} + x = \dfrac{1}{5}$

**K.** $1040 = x \cdot 884$

**L.** $\dfrac{250}{16.25} = \dfrac{x}{1000}$

**M.** $2.875 = x \cdot 5$

**N.** $x \cdot 884 = 156$

**O.** $x = 16.25 \cdot 250$

*Answers on page A-12*

6. *Land Sale.* Cole sold $2\frac{1}{8}$ acres from the 5 acres he inherited from his uncle. What percent did he sell?

7. *Travel Expenses.* A magazine photographer is reimbursed 16.25¢ per mile for business travel up to 1000 mi per week. In a recent week, he traveled 250 mi. What was the total reimbursement for travel?

8. *Trip Expenses.* The total expenses for Claire's recent business trip were $1040. She put $884 on her credit card and paid the balance in cash. What percent did she place on her credit card?

9. *Cost of Copies.* During the first summer session at a community college, the campus copy center advertised 250 copies for $16.25. At this rate, what is the cost of 1000 copies?

10. *Cost of Insurance.* Following a raise in the cost of health insurance, 250 of a company's 1040 employees dropped their health coverage. What percent of the employees canceled their insurance?

**7.5** **Exercise Set**

For Extra Help

*MyMathLab*

Math XL
PRACTICE

WATCH

DOWNLOAD

READ

REVIEW

**a** Solve.

1. *Foreign Students.* In the 2006–2007 school year, of the 15 million college students in the United States, 583,000 were from foreign countries. Approximately 10.70% of the foreign students were from South Korea and 6.05% were from Japan. How many foreign students were from South Korea? from Japan?

**Source:** Institute of International Education

2. *Mississippi River.* The Mississippi River, which extends from its source, at Lake Itasca in Minnesota, to the Gulf of Mexico, is 2348 mi long. Approximately 77% of the river is navigable. How many miles of the river are navigable?

**Source:** National Oceanic and Atmospheric Administration

3. A person earns $43,200 one year and receives an 8% raise in salary. What is the new salary?

4. A person earns $28,600 one year and receives a 5% raise in salary. What is the new salary?

5. *Test Results.* On a test, Juan got 85%, or 119, of the items correct. How many items were on the test?

6. *Test Results.* On a test, Maj Ling got 80%, or 76, of the items correct. How many items were on the test?

7. *Farmland.* In Kansas, 47,000,000 acres are farmland. About 5% of all the farm acreage in the United States is in Kansas. What is the total number of acres of farmland in the United States?

**Sources:** U.S. Department of Agriculture; National Agricultural Statistics Service

8. *World Population.* World population is increasing by 1.2% each year. In 2008, it was 6.68 billion. What will the population be in 2015?

**Sources:** U.S. Census Bureau; International Data Base

**9.** *Car Depreciation.* A car generally depreciates 25% of its original value in the first year. A car is worth $27,300 after the first year. What was its original cost?

**10.** *Car Depreciation.* Given normal use, an American-made car will depreciate 25% of its original cost the first year and 14% of its remaining value in the second year. What is the value of a car at the end of the second year if its original cost was $36,400? $28,400? $26,800?

**11.** *Test Results.* On a test of 80 items, Pedro got 95% correct. How many items did he get correct? incorrect?

**12.** *Test Results.* On a test of 60 items, Christina got 75% correct. How many items did she get correct? incorrect?

**13.** *Under 15 Years Old.* In Egypt, 32.6% of the population is under 15 years old. In the United States, 20.4% of the population is under 15. The population of Egypt is 75,449,000, and the population of the United States is 305,468,000. How many are under 15 years old in Egypt? in the United States?

Sources: U.S. Census Bureau; International Data Base

**14.** *Age 65 and Older.* In Egypt, 4.5% of the population is age 65 and older. In the United States, 12.5% of the population is age 65 and older. The population of Egypt is 75,449,000, and the population of the United States is 305,468,000. How many are age 65 and older in Egypt? in the United States?

Sources: U.S. Census Bureau; International Data Base

**15.** *Transplant Waiting List.* The total number of patients waiting for transplants as of September 27, 2007, was 96,749. Of this number, 16,737 were waiting for a liver transplant. What percent were waiting for a liver transplant?

Source: United Network for Organ Sharing

**16.** *Doctors' Salaries.* In 2007, the starting salary for a neurologist was approximately 64.5% of that of an anesthesiologist. The beginning salary for an anesthesiologist was $275,000. What was the beginning salary for a neurologist?

Source: *Journal of the American Medical Association*

**17.** *Tipping.* For a party of 8 or more, some restaurants add an 18% tip to the bill. What is the total amount charged for a party of 10 if the cost of the meal, without tip, is $195?

**18.** *Tipping.* Diners frequently add a 15% tip when charging a meal to a credit card. What is the total amount charged if the cost of the meal, without tip, is $18? $34? $49?

**19.** *Fast-Food Cooks.* The United States has 392,850 full-time farmers. This number is about 64.2% of the number of fast-food cooks. How many fast-food cooks are there in the United States?
Source: U.S. Department of Agriculture

**20.** *Spending in Restaurants.* Americans spent $364 billion in grocery stores in 2007. This amount is about $93\frac{1}{3}$% of the amount spent in restaurants. How much was spent in restaurants?
Source: U.S. Department of Agriculture

**21.** A lab technician has 540 mL of a solution of alcohol and water; 8% is alcohol. How many milliliters are alcohol? water?

**22.** A lab technician has 680 mL of a solution of water and acid; 3% is acid. How many milliliters are acid? water?

**23.** *U.S. Armed Forces.* There were 1,385,000 people in the United States in active military service in 2006. The numbers in the four armed services are listed in the table below. What percent of the total does each branch represent? Round the answers to the nearest tenth of a percent.

### U.S. ARMED FORCES: 2006

| TOTAL | 1,385,000* |
| --- | --- |
| AIR FORCE | 349,000 |
| ARMY | 505,000 |
| NAVY | 350,000 |
| MARINES | 180,000 |

*Includes National Guard, Reserve, and retired regular personnel on extended or continuous active duty. Excludes Coast Guard.

SOURCES: U.S. Department of Defense; U.S. Census Bureau

**24.** *Living Veterans.* There were 23,977,000 living veterans in the United States in 2006. Numbers in selected age groups are listed in the table below. What percent of the total does each age group represent? Round the answers to the nearest tenth of a percent.

### LIVING VETERANS BY AGE: 2006

| TOTAL | 23,977,000 |
| --- | --- |
| UNDER 35 YEARS OLD | 1,949,000 |
| 35–44 YEARS OLD | 2,901,000 |
| 45–54 YEARS OLD | 3,846,000 |
| 55–64 YEARS OLD | 6,081,000 |
| 65 YEARS OLD AND OLDER | 9,200,000 |

SOURCES: U.S. Department of Defense; U.S. Census Bureau

**b** Solve.

**25.** *Mortgage Payment Increase.* A monthly mortgage payment increases from $840 to $882. What is the percent of increase?

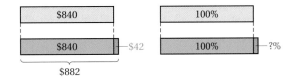

**26.** *Savings Increase.* The amount in a savings account increased from $200 to $216. What was the percent of increase?

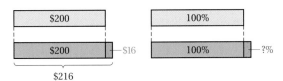

**27.** A person on a diet goes from a weight of 160 lb to a weight of 136 lb. What is the percent of decrease?

**28.** During a sale, a dress decreased in price from $90 to $72. What was the percent of decrease?

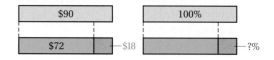

**29.** *Insulation.* A 15" roll of unfaced fiberglass insulation has a retail price of $23.43. For two weeks, it is on sale for $15.31. What is the percent of decrease?

**30.** *Set of Weights.* A 300-lb weight set retails for $199.95. For its grand opening, a sporting goods store reduced the price to $154.95. What is the percent of decrease?

**31.** *Mass Transit.* From April through June of 2008, people took 2.8 billion rides on public transit in the United States. The ridership was up from 2.1 billion rides in 1998. What is the percent of increase?

**Source:** American Public Transportation Association

**32.** *Miles of Railroad Track.* The greatest combined length of U.S.-owned operating railroad track was 254,037 mi in 1916, when industrial activity increased during World War I. The total length has decreased ever since. By 2006, the number of miles of track had decreased to 140,490 mi. What is the percent of decrease from 1916 to 2006?

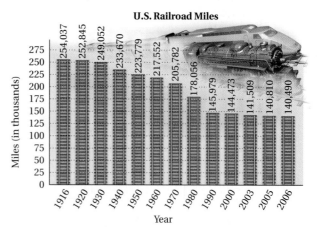

NOTE: The lengths exclude yard tracks, sidings, and parallel tracks.
SOURCE: Association of American Railroads

**33.** *Overdraft Fees.* Consumers are paying record amounts of fees for overdrawing their bank accounts. In 2007, banks, thrifts, and credit unions collected $45.6 billion in overdraft fees, which is $15.1 billion more than in 2001. What is the percent of increase?

Source: Moebs Services

**34.** *Credit-Card Debt.* In 2007, the average credit-card debt per household in the United States was $9840. In 1990, the average credit-card debt was $2966. What is the percent of increase?

Sources: CardWeb.com; *USA TODAY*

**35.** *Immigrant Applications.* From January through June of 2008, 46,866 immigrants applied for citizenship each month. During this same period in 2007, 114,469 immigrants applied each month. What is the percent of decrease?

Source: U.S. Citizenship and Immigration Services (USCIS)

**36.** *Pharmacists.* It is projected that there will be 296,000 people employed as pharmacists in 2016. In 2006, 243,000 members of the labor force were pharmacists. What is the percent of increase?

Source: EarnMyDegree.com

**37.** *Patents Issued.* The U.S. Patent and Trademark Office (USPTO) issued a total of 157,284 utility patents in 2007. This number of patents is down from 173,794 in 2006. What is the percent of decrease?

Source: IFI Patent Intelligence

**38.** *Highway Fatalities.* In 2007, there were 41,059 highway fatalities, which was 1649 fewer deaths than in 2006. What is the percent of decrease?

Source: National Highway Traffic Safety Administration

**39.** *Two-by-Four.* A cross-section of a standard or nominal "two-by-four" board actually measures $1\frac{1}{2}$ in. by $3\frac{1}{2}$ in. The rough board is 2 in. by 4 in. but is planed and dried to the finished size. What percent of the wood is removed in planing and drying?

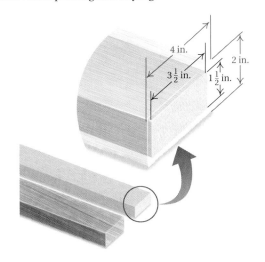

**40.** *Strike Zone.* In baseball, the *strike zone* is normally a 17-in. by 30-in. rectangle. Some batters give the pitcher an advantage by swinging at pitches thrown out of the strike zone. By what percent is the area of the strike zone increased if a 2-in. border is added to the outside?

Source: Major League Baseball

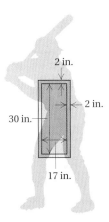

*Population Increase.* The table below provides data showing how the populations of various states increased from 2000 to 2006. Complete the table by filling in the missing numbers. Round percents to the nearest tenth of a percent.

| | STATE | POPULATION IN 2000 | POPULATION IN 2006 | CHANGE | PERCENT CHANGE |
|---|---|---|---|---|---|
| **41.** | Vermont | 608,827 | 623,908 | | |
| **42.** | Wisconsin | 5,363,675 | | 192,831 | |
| **43.** | Arizona | | 6,166,318 | 1,035,686 | |
| **44.** | Virginia | | 7,642,884 | 564,369 | |
| **45.** | Idaho | 1,293,953 | | 172,512 | |
| **46.** | Georgia | 8,186,453 | 9,363,941 | | |

SOURCE: U.S. Census Bureau

**47.** *Decrease in Population.* Between 2000 and 2006, the population of North Dakota decreased from 642,200 to 635,867. What was the percent of decrease?
**Sources:** U.S. Census Bureau; U.S. Department of Commerce

**48.** *Decrease in Population.* Between 2000 and 2006, the population of Louisiana decreased from 4,468,976 to 4,287,768. What was the percent of decrease?
**Sources:** U.S. Census Bureau; U.S. Department of Commerce

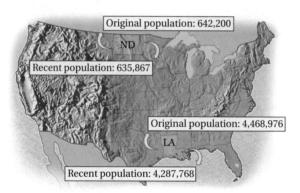

Original population: 642,200
ND
Recent population: 635,867
Original population: 4,468,976
LA
Recent population: 4,287,768

## Skill Maintenance

Convert to decimal notation. [5.1b], [5.5a]

**49.** $\dfrac{25}{11}$

**50.** $-\dfrac{11}{25}$

**51.** $-\dfrac{27}{8}$

**52.** $\dfrac{43}{9}$

**53.** $\dfrac{23}{25}$

**54.** $\dfrac{20}{24}$

**55.** $\dfrac{14}{32}$

**56.** $\dfrac{2317}{1000}$

**57.** $-\dfrac{34,809}{10,000}$

**58.** $-\dfrac{27}{40}$

## Synthesis

**59.** A coupon allows a couple to have dinner and then have $10 subtracted from the bill. Before subtracting $10, however, the restaurant adds a tip of 20%. If the couple is presented with a bill for $40.40, how much would the dinner (without tip) have cost without the coupon?

**60.** If $p$ is 120% of $q$, then $q$ is what percent of $p$?

# 7.6

# Sales Tax, Commission, and Discount

## (a) Sales Tax

**OBJECTIVES**

**a** Solve applied problems involving sales tax and percent.

**b** Solve applied problems involving commission and percent.

**c** Solve applied problems involving discount and percent.

Sales tax computations represent a special type of percent of increase problem. The sales tax rate in Pennsylvania is 6%. This means that the tax is 6% of the purchase price. Suppose the purchase price of a guitar is $839.95. The sales tax is then 6% of $839.95, or 0.06 × $839.95, or $50.397, or about $50.40.

**BILL:**

| | | |
|---|---|---|
| Purchase price | = | $839.95 |
| Sales tax (6% of $839.95) | = | + 50.40 |
| Total price | | $890.35 |

The total that you pay is the purchase price plus the sales tax:

$839.95 + $50.40,   or   $890.35.

---

**SALES TAX**

**Sales tax** = Sales tax rate × Purchase price

**Total price** = Purchase price + Sales tax

---

**EXAMPLE 1**  *Wisconsin Sales Tax.*   The sales tax rate in Wisconsin is 5%. How much tax is charged on the purchase of 3 gal of paint at $42.99 each? What is the total price?

a) We first find the cost of the paint. It is

3 × $42.99 = $128.97.

b) The sales tax on items costing $128.97 is

$$\underbrace{\text{Sales tax rate}}_{5\%} \times \underbrace{\text{Purchase price}}_{\$128.97},$$

or 0.05 × 128.97, or 6.4485. Thus the tax is $6.45 (rounded to the nearest cent).

c) The total price is given by the purchase price plus the sales tax:

$128.97 + $6.45,   or   $135.42.

To check, note that the total price is the purchase price plus 5% of the purchase price. Thus the total price is 105% of the purchase price. Since 1.05 × 128.97 ≈ 135.42, we have a check. The sales tax is $6.45, and the total price is $135.42.

1. **Texas Sales Tax.** The sales tax rate in Texas is 6.25%. In Texas, how much tax is charged on the purchase of an ultrasound toothbrush that sells for $139.95? What is the total price?

2. **Wyoming Sales Tax.** Samantha buys 7 copies of *The Last Lecture* by Randy Pausch with Jeffrey Zaslow for $14.95 each. The sales tax rate in Wyoming is 4%. In Wyoming, how much sales tax will be charged? What is the total price?

3. The sales tax on the purchase of a set of holiday dishes that costs $449 is $26.94. What is the sales tax rate?

**Price: ?**
**$12.74 tax @ 8%**

4. The sales tax on the purchase of a pair of designer jeans is $4.84 and the sales tax rate is 5.5%. Find the purchase price (the price before sales tax is added).

**Answers**

1. $8.75; $148.70    2. $4.19; $108.84
3. 6%    4. $88

---

Do Exercises 1 and 2.

**EXAMPLE 2**  The sales tax on the purchase of this GPS navigator, which costs $799, is $55.93. What is the sales tax rate?

GPS Navigator 4.3"-screen with speech recognition
**$799**
+ $55.93 sales tax

We rephrase and translate as follows:

*Rephrase:* Sales tax  is  what percent  of  purchase price?

*Translate:*  55.93  =  $r$  ×  799.

To solve the equation, we divide by 799 on both sides:

$$\frac{55.93}{799} = \frac{r \times 799}{799}$$

$$\frac{55.93}{799} = r$$

$$0.07 = r$$

$$7\% = r.$$

The sales tax rate is 7%.

Do Exercise 3.

**EXAMPLE 3**  The sales tax on the purchase of a stone-top firepit is $12.74 and the sales tax rate is 8%. Find the purchase price (the price before sales tax is added).

We rephrase and translate as follows:

*Rephrase:* Sales tax  is  8%  of  what?

*Translate:*  12.74  =  8%  ×  $b$,  or

  12.74 = 0.08 × $b$.

To solve, we divide by 0.08 on both sides:

$$\frac{12.74}{0.08} = \frac{0.08 \times b}{0.08}$$

$$\frac{12.74}{0.08} = b$$

$$159.25 = b.$$

The purchase price is $159.25.

Do Exercise 4.

## b Commission

When you work for a **salary**, you receive the same amount of money each week or month. When you work for a **commission**, you are paid a percentage of the total sales for which you are responsible.

> **COMMISSION**
>
> **Commission** = Commission rate × Sales

**EXAMPLE 4** *Appliance Sales.* A salesperson's commission rate is 3%. What is the commission from the sale of $8300 worth of appliances?

Commission rate **3%**

$$\text{Commission} = \text{Commission rate} \times \text{Sales}$$
$$C \quad = \quad 3\% \quad \times \quad 8300$$
$$C \quad = \quad 0.03 \quad \times \quad 8300$$
$$C = 249$$

The commission is $249.

Do Exercise 5.

**EXAMPLE 5** *Earth-Moving Equipment Sales.* Gavin earns a commission of $20,800 selling $320,000 worth of earth-moving equipment. What is the commission rate?

Commission $20,800
Total sales $320,000

$$\text{Commission} = \text{Commission rate} \times \text{Sales}$$
$$20,800 \quad = \quad r \quad \times \quad 320,000$$

*Answer*

**5.** $1389

To solve this equation, we divide by 320,000 on both sides:

$$\frac{20,800}{320,000} = \frac{r \times 320,000}{320,000}$$

$$0.065 = r$$

$$6.5\% = r.$$

The commission rate is 6.5%.

Do Exercise 6.

**6.** William earns a commission of $2040 selling $17,000 worth of concert tickets. What is the commission rate?

**EXAMPLE 6** *Cruise Vacations.* Mia's commission rate is 5.6%. She received a commission of $2457 on cruise vacation packages that she sold in November. How many dollars worth of cruise vacations did she sell?

$$Commission = Commission\ rate \times Sales$$
$$2457 \quad = \quad 5.6\% \quad \times \quad S, \quad or$$
$$2457 = 0.056 \times S$$

To solve this equation, we divide by 0.056 on both sides:

$$\frac{2457}{0.056} = \frac{0.056 \times S}{0.056}$$

$$\frac{2457}{0.056} = S$$

$$43,875 = S.$$

Mia sold $43,875 worth of cruise vacation packages.

Do Exercise 7.

**7.** Dylan's commission rate is 7.5%. He receives a commission of $2970 from the sale of winter ski passes. How many dollars worth of ski passes did he sell?

## c Discount

Suppose that the regular price of a rug is $60, and the rug is on sale at 25% off. Since 25% of $60 is $15, the sale price is $60 − $15, or $45. We call $60 the **original**, or **marked**, **price**, 25% the **rate of discount**, $15 the **discount**, and $45 the **sale price**. Note that discount problems are a type of percent of decrease problem.

**DISCOUNT AND SALE PRICE**

**Discount** = Rate of discount × Original price

**Sale price** = Original price − Discount

*Answers*

**6.** 12%     **7.** $39,600

**EXAMPLE 7** A leather sofa marked $2379 is on sale at $33\frac{1}{3}\%$ off. What is the discount? the sale price?

Leather sofa
$2379 original price
Save $33\frac{1}{3}\%$

**a)** Discount = Rate of discount × Original price

$$D = 33\frac{1}{3}\% \times 2379$$

$$D = \frac{1}{3} \times 2379$$

$$D = \frac{2379}{3} = 793$$

**b)** Sale price = Original price − Discount

$$S = 2379 - 793$$

$$S = 1586$$

The discount is $793, and the sale price is $1586.

Do Exercise 8.

**8.** A computer marked $660 is on sale at $16\frac{2}{3}\%$ off. What is the discount? the sale price?

**EXAMPLE 8** The price of a snowblower is marked down from $950 to $779. What is the rate of discount?

We first find the discount by subtracting the sale price from the original price:

$$950 - 779 = 171.$$

The discount is $171.

Next, we use the equation for discount:

Discount = Rate of discount × Original price

$$171 = r \times 950.$$

To solve, we divide by 950 on both sides:

$$\frac{171}{950} = \frac{r \times 950}{950}$$

$$\frac{171}{950} = r$$

$$0.18 = r$$

$$18\% = r.$$

The discount rate is 18%.

To check, note that an 18% discount rate means that 82% of the original price is paid:

$$0.82 \times \$950 = \$779.$$

**9.** The price of a winter coat is reduced from $75 to $60. Find the rate of discount.

Do Exercise 9.

*Answers*

**8.** $110; $550  **9.** 20%

Solve.

1. *Wyoming Sales Tax.*  The sales tax rate in Wyoming is 4%. How much sales tax would be charged on a fireplace screen with doors that costs $239?

2. *Kansas Sales Tax.*  The sales tax rate in Kansas is 5.3%. How much sales tax would be charged on a fireplace screen with doors that costs $239?

3. *Ohio Sales Tax.*  The sales tax rate in Ohio is 5.5%. How much sales tax would be charged on a dog jacket that sells for $29.50?

4. *New Mexico Sales Tax.*  The sales tax rate in New Mexico is 5%. How much sales tax would be charged on a pair of sunglasses that sells for $129.95?

5. *California Sales Tax.*  The sales tax rate in California is 7.25%. How much sales tax is charged on a purchase of 4 travel contour foam pillows at $39.95 each? What is the total price?

6. *Illinois Sales Tax.*  The sales tax rate in Illinois is 6.25%. How much sales tax is charged on a purchase of 3 wet–dry vacs at $60.99 each? What is the total price?

7. The sales tax is $30 on the purchase of a diamond ring that sells for $750. What is the sales tax rate?

8. The sales tax is $48 on the purchase of a dining room set that sells for $960. What is the sales tax rate?

9. The sales tax is $9.12 on the purchase of a patio set that sells for $456. What is the sales tax rate?

10. The sales tax is $35.80 on the purchase of a refrigerator–freezer that sells for $895. What is the sales tax rate?

11. The sales tax on the purchase of a new fishing boat is $112 and the sales tax rate is 2%. What is the purchase price (the price before tax is added)?

12. The sales tax on the purchase of a used car is $100 and the sales tax rate is 5%. What is the purchase price?

**13.** The sales tax rate in New York City, New York, is 4.375% for the city and 4% for the state. Find the total amount paid for 6 boxes of chocolates at $17.95 each.

**14.** The sales tax rate in Nashville, Tennessee, is 2.25% for Davidson County and 7% for the state. Find the total amount paid for 2 ladders at $39 each.

**15.** The sales tax rate in Seattle, Washington, is 2.5% for King County and 6.5% for the state. Find the total amount paid for 3 ceiling fans at $84.49 each.

**16.** The sales tax rate in Miami, Florida, is 1% for Dade County and 6% for the state. Find the total amount paid for 2 tires at $49.95 each.

**17.** The sales tax rate in Atlanta, Georgia, is 1% for the city, 3% for Fulton County, and 4% for the state. Find the total amount paid for 6 basketballs at $29.95 each.

**18.** The sales tax rate in Dallas, Texas, is 1% for the city, 1% for Dallas County, and 6.25% for the state. Find the total amount paid for 5 shrubs at $19.95 each.

 Solve.

**19.** Jose's commission rate is 21%. What is the commission from the sale of $12,500 worth of windows?

**20.** Jasmine's commission rate is 6%. What is the commission from the sale of $45,000 worth of lawn irrigation systems?

**21.** Olivia earns $408 selling $3400 worth of shoes. What is the commission rate?

**22.** Mitchell earns $120 selling $2400 worth of television sets. What is the commission rate?

**23.** *Real Estate Commission.* A real estate agent's commission rate is 7%. She receives a commission of $12,950 from the sale of a home. How much did the home sell for?

**24.** *Clothing Consignment Commission.* A clothing consignment shop's commission rate is 40%. The shop receives a commission of $552. How many dollars worth of clothing were sold?

**25.** A real estate commission rate is 8%. What is the commission from the sale of a piece of land for $68,000?

**26.** A real estate commission rate is 6%. What is the commission from the sale of a $98,000 home?

**27.** David earns $1147.50 selling $7650 worth of car parts. What is the commission rate?

**28.** Isabella earns $280.80 selling $2340 worth of tee shirts. What is the commission rate?

**29.** Sabrina's commission is increased according to how much she sells. She receives a commission of 4% for the first $1000 of sales and 7% for the amount over $1000. What is the total commission on sales of $5500?

**30.** Miguel's commission is increased according to how much he sells. He receives a commission of 5% for the first $2000 of sales and 8% for the amount over $2000. What is the total commission on sales of $6200?

**c** Complete the table below by filling in the missing numbers.

|  | MARKED PRICE | RATE OF DISCOUNT | DISCOUNT | SALE PRICE |
|---|---|---|---|---|
| **31.** | $300 | 10% |  |  |
| **32.** | $2000 | 40% |  |  |
| **33.** | $17 | 15% |  |  |
| **34.** | $20 | 25% |  |  |
| **35.** |  | 10% | $12.50 |  |
| **36.** |  | 15% | $65.70 |  |
| **37.** | $600 |  | $240 |  |
| **38.** | $12,800 |  | $1920 |  |

**39.** Find the marked price and the rate of discount for the steel log rack in this ad.

**40.** Find the marked price and the rate of discount for the digital photo frame in this ad.

**41.** Find the discount and the rate of discount for the pinball machine in this ad.

**42.** Find the discount and the rate of discount for the amaryllis in this ad.

## Skill Maintenance

Solve. [6.3b]

**43.** $\dfrac{x}{12} = \dfrac{24}{16}$

**44.** $\dfrac{7}{2} = \dfrac{11}{x}$

Solve. [5.4b]

**45.** $0.64 \cdot x = 170$

**46.** $-38.4 = 25.6 \times y$

Find decimal notation. [5.5a]

**47.** $\dfrac{5}{9}$

**48.** $\dfrac{23}{11}$

**49.** $\dfrac{11}{12}$

**50.** $\dfrac{13}{7}$

**51.** $-\dfrac{15}{7}$

**52.** $-\dfrac{19}{12}$

Convert to standard notation. [5.3b]

**53.** 4.03 trillion

**54.** 5.8 million

**55.** 42.7 million

**56.** 6.09 trillion

## Synthesis

**57.** *Magazine Subscriptions.* In a recent subscription drive, *Car and Driver* offered a 1-year subscription of 12 issues for the price of $1.50 per issue. They advertised that this was a savings of 69.94% off the cover price. What was the cover price?
**Source:** *Car and Driver*

**58.** Elijah collects baseball memorabilia. He bought two autographed plaques, but became short of funds and had to sell them quickly for $200 each. On one, he made a 20% profit, and on the other, he lost 20%. Did he make or lose money on the sale?

**59.** ▦ John receives a 10% commission on the first $5000 in sales and 15% on all sales beyond $5000. If John receives a commission of $2405, how much did he sell? Use a calculator and trial and error if you wish.

**60.** Tee shirts are being sold at the mall for $5 each, or 3 for $10. If you buy three tee shirts, what is the rate of discount?

# 7.7

# Simple Interest and Compound Interest; Credit Cards

## OBJECTIVES

**a** Solve applied problems involving simple interest.

**b** Solve applied problems involving compound interest.

**c** Solve applied problems involving interest rates on credit cards.

**SKILL TO REVIEW**

Objective 7.1b: Convert between percent notation and decimal notation.

Find decimal notation.

1. $34\frac{5}{8}\%$    2. $5\frac{1}{4}\%$

## a Simple Interest

Suppose you put $1000 into an investment for 1 year. The $1000 is called the **principal**. If the **interest rate** is 5%, in addition to the principal, you get back 5% of the principal, which is

$$5\% \text{ of } \$1000, \quad \text{or} \quad 0.05 \times \$1000, \quad \text{or} \quad \$50.00.$$

The $50.00 is called **simple interest**. It is, in effect, the price that a financial institution pays for the use of the money over time.

> ### SIMPLE INTEREST FORMULA
>
> The **simple interest** $I$ on principal $P$, invested for $t$ years at interest rate $r$, is given by
>
> $$I = P \cdot r \cdot t.$$

**EXAMPLE 1** What is the simple interest on $2500 invested at an interest rate of 6% for 1 year?

We use the formula $I = P \cdot r \cdot t$:

$$I = P \cdot r \cdot t = \$2500 \times 6\% \times 1$$
$$= \$2500 \times 0.06$$
$$= \$150.$$

The simple interest for 1 year is $150.

**Do Margin Exercise 1.**

1. What is the simple interest on $4300 invested at an interest rate of 4% for 1 year?

**EXAMPLE 2** What is the simple interest on a principal of $2500 invested at an interest rate of 6% for 3 months?

We use the formula $I = P \cdot r \cdot t$ and express 3 months as a fraction of a year:

$$I = P \cdot r \cdot t = \$2500 \times 6\% \times \frac{3}{12} = \$2500 \times 6\% \times \frac{1}{4}$$
$$= \frac{\$2500 \times 0.06}{4} = \$37.50.$$

The simple interest for 3 months is $37.50.

**Do Margin Exercise 2.**

2. What is the simple interest on a principal of $4300 invested at an interest rate of 4% for 9 months?

When time is given in days, we generally divide it by 365 to express the time as a fractional part of a year.

**EXAMPLE 3** To pay for a shipment of lawn furniture, Patio by Design borrows $8000 at $9\frac{3}{4}\%$ for 60 days. Find **(a)** the amount of simple interest that is due and **(b)** the total amount that must be paid after 60 days.

*Answers*

*Skill to Review:*
1. 0.34625    2. 0.0525

*Margin Exercises:*
1. $172    2. $129

**a)** We express 60 days as a fractional part of a year:

$$I = P \cdot r \cdot t = \$8000 \times 9\frac{3}{4}\% \times \frac{60}{365}$$

$$= \$8000 \times 0.0975 \times \frac{60}{365}$$

$$\approx \$128.22.$$

The interest due for 60 days is $128.22.

**b)** The total amount to be paid after 60 days is the principal plus the interest:

$$\$8000 + \$128.22 = \$8128.22.$$

The total amount due is $8128.22.

Do Exercise 3.

**3.** The Glass Nook borrows $4800 at $8\frac{1}{2}\%$ for 30 days. Find **(a)** the amount of simple interest due and **(b)** the total amount that must be paid after 30 days.

## b Compound Interest

When interest is paid *on interest*, we call it **compound interest**. This is the type of interest usually paid on investments. Suppose you have $5000 in a savings account at 6%. In 1 year, the account will contain the original $5000 plus 6% of $5000. Thus the total in the account after 1 year will be

106% of $5000,   or   $1.06 \times \$5000$,   or   $5300.

Now suppose that the total of $5300 remains in the account for another year. At the end of this second year, the account will contain the $5300 plus 6% of $5300. The total in the account would thus be

106% of $5300,   or   $1.06 \times \$5300$,   or   $5618.

Note that in the second year, interest is also earned on the first year's interest. When this happens, we say that interest is **compounded annually**.

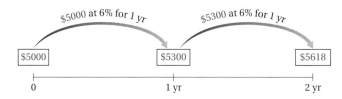

**EXAMPLE 4**   Find the amount in an account if $2000 is invested at 8%, compounded annually, for 2 years.

**a)** After 1 year, the account will contain 108% of $2000:

$$1.08 \times \$2000 = \$2160.$$

**b)** At the end of the second year, the account will contain 108% of $2160:

$$1.08 \times \$2160 = \$2332.80.$$

The amount in the account after 2 years is $2332.80.

Do Exercise 4.

**4.** Find the amount in an account if $2000 is invested at 9%, compounded annually, for 2 years.

*Answers*

**3. (a)** $33.53; **(b)** $4833.53    **4.** $2376.20

Suppose that the interest in Example 4 were **compounded semi-annually**—that is, every half year. Interest would then be calculated twice a year at a rate of 8% ÷ 2, or 4% each time. The approach used in Example 4 can then be adapted, as follows.

After the first $\frac{1}{2}$ year, the account will contain 104% of $2000:

$$1.04 \times \$2000 = \$2080.$$

After a second $\frac{1}{2}$ year (1 full year), the account will contain 104% of $2080:

$$1.04 \times \$2080 = \$2163.20.$$

After a third $\frac{1}{2}$ year $\left(1\frac{1}{2} \text{ full years}\right)$, the account will contain 104% of $2163.20:

$$1.04 \times \$2163.20 = \$2249.728$$
$$\approx \$2249.73. \qquad \text{Rounding to the nearest cent}$$

Finally, after a fourth $\frac{1}{2}$ year (2 full years), the account will contain 104% of $2249.73:

$$1.04 \times \$2249.73 = \$2339.7192$$
$$\approx \$2339.72. \qquad \text{Rounding to the nearest cent}$$

Let's summarize our results and look at them another way:

End of 1st $\frac{1}{2}$ year $\rightarrow 1.04 \times 2000 = 2000 \times (1.04)^1$;

End of 2nd $\frac{1}{2}$ year $\rightarrow 1.04 \times (1.04 \times 2000) = 2000 \times (1.04)^2$;

End of 3rd $\frac{1}{2}$ year $\rightarrow 1.04 \times (1.04 \times 1.04 \times 2000) = 2000 \times (1.04)^3$;

End of 4th $\frac{1}{2}$ year $\rightarrow 1.04 \times (1.04 \times 1.04 \times 1.04 \times 2000) = 2000 \times (1.04)^4$.

Note that each multiplication was by 1.04 and that

$$\$2000 \times 1.04^4 \approx \$2339.72. \qquad \begin{array}{l}\text{Using a calculator and rounding to the} \\ \text{nearest cent}\end{array}$$

We have illustrated the following result.

---

**COMPOUND INTEREST FORMULA**

If a principal $P$ has been invested at interest rate $r$, compounded $n$ times a year, in $t$ years it will grow to an amount $A$ given by

$$A = P \cdot \left(1 + \frac{r}{n}\right)^{n \cdot t}.$$

---

Let's apply this formula to confirm our preceding discussion, where the amount invested is $P = \$2000$, the number of years is $t = 2$, and the number of compounding periods each year is $n = 2$. Substituting into the compound interest formula, we have

$$A = P \cdot \left(1 + \frac{r}{n}\right)^{n \cdot t} = 2000 \cdot \left(1 + \frac{8\%}{2}\right)^{2 \cdot 2}$$

$$= \$2000 \cdot \left(1 + \frac{0.08}{2}\right)^4 = \$2000(1.04)^4$$

$$= \$2000 \times 1.16985856 \approx \$2339.72.$$

If you were using a calculator, you could perform this computation in one step.

**EXAMPLE 5** The Ibsens invest $4000 in an account paying $5\frac{5}{8}\%$, compounded quarterly. Find the amount in the account after $2\frac{1}{2}$ years.

The compounding is quarterly, so $n$ is 4. We substitute $4000 for $P$, $5\frac{5}{8}\%$, or 0.05625, for $r$, 4 for $n$, and $2\frac{1}{2}$, or $\frac{5}{2}$, for $t$ and compute $A$:

$$A = P \cdot \left(1 + \frac{r}{n}\right)^{n \cdot t} = \$4000 \cdot \left(1 + \frac{5\frac{5}{8}\%}{4}\right)^{4 \cdot 5/2}$$

$$= \$4000 \cdot \left(1 + \frac{0.05625}{4}\right)^{10}$$

$$= \$4000(1.0140625)^{10}$$

$$\approx \$4599.46.$$

The amount in the account after $2\frac{1}{2}$ years is $4599.46.

Do Exercise 5.

**5.** A couple invests $7000 in an account paying $6\frac{3}{8}\%$, compounded semiannually. Find the amount in the account after $1\frac{1}{2}$ years.

## Calculator Corner

**Compound Interest**  A calculator is useful in computing compound interest. Not only does it perform computations quickly but it also eliminates the need to round until the computation is completed. This minimizes "round-off errors" that occur when rounding is done at each stage of the computation. We must keep order of operations in mind when computing compound interest.

To find the amount due on a $20,000 loan made for 25 days at 11% interest, compounded daily, we would compute $20,000\left(1 + \frac{0.11}{365}\right)^{25}$. To do this on a calculator, we press $\boxed{2}\boxed{0}\boxed{0}\boxed{0}\boxed{0}\boxed{\times}\boxed{(}\boxed{1}\boxed{+}\boxed{.}\boxed{1}\boxed{1}\boxed{\div}\boxed{3}\boxed{6}\boxed{5}\boxed{)}\boxed{y^x}$

(or $\boxed{\wedge}$) $\boxed{2}\boxed{5}\boxed{=}$. Using a calculator that does not have parentheses, we would first find $1 + \frac{0.11}{365}$, raise this result to the 25th power, and then multiply by 20,000. To do this, we press $\boxed{1}\boxed{+}\boxed{.}\boxed{1}\boxed{1}\boxed{\div}\boxed{3}\boxed{6}\boxed{5}\boxed{=}\boxed{y^x}$ (or $\boxed{\wedge}$) $\boxed{2}\boxed{5}\boxed{=}\boxed{\times}\boxed{2}\boxed{0}\boxed{0}\boxed{0}\boxed{0}\boxed{=}$. In either case, the result is 20,151.23, rounded to the nearest cent.

Some calculators have business keys that allow such computations to be done more quickly.

**Exercises:**

**1.** Find the amount due on a $16,000 loan made for 62 days at 13% interest, compounded daily.

**2.** An investment of $12,500 is made for 90 days at 8.5% interest, compounded daily. How much is the investment worth after 90 days?

## c Credit Cards

According to CardWeb.com, the average credit-card debt per household in the United States was $9840 in 2007. According to a 2007 survey by student lender Nellie Mae, the average college graduate has a $2748 credit-card debt.

The money you obtain through the use of a credit card is not "free" money. There is a price (interest) to be paid for the privilege. A balance carried on a credit card is a type of loan. Comparing interest rates is essential if one is to become financially responsible. A small change in an interest rate can make a large difference in the cost of a loan. When you make a payment on a credit card, do you know how much of that payment is interest and how much is applied to reducing the principal?

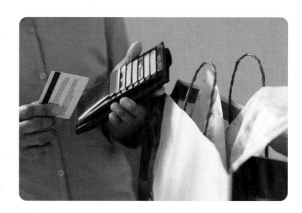

*Answer*

**5.** $7690.94

**EXAMPLE 6** *Credit Cards.* After the holidays, Sarah has a balance of $3216.28 on a credit card with an annual percentage rate (APR) of 19.7%. She decides not to make additional purchases with this card until she has paid off the balance.

a) Many credit cards require a minimum monthly payment of 2% of the balance. At this rate, what is Sarah's minimum payment on a balance of $3216.28? Round the answer to the nearest dollar.

b) Find the amount of interest and the amount applied to reduce the principal in the minimum payment found in part (a).

c) If Sarah had transferred her balance to a card with an APR of 12.5%, how much of her first payment would be interest and how much would be applied to reduce the principal?

d) Compare the amounts for 12.5% from part (c) with the amounts for 19.7% from part (b).

We solve as follows.

a) We multiply the balance of $3216.28 by 2%:

$$0.02 \times \$3216.28 = \$64.3256.$$

Sarah's minimum payment, rounded to the nearest dollar, is $64.

b) The amount of interest on $3216.28 at 19.7% for one month* is given by

$$I = P \cdot r \cdot t = \$3216.28 \times 0.197 \times \frac{1}{12} \approx \$52.80.$$

We subtract to find the amount applied to reduce the principal in the first payment:

$$\begin{aligned} \text{Amount applied to} \atop \text{reduce the principal} &= \text{Minimum payment} - {\text{Interest for} \atop \text{the month}} \\ &= \$64 - \$52.80 \\ &= \$11.20. \end{aligned}$$

Thus the principal of $3216.28 is decreased by only $11.20 with the first payment. (Sarah still owes $3205.08.)

c) The amount of interest on $3216.28 at 12.5% for one month is

$$I = P \cdot r \cdot t = \$3216.28 \times 0.125 \times \frac{1}{12} \approx \$33.50.$$

We subtract to find the amount applied to reduce the principal in the first payment:

$$\begin{aligned} \text{Amount applied to} \atop \text{reduce the principal} &= \text{Minimum payment} - {\text{Interest for} \atop \text{the month}} \\ &= \$64 - \$33.50 \\ &= \$30.50. \end{aligned}$$

Thus the principal of $3216.28 is decreased by $30.50 with the first payment. (Sarah still owes $3185.78.)

---

**6. Credit Cards.** After the holidays, Jamal has a balance of $4867.59 on a credit card with an annual percentage rate (APR) of 21.3%. He decides not to make additional purchases with this card until he has paid off the balance.

a) Many credit cards require a minimum monthly payment of 2% of the balance. What is Jamal's minimum payment on a balance of $4867.59? Round the answer to the nearest dollar.

b) Find the amount of interest and the amount applied to reduce the principal in the minimum payment found in part (a).

c) If Jamal had transferred his balance to a card with an APR of 13.6%, how much of his first payment would be interest and how much would be applied to reduce the principal?

d) Compare the amounts for 13.6% from part (c) with the amounts for 21.3% from part (b).

*Answers*

**6. (a)** $97; **(b)** interest: $86.40; amount applied to principal: $10.60; **(c)** interest: $55.17; amount applied to principal: $41.83; **(d)** At 13.6%, the principal was reduced by $31.23 more than at the 21.3% rate. The interest at 13.6% is $31.23 less than at 21.3%.

---

*Actually, the interest on a credit card is computed daily with a rate called a daily percentage rate (DPR). The DPR for Example 6 would be 19.7%/365 = 0.054%. When no payments or additional purchases are made during the month, the difference in total interest for the month is minimal and we will not deal with it here.

**d)** Let's organize the information for both rates in the following table.

| BALANCE BEFORE FIRST PAYMENT | FIRST MONTH'S PAYMENT | % APR | AMOUNT OF INTEREST | AMOUNT APPLIED TO PRINCIPAL | BALANCE AFTER FIRST PAYMENT |
|---|---|---|---|---|---|
| $3216.28 | $64 | 19.7% | $52.80 | $11.20 | $3205.08 |
| 3216.28 | 64 | 12.5 | 33.50 | 30.50 | 3185.78 |

Difference in balance after first payment → $19.30

At 19.7%, the interest is $52.80 and the principal is decreased by $11.20. At 12.5%, the interest is $33.50 and the principal is decreased by $30.50. Thus the interest at 19.7% is $52.80 − $33.50, or $19.30, greater than the interest at 12.5%. Thus the principal is decreased by $30.50 − $11.20, or $19.30, more with the 12.5% rate than with the 19.7% rate.

Do Exercise 6 on the preceding page.

Even though the mathematics of the information in the table below is beyond the scope of this text, it is interesting to compare how long it takes to pay off the balance of Example 6 if Sarah continues to pay $64 for each payment with how long it takes if she pays double that amount, $128, for each payment. Financial consultants frequently tell clients that if they want to take control of their debt, they should pay double the minimum payment.

| RATE | PAYMENT PER MONTH | NUMBER OF PAYMENTS TO PAY OFF DEBT | TOTAL PAID BACK | ADDITIONAL COST OF PURCHASES |
|---|---|---|---|---|
| 19.7% | $64 | 107, or 8 yr 11 mo | $6848 | $3631.72 |
| 19.7 | 128 | 33, or 2 yr 9 mo | 4224 | 1007.72 |
| 12.5 | 64 | 72, or 6 yr | 4608 | 1391.72 |
| 12.5 | 128 | 29, or 2 yr 5 mo | 3712 | 495.72 |

As with most loans, if you pay an extra amount toward the principal with each payment, the length of the loan can be greatly reduced. Note that at the rate of 19.7%, it will take Sarah almost 9 yr to pay off her debt if she pays only $64 per month and does not make additional purchases. If she transfers her balance to a card with a 12.5% rate and pays $128 per month, she can eliminate her debt in approximately $2\frac{1}{2}$ yr. You can see how debt can get out of control if you continue to make purchases and pay only the minimum payment each month. The debt will never be eliminated.

**a**　Find the simple interest.

| | PRINCIPAL | RATE OF INTEREST | TIME | SIMPLE INTEREST |
|---|---|---|---|---|
| **1.** | $200 | 4% | 1 year | |
| **2.** | $200 | 7.7% | $\frac{1}{2}$ year | |
| **3.** | $4300 | 10.56% | $\frac{1}{4}$ year | |
| **4.** | $80,000 | $6\frac{3}{4}$% | $\frac{1}{12}$ year | |
| **5.** | $20,000 | $4\frac{5}{8}$% | 1 year | |
| **6.** | $8000 | 9.42% | 2 months | |
| **7.** | $50,000 | $5\frac{3}{8}$% | 3 months | |
| **8.** | $100,000 | $3\frac{1}{4}$% | 1 year | |

Solve. Assume that simple interest is being calculated in each case.

**9.** CopiPix, Inc. borrows $10,000 at 9% for 60 days. Find **(a)** the amount of interest due and **(b)** the total amount that must be paid after 60 days.

**10.** Sal's Laundry borrows $8000 at 10% for 90 days. Find **(a)** the amount of interest due and **(b)** the total amount that must be paid after 90 days.

**11.** Animal Instinct, a pet supply shop, borrows $6500 at $5\frac{1}{4}$% for 90 days. Find **(a)** the amount of interest due and **(b)** the total amount that must be paid after 90 days.

**12.** Andante's Cafe borrows $4500 at $12\frac{1}{2}$% for 60 days. Find **(a)** the amount of interest due and **(b)** the total amount that must be paid after 60 days.

**13.** Jean's Garage borrows $5600 at 10% for 30 days. Find **(a)** the amount of interest due and **(b)** the total amount that must be paid after 30 days.

**14.** Shear Delights Hair Salon borrows $3600 at 4% for 30 days. Find **(a)** the amount of interest due and **(b)** the total amount that must be paid after 30 days.

**b** Interest is compounded annually. Find the amount in the account after the given length of time. Round to the nearest cent.

|  | PRINCIPAL | RATE OF INTEREST | TIME | AMOUNT IN THE ACCOUNT |
|---|---|---|---|---|
| **15.** | $400 | 5% | 2 years | |
| **16.** | $450 | 4% | 2 years | |
| **17.** | $2000 | 8.8% | 4 years | |
| **18.** | $4000 | 7.7% | 4 years | |
| **19.** | $4300 | 10.56% | 6 years | |
| **20.** | $8000 | 9.42% | 6 years | |
| **21.** | $20,000 | $6\frac{5}{8}\%$ | 25 years | |
| **22.** | $100,000 | $5\frac{7}{8}\%$ | 30 years | |

Interest is compounded semiannually. Find the amount in the account after the given length of time. Round to the nearest cent.

|  | PRINCIPAL | RATE OF INTEREST | TIME | AMOUNT IN THE ACCOUNT |
|---|---|---|---|---|
| **23.** | $4000 | 6% | 1 year | |
| **24.** | $1000 | 5% | 1 year | |
| **25.** | $20,000 | 8.8% | 4 years | |
| **26.** | $40,000 | 7.7% | 4 years | |
| **27.** | $5000 | 10.56% | 6 years | |
| **28.** | $8000 | 9.42% | 8 years | |
| **29.** | $20,000 | $7\frac{5}{8}\%$ | 25 years | |
| **30.** | $100,000 | $4\frac{7}{8}\%$ | 30 years | |

Solve.

**31.** A family invests $4000 in an account paying 6%, compounded monthly. How much is in the account after 5 months?

**32.** A couple invests $2500 in an account paying 3%, compounded monthly. How much is in the account after 6 months?

**33.** A couple invests $1200 in an account paying 10%, compounded quarterly. How much is in the account after 1 year?

**34.** The O'Hares invest $6000 in an account paying 8%, compounded quarterly. How much is in the account after 18 months?

C  Solve.

**35.** *Credit Cards.*   Amelia has a balance of $1278.56 on a credit card with an annual percentage rate (APR) of 19.6%. The minimum payment required in the current statement is $25.57. Find the amount of interest and the amount applied to reduce the principal in this payment and the balance after this payment.

**36.** *Credit Cards.*   Lawson has a balance of $1834.90 on a credit card with an annual percentage rate (APR) of 22.4%. The minimum payment required in the current statement is $36.70. Find the amount of interest and the amount applied to reduce the principal in this payment and the balance after this payment.

**37.** *Credit Cards.*   Antonio has a balance of $4876.54 on a credit card with an annual percentage rate (APR) of 21.3%.

   **a)** Many credit cards require a minimum monthly payment of 2% of the balance. What is Antonio's minimum payment on a balance of $4876.54? Round the answer to the nearest dollar.

   **b)** Find the amount of interest and the amount applied to reduce the principal in the minimum payment found in part (a).

   **c)** If Antonio had transferred his balance to a card with an APR of 12.6%, how much of his payment would be interest and how much would be applied to reduce the principal?

   **d)** Compare the amounts for 12.6% from part (c) with the amounts for 21.3% from part (b).

**38.** *Credit Cards.*   Becky has a balance of $5328.88 on a credit card with an annual percentage rate (APR) of 18.7%.

   **a)** Many credit cards require a minimum monthly payment of 2% of the balance. What is Becky's minimum payment on a balance of $5328.88? Round the answer to the nearest dollar.

   **b)** Find the amount of interest and the amount applied to reduce the principal in the minimum payment found in part (a).

   **c)** If Becky had transferred her balance to a card with an APR of 13.2%, how much of her payment would be interest and how much would be applied to reduce the principal?

   **d)** Compare the amounts for 13.2% from part (c) with the amounts for 18.7% from part (b).

## Skill Maintenance

In each of Exercises 39–46, fill in the blank with the correct term from the given list. Some of the choices may not be used.

**39.** If the product of two numbers is 1, they are _____ of each other.   [3.7a]

**40.** A number is _____ if its ones digit is even and the sum of its digits is divisible by 3.   [3.2a]

**41.** The number 0 is the _____ identity.   [1.2a]

**42.** A(n) _____ is the ratio of price to the number of units. [6.2b]

**43.** The distance around an object is its _____.   [1.2b]

**44.** A number is _____ if the sum of its digits is divisible by 3.   [3.2a]

**45.** A natural number that has exactly two different factors, only itself and 1, is called a _____ number.   [3.1c]

**46.** When two pairs of numbers have the same ratio, they are _____.   [6.3a]

divisible by 3

divisible by 4

divisible by 6

divisible by 9

perimeter

area

unit price

reciprocals

proportional

composite

prime

additive

multiplicative

## Synthesis

*Effective Yield.*   The *effective yield* is the yearly rate of simple interest that corresponds to a rate for which interest is compounded two or more times a year. For example, if $P$ is invested at 12%, compounded quarterly, we would multiply $P$ by $(1 + 0.12/4)^4$, or $1.03^4$. Since $1.03^4 \approx 1.126$, the 12% compounded quarterly corresponds to an effective yield of approximately 12.6%. In Exercises 47 and 48, find the effective yield for the indicated account.

**47.** ▦ The account pays 9% compounded monthly.

**48.** ▦ The account pays 10% compounded daily.

# Summary and Review

## Key Terms and Formulas

percent notation, $n\%$, p. 392
percent of decrease, p. 424
percent of increase, p. 426
purchase price, p. 435
sales tax, p. 435
total price, p. 435

commission, p. 437
original price, p. 438
marked price, p. 438
rate of discount, p. 438
discount, p. 438
sale price, p. 438

principal, p. 444
interest rate, p. 444
simple interest, p. 444
compound interest, p. 445
compounded annually, p. 445
compounded semi-annually, p. 446

Commission = Commission rate × Sales

Discount = Rate of discount × Original price

Sale price = Original price − Discount

*Simple Interest:*    $I = P \cdot r \cdot t$

*Compound Interest:*    $A = P \cdot \left(1 + \dfrac{r}{n}\right)^{n \cdot t}$

## Concept Reinforcement

Determine whether each statement is true or false.

_____ 1. A fixed principal invested for 4 years will earn more interest when interest is compounded quarterly than when interest is compounded semiannually.   [7.7b]

_____ 2. Of the numbers 0.5%, $\dfrac{5}{1000}\%$, $\dfrac{1}{2}\%$, $\dfrac{1}{5}$, and $0.\overline{1}$, the largest number is $0.\overline{1}$.   [7.1b], [7.2a, b]

_____ 3. If principal A equals principal B and principal A is invested for 2 years at 4%, compounded quarterly, while principal B is invested for 4 years at 2%, compounded semiannually, the interest earned from each investment is the same.   [7.7b]

## Important Concepts

**Objective 7.1b**   Convert between percent notation and decimal notation.

**Example**   Find decimal notation for $12\dfrac{3}{4}\%$.

$$12\dfrac{3}{4}\% = 12.75 \times 0.01 = 0.1275$$

**Practice Exercise**

1. Find decimal notation for $62\dfrac{5}{8}\%$.

**Objective 7.2a**   Convert from fraction notation to percent notation.

**Example**   Find percent notation for $\dfrac{5}{12}$.

$$
\begin{array}{r}
0.4\ 1\ 6 \\
1\ 2\ \overline{)\ 5.0\ 0\ 0} \\
\underline{4\ 8}\phantom{0000} \\
2\ 0\phantom{00} \\
\underline{1\ 2}\phantom{00} \\
8\ 0 \\
\underline{7\ 2} \\
8
\end{array}
$$

$\dfrac{5}{12} = 0.41\overline{6} = 41.\overline{6}\%$, or $41\dfrac{2}{3}\%$

**Practice Exercise**

2. Find percent notation for $\dfrac{7}{11}$.

**Objective 7.2b** Convert from percent notation to fraction notation.

**Example** Find fraction notation for 9.5%.

$$9.5\% = \frac{9.5}{100} = \frac{95}{1000} = \frac{5 \cdot 19}{5 \cdot 200}$$

$$= \frac{5}{5} \cdot \frac{19}{200} = \frac{19}{200}$$

**Practice Exercise**

**3.** Find fraction notation for 6.8%.

**Objective 7.3b** Solve basic percent problems.

**Example** Translate to a percent equation. Then solve.

165 is what percent of 3300?

We have

$$165 = p \cdot 3300 \qquad \text{Translating to a percent equation}$$

$$\frac{165}{3300} = \frac{p \cdot 3300}{3300}$$

$$\frac{165}{3300} = p$$

$$0.05 = p$$

$$5\% = p.$$

Thus, 165 is 5% of 3300.

**Practice Exercise**

**4.** Translate to a percent equation. Then solve.

12 is what percent of 288?

**Objective 7.4b** Solve basic percent problems.

**Example** Translate to a proportion. Then solve.

18% of what is 1296?

$$\frac{18}{100} = \frac{1296}{b} \qquad \text{Translating to a proportion}$$

$$18 \cdot b = 100 \cdot 1296$$

$$\frac{18 \cdot b}{18} = \frac{129{,}600}{18}$$

$$b = 7200.$$

Thus, 18% of 7200 is 1296.

**Practice Exercise**

**5.** Translate to a proportion. Then solve.

3% of what is 300?

**Objective 7.5b** Solve applied problems involving percent of increase or percent of decrease.

**Example** The total cost for 16 basic grocery items in the second quarter of 2008 averaged $46.67 nationally. The total cost of these 16 items in the second quarter of 2007 averaged $42.95. What was the percent of increase?

Source: American Farm Bureau Federation

We first determine the amount of increase: $46.67 − $42.95 = $3.72. Then we translate to a percent equation or a proportion and solve.

*Rewording:* $3.72 is what percent of $42.95?

**Practice Exercise**

**6.** In Indiana, the cost for 16 basic grocery items increased from $40.07 in the second quarter of 2007 to $46.20 in the second quarter of 2008. What was the percent of increase from 2007 to 2008?

**Objective 7.5b** *(continued)*

*Percent Equation:* $\qquad$ *Proportion:*

$$3.72 = p \cdot 42.95 \qquad \frac{N}{100} = \frac{3.72}{42.95}$$

$$\frac{3.72}{42.95} = \frac{p \cdot 42.95}{42.95} \qquad 42.95N = 100 \cdot 3.72$$

$$\frac{3.72}{42.95} = p \qquad \frac{42.95N}{42.95} = \frac{100 \cdot 3.72}{42.95}$$

$$0.087 \approx p \qquad N = \frac{372}{42.95}$$

$$8.7\% \approx p \qquad N \approx 8.7$$

The percent of increase was 8.7%.

---

**Objective 7.6a** Solve applied problems involving sales tax and percent.

**Example** The sales tax is $34.23 on the purchase of a flat-screen high-definition television that costs $489. What is the sales tax rate?

*Rephrase:* Sales tax is what percent of purchase price?

*Translate:* $34.23 = r \times 489$

*Solve:* $\dfrac{34.23}{489} = \dfrac{r \times 489}{489}$

$$\frac{34.23}{489} = r$$

$$0.07 = r$$

$$7\% = r$$

The sales tax rate is 7%.

**Practice Exercise**

7. The sales tax is $1102.20 on the purchase of a new car that costs $18,370. What is the sales tax rate?

---

**Objective 7.6b** Solve applied problems involving commission and percent.

**Example** A real estate agent's commission rate is $6\frac{1}{2}\%$. She received a commission of $17,160 on the sale of a home. For how much did the home sell?

*Rephrase:* Commission is $6\frac{1}{2}\%$ of what selling price?

*Translate:* $17,160 = 6\frac{1}{2}\% \times S$

*Solve:* $17,160 = 0.065 \times S$

$$\frac{17,160}{0.065} = \frac{0.065 \times S}{0.065}$$

$$264,000 = S$$

The home sold for $264,000.

**Practice Exercise**

8. A real estate agent's commission rate is 7%. He received a commission of $12,950 on the sale of a home. For how much did the home sell?

**Objective 7.7a**  Solve applied problems involving simple interest.

**Example**  To meet its payroll, a business borrows $5200 at $4\frac{1}{4}\%$ for 90 days. Find the amount of simple interest that is due and the total amount that must be paid after 90 days.

$$I = P \cdot r \cdot t = \$5200 \times 4\tfrac{1}{4}\% \times \frac{90}{365}$$

$$= \$5200 \times 0.0425 \times \frac{90}{365}$$

$$\approx \$54.49$$

The interest due for 90 days $= \$54.49$.

The total amount due $= \$5200 + \$54.49 = \$5254.49$.

**Practice Exercise**

9. A student borrows $2500 for tuition at $5\frac{1}{2}\%$ for 60 days. Find the amount of simple interest that is due and the total amount that must be paid after 60 days.

---

**Objective 7.7b**  Solve applied problems involving compound interest.

**Example**  Find the amount in an account if $3200 is invested at 5%, compounded semiannually, for $1\frac{1}{2}$ years.

$$A = P \cdot \left(1 + \frac{r}{n}\right)^{n \cdot t}$$

$$= \$3200\left(1 + \frac{0.05}{2}\right)^{2 \cdot \frac{3}{2}}$$

$$= \$3200(1.025)^3$$

$$= \$3446.05$$

The amount in the account after $1\frac{1}{2}$ years is $3446.05.

**Practice Exercise**

10. Find the amount in an account if $6000 is invested at $4\frac{3}{4}\%$, compounded quarterly, for 2 years.

---

# Review Exercises

Find decimal notation for the percent notations in each sentence.  [7.1b]

1. In the 2006–2007 school year, about 4% of the 15 million college students in the United States were foreign students. Approximately 14.4% of the foreign students were from India.
   **Source:** Institute of International Education

2. Poland is 62.1% urban; Sweden is 84.2% urban.
   **Source:** The World Almanac, 2008

Find percent notation   [7.2a]

3. $\dfrac{3}{8}$

4. $\dfrac{1}{3}$

Find percent notation.   [7.1b]

5. 1.7

6. 0.065

Find fraction notation.   [7.2b]

7. 24%

8. 6.3%

Translate to a percent equation. Then solve.   [7.3a, b]

9. 30.6 is what percent of 90?

10. 63 is 84% of what?

11. What is $38\frac{1}{2}\%$ of 168?

Translate to a proportion. Then solve.    [7.4a, b]

**12.** 24 percent of what is 16.8?

**13.** 42 is what percent of 30?

**14.** What is 10.5% of 84?

Solve.    [7.5a, b]

**15.** *Favorite Ice Creams.*    According to a survey, 8.9% of those interviewed chose chocolate as their favorite ice cream flavor and 4.2% chose butter pecan. At this rate, of the 2000 students in a freshman class, how many would choose chocolate as their favorite ice cream? butter pecan?
    Source: International Ice Cream Association

**16.** *Prescriptions.*    Of the 305 million people in the United States, 140.3 million take at least one type of prescription drug per day. What percent take at least one type of prescription drug per day?
    Source: William N. Kelly, *Pharmacy: What It Is and How It Works*, 2nd ed., CRC Press Pharmaceutical Education, 2006

**17.** *Water Output.*    The average person expels 200 mL of water per day by sweating. This is 8% of the total output of water from the body. How much is the total output of water?
    Source: Elaine N. Marieb, *Essentials of Human Anatomy and Physiology*, 6th ed. Boston: Addison Wesley Longman, Inc., 2000

**18.** *Test Scores.*    After Sheila got a 75 on a math test, she was allowed to go to the math lab and take a retest. She increased her score to 84. What was the percent of increase?

**19.** *Test Scores.*    James got an 80 on a math test. By taking a retest in the math lab, he increased his score by 15%. What was his new score?

Solve.    [7.6a, b, c]

**20.** A state charges a meals tax of $7\frac{1}{2}$%. What is the meals tax charged on a dinner party costing $320?

**21.** In a certain state, a sales tax of $453.60 is collected on the purchase of a used car for $7560. What is the sales tax rate?

**22.** Kim earns $753.50 selling $6850 worth of televisions. What is the commission rate?

**23.** An air conditioner has a marked price of $350. It is placed on sale at 12% off. What are the discount and the sale price?

**24.** The price of a fax machine is marked down from $305 to $262.30. What is the rate of discount?

**25.** An insurance salesperson receives a 7% commission. If $42,000 worth of life insurance is sold, what is the commission?

**26.** What is the rate of discount of this stepladder?

SPECIAL VALUE!
Now $67 Was $82
8' Aluminum Stepladder

Solve.  [7.7a, b, c]

**27.** What is the simple interest on $1800 at 6% for $\frac{1}{3}$ year?

**28.** The Dress Shack borrows $24,000 at 10% simple interest for 60 days. Find **(a)** the amount of interest due and **(b)** the total amount that must be paid after 60 days.

**29.** What is the simple interest on $2200 principal at an interest rate of 5.5% for 1 year?

**30.** The Armstrongs invest $7500 in an investment account paying an annual interest rate of 4%, compounded monthly. How much is in the account after 3 months?

**31.** Find the amount in an investment account if $8000 is invested at 9%, compounded annually, for 2 years.

**32.** *Credit Cards.* At the end of her junior year of college, Kasha has a balance of $6428.74 on a credit card with an annual percentage rate (APR) of 18.7%. She decides not to make additional purchases with this card until she has paid off the balance.

**a)** Many credit cards require a minimum payment of 2% of the balance. At this rate, what is Kasha's minimum payment on a balance of $6428.74? Round the answer to the nearest dollar.

**b)** Find the amount of interest and the amount applied to reduce the principal in the minimum payment found in part (a).

**c)** If Kasha had transferred her balance to a card with an APR of 13.2%, how much of her payment would be interest and how much would be applied to reduce the principal?

**d)** Compare the amounts for 13.2% from part (c) with the amounts for 18.7% from part (b).

**33.** A fishing boat listed at $16,500 is on sale at 15% off. What is the sale price?  [7.6c]

**A.** $14,025
**B.** $2475
**C.** 85%
**D.** $14,225

**34.** Find the amount in a money market account if $10,500 is invested at 6%, compounded semiannually, for $1\frac{1}{2}$ years.  [7.7b]

**A.** $11,139.45
**B.** $12,505.67
**C.** $11,473.63
**D.** $10,976.03

### Synthesis

**35.** Mike's Bike Shop reduces the price of a bicycle by 40% during a sale. By what percent must the store increase the sale price, after the sale, to get back to the original price?  [7.6c]

**36.** A worker receives raises of 3%, 6%, and then 9%. By what percent has the original salary increased?  [7.5a]

## Understanding Through Discussion and Writing

**1.** Which is the better deal for a consumer and why: a discount of 40% or a discount of 20% followed by another of 22%?  [7.6c]

**2.** Which is better for a wage earner, and why: a 10% raise followed by a 5% raise a year later, or a 5% raise followed by a 10% raise a year later?  [7.5a]

**3.** Ollie buys a microwave oven during a 10%-off sale. The sale price that Ollie paid was $162. To find the original price, Ollie calculates 10% of $162 and adds that to $162. Is this correct? Why or why not?  [7.6c]

**4.** You take 40% of 50% of a number. What percent of the number could you take to obtain the same result making only one multiplication? Explain your answer.  [7.5a]

**5.** A firm must choose between borrowing $5000 at 10% for 30 days and borrowing $10,000 at 8% for 60 days. Give arguments in favor of and against each option.  [7.7a]

**6.** On the basis of the mathematics presented in Section 7.7, discuss what you have learned about interest rates and credit cards.  [7.7c]

**Test**

**For Extra Help**

Step-by-step test solutions are found on the Chapter Test Prep Videos available via the Video Resources on DVD, in *MyMathLab*, and on YouTube (search "BittingerBasicMathEI" and click on "Channels").

**1.** *Land-Line Phone Service.* About 14.7% of U.S. households were projected to cut land-line phone service in 2009. Find decimal notation for 14.7%.
Source: Yankee Group 2006, *USA TODAY*, January 15, 2007

**2.** *Gravity.* The gravity of Mars is 0.38 as strong as Earth's. Find percent notation for 0.38.
Source: www.marsinstitute.info/epo/mermarsfacts.html

**3.** Find percent notation for $\frac{11}{8}$.

**4.** Find fraction notation for 65%.

**5.** Translate to a percent equation. Then solve.
What is 40% of 55?

**6.** Translate to a proportion. Then solve.
What percent of 80 is 65?

Solve.

**7.** *Organ Transplants.* In 2006, there were 28,291 organ transplants in the United States. The pie chart below shows the percentages for the main transplants. How many kidney transplants were there in 2006? liver transplants? heart transplants?

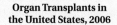

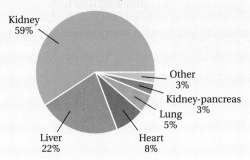

**Organ Transplants in the United States, 2006**

Kidney 59%
Other 3%
Kidney-pancreas 3%
Lung 5%
Liver 22%
Heart 8%

SOURCE: 2007 OPTN/SRTR Annual Report, Table 1.7; United Network for Organ Sharing

**8.** *Batting Average.* Garrett Atkins, third baseman for the Colorado Rockies, got 175 hits during the 2008 baseball season. This was about 28.64% of his at-bats. How many at-bats did he have?
Source: Major League Baseball

**9.** *Foreign Adoptions.* The number of foreign children adopted by Americans declined from 20,679 in 2006 to 19,292 in 2007. Find the percent of decrease.
Source: U.S. State Department, *USA TODAY*, August 13, 2008

**10.** There are about 6,603,000,000 people living in the world today, and approximately 4,002,000,000 live in Asia. What percent of people live in Asia?
Source: Population Division/International Programs Center, U.S. Census Bureau, U.S. Dept. of Commerce

**11.** *Oklahoma Sales Tax.* The sales tax rate in Oklahoma is 4.5%. How much tax is charged on a purchase of $560? What is the total price?

**12.** Noah's commission rate is 15%. What is the commission from the sale of $4200 worth of merchandise?

**13.** The marked price of a DVD player is $200 and the item is on sale at 20% off. What are the discount and the sale price?

**14.** What is the simple interest on a principal of $120 at the interest rate of 7.1% for 1 year?

**15.** A city orchestra invests $5200 at 6% simple interest. How much is in the account after $\frac{1}{2}$ year?

**16.** Find the amount in an account if $1000 is invested at $5\frac{3}{8}$% compounded annually, for 2 years.

**17.** The Suarez family invests $10,000 at an annual interest rate of 4.9%, compounded monthly. How much is in the account after 3 years?

**18.** *Job Opportunities.* The table below lists job opportunities in 2006 and projected increases for 2016. Complete the table by filling in the missing numbers.

| OCCUPATION | TOTAL EMPLOYMENT IN 2006 | PROJECTED EMPLOYMENT IN 2016 | CHANGE | PERCENT OF INCREASE |
|---|---|---|---|---|
| Dental assistant | 280,000 | 362,000 | 82,000 | 29.3% |
| Plumber | 705,000 | | 52,000 | |
| Veterinary assistant | 71,000 | 100,000 | | |
| Motorcycle repair technician | | 24,000 | 3000 | |
| Fitness professional | | 298,000 | | 26.8% |

SOURCE: EarnMyDegree.com

**19.** Find the discount and the discount rate of the television in this ad.

19" LCD HDTV

$299⁹⁹

was $349⁹⁹

**20.** *Credit Cards.* Jayden has a balance of $2704.27 on a credit card with an annual percentage rate of 16.3%. The minimum payment required on the current statement is $54. Find the amount of interest and the amount applied to reduce the principal in this payment and the balance after this payment.

**21.** 0.75% of what number is 300?

**A.** 2.25     **B.** 40,000     **C.** 400     **D.** 225

# Synthesis

**22.** By selling a home without using a realtor, Juan and Marie can avoid paying a 7.5% commission. They receive an offer of $180,000 from a potential buyer. In order to give a comparable offer, for what price would a realtor need to sell the house? Round to the nearest hundred.

**23.** Karen's commission rate is 16%. She invests her commission from the sale of $15,000 worth of merchandise at an interest rate of 12%, compounded quarterly. How much is Karen's investment worth after 6 months?

# Cumulative Review

1. *Passports.* In 2008, 44% of Canadians had passports while only 30% of Americans had passports. Find decimal notation for 44% and for 30%.

Sources: Government of Canada; U.S. Department of Homeland Security, U.S. Customs and Border Protection docket number USCBP 2007-0061

2. Find percent notation: 0.269.

3. Find percent notation: $\dfrac{9}{8}$.

4. Find decimal notation: $-\dfrac{13}{6}$.

5. Write fraction notation for the ratio 5 to 0.5.

6. Find the rate in kilometers per hour: 350 km, 15 hr.

Use $<$, $>$, or $=$ for $\square$ to write a true sentence.

7. $\dfrac{5}{7} \ \square \ \dfrac{6}{8}$

8. $\dfrac{-15}{25} \ \square \ \dfrac{-6}{14}$

Estimate the sum or the difference by first rounding to the nearest hundred.

9. $263{,}961 + 32{,}090 + 127.89$

10. $73{,}510 - 23{,}450$

Calculate.

11. $46 - [4(6 + 4 \div 2) + 2 \times 3 - 5]$

12. $\left[-0.8(1.5 - 9.8 \div 49) - (1 + 0.1)^2\right] \div 1.5$

Compute and simplify.

13. $\dfrac{6}{5} + 1\dfrac{5}{6}$

14. $-46.9 + 2.84$

15.
$$\begin{array}{r} 487{,}094 \\ 6{,}936 \\ +\ \ 21{,}120 \\ \hline \end{array}$$

16. $35 - 34.98$

17. $3\dfrac{1}{3} - 2\dfrac{2}{3}$

18. $-\dfrac{8}{9} - \dfrac{6}{7}$

19. $-\dfrac{7}{9} \cdot \left(-\dfrac{3}{14}\right)$

20.
$$\begin{array}{r} 236{,}984 \\ \times\ \ \ \ \ 3{,}600 \\ \hline \end{array}$$

21.
$$\begin{array}{r} 46.012 \\ \times\ \ \ \ 0.03 \\ \hline \end{array}$$

22. $-6\dfrac{3}{5} \div 4\dfrac{2}{5}$

23. $431.2 \div 35.2$

24. $15 \overline{)\,1\,8\,5\,0}$

Solve.

25. $36 \cdot x = 3420$

26. $y + 142.87 = -151$

27. $-\dfrac{2}{15} \cdot t = \dfrac{6}{5}$

28. $\dfrac{3}{4} + x = \dfrac{5}{6}$

29. $\dfrac{y}{25} = \dfrac{24}{15}$

30. $\dfrac{16}{n} = \dfrac{21}{11}$

**31. Museum Attendance.** Among leading art museums, the Indianapolis Museum of Art ranks 6th in the nation in the number of visitors as a percentage of the metropolitan population. In 2007, the number of visitors was 30.31% of the metropolitan population, which was 1,525,104. What was the total attendance?

Source: dashboard.imamuseum.org/series/2007+MSA+Attendance

**32. Salary of Medical Doctors.** In 2007, the starting salary for a doctor in radiology was $172,500 higher than the starting salary of a doctor in neurology. If the starting salary for a radiologist was $350,000, what was the starting salary of a neurologist?

Source: Journal of the American Medical Association

**33.** At one point during the 2008–2009 NBA season, the Cleveland Cavaliers had won 39 out of 49 games. At this rate, how many games would they win in the entire season of 82 games?

Source: National Basketball Association

**34. Shirts.** A total of $424.75 was paid for 5 shirts at an upscale men's store. How much did each shirt cost?

**35. Unit Price.** A 200-oz bottle of liquid laundry detergent costs $14.99. What is the unit price?

**36.** Patty walked $\frac{7}{10}$ mi to school and then $\frac{8}{10}$ mi to the library. How far did she walk?

**37.** On a map, 1 in. represents 80 mi. How much does $\frac{3}{4}$ in. represent?

**38. Compound Interest.** The Bakers invest $8500 in an investment account paying 8%, compounded monthly. How much is in the account after 5 years?

**39. Ribbons.** How many pieces of ribbon $1\frac{4}{5}$ yd long can be cut from a length of ribbon 9 yd long?

**40. Auto Repair Technicians.** In 2006, 773,000 people were employed as auto repair technicians. It is projected that by 2016, this number will increase by 110,000. Find the percent of increase in the number of auto repair technicians and the number of technicians in 2016.

Source: EarnMyDegree.com

**41.** Add and simplify: $-\dfrac{14}{25} + \dfrac{3}{20}$.

**A.** $-\dfrac{11}{500}$      **B.** $-\dfrac{11}{5}$

**C.** $-\dfrac{41}{100}$      **D.** $-\dfrac{71}{100}$

**42.** The population of the state of Louisiana decreased from 4,468,976 in 2000 to 4,287,768 in 2006. What was the percent of decrease?

Source: U.S. Census Bureau

**A.** 4.1%      **B.** 4.2%

**C.** 104%      **D.** 95.9%

## Synthesis

**43.** If $a$ is 50% of $b$, then $b$ is what percent of $a$?

# Data, Graphs, and Statistics

## Real-World Application

A gardener planted six varieties of bearded iris in a new garden on campus. Students from the horticulture department were assigned to record data on the range of heights for each variety. The vertical bar graph on p. 489 shows the results of their study. The length of the light green shaded portion and the blossom of each bar illustrates the range of heights. What is the range of heights for the border bearded iris?

*Source:* www.irises.org/classification.htm

*This problem appears as Exercise 5 in Exercise Set 8.3.*

# Averages, Medians, and Modes

**a** Find the average of a set of numbers and solve applied problems involving averages.

**b** Find the median of a set of numbers and solve applied problems involving medians.

**c** Find the mode of a set of numbers and solve applied problems involving modes.

**d** Compare two sets of data using their means.

---

**SKILL TO REVIEW**

Objectives 1.9c and 5.4c: Simplify expressions using the rules for order of operations.

1. Find the average of 282, 137, 5280, and 193.

2. Find the average of $23.40, $89.15, and $148.17 to the nearest cent.

---

In real-world applications, we can use tables and graphs of various kinds to show information about data and to extract information from data that can lead us to make analyses and predictions. Graphs allow us to communicate a message from the data.

For example, the following two pages show data regarding the number of prescriptions filled in supermarkets in recent years. Examine each method of presentation. Which method, if any, do you like the best and why?

## Paragraph

IMS HEALTH and the NACDS Economics Department have released data regarding the number of prescriptions filled in supermarkets for various years. In 1997, 269 million prescriptions were filled; in 1999, 357 million were filled; in 2001, 418 million were filled; in 2003, 462 million were filled; in 2005, 465 million were filled; and finally, in 2007, 478 million were filled.

## Table

| YEAR | NUMBER OF PRESCRIPTIONS FILLED IN SUPERMARKETS (in millions) |
|------|------|
| 1997 | 269 |
| 1999 | 357 |
| 2001 | 418 |
| 2003 | 462 |
| 2005 | 465 |
| 2007 | 478 |

SOURCE: IMS HEALTH and NACDS Economics Department

## Pictograph

**Number of Prescriptions Filled in Supermarkets**

| 1997 | | = 50 million prescriptions |
| 1999 | |
| 2001 | |
| 2003 | |
| 2005 | |
| 2007 | |

SOURCE: IMS HEALTH and NACDS Economics Department

---

**Answers**

*Skill to Review:*

1. 1473   2. $86.91

**Bar Graph**

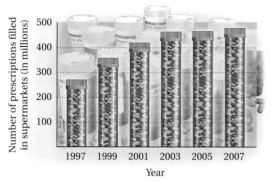

SOURCE: IMS HEALTH and NACDS Economics Department

**Line Graph**

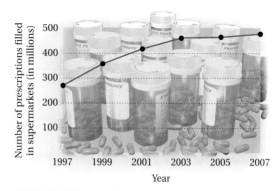

SOURCE: IMS HEALTH and NACDS Economics Department

Most people would not find the paragraph method for displaying the data most useful. It takes time to read the paragraph, and it is hard to look for a trend. The bar graph and the line graph are more helpful if we want to see the overall trend of the number of prescriptions filled and to make a prediction about years to come. The exact data values are most easily read in a table.

In this chapter, you will learn not only how to extract information from various kinds of tables and graphs, but also how to create bar, line, and circle graphs.

## a  Averages

A **statistic** is a number describing a set of data. One statistic is a *center point,* or *measure of central tendency,* that characterizes the data. The most common kind of center point is the *mean,* or *average,* of a set of numbers. We first considered averages in Section 1.9 and extended the coverage in Sections 4.7 and 5.4.

Let's consider the number of prescriptions, in millions, filled in supermarkets in the years 1997, 1999, 2001, 2003, 2005, and 2007:

  269,   357,   418,   462,   465,   478.

What is the average of this set of numbers? First, we add the numbers:

  269 + 357 + 418 + 462 + 465 + 478 = 2449.

Next, we divide by the number of data items, 6:

$$\frac{2449}{6} \approx 408. \qquad \text{Rounding to the nearest one}$$

Note that if the number of prescriptions had been the average (same) for each of the 6 yr, we would have

  408 + 408 + 408 + 408 + 408 + 408 = 2448 ≈ 2449.

The number 408 is called the *average* of the set of numbers. It is also called the **arithmetic** (pronounced ăr´ĭth-mĕt´-ĭk) **mean** or simply the **mean**.

---

**AVERAGE**

To find the **average** of a set of numbers, add the numbers and then divide by the number of items of data.

---

Find the average.

1. 14, 175, 36

2. 75, 36.8, 95.7, 12.1

3. A student earned the following scores on five tests: 68, 85, 82, 74, 96. What was the average score?

4. In the first five games of the season, a basketball player scored points as follows: 26, 21, 13, 14, 23. Find the average number of points scored per game.

5. **Home-Run Batting Average.**
Babe Ruth hit 714 home runs in 22 seasons played in the major leagues. What was his average number of home runs per season? Round to the nearest tenth.
**Source:** Major League Baseball

**EXAMPLE 1** On a 4-day trip, a car was driven the following number of miles: 240, 302, 280, 320. What was the average number of miles per day?

$$\frac{240 + 302 + 280 + 320}{4} = \frac{1142}{4}, \text{ or } 285.5$$

The car was driven an average of 285.5 mi per day. Had the car been driven exactly 285.5 mi each day, the same total distance (1142 mi) would have been traveled.

Do Exercises 1-4.

**EXAMPLE 2** *Scoring Average.* Michael Jordan is the third highest scorer in the history of the National Basketball Association. He scored 32,292 points in 1072 games. What was the average number of points scored per game? Round to the nearest tenth.
**Source:** National Basketball Association

We know the total number of points, 32,292, and the number of games, 1072. We divide and round to the nearest tenth:

$$\frac{32,292}{1072} = 30.123134\ldots \approx 30.1.$$

Michael Jordan's average was about 30.1 points per game.

Do Exercise 5.

**EXAMPLE 3** *Gas Mileage.* The 2009 Volkswagen Jetta TDI, shown here, is estimated to travel 522 mi in the city on 18 gal of diesel fuel. What is the expected average number of miles per gallon (mpg)—that is, what is its fuel mileage for city driving?
**Source:** Car and Driver, January 2009

*Answers*

1. 75    2. 54.9    3. 81    4. 19.4
5. 32.5 home runs per season

We divide the total number of miles, 522, by the total number of gallons, 18:

$$\frac{522 \text{ mi}}{18 \text{ gal}} = 29 \text{ mpg}.$$

The Volkswagen Jetta's expected average is 29 mi per gallon for city driving.

Do Exercise 6.

**6. Gas Mileage.** The 2009 Ford Flex SUV Crossover gets 384 mi of highway driving on 16 gal of gasoline. What is the average number of miles per gallon— that is, what is its gas mileage for highway driving?

**Source:** *Motor Trend*, November 2008

**EXAMPLE 4** *Grade Point Average.* In most colleges, students are assigned grade point values for grades obtained. The **grade point average**, or **GPA**, is the average of the grade point values for each credit hour taken. At most colleges, grade point values are assigned as follows:

A: 4.0   B: 3.0   C: 2.0   D: 1.0   F: 0.0

Meg earned the following grades for one semester. What was her grade point average?

| COURSE | GRADE | NUMBER OF CREDIT HOURS IN COURSE |
|---|---|---|
| Colonial History | B | 3 |
| Basic Mathematics | A | 4 |
| English Literature | A | 3 |
| French | C | 4 |
| Time Management | D | 1 |

To find the GPA, we first multiply the grade point value (in color below) by the number of credit hours in the course to determine the number of *quality points,* and then add, as follows:

Colonial History     $3.0 \cdot 3 =$   9
Basic Mathematics   $4.0 \cdot 4 =$ 16
English Literature    $4.0 \cdot 3 =$ 12
French                    $2.0 \cdot 4 =$   8
Time Management   $1.0 \cdot 1 =$   1
                                              46 (Total)

The total number of credit hours taken is $3 + 4 + 3 + 4 + 1$, or 15. We divide the number of quality points, 46, by the number of credit hours, 15, and round to the nearest tenth:

$$\text{GPA} = \frac{46}{15} \approx 3.1.$$

Meg's grade point average was 3.1.

Do Exercise 7.

**7. Grade Point Average.** Alex earned the following grades one semester.

| GRADE | NUMBER OF CREDIT HOURS IN COURSE |
|---|---|
| B | 3 |
| C | 4 |
| C | 4 |
| A | 2 |

What was Alex's grade point average? Assume that the grade point values are 4.0 for an A, 3.0 for a B, and so on. Round to the nearest tenth.

*Answers*

**6.** 24 mpg   **7.** 2.5

**EXAMPLE 5**  *Grading.*  To get a B in math, Geraldo must score an average of 80 on five tests. On the first four tests, his scores were 79, 88, 64, and 78. What is the lowest score that Geraldo can get on the last test and still get a B?

We can find the total of the five scores needed as follows:

$$80 + 80 + 80 + 80 + 80 = 5 \cdot 80, \quad \text{or} \quad 400.$$

The total of the scores on the first four tests is

$$79 + 88 + 64 + 78 = 309.$$

Thus Geraldo needs to get at least

$$400 - 309, \quad \text{or} \quad 91$$

in order to get a B. We can check this as follows:

$$\frac{79 + 88 + 64 + 78 + 91}{5} = \frac{400}{5}, \quad \text{or} \quad 80.$$

> **8. Grading.**  To get an A in math, Rosa must score an average of 90 on four tests. On the first three tests, her scores were 80, 100, and 86. What is the lowest score that Rosa can get on the last test and still get an A?

Do Exercise 8.

## b  Medians

Another type of center-point statistic is the *median.* Medians are useful when we wish to de-emphasize unusually extreme numbers. For example, suppose a small class scored as follows on an exam.

| Phil: | 78 | Pat: | 56 |
| Jill: | 81 | Olga: | 84 |
| Matt: | 82 | | |

Let's first list the scores in order from smallest to largest:

56, 78, 81, 82, 84.
       ↑
   Middle score

The middle score—in this case, 81—is called the **median**. Note that because of the extremely low score of 56, the average of the scores is 76.2. In this example, the median may be a more appropriate center-point statistic.

**EXAMPLE 6**  What is the median of this set of numbers?

99,  870,  91,  98,  106,  90,  98

We first rearrange the numbers in order from smallest to largest. Then we locate the middle number.

90,  91,  98,  98,  99,  106,  870
                  ↑
            Middle number

The median is 98.

> Find the median.
> **9.** 17, 13, 18, 14, 19
>
> **10.** 20, 14, 13, 19, 16, 18, 17
>
> **11.** 78, 81, 83, 91, 103, 102, 122, 119, 88

Do Exercises 9–11.

## MEDIAN

Once a set of data is listed in order, from smallest to largest, the **median** is the middle number if there is an odd number of data items. If there is an even number of items, the median is the number that is the average of the two middle numbers.

**EXAMPLE 7** What is the median of this set of numbers?

69, 80, 61, 63, 62, 65

We first rearrange the numbers in order from smallest to largest. There is an even number of numbers. We look for the middle two, which are 63 and 65. The median is halfway between 63 and 65, the number 64.

61, 62, 63, 65, 69, 80    The average of the middle numbers is

$$\frac{63 + 65}{2} = \frac{128}{2}, \text{ or } 64.$$

The median is 64.

**EXAMPLE 8** *Salaries.* The following are the salaries of the top four employees of Verducci's Dress Company. What is the median of the salaries?

$85,000,  $100,000,  $78,000,  $84,000

We rearrange the numbers in order from smallest to largest. The two middle numbers are $84,000 and $85,000. Thus the median is halfway between $84,000 and $85,000 (the average of $84,000 and $85,000):

$78,000,  $84,000,  $85,000,  $100,000

$$\text{Median} = \frac{\$84,000 + \$85,000}{2} = \frac{\$169,000}{2} = \$84,500.$$

Do Exercises 12 and 13.

## c Modes

The final type of center-point statistic is the *mode.*

## MODE

The **mode** of a set of data is the number or numbers that occur most often. If each number occurs the same number of times, there is *no* mode.

**EXAMPLE 9** Find the mode of these data.

13, 14, 17, 17, 18, 19

The number that occurs most often is 17. Thus the mode is 17.

**STUDY TIPS**

### REGISTERING FOR FUTURE COURSES

As you sign up for next semester's courses, evaluate your work and family commitments. Talk with instructors and other students to estimate how demanding the course is before registering. It is better to take one fewer course than one too many.

Find the median.

**12. Salaries of Part-Time Typists.**
$3300, $4000, $3900, $3600, $3800, $3400

**13.** 68, 34, 67, 69, 34, 70

*Answers*

**12.** $3700    **13.** 67.5

A set of data has just one average (mean) and just one median, but it can have more than one mode. It is also possible for a set of data to have no mode—when all numbers are equally represented. For example, the set of data 5, 7, 11, 13, 19 has no mode.

**EXAMPLE 10**   Find the modes of these data.

33,  34,  34,  34,  35,  36,  37,  37,  37,  38,  39,  40

There are two numbers that occur most often, 34 and 37. Thus the modes are 34 and 37.

Do Exercises 14–17.

Which statistic is best for a particular situation? If someone is bowling, the *average* from several games is a good indicator of that person's ability. If someone is applying for a job, the *median* salary at that business is often most indicative of what people are earning there because although executives tend to make a lot more money, there are fewer of them. Finally, if someone is reordering for a clothing store, the *mode* of the sizes sold is probably the most important statistic.

## d   Comparing Two Sets of Data

We have seen how to calculate averages, medians, and modes from data. A way to analyze two sets of data is to make a determination about which of two groups is "better." One way to do so is by comparing the averages.

**EXAMPLE 11**   *Growth of Wheat.*
A university agriculture department experiments to see which of two kinds of wheat is better. (In this situation, the shorter wheat is considered "better.") The researchers grow both kinds under similar conditions and measure stalk heights, in inches, as follows. Which kind is better?

| WHEAT A STALK HEIGHTS (in inches) | | | | WHEAT B STALK HEIGHTS (in inches) | | | |
|---|---|---|---|---|---|---|---|
| 16.2 | 42.3 | 19.5 | 25.7 | 19.7 | 18.4 | 19.7 | 17.2 |
| 25.6 | 18.0 | 15.6 | 41.7 | 19.7 | 14.6 | 32.0 | 25.7 |
| 22.6 | 26.4 | 18.4 | 12.6 | 14.0 | 21.6 | 42.5 | 32.6 |
| 41.5 | 13.7 | 42.0 | 21.6 | 22.6 | 10.9 | 26.7 | 22.8 |

Note that it is difficult to analyze the data at a glance because the numbers are close together. We need a way to compare the two groups. Let's compute the average of each set of data.

Wheat A: Average

$$= \frac{\begin{array}{c}16.2 + 25.6 + 22.6 + 41.5 + 42.3 + 18.0 + 26.4 + 13.7 + \\ 19.5 + 15.6 + 18.4 + 42.0 + 25.7 + 41.7 + 12.6 + 21.6\end{array}}{16}$$

$$\approx 25.21 \text{ in.}$$

Wheat B: Average

$$= \frac{\begin{array}{c}19.7 + 19.7 + 14.0 + 22.6 + 18.4 + 14.6 + 21.6 + 10.9 + \\ 19.7 + 32.0 + 42.5 + 26.7 + 17.2 + 25.7 + 32.6 + 22.8\end{array}}{16}$$

$$\approx 22.54 \text{ in.}$$

We see that the average stalk height of wheat B is less than that of wheat A. Thus wheat B is "better."

Do Exercise 18.

**18. Battery Testing.** Two kinds of battery were tested to see how long, in hours, they kept an emergency light running. On the basis of this test, which battery is better?

| BATTERY A (in hours) | | | |
|---|---|---|---|
| 26.6 | 28.3 | 27.1 | 27.6 |
| 27.6 | 28.0 | 26.8 | 27.4 |
| 28.1 | 28.2 | 26.9 | 27.3 |

| BATTERY B (in hours) | | | |
|---|---|---|---|
| 28.5 | 27.6 | 28.6 | 27.5 |
| 27.7 | 27.9 | 26.9 | 27.8 |
| 28.3 | 27.9 | 28.7 | 27.6 |

*Answers*

**18.** Battery A: average hours $\approx$ 27.49; battery B: average hours $\approx$ 27.92; battery B is better

## 8.1 Exercise Set

For Extra Help

MyMathLab
Math XL PRACTICE | WATCH | DOWNLOAD | READ | REVIEW

**a**, **b**, **c**   For each set of numbers, find the average, the median, and any modes that exist.

**1.** *Atlantic Storms and Hurricanes.*   The following bar graph shows the number of Atlantic storms and hurricanes that formed in various months from 1980 to 2007. What is the average number for the 9 months given? the median? the mode?

**Atlantic Storms and Hurricanes**

Tropical storm and hurricane formation in 1980–2007, by month

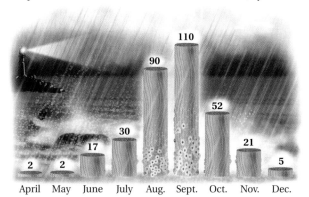

SOURCE: Colorado State University, Department of Atmospheric Science, Phil Klotzbach, Ph.D., Research Scientist

**2.** *Tornadoes.*   The following bar graph shows the average number of tornado deaths by month since 1950. What is the average number of tornado deaths for the 12 months? the median? the mode?

**Average Number of Deaths by Tornado by Month**

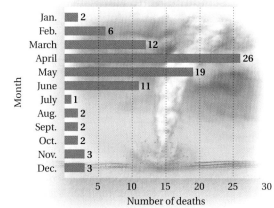

SOURCE: National Weather Service's Storm Prediction Center

**3.** 17, 19, 29, 18, 14, 29

**4.** 72, 83, 85, 88, 92

**5.** 5, 37, 20, 20, 35, 5, 25

**6.** 13, 32, 25, 27, 13

**7.** 4.3, 7.4, 1.2, 5.7, 8.3

**8.** 13.4, 13.4, 12.6, 42.9

**9.** 234, 228, 234, 229, 234, 278

**10.** $29.95, $28.79, $30.95, $28.79

**11.** *Gas Mileage.*  The 2009 Kia Sedona gets 253 mi of highway driving on 11 gal of gasoline. What is the average number of miles expected per gallon—that is, what is its gas mileage?
**Source:** *Motor Trend*, November 2008

**12.** *Gas Mileage.*  The 2009 Chevrolet Cobalt SS gets 330 mi of city driving on 15 gal of gasoline. What is the average number of miles expected per gallon—that is, what is its gas mileage?
**Source:** *Road & Track*, November 2008

*Grade Point Average.*  The tables in Exercises 13 and 14 show the grades of a student for one semester. In each case, find the grade point average. Assume that the grade point values are 4.0 for an A, 3.0 for a B, and so on. Round to the nearest tenth.

**13.**

| GRADE | NUMBER OF CREDIT HOURS IN COURSE |
|-------|----------------------------------|
| B | 4 |
| A | 5 |
| D | 3 |
| C | 4 |

**14.**

| GRADE | NUMBER OF CREDIT HOURS IN COURSE |
|-------|----------------------------------|
| A | 5 |
| C | 4 |
| F | 3 |
| B | 5 |

**15.** *Brussels Sprouts.*  The following prices per stalk of Brussels sprouts were found at five supermarkets:

$3.99,   $4.49,   $4.99,   $3.99,   $3.49.

What was the average price per stalk? the median price? the mode?

**16.** *Cheddar Cheese Prices.*  The following prices per pound of sharp cheddar cheese were found at five supermarkets:

$5.99,   $6.79,   $5.99,   $6.99,   $6.79.

What was the average price per pound? the median price? the mode?

**17.** *Grading.*  To get a B in math, Rich must score an average of 80 on five tests. His scores on the first four tests were 80, 74, 81, and 75. What is the lowest score that Rich can get on the last test and still receive a B?

**18.** *Grading.*  To get an A in math, Cybil must score an average of 90 on five tests. Her scores on the first four tests were 90, 91, 81, and 92. What is the lowest score that Cybil can get on the last test and still receive an A?

**19.** *Length of Pregnancy.* Marta was pregnant 270 days, 259 days, and 272 days for her first three pregnancies. In order for Marta's average pregnancy to equal the worldwide average of 266 days, how long must her fourth pregnancy last?

**Source:** Vardaan Hospital, Dr. Rekha Khandelwal, M.S.

**20.** *Male Height.* Jason's brothers are 174 cm, 180 cm, 179 cm, and 172 cm tall. The average male is 176.5 cm tall. How tall is Jason if he and his brothers have an average height of 176.5 cm?

 Solve.

**21.** *Light-Bulb Testing.* An experiment was performed to compare the lives of two types of light bulbs. Several bulbs of each type were tested, and the results are listed in the following table. On the basis of this test, which bulb is better?

| BULB A: HOTLIGHT TIME (in hours) | | | BULB B: BRIGHTBULB TIME (in hours) | | |
|---|---|---|---|---|---|
| 983 | 964 | 1214 | 979 | 1083 | 1344 |
| 1417 | 1211 | 1521 | 984 | 1445 | 975 |
| 1084 | 1075 | 892 | 1492 | 1325 | 1283 |
| 1423 | 949 | 1322 | 1325 | 1352 | 1432 |

**22.** *Cola Testing.* An experiment was conducted to determine which of two colas tastes better. Students drank the colas and gave each a rating from 1 to 10, with 10 indicating the best taste. The results are given in the following table. On the basis of this test, which cola tastes better?

| COLA A: VERYCOLA | | | | COLA B: COLA-COLA | | | |
|---|---|---|---|---|---|---|---|
| 6 | 8 | 10 | 7 | 10 | 9 | 9 | 6 |
| 7 | 9 | 9 | 8 | 8 | 8 | 10 | 7 |
| 5 | 10 | 9 | 10 | 8 | 7 | 4 | 3 |
| 9 | 4 | 7 | 6 | 7 | 8 | 10 | 9 |

## Skill Maintenance

Multiply.

**23.** $12.86 \times 17.5$   [5.3a]

**24.** $-222 \times (-0.5678)$   [5.3a]

**25.** $\dfrac{4}{5} \cdot \dfrac{3}{28}$   [3.6a]

**26.** $-\dfrac{28}{45} \cdot \dfrac{3}{2}$   [3.6a]

Convert to percent notation.   [7.2a]

**27.** $\dfrac{19}{16}$

**28.** $\dfrac{11}{16}$

**29.** $\dfrac{64}{125}$

**30.** $\dfrac{9781}{10,000}$

## Synthesis

**31.** The ordered set of data 18, 21, 24, *a*, 36, 37, *b* has a median of 30 and an average of 32. Find *a* and *b*.

**32.** *Hank Aaron.* Hank Aaron averaged $34\frac{7}{22}$ home runs per year over a 22-yr career. After 21 yr, Aaron had averaged $35\frac{10}{21}$ home runs per year. How many home runs did Aaron hit in his final year?

**33.** *Price Negotiations.* Amy offers $3200 for a used Ford Taurus advertised at $4000. The first offer from Jim, the car's owner, is to "split the difference" and sell the car for $(3200 + 4000) \div 2$, or $3600. Amy's second offer is to split the difference between Jim's offer and her first offer. Jim's second offer is to split the difference between Amy's second offer and his first offer. If this pattern continues and Amy accepts Jim's third (and final) offer, how much will she pay for the car?

# 8.2

# Tables and Pictographs

## OBJECTIVES

**a** Extract and interpret data from tables.

**b** Extract and interpret data from pictographs.

**SKILL TO REVIEW**

Objective 4.6a: Multiply using mixed numerals.

Multiply.

**1.** $3\frac{1}{2} \times 800$

**2.** $1\frac{3}{4} \times 3020$

Use the table in Example 1 to answer Margin Exercises 1–5.

**1.** Which country has the smallest amount of land area?

**2.** Which country has the greatest population density?

**3.** What is the population of Mexico?

**4.** Find the average of the land areas of the countries listed.

**5.** Find the median of the populations of these countries.

## **a** Reading and Interpreting Tables

A **table** is often used to present data in rows and columns.

**EXAMPLE 1** *Population Density.*  The following table lists data regarding 11 countries around the world.

| COUNTRY | POPULATION | LAND AREA (in square miles) | POPULATION DENSITY (per square mile) | PERCENT OF POPULATION THAT IS URBAN |
|---|---|---|---|---|
| Australia | 20,434,176 | 2,941,299 | 7 | 88.2 |
| Brazil | 190,010,647 | 3,265,077 | 58 | 84.2 |
| China | 1,321,851,888 | 3,600,947 | 367 | 40.4 |
| Finland | 5,238,460 | 117,558 | 45 | 61.1 |
| France | 63,718,187 | 210,669 | 258 | 76.7 |
| Germany | 82,400,996 | 134,836 | 611 | 75.2 |
| India | 1,129,866,154 | 1,147,955 | 984 | 28.7 |
| Japan | 127,433,494 | 144,689 | 881 | 65.8 |
| Kenya | 36,913,721 | 219,789 | 168 | 20.7 |
| Mexico | 108,700,891 | 742,490 | 146 | 76.0 |
| United States | 301,139,947 | 3,537,439 | 85 | 80.8 |

SOURCE: *The World Almanac*, 2008

**a)** Which country has the largest population?

**b)** Which country has the smallest population density?

**c)** What percent of the population in Kenya is urban?

**d)** Find the average of the population densities in the countries listed.

   Careful examination of the table will give the answers.

**a)** To determine which country has the largest population, we look down the column headed "Population" and find the largest number. That number is 1,321,851,888. Then we look across that row to find the name of the country, China.

**b)** To determine which country has the smallest population density, we look down the column headed "Population Density" and find the smallest number. That number is 7. Then we look across that row to find the country, Australia.

**c)** To determine what percent of the population in Kenya is urban, we look down the column headed "Country" and find the name Kenya. Then we look across that row to the column headed "Percent of Population That Is Urban" to find the percent. It is 20.7%.

*Answers*

*Skill to Review:*
**1.** 2800   **2.** 5285

*Margin Exercises:*
**1.** Finland   **2.** India   **3.** 108,700,891
**4.** About 1,460,250 sq mi   **5.** 108,700,891

**d)** We determine the average of all the numbers in the column headed "Population Density":

$$\frac{7 + 58 + 367 + 45 + 258 + 611 + 984 + 881 + 168 + 146 + 85}{11}$$

$$= \frac{3610}{11} \approx 328.2 \text{ per sq mi.}$$

Do Margin Exercises 1–5 on the preceding page.

**EXAMPLE 2** *Frosted Flakes Nutrition Facts.* Most foods are required by law to provide factual information regarding nutrition, as shown in the following table of nutrition facts from a box of Frosted Flakes cereal. Although this can be very helpful to the consumer, one must be careful in interpreting the data.

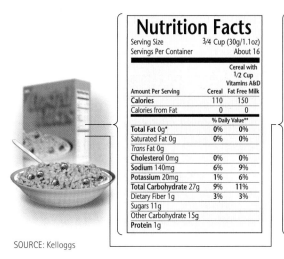

SOURCE: Kelloggs

Suppose your morning bowl of cereal consists of $1\frac{1}{2}$ cups of Frosted Flakes with 1 cup of fat-free milk.

**a)** How many calories have you consumed?

**b)** What percent of the daily value for dietary fiber have you consumed?

**c)** A nutritionist recommends that you look for foods that provide 10% or more of the daily value for vitamin C. Do you get that with your bowl of Frosted Flakes?

**d)** Suppose you are trying to limit your daily caloric intake to 2000 calories. How many bowls of cereal would it take to exceed the 2000 calories?

Careful examination of the table of nutrition facts will give the answers.

**a)** We look at the column marked "with $\frac{1}{2}$ Cup Fat Free Milk" and note that a serving size of $\frac{3}{4}$ cup of cereal with $\frac{1}{2}$ cup of fat-free milk contains 150 calories. Since you are having twice that amount, you are consuming

$$2 \times 150 \text{ calories, or } 300 \text{ calories.}$$

**b)** We read across from "Dietary Fiber" and note that in $\frac{3}{4}$ cup of cereal with $\frac{1}{2}$ cup of fat-free milk, you get 3% of the daily value of dietary fiber. Since you are doubling that, you get 6% of the daily value of dietary fiber.

**c)** We find the row labeled "Vitamin C" on the left and look under the column labeled "with $\frac{1}{2}$ Cup Fat Free Milk." Note that you get 10% of the daily value for $\frac{3}{4}$ cup of cereal with $\frac{1}{2}$ cup of fat-free milk, and since you are doubling that, you are more than satisfying the 10% requirement.

Use the nutrition facts data shown in the figure and the bowl of cereal described in Example 2 to answer Margin Exercises 6–10.

**6.** How many calories from fat are in your bowl of cereal?

**7.** A nutritionist recommends that you look for foods that provide 12% or more of the daily value for iron. Do you get that with your bowl of Frosted Flakes?

**8.** How much sodium have you consumed? (*Hint*: Use the data listed in the first footnote below the table of nutrition facts.)

**9.** What percent of the daily value for sodium have you consumed?

**10.** How much protein have you consumed? (*Hint*: Use the data listed in the first footnote below the table of nutrition facts.)

*Answers*

**6.** 0 calories   **7.** Yes   **8.** 410 mg   **9.** 18%
**10.** 10 g

**d)** From part (a), we know that you are consuming 300 calories per bowl. Dividing 2000 by 300 gives $\frac{2000}{300} \approx 6.67$. Thus if you eat 7 bowls of cereal in this manner, you will exceed the 2000 calories.

Do Exercises 6–10 on the preceding page.

## **b** Reading and Interpreting Pictographs

*Pictographs* (or *picture graphs*) are another way to show information. Instead of actually listing the amounts to be considered, a **pictograph** uses symbols to represent the amounts. In addition, a *key* is given telling what each symbol represents.

**EXAMPLE 3** *Touchdown Passes.* The following pictograph shows the career-high number of touchdown passes in a season for seven quarterbacks in the National Football League. Located below the graph is a key that tells you that each  symbol represents 10 touchdown passes.

**Touchdown Passes (Career high for quarterback)**

| | PLAYER | YEAR | TOUCHDOWN PASSES |
|---|---|---|---|
| **Retired** | John Elway | 1997 | |
| | Dan Marino | 1984 | |
| | Joe Montana | 1987 | |
| | Roger Staubach | 1978 | |
| **Active** | Tom Brady | 2007 | |
| | Brett Favre | 2001 and 2003 | |
| | Peyton Manning | 2004 | |

SOURCE: National Football League

 = 10 touchdown passes

**a)** Which quarterback has the greatest number of touchdown passes?

**b)** Which quarterback has the least number of touchdown passes?

**c)** How many more touchdown passes does Peyton Manning have in his career high than John Elway?

We can compute the answers by first reading the pictograph.

**a)** The quarterback with the most symbols has the greatest number of touchdown passes: Tom Brady, with 5 (symbols) × 10, or 50, touchdown passes.

**b)** The quarterback with the fewest symbols has the least number of touchdown passes: Roger Staubach, with about $2\frac{1}{2}$ (symbols) × 10, or 25, touchdown passes.

**c)** From the graph, we see that there are about $4\frac{9}{10}$ × 10, or 49, touchdown passes for Peyton Manning, and about $2\frac{7}{10}$ × 10, or 27, for John Elway. Thus Peyton Manning has 49−27, or 22, more touchdown passes in his career high than John Elway.

Do Exercises 11–13.

Use the pictograph in Example 3 to answer Margin Exercises 11–13.

**11.** How many more touchdown passes did Dan Marino have for his career high in a season than Brett Favre?

**12.** How many more touchdown passes did Joe Montana have in his career high than Roger Staubach?

**13.** What is the average number of career-high touchdown passes in a season for the seven quarterbacks?

*Answers*

**11.** 16 touchdown passes    **12.** Joe Montana had 6 more than Roger Staubach.    **13.** About 37 touchdown passes

**a** *Veggie Burgers.* Veggie burgers tend to be higher in fiber and lower in calories than conventional burgers. Their weakness is high sodium content. Comparative information on veggie burgers from eight companies is listed in the following table. Use the data for Exercises 1–10.

| COMPANY AND PRODUCT | PER PATTY (ABOUT 2.5 OZ) | | | | | |
| --- | --- | --- | --- | --- | --- | --- |
| | Cost | Calories | Fat (g) | Protein (g) | Fiber (g) | Sodium (mg) |
| Amy's All American The Classic | $1.12 | 120 | 3.0 | 10 | 3 | 390 |
| Boca All American Flame Grilled Meatless | 0.96 | 90 | 3.0 | 14 | 3 | 280 |
| Dr. Praeger's Sensible Foods California | 1.08 | 110 | 4.5 | 5 | 4 | 250 |
| Franklin Farms Portabella Fresh | 1.00 | 100 | 1.5 | 14 | 3 | 390 |
| Gardenburger Portabella | 0.98 | 90 | 2.5 | 5 | 5 | 360 |
| Lightlife Meatless Light | 1.48 | 120 | 1.5 | 16 | 3 | 500 |
| MorningStar Farms Garden | 1.07 | 110 | 3.5 | 10 | 3 | 350 |
| Veggie Patch Garlic Portabella | 1.00 | 120 | 6.0 | 11 | 4 | 320 |

SOURCE: *Consumer Reports*, June 2008

1. How many calories are in a Franklin Farms Portabella Fresh veggie burger?

2. Which veggie burger contains the most protein?

3. Which veggie burgers have less than 110 calories?

4. How many grams of fat are in a MorningStar Farms Garden veggie burger?

5. What are the average, the median, and the mode of the fiber content in these 8 veggie burgers?

6. What is the average number of calories in these 8 veggie burgers?

7. Which veggie burger is the most expensive? the least expensive?

8. Which veggie burger contains the most sodium? the least sodium?

9. Which veggie burger contains the greatest number of grams of fat? the least number of grams of fat?

10. What is the median cost of these veggie burgers?

*Heat Index.*   In warm weather, a person can feel hot because of reduced heat loss from the skin caused by higher humidity. The **temperature–humidity index**, or **apparent temperature**, is what the temperature would have to be with no humidity in order to give the same heat effect. The following table lists the apparent temperatures for various actual temperatures and relative humidities. Use this table for Exercises 11–22.

| ACTUAL TEMPERATURE (°F) | RELATIVE HUMIDITY | | | | | | | | | |
|---|---|---|---|---|---|---|---|---|---|---|
| | 10% | 20% | 30% | 40% | 50% | 60% | 70% | 80% | 90% | 100% |
| | APPARENT TEMPERATURE (°F) | | | | | | | | | |
| 75° | 75 | 77 | 79 | 80 | 82 | 84 | 86 | 88 | 90 | 92 |
| 80° | 80 | 82 | 85 | 87 | 90 | 92 | 94 | 97 | 99 | 102 |
| 85° | 85 | 88 | 91 | 94 | 97 | 100 | 103 | 106 | 108 | 111 |
| 90° | 90 | 93 | 97 | 100 | 104 | 107 | 111 | 114 | 118 | 121 |
| 95° | 95 | 99 | 103 | 107 | 111 | 115 | 119 | 123 | 127 | 131 |
| 100° | 100 | 105 | 109 | 114 | 118 | 123 | 127 | 132 | 137 | 141 |
| 105° | 105 | 110 | 115 | 120 | 125 | 131 | 136 | 141 | 146 | 151 |

In Exercises 11–14, find the apparent temperature for the given actual temperature and humidity combinations.

**11.** 80°, 60%

**12.** 90°, 70%

**13.** 85°, 90%

**14.** 95°, 80%

**15.** Which temperature–humidity combinations give an apparent temperature of 100°?

**16.** Which temperature–humidity combinations give an apparent temperature of 111°?

**17.** At a relative humidity of 50%, what actual temperatures give an apparent temperature above 100°?

**18.** At a relative humidity of 90%, what actual temperatures give an apparent temperature above 100°?

**19.** At an actual temperature of 95°, what relative humidities give an apparent temperature above 100°?

**20.** At an actual temperature of 85°, what relative humidities give an apparent temperature above 100°?

**21.** At an actual temperature of 85°, what is the difference in humidities required to raise the apparent temperature from 94° to 108°?

**22.** At an actual temperature of 80°, what is the difference in humidities required to raise the apparent temperature from 87° to 102°?

*Planets.* Use the following table, which lists information about the planets, for Exercises 23–32.

| PLANET | AVERAGE DISTANCE FROM SUN (in miles) | DIAMETER (in miles) | LENGTH OF PLANET'S DAY IN EARTH TIME (in days) | TIME OF REVOLUTION IN EARTH TIME (in years) |
|---|---|---|---|---|
| Mercury | 35,983,000 | 3,031 | 58.82 | 0.24 |
| Venus | 67,237,700 | 7,520 | 224.59 | 0.62 |
| Earth | 92,955,900 | 7,926 | 1.00 | 1.00 |
| Mars | 141,634,800 | 4,221 | 1.03 | 1.88 |
| Jupiter | 483,612,200 | 88,846 | 0.41 | 11.86 |
| Saturn | 888,184,000 | 74,898 | 0.43 | 29.46 |
| Uranus | 1,782,000,000 | 31,763 | 0.45 | 84.01 |
| Neptune | 2,794,000,000 | 31,329 | 0.66 | 164.78 |

SOURCE: *The Handy Science Answer Book*, Gale Research, Inc.

**23.** Find the average distance from the sun to Jupiter.

**24.** How long is a day on Venus?

**25.** Which planet has a time of revolution of 164.78 yr?

**26.** Which planet has a diameter of 4221 mi?

**27.** Which planets have an average distance from the sun that is greater than 1,000,000 mi?

**28.** Which planets have a diameter that is less than 100,000 mi?

**29.** About how many Earth diameters would it take to equal one Jupiter diameter?

**30.** How much longer is the longest time of revolution than the shortest?

**31.** What are the average, the median, and the mode of the diameters of the planets?

**32.** What are the average, the median, and the mode of the average distances from the sun of the planets?

*Rhino Population.* The rhinoceros is considered one of the world's most endangered animals. The worldwide total number of rhinoceroses is approximately 20,700. The following pictograph shows the population of the five remaining rhino species. Located in the graph is a key that tells you that each symbol 🦏 represents 300 rhinos. Use the pictograph for Exercises 33–38.

**Rhino Population (Remaining Species)**

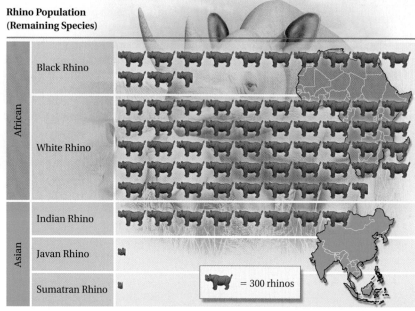

SOURCE: World Wildlife Fund, 2008

On July 5, 2004, at the Cincinnati Zoo and Botanical Gardens, Emi, a critically endangered Sumatran rhinoceros, became the first rhino to produce two calves in captivity.

**33.** Which species has the greatest number of rhinos?

**34.** Which species has the least number of rhinos?

**35.** How many more black rhinos are there than Indian rhinos?

**36.** How many more white rhinos are there than black rhinos?

**37.** What is the average number of rhinos in these five species?

**38.** How does the white rhino population compare with the Indian rhino population?

## Skill Maintenance

Solve. [7.5a]

**39.** *Kitchen Costs.* The average cost of a kitchen is $26,888. Some of the cost percentages are as follows.

|  |  |
|---|---|
| Cabinets: | 50% |
| Countertops: | 15% |
| Appliances: | 8% |
| Fixtures: | 3% |

Find the costs for each of these parts of a kitchen.

**Source:** National Kitchen and Bath Association

**40.** *Bathroom Costs.* The average cost of a bathroom is $11,605. Some of the cost percentages are as follows.

|  |  |
|---|---|
| Cabinets: | 31% |
| Countertops: | 11% |
| Labor: | 25% |
| Flooring: | 6% |

Find the costs for each of these parts of a bathroom.

**Source:** National Kitchen and Bath Association

# Mid-Chapter Review

## Concept Reinforcement

Determine whether each statement is true or false.

_____ **1.** A set of data has just one average and just one median, but it can have more than one mode.
[8.1a, b, c]

_____ **2.** It is possible for the average, the median, and the mode of a set of data to be the same number.
[8.1a, b, c]

_____ **3.** If there is an even number of items in a set of data, the middle number is the median. [8.1b]

## Guided Solutions

Fill in each blank with the number that creates a correct solution.

**4.** The average of 60, 45, 115, 15, and 35 is

$$\frac{60 + 45 + \square + 15 + 35}{\square} = \frac{\square}{5} = \square. \quad [8.1a]$$

**5.** Find the median of this set of numbers:

2.1,  11.3,  8.7,  6.3,  14.5,  4.8. [8.1b]

We first arrange the numbers from smallest to largest:

$\square$, $\square$, 6.3, $\square$, 11.3, $\square$.

There is an even number of items. The median is the average of

$\square$ and $\square$.

We find that average:

$$\frac{\square + \square}{2} = \frac{\square}{2} = \square.$$

The median is $\square$.

## Mixed Review

For each set of numbers, find the average, the median, and any modes that exist. [8.1a, b, c]

**6.** 56, 29, 45, 240, 175, 7, 29

**7.** 2.12, 18.42, 9.37, 43.89

**8.** $\frac{5}{10}, \frac{1}{10}, \frac{7}{10}, \frac{9}{10}, \frac{3}{10}$

**9.** 160, 102, 102, 116, 160, 116

**10.** $4.96, $5.24, $4.96, $10.05, $5.24

**11.** $\frac{1}{2}, \frac{3}{4}, \frac{7}{8}, \frac{5}{4}$

**12.** 2, 5, 7, 7, 8, 5, 5, 7, 8

**13.** 38.2, 38.2, 38.2, 38.2

*Downsizing.*   In tight economic times, companies sometimes downsize their products. That is, they charge the same price for a package that contains less product. The following table lists products that have been downsized recently. Use this table for Exercises 14–18.   [8.2a]

| PRODUCT | SIZE (in ounces) OLD | NEW | PERCENT SMALLER |
|---------|------|-----|-----------------|
| Breyer's ice cream | 56 | 48 | 14 |
| Hellmann's mayonnaise | 32 | 30 | 6 |
| Hershey's Special Dark chocolate bar | 8 | 6.8 | 15 |
| Iams cat food | 6 | 5.5 | 8 |
| Nabisco Chips Ahoy cookies | 16 | 15.25 | 5 |
| Skippy creamy peanut butter | 18 | 16.3 | 9 |
| Tropicana orange juice | 96 | 89 | 7 |

SOURCE: *Consumer Reports*, October 2008

**14.** How much less ice cream is in the new Breyer's ice cream package than in the old package?

**15.** By what percent has the size of Hellmann's mayonnaise changed in the downsizing process?

**16.** Which product in the table has the greatest percent of decrease?

**17.** How much less orange juice is in the new Tropicana orange juice package than in the old package?

**18.** Which product in the table has the smallest percent of decrease?

*Gun Ownership.*   According to the Small Arms Survey 2007 by the Graduate Institute of International Studies in Geneva, Switzerland, the United States is the most heavily armed society in the world. The following pictograph shows gun ownership per 100 civilians in six countries. Use the pictograph for Exercises 19–24.   [8.2b]

**Gun Ownership Rates (per 100 civilians)**

SOURCE: Reuters, August 28, 2007     = 5 guns

**19.** What is the gun ownership rate in the United States?

**20.** How much larger is the gun ownership rate in the United States than in Canada?

**21.** What is the gun ownership rate in Switzerland? in Finland?

**22.** Which country has the lowest gun ownership rate?

**23.** What is the average gun ownership rate for the six countries listed in the pictograph?

**24.** What is the median gun ownership rate for these six countries?

# Understanding Through Discussion and Writing

**25.** Is it possible for a driver to average 20 mph on a 30-mi trip and still receive a ticket for driving 75 mph? Why or why not?   [8.1a]

**26.** You are applying for an entry-level job at a large firm. You can be informed of the mean, median, or mode salary. Which of the three figures would you request? Why?   [8.1a, b, c]

# 8.3 Bar Graphs and Line Graphs

## (a) Reading and Interpreting Bar Graphs

A **bar graph** is convenient for showing comparisons because you can tell at a glance which amount represents the largest or smallest quantity. Since a bar graph is a more abstract form of pictograph, this is true of pictographs as well. However, with bar graphs, a *second scale* is usually included so that a more accurate determination of the amount can be made.

**EXAMPLE 1** *Foreign Visits by Presidents.* Theodore Roosevelt's trip to Panama in 1904 was the first presidential foreign visit. The 42nd and 43rd presidents of the United States made more than twice as many foreign visits while in office than previous presidents. The following horizontal bar graph shows the number of foreign visits by the 34th–43rd presidents.

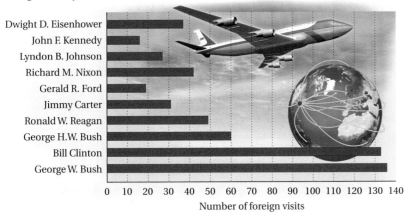

**Foreign Visits by U.S. Presidents While in Office**

Number of foreign visits

SOURCE: State Department, White House

a) Which president in this list made the fewest foreign visits while in office?

b) How many foreign visits did Ronald Reagan make?

c) Which president made approximately 130 foreign visits?

We look at the graph to answer the questions.

a) The shortest bar is for John F. Kennedy. Thus John F. Kennedy made the fewest foreign visits.

b) We move to the right along the bar that represents Ronald Reagan. We can read, fairly accurately, that he made approximately 49 foreign visits.

c) We locate the line representing 130 visits and then go up until we reach a bar that ends close to 130. We can go to the left and read the name of the president, Bill Clinton.

Do Exercises 1–3.

Use the bar graph in Example 1 to answer Margin Exercises 1–3.

1. How many foreign visits did George H. W. Bush make?

2. Which presidents made fewer than 30 foreign visits?

3. Which president made approximately 40 foreign visits?

*Answers*

**1.** 60 foreign visits  **2.** John F. Kennedy, Lyndon B. Johnson, Gerald R. Ford
**3.** Richard M. Nixon

Bar graphs are often drawn vertically, and sometimes a double bar graph is used to make comparisons.

**EXAMPLE 2** *Childhood Cancer.* The following graph indicates the survival rates for various childhood cancers, comparing the rates for 1962 and the present.

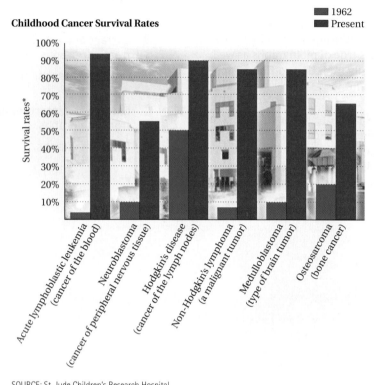

**Childhood Cancer Survival Rates**

■ 1962
■ Present

SOURCE: St. Jude Children's Research Hospital
*Cancer survival of 5 years or more based on national averages over the past 10 years.

**a)** What was the survival rate for osteosarcoma in 1962? in the present?

**b)** For which cancers is the present survival rate more than 85%?

**c)** How many of the cancers had survival rates lower than 50% in 1962? in the present?

**d)** The difference in survival rates from 1962 to the present is 75% for which type of cancer?

We look at the graph to answer the questions.

**a)** We go to the right, across the bottom, to the blue bar (1962) above osteosarcoma. Next, we go to the top of that bar and, from there, back to the left to read 20% on the vertical scale. We follow the same procedure for the red bar (the present) and read 65% on the vertical scale. The survival rate for osteosarcoma in 1962 was 20% and presently is 65%.

**b)** We move up the vertical scale to 85%. From there, we move to the right across all red bars, noting any that are above 85%. We see that the survival rates for acute lymphoblastic leukemia and Hodgkin's disease are greater than 85%.

**c)** We move up the vertical scale to 50%. From there, we move to the right across all blue bars, noting any that are below 50%. We see that five of the six blue bars are below 50%. We also see that all six red bars are above 50%. Thus, in 1962 five of the six cancers had survival rates below 50%. Today, none of the six cancers has a survival rate below 50%.

**d)** Looking at the heights of each set of bars, we see that the survival rate for medulloblastoma has increased from 10% in 1962 to 85% at the present. Thus the difference in survival rates is 75% for medulloblastoma.

Do Exercises 4–6.

Use the bar graph in Example 2 to answer Margin Exercises 4–6.

**4.** For which cancer(s) is the present survival rate between 50% and 70%?

**5.** What is the present survival rate for acute lymphoblastic leukemia?

**6.** What is the difference in survival rates for non-Hodgkin's lymphoma from 1962 to the present?

## b  Drawing Bar Graphs

**EXAMPLE 3** *Population by Age.* Listed below are U.S. population data for selected age groups. Make a vertical bar graph of the data.

| AGE GROUP | PERCENT OF POPULATION |
|---|---|
| 5–17 years | 17% |
| 18 years and older | 76 |
| 10–49 years | 54 |
| 16–64 years | 66 |
| 55 years and older | 25 |
| 65 years and older | 13 |
| 85 years and older | 2 |

SOURCE: U.S. Census Bureau

First, we indicate the age groups in seven equally spaced intervals on the horizontal scale and give the horizontal scale the title "Age category." (See Figure 1 below.)

Next, we scale the vertical axis. To do so, we look over the data and note that it ranges from 2% to 76%. We start the vertical scaling at 0, labeling the marks by 10's from 0 to 80. We give the vertical scale the title "Percent of population."

Finally, we draw vertical bars to show the various percents, as shown in Figure 2. We give the graph the overall title "U.S. Population by Age."

**7. Planetary Moons.** Make a horizontal bar graph to show the number of moons orbiting the various planets.

| PLANET | MOONS |
|---|---|
| Earth | 1 |
| Mars | 2 |
| Jupiter | 63 |
| Saturn | 60 |
| Uranus | 27 |
| Neptune | 13 |

SOURCE: National Aeronautics and Space Administration

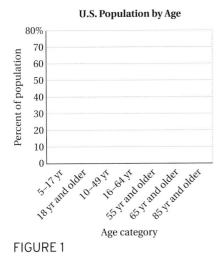

FIGURE 1

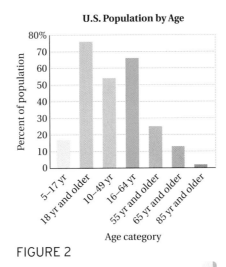

FIGURE 2

Do Exercise 7.

*Answers*

**4.** Neuroblastoma and osteosarcoma
**5.** About 94%   **6.** About 85% − 7% = 78%
**7.**

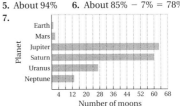

## (c) Reading and Interpreting Line Graphs

**Line graphs** are often used to show a change over time as well as to indicate patterns or trends.

**EXAMPLE 4** *Popcorn.* Americans consume 16 billion quarts of popcorn annually. Indiana ranks second in the United States in the production of popcorn. The following line graph shows the production of popcorn, in millions of pounds per year, in Indiana from 2002 to 2008.

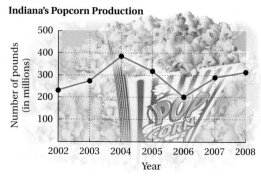

**Indiana's Popcorn Production**

SOURCE: *The Hoosier Farmer*, Spring 2008; Indiana Corn Growers Association

a) For which year was the production of popcorn the highest?

b) Between which years did production increase?

c) For which years was popcorn production about 315 million lb?

We look at the graph to answer the questions.

a) The greatest number of pounds was about 385 million in 2004.

b) Reading the graph from left to right, we see that popcorn production increased from 2002 to 2003, from 2003 to 2004, from 2006 to 2007, and from 2007 to 2008.

c) We look left to right along a line at 315. We see that there are points close to 315 million at 2005 and at 2008.

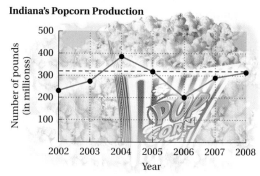

**Indiana's Popcorn Production**

SOURCE: *The Hoosier Farmer*, Spring 2008; Indiana Corn Growers Association

Use the line graph in Example 4 to answer Margin Exercises 8–10.

**8.** For which year was the production of popcorn the lowest?

**9.** Between which years did production decrease?

**10.** For which year was popcorn production about 230 million lb?

Do Exercises 8–10.

*Answers*

**8.** 2006  **9.** 2004–2005 and 2005–2006  **10.** 2002

**EXAMPLE 5** *Monthly Loan Payment.* Suppose you borrow $110,000 at an interest rate of $5\frac{1}{2}\%$ to buy a condominium. The following graph shows the monthly payment required to pay off the loan, depending on the length of the loan.

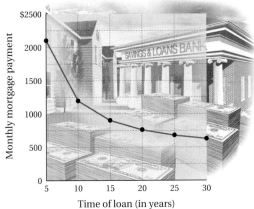

**$110,000 Loan Repayment**

a) Estimate the monthly payment for a loan of 15 yr.

b) What time period corresponds to a monthly payment of about $760?

c) By how much does the monthly payment decrease when the loan period is increased from 10 yr to 20 yr?

We look at the graph to answer the questions.

a) We find the time period labeled "15" on the bottom scale and move up from that point to the line. We then go straight across to the left and find that the monthly payment is about $900.

b) We locate $760 on the vertical axis. Then we move to the right until we come to the line. The point $760 is on the line at the 20-yr time period.

c) The graph shows that the monthly payment for 10 yr is about $1200; for 20 yr, it is about $760. Thus the monthly payment is decreased by $1200 − $760, or $440. (It should be noted that you will pay back $760 · 20 · 12 − $1200 · 10 · 12, or $38,400, more in interest for a 20-yr loan.)

Do Exercises 11–13.

Use the line graph in Example 5 to answer Margin Exercises 11–13.

**11.** Estimate the monthly payment for a loan of 25 yr.

**12.** What time period corresponds to a monthly payment of about $625?

**13.** By how much does the monthly payment decrease when the loan period is increased from 5 yr to 20 yr?

*Answers*
**11.** $675  **12.** 30 yr  **13.** About $1340

## d Drawing Line Graphs

**EXAMPLE 6** *Temperature in Enclosed Vehicle.* The temperature inside an enclosed vehicle increases rapidly with time. Listed in the table below are the inside temperatures of an enclosed vehicle for specified elapsed times when the outside temperature is 80°F. Make a line graph of the data.

| ELAPSED TIME | TEMPERATURE IN ENCLOSED VEHICLE WITH OUTSIDE TEMPERATURE 80°F |
|---|---|
| 10 min | 99° |
| 20 min | 109° |
| 30 min | 114° |
| 40 min | 118° |
| 50 min | 120° |
| 60 min | 123° |

SOURCES: General Motors; Jan Null, Golden Gate Weather Services

First, we indicate the 10-min elapsed time intervals on the horizontal scale and give the horizontal scale the title "Elapsed time (in minutes)." See the figure on the left below. Next, we scale the vertical axis by 10's beginning with 80 to show the number of degrees and give the vertical scale the title "Temperature (in degrees)." The jagged line at the base of the vertical scale indicates that an unnecessary portion of the scale has been omitted. We also give the graph the overall title "Temperature in Enclosed Vehicle with Outside Temperature 80°F."

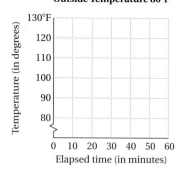

Temperature in Enclosed Vehicle with Outside Temperature 80°F

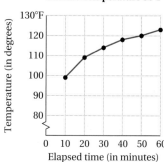

Temperature in Enclosed Vehicle with Outside Temperature 80°F

Next, we mark the temperature at the appropriate level above each elapsed time. (See the figure on the right above.) Then we draw line segments connecting the points. The rapid change in temperature can be observed easily from the graph.

Do Exercise 14.

**14. Touchdown Passes.** Listed below are the numbers of touchdown passes for Peyton Manning of the Indianapolis Colts for the years 2003 to 2008. Make a line graph of the data.

| YEAR (Season) | TOUCHDOWN PASSES |
|---|---|
| 2003 | 29 |
| 2004 | 49 |
| 2005 | 28 |
| 2006 | 31 |
| 2007 | 31 |
| 2008 | 27 |

SOURCE: National Football League

*Answer*

14.

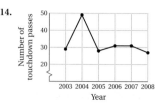

**a**   *Bearded Irises.*   A gardener planted six varieties of bearded iris in a new garden on campus. Students from the horticulture department were assigned to record data on the range of heights for each variety. The following vertical bar graph shows the results of their study. The length of the light green shaded portion and the blossom of each bar illustrates the range of heights. For example, the range of heights for the miniature dwarf bearded iris is 2 in. to 9 in.

**Bearded Irises**

SOURCE: www.irises.org/classification.htm

**1.** Which variety of iris has a minimum height of 17 in.?

**2.** What is the range of heights for the standard dwarf bearded iris?

**3.** Find the average of the maximum heights of all the varieties of irises shown in the graph.

**4.** Which variety of iris has a maximum height of 28 in.?

**5.** What is the range of heights for the border bearded iris?

**6.** Find the median maximum height of all varieties of irises shown in the graph.

**7.** Which variety of iris has the smallest range in heights?

**8.** Which irises have a maximum height less than 16 in.?

**9.** What is the difference between the maximum heights of the tallest iris and the shortest iris?

**10.** Which irises have a range in heights less than 10 in.?

*Chocolate Desserts.* The following horizontal bar graph shows the average caloric content of various kinds of chocolate desserts. Use the bar graph for Exercises 11–18.

11. Estimate how many calories there are in 1 cup of hot cocoa with skim milk.

12. Estimate how many calories there are in a 2-oz candy bar with peanuts.

13. Which dessert has the highest caloric content?

14. Which dessert has the lowest caloric content?

15. Which dessert contains about 460 calories?

16. Which desserts contain about 300 calories?

17. How many more calories are there in 1 cup of hot cocoa made with whole milk than in 1 cup of hot cocoa made with skim milk?

18. If Emily drinks a 4-cup chocolate milkshake, how many calories does she consume?

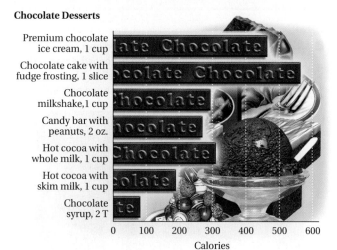

**Chocolate Desserts**

Premium chocolate ice cream, 1 cup
Chocolate cake with fudge frosting, 1 slice
Chocolate milkshake,1 cup
Candy bar with peanuts, 2 oz.
Hot cocoa with whole milk, 1 cup
Hot cocoa with skim milk, 1 cup
Chocolate syrup, 2 T

0   100   200   300   400   500   600
Calories

*Bachelor's Degrees.* The graph shown at right provides data on the number of bachelor's degrees conferred on men and on women in selected years. Use the bar graph for Exercises 19–22.

19. In which years were more bachelor's degrees conferred on men than on women?

20. How many more bachelor's degrees were conferred on women in 2007 than in 1970?

21. How many more bachelor's degrees were conferred on women than on men in 2000?

22. In which years were the numbers of bachelor's degrees conferred on men and on women each greater than 500,000?

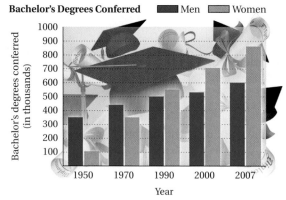

**Bachelor's Degrees Conferred**   ■ Men   ▨ Women

Bachelor's degrees conferred (in thousands)

1000
900
800
700
600
500
400
300
200
100

1950   1970   1990   2000   2007
Year

SOURCE: National Center for Education Statistics, U.S. Department of Education

**23.** *Cost of Adult Day Care.* The following table lists the average daily cost of adult day care in six cities in the United States, along with the national average. Make a horizontal bar graph to illustrate the data.

| CITY | AVERAGE DAILY COST OF ADULT DAY CARE |
|---|---|
| Chicago | $50 |
| Denver | 61 |
| New York City | 107 |
| Pittsburgh | 53 |
| San Antonio | 32 |
| San Diego | 77 |
| U.S. Average | 64 |

SOURCES: www.consumerhealthratings.com; MetLife Mature Market Institute survey, 2008

Use the data and the bar graph in Exercise 23 to do Exercises 24–27.

**24.** Which city has the highest average daily cost of adult day care?

**25.** In which cities is the average daily cost of adult day care less than $75?

**26.** What is the average daily cost of adult day care in the cities listed in the table? How does it compare with the national average?

**27.** How much higher is the average daily cost of adult day care in New York City than the national average?

**28.** *Commuting Time.* The following table lists the average commuting time in six metropolitan areas with more than 1 million people. Make a vertical bar graph to illustrate the data.

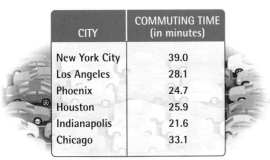

| CITY | COMMUTING TIME (in minutes) |
|---|---|
| New York City | 39.0 |
| Los Angeles | 28.1 |
| Phoenix | 24.7 |
| Houston | 25.9 |
| Indianapolis | 21.6 |
| Chicago | 33.1 |

SOURCE: U.S. Census Bureau

Use the data and the bar graph in Exercise 28 to do Exercises 29–32.

**29.** Which city has the longest commuting time?

**30.** Which city has the shortest commuting time?

**31.** What was the median commuting time for all six cities?

**32.** What was the average commuting time for the six cities?

*International Tourists to Beijing.* The line graph below illustrates the number of international tourists, in millions, who visited Beijing in recent years. Use the graph for Exercises 33–36.

**International Tourists Visiting Beijing**

International tourists (in millions) vs Year (2003–2008)

SOURCE: *Beijing Statistics Year Book*

**33.** Estimate how many international tourists visited Beijing in 2004; in 2007.

**34.** How many more international tourists visited Beijing in 2008 than in 2003?

**35.** Between which years did the number of international tourists decrease?

**36.** Between which two years was the greatest increase in the number of international tourists?

*Prime Time for Bank Crimes.* In 2006, 7272 bank crimes occurred in the United States. The line graph below shows the number of those incidents that occurred by time of day. Use the graph for Exercises 37–40.

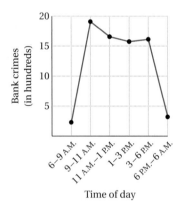

**37.** In which time interval did the most bank crimes occur?

**38.** In which time interval did the smallest number of bank crimes occur?

**39.** Estimate how many more bank crimes occurred between 9 A.M. and 11 A.M. than between 3 P.M. and 6 P.M.

**40.** Estimate how many bank crimes occurred between 11 A.M. and 1 P.M.

**41.** *Longevity Beyond Age 65.* The data in the table below tell us the number of years that a 65-year-old male in the given year can expect to live beyond age 65. Draw a line graph using the horizontal axis to scale "Year."

| YEAR | AVERAGE NUMBER OF YEARS MEN ARE ESTIMATED TO LIVE BEYOND AGE 65 |
|------|------|
| 1980 | 14 |
| 1990 | 15 |
| 2000 | 15.9 |
| 2010 | 16.4 |
| 2020 | 16.9 |
| 2030 | 17.5 |

SOURCE: 2000 Social Security Report

**42.** What was the percent of increase in longevity (years beyond 65) between 1980 and 2000?

**43.** What is the expected percent of increase in longevity between 1980 and 2030?

**44.** What is the expected percent of increase in longevity between 2020 and 2030?

**45.** What is the expected percent of increase in longevity between 2000 and 2030?

## Skill Maintenance

In each of Exercises 46–53, fill in the blank with the correct term from the given list. Some of the choices may not be used.

**46.** The set of numbers 1, 2, 3, 4, 5, ... is called the set of _____ numbers. [1.1b]

**47.** To find the average of a set of numbers, _____ the numbers and then _____ by the number of items of data. [8.1a]

**48.** The _____ interest $I$ on principal $P$, invested for $t$ years at interest rate $r$, is given by $I = P \cdot r \cdot t$. [7.7a]

**49.** If an item has a regular price of $100 and is on sale for 25% off, $100 is called the _____, 25% the _____, 25% of $100, or $25, the _____, and $100 − $25, or $75, the _____. [7.6c]

**50.** When interest is paid on interest, it is called _____ interest. [7.7b]

**51.** The decimal $0.\overline{1518}$ is an example of a(n) _____ decimal. [5.5a]

**52.** The statement $a(b + c) = ab + ac$ illustrates the _____ law. [1.4a]

**53.** The statement $x + t = t + x$ illustrates the _____ law of addition. [1.2a]

add
subtract
multiply
divide
simple
compound
discount
rate of discount
sale price
marked price
repeating
terminating
natural
whole
distributive
associative
commutative

# 8.4

# Circle Graphs

## OBJECTIVES

**a** Extract and interpret data from circle graphs.

**b** Draw circle graphs.

We often use **circle graphs**, also called **pie charts**, to show the percent of a quantity in each of several categories. Circle graphs can also be used very effectively to show visually the *ratio* of one category to another. In either case, it is quite often necessary to use mathematics to find the actual amounts represented for each specific category.

## **a** Reading and Interpreting Circle Graphs

**EXAMPLE 1** *Children in Foster Care.* There are over one half million children in publicly supported foster care in the United States. This circle graph shows the percentage of foster children by age.

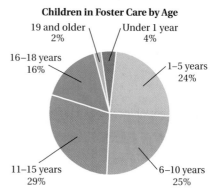

**Children in Foster Care by Age**

- 19 and older 2%
- Under 1 year 4%
- 16–18 years 16%
- 1–5 years 24%
- 11–15 years 29%
- 6–10 years 25%

SOURCE: Adoption and Foster Care Analysis and Reporting System (AFCARS)

a) Which age group has the most children?

b) What percent of the children are 6–10 years old?

c) Which age group includes 4% of the total number of children?

d) In 2008, there were 580,000 children in foster care. How many were 16–18 years old?

e) What percent of children in foster care are less than 11 years old?

We look at the sections of the graph to find the answers.

a) The largest section (or sector) of the graph, 29%, represents 11–15 years old.

b) We see that those 6–10 years old are 25% of all children in foster care.

c) Those under 1 year old account for 4% of the total number of children in foster care.

d) The section of the graph representing 16–18 years old is 16%; 16% of 580,000 is 92,800. Thus, 92,800 children in foster care in 2008 were 16–18 years old.

e) We add the percents corresponding to 6–10 years old, 1–5 years old, and under 1 year. We have

$$25\% + 24\% + 4\% = 53\%.$$

Do Margin Exercises 1–3.

Use the circle graph in Example 1 to answer Margin Exercises 1–3.

**1.** Which age group includes 24% of all children in foster care?

**2.** What percent of children in foster care are over 15 years old?

**3.** In 2008, there were 580,000 children in foster care. How many were 11–15 years old?

*Answers*

*Skill to Review:*
**1.** 7%   **2.** 81%

*Margin Exercises:*
**1.** 1–5 years old   **2.** 18%
**3.** 168,200 children

## b Drawing Circle Graphs

To draw a circle graph, or pie chart, like the one in Example 1, think of a pie cut into 100 equally sized pieces. We then shade in a wedge equal in size to 4 of these pieces to represent 4% under 1 year. We shade a wedge equal in size to 24 of these pieces to represent 24% for 1–5 years, and so on.

**EXAMPLE 2** *Fruit Juice Sales.* The percents of various kinds of fruit juice sold are given in the list at right. Use this information to draw a circle graph.

**Source:** Beverage Marketing Corporation

| | |
|---|---|
| Apple: | 14% |
| Orange: | 56% |
| Blends: | 6% |
| Grape: | 5% |
| Grapefruit: | 4% |
| Prune: | 1% |
| Other: | 14% |

Using a circle with 100 equally spaced tick marks, we start with the 14% given for apple juice. We draw a line from the center to any tick mark. Then we count off 14 ticks and draw another line. We shade the wedge with a color—in this case, pink—and label the wedge as shown in the figure on the left below.

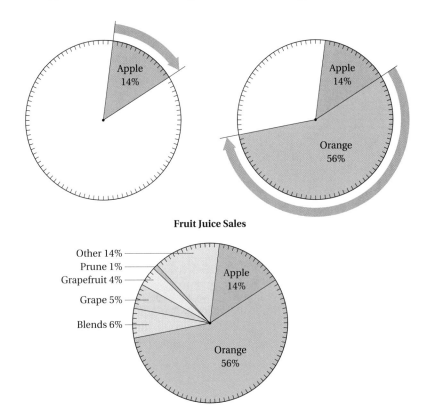

**Fruit Juice Sales**

To shade a wedge for orange juice, at 56%, we start at one side of the apple wedge, count off 56 ticks, and draw another line. We shade the wedge with a different color—in this case, orange—and label the wedge as shown in the figure on the right above. Continuing in this manner and choosing different colors, we obtain the final graph shown above. Finally, we give the graph the overall title "Fruit Juice Sales."

Do Exercise 4.

4. **Lengths of Engagement of Married Couples.** The data below relate the percent of married couples who were engaged for certain periods of time before marriage. Use this information to draw a circle graph.

| ENGAGEMENT PERIOD | PERCENT |
|---|---|
| Less than 1 year | 24% |
| 1–2 years | 21 |
| More than 2 years | 35 |
| Never engaged | 20 |

SOURCE: Bruskin Goldring Research

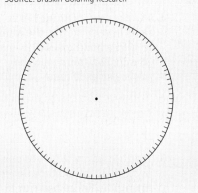

*Answer*

4.

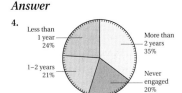

# Translating for Success

1. *Vacation Miles.* The Saenz family drove their new van 13,640.8 mi in the first year. Of this total, 2018.2 mi were driven while on vacation. How many nonvacation miles did they drive?

2. *Rail Miles.* Of the recent $15\frac{1}{2}$ million passenger miles on a rail passenger line, 80% were transportation-to-work miles. How many rail miles, in millions, were to and from work?

3. *Sales Tax Rate.* The sales tax on the purchase of 10 bath towels that cost $129.50 is $8.42. What is the sales tax rate?

4. *Water Level.* During heavy rains in early spring, the water level in a pond rose 0.5 in. every 35 min. How much did the water rise in 90 min?

5. *Marathon Training.* At one point in his daily training routine for a marathon, Rocco had run $15\frac{1}{2}$ mi. This was 80% of the distance he intended to run that day. How far did Rocco plan to run?

The goal of these matching questions is to practice step (2), *Translate*, of the five-step problem-solving process. Translate each problem to an equation and select a correct translation from equations A–O.

**A.** $8.42 \cdot x = 129.50$

**B.** $x = 80\% \cdot 15\frac{1}{2}$

**C.** $x = \dfrac{84 - 68}{84}$

**D.** $2018.2 + x = 13{,}640.8$

**E.** $\dfrac{5}{100} = \dfrac{x}{3875}$

**F.** $2018.2 = x \cdot 13{,}640.8$

**G.** $4\frac{1}{6} \cdot 73 = x$

**H.** $\dfrac{x}{5} = \dfrac{100}{3875}$

**I.** $15\frac{1}{2} = 80\% \cdot x$

**J.** $8.42 = x \cdot 129.50$

**K.** $\dfrac{0.5}{35} = \dfrac{x}{90}$

**L.** $x \cdot 4\frac{1}{6} = 73$

**M.** $x = \dfrac{84 - 68}{68}$

**N.** $x = 8.42\% \cdot 129.50$

**O.** $0.5 \times 35 = 90 \cdot x$

*Answers on page A-13*

6. *Vacation Miles.* The Ning family drove 2018.2 mi on their summer vacation. If they put a total of 13,640.8 mi on their new van during that year, what percent were vacation miles?

7. *Sales Tax.* The sales tax rate is 8.42%. Salena purchased 10 pillows at $12.95 each. How much tax was charged on this purchase?

8. *Charity Donations.* Rachel donated $5 to her favorite charity for each $100 she earned. One month, she earned $3875. How much did she donate that month?

9. *Tuxedos.* Emil Tailoring Company purchased 73 yd of fabric for a new line of tuxedos. How many tuxedos can be produced if it takes $4\frac{1}{6}$ yd of fabric for each tuxedo?

10. *Percent of Increase.* In a calculus-based physics course, Mime got 68% on the first exam and 84% on the second. What was the percent of increase in her score?

**a** *Foreign Students.* The circle graph below shows the foreign countries sending the most students to the United States to attend colleges and universities. Use this graph for Exercises 1–6.

1. What percent of foreign students are from South Korea?

2. Together, what percent of foreign students are from China and Taiwan?

3. In 2008, there were approximately 625,000 foreign students studying at colleges and universities in the United States. According to the data in the graph, how many were from India?

4. In 2008, there were approximately 625,000 foreign students studying in the United States. How many were from Nepal?

5. Which country accounted for 6% of the foreign students?

6. Which country accounted for 11% of the foreign students?

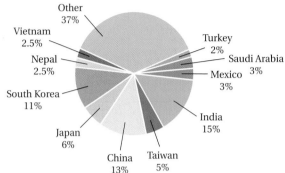

**Foreign Students Enrolled in U.S. Colleges and Universities, 2008**

SOURCE: Institute of International Education

*Outdoor Recreation.* The circle graph below shows what people look forward to the most when participating in outdoor recreational activities. The data shown in the graph are from a survey of 1027 adults. Use the graph for Exercises 7–10.

7. What percent of those who participate in outdoor activities look forward the most to forgetting about work?

8. What percent of those who participate in outdoor activities look forward the most to solitude?

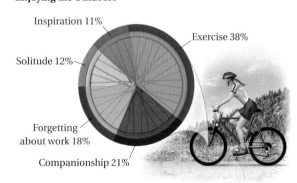

**Enjoying the Outdoors**

SOURCES: *Open Air Magazine*; International Communications Research

9. How many participants in the survey listed exercise as the main reason they look forward to outdoor recreation?

10. How many participants in the survey listed inspiration as the main reason they look forward to outdoor recreation?

**b** In Exercises 11–14, use the given information to complete a circle graph. Note that each circle is divided into 100 sections.

**11.** *College Loans.* The table below lists where families first look for college loans.

| WHERE | PERCENT |
|---|---|
| College | 57% |
| Local bank | 18 |
| Internet | 15 |
| Family/friends | 5 |
| Other | 5 |

SOURCES: Survey.com; TuitionBids.com

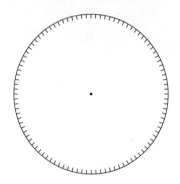

**12.** *Population of Continents.* The table below lists the percentage of the world population on each continent.

| CONTINENT | PERCENT |
|---|---|
| Africa | 14% |
| Asia | 61 |
| Europe | 11 |
| Oceania, includes Australia | 1 |
| North America | 8 |
| South America | 5 |

SOURCE: Population Division/International Programs Center, U.S. Census Bureau

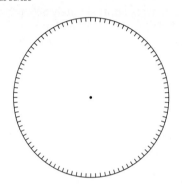

**13.** *Addiction Help.* About 5 million people in the United States participated in self-help programs to fight addiction in 2008. The table below lists the participants by age.

| AGE | PERCENT |
|---|---|
| 12–17 | 5% |
| 18–25 | 15 |
| 26–49 | 57 |
| 50 and older | 23 |

SOURCE: Substance Abuse and Mental Health Services Administration

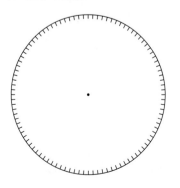

**14.** *Causes of Spinal Cord Injuries.* The table below lists the causes of spinal cord injury.

| CAUSES | PERCENT |
|---|---|
| Motor vehicle accidents | 44% |
| Acts of violence | 24 |
| Falls | 22 |
| Sports | 8 |
| Other | 2 |

SOURCE: National Spinal Cord Injury Association

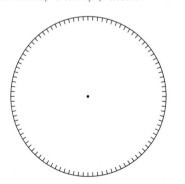

# Summary and Review

## Key Terms

statistic, p. 465
average, p. 465
arithmetic mean, p. 465
mean, p. 465
grade point average (GPA), p. 467

median, p. 468
mode, p. 469
table, p. 474
pictograph, p. 476

bar graph, p. 483
line graph, p. 486
circle graph, p. 494
pie chart, p. 494

## Concept Reinforcement

Determine whether each statement is true or false.

_____ **1.** To find the average of a set of numbers, add the numbers and then multiply by the number of items of data.   [8.1a]

_____ **2.** If each number in a set of data occurs the same number of times, there is no mode.   [8.1c]

_____ **3.** If there is an odd number of items in a set of data, the middle number is the median.   [8.1b]

## Important Concepts

**Objectives 8.la, b, c**   Find the average, the median, and the mode of a set of numbers.

**Example**   Find the average, the median, and the mode of this set of numbers:

2.6,   3.5,   61.8,   10.4,   3.5,   21.6,   10.4,   3.5.

*Average:* We add the numbers and divide by the number of data items:

$$\frac{2.6 + 3.5 + 61.8 + 10.4 + 3.5 + 21.6 + 10.4 + 3.5}{8}$$

$$= \frac{117.3}{8} = 14.6625.$$

The *average* is 14.6625.

*Median:* We first rearrange the numbers from the smallest to the largest. There is an even number of numbers. We look for the middle two, which are 3.5 and 10.4. The median is halfway between 3.5 and 10.4. (If there is an odd number of numbers, the median is the middle number.) Thus we have

2.6,   3.5,   3.5,   3.5,   10.4,   10.4,   21.6,   61.8.

The average of the middle numbers is

$$\frac{3.5 + 10.4}{2} = \frac{13.9}{2} = 6.95.$$

The *median* is 6.95.

*Mode:* The number that occurs most often is 3.5, so it is the mode.

The *mode* is 3.5.

**Practice Exercise**

**1.** Find the average, the median, and the mode of this set of numbers:

8,   13,   1,   4,   8,   7,   15.

**Objective 8.2a**  Extract and interpret data from tables.

**Example**  The table below lists comparative information for oatmeal sold by six companies.

| PRODUCT | PER PACKET (instant) OR SERVING (longer-cooking) | | | | | |
|---|---|---|---|---|---|---|
| | Cost | Calories | Fat (g) | Fiber (g) | Sugars (g) | Sodium (mg) |
| Quaker Quick-1 Minute | 0.19 | 150 | 3.0 | 4 | 1 | 0 |
| Market Pantry Maple & Brown Sugar | 0.17 | 160 | 2.0 | 3 | 13 | 240 |
| 365 Organic Maple Spice | 0.42 | 150 | 1.5 | 3 | 13 | 200 |
| Kashi Heart to Heart Golden Brown Maple | 0.44 | 160 | 2.0 | 5 | 12 | 100 |
| McCann's Irish Maple & Brown Sugar | 0.45 | 160 | 2.0 | 3 | 13 | 240 |
| Quaker Organic Maple & Brown Sugar | 0.54 | 150 | 2.0 | 3 | 12 | 95 |
| Nature's Path Organic Maple Nut | 0.47 | 200 | 4.0 | 4 | 12 | 105 |

SOURCE: *Consumer Reports*, November 2008

**a)** Which oatmeal has the greatest number of calories?
**b)** How much sodium is in a serving of Market Pantry Maple & Brown Sugar?

An examination of the table will give the answers.

**a)** We look down the column headed "Calories" and find the largest number. That number is 200. Then we look left across that row to find the name of the oatmeal, Nature's Path Organic Maple Nut.
**b)** We look down the column of products and find Market Pantry. Then we move right across that row to the column headed "Sodium" and find the amount of sodium: 240 mg.

---

**Objective 8.3a**  Extract and interpret data from bar graphs.

**Example**  The horizontal bar graph below shows the building costs of selected stadiums. When comparing the costs, note the year in which each stadium was built.

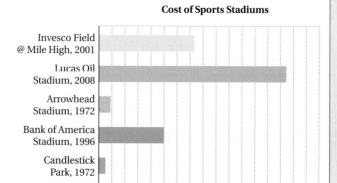

SOURCE: National Football League

**a)** Estimate how much more Lucas Oil Stadium cost than Invesco Field @ Mile High.
**b)** Which stadium cost approximately $250 million to build?

We look at the graph to answer the questions.

**a)** We move to the right along the bars for Lucas Oil Stadium and Invesco Field @ Mile High and move down to the horizontal scale to estimate the costs: about $720 million for Lucas Oil Stadium and $360 million for Invesco Field @ Mile High. The difference in cost is about $720 million − $360 million, or $360 million. Thus Lucas Oil cost about $360 million more than Invesco Field @ Mile High.
**b)** We locate the line representing $250 million and then go up until we reach a bar that ends close to $250 million. Then we go to the left and read the name of the stadium, Bank of America Stadium.

---

**Practice Exercises**

Use the table in the example shown above for Exercises 2 and 3.

2. Which oatmeal has the greatest cost per serving? What is that cost?

3. How many grams of sugar are in the Kashi oatmeal?

**Practice Exercises**

Use the bar graph in the example shown above for Exercises 4 and 5.

4. Which stadiums cost less than $100 million?

5. Estimate how much more Invesco Field @ Mile High cost than Bank of America Stadium.

**Objective 8.4a**  Extract and interpret data from circle graphs.

**Example**  The circle graph below shows the percentages of the population of the United States in various age groups.

**Population of United States by Age**

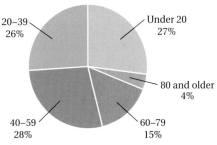

20–39
26%

Under 20
27%

80 and older
4%

40–59
28%

60–79
15%

a) What percent of the population is 40–59 years old?
b) How much more of the population is in the 20–39 years age group than in the 60–79 years age group?

The graph gives us the answers.

a) The graph shows us that the segment of the population that is 40–59 years old is 28% of the total population.
b) From the graph, we see that 26% of the population is 20–39 years old and only 15% is 60–79 years old. We subtract: 26% − 15% = 11%. Thus 11% more of the population is in the 20–39 years age group than in the 60–79 years age group.

**Practice Exercises**

Use the circle graph at left to answer Exercises 6 and 7.

**6.** Which age group has the fewest people?

**7.** What percent of the population is under 20 years old?

# Review Exercises

Find the average.  [8.1a]

**1.** 26, 34, 43, 51

**2.** 11, 14, 17, 18, 7

**3.** 0.2, 1.7, 1.9, 2.4

**4.** 700, 2700, 3000, 900, 1900

**5.** $2, $14, $17, $17, $21, $29

**6.** 20, 190, 280, 470, 470, 500

**7.** To get an A in math, Naomi must score an average of 90 on four tests. Her scores on the first three tests were 94, 78, and 92. What is the lowest score she can make on the last test and still get an A?  [8.1a]

**8.** *Gas Mileage.*  A 2009 Pontiac Solstice GXP gets 532 mi of highway driving on 19 gal of gasoline. What is the gas mileage?  [8.1a]
Source: *Car and Driver,* December 2008

**9.** *Grade Point Average.*  Find the grade point average for one semester given the following grades. Assume the grade point values are 4.0 for A, 3.0 for B, and so on. Round to the nearest tenth.  [8.1a]

| COURSE | GRADE | NUMBER OF CREDIT HOURS IN COURSE |
|---|---|---|
| Math | A | 5 |
| English | B | 3 |
| Computer Science | C | 4 |
| Spanish | B | 3 |
| College Skills | B | 1 |

Find the median.  [8.1b]

**10.** 26, 34, 43, 51

**11.** 7, 11, 14, 17, 18

**12.** 0.2, 1.7, 1.9, 2.4

**13.** 700, 900, 1900, 2700, 3000

**14.** $2, $17, $21, $29, $14, $17

**15.** 470, 20, 190, 280, 470, 500

**16.** One summer, a student worked part time as a veterinary assistant. She earned the following weekly amounts over a six-week period: $360, $192, $240, $216, $420, and $132. What was the average amount earned per week? the median?  [8.1a, b]

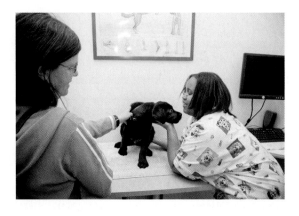

Find the mode.  [8.1c]

**17.** 26, 34, 43, 26, 51

**18.** 17, 7, 11, 11, 14, 17, 18

**19.** 0.2, 0.2, 1.7, 1.9, 2.4, 0.2

**20.** 700, 700, 800, 2700, 800

**21.** $14, $17, $21, $29, $17, $2

**22.** 20, 20, 20, 20, 20, 500

**23.** *Battery Testing.*  An experiment is performed to compare battery quality. Two kinds of battery were tested to see how long, in hours, they kept a hand radio running. On the basis of this test, which battery is better?  [8.1d]

| BATTERY A: TIMES (in hours) | | | BATTERY B: TIMES (in hours) | | |
|---|---|---|---|---|---|
| 38.9 | 39.3 | 40.4 | 39.3 | 38.6 | 38.8 |
| 53.1 | 41.7 | 38.0 | 37.4 | 47.6 | 37.9 |
| 36.8 | 47.7 | 48.1 | 46.9 | 37.8 | 38.1 |
| 38.2 | 46.9 | 47.4 | 47.9 | 50.1 | 38.2 |

*UPS Mailing Costs.*  The table below lists the charges for three types of UPS delivery service of packages of various weights from zip code 46143 to zip code 60614. Use this table for Exercises 24–26.  [8.2a]

| UPS PACKAGE | UPS GROUND | UPS NEXT DAY SAVER DELIVERY BY END OF DAY | UPS NEXT DAY AIR DELIVERY BY 10:30 A.M. |
|---|---|---|---|
| 1 lb | $ 9.13 | $21.29 | $24.30 |
| 2 | 9.25 | 22.90 | 25.21 |
| 3 | 9.32 | 24.03 | 28.06 |
| 4 | 9.52 | 25.75 | 30.05 |
| 5 | 9.88 | 26.61 | 30.37 |
| 6 | 10.11 | 27.79 | 32.79 |
| 7 | 10.52 | 28.17 | 34.24 |
| 8 | 10.85 | 29.51 | 34.78 |
| 9 | 11.10 | 30.91 | 35.48 |
| 10 | 11.41 | 31.77 | 35.53 |

SOURCE: United Parcel Service

**24.** Find the cost of a 5-lb UPS Next Day Air delivery.

**25.** Find the cost of an 8-lb UPS Ground delivery.

**26.** How much would you save by sending the package in Exercise 24 by UPS Next Day Saver delivery?

*PGA Champions by Age.* The pictograph below shows the number of Professional Golf Association (PGA) champions by age. Use this graph for Exercises 27–29.  [8.2b]

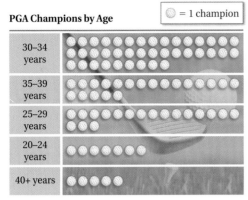

**PGA Champions by Age**
= 1 champion

SOURCE: Professional Golf Association

**27.** How many PGA champions were 25–29 years old?

**28.** Of the age categories listed, which one has the most champions?

**29.** How many more champions were 30–34 years old than 35–39 years old?

*Melting Snow Runoff.* The amount of snow that accumulates during the winter in California is vital to the state's water supply. The bar graph below shows the state's runoff as a percentage of normal for 2004–2008. Use this graph for Exercises 30–33.  [8.3a]

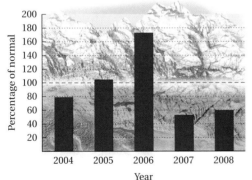

**California's Snow Runoff**

SOURCE: California Department of Water Resources

**30.** Which years had less than normal runoff?

**31.** What was the percentage of runoff in 2008?

**32.** What was the difference in the runoffs for 2005 and 2007?

**33.** What year had over 170% runoff?

*Masters Tournament Scores.* The line graph below shows the total scores that Tiger Woods shot in the Masters Tournaments from 2000 to 2008. Use this graph for Exercises 34–40.

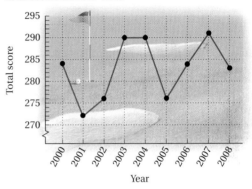

**Tiger Woods' Masters Tournament, Total Scores for 2000–2008**

SOURCE: Professional Golf Association

**34.** In which year did Tiger Woods shoot his lowest score? [8.3c]

**35.** In which year did Tiger Woods shoot his highest score? [8.3c]

**36.** In which years did Tiger Woods score less than 280? [8.3c]

**37.** In which years was Tiger Woods' total score for the Masters 276?  [8.3c]

**38.** How much higher was Tiger Woods' highest total score than his lowest total score for the tournaments from 2000 to 2008?  [8.3c]

**39.** What is the median score of the total scores for 2000–2008?  [8.1b], [8.3c]

**40.** What is the average of Tiger Woods' scores for the years shown in the graph?  [8.1a], [8.3c]

*Armed Forces.* The circle graph below shows the percentage of defense personnel for each branch of the U.S. Armed Forces. Use this graph for Exercises 41–44.   [8.4a]

**Total Armed Forces**

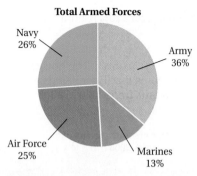

SOURCE: U.S. Department of Defense

**41.** What percent of the Armed Forces is the Army?

**42.** Which branch represents 13% of the Armed Forces?

**43.** In 2008, there were approximately 1,400,000 members of the Armed Forces. How many were in the Air Force?

**44.** The Army represents how much more of the total Armed Forces than the Navy does?

**45.** Find the mode(s) of this set of data.

6, 9, 6, 8, 8, 5, 10, 5, 9, 10   [8.1c]

**A.** 8
**C.** 9
**B.** 5, 6, 8, 9, 10
**D.** No mode exists.

**46.** What is the average of this set of data?

$$\frac{1}{2}, \quad \frac{1}{3}, \quad \frac{1}{4}, \quad \frac{1}{5} \quad \text{[8.1a]}$$

**A.** $\dfrac{77}{240}$    **B.** $\dfrac{1}{3}$

**C.** $\dfrac{7}{24}$    **D.** $\dfrac{1}{4}$

*First-Class Postage.* The table below lists the cost of first-class postage in various years. Use the table for Exercises 47 and 48.

| YEAR | FIRST-CLASS POSTAGE |
|------|---------------------|
| 1995 | 32¢ |
| 1999 | 33 |
| 2001 | 34 |
| 2002 | 37 |
| 2006 | 39 |
| 2007 | 41 |
| 2008 | 42 |
| 2009 | 44 |

SOURCE: U.S. Postal Service

**47.** Make a vertical bar graph of the data.   [8.3b]

**48.** Make a line graph of the data.   [8.3d]

**49.** Construct a circle graph for the PGA Champions by Age data given in the pictograph for Exercises 27–29: 20–24 years, 8%; 25–29 years, 20%; 30–34 years, 44%; 35–39 years, 22%; 40+ years, 6%.   [8.4b]

## Synthesis

**50.** The ordered set of data 298, 301, 305, *a*, 323, *b*, 390 has a median of 316 and an average of 326. Find *a* and *b*.   [8.1a, b]

# Understanding Through Discussion and Writing

**1.** Find a real-world situation that fits this equation:

$$T = \frac{20{,}500 + 22{,}800 + 23{,}400 + 26{,}000}{4}. \quad \text{[8.1a]}$$

**2.** Can bar graphs always, sometimes, or never be converted to line graphs? Why?   [8.3b, d]

**3.** Discuss the advantages of being able to read a circle graph.   [8.4a]

**4.** Compare bar graphs and line graphs. Discuss why you might use one rather than the other to graph a particular set of data.   [8.3b, d]

**5.** Compare and contrast averages, medians, and modes. Discuss why you might use one over the others to analyze a set of data.   [8.1a, b, c]

**6.** Compare circle graphs to bar graphs.   [8.3a], [8.4a]

# Test

**For Extra Help**

CHAPTER
Test Prep
VIDEOS

Step-by-step test solutions are found on the Chapter Test Prep Videos available via the Video Resources on DVD, in *MyMathLab* , and on You[Tube] (search "BittingerBasicMathEl" and click on "Channels").

Find the average.

**1.** 45, 49, 52, 52

**2.** 1, 1, 3, 5, 3

**3.** 3, 17, 17, 18, 18, 20

Find the median and the mode.

**4.** 45, 49, 52, 53

**5.** 1, 1, 3, 5, 3

**6.** 3, 17, 17, 18, 18, 20

**7.** *Gas Mileage.* A 2009 Honda Fit gets 462 mi of highway driving on 14 gal of gasoline. What is the gas mileage?
**Source:** *Car and Driver*, December 2008

**8.** *Grades.* To get a C in chemistry, Ted must score an average of 70 on four tests. His scores on the first three tests were 68, 71, and 65. What is the lowest score he can make on the last test and still get a C?

**9.** *Grade Point Average.* Find the grade point average for one semester given the following grades. Assume the grade point values are 4.0 for A, 3.0 for B, and so on. Round to the nearest tenth.

| COURSE | GRADE | NUMBER OF CREDIT HOURS IN COURSE |
|--------|-------|----------------------------------|
| Introductory Algebra | B | 3 |
| English | A | 3 |
| Business | C | 4 |
| Spanish | B | 3 |
| Typing | B | 2 |

**10.** *Chocolate Bars.* An experiment was performed to compare the quality of new Swiss chocolate bars being introduced in the United States. People were asked to taste the candies and rate them on a scale of 1 to 10, with 10 being the best. On the basis of this test, which chocolate bar is better?

| BAR A: SWISS PECAN | | | BAR B: SWISS HAZELNUT | | |
|---|---|---|---|---|---|
| 9 | 10 | 8 | 10 | 6 | 8 |
| 10 | 9 | 7 | 9 | 10 | 10 |
| 6 | 9 | 10 | 8 | 7 | 6 |
| 7 | 8 | 8 | 9 | 10 | 8 |

*Desirable Body Weights.* The following tables list the desirable body weights for men and women over age 25. Use the tables for Exercises 11–14.

| DESIRABLE WEIGHT OF MEN | | | |
|---|---|---|---|
| HEIGHT | SMALL FRAME (in pounds) | MEDIUM FRAME (in pounds) | LARGE FRAME (in pounds) |
| 5 ft 7 in. | 138 | 152 | 166 |
| 5 ft 9 in. | 146 | 160 | 174 |
| 5 ft 11 in. | 154 | 169 | 184 |
| 6 ft 1 in. | 163 | 179 | 194 |
| 6 ft 3 in. | 172 | 188 | 204 |

| DESIRABLE WEIGHT OF WOMEN | | | |
|---|---|---|---|
| HEIGHT | SMALL FRAME (in pounds) | MEDIUM FRAME (in pounds) | LARGE FRAME (in pounds) |
| 5 ft 1 in. | 105 | 113 | 122 |
| 5 ft 3 in. | 111 | 120 | 130 |
| 5 ft 5 in. | 118 | 128 | 139 |
| 5 ft 7 in. | 126 | 137 | 147 |
| 5 ft 9 in. | 134 | 144 | 155 |

SOURCE: U.S. Department of Agriculture

**11.** What is the desirable weight for a 6 ft 1 in. man with a medium frame?

**12.** What size woman has a desirable weight of 120 lb?

**13.** How much more should a 5 ft 3 in. woman with a medium frame weigh than one with a small frame?

**14.** How much more should a 6 ft 3 in. man with a large frame weigh than one with a small frame?

*Waste Generated.* The number of pounds of waste generated per person per year varies greatly among countries around the world. In the pictograph at right, each symbol represents approximately 100 lb of waste. Use the pictograph for Exercises 15–18.

**15.** In which country does each person generate 1300 lb of waste per year?

**16.** In which countries does each person generate more than 1500 lb of waste per year?

**17.** How many pounds of waste per person per year are generated in Canada?

**18.** How many more pounds of waste per person per year are generated in the United States than in Mexico?

**Amount of Waste Generated (per person per year)**

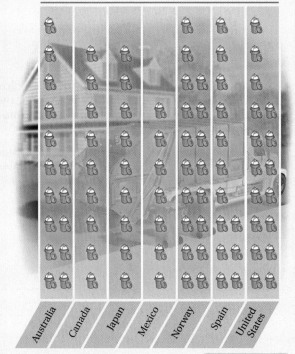

SOURCE: OECD, Key Environmental Indicators 2008

 = 100 pounds

**19.** *Animal Speeds.* The following table lists maximum speeds of movement for various animals, in miles per hour, compared to the speed of the fastest human. Make a vertical bar graph of the data.

| ANIMAL | SPEED (in miles per hour) |
|---|---|
| Human | 28 |
| Chicken | 9 |
| Elephant | 25 |
| Cheetah | 70 |
| Zebra | 40 |
| Lion | 50 |
| Pronghorn antelope | 61 |
| Elk | 45 |

SOURCE: *Natural History Magazine*, © American Museum of Natural History

Refer to the table and the graph in Exercise 19 for Exercises 20–23.

**20.** By how much does the fastest speed exceed the slowest speed?

**21.** Does a human have a chance of outrunning a zebra? Explain.

**22.** Find the average of all the speeds.

**23.** Find the median of all the speeds.

*Food Dollars Spent Away from Home.* The line graph below shows the percent of food dollars spent away from home for various years, and projected to 2010. Use the graph for Exercises 24–27.

**Food Dollars Spent Away from Home**

SOURCES: U.S. Bureau of Labor Statistics; National Restaurant Association

**24.** What percent of food dollars will be spent away from home in 2010?

**25.** What percent of food dollars were spent away from home in 1985?

**26.** In what year was the percent of food dollars spent away from home about 30%?

**27.** In what year was the percent of food dollars spent away from home about 50%?

*Book Circulation.* The table below lists the average number of books checked out per day of the week for a branch library. Use this table for Exercises 28 and 29.

| DAY | NUMBER OF BOOKS CHECKED OUT |
|---|---|
| Sunday | 210 |
| Monday | 160 |
| Tuesday | 240 |
| Wednesday | 270 |
| Thursday | 310 |
| Friday | 275 |
| Saturday | 420 |

**28.** Make a vertical bar graph of the data.

**29.** Make a line graph of the data.

**30.** *Food Budget.* The following table lists the percent of a family's food budget spent on selected food categories. Construct a circle graph representing these data.

| FOOD CATEGORY | PERCENT OF BUDGET |
|---|---|
| Meat, poultry, fish, and eggs | 23 |
| Fruits and vegetables | 17 |
| Cereals and bakery products | 13 |
| Dairy products | 11 |
| Other | 36 |

SOURCE: Consumer Expenditure Survey

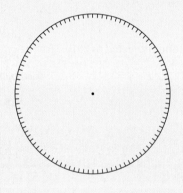

**31.** Referring to Exercise 30, consider a family that spends $664 per month on food. Using the percents from the table and the circle graph, find the amount of money spent on cereals and bakery products.
**A.** $8.63 **B.** $21.58 **C.** $86.32 **D.** $577.68

# Synthesis

**32.** The ordered set of data 69, 71, 73, $a$, 78, 98, $b$ has a median of 74 and a mean of 82. Find $a$ and $b$.

1. *Net Worth.* In 2009, Warren Buffett of the United States was worth $62.1 billion. Write standard notation for 62.1 billion.

2. *Gas Mileage.* A 2009 Saturn Outlook gets 312 mi of highway driving on 13 gal of gasoline. What is the gas mileage?

**Source:** *Motor Trend*, November 2008

3. In 402,513, what does the digit 5 mean?

4. Evaluate: $3 + 5^3$.

5. Find all the factors of 60.

6. Round 52.045 to the nearest tenth.

7. Convert to fraction notation: $3\frac{3}{10}$.

8. Convert from cents to dollars: 210¢.

9. Find percent notation for $\frac{7}{20}$.

10. Determine whether 11, 30 and 4, 12 are proportional.

Compute and simplify.

11. $2\frac{2}{5} + 4\frac{3}{10}$

12. $41.063 + (-43.5721)$

13. $-\frac{11}{15} - \frac{3}{5}$

14. $350 - 24.57$

15. $3\frac{3}{7} \cdot 4\frac{3}{8}$

16. $12{,}456 \times 220$

17. $-\frac{13}{15} \div \left(-\frac{26}{27}\right)$

18. $104{,}676 \div 24$

Solve.

19. $\frac{5}{8} = \frac{6}{x}$

20. $\frac{2}{5} \cdot y = \frac{3}{10}$

21. $21.5 \cdot y = -146.2$

22. $x = 398{,}112 \div 26$

Solve.

23. Tortilla chips cost $2.99 for 14.5 oz. Find the unit price in cents per ounce, rounded to the nearest tenth of a cent.

24. A college has a student body of 6000 students. Of these, 55.4% own a car. How many students own a car?

25. A piece of fabric $1\frac{3}{4}$ yd long is cut into 7 equal strips. What is the length of each strip?

26. A recipe calls for $\frac{3}{4}$ cup of sugar. How much sugar should be used for $\frac{1}{2}$ of the recipe?

**27. Peanut Products.** In any given year, the average American eats 2.7 lb of peanut butter, 1.5 lb of salted peanuts, 1.2 lb of peanut candy, 0.7 lb of in-shell peanuts, and 0.1 lb of peanuts in other forms. How many pounds of peanuts and products containing peanuts does the average American eat in one year?

**28. Energy Consumption.** In a recent year, American utility companies generated 1464 billion kilowatt-hours of electricity using coal, 455 billion using nuclear power, 273 billion using natural gas, 250 billion using hydroelectric plants, 118 billion using petroleum, and 12 billion using geothermal technology and other methods. How many kilowatt-hours of electricity were produced that year?

**29. Heart Disease.** Of the 301 million people in the United States, about 7.5 million have coronary heart disease and about 509,000 die of heart attacks each year. What percent have coronary heart disease? What percent die of heart attacks? Round your answers to the nearest tenth of a percent.
Source: U.S. Centers for Disease Control

**30. Billionaires.** In 2003, the mean net worth of U.S. billionaires was $3.18 billion. By 2004, this figure had increased to $3.31 billion. What was the percent of increase?
Source: Forbes

**31.** A business is owned by four people. One owns $\frac{1}{3}$, the second owns $\frac{1}{4}$, and the third owns $\frac{1}{6}$. How much does the fourth person own?

**32.** In manufacturing valves for engines, a factory was discovered to have made 4 defective valves in a lot of 18 valves. At this rate, how many defective valves can be expected in a lot of 5049 valves?

**33.** A landscaper bought 22 evergreen trees for $210. What was the cost of each tree? Round to the nearest cent.

**34.** A salesperson earns $182 selling $2600 worth of electronic equipment. What is the commission rate?

**FedEx.** The following table lists the cost of delivesring a package by FedEx Priority Overnight shipping from zip code 46143 to zip code 80403. Use the table for Exercises 35–37.

| WEIGHT (in pounds) | COST |
|---|---|
| 1 | $38.08 |
| 2 | 41.86 |
| 3 | 46.11 |
| 4 | 50.25 |
| 5 | 54.69 |
| 6 | 58.83 |
| 7 | 62.97 |
| 8 | 66.86 |
| 9 | 71.00 |
| 10 | 74.59 |

SOURCE: Federal Express Corporation

**35.** Find the average and the median of these costs.

**36.** Make a vertical bar graph of the data.

**37.** Make a line graph of the data.

## Synthesis

**38.** A photography club meets four times a month. In September, the attendance figures were 28, 23, 26, and 23. In October, the attendance figures were 26, 20, 14, and 28. What was the percent of increase or decrease in average monthly attendance from September to October?

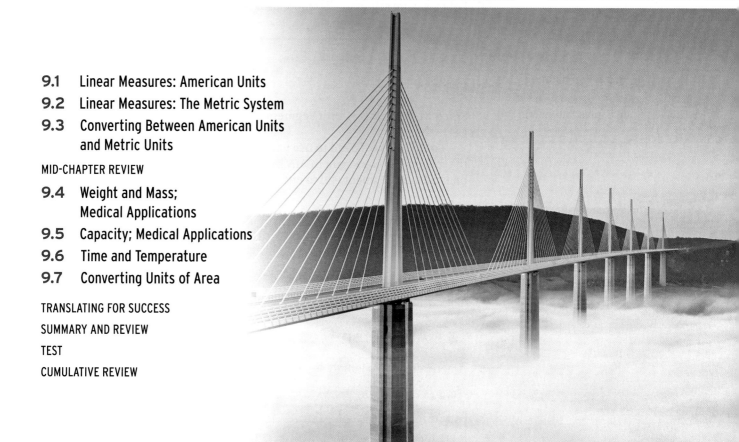

# Measurement

## Real-World Application

The Millau viaduct is part of the E11 expressway connecting Paris, France, and Barcelona, Spain. The viaduct has the highest bridge piers ever constructed. The tallest pier is 804 ft high and the overall height including the pylon is 1122 ft, making this the highest bridge in the world. Convert 804 feet and 1122 feet to meters.

*Source:* www.abelard.org/france/viaduct-de-millau.php

***This problem appears as Example 5 in Section 9.3.***

# 9.1

# Linear Measures: American Units

## OBJECTIVE

**a** Convert from one American unit of length to another.

**SKILL TO REVIEW**
Objective 4.4a: Convert between mixed numerals and fraction notation.

**1.** Convert $6\frac{3}{8}$ to fraction notation.

**2.** Convert $\frac{96}{5}$ to a mixed numeral.

Use the unit below to measure the length of each segment or object.

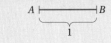

**1.** ⊢—————————⊣

**2.** ⊢———————————————⊣

**3.**

**4.**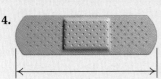

Length, or distance, is one kind of measure. To find lengths, we start with some **unit segment** and assign to it a measure of 1. Suppose $\overline{AB}$ below is a unit segment.

Let's measure segment $\overline{CD}$ below, using $\overline{AB}$ as our unit segment.

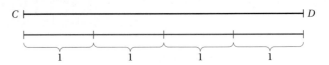

Since we can place 4 unit segments end to end along $\overline{CD}$, the measure of $\overline{CD}$ is 4.

Sometimes we need to use parts of units, called **subunits**. For example, the measure of the segment $\overline{MN}$ below is $1\frac{1}{2}$. We place one unit segment and one half-unit segment end to end.

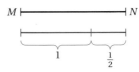

Do Margin Exercises 1–4.

## **a** American Measures

American units of length are related as follows.

| **AMERICAN UNITS OF LENGTH** | |
|---|---|
| 12 inches (in.) = 1 foot (ft) | 3 feet = 1 yard (yd) |
| 36 inches = 1 yard | 5280 feet = 1 mile (mi) |

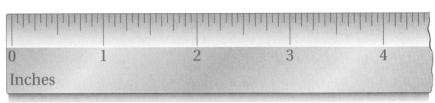

(Actual size, in inches)

*Answers*

*Skill to Review:*

1. $\frac{51}{8}$   2. $19\frac{1}{5}$

*Margin Exercises:*

1. 2   2. 3   3. $1\frac{1}{2}$   4. $2\frac{1}{2}$

We can visualize comparisons of the units as follows:

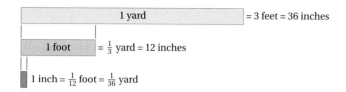

The symbolism 13 in. = 13'' and 27 ft = 27' is also used for inches and feet. American units have also been called "English," or "British–American," because at one time they were used by both countries. Today, both Canada and England have officially converted to the metric system.

To change from certain American units to others, we make substitutions. Such a substitution is usually helpful when we are converting from a *larger* unit to a *smaller* one.

**EXAMPLE 1**  Complete:  $7\frac{1}{3}$ yd = _____ in.

$$7\frac{1}{3}\text{ yd} = 7\frac{1}{3} \times 1\text{ yd} \qquad \text{We think of } 7\frac{1}{3}\text{ yd as } 7\frac{1}{3} \times \text{ yd, or } 7\frac{1}{3} \times 1\text{ yd.}$$

$$= 7\frac{1}{3} \times 36\text{ in.} \qquad \text{Substituting 36 in. for 1 yd}$$

$$= \frac{22}{3} \times 36\text{ in.}$$

$$= 264\text{ in.}$$

Do Exercises 5–7.

Complete.
**5.** 8 yd = _____ in.

**6.** $2\frac{5}{6}$ yd = _____ ft

**7.** 3.8 mi = _____ in.

Sometimes it helps to use multiplying by 1 in making conversions. For example, 12 in. = 1 ft, so

$$\frac{12\text{ in.}}{1\text{ ft}} = 1 \quad \text{and} \quad \frac{1\text{ ft}}{12\text{ in.}} = 1.$$

If we divide 12 in. by 1 ft or 1 ft by 12 in., we get 1 because the lengths are the same. Let's first convert from *smaller* units to *larger* units.

**EXAMPLE 2**  Complete:  48 in. = _____ ft.

We want to convert from "in." to "ft." We multiply by 1 using a symbol for 1 with "in." on the bottom and "ft" on the top to eliminate inches and to convert to feet:

$$48\text{ in.} = \frac{48\text{ in.}}{1} \times \frac{1\text{ ft}}{12\text{ in.}} \qquad \text{Multiplying by 1 using } \frac{1\text{ ft}}{12\text{ in.}} \text{ to eliminate in.}$$

$$= \frac{48\text{ in.}}{12\text{ in.}} \times 1\text{ ft}$$

$$= \frac{48}{12} \times \frac{\text{in.}}{\text{in.}} \times 1\text{ ft}$$

$$= 4 \times 1\text{ ft} \qquad \text{The } \frac{\text{in.}}{\text{in.}} \text{ acts like 1, so we can omit it.}$$

$$= 4\text{ ft.}$$

*Answers*

**5.** 288   **6.** $8\frac{1}{2}$   **7.** 240,768

We can also look at this conversion as "canceling" units:

$$48 \text{ in.} = \frac{48 \text{ in.}}{1} \times \frac{1 \text{ ft}}{12 \text{ in.}} = \frac{48}{12} \times 1 \text{ ft} = 4 \text{ ft.}$$

This method is used not only in mathematics, as here, but also in fields such as medicine, chemistry, and physics.

Do Exercises 8 and 9.

**EXAMPLE 3**  Complete: 25 ft = _____ yd.

Since we are converting from "ft" to "yd," we choose a symbol for 1 with "yd" on the top and "ft" on the bottom:

$$25 \text{ ft} = 25 \text{ ft} \times \frac{1 \text{ yd}}{3 \text{ ft}} \qquad 3 \text{ ft} = 1 \text{ yd, so } \frac{3 \text{ ft}}{1 \text{ yd}} = 1 \text{ and } \frac{1 \text{ yd}}{3 \text{ ft}} = 1.$$

We use $\frac{1 \text{ yd}}{3 \text{ ft}}$ to eliminate ft.

$$= \frac{25}{3} \times \frac{\text{ft}}{\text{ft}} \times 1 \text{ yd}$$

$$= 8\frac{1}{3} \times 1 \text{ yd} \qquad \text{The } \frac{\text{ft}}{\text{ft}} \text{ acts like 1, so we can omit it.}$$

$$= 8\frac{1}{3} \text{ yd, or } 8.\overline{3} \text{ yd.}$$

Again, in this example, we can consider conversion from the point of view of canceling:

$$25 \text{ ft} = 25 \text{ ft} \times \frac{1 \text{ yd}}{3 \text{ ft}} = \frac{25}{3} \times 1 \text{ yd} = 8\frac{1}{3} \text{ yd, or } 8.\overline{3} \text{ yd.}$$

Do Exercises 10 and 11.

**EXAMPLE 4**  Complete: 23,760 ft = _____ mi.

We choose a symbol for 1 with "mi" on the top and "ft" on the bottom:

$$23,760 \text{ ft} = 23,760 \text{ ft} \times \frac{1 \text{ mi}}{5280 \text{ ft}} \qquad 5280 \text{ ft} = 1 \text{ mi, so } \frac{1 \text{ mi}}{5280 \text{ ft}} = 1.$$

$$= \frac{23,760}{5280} \times \frac{\text{ft}}{\text{ft}} \times 1 \text{ mi}$$

$$= 4.5 \times 1 \text{ mi} \qquad\qquad \text{Dividing}$$

$$= 4.5 \text{ mi.}$$

Let's also consider this example using canceling:

$$23,760 \text{ ft} = 23,760 \text{ ft} \times \frac{1 \text{ mi}}{5280 \text{ ft}}$$

$$= \frac{23,760}{5280} \times 1 \text{ mi}$$

$$= 4.5 \times 1 \text{ mi} = 4.5 \text{ mi.}$$

Do Exercises 12 and 13.

---

Complete.
**8.** 72 in. = _____ ft

**9.** 17 in. = _____ ft

Complete.
**10.** 24 ft = _____ yd

**11.** 35 ft = _____ yd

Complete.
**12.** 26,400 ft = _____ mi

**13.** 2640 ft = _____ mi

*Answers*

**8.** 6    **9.** $1\frac{5}{12}$    **10.** 8    **11.** $11\frac{2}{3}$, or $11.\overline{6}$

**12.** 5    **13.** $\frac{1}{2}$, or 0.5

We can also use multiplying by 1 to convert from larger units to smaller units. Let's redo Example 1.

**EXAMPLE 5**  Complete: $7\frac{1}{3}$ yd = _____ in.

$$7\frac{1}{3}\text{ yd} = \frac{22\text{ yd}}{3} \times \frac{36\text{ in.}}{1\text{ yd}} = \frac{22 \times 36}{3} \times 1\text{ in.} = 264\text{ in.}$$

Do Exercise 14.

**14.** Complete. Use multiplying by 1.

$$2\frac{2}{3}\text{ yd} = \text{_____ in.}$$

**EXAMPLE 6**  *Rope Sculpture.*  New York–based artist Orly Genger is known for transforming common nylon ropes into elaborate, monumental sculptures. For her work titled *Whole*, Genger hand-knotted and piled an estimated 113.6 mi of painted rope into nine separate sculptures that were on display at the Indianapolis Museum of Art from November 21, 2008, to June 14, 2009. Convert 113.6 miles to yards.

**Sources:** www.art-directory.cn; *IMA Previews*, Winter 2008–2009

We have

$$113.6\text{ mi} = 113.6\text{ mi} \times \frac{5280\text{ ft}}{1\text{ mi}} \times \frac{1\text{ yd}}{3\text{ ft}}$$

$$= \frac{113.6 \times 5280}{1 \times 3} \times 1\text{ yd}$$

$$= 199{,}936\text{ yd.}$$

The nine sculptures of *Whole* contain 199,936 yd of rope.

Do Exercise 15.

**15. Pedestrian Paths.**  There are 23 mi of pedestrian paths in Central Park in New York City. Convert 23 miles to yards.

*Answers*
**14.** 96    **15.** 40,480 yd

**a** Complete.

**1.** 1 ft = _____ in.

**2.** 1 yd = _____ ft

**3.** 1 in. = _____ ft

**4.** 1 mi = _____ yd

**5.** 1 mi = _____ ft

**6.** 1 ft = _____ yd

**7.** 3 yd = _____ in.

**8.** 10 yd = _____ ft

**9.** 84 in. = _____ ft

**10.** 48 ft = _____ yd

**11.** 18 in. = _____ ft

**12.** 29 ft = _____ yd

**13.** 5 mi = _____ ft

**14.** 5 mi = _____ yd

**15.** 63 in. = _____ ft

**16.** 11,616 ft = _____ mi

**17.** 10 ft = _____ yd

**18.** 9.6 yd = _____ ft

**19.** 7.1 mi = _____ ft

**20.** 31,680 ft = _____ mi

**21.** $4\frac{1}{2}$ ft = _____ yd

**22.** 48 in. = _____ ft

**23.** 45 in. = _____ yd

**24.** $6\frac{1}{3}$ yd = _____ in.

**25.** 330 ft = _____ yd

**26.** 5280 yd = _____ mi

**27.** 3520 yd = _____ mi

**28.** 25 mi = _____ ft

**29.** 100 yd = _____ ft

**30.** 480 in. = _____ ft

**31.** 360 in. = _____ ft

**32.** 720 in. = _____ yd

**33.** 1 in. = _____ yd

**34.** 25 in. = _____ ft

**35.** 2 mi = _____ in.

**36.** 63,360 in. = _____ mi

**37.** 83 yd = _____ in.

**38.** 450 in. = _____ yd

## Skill Maintenance

Convert to fraction notation.   [7.2b]

**39.** 9.25%

**40.** $87\frac{1}{2}\%$

Find fraction notation for the percent notation in each sentence.   [7.2b]

**41.** Of all 18-year-olds, 27.5% are registered to vote.

**42.** Of all those who buy CDs, 57% are in the 20–39 age group.

Convert to percent notation.   [7.2a]

**43.** $\frac{11}{8}$

**44.** $\frac{2}{3}$

**45.** $\frac{1}{4}$

**46.** $\frac{7}{16}$

Solve.   [6.1a, b]

**47.** *Trout and Catfish Production.*   In the United States in 2006, 49.2 million trout (food-size) were produced and sold, and 368.7 million catfish (food-size) were produced and sold. What is the ratio of the number of trout to the number of catfish? What is the ratio of the number of catfish to the number of trout?
Source: U.S. Department of Agriculture

**48.** *Wildfires.*   In 2006, 96,385 wildland forest fires burned 9.87 million acres in the United States. What was the ratio of number of fires to number of acres? What was the ratio of number of acres to number of fires?
Source: National Interagency Coordination Center, *Wildland Fires and Acres* (1960–2006)

## Synthesis

**49.** *Noah's Ark.*   In biblical measures, it is thought that 1 cubit ≈ 18 in. The dimensions of Noah's ark are given as follows: "The length of the ark shall be three hundred cubits, the breadth of it fifty cubits, and the height of it thirty cubits." What were the dimensions of Noah's ark, in inches? in feet?
Source: *Holy Bible, King James Version*, Gen. 6:15

**50.** *Goliath's Height.*   In biblical measures, a span was considered to be half of a cubit (1 cubit ≈ 18 in.; see Exercise 49). The giant Goliath's height "was six cubits and a span." What was the height of Goliath, in inches? in feet?
Source: *Holy Bible, King James Version*, 1 Sam. 17:4

# 9.2

## Linear Measures: The Metric System

### OBJECTIVE

**a** Convert from one metric unit of length to another.

---

**SKILL TO REVIEW**

Objective 5.3a: Multiply using decimal notation.

Multiply.

1. $0.5603 \times 1000$
2. $18.7 \times 100$

---

Although the **metric system** is used in most countries of the world, it is used very little in the United States. The metric system does not use inches, feet, pounds, and so on, but units for time and electricity are the same as those used now in the United States.

An advantage of the metric system is that it is easier to convert from one unit to another within this system than within the American system. That is because the metric system is based on the number 10.

The basic unit of length is the **meter**. It is just over a yard. In fact, 1 meter $\approx$ 1.1 yd.

(Comparative sizes are shown.)

| 1 Meter |
|---|

| 1 Yard |
|---|

The other units of length are multiples of the length of a meter:

10 times a meter,   100 times a meter,   1000 times a meter,   and so on,

or fractions of a meter:

$\frac{1}{10}$ of a meter,   $\frac{1}{100}$ of a meter,   $\frac{1}{1000}$ of a meter,   and so on.

### METRIC UNITS OF LENGTH

1 *kilo*meter (km) = 1000 meters (m)

1 *hecto*meter (hm) = 100 meters (m)

1 *deka*meter (dam) = 10 meters (m)

1 meter (m)

1 *deci*meter (dm) = $\frac{1}{10}$ meter (m)

1 *centi*meter (cm) = $\frac{1}{100}$ meter (m)

1 *milli*meter (mm) = $\frac{1}{1000}$ meter (m)

You should memorize these names and abbreviations. Think of *kilo-* for 1000, *hecto-* for 100, *deka-* for 10, *deci-* for $\frac{1}{10}$, *centi-* for $\frac{1}{100}$, and *milli-* for $\frac{1}{1000}$. (The units dekameter and decimeter are not used often.) We will also use these prefixes when considering units of area, capacity, and mass.

### Thinking Metric

To familiarize yourself with metric units, consider the following.

| | |
|---|---|
| 1 kilometer (1000 meters) | is slightly more than $\frac{1}{2}$ mile (0.6 mi). |
| 1 meter | is just over a yard (1.1 yd). |
| 1 centimeter (0.01 meter) | is a little more than the width of a paperclip (about 0.3937 inch). |

1 cm

1 cm

---

1 inch is about 2.54 centimeters.

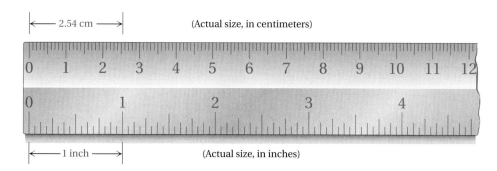

(Actual size, in centimeters)

1 inch (Actual size, in inches)

1 millimeter is about the diameter of paperclip wire.

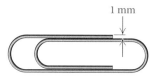

1 mm

The millimeter (mm) is used to measure small distances, especially in industry.

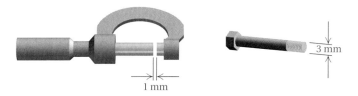

3 mm

1 mm

In many countries, the centimeter (cm) is used for body dimensions and clothing sizes.

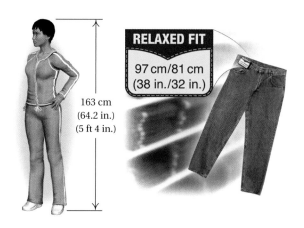

163 cm
(64.2 in.)
(5 ft 4 in.)

**RELAXED FIT**
97 cm/81 cm
(38 in./32 in.)

Do Exercises 1–3.

Using a centimeter ruler, measure each object.

**1.**

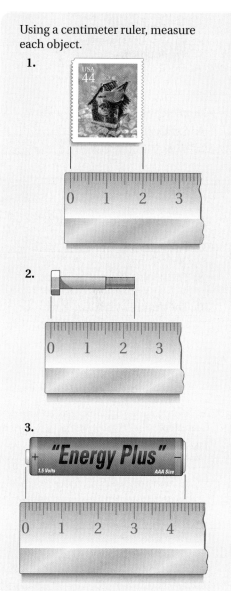

**2.**

**3.**

"Energy Plus"
1.5 Volts          AAA Size

*Answers*

**1.** 2 cm, or 20 mm    **2.** 2.3 cm, or 23 mm
**3.** 4.4 cm, or 44 mm

The meter (m) is used for expressing dimensions of larger objects—say, the height of a diving board—and for shorter distances, like the length of a rug.

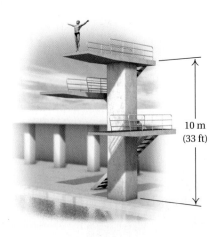

10 m
(33 ft)

3.7 m
(12 ft)

2.7 m
(9 ft)

Chicago
30 mi
48 km

The kilometer (km) is used for longer distances, mostly in cases in which miles would be used.

1 mile is about 1.6 km.

| 1 km |
|---|
| 1 mi |

Do Exercises 4–9.

Do Exercises 4–9.

**Complete with mm, cm, m, or km.**

**4.** A stick of gum is 7 _____ long.

**5.** New Orleans is 2941 _____ from San Diego.

**6.** A penny is 1 _____ thick.

**7.** The halfback ran 7 _____.

**8.** The book is 3 _____ thick.

**9.** The desk is 2 _____ long.

## (a) Changing Metric Units

As with American units, when changing from a *larger* unit to a *smaller* unit, we usually make substitutions.

**EXAMPLE 1** Complete: 4 km = _____ m.

Since we are converting from a *larger* unit to a *smaller* unit, we use substitution.

$$4 \text{ km} = 4 \times 1 \text{ km}$$
$$= 4 \times 1000 \text{ m} \quad \text{Substituting 1000 m for 1 km}$$
$$= 4000 \text{ m}$$

Do Exercises 10 and 11.

**Complete.**

**10.** 23 km = _____ m

**11.** 4 hm = _____ m

Since

$$\frac{1}{10} \text{ m} = 1 \text{ dm}, \quad \frac{1}{100} \text{ m} = 1 \text{ cm}, \text{ and } \frac{1}{1000} \text{ m} = 1 \text{ mm},$$

it follows that

$$1 \text{ m} = 10 \text{ dm}, \quad 1 \text{ m} = 100 \text{ cm}, \text{ and } 1 \text{ m} = 1000 \text{ mm}.$$

**EXAMPLE 2** Complete: 93.4 m = _____ cm.

Since we are converting from a *larger* unit to a *smaller* unit, we use substitution. We substitute 100 cm for 1 m:

$$93.4 \text{ m} = 93.4 \times 1 \text{ m} = 93.4 \times 100 \text{ cm} = 9340 \text{ cm}.$$

*Answers*

**4.** cm   **5.** km   **6.** mm   **7.** m
**8.** cm   **9.** m   **10.** 23,000   **11.** 400

**EXAMPLE 3** Complete: 0.248 m = _____ mm.

Since we are converting from a *larger* unit to a *smaller* unit, we use substitution.

$$0.248 \text{ m} = 0.248 \times 1 \text{ m}$$
$$= 0.248 \times 1000 \text{ mm} \quad \text{Substituting 1000 mm for 1 m}$$
$$= 248 \text{ mm}$$

Do Exercises 12 and 13.

Complete.

**12.** 1.78 m = _____ cm

**13.** 9.04 m = _____ mm

We now convert from "m" to "km." Since we are converting from a *smaller* unit to a *larger* unit, we use multiplying by 1. We choose a symbol for 1 with "km" in the numerator and "m" in the denominator.

**EXAMPLE 4** Complete: 2347 m = _____ km.

$$2347 \text{ m} = 2347 \text{ m} \times \frac{1 \text{ km}}{1000 \text{ m}} \quad \text{Multiplying by 1 using } \frac{1 \text{ km}}{1000 \text{ m}}$$
$$= \frac{2347}{1000} \times \frac{\text{m}}{\text{m}} \times 1 \text{ km} \quad \text{The } \frac{\text{m}}{\text{m}} \text{ acts like 1, so we will omit it.}$$
$$= 2.347 \text{ km} \quad \text{Dividing by 1000 moves the decimal point three places to the left.}$$

Using canceling, we can work this example as follows:

$$2347 \text{ m} = 2347 \text{ m̸} \times \frac{1 \text{ km}}{1000 \text{ m̸}} = \frac{2347}{1000} \times 1 \text{ km} = 2.347 \text{ km}.$$

Sometimes we multiply by 1 more than once.

**EXAMPLE 5** Complete: 8.42 mm = _____ cm.

$$8.42 \text{ mm} = 8.42 \text{ mm} \times \frac{1 \text{ m}}{1000 \text{ mm}} \times \frac{100 \text{ cm}}{1 \text{ m}} \quad \text{Multiplying by 1 using } \frac{1 \text{ m}}{1000 \text{ mm}} \text{ and } \frac{100 \text{ cm}}{1 \text{ m}}$$

$$= \frac{8.42 \times 100}{1000} \times \frac{\text{mm}}{\text{mm}} \times \frac{\text{m}}{\text{m}} \times 1 \text{ cm}$$

$$= \frac{842}{1000} \text{ cm} = 0.842 \text{ cm}$$

Do Exercises 14–17.

Complete.

**14.** 7814 m = _____ km

**15.** 7814 m = _____ dam

**16.** 9.67 mm = _____ cm

**17.** 89 km = _____ cm

## Mental Conversion

Changing from one unit to another in the metric system amounts only to the movement of a decimal point. That is because the metric system is based on 10. Let's find a faster way to convert. Look at the following table.

| 1000 m | 100 m | 10 m | 1 m | 0.1 m | 0.01 m | 0.001 m |
|--------|-------|------|-----|-------|--------|---------|
| 1 km | 1 hm | 1 dam | 1 m | 1 dm | 1 cm | 1 mm |

Each place in the table has a value $\frac{1}{10}$ that to the left or 10 times that to the right. Thus moving one place in the table corresponds to moving one decimal place.

*Answers*

**12.** 178    **13.** 9040    **14.** 7.814
**15.** 781.4    **16.** 0.967    **17.** 8,900,000

Let's convert mentally.

**EXAMPLE 6** Complete: 8.42 mm = _____ cm.

*Think:* To go from mm to cm in the table is a move of one place to the left. Thus we move the decimal point one place to the left.

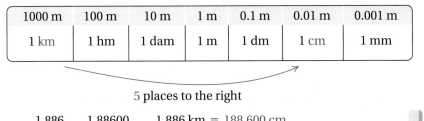

| 1000 m | 100 m | 10 m | 1 m | 0.1 m | 0.01 m | 0.001 m |
|--------|-------|------|-----|-------|--------|---------|
| 1 km | 1 hm | 1 dam | 1 m | 1 dm | 1 cm | 1 mm |

1 place to the left

8.42     0.8.42     8.42 mm = 0.842 cm

**EXAMPLE 7** Complete: 1.886 km = _____ cm.

*Think:* To go from km to cm in the table is a move of five places to the right. Thus we move the decimal point five places to the right.

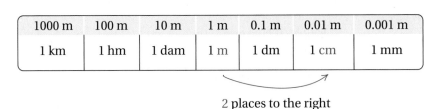

| 1000 m | 100 m | 10 m | 1 m | 0.1 m | 0.01 m | 0.001 m |
|--------|-------|------|-----|-------|--------|---------|
| 1 km | 1 hm | 1 dam | 1 m | 1 dm | 1 cm | 1 mm |

5 places to the right

1.886     1.88600.     1.886 km = 188,600 cm

**EXAMPLE 8** Complete: 3 m = _____ cm.

*Think:* To go from m to cm in the table is a move of two places to the right. Thus we move the decimal point two places to the right.

| 1000 m | 100 m | 10 m | 1 m | 0.1 m | 0.01 m | 0.001 m |
|--------|-------|------|-----|-------|--------|---------|
| 1 km | 1 hm | 1 dam | 1 m | 1 dm | 1 cm | 1 mm |

2 places to the right

3     3.00.     3 m = 300 cm

You should try to make metric conversions mentally as much as possible. The fact that conversions can be done so easily is an important advantage of the metric system. The most commonly used metric units of length are km, m, cm, and mm. We have purposely used these more often than the others in the exercises.

Do Exercises 18–21.

Complete. Try to do this mentally using the table.

**18.** 6780 m = _____ km

**19.** 9.74 cm = _____ mm

**20.** 1 mm = _____ cm

**21.** 845.1 mm = _____ dm

*Answers*

**18.** 6.78   **19.** 97.4   **20.** 0.1   **21.** 8.451

**a** Complete. Do as much as possible mentally.

**1. a)** 1 km = _____ m
   **b)** 1 m = _____ km

**2. a)** 1 hm = _____ m
   **b)** 1 m = _____ hm

**3. a)** 1 dam = _____ m
   **b)** 1 m = _____ dam

**4. a)** 1 dm = _____ m
   **b)** 1 m = _____ dm

**5. a)** 1 cm = _____ m
   **b)** 1 m = _____ cm

**6. a)** 1 mm = _____ m
   **b)** 1 m = _____ mm

**7.** 6.7 km = _____ m

**8.** 27 km = _____ m

**9.** 98 cm = _____ m

**10.** 0.789 cm = _____ m

**11.** 8921 m = _____ km

**12.** 8664 m = _____ km

**13.** 56.66 m = _____ km

**14.** 4.733 m = _____ km

**15.** 5666 m = _____ cm

**16.** 869 m = _____ cm

**17.** 477 cm = _____ m

**18.** 6.27 mm = _____ m

**19.** 6.88 m = _____ cm

**20.** 6.88 m = _____ dm

**21.** 1 mm = _____ cm

**22.** 1 cm = _____ km

**23.** 1 km = _____ cm

**24.** 2 km = _____ cm

**25.** 14.2 cm = _____ mm

**26.** 25.3 cm = _____ mm

**27.** 8.2 mm = _____ cm

**28.** 9.7 mm = _____ cm

**29.** 4500 mm = _____ cm

**30.** 8,000,000 m = _____ km

**31.** 0.024 mm = _____ m

**32.** 60,000 mm = _____ dam

**33.** 6.88 m = _____ dam

**34.** 7.44 m = _____ hm

**35.** 2.3 dam = _____ dm

**36.** 9 km = _____ hm

**37.** 392 dam = _____ km

**38.** 0.056 mm = _____ dm

Complete the following table.

| | OBJECT | MILLIMETERS (mm) | CENTIMETERS (cm) | METERS (m) |
|---|---|---|---|---|
| **39.** | Length of a calculator | | 18 | |
| **40.** | Width of a calculator | 85 | | |
| **41.** | Length of a piece of typing paper | | | 0.278 |
| **42.** | Length of a football field | | | 109.09 |
| **43.** | Width of a football field | | 4844 | |
| **44.** | Width of a credit card | 56 | | |
| **45.** | Length of 4 meter sticks | | | 4 |
| **46.** | Length of 3 meter sticks | | 300 | |
| **47.** | Thickness of an index card | 0.27 | | |
| **48.** | Thickness of a piece of cardboard | | 0.23 | |
| **49.** | Height of the Willis Tower | | | 442 |
| **50.** | Height of the CN Tower (Toronto) | 553,000 | | |

## Skill Maintenance

Divide.   [5.4a]

**51.** $23.4 \div 100$

**52.** $23.4 \div 1000$

Multiply.

**53.** $3.14 \times 4.41$   [5.3a]

**54.** $-4 \times 20\frac{1}{8}$   [4.6a]

Convert to percent notation.   [7.2a]

**55.** $\frac{3}{8}$

**56.** $\frac{5}{8}$

**57.** $\frac{2}{3}$

**58.** $\frac{1}{5}$

Find decimal notation for the percent notation in each sentence.   [7.1b]

**59.** Blood is 90% water.

**60.** Of those accidents requiring medical attention, 10.8% occur on roads.

**61.** Add: $\frac{7}{15} + \frac{4}{25}$.   [4.2a]

**62.** Subtract: $\frac{11}{18} - \left(-\frac{5}{24}\right)$.   [4.3a]

## Synthesis

Each sentence is incorrect. Insert or move a decimal point to make the sentence correct.

**63.** When my right arm is extended, the distance from my left shoulder to the end of my right hand is 10 m.

**64.** The height of the Shanghai World Financial Center is 49.2 m.

**65.** A stack of ten quarters is 140 cm high.

**66.** The width of an adult's hand is 112 cm.

# 9.3 Converting Between American Units and Metric Units

## (a) Converting Units

We can make conversions between American units and metric units by using the following table. These listings are rounded approximations. Again, we either make a substitution or multiply by 1 appropriately.

| AMERICAN | METRIC |
|----------|--------|
| 1 in. | 2.540 cm |
| 1 ft | 0.305 m |
| 1 yd | 0.914 m |
| 1 mi | 1.609 km |
| 0.621 mi | 1 km |
| 1.094 yd | 1 m |
| 3.281 ft | 1 m |
| 39.370 in. | 1 m |

This table is set up to enable us to make conversions by substitution.

**EXAMPLE 1** Complete: $26.2 \text{ mi} =$ _____ km. (The length of the Olympic marathon)

$$26.2 \text{ mi} = 26.2 \times 1 \text{ mi}$$
$$\approx 26.2 \times 1.609 \text{ km} \qquad \text{Substituting } 1.609 \text{ km for } 1 \text{ mi}$$
$$= 42.1558 \text{ km}$$

**EXAMPLE 2** Complete: $2.36 \text{ m} =$ _____ in. (The height of Bao Xishun, the world's tallest man)

$$2.36 \text{ m} = 2.36 \times 1 \text{ m}$$
$$\approx 2.36 \times 39.37 \text{ in.} \qquad \text{Substituting } 39.37 \text{ in. for } 1 \text{ m}$$
$$= 92.9132 \text{ in.}$$

In an application like this one, the answer would probably be rounded to the nearest one, as 93 in.

**EXAMPLE 3** Complete: $100 \text{ m} =$ _____ ft. (The length of the 100-meter dash)

$$100 \text{ m} = 100 \times 1 \text{ m}$$
$$\approx 100 \times 3.281 \text{ ft} \qquad \text{Substituting } 3.281 \text{ ft for } 1 \text{ m}$$
$$= 328.1 \text{ ft}$$

## OBJECTIVE

(a) Convert between American units of length and metric units of length.

**SKILL TO REVIEW**
Objective 5.3a: Multiply using decimal notation.

Multiply.
1. $3.89 \times 1.609$
2. $5012 \times 0.621$

## STUDY TIPS

### TAKE THE TIME!

The foundation of all your study skills is making time to study! If you invest your time wisely, you increase your likelihood of succeeding.

*Answers*

*Skill to Review:*
**1.** 6.25901 **2.** 3112.452

Complete.

**1.** 100 yd = _____ m
(The length of a football field excluding the end zones)

**2.** 2.5 mi = _____ km
(The length of the tri-oval track at Daytona International Speedway)

**3.** 2383 km = _____ mi
(The distance from St. Louis to Phoenix)

**4.** The distance from San Francisco to Las Vegas is 568 mi. Find the distance in kilometers.

**5.** The height of the Stratosphere Tower in Las Vegas, Nevada, is 1149 ft. Find the height in meters.

**6.** Complete: 3.175 mm = _____ in.
(The thickness of a quarter)

**EXAMPLE 4** Complete: 4544 km = _____ mi.
(The distance from New York to Los Angeles)

$$4544 \text{ km} = 4544 \times 1 \text{ km}$$
$$\approx 4544 \times 0.621 \text{ mi} \quad \text{Substituting 0.621 mi for 1 km}$$
$$= 2821.824 \text{ mi}$$

In practical situations, we would probably round this answer to 2822 mi.

Do Exercises 1–3.

**EXAMPLE 5** *Millau Viaduct.* The Millau viaduct is part of the E11 expressway connecting Paris, France, and Barcelona, Spain. The viaduct has the highest bridge piers ever constructed. The tallest pier is 804 ft high and the overall height including the pylon is 1122 ft, making this the highest bridge in the world. Convert 804 feet and 1122 feet to meters.

**Source:** www.abelard.org/france/viaduct-de-millau.php

We let $P$ = the height of the pier and $H$ = the overall height of the bridge. To convert feet to meters, we substitute 0.305 m for 1 ft.

$$P = 804 \text{ ft} \qquad\qquad H = 1122 \text{ ft}$$
$$= 804 \times 1 \text{ ft} \qquad\quad = 1122 \times 1 \text{ ft}$$
$$\approx 804 \times 0.305 \text{ m} \qquad \approx 1122 \times 0.305 \text{ m}$$
$$= 245.22 \text{ m} \qquad\quad = 342.21 \text{ m}$$

Do Exercises 4 and 5.

**EXAMPLE 6** Complete: 0.10414 mm = _____ in.
(The thickness of a $1 bill)

In this case, we must make two substitutions or multiply by two forms of 1 since the table on the preceding page does not provide an easy way to convert from millimeters to inches. Here we choose to multiply by forms of 1.

$$0.10414 \text{ mm} = 0.10414 \times 1 \text{ mm} \times \frac{1 \text{ cm}}{10 \text{ mm}}$$
$$= 0.010414 \text{ cm}$$
$$\approx 0.010414 \times 1 \text{ cm} \times \frac{1 \text{ in.}}{2.54 \text{ cm}}$$
$$= 0.0041 \text{ in.}$$

Do Exercise 6.

*Answers*

**1.** 91.4　**2.** 4.0225　**3.** 1479.843
**4.** 913.912　**5.** 350.445 m　**6.** 0.125

**a** Complete.

**1.** 330 ft = _____ m
(The length of most baseball foul lines)

**2.** 12 in. = _____ cm
(The length of a common ruler)

**3.** 1171.4 km = _____ mi
(The distance from Cleveland to Atlanta)

**4.** 2 m = _____ ft
(The length of a desk)

**5.** 65 mph = _____ km/h
(A common speed limit in the United States)

**6.** 100 km/h = _____ mph
(A common speed limit in Canada)

**7.** 180 mi = _____ km
(The distance from Indianapolis to Chicago)

**8.** 141,600,000 mi = _____ km
(The farthest distance of Mars from the sun)

**9.** 70 mph = _____ km/h
(An interstate speed limit in Arizona)

**10.** 60 km/h = _____ mph
(A city speed limit in Canada)

**11.** 10 yd = _____ m
(The length needed for a first down in football)

**12.** 450 ft = _____ m
(The length of a long home run in baseball)

**13.** 2.08 m = _____ in.
(The height of Amare Stoudemire of the Phoenix Suns)

**14.** 76 in. = _____ m
(The height of Jason Kidd of the Dallas Mavericks)

**15.** 381 m = _____ ft
(The height of the Empire State Building)

**16.** 1127 ft = _____ m
(The height of the John Hancock Center)

**17.** 15.7 cm = _____ in.
(The length of a $1 bill)

**18.** 7.5 in. = _____ cm
(The length of a pencil)

**19.** 2216 km = _____ mi
(The distance from Chicago to Miami)

**20.** 1862 mi = _____ km
(The distance from Seattle to Kansas City)

**21.** 13 mm = _____ in.
(The thickness of a plastic case for a DVD)

**22.** 0.25 in. = _____ mm
(The thickness of an eraser on a pencil)

Complete the following table. Answers may vary, depending on the conversion factor used.

| | OBJECT | YARDS (yd) | CENTIMETERS (cm) | INCHES (in.) | METERS (m) | MILLIMETERS (mm) |
|---|---|---|---|---|---|---|
| **23.** | Width of a piece of typing paper | | | $8\frac{1}{2}$ | | |
| **24.** | Length of a football field | 120 | | | | |
| **25.** | Width of a football field | | 4844 | | | |
| **26.** | Width of a credit card | | | | | 56 |
| **27.** | Length of 4 yardsticks | 4 | | | | |
| **28.** | Length of 3 meter sticks | | 300 | | | |
| **29.** | Thickness of an index card | | | | 0.00027 | |
| **30.** | Thickness of a piece of cardboard | | 0.23 | | | |
| **31.** | Height of the Willis Tower | | | | 442 | |
| **32.** | Height of Central Plaza, Hong Kong | 409 | | | | |

## Skill Maintenance

**33.** *Rail Travel.* In 2008, 28.7 million passengers traveled by rail on Amtrak. Convert 28.7 million to standard notation. [5.3b]
**Source:** National Railroad Passenger Corporation

**34.** *Domestic and Foreign Travel.* It is estimated that domestic and foreign travelers spent 763 billion dollars in 2008 while traveling in the United States. Convert 763 billion to standard notation. [5.3b]
**Sources:** TIA's Travel Forecast Model; Office of Travel and Tourism Industries

## Synthesis

**35.** Develop a formula to convert from inches to millimeters.

**36.** Develop a formula to convert from millimeters to inches. How does it relate to the answer for Exercise 35?

**37.** ▦ As of November 2008, the women's world record for the 100-m dash was 10.49 sec, set by Florence Griffith-Joyner in Indianapolis, Indiana, on July 16, 1988. How fast is this in miles per hour? Round to the nearest tenth of a mile per hour.
**Source:** International Association of Athletics Federations

**38.** ▦ As of November 2008, the men's world record for the 100-m dash was 9.69 sec, set by Usain Bolt of Jamaica in the 2008 Summer Olympics in Beijing. How fast is this in miles per hour? Round to the nearest hundredth of a mile per hour.
**Source:** en.wikipedia.com

# Mid-Chapter Review

## Concept Reinforcement

Determine whether each statement is true or false.

_____ **1.** Distances that are measured in miles in the American system would probably be measured in meters in the metric system.   [9.2a], [9.3a]

_____ **2.** One meter is slightly more than one yard.   [9.2a], [9.3a]

_____ **3.** One kilometer is longer than one mile.   [9.2a], [9.3a]

_____ **4.** When converting from meters to centimeters, move the decimal point to the right.   [9.2a]

_____ **5.** One foot is approximately 30 centimeters.   [9.3a]

## Guided Solutions

Fill in each blank with the unit of length that creates a correct solution.

**6.** Complete: $16\frac{2}{3}$ yd = _____ ft.   [9.1a]

$$16\frac{2}{3} \text{ yd} = 16\frac{2}{3} \times 1\,\square = \frac{50}{3} \times 3\,\square = 50\,\square$$

**7.** Complete: 13,200 ft = _____ mi.   [9.1a]

$$13,200 \text{ ft} = 13,200 \text{ ft} \times \frac{1\,\square}{5280\,\square} = 2.5\,\square$$

**8.** Complete: 520 mm = _____ km.   [9.2a]

$$520 \text{ mm} = 520 \text{ mm} \times \frac{1\,\square}{1000\,\square} = 0.52\,\square \times \frac{1\,\square}{1000\,\square} = 0.00052\,\square$$

**9.** Complete: 10,200 mm = _____ ft.   [9.3a]

$$10,200 \text{ mm} = 10,200 \text{ mm} \times \frac{1\,\square}{1000\,\square} \approx 10.2\,\square \times \frac{3.281\,\square}{1\,\square} = 33.4662\,\square$$

## Mixed Review

Complete.   [9.1a], [9.2a], [9.3a]

**10.** $5\frac{1}{2}$ mi = _____ yd

**11.** 840 in. = _____ ft

**12.** 24.05 cm = _____ dm

**13.** 0.15 m = _____ km

**14.** 630 yd = _____ in.

**15.** 100 ft = _____ in.

**16.** 6000 dam = _____ m

**17.** 85,000 mm = _____ dm

**18.** 26,400 ft = _____ mi

**19.** 3753 ft = _____ yd

**20.** 10 mi = _____ ft

**21.** 1800 m = _____ cm

**22.** 8.4 km = _____ dm

**23.** 0.007 km = _____ cm

**24.** 40 dm = _____ dam

**25.** 80.09 cm = _____ m

**26.** 360 in. = _____ yd

**27.** 19.2 m = _____ mm

**28.** 1200 in. = _____ ft

**29.** 0.0001 mm = _____ hm

**30.** 4 km = _____ cm

**31.** 12 mi = _____ in.

**32.** 36 m = _____ ft

**33.** 80 dm = _____ dam

**34.** 2.5 yd = _____ m

**35.** 6000 mm = _____ dm

**36.** 0.0635 mm = _____ in.

**37.** Match each measure in the first column with an equivalent measure in the second column by drawing connecting lines.   [9.1a], [9.2a]

| | |
|---|---|
| $\frac{1}{4}$ yd | 24,000 dm |
| 144 in. | 1320 yd |
| 2400 m | 2400 mm |
| 0.75 mi | 9 in. |
| 24 m | 0.024 km |
| 240 cm | 12 ft |

**38.** Arrange from smallest to largest:

      100 in.,   430 ft,   $\frac{1}{100}$ mi,   3.5 ft,   6000 ft,   1000 in.,   2 yd.   [9.1a]

**39.** Arrange from largest to smallest:

      3240 cm,   300 m,   250 dm,   150 hm,   33,000 mm,   310 dam,   13 km.   [9.2a]

**40.** Arrange from smallest to largest:

      2 yd,   1.5 mi,   65 cm,   $\frac{1}{2}$ ft,   3 km,   2.5 m.   [9.3a]

# Understanding Through Discussion and Writing

**41.** A student makes the following error:

      23 in. = 23 · (12 ft) = 276 ft.

Explain the error.   [9.1a]

**42.** Explain in your own words why metric units are easier to work with than American units.   [9.2a]

**43.** Recall the guidelines for conversion: (1) "If the conversion is from a larger unit to a smaller unit, substitute." (2) "If the conversion is from a smaller unit to a larger unit, multiply by 1." Explain why each is the easier way to convert in that situation.   [9.1a], [9.2a]

**44.** Do some research in a library or on the Internet about the metric system versus the American system. Why do you think the United States has not converted to the metric system?   [9.3a]

# 9.4 Weight and Mass; Medical Applications

There is a difference between **mass** and **weight**, but the terms are often used interchangeably. People sometimes use the word "weight" when, technically, they are referring to "mass." Weight is related to the force of gravity. The farther you are from the center of the earth, the less you weigh. Your mass stays the same no matter where you are.

## a Weight: The American System

| AMERICAN UNITS OF WEIGHT | |
|---|---|
| 1 ton (T) = 2000 pounds (lb) | 1 lb = 16 ounces (oz) |

The term "ounce" used here for weight is different from the "ounce" we will use for capacity in Section 9.5. We convert units of weight using the same techniques that we use with linear measures.

**EXAMPLE 1** A well-known hamburger is called a "quarter-pounder." Find its name in ounces: a "_____ ouncer."

Since we are converting from a larger unit to a smaller unit, we use substitution.

$$\frac{1}{4} \text{ lb} = \frac{1}{4} \cdot 1 \text{ lb} = \frac{1}{4} \cdot 16 \text{ oz} \qquad \text{Substituting 16 oz for 1 lb}$$

$$= 4 \text{ oz}$$

A "quarter-pounder" can also be called a "four-ouncer."

**EXAMPLE 2** Complete: 15,360 lb = _____ T.

Since we are converting from a smaller unit to a larger unit, we use multiplying by 1.

$$15,360 \text{ lb} = 15,360 \text{ lb} \times \frac{1 \text{ T}}{2000 \text{ lb}} \qquad \text{Multiplying by 1}$$

$$= \frac{15,360}{2000} \text{ T} = 7.68 \text{ T}$$

Do Margin Exercises 1–3.

## b Mass: The Metric System

The basic unit of mass is the **gram** (g), which is the mass of 1 cubic centimeter (1 cm³) of water. Since a cubic centimeter is small, a gram is a small unit of mass.

$$1 \text{ g} = 1 \text{ gram} = \text{the mass of 1 cm}^3 \text{ of water}$$

## OBJECTIVES

**a** Convert from one American unit of weight to another.

**b** Convert from one metric unit of mass to another.

**c** Make conversions and solve applied problems concerning medical dosages.

**SKILL TO REVIEW**
Objective 3.4a: Multiply an integer and a fraction.

Multiply.

**1.** $\frac{3}{4} \cdot 16$

**2.** $2480 \times \frac{1}{2000}$

Complete.

**1.** 5 lb = _____ oz

**2.** 8640 lb = _____ T

**3.** 1 T = _____ oz

1 g = 1 cm³ of water

*Answers*

*Skill to Review:*
**1.** 12  **2.** $\frac{31}{25}$, or 1.24

*Margin Exercises:*
**1.** 80  **2.** 4.32  **3.** 32,000

1 metric ton (t) = 1000 kilograms (kg)

1 *kilo*gram (kg) = 1000 grams (g)

1 *hecto*gram (hg) = 100 grams (g)

1 *deka*gram (dag) = 10 grams (g)

1 gram (g)

1 *deci*gram (dg) = $\frac{1}{10}$ gram (g)

1 *centi*gram (cg) = $\frac{1}{100}$ gram (g)

1 *milli*gram (mg) = $\frac{1}{1000}$ gram (g)

The table at left lists the metric units of mass. The prefixes are the same as those for length.

## Thinking Metric

One gram is about the mass of 1 raisin or 1 package of artificial sweetener. Since 1 kg is about 2.2 lb, 1000 kg is about 2200 lb, or 1 metric ton (t), which is just a little more than 1 American ton (T), which is 2000 lb.

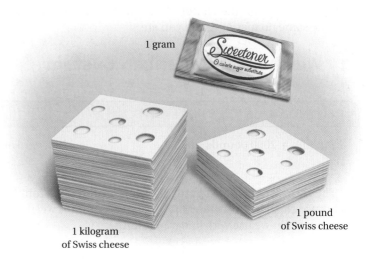

1 gram

1 kilogram
of Swiss cheese

1 pound
of Swiss cheese

Small masses, such as dosages of medicine and vitamins, may be measured in milligrams (mg). The gram (g) is used for objects ordinarily measured in ounces, such as the mass of a letter, a piece of candy, a coin, or a small package of food.

Complete with mg, g, kg, or t.

**4.** A laptop computer has a mass of 6 _____.

**5.** Eric has a body mass of 85.4 _____.

**6.** This is a 3-_____ vitamin.

**7.** A pen has a mass of 12 _____.

**8.** A sport utility vehicle has a mass of 3 _____.

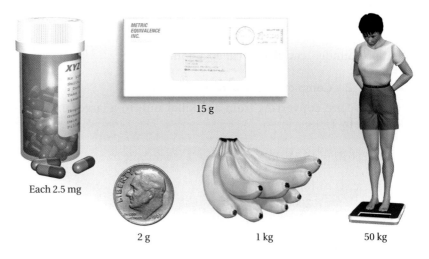

Each 2.5 mg

15 g

2 g

1 kg

50 kg

The kilogram (kg) is used for larger food packages, such as fruit, or for human body mass. The metric ton (t) is used for very large masses, such as the mass of an automobile, a truckload of gravel, or an airplane.

Do Exercises 4–8.

*Answers*

**4.** kg    **5.** kg    **6.** mg    **7.** g    **8.** t

## Changing Units Mentally

As before, changing from one metric unit to another amounts only to the movement of a decimal point. We use this table.

| 1000 g | 100 g | 10 g | 1 g | 0.1 g | 0.01 g | 0.001 g |
|--------|-------|------|-----|-------|--------|---------|
| 1 kg | 1 hg | 1 dag | 1 g | 1 dg | 1 cg | 1 mg |

**EXAMPLE 3**  Complete: 8 kg = _____ g.

*Think:* To go from kg to g in the table is a move of three places to the right. Thus we move the decimal point three places to the right.

| 1000 g | 100 g | 10 g | 1 g | 0.1 g | 0.01 g | 0.001 g |
|--------|-------|------|-----|-------|--------|---------|
| 1 kg | 1 hg | 1 dag | 1 g | 1 dg | 1 cg | 1 mg |

3 places to the right

8.0    8.000.    8 kg = 8000 g

**EXAMPLE 4**  Complete: 4235 g = _____ kg.

*Think:* To go from g to kg in the table is a move of three places to the left. Thus we move the decimal point three places to the left.

| 1000 g | 100 g | 10 g | 1 g | 0.1 g | 0.01 g | 0.001 g |
|--------|-------|------|-----|-------|--------|---------|
| 1 kg | 1 hg | 1 dag | 1 g | 1 dg | 1 cg | 1 mg |

3 places to the left

4235.0    4.235.0    4235 g = 4.235 kg

Do Exercises 9 and 10.

**EXAMPLE 5**  Complete: 6.98 cg = _____ mg.

*Think:* To go from cg to mg is a move of one place to the right. Thus we move the decimal point one place to the right.

| 1000 g | 100 g | 10 g | 1 g | 0.1 g | 0.01 g | 0.001 g |
|--------|-------|------|-----|-------|--------|---------|
| 1 kg | 1 hg | 1 dag | 1 g | 1 dg | 1 cg | 1 mg |

1 place to the right

6.98    6.9.8    6.98 cg = 69.8 mg

Complete.

**9.** 6.2 kg = _____ g

**10.** 304.8 cg = _____ g

*Answers*

**9.** 6200    **10.** 3.048

The most commonly used metric units of mass are kg, g, cg, and mg. We have purposely used those more than the others in the exercises.

**EXAMPLE 6** Complete: 89.21 mg = _____ g.

*Think:* To go from mg to g is a move of three places to the left. Thus we move the decimal point three places to the left.

| 1000 g | 100 g | 10 g | 1 g | 0.1 g | 0.01 g | 0.001 g |
|--------|-------|------|-----|-------|--------|---------|
| 1 kg | 1 hg | 1 dag | 1 g | 1 dg | 1 cg | 1 mg |

3 places to the left

89.21    0.089.21    89.21 mg = 0.08921 g

Do Exercises 11–13.

### (c) Medical Applications

Another metric unit that is used in medicine is the microgram (mcg). It is defined as follows.

> **MICROGRAM**
>
> $1 \text{ microgram} = 1 \text{ mcg} = \dfrac{1}{1,000,000} \text{ g} = 0.000001 \text{ g}$
>
> $1,000,000 \text{ mcg} = 1 \text{ g}$

**EXAMPLE 7** Complete: 1 mg = _____ mcg.

We convert to grams and then to micrograms:

$$1 \text{ mg} = 0.001 \text{ g}$$
$$= 0.001 \times 1 \text{ g}$$
$$= 0.001 \times 1,000,000 \text{ mcg} \quad \text{Substituting 1,000,000 mcg for 1 g}$$
$$= 1000 \text{ mcg}.$$

Do Exercise 14.

**EXAMPLE 8** *Medical Dosage.* Nitroglycerin sublingual tablets come in 0.4-mg tablets. How many micrograms are in each tablet?
Source: Steven R. Smith, M.D.

We are to complete: 0.4 mg = _____ mcg. Thus,

$$0.4 \text{ mg} = 0.4 \times 1 \text{ mg}$$
$$= 0.4 \times 1000 \text{ mcg} \quad \text{From Example 7, substituting 1000 mcg for 1 mg}$$
$$= 400 \text{ mcg}.$$

We can also do this problem in a manner similar to Example 7.

Do Exercise 15.

---

Complete.

**11.** 7.7 cg = _____ mg

**12.** 2344 mg = _____ cg

**13.** 67 dg = _____ mg

**14.** Complete:
$$1 \text{ mcg} = \text{_____ mg}.$$

**15. Medical Dosage.** A physician prescribes 500 mcg of alprazolam, an antianxiety medication. How many milligrams is this dosage?
**Source:** Steven R. Smith, M.D.

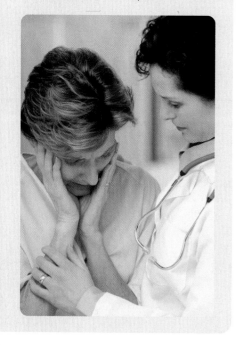

*Answers*

**11.** 77    **12.** 234.4    **13.** 6700    **14.** 0.001
**15.** 0.5 mg

**9.4** Exercise Set

For Extra Help

*MyMathLab*

Math XL
PRACTICE

WATCH

DOWNLOAD

READ

REVIEW

**a** Complete.

**1.** 1 T = _____ lb

**2.** 1 lb = _____ oz

**3.** 6000 lb = _____ T

**4.** 8 T = _____ lb

**5.** 4 lb = _____ oz

**6.** 10 lb = _____ oz

**7.** 6.32 T = _____ lb

**8.** 8.07 T = _____ lb

**9.** 3200 oz = _____ T

**10.** 6400 oz = _____ T

**11.** 80 oz = _____ lb

**12.** 960 oz = _____ lb

**13.** *Excelsior.* Western Excelsior is a company that makes a packing material called excelsior. Excelsior is produced from the wood of aspen trees. The largest grove of aspen trees on record contained 13,000,000 tons of aspen. How many pounds of aspen were there?
Source: Western Excelsior Corporation, Mancos, Colorado

**14.** *Excelsior.* Western Excelsior buys 44,800 tons of aspen each year to make excelsior. How many pounds of aspen does it buy each year?
Source: Western Excelsior Corporation, Mancos, Colorado

**b** Complete.

**15.** 1 kg = _____ g

**16.** 1 hg = _____ g

**17.** 1 dag = _____ g

**18.** 1 dg = _____ g

**19.** 1 cg = _____ g

**20.** 1 mg = _____ g

**21.** 1 g = _____ mg

**22.** 1 g = _____ cg

**23.** 1 g = _____ dg

**24.** 25 kg = _____ g

**25.** 234 kg = _____ g

**26.** 9403 g = _____ kg

**27.** 5200 g = _____ kg

**28.** 1.506 kg = _____ g

**29.** 67 hg = _____ kg

**30.** 45 cg = _____ g

**31.** 0.502 dg = _____ g

**32.** 0.0025 cg = _____ mg

**33.** 8492 g = _____ kg

**34.** 9466 g = _____ kg

**35.** 585 mg = _____ cg

**36.** 96.1 mg = _____ cg

**37.** 8 kg = _____ cg

**38.** 0.06 kg = _____ mg

**39.** 1 t = _____ kg

**40.** 2 t = _____ kg

**41.** 3.4 cg = _____ dag

**42.** 115 mg = _____ g

**43.** 60.3 kg = _____ t

**44.** 15.68 kg = _____ t

**c** Complete.

**45.** 1 mg = _____ mcg

**46.** 1 mcg = _____ mg

**47.** 325 mcg = _____ mg

**48.** 0.45 mg = _____ mcg

**49.** 210.6 mg = _____ mcg

**50.** 8000 mcg = _____ mg

**51.** 4.9 mcg = _____ mg

**52.** 0.075 mg = _____ mcg

*Medical Dosage.*    Solve each of the following. (None of these medications should be taken without consulting your own physician.)
**Source:** Steven R. Smith, M.D.

**53.** Digoxin is a medication used to treat heart problems. A physician orders 0.125 mg of digoxin to be taken once daily. How many micrograms of digoxin are there in the daily dosage?

**54.** Digoxin is a medication used to treat heart problems. A physician orders 0.25 mg of digoxin to be taken once a day. How many micrograms of digoxin are there in the daily dosage?

**55.** Triazolam is a medication used for the short-term treatment of insomnia. A physician advises her patient to take one of the 0.125-mg tablets each night for 7 nights. How many milligrams of triazolam will the patient have ingested over that 7-day period? How many micrograms?

**56.** Clonidine is a medication used to treat high blood pressure. The usual starting dose of clonidine is one 0.1-mg tablet twice a day. If a patient is started on this dose by his physician, how many total milligrams of clonidine will the patient have taken before he returns to see his physician 14 days later? How many micrograms?

**57.** Cephalexin is an antibiotic that frequently is prescribed in a 500-mg tablet form. A physician prescribes 2 g of cephalexin per day for a patient with a skin sore. How many 500-mg tablets would have to be taken in order to achieve this daily dosage?

**58.** Quinidine gluconate is a liquid mixture, part medicine and part water, that is administered intravenously. There are 80 mg of quinidine gluconate in each cubic centimeter (cc) of the liquid mixture. A physician orders 900 mg of quinidine gluconate to be administered daily to a patient with malaria. How much of the solution would have to be administered in order to achieve the recommended daily dosage?

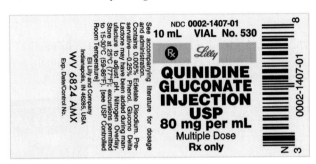

**59.** Amoxicillin is a common antibiotic prescribed for children. It is a liquid suspension composed of part amoxicillin and part water. In one formulation of amoxicillin suspension, there are 250 mg of amoxicillin in 5 cc of the liquid suspension. A physician prescribes 400 mg per day for a 2-year-old child with an ear infection. How much of the amoxicillin liquid suspension would the child's parent need to administer in order to achieve the recommended daily dosage of amoxicillin?

**60.** Albuterol is a medication used for the treatment of asthma. It comes in an inhaler that contains 17 mg of albuterol mixed with a liquid. One actuation (inhalation) from the mouthpiece delivers a 90-mcg dose of albuterol.

**a)** A physician orders 2 inhalations 4 times per day. How many micrograms of albuterol does the patient inhale per day?

**b)** How many actuations/inhalations are contained in one inhaler?

**c)** Danielle is leaving for 4 months of college and wants to take enough albuterol to last for that time. Her physician has prescribed 2 inhalations 4 times per day. How many inhalers will Danielle need to take with her for the 4-month period?

## Skill Maintenance

Convert to fraction notation.  [7.2b]

**61.** 35%

**62.** 99%

**63.** 85.5%

**64.** 34.2%

**65.** $37\frac{1}{2}\%$

**66.** $66.\overline{6}\%$

**67.** $83.\overline{3}\%$

**68.** $16\frac{2}{3}\%$

Solve.

**69.** A country has a population that is increasing by 4% each year. This year the population is 180,000. What will it be next year?    [7.5b]

**70.** A college has a student body of 1850 students. Of these, 17.5% are seniors. How many students are seniors? [7.5a]

**71.** A state charges a meals tax of $4\frac{1}{2}$%. What is the meals tax charged on a dinner party costing $540?    [7.6a]

**72.** The price of a microwave oven was reduced from $350 to $308. Find the percent of decrease in price.    [7.5b]

**73.** A ream of paper contains 500 sheets. How many sheets are there in 15 reams?    [1.8a]

**74.** A lab technician separates a vial containing 140 cc of blood into test tubes, each of which contains 3 cc of blood. How many test tubes can be filled? How much blood is left over?    [1.8a]

# Synthesis

**75.** A box of gelatin-mix packages weighs $15\frac{3}{4}$ lb. Each package weighs $1\frac{3}{4}$ oz. How many packages are in the box?

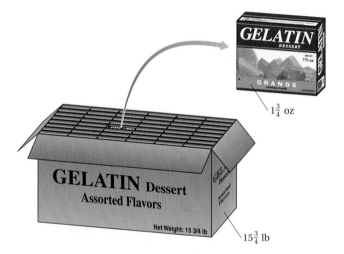

$1\frac{3}{4}$ oz

$15\frac{3}{4}$ lb

**76.** At $1.59 a dozen, the cost of eggs is $1.06 per pound. How much does an egg weigh?

**77.** *Large Diamonds.*   A **carat** (also spelled **karat**) is a unit of weight for precious stones; 1 carat = 200 mg. The Golden Jubilee Diamond weighs 545.67 carats and is the largest cut diamond in the world. The Hope Diamond, located at the Smithsonian Institution Museum of Natural History, weighs 45.52 carats.

Source: *National Geographic*, February 2001

**a)** How many grams does the Golden Jubilee Diamond weigh?

**b)** How many grams does the Hope Diamond weigh?

**c)** ▦ Given that 1 lb = 453.6 g, how many ounces does each diamond weigh?

# 9.5 Capacity; Medical Applications

## a Capacity

### American Units

To answer a question like "How much soda is in the can?" we need measures of **capacity**. American units of capacity are fluid ounces, cups, pints, quarts, and gallons. These units are related as follows.

> **AMERICAN UNITS OF CAPACITY**
>
> 1 gallon (gal) = 4 quarts (qt)    1 pt = 2 cups
> = 16 fluid ounces (fl oz)
> 1 qt = 2 pints (pt)    1 cup = 8 fluid oz

Fluid ounces, abbreviated fl oz, are often referred to as ounces, or oz.

**EXAMPLE 1**  Complete:  9 gal = _____ oz.

Since we are converting from a *larger* unit to a *smaller* unit, we use substitution:

$$9 \text{ gal} = 9 \cdot 1 \text{ gal} = 9 \cdot 4 \text{ qt} \qquad \text{Substituting 4 qt for 1 gal}$$
$$= 9 \cdot 4 \cdot 1 \text{ qt} = 9 \cdot 4 \cdot 2 \text{ pt} \qquad \text{Substituting 2 pt for 1 qt}$$
$$= 9 \cdot 4 \cdot 2 \cdot 1 \text{ pt} = 9 \cdot 4 \cdot 2 \cdot 16 \text{ oz} \qquad \text{Substituting 16 oz for 1 pt}$$
$$= 1152 \text{ oz}.$$

**EXAMPLE 2**  Complete:  24 qt = _____ gal.

Since we are converting from a *smaller* unit to a *larger* unit, we multiply by 1 using 1 gal in the numerator and 4 qt in the denominator:

$$24 \text{ qt} = 24 \text{ qt} \cdot \frac{1 \text{ gal}}{4 \text{ qt}} = \frac{24}{4} \cdot 1 \text{ gal} = 6 \text{ gal}.$$

Do Exercises 1 and 2.

### Metric Units

One unit of capacity in the metric system is a **liter**. A liter is just a bit more than a quart. It is defined as follows.

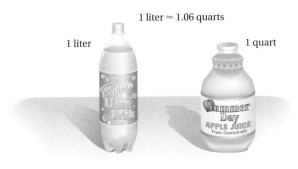

1 liter ≈ 1.06 quarts

1 liter      1 quart

## OBJECTIVES

a Convert from one unit of capacity to another.

b Solve applied problems concerning medical dosages.

Complete.

1. 5 gal = _____ pt

2. 80 qt = _____ gal

**Answers**

1. 40    2. 20

## METRIC UNITS OF CAPACITY

1 liter (L) = 1000 cubic centimeters (1000 cm$^3$)

The script letter $\ell$ is also used for "liter."

The metric prefixes are also used with liters. The most common is **milli-**. The milliliter (mL) is, then, $\frac{1}{1000}$ liter. Thus,

$$1\,L = 1000\,mL = 1000\,cm^3;$$
$$0.001\,L = 1\,mL = 1\,cm^3.$$

Although the other metric prefixes are rarely used for capacity, we display them in the following table as we did for linear measure.

| 1000 L | 100 L | 10 L | 1 L | 0.1 L | 0.01 L | 0.001 L |
|--------|-------|------|-----|-------|--------|---------|
| 1 kL | 1 hL | 1 daL | 1 L | 1 dL | 1 cL | 1 mL (cc) |

A preferred unit for drug dosage is the milliliter (mL) or the cubic centimeter (cm$^3$). The notation "cc" is also used for cubic centimeter, especially in medicine. The milliliter and the cubic centimeter represent the same measure of capacity. A milliliter is about $\frac{1}{5}$ of a teaspoon.

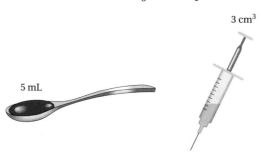

3 cm$^3$

5 mL

$$1\,mL = 1\,cm^3 = 1\,cc$$

Volumes for which quarts and gallons are used are expressed in liters. Large volumes in business and industry are expressed using measures of cubic meters (m$^3$).

Do Exercises 3–6.

Complete with mL or L.

**3.** The patient received an injection of 2 _____ of penicillin.

**4.** There are 250 _____ in a coffee cup.

**5.** The gas tank holds 80 _____.

**6.** Bring home 8 _____ of milk.

**EXAMPLE 3** Complete: 4.5 L = _____ mL.

$$4.5\,L = 4.5 \times 1\,L = 4.5 \times 1000\,mL \qquad \text{Substituting 1000 mL for 1 L}$$
$$= 4500\,mL$$

| 1000 L | 100 L | 10 L | 1 L | 0.1 L | 0.01 L | 0.001 L |
|--------|-------|------|-----|-------|--------|---------|
| 1 kL | 1 hL | 1 daL | 1 L | 1 dL | 1 cL | 1 mL (cc) |

3 places to the right

*Answers*

**3.** mL   **4.** mL   **5.** L   **6.** L

**EXAMPLE 4**  Complete: 280 mL = _____ L.

$$280 \text{ mL} = 280 \times 1 \text{ mL}$$
$$= 280 \times 0.001 \text{ L} \qquad \text{Substituting 0.001 L for 1 mL}$$
$$= 0.28 \text{ L}$$

| 1000 L | 100 L | 10 L | 1 L | 0.1 L | 0.01 L | 0.001 L |
|--------|-------|------|-----|-------|--------|---------|
| 1 kL | 1 hL | 1 daL | 1 L | 1 dL | 1 cL | 1 mL (cc) |

3 places to the left

We do find metric units of capacity in frequent use in the United States—for example, in sizes of soda bottles and automobile engines.

Do Exercises 7 and 8.

Complete.

**7.** 0.97 L = _____ mL

**8.** 8990 mL = _____ L

## b  Medical Applications

The metric system is used extensively in medicine.

**EXAMPLE 5**  *Medical Dosage.*  A physician orders 3.5 L of 5% dextrose in water (abbrev. D5W) to be administered over a 24-hr period. How many milliliters were ordered?

We convert 3.5 L to milliliters:

$$3.5 \text{ L} = 3.5 \times 1 \text{ L} = 3.5 \times 1000 \text{ mL} = 3500 \text{ mL}.$$

The physician ordered 3500 mL of D5W.

Do Exercise 9.

**9. Medical Dosage.**  A physician orders 2400 mL of 0.9% saline solution to be administered intravenously over a 24-hr period. How many liters were ordered?

**EXAMPLE 6**  *Medical Dosage.*  Liquids at a pharmacy are often labeled in liters or milliliters. Thus if a physician's prescription is given in ounces, it must be converted. For conversion, a pharmacist knows that 1 fluid oz ≈ 29.57 mL.* A prescription calls for 3 fluid oz of theophylline. For how many milliliters is the prescription?

We convert as follows:

$$3 \text{ oz} = 3 \times 1 \text{ oz} \approx 3 \times 29.57 \text{ mL} = 88.71 \text{ mL}.$$

The prescription calls for 88.71 mL of theophylline.

Do Exercise 10.

**10. Medical Dosage.**  A prescription calls for 2 oz of theophylline.

**a)** For how many milliliters is the prescription?

**b)** For how many liters is the prescription?

---

*In practice, most pharmacists use 30 mL as an approximation to 1 oz.

**a** Complete.

**1.** 1 L = _____ mL = _____ cm³

**2.** _____ L = 1 mL = _____ cm³

**3.** 87 L = _____ mL

**4.** 806 L = _____ mL

**5.** 49 mL = _____ L

**6.** 19 mL = _____ L

**7.** 0.401 mL = _____ L

**8.** 0.816 mL = _____ L

**9.** 78.1 L = _____ cm³

**10.** 99.6 L = _____ cm³

**11.** 10 qt = _____ oz

**12.** 9.6 oz = _____ pt

**13.** 20 cups = _____ pt

**14.** 1 gal = _____ oz

**15.** 8 gal = _____ qt

**16.** 1 gal = _____ cups

**17.** 5 gal = _____ qt

**18.** 11 gal = _____ qt

**19.** 56 qt = _____ gal

**20.** 84 qt = _____ gal

**21.** 11 gal = _____ pt

**22.** 5 gal = _____ pt

Complete.

| | OBJECT | GALLONS (gal) | QUARTS (qt) | PINTS (pt) | CUPS | OUNCES (oz) |
|---|---|---|---|---|---|---|
| **23.** | 12-can package of 12-oz sodas | | | | | 144 |
| **24.** | 6-bottle package of 16-oz sodas | | 3 | | | |
| **25.** | Full tank of gasoline | 16 | | | | |
| **26.** | Container of milk | | | 8 | | |
| **27.** | Tropicana Punch | | | | 4 | |
| **28.** | Dove shampoo | | | | | 12 |
| **29.** | Downy fabric softener | | | | | 51 |
| **30.** | Williams Electric Shave | | | | | 7 |

Complete.

| | OBJECT | LITERS (L) | MILLILITERS (mL) | CUBIC CENTIMETERS (cc) | CUBIC CENTIMETERS (cm³) |
|---|---|---|---|---|---|
| 31. | 2-L bottle of soda | 2 | | | |
| 32. | Heinz Vinegar | | 3755 | | |
| 33. | Full tank of gasoline in Europe | 64 | | | |
| 34. | Williams Electric Shave | | | | 207 |
| 35. | Dove shampoo | | | 355 | |
| 36. | Newman's Own Salad Dressing | | 473 | | |

**b** *Medical Dosage.* Solve each of the following.
**Source:** Steven R. Smith, M.D.

37. An emergency-room physician orders 2.0 L of Ringer's lactate to be administered over 2 hr for a patient in shock. How many milliliters is this?

38. An emergency-room physician orders 2.5 L of 0.9% saline solution over 4 hr for a patient suffering from dehydration. How many milliliters is this?

39. A physician orders 320 mL of 5% dextrose in water (D5W) solution to be administered intravenously over 4 hr. How many liters of D5W is this?

40. A physician orders 40 mL of 5% dextrose in water (D5W) solution to be administered intravenously over 2 hr to an elderly patient. How many liters of D5W is this?

41. A physician orders 0.5 oz of magnesia and alumina oral suspension antacid 4 times per day for a patient with indigestion. How many milliliters of the antacid is the patient to ingest in a day?

42. A physician orders 0.25 oz of magnesia and alumina oral suspension antacid 3 times per day for a child with upper abdominal discomfort. How many milliliters of the antacid is the child to ingest in a day?

43. A physician orders 0.5 L of normal saline solution. How many milliliters are ordered?

44. A physician has ordered that his patient receive 60 mL per hour of normal saline solution intravenously. How many liters of the saline solution is the patient to receive in a 24-hr period?

45. A physician wants her patient to receive 3.0 L of normal saline intravenously over a 24-hr period. How many milliliters per hour must the nurse administer?

46. A physician tells a patient to purchase 0.5 L of hydrogen peroxide. Commercially, hydrogen peroxide is found on the shelf in bottles that hold 4 oz, 8 oz, and 16 oz. Which bottle comes closest to filling the prescription?

*Medical Dosage.* Because patients do not always have a working knowledge of the metric system, physicians often prescribe dosages in teaspoons (t or tsp) and tablespoons (T or Tbsp). The units are related to the metric system and to each other as follows:

$$5 \text{ mL} \approx 1 \text{ tsp}, \qquad 3 \text{ tsp} = 1 \text{ T}.$$

Complete.

**47.** 45 mL = _____ tsp

**48.** 3 T = _____ tsp

**49.** 1 mL = _____ tsp

**50.** 18.5 mL = _____ tsp

**51.** 2 T = _____ tsp

**52.** 8.5 tsp = _____ T

**53.** 1 T = _____ mL

**54.** 18.5 mL = _____ T

## Skill Maintenance

Convert to percent notation.   [7.1b], [7.2a]

**55.** 0.452

**56.** 0.999

**57.** $\dfrac{1}{3}$

**58.** $\dfrac{2}{3}$

**59.** $\dfrac{11}{20}$

**60.** $\dfrac{21}{20}$

**61.** $\dfrac{22}{25}$

**62.** $\dfrac{2}{25}$

**63.** *Lumber Consumption.*   Of the 71.3 billion board feet of lumber consumed in 2006 in the United States, 60.7 billion board feet were softwood and 10.6 billion board feet were hardwood. What is the ratio of hardwood board feet to total amount of board feet? What is the ratio of softwood board feet to hardwood board feet?   [6.1a]

**Source:** U.S. Forest Service

**64.** One person in four plays a musical instrument. In a given group of people, what is the ratio of those who play an instrument to total number of people? What is the ratio of those who do not play an instrument to total number of people? What is the ratio of the number of those who play to the number of those who do not play?   [6.1a]

## Synthesis

**65.** *Bees and Honey.*   The average bee produces only $\frac{1}{8}$ teaspoon of honey in its lifetime. It takes 60,000 honeybees to produce 100 lb of honey. How much does a teaspoon of honey weigh? Express the answer in ounces.

**67.** *Cost of Gasoline.*   Suppose that premium gasoline is selling for about \$2.98/gal. Using the fact that 1 L = 1.057 qt, determine the price of the gasoline in dollars per liter.

**66.** *Wasting Water.*   Many people leave the water running while they are brushing their teeth. Suppose that 32 oz of water is wasted in such a way each day by one person. How much water, in gallons, is wasted in a week? in a month (30 days)? in a year (365 days)? Assuming each of the 306 million people in the United States wastes water in this way, estimate how much water is wasted in the United States in a day; in a year.

# 9.6 Time and Temperature

## a Time

A table of units of time is shown below. The metric system sometimes uses "h" for hour and "s" for second, but we will use the more familiar "hr" and "sec."

> **UNITS OF TIME**
>
> 1 day  = 24 hours (hr)
> 1 hr  = 60 minutes (min)
> 1 min = 60 seconds (sec)
>
> 1 year (yr) = $365\frac{1}{4}$ days
> 1 week (wk) = 7 days

The earth revolves completely around the sun in $365\frac{1}{4}$ days. Since we cannot have $\frac{1}{4}$ day on the calendar, we give each year 365 days and every fourth year 366 days (a leap year), unless it is a year at the beginning of a century not divisible by 400.

**EXAMPLE 1**  Complete: 1 hr = _____ sec.

$$1\text{ hr} = 60\text{ min}$$
$$= 60 \cdot 1\text{ min}$$
$$= 60 \cdot 60\text{ sec} \qquad \text{Substituting 60 sec for 1 min}$$
$$= 3600\text{ sec}$$

**EXAMPLE 2**  Complete: 5 yr = _____ days.

$$5\text{ yr} = 5 \cdot 1\text{ yr}$$
$$= 5 \cdot 365\frac{1}{4}\text{ days} \qquad \text{Substituting } 365\frac{1}{4} \text{ days for 1 yr}$$
$$= 5 \cdot \frac{1461}{4}\text{ days}$$
$$= \frac{7305}{4}\text{ days}$$
$$= 1826\frac{1}{4}\text{ days}$$

**EXAMPLE 3**  Complete: 4320 min = _____ days.

$$4320\text{ min} = 4320\ \cancel{\text{min}} \cdot \frac{1\ \cancel{\text{hr}}}{60\ \cancel{\text{min}}} \cdot \frac{1\text{ day}}{24\ \cancel{\text{hr}}} = \frac{4320}{60 \cdot 24}\text{ days} = 3\text{ days}$$

> Do Margin Exercises 1–4.

Complete.

1. 2 hr = _____ min

2. 4 yr = _____ days

3. 1 day = _____ min

4. 168 hr = _____ wk

## b Temperature

Below are two temperature scales: **Fahrenheit** for American measure and **Celsius** for metric measure.

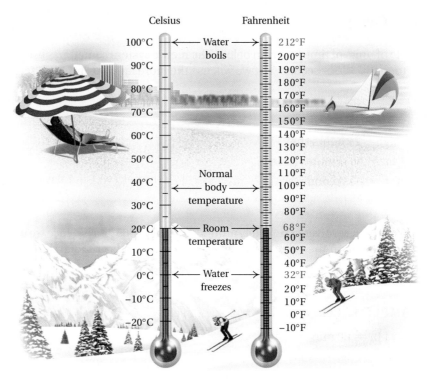

By laying a straight edge horizontally between the scales, we can make an approximate conversion from one measure of temperature to the other and get an idea of how the temperature scales compare.

| **EXAMPLES** Convert to Celsius using the scales shown above. Approximate to the nearest ten degrees.

**4.** 212°F (Boiling point of water)    100°C    This is exact.

**5.** 32°F (Freezing point of water)    0°C    This is exact.

**6.** 105°F                                        40°C    This is approximate.

Do Exercises 5–7.

| **EXAMPLES** Make an approximate conversion to Fahrenheit using the scales shown above.

**7.** 44°C (Hot bath)                        110°F    This is approximate.

**8.** 20°C (Room temperature)          68°F    This is exact.

**9.** 83°C                                        180°F    This is approximate.

Do Exercises 8–10.

---

Convert to Celsius. Approximate to the nearest ten degrees.

**5.** 180°F (Brewing coffee)

**6.** 25°F (Cold day)

**7.** −10°F (Miserably cold day)

---

Convert to Fahrenheit. Approximate to the nearest ten degrees.

**8.** 25°C (Warm day at the beach)

**9.** 40°C (Temperature of a patient with a high fever)

**10.** 10°C (Cold bath)

---

*Answers*

**5.** 80°C    **6.** 0°C    **7.** −20°C    **8.** 80°F
**9.** 100°F    **10.** 50°F

The following formula allows us to make exact conversions from Celsius to Fahrenheit.

**CELSIUS TO FAHRENHEIT**

$$F = \frac{9}{5} \cdot C + 32, \quad \text{or} \quad F = 1.8 \cdot C + 32$$

$\left(\text{Multiply the Celsius temperature by } \frac{9}{5}, \text{ or } 1.8, \text{ and add } 32.\right)$

**EXAMPLES** Convert to Fahrenheit.

**10.** 0°C (Freezing point of water)

$$F = \frac{9}{5} \cdot C + 32 = \frac{9}{5} \cdot 0 + 32 = 0 + 32 = 32$$

Thus, 0°C = 32°F.

**11.** 37°C (Normal body temperature)

$$F = 1.8 \cdot C + 32 = 1.8 \cdot 37 + 32 = 66.6 + 32 = 98.6$$

Thus, 37°C = 98.6°F.

Check the answers to Examples 10 and 11 using the scales on p. 546.

Do Exercises 11 and 12.

The following formula allows us to make exact conversions from Fahrenheit to Celsius.

**FAHRENHEIT TO CELSIUS**

$$C = \frac{5}{9} \cdot (F - 32), \quad \text{or} \quad C = \frac{(F - 32)}{1.8}$$

$\left(\text{Subtract } 32 \text{ from the Fahrenheit temperature and multiply by } \frac{5}{9} \text{ or divide by } 1.8.\right)$

**EXAMPLES** Convert to Celsius.

**12.** 212°F (Boiling point of water)

$$C = \frac{5}{9} \cdot (F - 32)$$

$$= \frac{5}{9} \cdot (212 - 32)$$

$$= \frac{5}{9} \cdot 180 = 100$$

Thus, 212°F = 100°C.

**13.** 5°F

$$C = \frac{F - 32}{1.8}$$

$$= \frac{5 - 32}{1.8}$$

$$= \frac{-27}{1.8} = -15$$

Thus, 5°F = −15°C.

Check the answers to Examples 12 and 13 using the scales on p. 546.

Do Exercises 13 and 14.

---

Convert to Fahrenheit.

**11.** 80°C

**12.** 35°C

**Calculator Corner**

**Temperature Conversions** Temperature conversions can be done quickly using a calculator. To convert 37°C to Fahrenheit, for example, we press $\boxed{1}\boxed{.}\boxed{8}\boxed{\times}\boxed{3}\boxed{7}\boxed{+}$ $\boxed{3}\boxed{2}\boxed{=}$. The calculator displays $\boxed{98.6}$, so 37°C = 98.6°F. We can convert 212°F to Celsius by pressing $\boxed{(}\boxed{2}\boxed{1}\boxed{2}\boxed{-}\boxed{3}\boxed{2}\boxed{)}$ $\boxed{\div}\boxed{1}\boxed{.}\boxed{8}\boxed{=}$. The display reads $\boxed{100}$, so 212°F = 100°C. Note that we must use parentheses when converting from Fahrenheit to Celsius in order to get the correct result.

**Exercises:** Use a calculator to convert each temperature to Fahrenheit.

**1.** 5°C

**2.** 50°C

Use a calculator to convert each temperature to Celsius.

**3.** 68°F

**4.** 113°F

Convert to Celsius.

**13.** 95°F

**14.** 14°F

*Answers*

**11.** 176°F   **12.** 95°F   **13.** 35°C
**14.** −10°C

**a** Complete.

**1.** 1 day = _____ hr

**2.** 1 hr = _____ min

**3.** 1 min = _____ sec

**4.** 1 wk = _____ days

**5.** 1 yr = _____ days

**6.** 2 yr = _____ days

**7.** 180 sec = _____ hr

**8.** 60 sec = _____ hr

**9.** 492 sec = _____ min
(The amount of time it takes for the rays of the sun to reach the earth)

**10.** 18,000 sec = _____ hr

**11.** 156 hr = _____ days

**12.** 444 hr = _____ days

**13.** 645 min = _____ hr

**14.** 375 min = _____ hr

**15.** 2 wk = _____ hr

**16.** 4 hr = _____ sec

**17.** 756 hr = _____ wk

**18.** 166,320 min = _____ wk

**19.** 2922 wk = _____ yr

**20.** 623 days = _____ wk

**21.** *Actual Time in a Day.* Although we round it to 24 hr, the actual length of a day is 23 hr, 56 min, and 4.2 sec. How many seconds are there in an actual day?
**Source:** *The Handy Geography Answer Book*

**22.** *Time Length.* What length of time is 86,400 sec? Is it 1 hr, 1 day, 1 week, or 1 month?

**b** Convert to Fahrenheit. Use the formula $F = \frac{9}{5} \cdot C + 32$ or $F = 1.8 \cdot C + 32$.

**23.** 25°C

**24.** 85°C

**25.** 40°C

**26.** 90°C

**27.** 86°C

**28.** 93°C

**29.** −20°C

**30.** −25°C

**31.** 2°C

**32.** 78°C

**33.** −24°C

**34.** −28°C

**35.** 3000°C
(The melting point of iron)

**36.** 1000°C
(The melting point of gold)

Convert to Celsius. Use the formula $C = \dfrac{5}{9} \cdot (F - 32)$ or $C = \dfrac{F - 32}{1.8}$.

**37.** 86°F

**38.** 59°F

**39.** −13°F

**40.** −4°F

**41.** 178°F

**42.** 195°F

**43.** 140°F

**44.** 107°F

**45.** 68°F

**46.** 50°F

**47.** 10°F

**48.** 0°F

**49.** 98.6°F
(Normal body temperature)

**50.** 104°F
(High-fevered body temperature)

**51.** *Highest Temperatures.* The highest temperature ever recorded in the world is 136°F in the desert of Libya in 1922. The highest temperature ever recorded in the United States is $56\frac{2}{3}$°C in California's Death Valley in 1913.
**Source:** *The Handy Geography Answer Book*

**a)** Convert each temperature to the other scale.
**b)** How much higher in degrees Fahrenheit was the world record than the U.S. record?

**52.** *Boiling Point and Altitude.* The boiling point of water actually changes with altitude. The boiling point is 212°F at sea level, but lowers about 1°F for every 500 ft that the altitude increases above sea level.
**Sources:** *The Handy Geography Answer Book; The New York Times Almanac*

**a)** What is the boiling point at an elevation of 1500 ft above sea level?
**b)** The elevation of Tucson is 2564 ft above sea level and that of Phoenix is 1117 ft. What is the boiling point in each city?
**c)** How much lower is the boiling point in Denver, whose elevation is 5280 ft, than in Tucson?
**d)** What is the boiling point at the top of Mt. McKinley in Alaska, the highest point in the United States, at 20,320 ft?

**53.** *Record Low Temperature.* The record low temperature in Hawaii as of August 2006 occurred on May 17, 1979. The record was 12°F at Mauna Kea Observatory. Convert 12°F to Celsius.

Source: National Climatic Data Center, NESDIS, NOAA, U.S. Department of Commerce

**54.** *Record High Temperature.* The record high temperature in Utah as of August 2006 occurred on July 5, 1985. The record was 117°F at Saint George. Convert 117°F to Celsius.

Source: National Climatic Data Center, NESDIS, NOAA, U.S. Department of Commerce

## Skill Maintenance

In each of Exercises 55–62, fill in the blank with the correct term from the given list. Some of the choices may not be used.

**55.** When interest is paid on interest, it is called _____ interest.   [7.7b]

**56.** A _____ is the quotient of two quantities.   [6.1a]

**57.** The _____ of a set of data is the middle number if there is an odd number of data items.   [8.1b]

**58.** When you work for a _____, you are paid a percentage of the total sales for which you are responsible.   [7.6b]

**59.** In _____ triangles, the lengths of their corresponding sides have the same ratio.   [6.5a]

**60.** To find the _____ of a set of data, add the numbers and then divide by the number of items of data.   [8.1a]

**61.** A natural number, other than 1, that is not prime is _____.   [3.1c]

**62.** In the sentence 35 − 11 = 24, 11 is called the _____.   [1.3a]

mean

median

mode

complementary

composite

commission

salary

ratio

difference

minuend

subtrahend

simple

compound

isosceles

similar

## Synthesis

**63.** Estimate the number of years in one million seconds.

**64.** Estimate the number of years in one billion seconds.

**65.** Estimate the number of years in one trillion seconds.

Complete.

**66.** $88 \dfrac{\text{ft}}{\text{sec}} = \underline{\hspace{1cm}} \dfrac{\text{mi}}{\text{hr}}$

**67.** $0.9 \dfrac{\text{L}}{\text{hr}} = \underline{\hspace{1cm}} \dfrac{\text{mL}}{\text{sec}}$

# 9.7 Converting Units of Area

## a American Units

Let's do some conversions from one American unit of area to another.

**EXAMPLE 1** Complete: $1\,\text{ft}^2 =$ _____ $\text{in}^2$.

$$1\,\text{ft}^2 = 1 \cdot (12\,\text{in.})^2 \qquad \text{Substituting 12 in. for 1 ft}$$
$$= 12\,\text{in.} \cdot 12\,\text{in.} = 144\,\text{in}^2$$

**EXAMPLE 2** Complete: $8\,\text{yd}^2 =$ _____ $\text{ft}^2$.

$$8\,\text{yd}^2 = 8 \cdot (3\,\text{ft})^2 \qquad \text{Substituting 3 ft for 1 yd}$$
$$= 8 \cdot 3\,\text{ft} \cdot 3\,\text{ft} = 8 \cdot 3 \cdot 3 \cdot \text{ft} \cdot \text{ft} = 72\,\text{ft}^2$$

Do Margin Exercises 1–3.

### AMERICAN UNITS OF AREA

1 square yard $(\text{yd}^2) = 9$ square feet $(\text{ft}^2)$
1 square foot $(\text{ft}^2) = 144$ square inches $(\text{in}^2)$
1 square mile $(\text{mi}^2) = 640$ acres
1 acre $= 43{,}560\,\text{ft}^2$

**EXAMPLE 3** Complete: $36\,\text{ft}^2 =$ _____ $\text{yd}^2$.

We are converting from "$\text{ft}^2$" to "$\text{yd}^2$." Thus we choose a symbol for 1 with $\text{yd}^2$ on top and $\text{ft}^2$ on the bottom.

$$36\,\text{ft}^2 = 36\,\text{ft}^2 \times \frac{1\,\text{yd}^2}{9\,\text{ft}^2} \qquad \text{Multiplying by 1 using } \frac{1\,\text{yd}^2}{9\,\text{ft}^2}$$
$$= \frac{36}{9} \times \frac{\text{ft}^2}{\text{ft}^2} \times 1\,\text{yd}^2 = 4\,\text{yd}^2$$

**EXAMPLE 4** Complete: $7\,\text{mi}^2 =$ _____ acres.

$$7\,\text{mi}^2 = 7 \cdot 1\,\text{mi}^2$$
$$= 7 \cdot 640\,\text{acres} \qquad \text{Substituting 640 acres for 1 mi}^2$$
$$= 4480\,\text{acres}$$

Do Exercises 4 and 5.

## b Metric Units

Let's now convert from one metric unit of area to another.

**EXAMPLE 5** Complete: $1\,\text{km}^2 =$ _____ $\text{m}^2$.

$$1\,\text{km}^2 = 1 \cdot (1000\,\text{m})^2 \qquad \text{Substituting 1000 m for 1 km}$$
$$= 1000\,\text{m} \cdot 1000\,\text{m} = 1{,}000{,}000\,\text{m}^2$$

---

### OBJECTIVES

**a** Convert from one American unit of area to another.

**b** Convert from one metric unit of area to another.

**SKILL TO REVIEW**
Objective 1.9b: Evaluate exponential notation.

Evaluate.
1. $9^2$
2. $10^3$

Complete.
1. $1\,\text{yd}^2 =$ _____ $\text{ft}^2$
2. $5\,\text{yd}^2 =$ _____ $\text{ft}^2$
3. $20\,\text{ft}^2 =$ _____ $\text{in}^2$

Complete.
4. $360\,\text{in}^2 =$ _____ $\text{ft}^2$
5. $5\,\text{mi}^2 =$ _____ acres

*Answers*

*Skill to Review:*
1. 81    2. 1000

*Margin Exercises:*
1. 9    2. 45    3. 2880
4. 2.5    5. 3200

**STUDY TIPS**

**LEARN FROM YOUR MISTAKES**

Immediately after each quiz or test, write out a step-by-step solution to any questions you missed. Visit your instructor during office hours or consult with a tutor for help with questions that are giving you trouble. These corrected solutions provide excellent review material for the final examination.

**EXAMPLE 6** Complete: $10,000 \text{ cm}^2 = \underline{\hspace{1.5cm}} \text{ m}^2$.

$$10,000 \text{ cm}^2 = 10,000 \text{ cm}^2 \cdot \frac{1 \text{ m}}{100 \text{ cm}} \cdot \frac{1 \text{ m}}{100 \text{ cm}}$$

$$= 10,000 \text{ cm}^2 \cdot \frac{1 \text{ m}^2}{10,000 \text{ cm}^2}$$

$$= 1 \text{ m}^2$$

Do Exercises 6 and 7.

## Mental Conversion

To convert mentally, we first note that $10^2 = 100$, $100^2 = 10,000$, and $0.1^2 = 0.01$. We use the table as before and multiply the number of places we move by 2 to determine the number of places to move the decimal point.

| 1000 m | 100 m | 10 m | 1 m | 0.1 m | 0.01 m | 0.001 m |
|--------|-------|------|-----|-------|--------|---------|
| 1 km | 1 hm | 1 dam | 1 m | 1 dm | 1 cm | 1 mm |

**EXAMPLE 7** Complete: $3.48 \text{ km}^2 = \underline{\hspace{1.5cm}} \text{ m}^2$.

*Think:* To go from km to m in the table is a move of 3 places to the right.

| 1000 m | 100 m | 10 m | 1 m | 0.1 m | 0.01 m | 0.001 m |
|--------|-------|------|-----|-------|--------|---------|
| 1 km | 1 hm | 1 dam | 1 m | 1 dm | 1 cm | 1 mm |

3 moves to the right

So we move the decimal point $2 \cdot 3$, or 6, places to the right.

$3.48 \qquad 3.480000. \qquad 3.48 \text{ km}^2 = 3,480,000 \text{ m}^2$

6 places to the right

**EXAMPLE 8** Complete: $586.78 \text{ cm}^2 = \underline{\hspace{1.5cm}} \text{ m}^2$.

*Think:* To go from cm to m in the table is a move of 2 places to the left.

| 1000 m | 100 m | 10 m | 1 m | 0.1 m | 0.01 m | 0.001 m |
|--------|-------|------|-----|-------|--------|---------|
| 1 km | 1 hm | 1 dam | 1 m | 1 dm | 1 cm | 1 mm |

2 moves to the left

So we move the decimal point $2 \cdot 2$, or 4, places to the left.

$586.78 \qquad 0.0586.78 \qquad 586.78 \text{ cm}^2 = 0.058678 \text{ m}^2$

4 places to the left

Do Exercises 8–10.

# Translating
# for Success

**1.** *Test Items.* On a test of 90 items, Sally got 80% correct. How many items did she get correct?

**2.** *Suspension Bridge.* The San Francisco/Oakland Bay suspension bridge is 0.4375 mi long. Convert this distance to yards.

**3.** *Population Growth.* City A's growth rate per year is 0.9%. If the population was 1,500,000 in 2005, what was the population in 2006?

**4.** *Roller Coaster Drop.* The Manhattan Express Roller Coaster at the New York–New York Hotel and Casino, Las Vegas, Nevada, has a 144-ft drop. The California Screamin' Roller Coaster at Disney's California Adventure, Anaheim, California, has a 32.635-m drop. How much larger, in meters, is the drop of the Manhattan Express than the drop of the California Screamin'?

**5.** *Driving Distance.* Nate drives the company car 675 mi in 15 days. At this rate, how far will he drive in 20 days?

---

The goal of these matching questions is to practice step (2), *Translate,* of the five-step problem-solving process. Translate each word problem to an equation and select a correct translation from equations A–O.

**A.** $32.635 \text{ m} + x = 144 \text{ ft} \cdot \dfrac{0.305 \text{ m}}{1 \text{ ft}}$

**B.** $x = 0.4375 \text{ mi} \times \dfrac{5280 \text{ ft}}{1 \text{ mi}} \times \dfrac{12 \text{ in.}}{1 \text{ ft}}$

**C.** $80\% \cdot x = 90$

**D.** $x = 0.89 \text{ km} \cdot \dfrac{1000 \text{ m}}{1 \text{ km}} \cdot \dfrac{1 \text{ m}}{3.281 \text{ ft}}$

**E.** $x = 80\% \cdot 90$

**F.** $x = 420 \text{ m} + 75 \text{ ft} \cdot \dfrac{0.305 \text{ m}}{1 \text{ ft}}$

**G.** $\dfrac{x}{20} = \dfrac{675}{15}$

**H.** $x = 0.4375 \text{ mi} \times \dfrac{5280 \text{ ft}}{1 \text{ mi}} \times \dfrac{1 \text{ yd}}{3 \text{ ft}}$

**I.** $x = 0.89 \text{ km} \cdot \dfrac{0.621 \text{ mi}}{1 \text{ km}} \cdot \dfrac{5280 \text{ ft}}{1 \text{ mi}}$

**J.** $\dfrac{x}{15} = \dfrac{675}{20}$

**K.** $x = 420 \text{ m} \cdot \dfrac{3.281 \text{ ft}}{1 \text{ m}} + 75 \text{ ft}$

**L.** $144 \text{ ft} + x = 32.635 \text{ m} \cdot \dfrac{1 \text{ ft}}{0.305 \text{ m}}$

**M.** $x = 1,500,000 - 0.9\%(1,500,000)$

**N.** $20 \cdot x = 675$

**O.** $x = 1,500,000 + 0.9\%(1,500,000)$

*Answers on page A-16*

---

**6.** *Test Items.* Jason correctly answered 90 items on a recent test. These items represented 80% of the total number of questions. How many items were on the test?

**7.** *Population Decline.* City B's growth rate per year is −0.9%. If the population was 1,500,000 in 2005, what was the population in 2006?

**8.** *Bridge Length.* The Tatara Bridge in Onomichi-Imabari, Japan, is 0.89 km long. Convert this distance to feet.

**9.** *Height of Tower.* The Willis Tower in Chicago is 75 ft taller than the Jin Mao Building in Shanghai. The height of the Jin Mao Building is 420 m. What is the height of the Willis Tower in feet?

**10.** *Gasoline Usage.* Nate's company car gets 20 mi to the gallon in city driving. How many gallons will it use in 675 mi of city driving?

**a** Complete.

**1.** $1 \text{ ft}^2 = \underline{\hspace{1cm}} \text{ in}^2$

**2.** $1 \text{ yd}^2 = \underline{\hspace{1cm}} \text{ ft}^2$

**3.** $1 \text{ mi}^2 = \underline{\hspace{1cm}} \text{ acres}$

**4.** $1 \text{ acre} = \underline{\hspace{1cm}} \text{ ft}^2$

**5.** $1 \text{ in}^2 = \underline{\hspace{1cm}} \text{ ft}^2$

**6.** $1 \text{ ft}^2 = \underline{\hspace{1cm}} \text{ yd}^2$

**7.** $22 \text{ yd}^2 = \underline{\hspace{1cm}} \text{ ft}^2$

**8.** $40 \text{ ft}^2 = \underline{\hspace{1cm}} \text{ in}^2$

**9.** $44 \text{ yd}^2 = \underline{\hspace{1cm}} \text{ ft}^2$

**10.** $144 \text{ ft}^2 = \underline{\hspace{1cm}} \text{ yd}^2$

**11.** $20 \text{ mi}^2 = \underline{\hspace{1cm}} \text{ acres}$

**12.** $576 \text{ in}^2 = \underline{\hspace{1cm}} \text{ ft}^2$

**13.** $1 \text{ mi}^2 = \underline{\hspace{1cm}} \text{ ft}^2$

**14.** $1 \text{ mi}^2 = \underline{\hspace{1cm}} \text{ yd}^2$

**15.** $720 \text{ in}^2 = \underline{\hspace{1cm}} \text{ ft}^2$

**16.** $27 \text{ ft}^2 = \underline{\hspace{1cm}} \text{ yd}^2$

**17.** $144 \text{ in}^2 = \underline{\hspace{1cm}} \text{ ft}^2$

**18.** $72 \text{ in}^2 = \underline{\hspace{1cm}} \text{ ft}^2$

**19.** $1 \text{ acre} = \underline{\hspace{1cm}} \text{ mi}^2$

**20.** $4 \text{ acres} = \underline{\hspace{1cm}} \text{ ft}^2$

**21.** $40.3 \text{ mi}^2 = \underline{\hspace{1cm}} \text{ acres}$

**22.** $1080 \text{ in}^2 = \underline{\hspace{1cm}} \text{ ft}^2$

**b** Complete.

**23.** $5.21 \text{ km}^2 = \underline{\hspace{1cm}} \text{ m}^2$

**24.** $65 \text{ km}^2 = \underline{\hspace{1cm}} \text{ m}^2$

**25.** $0.014 \text{ m}^2 = \underline{\hspace{1cm}} \text{ cm}^2$

**26.** $0.028 \text{ m}^2 = \underline{\hspace{1cm}} \text{ mm}^2$

**27.** $2345.6 \, \text{mm}^2 =$ _____ $\text{cm}^2$

**28.** $8.38 \, \text{cm}^2 =$ _____ $\text{mm}^2$

**29.** $852.14 \, \text{cm}^2 =$ _____ $\text{m}^2$

**30.** $125 \, \text{mm}^2 =$ _____ $\text{m}^2$

**31.** $250{,}000 \, \text{mm}^2 =$ _____ $\text{cm}^2$

**32.** $2400 \, \text{mm}^2 =$ _____ $\text{cm}^2$

**33.** $472{,}800 \, \text{m}^2 =$ _____ $\text{km}^2$

**34.** $1.37 \, \text{cm}^2 =$ _____ $\text{mm}^2$

## Skill Maintenance

In Exercises 35 and 36, find the simple interest.   [7.7a]

**35.** On $2000 at an interest rate of 8% for 1.5 years

**36.** On $2000 at an interest rate of 5.3% for 2 years

In each of Exercises 37–40, find **(a)** the amount of simple interest due and **(b)** the total amount that must be paid back.   [7.7a]

**37.** A firm borrows $15,500 at 9.5% for 120 days.

**38.** A firm borrows $8500 at 10% for 90 days.

**39.** A firm borrows $6400 at 8.4% for 150 days.

**40.** A firm borrows $4200 at 11% for 30 days.

## Synthesis

Complete.

**41.** $1 \, \text{m}^2 =$ _____ $\text{ft}^2$

**42.** $1 \, \text{in}^2 =$ _____ $\text{cm}^2$

**43.** $2 \, \text{yd}^2 =$ _____ $\text{m}^2$

**44.** 1 acre = _____ $\text{m}^2$

**45.** *Ronald Reagan Library.*   The largest of the presidential libraries is the Ronald Reagan Library in Simi Valley, California. It covers approximately $153{,}000 \, \text{ft}^2$. Convert $153{,}000 \, \text{ft}^2$ to square meters.

# Summary and Review

## Key Terms and Formulas

Fahrenheit, p. 546       Celsius, p. 546

*American Units of Length:*    12 in. = 1 ft; 3 ft = 1 yd; 36 in. = 1 yd; 5280 ft = 1 mi

*Metric Units of Length:*    1 km = 1000 m; 1 hm = 100 m; 1 dam = 10 m; 1 dm = 0.1 m; 1 cm = 0.01 m; 1 mm = 0.001 m

*American–Metric Conversion:*    1 m = 39.370 in.; 1 m = 3.281 ft; 1 ft = 0.305 m; 1 in. = 2.540 cm; 1 km = 0.621 mi; 1 mi = 1.609 km; 1 yd = 0.914 m; 1 m = 1.094 yd

*American Units of Weight:*    1 T = 2000 lb; 1 lb = 16 oz

*Metric Units of Mass:*    1 t = 1000 kg; 1 kg = 1000 g; 1 hg = 100 g; 1 dag = 10 g; 1 dg = 0.1 g; 1 cg = 0.01 g; 1 mg = 0.001 g; 1 mcg = 0.000001 g

*American Units of Capacity:*    1 gal = 4 qt; 1 qt = 2 pt; 1 pt = 16 oz; 1 pt = 2 cups; 1 cup = 8 oz

*Metric Units of Capacity:*    1 L = 1000 mL = 1000 cm$^3$ = 1000 cc

*American–Metric Conversion:*    1 oz = 29.57 mL; 1 L = 1.06 qt

*Units of Time:*    1 min = 60 sec; 1 hr = 60 min; 1 day = 24 hr; 1 wk = 7 days; 1 yr = $365\frac{1}{4}$ days

*Temperature Conversion:*    $F = \dfrac{9}{5} \cdot C + 32$, or $F = 1.8 \cdot C + 32$;

$C = \dfrac{5}{9} \cdot (F - 32)$, or $C = \dfrac{F - 32}{1.8}$

## Concept Reinforcement

Determine whether each statement is true or false.

_____ **1.** Distances measured in feet in the American system would probably be measured in meters in the metric system.   [9.2a]

_____ **2.** When converting from grams to milligrams, move the decimal point two places to the left.   [9.4b]

_____ **3.** To convert mm$^2$ to cm$^2$, move the decimal point two places to the left.   [9.7b]

_____ **4.** Since 1 yd = 3 ft, multiply by 3 to convert square yards to square feet.   [9.7a]

_____ **5.** You would probably use your furnace when the temperature outside was 40°C.   [9.6b]

_____ **6.** You could go ice fishing when the temperature outside was 10°C.   [9.6b]

# Important Concepts

**Objective 9.1a**  Convert from one American unit of length to another.

**Examples**  Complete: 126 in. = _____ yd and $5\frac{2}{3}$ yd = _____ ft.

$$126 \text{ in.} = \frac{126 \text{ in.}}{1} \times \frac{1 \text{ yd}}{36 \text{ in.}}$$

$$= \frac{126 \text{ in.}}{36 \text{ in.}} \times 1 \text{ yd}$$

$$= \frac{126}{36} \times \frac{\text{in.}}{\text{in.}} \times 1 \text{ yd}$$

$$= 3.5 \times 1 \text{ yd} = 3.5 \text{ yd};$$

$$5\frac{2}{3} \text{ yd} = 5\frac{2}{3} \times 1 \text{ yd} = \frac{17}{3} \times 3 \text{ ft} = 17 \text{ ft}$$

**Practice Exercises**

Complete.

**1.** 7 ft = _____ yd

**2.** $2\frac{1}{2}$ mi = _____ ft

---

**Objective 9.2a**  Convert from one metric unit of length to another.

**Example**  Complete: 38 km = _____ cm and 2.9 mm = _____ m.

To go from km to cm, we move the decimal point 5 places to the right.

38      38.00000.      38 km = 3,800,000 cm

To go from mm to m, we move the decimal point 3 places to the left.

2.9      0.002.9      2.9 mm = 0.0029 m

**Practice Exercises**

Complete.

**3.** 12 hm = _____ m

**4.** 4.6 cm = _____ km

---

**Objective 9.3a**  Convert between American units of length and metric units of length.

**Example**  Complete: 42 ft = _____ m. (Note: 1 ft ≈ 0.305 m.)

$$42 \text{ ft} = 42 \times 1 \text{ ft} \approx 42 \times 0.305 \text{ m}$$

$$= 12.81 \text{ m}$$

**Practice Exercise**

**5.** Complete: 10 m = _____ yd. (Note: 1 m ≈ 1.094 yd.)

---

**Objective 9.4a**  Convert from one American unit of weight to another.

**Example**  Complete: 4020 oz = _____ lb.

$$4020 \text{ oz} = 4020 \text{ oz} \times \frac{1 \text{ lb}}{16 \text{ oz}} = \frac{4020}{16} \text{ lb} = 251.25 \text{ lb}$$

**Practice Exercise**

**6.** Complete: 10,280 lb = _____ T.

---

**Objective 9.4b**  Convert from one metric unit of mass to another.

**Example**  Complete: 5.62 cg = _____ g.

To go from cg to g, we move the decimal point 2 places to the left.

5.62      0.05.62      5.62 cg = 0.0562 g

**Practice Exercise**

**7.** Complete: 9.78 mg = _____ g.

**Objective 9.5a**   Convert from one unit of capacity to another.

**Examples**   Complete: 6 gal = _____ pt and 3800 mL = _____ L.

$$6 \text{ gal} = 6 \times 1 \text{ gal}$$
$$= 6 \times 4 \text{ qt}$$
$$= 6 \times 4 \times 1 \text{ qt}$$
$$= 6 \times 4 \times 2 \text{ pt} = 48 \text{ pt}$$

To go from mL to L, we move the decimal point 3 places to the left.

3800     3.800.     3800 mL = 3.8 L

**Practice Exercises**

Complete.

**8.** 16 qt = _____ cups

**9.** 42,670 mL = _____ L

---

**Objective 9.6a**   Convert from one unit of time to another.

**Example**   Complete: 7200 min = _____ days.

$$7200 \text{ min} = 7200 \text{ min} \cdot \frac{1 \text{ hr}}{60 \text{ min}} \cdot \frac{1 \text{ day}}{24 \text{ hr}}$$

$$= \frac{7200}{60 \cdot 24} \text{ days} = 5 \text{ days}$$

**Practice Exercise**

**10.** Complete: 3600 sec = _____ hr.

---

**Objective 9.6b**   Convert between Celsius and Fahrenheit temperatures using the formulas $F = \frac{9}{5} \cdot C + 32$ and $C = \frac{5}{9} \cdot (F - 32)$.

**Examples**   Convert 18°C to Fahrenheit and 95°F to Celsius.

$$F = \frac{9}{5}C + 32 = 1.8 \cdot 18 + 32$$
$$= 32.4 + 32 = 64.4$$

Thus, 18°C = 64.4°F.

$$C = \frac{5}{9} \cdot (F - 32) = \frac{5}{9} \cdot (95 - 32)$$

$$= \frac{5}{9} \cdot 63 = 35$$

Thus, 95°F = 35°C.

**Practice Exercises**

**11.** Convert 68°C to Fahrenheit.

**12.** Convert 104°F to Celsius.

---

**Objective 9.7a**   Convert from one American unit of area to another.

**Example**   Complete: 14,400 in² = _____ ft².

$$14{,}400 \text{ in}^2 = 14{,}400 \text{ in}^2 \times \frac{1 \text{ ft}^2}{144 \text{ in}^2}$$

$$= \frac{14{,}400}{144} \times \frac{\text{in}^2}{\text{in}^2} \times 1 \text{ ft}^2$$

$$= 100 \text{ ft}^2$$

**Practice Exercise**

**13.** Complete: 81 ft² = _____ yd².

---

**Objective 9.7b**   Convert from one metric unit of area to another.

**Example**   Complete: 9.6 m² = _____ cm².

To go from m² to cm², we move the decimal point 2 × 2, or 4, places to the right.

9.6     9.6000.     9.6 m² = 96,000 cm²

**Practice Exercise**

**14.** Complete: 52.4 cm² = _____ mm².

# Review Exercises

Complete.  [9.1a], [9.2a], [9.3a]

**1.** 8 ft = _____ yd

**2.** $\frac{5}{6}$ yd = _____ in.

**3.** 0.3 mm = _____ cm

**4.** 4 m = _____ km

**5.** 2 yd = _____ in.

**6.** 4 km = _____ cm

**7.** 14 in. = _____ ft

**8.** 15 cm = _____ m

**9.** 200 m = _____ yd

**10.** 20 mi = _____ km

Complete the table below.  [9.2a]

|  | MILLIMETERS (mm) | CENTIMETERS (cm) | METERS (m) |
|---|---|---|---|
| **11.** |  | 1 |  |
| **12.** |  |  | 305 |

Complete.  [9.4a,b], [9.5a], [9.6a]

**13.** 7 lb = _____ oz

**14.** 4 g = _____ kg

**15.** 16 min = _____ hr

**16.** 464 mL = _____ L

**17.** 3 min = _____ sec

**18.** 4.7 kg = _____ g

**19.** 8.07 T = _____ lb

**20.** 0.83 L = _____ mL

**21.** 6 hr = _____ days

**22.** 4 cg = _____ g

**23.** 0.2 g = _____ mg

**24.** 0.0003 kg = _____ cg

**25.** 0.7 mL = _____ L

**26.** 60 mL = _____ L

**27.** 0.8 T = _____ lb

**28.** 0.4 L = _____ mL

**29.** 20 oz = _____ lb

**30.** $\frac{5}{6}$ min = _____ sec

**31.** 20 gal = _____ pt

**32.** 960 oz = _____ gal

**33.** 54 qt = _____ gal

**34.** 2.5 days = _____ hr

**35.** 3020 cg = _____ kg

**36.** 10,500 lb = _____ T

*Medical Dosage.*  Solve.

**37.** Amoxicillin is an antibiotic obtainable in a liquid suspension form, part medication and part water, and is frequently used to treat infections in infants. One formulation of the drug contains 125 mg of amoxicillin per 5 mL of liquid. A pediatrician orders 150 mg per day for a 4-month-old child with an ear infection. How much of the amoxicillin suspension would the parent need to administer to the infant in order to achieve the recommended daily dose?  [9.4c], [9.5b]

**38.** An emergency-room physician orders 3 L of Ringer's lactate to be administered over 4 hr for a patient suffering from shock and severe low blood pressure. How many milliliters is this?   [9.5b]

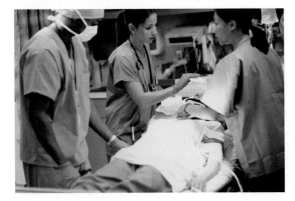

**39.** A physician prescribes 0.25 mg of alprazolam, an antianxiety medication. How many micrograms are in this dose?   [9.4c]

**40.** Convert −6°C to Fahrenheit.   [9.6b]

**41.** Convert 45°C to Fahrenheit.   [9.6b]

**42.** Convert 68°F to Celsius.   [9.6b]

**43.** Convert −20°F to Celsius.   [9.6b]

Complete.   [9.7a, b]

**44.** $4 \text{ yd}^2 =$ _____ $\text{ft}^2$

**45.** $0.3 \text{ km}^2 =$ _____ $\text{m}^2$

**46.** $2070 \text{ in}^2 =$ _____ $\text{ft}^2$

**47.** $600 \text{ cm}^2 =$ _____ $\text{m}^2$

**48.** Complete: $172.6 \text{ cm} =$ _____ hm.   [9.2a]
   **A.** 0.1726             **B.** 1,726,000
   **C.** 0.01726          **D.** 17.26

**49.** Complete: $0.16 \text{ gal} =$ _____ cups.   [9.5a]
   **A.** 1.28               **B.** 2.56
   **C.** 0.64               **D.** 160

## Synthesis

**50.** *Running Record.*   The world's record for running the 200-m dash is 19.30 sec, set by Usain Bolt of Jamaica in the 2008 Summer Olympics in Beijing. How should the record be changed if the run were a 200-yd dash?   [9.3a]
**Source:** *The New York Times*, August 17, 2008

**51.** It is known that 1 gal of water weighs 8.3453 lb. Which weighs more: an ounce of pennies or an ounce (as capacity) of water? Explain.   [9.4a], [9.5a]

# Understanding Through Discussion and Writing

**1.** Give at least two reasons why someone might prefer the use of grams to the use of ounces.   [9.4a, b]

**2.** Why do you think most liquid containers list both metric and American units of measure?   [9.5a]

**3.** Explain the difference between the way we move the decimal point for area conversion and the way we do so for length conversion.   [9.2a], [9.7b]

**4.** What advantages does the use of metric units of capacity have over that of American units?   [9.5a]

**5.** **a)** The temperature is 23°C. Would you want to play golf? Explain.
   **b)** Your bathwater has a temperature of 10°C. Would you want to take a bath?
   **c)** The nearby lake has a temperature of −10°C. Would it be safe to go ice skating?   [9.6b]

**6.** Which is larger and why: one square meter or one square yard?   [9.3a], [9.7a, b]

Test

For Extra Help

CHAPTER
Test Prep
VIDEOS

Step-by-step test solutions are found on the Chapter Test Prep Videos available via the Video Resources on DVD, in *MyMathLab* , and on You Tube (search "BittingerBasicMathEI" and click on "Channels").

Complete.

**1.** 4 ft = _____ in.

**2.** 4 in. = _____ ft

**3.** 6 km = _____ m

**4.** 8.7 mm = _____ cm

**5.** 200 yd = _____ m

**6.** 2400 km = _____ mi

Complete the table below.

| | OBJECT | MILLIMETERS (mm) | CENTIMETERS (cm) | METERS (m) |
|---|---|---|---|---|
| **7.** | Width of a key on a calculator | | 0.5 | |
| **8.** | Height of your author | | | 1.8542 |

Complete.

**9.** 3080 mL = _____ L

**10.** 0.24 L = _____ mL

**11.** 4 lb = _____ oz

**12.** 4.11 T = _____ lb

**13.** 3.8 kg = _____ g

**14.** 4.325 mg = _____ cg

**15.** 2200 mg = _____ g

**16.** 5 hr = _____ min

**17.** 15 days = _____ hr

**18.** 64 pt = _____ qt

**19.** 10 gal = _____ oz

**20.** 5 cups = _____ oz

**21.** 0.37 mg = _____ mcg

**22.** Convert 95°F to Celsius.

**23.** Convert 59°C to Fahrenheit.

Complete the table below.

| | OBJECT | YARDS (yd) | CENTIMETERS (cm) | INCHES (in.) | METERS (m) | MILLIMETERS (mm) |
|---|---|---|---|---|---|---|
| **24.** | Length of a meter stick | | | | 1 | |
| **25.** | Height of Xiamen Posts and Telecommunications Building, Xiamen, China | 398 | | | | |

*Medical Dosage.* Solve.

**26.** An emergency-room physician prescribes 2.5 L of normal saline intravenously over 8 hr for a patient who is severely dehydrated. How many milliliters is this?

**27.** A physician prescribes 0.5 mg of alprazolam to be taken 3 times a day by a patient suffering from anxiety. How many micrograms of alprazolam is the patient to ingest each day?

**28.** A prescription calls for 4 oz of dextromethorphan, a cough-suppressant medication. For how many milliliters is the prescription? (Use 1 oz ≈ 29.57 mL.)

Complete.

**29.** $12 \text{ ft}^2 = $ _____ $\text{in}^2$

**30.** $3 \text{ cm}^2 = $ _____ $\text{m}^2$

**31.** Convert 45.5°C to Fahrenheit.
   **A.** 49.9°F
   **B.** 24.3°F
   **C.** 7.5°F
   **D.** 113.9°F

# Synthesis

**32.** *Running Record.* The world's record for running the 400-m run is 43.18 sec, set by Michael Johnson of the United States in Seville, Spain, on August 26, 1999. How should the record be changed if the run were a 400-yd run?
   Source: The *World Almanac,* 2008

Solve.

**1.** *Population 65 and Older.* The U.S. population 65 and older was about 37.3 million in 2006. It is projected to be 54.6 million in 2020. Find standard notation for 37.3 million and 54.6 million.

**Source:** Decennial Censuses, Annual Population Estimates, U.S. Interim Projections, U.S. Census Bureau, U.S. Department of Commerce

**2.** *Gas Mileage.* A Honda Fit travels 561 mi on the highway on 17 gal of gasoline. What is the gas mileage?

**Source:** *Car and Driver,* December 2008

**3.** *Court Reporters.* In 2006, approximately 19,000 people in the United States were employed as court reporters. It is projected that this number will increase to 24,000 by 2016. What is the percent of increase?

**Sources:** *Occupational Outlook Handbook* 2008–09 Edition, January 2008; U.S. Department of Labor, Bulletin 2700

**4.** *Jewelers.* The number of employed jewelers and precious-stone and metal workers was 52,000 in 2006. This number is projected to decrease to 51,000 in 2016. What is the percent of decrease?

**Sources:** *Occupational Outlook Handbook* 2008–09 Edition, January 2008; U.S. Department of Labor, Bulletin 2700

Perform the indicated operation and simplify.

**5.** $46{,}231 \times 1100$

**6.** $\dfrac{1}{10} \cdot \left(-\dfrac{5}{6}\right)$

**7.** $-14.5 + \dfrac{4}{5} - 0.1$

**8.** $-2\dfrac{3}{5} \div -3\dfrac{9}{10}$

**9.** $0.1\overline{)3.56}$

**10.** $3\dfrac{1}{2} - 2\dfrac{2}{3}$

**11.** Determine whether 1,298,032 is divisible by 8.

**12.** Determine whether 5,024,120 is divisible by 3.

**13.** Find the prime factorization of 99.

**14.** Find the LCM of 35 and 49.

**15.** Round $35.\overline{7}$ to the nearest tenth.

**16.** Write a word name for 103.064.

**17.** Find the average and the median of this set of numbers: 29, 21, 9, 13, 17, 18.

Find percent notation.

**18.** 0.08

**19.** $\dfrac{3}{5}$

Complete.

**20.** 2 yd = _____ ft

**21.** 6 oz = _____ lb

**22.** 15°C = _____ °F

**23.** 0.087 L = _____ mL

**24.** 9 sec = _____ min

**25.** 17 cm = _____ m

**26.** 2200 mi = _____ km

**27.** 2000 mL = _____ L

**28.** 0.23 mg = _____ mcg

**29.** 12 yd$^2$ = _____ ft$^2$

Solve.

**30.** $0.07 \cdot x = -10.535$          **31.** $x + 12{,}843 = 32{,}091$

**32.** $\frac{2}{3} \cdot y = 5$          **33.** $\frac{4}{5} + y = -\frac{6}{7}$

Solve.

**34.** A mechanic spent $\frac{1}{3}$ hr changing a car's oil, $\frac{1}{2}$ hr rotating the tires, $\frac{1}{10}$ hr changing the air filter, $\frac{1}{4}$ hr adjusting the idle speed, and $\frac{1}{15}$ hr checking the brake and transmission fluids. How many hours did the mechanic spend working on the car?

**35.** *Milk Production.*   There are 9.112 million milk cows in America, each producing, on average, 19,950 lb of milk per year. How many pounds of milk are produced each year in America?
**Source:** U.S. Department of Agriculture

**36.** A driver filled the gas tank when the odometer read 86,897.2. At the next gasoline purchase, the odometer read 87,153.0. How many miles had been driven? The tank was filled with 16 gal. What was the gas mileage?

**37.** *Real Estate Commission.*   A real estate commission rate is $7\frac{1}{2}\%$. What is the commission on the sale of a property for $215,000?

**38.** *Compound Interest.*   A student invests $2000 in an account paying 6%, compounded semiannually. How much is in the account after 3 years?

**39.** A man on a diet loses $3\frac{1}{2}$ lb in 2 weeks. At this rate, how many pounds will he lose in 5 weeks?

**40.** A family has an annual income of $52,800. Of this, $\frac{1}{4}$ is spent for food. How much does the family spend for food?

**41.** *Seed Production.*   The U.S. Department of Agriculture requires that 80% of the seeds that a company produces must sprout. To determine the quality of the seeds it has produced, a company plants 500 seeds. It finds that 417 of the seeds sprout. Do the seeds meet government standards?
**Source:** U.S. Department of Agriculture

*Sundae's Homemade Ice Cream & Coffee Co.*   This company in Indianapolis, Indiana, makes ice cream, sorbet, and frozen yogurt.

**42.** Ice cream is packaged in 15-lb tubs. How many ounces are in one tub?

**43.** Although the process is not perfect, Sundae's attempts to have about 4 oz in 1 dip of ice cream. How many dips are there in a tub of ice cream?

**44.** By weighing each tub, the owner can determine how many dips have been sold of that flavor. The weight of a tub changes from 15 lb to $8\frac{5}{8}$ lb over a busy weekend. How many dips of ice cream were served from the tub?

**45.** A 1-dip ice cream cone sells for $1.99. If the entire contents of a tub were used to make 1-dip cones, how much money would be taken in from the sale of a tub of ice cream?

**46.** A 2-dip ice cream cone sells for $2.99. If the entire contents of a tub were used to make 2-dip cones, how much money would be taken in from the sale of a tub of ice cream?

## Synthesis

**47.** If $r$ is $\frac{2}{5}$ of $q$, then $q$ is what fractional part of $r$?

# Geometry

## Real-World Application

Malaria, a disease transmitted by mosquitoes, is the leading cause of death among children in Africa. Bed nets prevent malaria transmission by creating a protective barrier against mosquitoes at night. In November 2006, the United Nations Foundation, the United Methodist Church, and the National Basketball Association launched the Nothing But Nets campaign to distribute mosquito netting in Africa. Two years later, 2,194,124 insecticide-treated bed nets had been sent to seven African countries. A medium-sized net measures approximately 9.843 ft by 8.2025 ft. A large-sized net measures approximately 13.124 ft by 8.2025 ft. Find the area of each net. How much larger is the area of the large net than that of the medium net?

*Source:* www.nothingbutnets.net

***This problem appears as Example 10 in Section 10.2.***

# 10.1

# Perimeter

**SKILL TO REVIEW**

Objective 4.6a: Multiply using mixed numerals.

Multiply.

**1.** $2 \times 8\frac{1}{3}$    **2.** $4 \times 6\frac{2}{5}$

Find the perimeter of each polygon.

**1.**

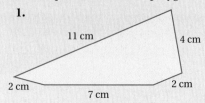

**2.**

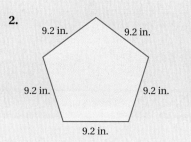

**3.** Find the perimeter of a rectangle that is 2 cm by 4 cm.

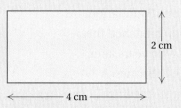

## **a** Finding Perimeters

> **PERIMETER OF A POLYGON**
>
> A **polygon** is a closed geometric figure with three or more sides. The **perimeter of a polygon** is the distance around it, or the sum of the lengths of its sides.

**EXAMPLE 1**  Find the perimeter of this polygon.

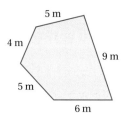

We add the lengths of the sides. Since all units are the same, we add the numbers, keeping meters (m) as the unit.

$$\text{Perimeter} = 6\,\text{m} + 5\,\text{m} + 4\,\text{m} + 5\,\text{m} + 9\,\text{m}$$
$$= (6 + 5 + 4 + 5 + 9)\,\text{m}$$
$$= 29\,\text{m}$$

Do Margin Exercises 1 and 2.

A **rectangle** is a polygon with four sides and four 90° angles, like the one shown in Example 2.

**EXAMPLE 2**  Find the perimeter of a rectangle that is 3 cm by 4 cm. The symbol ⌐ in the corner indicates 90°.

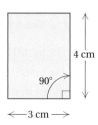

$$\text{Perimeter} = 3\,\text{cm} + 4\,\text{cm} + 3\,\text{cm} + 4\,\text{cm}$$
$$= (3 + 4 + 3 + 4)\,\text{cm}$$
$$= 14\,\text{cm}$$

Do Exercise 3.

*Answers*

*Skill to Review:*

**1.** $16\frac{2}{3}$   **2.** $25\frac{3}{5}$

*Margin Exercises:*

**1.** 26 cm   **2.** 46 in.   **3.** 12 cm

## PERIMETER OF A RECTANGLE

The **perimeter of a rectangle** is twice the sum of the length and the width, or 2 times the length plus 2 times the width:

$$P = 2 \cdot (l + w), \quad \text{or} \quad P = 2 \cdot l + 2 \cdot w.$$

**EXAMPLE 3** Find the perimeter of a rectangle that is 7.8 ft by 4.3 ft.

$$P = 2 \cdot (l + w)$$
$$= 2 \cdot (7.8\,\text{ft} + 4.3\,\text{ft})$$
$$= 2 \cdot (12.1\,\text{ft})$$
$$= 24.2\,\text{ft}$$

Do Exercises 4 and 5.

**4.** Find the perimeter of a rectangle that is 5.25 yd by 3.5 yd.

**5.** Find the perimeter of a rectangle that is $8\frac{1}{4}$ in. by $5\frac{2}{3}$ in.

A **square** is a rectangle with all sides the same length.

**EXAMPLE 4** Find the perimeter of a square whose sides are 9 mm long.

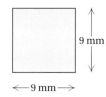

$$P = 9\,\text{mm} + 9\,\text{mm} + 9\,\text{mm} + 9\,\text{mm}$$
$$= (9 + 9 + 9 + 9)\,\text{mm}$$
$$= 36\,\text{mm}$$

Do Exercise 6.

**6.** Find the perimeter of a square with sides of length 10 km.

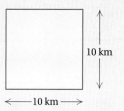

## PERIMETER OF A SQUARE

The **perimeter of a square** is four times the length of a side:

$$P = 4 \cdot s.$$

**EXAMPLE 5**  Find the perimeter of a square whose sides are $20\frac{1}{8}$ in. long.

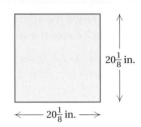

$20\frac{1}{8}$ in.

$20\frac{1}{8}$ in.

$$P = 4 \cdot s = 4 \cdot 20\frac{1}{8} \text{ in.}$$

$$= 4 \cdot \frac{161}{8} \text{ in.} = \frac{4 \cdot 161}{4 \cdot 2} \text{ in.}$$

$$= \frac{4}{4} \cdot \frac{161}{2} \text{ in.} = 80\frac{1}{2} \text{ in.}$$

**7.** Find the perimeter of a square with sides of length $5\frac{1}{4}$ yd.

**8.** Find the perimeter of a square with sides of length 7.8 km.

Do Exercises 7 and 8.

## b  Solving Applied Problems

**EXAMPLE 6**  Jaci is adding crown molding to the top edge of each wall in her rectangular dining room, which measures 14 ft by 12 ft. How many feet of trim will be needed? If the crown molding sells for $2.25 per foot, what will the trim cost?

1. **Familiarize.**  We make a drawing and let $P =$ the perimeter.

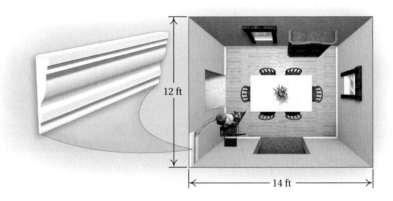

12 ft

14 ft

2. **Translate.**  The perimeter of the room is given by

$$P = 2 \cdot (l + w) = 2 \cdot (14 \text{ ft} + 12 \text{ ft}).$$

3. **Solve.**  We calculate the perimeter as follows:

$$P = 2 \cdot (14 \text{ ft} + 12 \text{ ft}) = 2 \cdot (26 \text{ ft}) = 52 \text{ ft}.$$

Then we multiply by $2.25 to find the cost of the crown molding:

$$\text{Cost} = \$2.25 \times \text{Perimeter} = \$2.25 \times 52 \text{ ft} = \$117.$$

4. **Check.**  The check is left to the student.

5. **State.**  The 52 ft of crown molding that is needed will cost $117.

**9.** A fence is to be built around a vegetable garden that measures 20 ft by 15 ft. How many feet of fence will be needed? If fencing sells for $2.95 per foot, what will the fencing cost?

Do Exercise 9.

*Answers*

**7.** 21 yd   **8.** 31.2 km   **9.** 70 ft; $206.50

**a**  Find the perimeter of each polygon.

**1.**

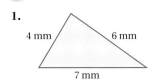

4 mm   6 mm
7 mm

**2.**

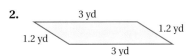

3 yd
1.2 yd   1.2 yd
3 yd

**3.**

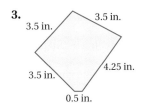

3.5 in.   3.5 in.
3.5 in.   4.25 in.
3.5 in.
0.5 in.

**4.**

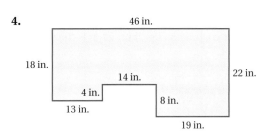

46 in.
18 in.   14 in.   22 in.
4 in.   8 in.
13 in.
19 in.

**5.**

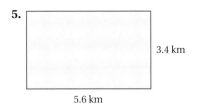

3.4 km
5.6 km

**6.**

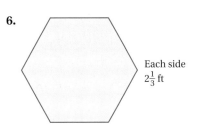

Each side
$2\frac{1}{3}$ ft

Find the perimeter of each rectangle.

**7.** 5 ft by 10 ft

**8.** 2.5 m by 100 m

**9.** $3\frac{1}{2}$ yd by $4\frac{1}{2}$ yd

**10.** 34.67 cm by 4.9 cm

Find the perimeter of each square.

**11.** 22 ft on a side

**12.** 56.9 km on a side

**13.** 45.5 mm on a side

**14.** $3\frac{1}{8}$ yd on a side

**b**  Solve.

**15.** Most billiard tables are twice as long as they are wide. What is the perimeter of a billiard table that measures 4.5 ft by 9 ft?

**16.** A security fence is to be built around a 173-m by 240-m rectangular lot. What is the perimeter of the lot? If 5-ft-high galvanized fence wire costs $7.29 per meter, what will the fencing cost?

**17.** A piece of flooring tile is a square with sides of length 30.5 cm. What is the perimeter of a piece of tile?

**18.** A rectangular posterboard is 61.8 cm by 87.9 cm. What is the perimeter of the board?

**19.** A rectangular glass backboard for a basketball goal measures 2 ft 8 in. by 4 ft 6 in. What is the perimeter of the backboard? (*Hint:* Convert 2 ft 8 in. to 32 in. and 4 ft 6 in. to 54 in.)

**20.** A standard license plate measures 12 in. by 6 in. What is the perimeter of the license plate?

**21.** A rain gutter is to be installed around the office building shown in the figure.

  **a)** Find the perimeter of the office building.

  **b)** If the gutter costs $4.59 per foot, what is the total cost of the gutter?

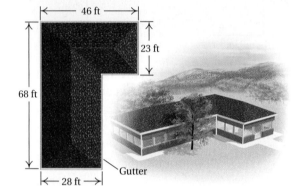

**22.** A carpenter is to build a fence around a 9-m by 12-m garden.

  **a)** The posts are 3 m apart. How many posts will be needed?

  **b)** The posts cost $8.65 each. How much will the posts cost?

  **c)** The fence will surround all but 3 m of the garden, which will be a gate. How long will the fence be?

  **d)** The fence costs $3.85 per meter. What will the cost of the fence be?

  **e)** The gate costs $69.95. What is the total cost of the materials?

## Skill Maintenance

**23.** Find the simple interest on $600 at 6.4% for $\frac{1}{2}$ year. [7.7a]

**24.** Find the simple interest on $600 at 8% for 2 years. [7.7a]

Evaluate. [1.9b]

**25.** $10^3$

**26.** $11^3$

**27.** $15^2$

**28.** $22^2$

**29.** $7^2$

**30.** $4^3$

Solve.

**31.** *Sales Tax.* In a certain state, a sales tax of $878 is collected on the purchase of a car for $17,560. What is the sales tax rate? [7.6a]

**32.** *Commission Rate.* Rich earns $1854.60 selling $16,860 worth of cell phones. What is the commission rate? [7.6b]

## Synthesis

**33.** If it takes 18 in. to make the bow, how much ribbon is needed for the entire package shown here?

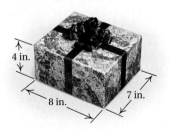

**34.** Find the perimeter, in feet, of the figure.

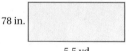

78 in.

5.5 yd

# 10.2 Area

## a) Rectangles and Squares

A polygon and its interior form a plane region. We can find the area of a *rectangular region*, or *rectangle*, by filling it in with square units. Two such units, a *square inch* and a *square centimeter*, are shown below.

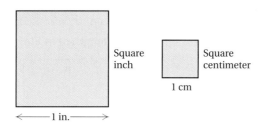

Square inch
1 in.

Square centimeter
1 cm

**EXAMPLE 1** What is the area of this region?

We have a rectangular array. Since the region is filled with 12 square centimeters, its area is 12 square centimeters (sq cm), or 12 cm². The number of units is 3 × 4, or 12.

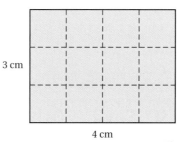

3 cm

4 cm

Do Margin Exercise 1.

### AREA OF A RECTANGLE

The **area of a rectangle** is the product of the length *l* and the width *w*:

$$A = l \cdot w.$$

w

l

**EXAMPLE 2** Find the area of a rectangle that is 7 yd by 4 yd.

We have

$$A = l \cdot w = 7 \text{ yd} \cdot 4 \text{ yd}$$
$$= 7 \cdot 4 \cdot \text{yd} \cdot \text{yd} = 28 \text{ yd}^2.$$

We think of yd · yd as (yd)² and denote it yd². Thus we read "28 yd²" as "28 square yards."

Do Exercises 2 and 3.

## OBJECTIVES

**a** Find the area of a rectangle and a square.

**b** Find the area of a parallelogram, a triangle, and a trapezoid.

**c** Solve applied problems involving areas of rectangles, squares, parallelograms, triangles, and trapezoids.

### SKILL TO REVIEW
Objective 5.5c: Calculate using fraction notation and decimal notation together.

Calculate.

**1.** $\frac{1}{2} \times 16.243$    **2.** $0.5 \times \frac{3}{8}$

**1.** What is the area of this region? Count the number of square centimeters.

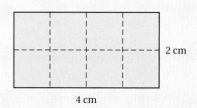

2 cm

4 cm

**2.** Find the area of a rectangle that is 7 km by 8 km.

**3.** Find the area of a rectangle that is $5\frac{1}{4}$ yd by $3\frac{1}{2}$ yd.

*Answers*

*Skill to Review:*

**1.** 8.1215    **2.** 0.1875, or $\frac{3}{16}$

*Margin Exercises:*

**1.** 8 cm²    **2.** 56 km²    **3.** $18\frac{3}{8}$ yd²

**4.** Find the area of a square with sides of length 12 km.

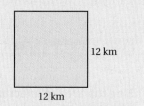
12 km
12 km

**EXAMPLE 3** Find the area of a square with sides of length 9 mm.

$$A = (9\,\text{mm}) \cdot (9\,\text{mm})$$
$$= 9 \cdot 9 \cdot \text{mm} \cdot \text{mm}$$
$$= 81\,\text{mm}^2$$

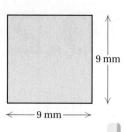
9 mm
9 mm

Do Exercise 4.

### AREA OF A SQUARE

The **area of a square** is the square of the length of a side:

$$A = s \cdot s, \quad \text{or} \quad A = s^2.$$

*s*
*s*

**5.** Find the area of a square with sides of length 10.9 m.

**6.** Find the area of a square with sides of length $3\frac{1}{2}$ yd.

**EXAMPLE 4** Find the area of a square with sides of length 20.3 m.

$$A = s \cdot s = 20.3\,\text{m} \times 20.3\,\text{m} = 20.3 \times 20.3 \times \text{m} \times \text{m} = 412.09\,\text{m}^2$$

Do Exercises 5 and 6.

### **b** Finding Other Areas

#### *Parallelograms*

A **parallelogram** is a four-sided figure with two pairs of parallel sides, as shown below.

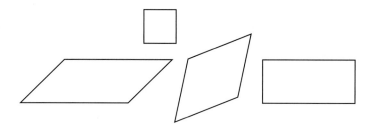

To find the area of a parallelogram, consider the one below.

If we cut off a piece and move it to the other end, we get a rectangle.

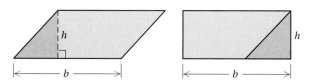
*h*
*b*
*h*
*b*

*Answers*

**4.** 144 km² **5.** 118.81 m² **6.** $12\frac{1}{4}$ yd²

We can find the area by multiplying the length *b*, called the **base**, by *h*, called the **height**.

## AREA OF A PARALLELOGRAM

The **area of a parallelogram** is the product of the length of the base $b$ and the height $h$:

$$A = b \cdot h.$$

**EXAMPLE 5** Find the area of this parallelogram.

$A = b \cdot h$

$\quad = 7\,\text{km} \cdot 5\,\text{km}$

$\quad = 35\,\text{km}^2$

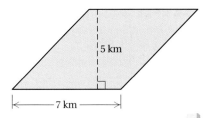

**EXAMPLE 6** Find the area of this parallelogram.

$A = b \cdot h$

$\quad = 1.2\,\text{m} \times 6\,\text{m}$

$\quad = 7.2\,\text{m}^2$

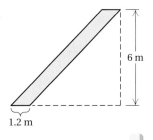

Do Exercises 7 and 8.

Find the area.

7.

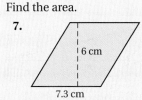

8.

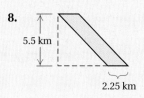

### Triangles

A **triangle** is a polygon with three sides. To find the area of a triangle like the one shown on the left below, think of cutting out another just like it and placing it as shown on the right below.

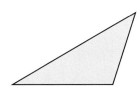
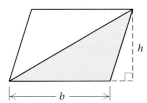

The resulting figure is a parallelogram whose area is

$$b \cdot h.$$

The triangle we began with has half the area of the parallelogram, or

$$\frac{1}{2} \cdot b \cdot h.$$

## AREA OF A TRIANGLE

The **area of a triangle** is half the length of the base times the height:

$$A = \frac{1}{2} \cdot b \cdot h.$$

*Answers*

**7.** $43.8\,\text{cm}^2$   **8.** $12.375\,\text{km}^2$

**EXAMPLE 7** Find the area of this triangle.

$$A = \frac{1}{2} \cdot b \cdot h$$

$$= \frac{1}{2} \cdot 9\,\text{m} \cdot 6\,\text{m}$$

$$= \frac{9 \cdot 6}{2}\,\text{m}^2$$

$$= 27\,\text{m}^2$$

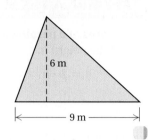

**EXAMPLE 8** Find the area of this triangle.

$$A = \frac{1}{2} \cdot b \cdot h$$

$$= \frac{1}{2} \times 6.25\,\text{cm} \times 5.5\,\text{cm}$$

$$= 0.5 \times 6.25 \times 5.5\,\text{cm}^2$$

$$= 17.1875\,\text{cm}^2$$

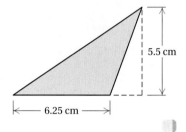

Do Exercises 9 and 10.

### Trapezoids

A **trapezoid** is a polygon with four sides, two of which, the **bases**, are parallel to each other.

To find the area of a trapezoid, think of cutting out another just like it.

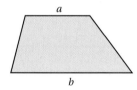

Then place the second one like this.

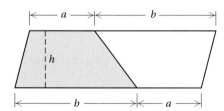

The resulting figure is a parallelogram whose area is

$$h \cdot (a + b). \qquad \text{The base is } a + b.$$

The trapezoid we began with has half the area of the parallelogram, or

$$\frac{1}{2} \cdot h \cdot (a + b).$$

Find the area.

**9.**

**10.**

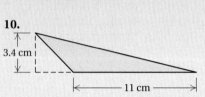

**Answers**

**9.** 96 m² **10.** 18.7 cm²

> **AREA OF A TRAPEZOID**
>
> The **area of a trapezoid** is half the product of the height and the sum of the lengths of the parallel sides (bases):
>
> $$A = \frac{1}{2} \cdot h \cdot (a + b), \quad \text{or} \quad A = \frac{a + b}{2} \cdot h.$$

**Find the area.**

**11.**

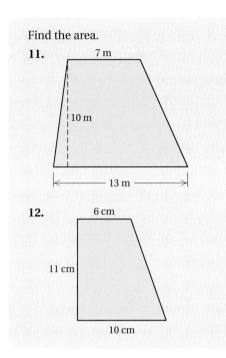

**EXAMPLE 9**   Find the area of this trapezoid.

$$A = \frac{1}{2} \cdot h \cdot (a + b)$$

$$= \frac{1}{2} \cdot 7 \text{ cm} \cdot (12 + 18) \text{ cm}$$

$$= \frac{1}{2} \cdot 7 \text{ cm} \cdot (30 \text{ cm})$$

$$= \frac{7 \cdot 30}{2} \cdot \text{cm}^2$$

$$= \frac{7 \cdot 15 \cdot 2}{1 \cdot 2} \text{ cm}^2$$

$$= \frac{2}{2} \cdot \frac{7 \cdot 15}{1} \text{ cm}^2$$

$$= 105 \text{ cm}^2$$

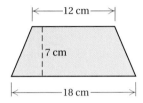

Do Exercises 11 and 12.

**12.**

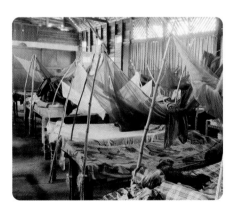

## (c) Solving Applied Problems

**EXAMPLE 10**   *Mosquito Netting.*   Malaria, a disease transmitted by mosquitoes, is the leading cause of death among children in Africa. Bed nets prevent malaria transmission by creating a protective barrier against mosquitoes at night. In November 2006, the United Nations Foundation, the United Methodist Church, and the National Basketball Association launched the Nothing But Nets campaign to distribute mosquito netting in Africa. Two years later, 2,194,124 insecticide-treated bed nets had been sent to seven African countries. A medium-sized net measures approximately 9.843 ft by 8.2025 ft. A large-sized net measures approximately 13.124 ft by 8.2025 ft. Find the area of each net. How much larger is the area of the large net than that of the medium net?
**Source:** www.nothingbutnets.net

We find the area of each net using the area formula $A = l \cdot w$ and substituting values for $l$ and $w$:

$$A = l \times w \qquad\qquad A = l \times w$$
$$A \approx 9.843 \text{ ft} \times 8.2025 \text{ ft} \qquad A \approx 13.124 \text{ ft} \times 8.2025 \text{ ft}$$
$$A \approx 80.74 \text{ ft}^2; \qquad\qquad A \approx 107.65 \text{ ft}^2.$$

The area of the medium bed net is about $80.74 \text{ ft}^2$; the area of the large bed net is about $107.65 \text{ ft}^2$. The large net is approximately $107.65 - 80.74$, or $26.91 \text{ ft}^2$, larger than the medium net.

*Answers*
**11.** $100 \text{ m}^2$   **12.** $88 \text{ cm}^2$

**EXAMPLE 11** *Lucas Oil Stadium.* The retractable roof of Lucas Oil Stadium, the home of the Indianapolis Colts football team, divides lengthwise. Each half measures 600 ft by 160 ft. The roof opens and closes in approximately 9–11 min. The opening measures 293 ft across. What is the total area of the rectangular roof? What is the area of the opening?

**Source:** HKS Sports and Entertainment

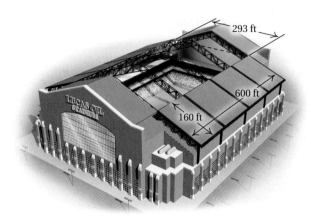

Each half of the retractable roof is a rectangle that measures 600 ft by 160 ft. The area of a rectangle is length times width, so we have

$$A = l \cdot w$$
$$= 600 \text{ ft} \times 160 \text{ ft}$$
$$= 96{,}000 \text{ ft}^2.$$

The total area of the two halves of the retractable roof is

$$\text{Total area} = 2 \times 96{,}000 \text{ ft}^2$$
$$= 192{,}000 \text{ ft}^2.$$

When the retractable roof is open, the dimensions of the opening are 600 ft by 293 ft. The area of this rectangle is

$$A = l \cdot w$$
$$= 600 \text{ ft} \times 293 \text{ ft}$$
$$= 175{,}800 \text{ ft}^2.$$

When the roof is open, the area of the opening is 175,800 ft$^2$.

Do Exercise 13.

**13.** Find the area of this kite.

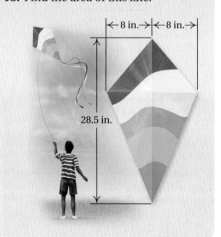

← 8 in. → ← 8 in. →

28.5 in.

**a** Find the area.

**1.**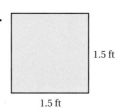
5 km, 3 km

**2.**
1.5 ft, 1.5 ft

**3.**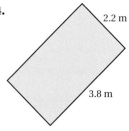
2 in., 0.7 in.

**4.**
2.2 m, 3.8 m

**5.**
$2\frac{1}{2}$ yd, $2\frac{1}{2}$ yd

**6.**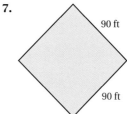
$3\frac{1}{2}$ mi, $3\frac{1}{2}$ mi

**7.**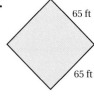
90 ft, 90 ft

**8.**
65 ft, 65 ft

Find the area of each rectangle.

**9.** 5 ft by 10 ft

**10.** 14 yd by 8 yd

**11.** 34.67 cm by 4.9 cm

**12.** 2.45 km by 100 km

**13.** $4\frac{2}{3}$ in. by $8\frac{5}{6}$ in.

**14.** $10\frac{1}{3}$ mi by $20\frac{2}{3}$ mi

Find the area of the square.

**15.** 22 ft on a side

**16.** 18 yd on a side

**17.** 56.9 km on a side

**18.** 45.5 m on a side

**19.** $5\frac{3}{8}$ yd on a side

**20.** $7\frac{2}{3}$ ft on a side

**b** Find the area.

**21.**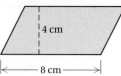

4 cm

8 cm

**22.**

4 cm

4 cm

**23.**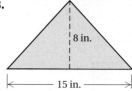

8 in.

15 in.

**24.**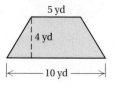

5 yd

4 yd

10 yd

**25.**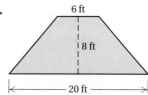

6 ft

8 ft

20 ft

**26.**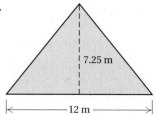

7.25 m

12 m

**27.**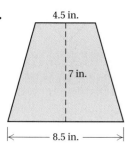

4.5 in.

7 in.

8.5 in.

**28.**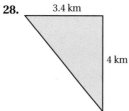

3.4 km

4 km

**29.**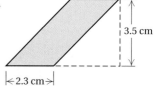

3.5 cm

2.3 cm

**30.**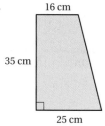

16 cm

35 cm

25 cm

**31.**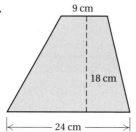

9 cm

18 cm

24 cm

**32.**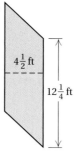

$4\frac{1}{2}$ ft

$12\frac{1}{4}$ ft

**33.**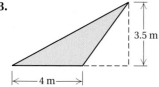

3.5 m

4 m

**34.**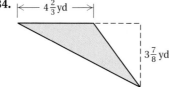

$4\frac{2}{3}$ yd

$3\frac{7}{8}$ yd

**35.** *Area of a Lawn.* A lot is 40 m by 36 m. A house 27 m by 9 m is built on the lot. How much area is left over for a lawn?

**36.** *Area of a Field.* A field is 240.8 m by 450.2 m. A rectangular area that measures 160.4 m by 90.6 m is paved for a parking lot. How much area is unpaved?

**37.** *Mowing Expense.* A square basketball court $19\frac{1}{2}$ ft on a side is placed on an 80-ft by $110\frac{2}{3}$-ft lawn.

  **a)** Find the area of the lawn that is not covered by the court.

  **b)** It costs $0.012 per square foot to have the lawn mowed. What is the total cost of the mowing?

**38.** *Mowing Expense.* A square flower bed 10.5 ft on a side is dug on a 90-ft by $67\frac{1}{4}$-ft lawn.

  **a)** Find the area of the lawn that is not covered by the flower bed.

  **b)** It costs $0.03 per square foot to have the lawn mowed. What is the total cost of the mowing?

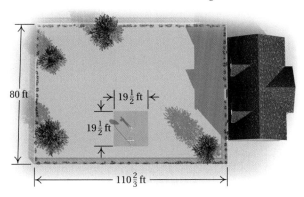

**39.** *Area of a Sidewalk.* Franklin Construction Company builds a sidewalk around two sides of a new library, as shown in the figure. What is the area of the sidewalk?

**40.** *Margin Area.* A standard sheet of typewriter paper is $8\frac{1}{2}$ in. by 11 in. For a quarterly report, Don types on a 7-in. by 9-in. area of the paper. What is the area of the margins?

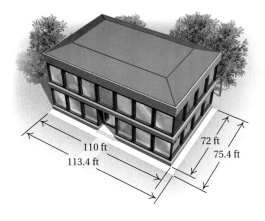

**41.** *Painting Costs.* A room is 15 ft by 20 ft. The ceiling is 8 ft above the floor. There are two windows in the room, each 3 ft by 4 ft. The door is $2\frac{1}{2}$ ft by $6\frac{1}{2}$ ft.

  **a)** What is the total area of the walls and the ceiling?

  **b)** A gallon of paint will cover 360.625 ft$^2$. How many gallons of paint are needed for the room, including the ceiling?

  **c)** Paint costs $24.95 a gallon. How much will it cost to paint the room?

**42.** *Carpeting Costs.* A restaurant owner wants to carpet a 15-yd by 20-yd room.

  **a)** How many square yards of carpeting are needed?

  **b)** The carpeting she wants is $18.50 per square yard. How much will it cost to carpet the room?

Find the area of the shaded region in each figure.

**43.**

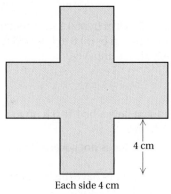

4 cm

Each side 4 cm

**44.**

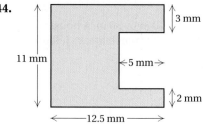

11 mm

3 mm

←5 mm→

2 mm

12.5 mm

**45.**

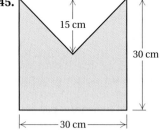

15 cm

30 cm

30 cm

**46.**

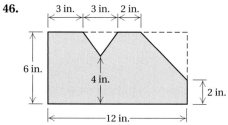

3 in.  3 in.  2 in.

6 in.

4 in.

12 in.

2 in.

**47.**

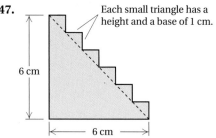

Each small triangle has a height and a base of 1 cm.

6 cm

6 cm

**48.**

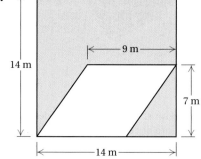

14 m

9 m

7 m

14 m

**49.** *Triangular Sail.* Jane's Custom Sails is making a custom sail for a laser sailboat. From a rectangular piece of dacron sailcloth that measures 18 ft by 12 ft, she cuts out a right triangular area plus a rectangular extension on each side for the hems, with the dimensions shown below. How much fabric (area) is left over?

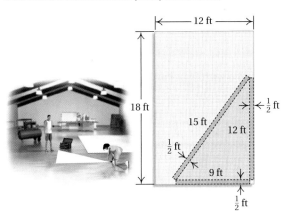

**50.** *Building Area.* Find the total area of the sides and the ends of the building.

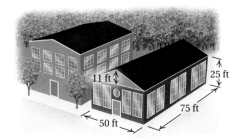

## Skill Maintenance

Complete.  [9.1a], [9.2a]

**51.** 23.4 cm = _____ mm

**52.** 0.23 km = _____ m

**53.** 28 ft = _____ in.

**54.** 72 ft = _____ yd

**55.** 72.4 cm = _____ m

**56.** 72.4 m = _____ km

**57.** 70 yd = _____ in.

**58.** 31,680 ft = _____ mi

**59.** 84 ft = _____ yd

**60.** $7\frac{1}{2}$ yd = _____ ft

**61.** 144 in. = _____ ft

**62.** 0.73 mi = _____ in.

Interest is compounded semiannually. Find the amount in the account after the given length of time. Round to the nearest cent.  [7.7b]

|  | PRINCIPAL | RATE OF INTEREST | TIME | AMOUNT IN THE ACCOUNT |
|---|---|---|---|---|
| **63.** | $25,000 | 4% | 5 years | |
| **64.** | $150,000 | $6\frac{7}{8}$% | 15 years | |
| **65.** | $150,000 | 7.4% | 20 years | |
| **66.** | $160,000 | 5% | 20 years | |

## Synthesis

**67.** Find the area, in square inches, of the shaded region.

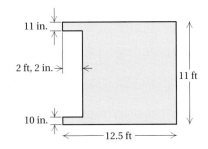

**68.** Find the area, in square feet, of the shaded region.

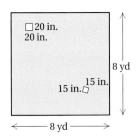

# 10.3

# Circles

## OBJECTIVES

**a** Find the length of a radius of a circle given the length of a diameter, and find the length of a diameter given the length of a radius.

**b** Find the circumference of a circle given the length of a diameter or a radius.

**c** Find the area of a circle given the length of a diameter or a radius.

**d** Solve applied problems involving circles.

**1.** Find the length of a radius.

18"

**2.** Find the length of a diameter.

$2\frac{1}{2}$ ft

## a Radius and Diameter

Shown below is a circle with center $O$. Segment $\overline{AC}$ is a *diameter*. A **diameter** is a segment that passes through the center of the circle and has endpoints on the circle. Segment $\overline{OB}$ is called a *radius*. A **radius** is a segment with one endpoint on the center and the other endpoint on the circle.

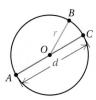

B

C

r

O

d

A

> ### DIAMETER AND RADIUS
>
> Suppose that $d$ is the length of the diameter of a circle and $r$ is the length of the radius. Then
>
> $$d = 2 \cdot r \quad \text{and} \quad r = \frac{d}{2}.$$

**EXAMPLE 1** Find the length of a radius of this circle.

$$r = \frac{d}{2}$$

$$= \frac{12 \text{ m}}{2}$$

$$= 6 \text{ m}$$

12 m

The radius is 6 m.

**EXAMPLE 2** Find the length of a diameter of this circle.

$$d = 2 \cdot r$$

$$= 2 \cdot \frac{1}{4} \text{ ft}$$

$$= \frac{1}{2} \text{ ft}$$

$\frac{1}{4}$ ft

The diameter is $\frac{1}{2}$ ft.

Do Exercises 1 and 2.

*Answers*

**1.** 9"    **2.** 5 ft

## b Circumference

The **circumference** of a circle is the distance around it. Calculating circumference is similar to finding the perimeter of a polygon.

To find a formula for the circumference of any circle given its diameter, we first need to consider the ratio $C/d$. Take a dinner plate and measure the circumference $C$ with a tape measure. Also measure the diameter $d$. The results for a specific plate are shown in the figure below.

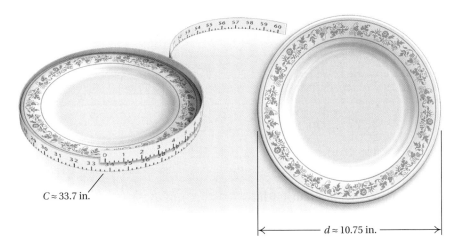

$C \approx 33.7$ in.

$d \approx 10.75$ in.

Then we have

$$\frac{C}{d} = \frac{33.7 \text{ in.}}{10.75 \text{ in.}} \approx 3.1.$$

Suppose we do this with plates and circles of several sizes. We get different values for $C$ and $d$, but always a number close to 3.1 for $C/d$. For any circle, if we divide the circumference $C$ by the diameter $d$, we get the same number. We call this number $\pi$ **(pi)**. The *exact* value of the ratio $C/d$ is $\pi$; 3.14 and 22/7 are approximations of $\pi$. If $C/d = \pi$, then $C = \pi \cdot d$.

---

### CIRCUMFERENCE AND DIAMETER

The circumference $C$ of a circle of diameter $d$ is given by

$$C = \pi \cdot d.$$

The number $\pi$ is about 3.14, or about $\dfrac{22}{7}$.

---

**EXAMPLE 3** Find the circumference of this circle. Use 3.14 for $\pi$.

$$
\begin{aligned}
C &= \pi \cdot d \\
&\approx 3.14 \times 6 \text{ cm} \\
&= 18.84 \text{ cm}
\end{aligned}
$$

6 cm

The circumference is about 18.84 cm.

Do Exercise 3.

**3.** Find the circumference of this circle. Use 3.14 for $\pi$.

20 m

*Answer*

**3.** 62.8 m

Since $d = 2 \cdot r$, where $r$ is the length of a radius, it follows that

$$C = \pi \cdot d = \pi \cdot (2 \cdot r), \text{ or } 2 \cdot \pi \cdot r.$$

### CIRCUMFERENCE AND RADIUS

The circumference $C$ of a circle of radius $r$ is given by

$$C = 2 \cdot \pi \cdot r.$$

**EXAMPLE 4** Find the circumference of this circle. Use $\frac{22}{7}$ for $\pi$.

$$C = 2 \cdot \pi \cdot r$$
$$\approx 2 \cdot \frac{22}{7} \cdot 70 \text{ in.}$$
$$= 2 \cdot 22 \cdot \frac{70}{7} \text{ in.}$$
$$= 44 \cdot 10 \text{ in.}$$
$$= 440 \text{ in.}$$

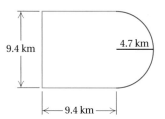
70 in.

The circumference is about 440 in.

**EXAMPLE 5** Find the perimeter of this figure. Use 3.14 for $\pi$.

9.4 km     4.7 km

|← 9.4 km →|

We let $P = $ the perimeter. We see that we have half a circle attached to a square. Thus we add half the circumference of the circle to the lengths of the three sides of the square.

$$P = \begin{array}{c}\text{Length of}\\\text{three sides}\\\text{of the square}\end{array} + \begin{array}{c}\text{Half of the}\\\text{circumference}\\\text{of the circle}\end{array}$$

$$= 3 \times 9.4 \text{ km} + \frac{1}{2} \times 2 \times \pi \times 4.7 \text{ km}$$
$$\approx 28.2 \text{ km} + 3.14 \times 4.7 \text{ km}$$
$$= 28.2 \text{ km} + 14.758 \text{ km}$$
$$= 42.958 \text{ km}$$

The perimeter is about 42.958 km.

Do Exercises 4 and 5.

**4.** Find the circumference of this circle. Use $\frac{22}{7}$ for $\pi$.

14 m

**5.** Find the perimeter of this figure. Use 3.14 for $\pi$.

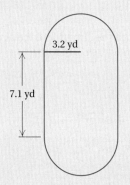
3.2 yd

7.1 yd

*Answers*

**4.** 88 m    **5.** 34.296 yd

## c  Area

Below is a circle of radius $r$.

Think of cutting half the circular region into small pieces and arranging them as shown below.

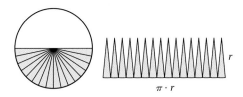

Then imagine cutting the other half of the circular region and arranging the pieces in with the others as shown below.

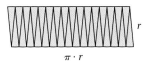

This is almost a parallelogram. The base has length $\frac{1}{2} \cdot 2 \cdot \pi \cdot r$, or $\pi \cdot r$ (half the circumference), and the height is $r$. Thus the area is

$$(\pi \cdot r) \cdot r.$$

This is the area of a circle.

---

**AREA OF A CIRCLE**

The **area of a circle** with radius of length $r$ is given by

$$A = \pi \cdot r \cdot r, \quad \text{or} \quad A = \pi \cdot r^2.$$

---

**EXAMPLE 6**  Find the area of this circle. Use $\frac{22}{7}$ for $\pi$.

$A = \pi \cdot r \cdot r$

$\approx \dfrac{22}{7} \cdot 14 \text{ cm} \cdot 14 \text{ cm}$

$= \dfrac{22}{7} \cdot 196 \text{ cm}^2$

$= 616 \text{ cm}^2$

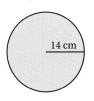

14 cm

The area is about $616 \text{ cm}^2$.

Do Exercise 6.

**6.** Find the area of this circle. Use $\frac{22}{7}$ for $\pi$.

5 km

*Answer*

**6.** $78\frac{4}{7} \text{ km}^2$

---

**STUDY TIPS**

**BEGINNING TO STUDY FOR THE FINAL EXAM**

It is never too soon to begin to study for the final examination. Take a few minutes each week to review the highlighted information, such as formulas, properties, and procedures. Make special use of the Mid-Chapter Reviews, Summary and Reviews, Chapter Tests, and Cumulative Reviews, as well as supplements such as the Video Resources on DVD Featuring Chapter Test Prep Videos. The Cumulative Review for Chapters 1–11 is a sample final exam.

**EXAMPLE 7** Find the area of this circle. Use 3.14 for $\pi$. Round to the nearest hundredth.

The diameter is 4.2 m; the radius is 4.2 m ÷ 2, or 2.1 m.

$$A = \pi \cdot r \cdot r$$
$$\approx 3.14 \times 2.1 \text{ m} \times 2.1 \text{ m}$$
$$= 3.14 \times 4.41 \text{ m}^2$$
$$= 13.8474 \text{ m}^2$$
$$\approx 13.85 \text{ m}^2$$

The area is about 13.85 m².

4.2 m

**7.** Find the area of this circle. Use 3.14 for $\pi$. Round to the nearest hundredth.

10.4 cm

-------------------------------------------- *Caution!* --------------------------------------------

Remember that circumference is always measured in linear units like ft, m, cm, yd, and so on. But area is measured in square units like ft², m², cm², yd², and so on.

Do Exercise 7.

## (d) Solving Applied Problems

**EXAMPLE 8** *Area of Pizza Pans.* How much larger is a pizza made in a 16-in.-square pizza pan than a pizza made in a 16-in.-diameter circular pan?

First, we make a drawing of each.

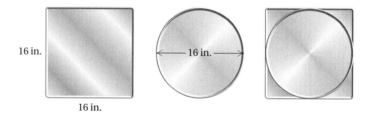

16 in.

16 in.

16 in.

Then we compute areas.
The area of the square is

$$A = s \cdot s$$
$$= 16 \text{ in.} \times 16 \text{ in.} = 256 \text{ in}^2.$$

The diameter of the circle is 16 in., so the radius is 16 in./2, or 8 in. The area of the circle is

$$A = \pi \cdot r \cdot r$$
$$\approx 3.14 \times 8 \text{ in.} \times 8 \text{ in.} = 200.96 \text{ in}^2.$$

The square pizza is larger by about

$$256 \text{ in}^2 - 200.96 \text{ in}^2, \quad \text{or} \quad 55.04 \text{ in}^2.$$

**8.** Which is larger and by how much: a 10-ft-square flower bed or a 12-ft-diameter flower bed?

Do Exercise 8.

*Answers*

**7.** 339.62 cm²   **8.** 12-ft-diameter flower bed, by about 13.04 ft²

**a** , **b** , **c**    For each circle, find the length of a diameter, the circumference, and the area. Use $\frac{22}{7}$ for $\pi$.

**1.**

**2.**
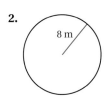
8 m

**3.**
$\frac{3}{4}$ in.

**4.**
$8\frac{2}{3}$ mi

7 cm

For each circle, find the length of a radius, the circumference, and the area. Use 3.14 for $\pi$.

**5.**

32 ft

**6.**

24 in.

**7.**

1.4 cm

**8.**

60.9 km

**d**    Solve. Use 3.14 for $\pi$.

**9.** *Treated Mosquito Net.*    The Adventure II treated mosquito net is perfect for the rural or tropic traveler. This net is circular with a radius of 6.37 ft. What is its diameter? its circumference? its area? Compare its area to that of the rectangular nets described in Example 10 of Section 10.2. How much larger is this net?

Source: www.travelhealthhelp.com/nets1.html

**10.** *Gypsy-Moth Tape.*    To protect an elm tree in your backyard, you need to attach gypsy moth caterpillar tape around the trunk. The tree has a 1.1-ft diameter. What length of tape is needed?

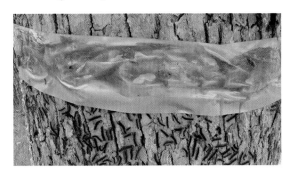

**11.** *Area of Pizza Pans.*    How much larger is a pizza made in a 12-in.-square pizza pan than a pizza made in a 12-in.-diameter circular pan?

**12.** *Penny.*    A penny has a 1-cm radius. What is its diameter? its circumference? its area?

1 cm

**13.** *Earth.* The diameter of the earth at the equator is 7926.41 mi. What is the circumference of the earth at the equator?

**14.** *Dimensions of a Quarter.* The circumference of a quarter is 7.85 cm. What is the diameter? the radius? the area?

**15.** *Circumference of a Baseball Bat.* In Major League Baseball, the diameter of the barrel of a bat cannot be more than $2\frac{3}{4}$ in., and the diameter of the bat handle can be no thinner than $\frac{16}{19}$ in. Find the maximum circumference of the barrel of a bat and the minimum circumference of the bat handle. Use $\frac{22}{7}$ for $\pi$.

Source: Major League Baseball

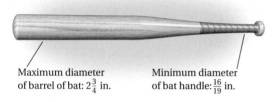

Maximum diameter of barrel of bat: $2\frac{3}{4}$ in.  Minimum diameter of bat handle: $\frac{16}{19}$ in.

**16.** *Trampoline.* The standard backyard trampoline has a diameter of 14 ft. What is its area?

Source: International Trampoline Industry Association, Inc.

14 ft

Frame height: 36 in.

**17.** *Swimming-Pool Walk.* You want to install a 1-yd-wide walk around a circular swimming pool. The diameter of the pool is 20 yd. What is the area of the walk?

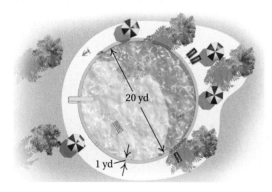

20 yd

1 yd

**18.** *Roller-Rink Floor.* A roller-rink floor is shown below. Each end is a semicircle. What is its area? If hardwood flooring costs $32.50 per square meter, how much will the flooring cost?

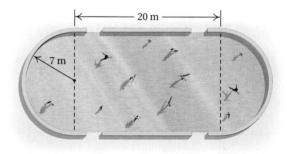

20 m

7 m

Find the perimeter of each figure. Use 3.14 for $\pi$.

**19.**

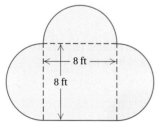

8 ft

8 ft

**20.**

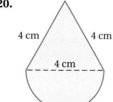

4 cm    4 cm

4 cm

**21.**

4 yd

4 yd

**22.**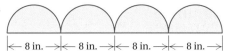

|← 8 in. →|← 8 in. →|← 8 in. →|← 8 in. →|

**23.**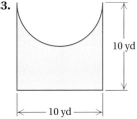

10 yd

|← 10 yd →|

**24.**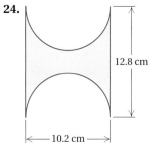

12.8 cm

|← 10.2 cm →|

Find the area of the shaded region in each figure. Use 3.14 for $\pi$.

**25.**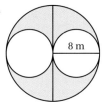

8 m

**26.**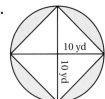

10 yd

10 yd

**27.** |← 2.8 cm →|

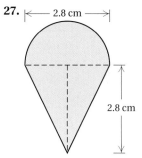

2.8 cm

**28.**

8 km

8 km

**29.**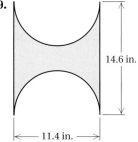

14.6 in.

|← 11.4 in. →|

**30.**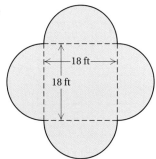

|←18 ft→|

18 ft

## Skill Maintenance

Complete.   [9.2a]

**31.** 5.43 m = _____ cm

**32.** 5.43 m = _____ km

**33.** The weight of a human brain is 2.5% of total body weight. A person weighs 200 lb. What does the brain weigh?   [7.5a]

**34.** Jack's commission is increased according to how much he sells. He receives a commission of 6% for the first $3000 and 10% on the amount over $3000. What is the total commission on sales of $8500?   [7.6b]

## Synthesis

*Comparing Perimeters and Fencing Costs.*   An **acre** is a unit of area that is defined to be 43,560 ft². A farmer needs to fence an acre of land. She is using 32-in. fencing that costs $149.99 for a 330-ft roll. Complete the following table for Exercises 35–39 and then use the data to answer Exercise 40. Use 3.14 for $\pi$.

| | FIGURE | AREA | PERIMETER OR CIRCUMFERENCE | COST OF FENCING |
|---|---|---|---|---|
| **35.** | 75 ft, 580.8 ft | | | |
| **36.** | 100 ft, 435.6 ft | | | |
| **37.** | 117.83 ft | | | |
| **38.** | 208.71 ft, 208.71 ft | | | |
| **39.** | 180 ft, 242 ft | | | |

**40.** Which dimensions of the acre yield the fence with **(a)** the shortest perimeter? **(b)** the least area? **(c)** the lowest cost and the largest area?

# Mid-Chapter Review

## Concept Reinforcement

Determine whether each statement is true or false.

_____ **1.** The area of a square is four times the length of a side. [10.2a]

_____ **2.** The area of a parallelogram with base 8 cm and height 5 cm is the same as the area of a rectangle with length 8 cm and width 5 cm. [10.2a, b]

_____ **3.** The area of a square that is 4 in. on a side is less than the area of a circle whose radius is 4 in. [10.2a], [10.3c]

_____ **4.** The perimeter of a rectangle that is 6 ft by 3 ft is greater than the circumference of a circle whose radius is 3 ft. [10.1a], [10.3b]

_____ **5.** The exact value of the ratio $C/d$ is $\pi$. [10.3b]

## Guided Solutions

Fill in each blank with the number that creates a correct solution.

**6.** Find the perimeter and the area. [10.1a], [10.2a]

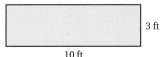

3 ft

10 ft

$P = 2 \cdot (l + w)$ $\qquad$ $A = l \cdot w$

$P = 2 \cdot (\Box \text{ ft} + \Box \text{ ft})$ $\qquad$ $A = \Box \text{ ft} \cdot \Box \text{ ft}$

$P = 2 \cdot (\Box \text{ ft})$ $\qquad$ $A = \Box \cdot \Box \cdot \text{ ft} \cdot \text{ ft}$

$P = \Box \text{ ft}$ $\qquad$ $A = \Box \text{ ft}^{\Box}$

**7.** Find the area. [10.2b]

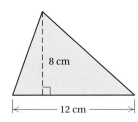

8 cm

12 cm

$A = \dfrac{1}{2} \cdot b \cdot h$

$A = \dfrac{1}{2} \cdot \Box \text{ cm} \cdot \Box \text{ cm}$

$A = \dfrac{\Box \cdot \Box}{2} \text{ cm}^{\Box}$

$A = \dfrac{\Box}{2} \text{ cm}^{\Box}, \text{ or } \Box \text{ cm}^{\Box}$

**8.** Find the circumference and the area. Use 3.14 for $\pi$. [10.3b, c]

10.2 in.

$C = \pi \cdot d$ $\qquad$ $A = \pi \cdot r \cdot r$

$C \approx \Box \cdot \Box \text{ in.}$ $\qquad$ $A \approx \Box \cdot \Box \text{ in.} \cdot \Box \text{ in.}$

$C = \Box \text{ in.}$ $\qquad$ $A = \Box \text{ in}^{\Box}$

## Mixed Review

**9.** Find the perimeter. [10.1a]

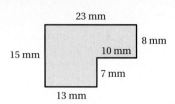

**10.** Find the perimeter and the area. [10.1a], [10.2a]

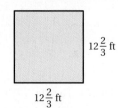

$12\frac{2}{3}$ ft

$12\frac{2}{3}$ ft

**11.** Find the area. [10.2b]

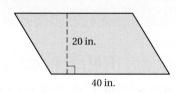

Find the area. [10.2b]

**12.**

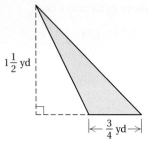

$1\frac{1}{2}$ yd

$\leftarrow \frac{3}{4}$ yd $\rightarrow$

**13.**

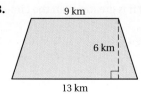

9 km

6 km

13 km

Find the circumference and the area. Use 3.14 for $\pi$. [10.3b, c]

**14.**

7 in.

**15.**

8.6 cm

**16.** *Matching.* Match each item in the first column with the appropriate item in the second column by drawing connecting lines. Some expressions in the second column might be used more than once. Some expressions might not be used. [10.1a], [10.2a, b], [10.3b, c]

| | |
|---|---|
| Area of a circle with radius 4 ft | 24 ft |
| Area of a square with side 4 ft | 16 ft |
| Circumference of a circle with radius 4 ft | $16 \cdot \pi$ ft$^2$ |
| Area of a rectangle with length 8 ft and width 4 ft | $8 \cdot \pi$ ft$^2$ |
| Area of a triangle with base 4 ft and height 8 ft | 32 ft$^2$ |
| Perimeter of a square with side 4 ft | $4 \cdot \pi$ ft |
| Perimeter of a rectangle with length 8 ft and width 4 ft | $8 \cdot \pi$ ft |
| | 64 ft |
| | 16 ft$^2$ |

## Understanding Through Discussion and Writing

**17.** Explain why a 16-in.-diameter pizza that costs $16.25 is a better buy than a 10-in.-diameter pizza that costs $7.85. [10.3d]

**18.** The length and the width of one rectangle are each three times the length and the width of another rectangle. Is the area of the first rectangle three times the area of the other rectangle? Why or why not? [10.2a]

**19.** The length of a side of a square is $\frac{1}{2}$ the length of a side of another square. Is the perimeter of the first square $\frac{1}{2}$ the perimeter of the other square? Why or why not? [10.1a]

**20.** Create for a fellow student a development of the formula
$$P = 2 \cdot (l + w) = 2 \cdot l + 2 \cdot w$$
for the perimeter of a rectangle. [10.1a]

**21.** Explain how the area of a triangle can be found by considering the area of a parallelogram. [10.2b]

**22.** The radius of one circle is twice the length of another circle's radius. Is the area of the first circle twice the area of the other circle? Why or why not? [10.3c]

# 10.4  Volume

## a  Rectangular Solids

The **volume** of a **rectangular solid** is the number of unit cubes needed to fill it.

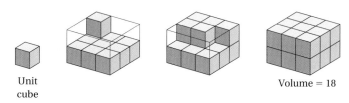

Unit cube

Volume = 18

Two other units are shown below.

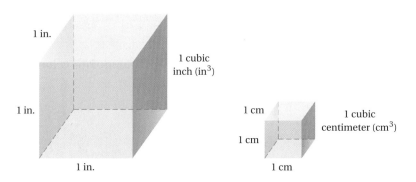

1 in.

1 in.

1 in.

1 cubic inch (in³)

1 cm

1 cm

1 cm

1 cubic centimeter (cm³)

**EXAMPLE 1**   Find the volume.

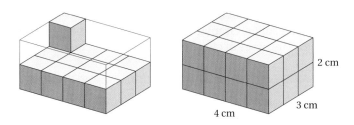

2 cm

4 cm

3 cm

The figure is made up of 2 layers of 12 cubes each, so its volume is 24 cubic centimeters (cm³).

> Do Margin Exercise 1.

---

### VOLUME OF A RECTANGULAR SOLID

The **volume of a rectangular solid** is found by multiplying length by width by height:

$$V = l \cdot w \cdot h.$$

h

w

l

---

### SKILL TO REVIEW
Objective 10.3c:  Find the area of a circle given the length of a diameter or a radius.

**1.** Find the area of a circle whose radius is 21 in. Use $\frac{22}{7}$ for $\pi$.

**2.** Find the area of a circle whose diameter is $2\frac{4}{5}$ ft. Use $\frac{22}{7}$ for $\pi$.

**1.** Find the volume.

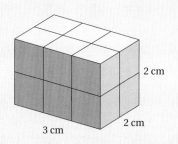

2 cm

3 cm

2 cm

*Answers*

*Skill to Review:*

**1.** 1386 in² **2.** $6\frac{4}{25}$ ft², or 6.16 ft²

*Margin Exercise:*

**1.** 12 cm³

**2. Carry-on Luggage.** The largest piece of luggage that you can carry on an airplane measures 23 in. by 10 in. by 13 in. Find the volume of this solid.

**3. Cord of Wood.** A cord of wood measures 4 ft by 4 ft by 8 ft. What is the volume of a cord of wood?

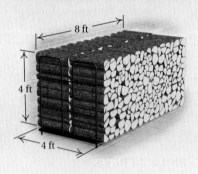

**EXAMPLE 2**  *Volume of a Safe.*  For office security, William purchases a safe whose dimensions are 29 in. × 25 in. × 53 in. Find the volume of this rectangular solid.

$$V = l \cdot w \cdot h$$
$$= 29 \text{ in.} \times 25 \text{ in.} \times 53 \text{ in.}$$
$$= 38{,}425 \text{ in}^3$$

Do Exercises 2 and 3.

## **b** Cylinders

A rectangular solid is shown below. Note that we can think of the volume as the product of the area of the base times the height:

$$V = l \cdot w \cdot h$$
$$= (l \cdot w) \cdot h$$
$$= (\text{Area of the base}) \cdot h$$
$$= B \cdot h,$$

Area of base $= B = l \cdot w$

where $B$ represents the area of the base.

Like rectangular solids, **circular cylinders** have bases of equal area that lie in parallel planes. The bases of circular cylinders are circular regions.

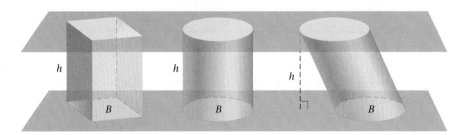

The volume of a circular cylinder is found in a manner similar to the way the volume of a rectangular solid is found. The volume is the product of the area of the base times the height. The height is always measured perpendicular to the base.

## VOLUME OF A CIRCULAR CYLINDER

The **volume of a circular cylinder** is the product of the area of the base $B$ and the height $h$:

$$V = B \cdot h, \quad \text{or} \quad V = \pi \cdot r^2 \cdot h.$$

**EXAMPLE 3** Find the volume of this circular cylinder. Use 3.14 for $\pi$.

$$V = B \cdot h = \pi \cdot r^2 \cdot h$$
$$\approx 3.14 \times 4 \text{ cm} \times 4 \text{ cm} \times 12 \text{ cm}$$
$$= 602.88 \text{ cm}^3$$

12 cm

4 cm

**EXAMPLE 4** Find the volume of this circular cylinder. Use $\dfrac{22}{7}$ for $\pi$.

$$V = B \cdot h = \pi \cdot r^2 \cdot h$$
$$\approx \frac{22}{7} \times 6.8 \text{ yd} \times 6.8 \text{ yd} \times 11.2 \text{ yd}$$
$$= 1627.648 \text{ yd}^3$$

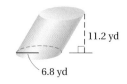

11.2 yd

6.8 yd

Do Exercises 4 and 5.

**4.** Find the volume of the cylinder. Use 3.14 for $\pi$.

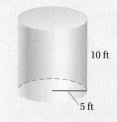

10 ft

5 ft

**5.** Find the volume of the cylinder. Use $\frac{22}{7}$ for $\pi$.

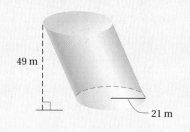

49 m

21 m

## c Spheres

A **sphere** is the three-dimensional counterpart of a circle. It is the set of all points in space that are a given distance (the radius) from a given point (the center).

We find the volume of a sphere as follows.

## VOLUME OF A SPHERE

The **volume of a sphere** of radius $r$ is given by

$$V = \frac{4}{3} \cdot \pi \cdot r^3.$$

*Answers*

**4.** 785 ft³   **5.** 67,914 m³

**EXAMPLE 5**  *Bowling Ball.*  The radius of a standard-sized bowling ball is 4.2915 in. Find the volume of a standard-sized bowling ball. Round to the nearest hundredth of a cubic inch. Use 3.14 for $\pi$.

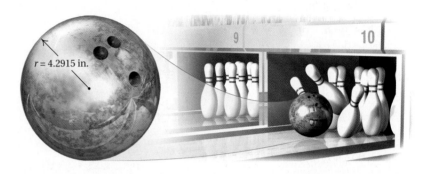

r = 4.2915 in.

6. Find the volume of the sphere. Use $\frac{22}{7}$ for $\pi$.

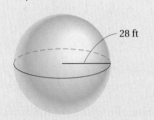

28 ft

7. The radius of a standard-sized golf ball is 2.1 cm. Find its volume. Use 3.14 for $\pi$.

We have

$$V = \frac{4}{3} \cdot \pi \cdot r^3 \approx \frac{4}{3} \times 3.14 \times (4.2915 \text{ in.})^3$$
$$\approx 330.90 \text{ in}^3. \quad \text{Using a calculator}$$

Do Exercises 6 and 7.

**d  Cones**

Consider a circle in a plane and choose any point $P$ not in the plane. The circular region, together with the set of all segments connecting $P$ to a point on the circle, is called a **circular cone**.

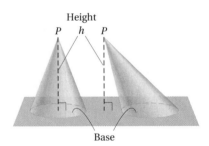

Height
P   h   P

Base

We find the volume of a cone as follows.

8. Find the volume of this cone. Use 3.14 for $\pi$.

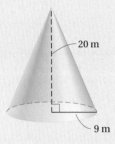

20 m

9 m

9. Find the volume of this cone. Use $\frac{22}{7}$ for $\pi$.

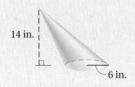

14 in.

6 in.

> **VOLUME OF A CIRCULAR CONE**
>
> The **volume of a circular cone** with base radius $r$ is one-third the product of the area of the base and the height:
>
> $$V = \frac{1}{3} \cdot B \cdot h = \frac{1}{3} \cdot \pi \cdot r^2 \cdot h.$$

**EXAMPLE 6**  Find the volume of this circular cone. Use 3.14 for $\pi$.

$$V = \frac{1}{3} \cdot \pi \cdot r^2 \cdot h$$
$$\approx \frac{1}{3} \times 3.14 \times 3 \text{ cm} \times 3 \text{ cm} \times 7 \text{ cm}$$
$$= 65.94 \text{ cm}^3$$

7 cm

3 cm

Do Exercises 8 and 9.

**Answers**

6. $91,989\frac{1}{3}$ ft$^3$    7. 38.77272 cm$^3$

8. 1695.6 m$^3$    9. 528 in$^3$

## (e) Solving Applied Problems

**EXAMPLE 7** *Propane Gas Tank.* A propane gas tank is shaped like a circular cylinder with half of a sphere at each end. Find the volume of the tank if the cylindrical section is 5 ft long with a 4-ft diameter. Use 3.14 for $\pi$.

1. **Familiarize.** We first make a drawing.

2. **Translate.** This is a two-step problem. We first find the volume of the cylindrical portion. Then we find the volume of the two ends and add. Note that together the two ends make a sphere with a radius of 2 ft. We let

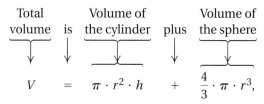

$$ V = \pi \cdot r^2 \cdot h + \frac{4}{3} \cdot \pi \cdot r^3, $$

where $V$ is the total volume. Then

$$ V \approx 3.14 \cdot (2\,\text{ft})^2 \cdot 5\,\text{ft} + \frac{4}{3} \cdot 3.14 \cdot (2\,\text{ft})^3. $$

3. **Solve.** The volume of the cylinder is approximately

$$ 3.14 \cdot (2\,\text{ft})^2 \cdot 5\,\text{ft} = 3.14 \cdot 2\,\text{ft} \cdot 2\,\text{ft} \cdot 5\,\text{ft} $$
$$ = 62.8\,\text{ft}^3. $$

The volume of the two ends is approximately

$$ \frac{4}{3} \cdot 3.14 \cdot (2\,\text{ft})^3 = \frac{4}{3} \cdot 3.14 \cdot 2\,\text{ft} \cdot 2\,\text{ft} \cdot 2\,\text{ft} $$
$$ \approx 33.5\,\text{ft}^3. $$

The total volume is about

$$ 62.8\,\text{ft}^3 + 33.5\,\text{ft}^3 = 96.3\,\text{ft}^3. $$

4. **Check.** We can repeat the calculations. The answer checks.
5. **State.** The volume of the tank is about $96.3\,\text{ft}^3$.

Do Exercise 10.

10. **Medicine Capsule.** A cold capsule is 8 mm long and 4 mm in diameter. Find the volume of the capsule. Use 3.14 for $\pi$. (*Hint*: First find the length of the cylindrical section.)

*Answer*
10. $83.7\overline{3}\,\text{mm}^3$

# 10.4

## Exercise Set

For Extra Help
*MathXL*

 Math XL
PRACTICE

WATCH

DOWNLOAD

READ

REVIEW

**a**    Find the volume of the rectangular solid.

**1.**

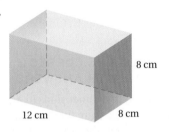

8 cm

12 cm    8 cm

**2.**

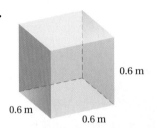

0.6 m

0.6 m    0.6 m

**3.**

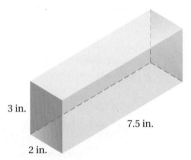

3 in.

7.5 in.

2 in.

**4.**

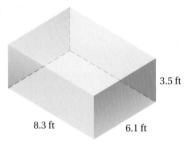

3.5 ft

8.3 ft    6.1 ft

**5.**

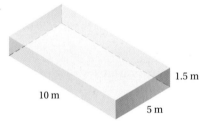

1.5 m

10 m

5 m

**6.**

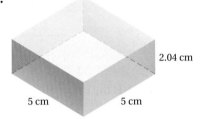

2.04 cm

5 cm    5 cm

**7.**

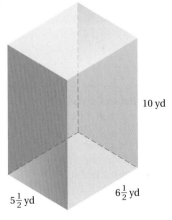

10 yd

$5\frac{1}{2}$ yd    $6\frac{1}{2}$ yd

**8.**

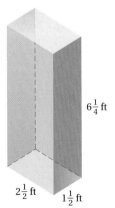

$6\frac{1}{4}$ ft

$2\frac{1}{2}$ ft    $1\frac{1}{2}$ ft

**b** Find the volume of the circular cylinder. Use 3.14 for $\pi$ in Exercises 9–12. Use $\frac{22}{7}$ for $\pi$ in Exercises 13 and 14.

**9.**

4 in.
8 in.

**10.**

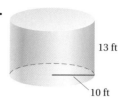
13 ft
10 ft

**11.**

4.5 cm
5 cm

**12.**

40 cm
4 cm

**13.**

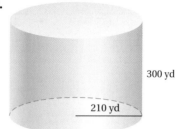
300 yd
210 yd

**14.**

28 m
4 m

**c** Find the volume of the sphere. Use 3.14 for $\pi$ in Exercises 15–18 and round to the nearest hundredth. Use $\frac{22}{7}$ for $\pi$ in Exercises 19 and 20.

**15.**

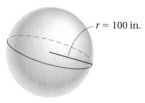
$r = 100$ in.

**16.**

$r = 200$ ft

**17.**

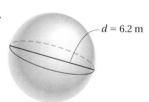
$d = 6.2$ m

**18.**

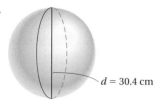
$d = 30.4$ cm

**19.**

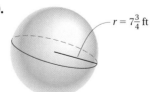
$r = 7\frac{3}{4}$ ft

**20.**

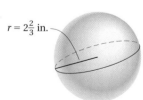
$r = 2\frac{2}{3}$ in.

**(d)** Find the volume of the circular cone. Use 3.14 for $\pi$ in Exercises 21, 22, and 26. Use $\frac{22}{7}$ for $\pi$ in Exercises 23, 24, and 25.

**21.**

100 ft

33 ft

**22.**

10 m

3 m

**23.**

12 cm

1.4 cm

**24.**

30 mm

35 mm

**25.**

$\frac{7}{5}$ yd

$\frac{3}{4}$ yd

**26.**

9.3 in.

4.6 in.

**(e)** Solve.

**27.** *Oak Log.* An oak log has a diameter of 12 cm and a length (height) of 42 cm. Find the volume. Use 3.14 for $\pi$.

**28.** *Ladder Rung.* A rung of a ladder is 2 in. in diameter and 16 in. long. Find the volume. Use 3.14 for $\pi$.

**29.** *Times Square Crystal Ball.* In 2009, a Waterford crystal ball, made up of more than 2500 pieces of crystal, was used to usher in the new year in Times Square. The ball will remain in place all year. The ball is 12 ft in diameter and weighs almost 6 tons. Find the volume of the crystal ball. Use 3.14 for $\pi$.

**30.** *Gas Pipeline.* The 638-mi Rockies Express–East pipeline from Colorado to Ohio is constructed with 80-ft sections of 42-in, or $3\frac{1}{2}$-ft, steel gas pipeline. Find the volume of one section. Use $\frac{22}{7}$ for $\pi$.

**Source:** Rockies Express Pipeline

**31.** *Tennis Ball.* The diameter of a tennis ball is 6.5 cm. Find the volume. Use 3.14 for $\pi$ and round the answer to the nearest hundredth.

**32.** *Volume of a Trash Can.* The diameter of the base of a cylindrical trash can is 0.7 yd. The height is 1.1 yd. Find the volume. Use 3.14 for $\pi$ and round the answer to the nearest hundredth.

**33.** *Volume of Earth.* The radius of the earth is about 3960 mi. Find the volume of the earth. Use 3.14 for $\pi$. Round to the nearest ten thousand cubic miles.

**34.** *Astronomy.* The radius of the largest moon of Uranus is about 789 km. Find the volume of this satellite. Use $\frac{22}{7}$ for $\pi$. Round to the nearest ten thousand cubic kilometers.

**35.** *Volume of a Candle.* Find the approximate volume of a candle that is a circular cone. The diameter of the base of the candle is 4.875 in., and the height is 12.5 in. Use 3.14 for $\pi$.

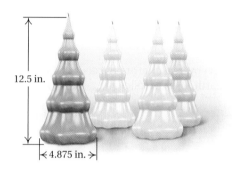

**36.** *Golf-Ball Packaging.* The box shown is just big enough to hold 3 golf balls. If the radius of a golf ball is 2.1 cm, how much air surrounds the three balls? Use 3.14 for $\pi$.

**37.** *Water Storage.* A water storage tank is a right circular cylinder with a radius of 14 m and a height of 100 m. What is the tank's volume? Use $\frac{22}{7}$ for $\pi$.

**38.** *Roof of a Turret.* The roof of a turret is often in the shape of a circular cone. Find the volume of this circular-cone structure if the radius is 2.5 m and the height is 4.6 m. Use 3.14 for $\pi$.

**39.** *Metallurgy.* If all the gold in the world could be gathered together, it would form a cube 18 yd on a side. Find the volume of the world's gold.

**40.** The volume of a ball is $36\pi$ cm$^3$. Find the dimensions of a rectangular box that is just large enough to hold the ball.

**41.** *Tennis-Ball Packaging.* Tennis balls are generally packaged in circular cylinders that hold 3 balls each. The diameter of a tennis ball is 6.5 cm. Find the volume of a can of tennis balls. Use 3.14 for $\pi$ and round the answer to the nearest hundredth.

**42.** *Oceanography.* A research submarine is capsule-shaped. Find the volume of the submarine if it has a length of 10 m and a diameter of 8 m. Use 3.14 for $\pi$ and round the answer to the nearest hundredth. (*Hint:* First find the length of the cylindrical section.)

## Skill Maintenance

Complete.  [9.1a]

**43.** 11 yd = _____ in.

**44.** 15,840 ft = _____ mi

**45.** 42 ft = _____ yd

**46.** 48 mi = _____ ft

**47.** 144 in. = _____ ft

**48.** 5.3 mi = _____ in.

Complete.  [9.5a]

**49.** 6 gal = _____ qt

**50.** 56 qt = _____ gal

**51.** 566 mL = _____ L

**52.** 13.4 L = _____ cm³

*Great Lakes.*  The Great Lakes contain about 5500 mi³ (23,000 km³) of water that covers a total area of about 94,000 mi² (244,000 km²). The Great Lakes are the largest system of fresh surface water on Earth. Use the following table for Exercises 53–58.

| FEATURE | UNITS | GREAT LAKE | | | | |
| | | Superior | Michigan | Huron | Erie | Ontario |
|---|---|---|---|---|---|---|
| Average depth | ft | 483 | 279 | 195 | 62 | 283 |
| Volume | mi³ | 2,900 | 1,180 | 850 | 116 | 393 |
| Water area | mi² | 31,700 | 22,300 | 23,000 | 9,910 | 7,340 |

SOURCES: http://www.epa.gov/glnpo/factsheet.html;
http://earth1.epa.gov/glnpo/statrefs.html

**53.** How much greater is the volume of water in Lake Michigan than in Lake Erie?  [1.8a], [8.2a]

**54.** How much less is the water area of Lake Ontario than of Lake Superior?  [1.8a], [8.2a]

**55.** Find the average of the average depths of the five Great Lakes.  [8.1a], [8.2a]

**56.** Find the average volume of water in the five Great Lakes.  [8.1a], [8.2a]

## Synthesis

Use the data in the table above for Exercises 57 and 58.

**57.** Convert the volume of water in Lake Huron from cubic miles to cubic kilometers. Round to the nearest hundredth of a cubic kilometer.

**58.** Convert the water area of Lake Superior from square miles to square kilometers. Round to the nearest hundredth of a square kilometer.

**59.** ▦ A sphere with diameter 1 m is circumscribed by a cube. How much greater is the volume of the cube than the volume of the sphere?

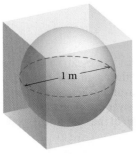

1 m

# 10.5 Angles and Triangles

## (a) Measuring Angles

We see a real-world application of *angles* of various types in the different back postures of the bicycle riders illustrated below.

**Style of Biking Determines Cycling Posture**

| **Road** About 180° flat | **Mountain** About 45° | **Comfort** About 90° |
|---|---|---|

Riders prefer a more aerodynamic flat-back position.

Riders prefer a semi-upright position to help lift the front wheel over obstacles.

Riders prefer an upright position that lessens stress on the lower back and neck.

SOURCE: USA TODAY research

An **angle** is a set of points consisting of two **rays**, or half-lines, with a common endpoint. The endpoint is called the **vertex** of the angle. The rays are called the **sides** of the angle.

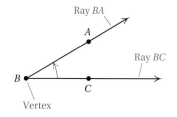

The angle above can be named

angle *ABC*,    angle *CBA*,    ∠*ABC*,    ∠*CBA*,    or    ∠*B*.

Note that the name of the vertex is either in the middle or, if no confusion results, listed by itself.

> Do Exercises 1 and 2.

**OBJECTIVES**

**(a)** Name an angle in five different ways, and given an angle, measure it with a protractor.

**(b)** Classify an angle as right, straight, acute, or obtuse.

**(c)** Identify complementary and supplementary angles and find the measure of a complement or a supplement of a given angle.

**(d)** Classify a triangle as equilateral, isosceles, or scalene, and as right, obtuse, or acute.

**(e)** Given two of the angle measures of a triangle, find the third.

Name the angle in five different ways.

1.

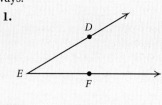

2.

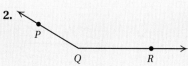

***Answers***

**1.** Angle *DEF*, angle *FED*, ∠*DEF*, ∠*FED*, or ∠*E*
**2.** Angle *PQR*, angle *RQP*, ∠*PQR*, ∠*RQP*, or ∠*Q*

Measuring angles is similar to measuring segments. To measure angles, we start with some arbitrary angle and assign to it a measure of 1. We call it a *unit angle*. Suppose that ∠U, shown below, is a unit angle. Let's measure ∠DEF. If we made 3 copies of ∠U, they would "fill up" ∠DEF. Thus the measure of ∠DEF would be 3.

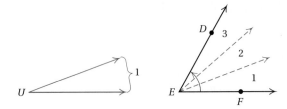

The unit most commonly used for angle measure is the degree. Below is such a unit. Its measure is 1 degree, or 1°.

A 1° angle:

Here are some other angles with their degree measures.

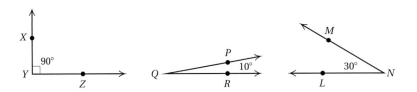

To indicate the *measure* of ∠XYZ, we write m ∠XYZ = 90°. The symbol ⌐ is sometimes drawn on a figure to indicate a 90° angle.

A device called a **protractor** is used to measure angles. Protractors have two scales. To measure an angle like ∠Q below, we place the protractor's ▲ at the vertex and line up one of the angle's sides at 0°. Then we check where the angle's other side crosses the scale. In the figure below, 0° is on the inside scale, so we check where the angle's other side crosses the inside scale. We see that m ∠Q = 145°. The notation m ∠Q is read "the measure of angle Q."

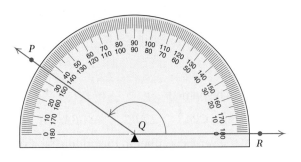

Do Exercise 3.

**3.** Use a protractor to measure this angle.

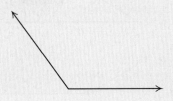

*Answer*

**3.** 127°

Let's find the measure of ∠ABC. This time we will use the 0° on the outside scale. We see that m ∠ABC = 42°.

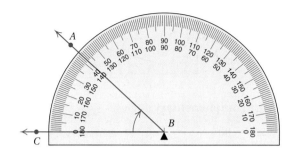

Do Exercise 4.

## b Classifying Angles

The following are ways in which we classify angles.

> **TYPES OF ANGLES**
>
> **Right angle:** An angle whose measure is 90°.
>
> **Straight angle:** An angle whose measure is 180°.
>
> **Acute angle:** An angle whose measure is greater than 0° and less than 90°.
>
> **Obtuse angle:** An angle whose measure is greater than 90° and less than 180°.

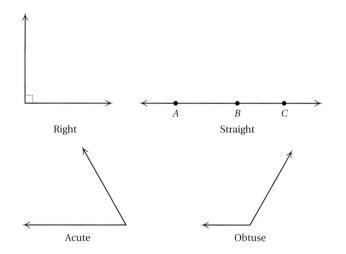

Do Exercises 5-8.

**4.** Use a protractor to measure this angle.

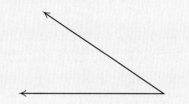

Classify each angle as right, straight, acute, or obtuse. Use a protractor if necessary.

**5.**

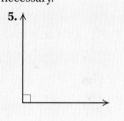

**6.**

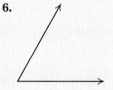

**7.**

**8.**

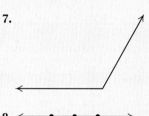

**Answers**

**4.** 33°   **5.** Right   **6.** Acute   **7.** Obtuse
**8.** Straight

## (c) Complementary and Supplementary Angles

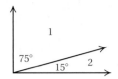

∠1 and ∠2 above are **complementary** angles.

$$m\angle 1 + m\angle 2 = 90°$$
$$75° \; + \; 15° \; = 90°$$

> ### COMPLEMENTARY ANGLES
>
> Two angles are **complementary** if the sum of their measures is 90°.
> Each angle is called a **complement** of the other.

If two angles are complementary, each is an acute angle. When complementary angles are adjacent to each other, they form a right angle.

**EXAMPLE 1** Identify each pair of complementary angles.

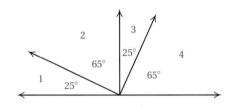

| ∠1 and ∠2 | 25° + 65° = 90° | ∠2 and ∠3 |
| ∠1 and ∠4 | | ∠3 and ∠4 |

**EXAMPLE 2** Find the measure of a complement of an angle of 39°.

$$90° - 39° = 51°$$

The measure of a complement is 51°.

Do Exercises 9–13.

Next, consider ∠1 and ∠2 as shown below. Because the sum of their measures is 180°, ∠1 and ∠2 are said to be **supplementary**. Note that when supplementary angles are adjacent, they form a straight angle.

$$m\angle 1 + m\angle 2 = 180°$$
$$30° \; + \; 150° \; = 180°$$

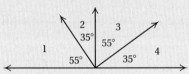

**9.** Identify each pair of complementary angles.

Find the measure of a complement of each angle.

**10.** 45°  **11.** 18°

**12.** 85°  **13.** 67°

*Answers*

**9.** ∠1 and ∠2; ∠1 and ∠4; ∠2 and ∠3;
∠3 and ∠4  **10.** 45°  **11.** 72°  **12.** 5°
**13.** 23°

> **SUPPLEMENTARY ANGLES**
>
> Two angles are **supplementary** if the sum of their measures is 180°.
> Each angle is called a **supplement** of the other.

**EXAMPLE 3** Identify each pair of supplementary angles.

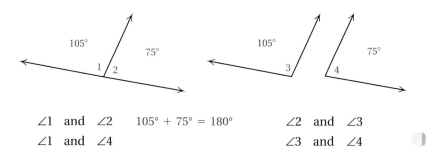

∠1 and ∠2    105° + 75° = 180°       ∠2 and ∠3

∠1 and ∠4                            ∠3 and ∠4

**EXAMPLE 4** Find the measure of a supplement of an angle of 112°.

180° − 112° = 68°

The measure of a supplement is 68°.

Do Exercises 14–18.

Do Exercises 14–18.

**14.** Identify each pair of supplementary angles.

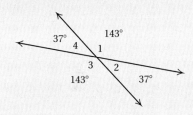

Find the measure of a supplement of each angle.

**15.** 38°               **16.** 157°

**17.** 90°               **18.** 71°

## (d) Triangles

A **triangle** is a polygon made up of three segments, or sides. Consider these triangles. The triangle with vertices *A*, *B*, and *C* can be named △*ABC*.

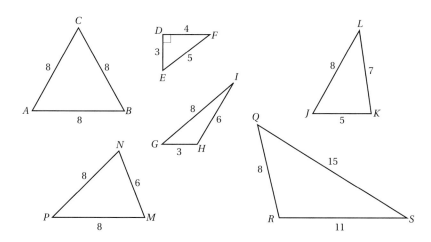

**19.** Which triangles on p. 607 are:

  **a)** equilateral?

  **b)** isosceles?

  **c)** scalene?

**20.** Are all equilateral triangles isosceles?

**21.** Are all isosceles triangles equilateral?

**22.** Which triangles on p. 607 are:

  **a)** right triangles?

  **b)** obtuse triangles?

  **c)** acute triangles?

**23.** Find $m \angle P + m \angle Q + m \angle R$.

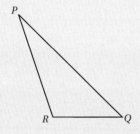

**24.** Find the missing angle measure.

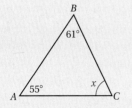

**Answers**

**19.** (a) △ABC; (b) △ABC, △MPN; (c) △DEF, △GHI, △JKL, △QRS   **20.** Yes   **21.** No
**22.** (a) △DEF; (b) △GHI, △QRS; (c) △ABC, △MPN, △JKL   **23.** 180°   **24.** 64°

---

We can classify triangles according to sides and according to angles.

> **TYPES OF TRIANGLES**
>
> **Equilateral triangle:** All sides are the same length.
>
> **Isosceles triangle:** Two or more sides are the same length.
>
> **Scalene triangle:** All sides are of different lengths.
>
> **Right triangle:** One angle is a right angle.
>
> **Obtuse triangle:** One angle is an obtuse angle.
>
> **Acute triangle:** All three angles are acute.

Do Exercises 19–22.

## (e) Sum of the Angle Measures of a Triangle

The sum of the angle measures of a triangle is 180°. To see this, note that we can think of cutting apart a triangle as shown on the left below. If we reassemble the pieces, we see that a straight angle is formed.

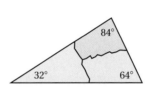
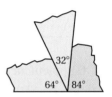

$$64° + 32° + 84° = 180°$$

> **SUM OF THE ANGLE MEASURES OF A TRIANGLE**
>
> In any △$ABC$, the sum of the measures of the angles is 180°:
>
> $$m \angle A + m \angle B + m \angle C = 180°.$$

Do Exercise 23.

If we know the measures of two angles of a triangle, we can calculate the measure of the third angle.

**EXAMPLE 5** Find the missing angle measure.

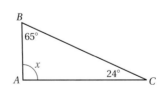

$$
\begin{aligned}
m \angle A + m \angle B + m \angle C &= 180° \\
x + 65° + 24° &= 180° \\
x + 89° &= 180° \\
x &= 180° - 89° \\
x &= 91°
\end{aligned}
$$

Thus, $m \angle A = 91°$.

Do Exercise 24.

**a**   Name each angle in five different ways.

**1.**

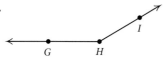

**2.**

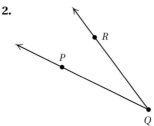

Use a protractor to measure each angle.

**3.**

**4.**

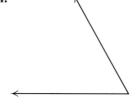

**5.**

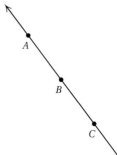

**6.**

**7.**

**8.**

**b**

**9.–16.** Classify each of the angles in Exercises 1–8 as right, straight, acute, or obtuse.

**17.–20.** Classify each of the angles in Margin Exercises 1–4 as right, straight, acute, or obtuse.

**c**   Find the measure of a complement of each angle.

**21.** 11°

**22.** 83°

**23.** 67°

**24.** 5°

**25.** 58°

**26.** 32°

**27.** 29°

**28.** 54°

Find the measure of a supplement of each angle.

**29.** 3°

**30.** 54°

**31.** 139°

**32.** 13°

**33.** 85°          **34.** 129°          **35.** 102°          **36.** 45°

---

**d**  Classify each triangle as equilateral, isosceles, or scalene. Then classify it as right, obtuse, or acute.

**37.**

**38.**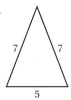

**39.**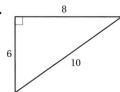

**40.**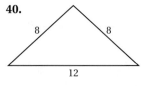

**41.**

**42.**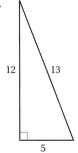

**43.**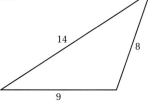

**44.**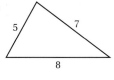

---

**e**  Find the missing angle measure.

**45.**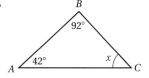

**46.**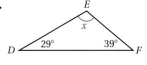

**47.**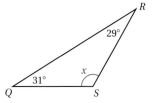

**48.**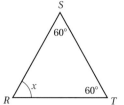

**49.**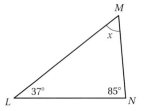

**50.**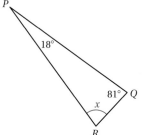

## Skill Maintenance

In each of Exercises 51–58, fill in the blank with the correct term from the given list. Some of the choices may not be used.

**51.** A number is divisible by 8 if the number named by the last _____ digits of the given number is divisible by 8. [3.2a]

**52.** A parallelogram is a four-sided figure with two pairs of _____ sides. [10.2b]

**53.** In the metric system, the _____ is the basic unit of mass. [9.4b]

**54.** A natural number, other than 1, that is not _____ is composite. [3.1c]

**55.** A(n) _____ is a set of points consisting of two rays with a common endpoint. [10.5a]

**56.** To convert from _____ to _____ , move the decimal point two places to the left and change the ¢ sign at the end to the $ sign in front. [5.3b]

**57.** The _____ of a polygon is the sum of the lengths of its sides. [10.1a]

**58.** The number 1 is known as the _____ identity, and the number 0 is known as the _____ identity. [1.2a], [1.4a]

prime

composite

two

three

dollars

cents

perimeter

parallel

perpendicular

additive

multiplicative

meter

gram

vertex

angle

area

## Synthesis

**59.** ⊞ In the figure, $m\angle 1 = 79.8°$ and $m\angle 6 = 33.07°$. Find $m\angle 2$, $m\angle 3$, $m\angle 4$, and $m\angle 5$.

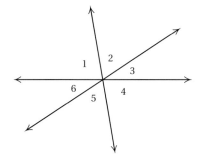

**60.** ⊞ In the figure, $m\angle 2 = 42.17°$ and $m\angle 3 = 81.9°$. Find $m\angle 1$, $m\angle 4$, $m\angle 5$, and $m\angle 6$.

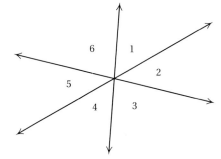

**61.** Find $m\angle ACB$, $m\angle CAB$, $m\angle EBC$, $m\angle EBA$, $m\angle AEB$, and $m\angle ADB$ in the rectangle shown below.

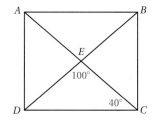

# 10.6

# Square Roots and the Pythagorean Theorem

**SKILL TO REVIEW**

Objective 1.9b: Evaluate exponential notation.

Evaluate.

**1.** $5^2$        **2.** $8^2$

Find the square.

**1.** $9^2$       **2.** $10^2$

**3.** $11^2$      **4.** $12^2$

**5.** $13^2$      **6.** $14^2$

**7.** $15^2$      **8.** $16^2$

**9.** $17^2$     **10.** $18^2$

**11.** $20^2$    **12.** $25^2$

Simplify. The results of Exercises 1–12 above may be helpful here.

**13.** $\sqrt{9}$      **14.** $\sqrt{16}$

**15.** $\sqrt{121}$    **16.** $\sqrt{100}$

**17.** $\sqrt{81}$     **18.** $\sqrt{64}$

**19.** $\sqrt{324}$   **20.** $\sqrt{400}$

**21.** $\sqrt{225}$   **22.** $\sqrt{169}$

**23.** $\sqrt{1}$      **24.** $\sqrt{0}$

Answers are on p. 613.

## a Square Roots

**SQUARE ROOT**

If a number is a product of two identical factors, then either factor is called a **square root** of the number. (If $a = c^2$, then $c$ is a square root of $a$.) The symbol $\sqrt{\phantom{x}}$ (called a **radical sign**) is used in naming square roots.

For example, $\sqrt{36}$ is the square root of 36. It follows that

$$\sqrt{36} = \sqrt{6 \cdot 6} = 6 \qquad \text{The square root of 36 is 6.}$$

because $6^2 = 36$.

**EXAMPLE 1** Simplify: $\sqrt{25}$.

$$\sqrt{25} = \sqrt{5 \cdot 5} = 5 \qquad \text{The square root of 25 is 5 because } 5^2 = 25.$$

**EXAMPLE 2** Simplify: $\sqrt{144}$.

$$\sqrt{144} = \sqrt{12 \cdot 12} = 12 \qquad \text{The square root of 144 is 12 because } 12^2 = 144.$$

---------------------------------------------- *Caution!* ----------------------------------------------

It is common to confuse squares and square roots. A number squared is that number multiplied by itself. For example, $16^2 = 16 \cdot 16 = 256$. A square root of a number is a number that when multiplied by itself gives the original number. For example, $\sqrt{16} = 4$, because $4 \cdot 4 = 16$.

---------------------------------------------------------------------------------------------------------

**EXAMPLES** Simplify.

**3.** $\sqrt{4} = 2$          **4.** $\sqrt{256} = 16$          **5.** $\sqrt{361} = 19$

Do Margin Exercises 1–24.

## b Approximating Square Roots

Many square roots cannot be written as whole numbers or fractions. For example, $\sqrt{2}$, $\sqrt{3}$, $\sqrt{39}$, and $\sqrt{70}$ cannot be precisely represented in decimal notation. To see this, consider the following decimal approximations for $\sqrt{2}$. Each gives a closer approximation, but none is exactly $\sqrt{2}$:

$$\sqrt{2} \approx 1.4 \qquad \text{because} \quad (1.4)^2 = 1.96;$$
$$\sqrt{2} \approx 1.41 \qquad \text{because} \quad (1.41)^2 = 1.9881;$$
$$\sqrt{2} \approx 1.414 \qquad \text{because} \quad (1.414)^2 = 1.999396;$$
$$\sqrt{2} \approx 1.4142 \quad \text{because} \quad (1.4142)^2 = 1.99996164.$$

Decimal approximations like these are commonly found by using a calculator.

**EXAMPLE 6** Use a calculator to approximate $\sqrt{3}$, $\sqrt{27}$, and $\sqrt{180}$ to three decimal places.

We use a calculator to find each square root. Since the calculator displays more than three decimal places, we round back to three places.

$$\sqrt{3} \approx 1.732, \qquad \sqrt{27} \approx 5.196, \qquad \sqrt{180} \approx 13.416$$

As a check, note that $1 \cdot 1 = 1$ and $2 \cdot 2 = 4$, so we expect $\sqrt{3}$ to be between 1 and 2. Similarly, we expect $\sqrt{27}$ to be between 5 and 6 and $\sqrt{180}$ to be between 13 and 14.

Do Exercises 25–28.

Use a calculator to approximate to three decimal places.

**25.** $\sqrt{5}$        **26.** $\sqrt{78}$

**27.** $\sqrt{168}$      **28.** $\sqrt{321}$

## c  The Pythagorean Theorem

A **right triangle** is a triangle with a 90° angle, as shown here. In a right triangle, the longest side is called the **hypotenuse**. It is the side opposite the right angle. The other two sides are called **legs**. We generally use the letters $a$ and $b$ for the lengths of the legs and $c$ for the length of the hypotenuse. They are related as follows.

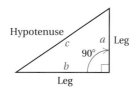

Hypotenuse $c$   $a$ Leg   90°   $b$   Leg

---

**THE PYTHAGOREAN THEOREM**

In any right triangle, if $a$ and $b$ are the lengths of the legs and $c$ is the length of the hypotenuse, then

$$a^2 + b^2 = c^2, \quad \text{or}$$

$$(\text{Leg})^2 + (\text{Other leg})^2 = (\text{Hypotenuse})^2.$$

The equation $a^2 + b^2 = c^2$ is called the **Pythagorean equation**.*

*Answers*

*Skill to Review:*
**1.** 25    **2.** 64

*Margin Exercises:*
**1.** 81    **2.** 100    **3.** 121    **4.** 144
**5.** 169    **6.** 196    **7.** 225    **8.** 256
**9.** 289    **10.** 324    **11.** 400    **12.** 625
**13.** 3    **14.** 4    **15.** 11    **16.** 10
**17.** 9    **18.** 8    **19.** 18    **20.** 20
**21.** 15    **22.** 13    **23.** 1    **24.** 0
**25.** 2.236    **26.** 8.832    **27.** 12.961
**28.** 17.916

---

### Calculator Corner

**Finding Square Roots** Many calculators have a square root key, $\boxed{\sqrt{\phantom{x}}}$. This is often the second function associated with the $\boxed{x^2}$ key and is accessed by pressing $\boxed{\text{2nd}}$ or $\boxed{\text{SHIFT}}$ followed by $\boxed{x^2}$. Often we enter the expression under the radical first followed by the $\boxed{\sqrt{\phantom{x}}}$ key. To find $\sqrt{30}$, for example, we press $\boxed{3}\ \boxed{0}\ \boxed{\text{2nd}}\ \boxed{\sqrt{\phantom{x}}}$ or $\boxed{3}\ \boxed{0}\ \boxed{\text{SHIFT}}\ \boxed{\sqrt{\phantom{x}}}$. We get 5.477225575.

It is always best to wait until calculations are complete before rounding. For example, to find $9 \cdot \sqrt{30}$ rounded to the nearest tenth, we do not first determine that $\sqrt{30} \approx 5.5$ and then multiply by 9 to get 49.5. Rather, we press $\boxed{9}\ \boxed{\times}\ \boxed{3}\ \boxed{0}\ \boxed{\text{2nd}}\ \boxed{\sqrt{\phantom{x}}}\ \boxed{=}$ or $\boxed{9}\ \boxed{\times}$ $\boxed{3}\ \boxed{0}\ \boxed{\text{SHIFT}}\ \boxed{\sqrt{\phantom{x}}}\ \boxed{=}$. The result is 49.29503018, so $9 \cdot \sqrt{30} \approx 49.3$.

**Exercises:** Use a calculator to find each of the following. Round to the nearest tenth.

**1.** $\sqrt{43}$           **4.** $5 \cdot \sqrt{12}$           **7.** $13\sqrt{68} + 14$

**2.** $\sqrt{94}$           **5.** $\sqrt{35} + 19$          **8.** $24 \cdot \sqrt{31} - 18$

**3.** $7 \cdot \sqrt{8}$         **6.** $17 + \sqrt{57}$          **9.** $7 \cdot \sqrt{90} + 3 \cdot \sqrt{40}$

---

*The *converse* of the Pythagorean theorem is also true. That is, if $a^2 + b^2 = c^2$, then the triangle is a right triangle.

The Pythagorean theorem is named for the Greek mathematician Pythagoras (569?–500? B.C.). We can think of this relationship as adding areas.

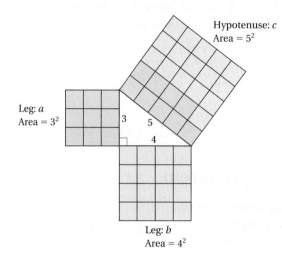

$$a^2 + b^2 = c^2$$
$$3^2 + 4^2 = 5^2$$
$$9 + 16 = 25$$

If we know the lengths of any two sides of a right triangle, we can use the Pythagorean equation to determine the length of the third side.

**EXAMPLE 7** Find the length of the hypotenuse of this right triangle.

We substitute in the Pythagorean equation:

$$a^2 + b^2 = c^2$$
$$6^2 + 8^2 = c^2 \quad \text{Substituting}$$
$$36 + 64 = c^2$$
$$100 = c^2.$$

The solution of this equation is the square root of 100, which is 10:

$$c = \sqrt{100} = 10.$$

Do Exercise 29.

**EXAMPLE 8** Find the length $b$ for the right triangle shown. Give an exact answer and an approximation to three decimal places.

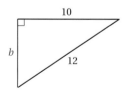

We substitute in the Pythagorean equation. Next, we solve for $b^2$ and then $b$, as follows:

$$a^2 + b^2 = c^2$$
$$10^2 + b^2 = 12^2 \quad \text{Substituting}$$
$$100 + b^2 = 144.$$

**29.** Find the length of the hypotenuse of this right triangle.

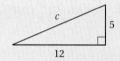

*Answer*

**29.** $c = 13$

Then

$$100 + b^2 - 100 = 144 - 100 \qquad \text{Subtracting 100 on both sides}$$
$$b^2 = 144 - 100$$
$$b^2 = 44$$

*Exact answer*:   $b = \sqrt{44}$

*Approximation*:   $b \approx 6.633.$ \qquad Using a calculator

Do Exercises 30-32.

## (d) Applications

**EXAMPLE 9**   *Height of Ladder.*   A 12-ft ladder leans against a building. The bottom of the ladder is 7 ft from the building. How high is the top of the ladder? Give an exact answer and an approximation to the nearest tenth of a foot.

1. **Familiarize.**   We first make a drawing. In it we see a right triangle. We let $h$ = the unknown height.

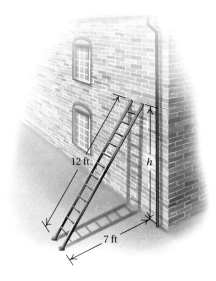

2. **Translate.**   We substitute 7 for $a$, $h$ for $b$, and 12 for $c$ in the Pythagorean equation:

$$a^2 + b^2 = c^2 \qquad \text{Pythagorean equation}$$
$$7^2 + h^2 = 12^2.$$

3. **Solve.**   We solve for $h^2$ and then $h$:

$$49 + h^2 = 144$$
$$49 + h^2 - 49 = 144 - 49$$
$$h^2 = 144 - 49$$
$$h^2 = 95$$

*Exact answer*:   $h = \sqrt{95}$

*Approximation*:   $h \approx 9.7$ ft.

4. **Check.**   $7^2 + (\sqrt{95})^2 = 49 + 95 = 144 = 12^2.$

5. **State.**   The top of the ladder is $\sqrt{95}$ ft, or about 9.7 ft, from the ground.

Do Exercise 33.

Find the length of the leg of each right triangle. Give an exact answer and an approximation to three decimal places.

30.

31.

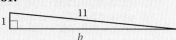

32.

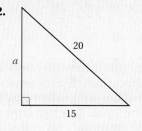

33. How long is a guy wire reaching from the top of an 18-ft pole to a point on the ground 10 ft from the pole? Give an exact answer and an approximation to the nearest tenth of a foot.

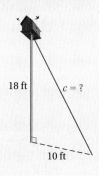

*Answers*

**30.** $a = \sqrt{75}$; $a \approx 8.660$   **31.** $b = \sqrt{120}$; $b \approx 10.954$   **32.** $a = \sqrt{175}$; $a \approx 13.229$
**33.** $\sqrt{424}$ ft $\approx 20.6$ ft

# Translating
# for Success

1. *Servings of Pork.* An 8-lb pork roast contains 37 servings of meat. How many pounds of pork would be needed for 55 servings?

2. *Height of a Ladder.* A 14.5-ft ladder leans against a house. The bottom of the ladder is 9.4 ft from the house. How high is the top of the ladder?

3. *Cruise Cost.* A group of 6 college students pays $4608 for a spring break cruise. What is each person's share?

4. *Sales Tax Rate.* The sales tax is $14.95 on the purchase of a new ladder that costs $299. What is the sales tax rate?

5. *Volume of a Sphere.* Find the volume of a sphere whose radius is 7.2 cm.

The goal of these matching questions is to practice step (2), *Translate*, of the five-step problem-solving process. Translate each problem to an equation and select a correct translation from equations A–O.

**A.** $x = \frac{1}{3} \pi \cdot 6^2 \cdot (7.2)$

**B.** $6 \cdot x = \$4608$

**C.** $x = \frac{4}{3} \cdot \pi \cdot 6^2 \cdot (7.2)$

**D.** $x = \pi \cdot \left(5\frac{1}{2} \div 2\right)^2 \cdot 7$

**E.** $x = 6\% \times 5 \times \$14.95$

**F.** $x = \frac{1}{3}\pi\left(5\frac{1}{2}\right)^2$

**G.** $(9.4)^2 + x^2 = (14.5)^2$

**H.** $\$14.95 = x \cdot \$299$

**I.** $x = 2(14.5 + 9.4)$

**J.** $(9.4 + 14.5)^2 = x$

**K.** $\frac{8}{37} = \frac{x}{55}$

**L.** $x = 4(14.5 + 9.4)$

**M.** $x = 6 \cdot \$4608$

**N.** $8 \cdot 37 = 55 \cdot x$

**O.** $x = \frac{4}{3}\pi(7.2)^3$

*Answers on page A-17*

6. *Inheritance.* Six children each inherit $4608 from their mother's estate. What is the total inheritance?

7. *Sales Tax.* Erica buys 5 pairs of earrings at $14.95 each. The sales tax rate is 6%. How much sales tax will be charged?

8. *Volume of a Cone.* Find the volume of a circular cone with a 6-cm base radius and a height of 7.2 cm.

9. *Volume of a Storage Tank.* The diameter of a cylindrical grain-storage tank is $5\frac{1}{2}$ yd. Its height is 7 yd. Find its volume.

10. *Perimeter of a Photo.* A rectangular photo is 14.5 cm by 9.4 cm. What is the perimeter of the photo?

**a** Simplify.

**1.** $\sqrt{100}$

**2.** $\sqrt{25}$

**3.** $\sqrt{441}$

**4.** $\sqrt{225}$

**5.** $\sqrt{625}$

**6.** $\sqrt{576}$

**7.** $\sqrt{361}$

**8.** $\sqrt{484}$

**9.** $\sqrt{529}$

**10.** $\sqrt{169}$

**11.** $\sqrt{10,000}$

**12.** $\sqrt{4,000,000}$

**b** Approximate to three decimal places.

**13.** $\sqrt{48}$

**14.** $\sqrt{17}$

**15.** $\sqrt{8}$

**16.** $\sqrt{3}$

**17.** $\sqrt{18}$

**18.** $\sqrt{7}$

**19.** $\sqrt{6}$

**20.** $\sqrt{61}$

**21.** $\sqrt{10}$

**22.** $\sqrt{21}$

**23.** $\sqrt{75}$

**24.** $\sqrt{220}$

**25.** $\sqrt{196}$

**26.** $\sqrt{123}$

**27.** $\sqrt{183}$

**28.** $\sqrt{300}$

**c** Find the length of the third side of each right triangle. Give an exact answer and, where appropriate, an approximation to three decimal places.

**29.**

**30.**

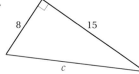

**31.**

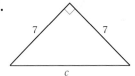

**32.**

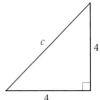

**33.**

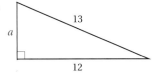

**34.**

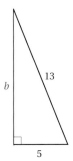

**35.**

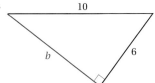

**36.**

For each right triangle, find the length of the side not given. Assume that *c* represents the length of the hypotenuse. Give an exact answer and, when appropriate, an approximation to three decimal places.

**37.** $a = 10, b = 24$

**38.** $a = 5, b = 12$

**39.** $a = 9, c = 15$

**40.** $a = 18, c = 30$

**41.** $a = 1, c = 32$

**42.** $b = 1, c = 20$

**43.** $a = 4, b = 3$

**44.** $a = 1, c = 15$

 In Exercises 45–54, give an exact answer and an approximation to the nearest tenth.

**45.** How long must a wire be in order to reach from the top of a 13-m telephone pole to a point on the ground 9 m from the base of the pole?

**46.** How long is a light cord reaching from the top of a 12-ft pole to a point on the ground 8 ft from the base of the pole?

**47.** *Softball Diamond.* A slow-pitch softball diamond is actually a square 65 ft on a side. How far is it from home plate to second base?

**48.** *Baseball Diamond.* A baseball diamond is actually a square 90 ft on a side. How far is it from home plate to second base?

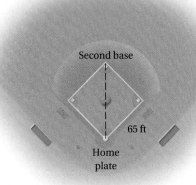

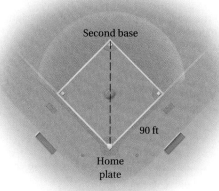

**49.** How tall is this tree?

**50.** How far is the base of the fence post from point *A*?

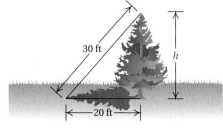

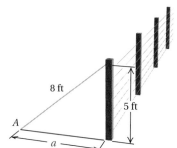

**51.** An airplane is flying at an altitude of 4100 ft. The slanted distance directly to the point where it will touch down is 15,100 ft. How far is the airplane horizontally from that point?

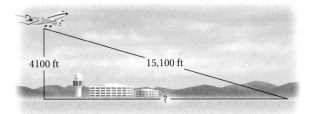

4100 ft    15,100 ft    ?

**52.** A surveyor had poles located at points $P$, $Q$, and $R$ around a lake. The distances that the surveyor was able to measure are marked on the drawing. What is the approximate distance from $P$ to $R$ across the lake?

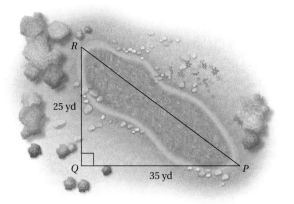

$R$    25 yd    $Q$    35 yd    $P$

**53.** How long is a string of lights reaching from the top of a 24-ft pole to a point on the ground 16 ft from the base of the pole?

**54.** How long must a wire be in order to reach from the top of a 26-m telephone pole to a point on the ground 18 m from the base of the pole?

## Skill Maintenance

Evaluate. [1.9b]

**55.** $10^3$

**56.** $10^2$

**57.** $10^5$

**58.** $10^4$

Interest is compounded annually. Find the amount in each account after the given length of time. Round to the nearest cent. [7.7b]

|  | PRINCIPAL | RATE OF INTEREST | TIME | AMOUNT IN THE ACCOUNT |
|---|---|---|---|---|
| **59.** | $200,000 | $6\frac{5}{8}\%$ | 25 years | |
| **60.** | $150,000 | 8.4% | 30 years | |
| **61.** | $45,000 | 4.5% | 15 years | |
| **62.** | $31,000 | $5\frac{1}{2}\%$ | 20 years | |

## Synthesis

**63.** Which of the triangles below has the larger area?

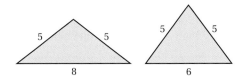

5    5    5    5    8    6

**64.** ▦ Find the area of the trapezoid shown. Round to the nearest hundredth.

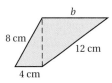

$b$    8 cm    12 cm    4 cm

# Summary and Review

## Key Terms and Formulas

polygon, p. 566
perimeter, p. 566
rectangle, p. 566
square, p. 567
area, p. 571
parallelogram, p. 572
triangle, p. 573
trapezoid, p. 574
bases, p. 574
diameter, p. 582
radius, p. 582
circumference, p. 583
pi ($\pi$), p. 583
volume, p. 593
rectangular solid, p. 593

circular cylinder, p. 594
sphere, p. 595
circular cone, p. 596
angle, p. 603
ray, p. 603
vertex, p. 603
sides, 603
protractor, p. 604
right angle, p. 605
straight angle, p. 605
acute angle, p. 605
obtuse angle, p. 605
complementary angles, p. 606
supplementary angles, p. 606

equilateral triangle, p. 608
isosceles triangle, p. 608
scalene triangle, p. 608
right triangle, p. 608
obtuse triangle, p. 608
acute triangle, p. 608
square root of a number, p. 612
radical sign, p. 612
square of a number, p. 612
hypotenuse, p. 613
legs, p. 613
Pythagorean theorem, p. 613
Pythagorean equation, p. 613

| | |
|---|---|
| *Perimeter of a Rectangle:* | $P = 2 \cdot (l + w)$, or $P = 2 \cdot l + 2 \cdot w$ |
| *Perimeter of a Square:* | $P = 4 \cdot s$ |
| *Area of a Rectangle:* | $A = l \cdot w$ |
| *Area of a Square:* | $A = s \cdot s$, or $A = s^2$ |
| *Area of a Parallelogram:* | $A = b \cdot h$ |
| *Area of a Triangle:* | $A = \frac{1}{2} \cdot b \cdot h$ |
| *Area of a Trapezoid:* | $A = \frac{1}{2} \cdot h \cdot (a + b)$ |
| *Radius and Diameter of a Circle:* | $d = 2 \cdot r; r = \frac{d}{2}$ |

| | |
|---|---|
| *Circumference of a Circle:* | $C = \pi \cdot d$, or $C = 2 \cdot \pi \cdot r$ |
| *Area of a Circle:* | $A = \pi \cdot r \cdot r$, or $A = \pi \cdot r^2$ |
| *Volume of a Rectangular Solid:* | $V = l \cdot w \cdot h$ |
| *Volume of a Circular Cylinder:* | $V = \pi \cdot r^2 \cdot h$ |
| *Volume of a Sphere:* | $V = \frac{4}{3} \cdot \pi \cdot r^3$ |
| *Volume of a Cone:* | $V = \frac{1}{3} \cdot \pi \cdot r^2 \cdot h$ |
| *Sum of Angle Measures of a Triangle:* | $m\angle A + m\angle B + m\angle C = 180°$ |
| *Pythagorean Equation:* | $a^2 + b^2 = c^2$ |

## Concept Reinforcement

Determine whether each statement is true or false.

_____ **1.** The acute angles of a right triangle are complementary.   [10.5c, d]

_____ **2.** Two angles are supplementary if the sum of their measures is between 90° and 180°.   [10.5c]

_____ **3.** The number $\pi$ is greater than 3.14 and $\frac{22}{7}$.   [10.3b]

_____ **4.** The volume of a sphere with diameter 6 ft is less than the volume of a rectangular solid that measures 6 ft by 6 ft by 6 ft.   [10.4a, c]

_____ **5.** The measure of any obtuse angle is larger than the measure of any acute angle.   [10.5b]

_____ **6.** The length of the hypotenuse of a right triangle is greater than the length of either of its legs.   [10.6c]

# Important Concepts

## Objectives 10.1a and 10.2a   Find the perimeter of a polygon; find the area of a rectangle and a square.

**Example**   Find the perimeter and the area of this rectangle.

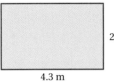

$$P = 2 \cdot (l + w)$$
$$= 2 \cdot (4.3\,\text{m} + 2.7\,\text{m})$$
$$= 2 \cdot (7\,\text{m}) = 14\,\text{m}$$
$$A = l \cdot w$$
$$= 4.3\,\text{m} \cdot 2.7\,\text{m} = 11.61\,\text{m}^2$$

**Practice Exercise**

1. Find the perimeter and the area of this rectangle.

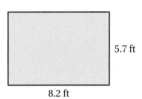

## Objective 10.2b   Find the area of a parallelogram, a triangle, and a trapezoid.

**Example**   Find the area of this parallelogram.

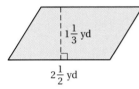

$$A = b \cdot h$$
$$= 2\frac{1}{2}\,\text{yd} \cdot 1\frac{1}{3}\,\text{yd}$$
$$= \frac{5}{2} \cdot \frac{4}{3} \cdot \text{yd} \cdot \text{yd}$$
$$= \frac{20}{6}\,\text{yd}^2 = \frac{10}{3}\,\text{yd}^2,$$
$$\text{or } 3\frac{1}{3}\,\text{yd}^2$$

**Practice Exercises**

2. Find the area of this parallelogram.

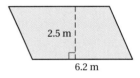

Find the area of this triangle.

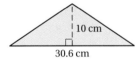

$$A = \frac{1}{2} \cdot b \cdot h$$
$$= \frac{1}{2} \cdot 30.6\,\text{cm} \cdot 10\,\text{cm}$$
$$= \frac{1}{2} \cdot 30.6 \cdot 10 \cdot \text{cm}^2$$
$$= 153\,\text{cm}^2$$

3. Find the area of this triangle.

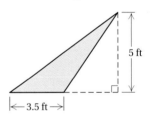

Find the area of this trapezoid.

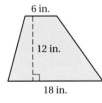

$$A = \frac{1}{2} \cdot h \cdot (a + b)$$
$$= \frac{1}{2} \times 12\,\text{in.} \times (6\,\text{in.} + 18\,\text{in.})$$
$$= \frac{1}{2} \times 12\,\text{in.} \times (24\,\text{in.})$$
$$= \frac{12 \times 24}{2}\,\text{in}^2 = 144\,\text{in}^2$$

4. Find the area of this trapezoid.

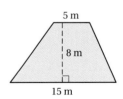

**Objective 10.3b** Find the circumference of a circle given the length of a diameter or a radius.

**Example** Find the circumference of this circle. Use 3.14 for $\pi$.

$$C = \pi \cdot d, \quad \text{or} \quad 2 \cdot \pi \cdot r$$
$$\approx 2 \times 3.14 \times 4\,\text{ft}$$
$$= 25.12\,\text{ft}$$

**Practice Exercise**

**5.** Find the circumference of this circle. Use 3.14 for $\pi$.

---

**Objective 10.3c** Find the area of a circle given the length of a diameter or a radius.

**Example** Find the area of this circle. Use $\dfrac{22}{7}$ for $\pi$.

$$A = \pi \cdot r \cdot r, \quad \text{or} \quad \pi \cdot r^2$$
$$\approx \frac{22}{7} \cdot 21\,\text{mm} \cdot 21\,\text{mm}$$
$$= \frac{22 \cdot 21 \cdot 21}{7}\,\text{mm}^2 = 1386\,\text{mm}^2$$

**Practice Exercise**

**6.** Find the area of this circle. Use $\dfrac{22}{7}$ for $\pi$.

---

**Objective 10.4a** Find the volume of a rectangular solid using the formula $V = l \cdot w \cdot h$.

**Example** Find the volume of this rectangular solid.

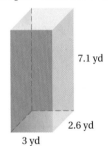

$$V = l \cdot w \cdot h$$
$$= 3\,\text{yd} \times 2.6\,\text{yd} \times 7.1\,\text{yd}$$
$$= 3 \times 2.6 \times 7.1\,\text{yd}^3$$
$$= 55.38\,\text{yd}^3$$

**Practice Exercise**

**7.** Find the volume of this rectangular solid.

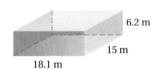

---

**Objective 10.4b** Given the radius and the height, find the volume of a circular cylinder.

**Example** Find the volume of this circular cylinder. Use 3.14 for $\pi$.

$$V = B \cdot h, \quad \text{or} \quad \pi \cdot r^2 \cdot h$$
$$\approx 3.14 \times 1.4\,\text{m} \times 1.4\,\text{m} \times 2.7\,\text{m}$$
$$= 16.61688\,\text{m}^3$$

**Practice Exercise**

**8.** Find the volume of this circular cylinder. Use $\dfrac{22}{7}$ for $\pi$.

**Objective 10.4c**   Given the radius, find the volume of a sphere.

**Example**   Find the volume of this sphere. Use $\frac{22}{7}$ for $\pi$.

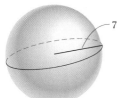

7 in.

$$V = \frac{4}{3} \cdot \pi \cdot r^3$$

$$\approx \frac{4}{3} \times \frac{22}{7} \times 7 \text{ in.} \times 7 \text{ in.} \times 7 \text{ in.}$$

$$= 1437\frac{1}{3} \text{ in}^3$$

**Practice Exercise**

**9.** Find the volume of this sphere. Use 3.14 for $\pi$.

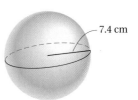

7.4 cm

---

**Objective 10.4d**   Given the radius and the height, find the volume of a circular cone.

**Example**   Find the volume of this circular cone. Use 3.14 for $\pi$.

20 cm

9 cm

$$V = \frac{1}{3} \cdot B \cdot h, \quad \text{or} \quad \frac{1}{3} \cdot \pi \cdot r^2 \cdot h$$

$$\approx \frac{1}{3} \times 3.14 \times 9 \text{ cm} \times 9 \text{ cm} \times 20 \text{ cm}$$

$$= \frac{3.14 \times 9 \times 9 \times 20}{3} \text{ cm}^3$$

$$= 1695.6 \text{ cm}^3$$

**Practice Exercise**

**10.** Find the volume of this circular cone. Use 3.14 for $\pi$.

5 ft

2.25 ft

---

**Objective 10.5c**   Find the measure of a complement or a supplement of a given angle.

**Example**   Find the measure of a complement and a supplement of an angle that measures 65°.

   The measure of the complement of an angle of 65° is 90° − 65°, or 25°.

   The measure of the supplement of an angle of 65° is 180° − 65°, or 115°.

**Practice Exercise**

**11.** Find the measure of a complement and a supplement of an angle that measures 38°.

---

**Objective 10.5e**   Given two of the angle measures of a triangle, find the third.

**Example**   Find the missing angle measure.

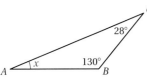

$$m\angle A + m\angle B + m\angle C = 180°$$

$$x + 130° + 28° = 180°$$

$$x + 158° = 180°$$

$$x = 180° - 158°$$

$$x = 22°$$

The measure of $\angle A$ is 22°.

**Practice Exercise**

**12.** Find the missing angle measure.

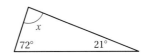

72°     $x$     21°

---

**Objective 10.6c**   Given the lengths of any two sides of a right triangle, find the length of the third side.

**Example**   Find the length of the third side of this triangle. Give an exact answer and an approximation to three decimal places.

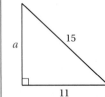

$$a^2 + b^2 = c^2 \quad \text{Pythagorean equation}$$
$$a^2 + 11^2 = 15^2$$
$$a^2 + 121 = 225$$
$$a^2 = 225 - 121$$
$$a^2 = 104$$
$$a = \sqrt{104} \approx 10.198$$

**Practice Exercise**

**13.** Find the length of the third side of this right triangle. Give an exact answer and an approximation to three decimal places.

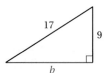

# Review Exercises

Find the perimeter.   [10.1a]

**1.**

**2.**

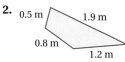

**3.** *Tennis Court.*   The dimensions of a standard-sized tennis court are 78 ft by 36 ft. Find the perimeter and the area of the tennis court.   [10.1b], [10.2c]

Find the perimeter and the area.   [10.1a], [10.2a]

**4.**

9 ft

9 ft

**5.**

1.8 cm

7 cm

Find the area.   [10.2b]

**6.**

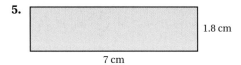

**7.**

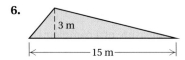

**8.**

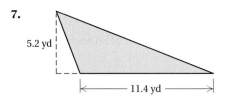

**9.**

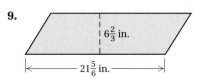

**10.** *Seeded Area.* A grassy area is to be seeded around three sides of a building and has equal width on the three sides, as shown below. What is the area of the seeded area?  [10.2c]

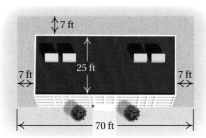

Find the length of a radius of each circle.  [10.3a]

**11.**

16 m

**12.**

$\frac{28}{11}$ in.

Find the length of a diameter of each circle.  [10.3a]

**13.**

7 ft

**14.**

10 cm

**15.** Find the circumference of the circle in Exercise 11. Use 3.14 for $\pi$.  [10.3b]

**16.** Find the circumference of the circle in Exercise 12. Use $\frac{22}{7}$ for $\pi$.  [10.3b]

**17.** Find the area of the circle in Exercise 11. Use 3.14 for $\pi$.  [10.3c]

**18.** Find the area of the circle in Exercise 12. Use $\frac{22}{7}$ for $\pi$.  [10.3c]

**19.** Find the area of the shaded region. Use 3.14 for $\pi$.  [10.3d]

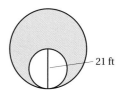

21 ft

Find the volume.  [10.4a]

**20.**

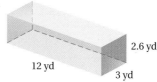

2.6 yd
12 yd
3 yd

**21.**

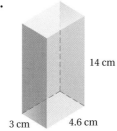

14 cm
3 cm
4.6 cm

Find the volume. Use 3.14 for $\pi$.  [10.4b, c, d]

**22.**

100 ft
20 ft

**23.**

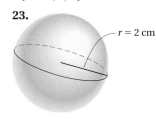

$r = 2$ cm

**24.**

4.5 in.
1 in.

**25.**

12 cm
5 cm

**26.** A "Norman" window is designed with dimensions as shown. Find its area and its perimeter. Use 3.14 for $\pi$. [10.2c], [10.3d]

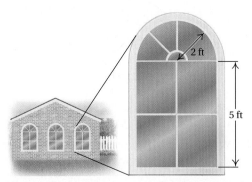

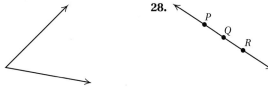

Use a protractor to measure each angle. [10.5a]

**27.**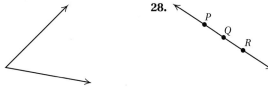

**28.**

**29.**

**30.**

**31.–34.** Classify each of the angles in Exercises 27–30 as right, straight, acute, or obtuse. [10.5b]

**35.** Find the measure of a complement of $\angle BAC$. [10.5c]

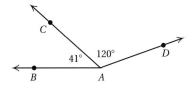

**36.** Find the measure of a supplement of a 44° angle. [10.5c]

Use the following triangle for Exercises 37–39.

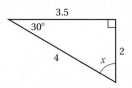

**37.** Find the missing angle measure. [10.5e]

**38.** Classify the triangle as equilateral, isosceles, or scalene. [10.5d]

**39.** Classify the triangle as right, obtuse, or acute. [10.5d]

**40.** Simplify: $\sqrt{64}$. [10.6a]

**41.** Use a calculator to approximate $\sqrt{83}$ to three decimal places. [10.6b]

In a right triangle, find the length of the side not given. Give an exact answer and an approximation to three decimal places. Assume that $c$ represents the length of the hypotenuse. [10.6c]

**42.** $a = 15, b = 25$        **43.** $a = 7, c = 10$

Find the length of the side not given. Give an exact answer and an approximation to three decimal places. [10.6c]

**44.**

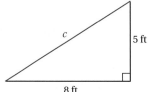

**45.**

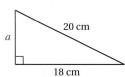

Solve. [10.6d]

**46.** How long is a wire reaching from the top of a 24-ft pole to a point on the ground 16 ft from the base of the pole? Round to the nearest tenth of a foot.

**47.** *Muir Woods.* On a tour of Muir Woods in Marin County, California, Eliza photographed the tallest redwood in the park. From 24 m from the base of the tree, she estimates that it is 83 m to the top of the tree. How tall is this redwood? Round to the nearest meter.

**Source:** National Register of Historic Places

**48.** Find the length of a diagonal from one corner to another of the tennis court in Exercise 3. Round to the nearest tenth of a foot.

**49.** Find the measure of a supplement of a $20\frac{3}{4}^{\circ}$ angle.
[10.5c]

**A.** $339\frac{1}{4}^{\circ}$        **B.** $159\frac{1}{4}^{\circ}$

**C.** $69\frac{1}{4}^{\circ}$        **D.** $70\frac{1}{4}^{\circ}$

**50.** Find the area of a circle whose diameter is $\frac{7}{9}$ in. Use $\frac{22}{7}$ for $\pi$. [10.3c]

**A.** $\frac{11}{9}$ in²        **B.** $\frac{77}{162}$ in²

**C.** $\frac{22}{9}$ in²        **D.** $\frac{154}{81}$ in²

# Synthesis

**51.** A square is cut in half so that the perimeter of the resulting rectangle is 30 ft. Find the area of the original square. [10.1a], [10.2a]

**52.** Find the area, in square meters, of the shaded region. [9.2a], [10.2c]

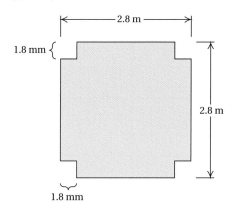

**53.** Find the area, in square centimeters, of the shaded region. [9.2a], [10.2c]

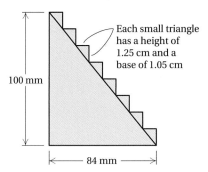

Each small triangle has a height of 1.25 cm and a base of 1.05 cm

# Understanding Through Discussion and Writing

**1.** Explain a procedure that could be used to determine the measure of an angle's supplement from the measure of the angle's complement. [10.5c]

**2.** How could you use the volume formulas given in this section to help estimate the volume of an egg? [10.4a, b, c, e]

**3.** Explain how the Pythagorean theorem can be used to prove that a triangle is a *right* triangle. [10.6c]

**4.** Explain how you might use triangles to find the sum of the angle measures of this figure. [10.5e]

**5.** The design of a home includes a cylindrical tower that will be capped with either a 10-ft-high dome (half of a sphere) or a 10-ft-high cone. Which type of cap will be more energy-efficient and why? [10.4c, d]

**6.** Which occupies more volume: two spheres, each with radius $r$, or one sphere with radius $2r$? Explain why. [10.4c]

## Test

For Extra Help

CHAPTER
Test Prep
VIDEOS

Step-by-step test solutions are found on the Chapter Test Prep Videos available via the Video Resources on DVD, in *MyMathLab*, and on YouTube (search "BittingerBasicMathEI" and click on "Channels").

Find the perimeter and the area.

**1.**

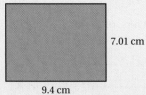

7.01 cm

9.4 cm

**2.**

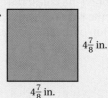

$4\frac{7}{8}$ in.

$4\frac{7}{8}$ in.

Find the area.

**3.**

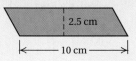

2.5 cm

10 cm

**4.**

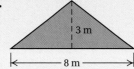

3 m

8 m

**5.**

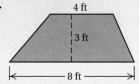

4 ft

3 ft

8 ft

**6.** Find the length of a diameter of this circle.

$\frac{1}{8}$ in.

**7.** Find the length of a radius of this circle.

18 cm

**8.** Find the circumference of the circle in Exercise 6. Use $\frac{22}{7}$ for $\pi$.

**9.** Find the area of the circle in Exercise 7. Use 3.14 for $\pi$.

**10.** Find the perimeter and the area of the shaded region. Use 3.14 for $\pi$.

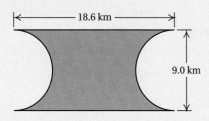

18.6 km

9.0 km

**11.** Find the volume.

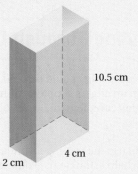

10.5 cm

2 cm

4 cm

**12.** A twelve-box rectangular carton of 12-oz juice boxes measures $10\frac{1}{2}$ in. by 8 in. by 5 in. What is the volume of the carton?

Find the volume. Use 3.14 for $\pi$.

**13.**

15 ft

5 ft

**14.**

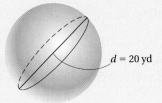

$d = 20$ yd

**15.**

12 cm

3 cm

Use a protractor to measure each angle.

**16.**

**17.**

**18.**

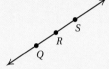

Q   R   S

**19.**

**20.–23.** Classify each of the angles in Exercises 16–19 as right, straight, acute, or obtuse.

Use the following triangle for Exercises 24–26.

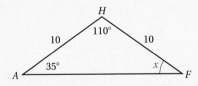

H
110°
10        10
35°              x
A                    F

**24.** Find the missing angle measure.

**25.** Classify the triangle as equilateral, isosceles, or scalene.

**26.** Classify the triangle as right, obtuse, or acute.

**27.** Find the measure of a complement and a supplement of $\angle CAD$.

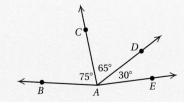

C
D
65°
75°    30°
B    A    E

**28.** Simplify: $\sqrt{225}$.

**29.** Approximate to three decimal places: $\sqrt{87}$

In a right triangle, find the length of the side not given. Give an exact answer and, where appropriate, an approximation to three decimal places. Assume that $c$ represents the length of the hypotenuse.

**30.** $a = 24, b = 32$

**31.** $a = 2, c = 8$

**32.**

**33.**

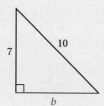

**34.** How long must a wire be in order to reach from the top of a 13-m antenna to a point on the ground 9 m from the base of the antenna? Round to the nearest tenth of a meter.

**35.** Find the volume of a sphere whose diameter is 42 cm. Use $\dfrac{22}{7}$ for $\pi$.

  **A.** $310,464 \text{ cm}^3$      **B.** $9702 \text{ cm}^3$

  **C.** $1848 \text{ cm}^3$      **D.** $38,808 \text{ cm}^3$

# Synthesis

Find the area of the shaded region. (Note that the figures are not drawn in perfect proportion.) Give the answer in square feet.

**36.**

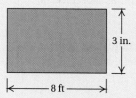

**37.**

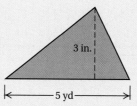

Find the volume of the solid. (Note that the solids are not drawn in perfect proportion.) Give the answer in cubic feet. Use 3.14 for $\pi$ and round to the nearest thousandth in Exercises 39 and 40.

**38.**

**39.**

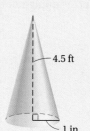

**40.**

# Cumulative Review

Solve.

**1.** *Neon Tubing.* The highest concentration of neon tubing in the world is found in Las Vegas. There are approximately 79.2 million ft of neon tubing in the city. Find standard notation for 79.2 million.

**Source:** *Las Vegas: An Unconventional History*, by Michelle Ferrari and Stephen Ives, Bulfinch Press

**2.** *Firefighting.* During a fire, the firefighters get a 1-ft layer of water on the 25-ft by 60-ft first floor of a 5-floor building. Water weighs $62\frac{1}{2}$ lb per cubic foot. What is the total weight of the water on the floor?

Calculate.

**3.** $1\frac{1}{2} + 2\frac{2}{3}$

**4.** $120.5 - 32.98$

**5.** $22\overline{)27{,}148}$

**6.** $8^3 - 45 \cdot 24 - 9^2 \div 3$

**7.** $\left(\frac{1}{4}\right)^2 \div \left(\frac{1}{2}\right)^3 \times 2^4 - (10.3)(4)$

**8.** $14 \div [33 \div 11 + 8 \times 2 - (15 - 3)]$

Find fraction notation.

**9.** $1.209$

**10.** $17\%$

Use $<$, $>$, or $=$ for $\square$ to write a true sentence.

**11.** $\dfrac{5}{6} \square \dfrac{7}{8}$

**12.** $\dfrac{-15}{18} \square \dfrac{-10}{12}$

Complete.

**13.** $42 \text{ in.} = \underline{\hspace{1cm}} \text{ ft}$

**14.** $9 \text{ sec} = \underline{\hspace{1cm}} \text{ min}$

**15.** $15°\text{C} = \underline{\hspace{1cm}} °\text{F}$

**16.** $0.087 \text{ L} = \underline{\hspace{1cm}} \text{ mL}$

**17.** $3 \text{ yd}^2 = \underline{\hspace{1cm}} \text{ ft}^2$

**18.** $17 \text{ cm} = \underline{\hspace{1cm}} \text{ m}$

Solve.

**19.** $x + \dfrac{3}{4} = -\dfrac{7}{8}$

**20.** $\dfrac{3}{x} = \dfrac{7}{10}$

**21.** $25 \cdot x = 2835$

**22.** $\dfrac{12}{15} = \dfrac{x}{18}$

**23.** Find the perimeter and the area.

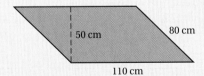

**24.** Find the diameter, the circumference, and the area of this circle. Use $\frac{22}{7}$ for $\pi$.

35 in.

**25.** Find the volume of this sphere. Use $\frac{22}{7}$ for $\pi$.

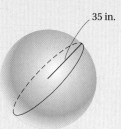

35 in.

Solve.

**26.** To get an A in math, a student must score an average of 90 on five tests. On the first four tests, the scores were 85, 92, 79, and 95. What is the lowest score that the student can get on the last test and still get an A?

**27.** What is the simple interest on $8000 at 4.2% for $\frac{1}{4}$ year?

**28.** What is the amount in an account after 25 years if $8000 is invested at 4.2%, compounded annually?

**29.** How long must a rope be in order to reach from the top of an 8-m tree to a point on the ground 15 m from the bottom of the tree?

**30.** The sales tax on an office supply purchase of $5.50 is $0.33. What is the sales tax rate?

**31.** A bolt of fabric in a fabric store has $10\frac{3}{4}$ yd on it. A customer purchases $8\frac{5}{8}$ yd. How many yards remain on the bolt?

**32.** What is the cost, in dollars, of 15.6 gal of gasoline at 239.9¢ per gallon? Round to the nearest cent.

**33.** A box of powdered milk that makes 20 qt costs $4.99. A box that makes 8 qt costs $1.99. Which size has the lower unit price?

**34.** It is $\frac{7}{10}$ km from Maria's dormitory to the library. Maria starts to walk from the dorm to the library, changes her mind after going $\frac{1}{4}$ of the distance, and returns to the dorm. How far did she walk?

**35.** Find the missing angle measure.

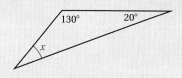

130°     20°

$x$

**36.** Classify the triangle as equilateral, isosceles, or scalene.

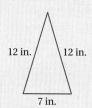

12 in.     12 in.

7 in.

**37.** Classify the triangle in Exercise 35 as right, obtuse, or acute.

## Synthesis

Find the volume in cubic feet. Use 3.14 for $\pi$.

**38.**

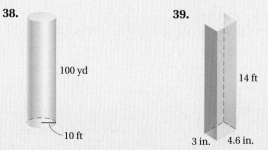

100 yd

10 ft

**39.**

14 ft

3 in.     4.6 in.

# Algebra: Solving Equations and Problems

## Real-World Application

The manatee, Florida's state marine mammal, is an endangered species. An aerial wintertime manatee census counted 2817 of these animals in 2007. This was 296 fewer than the number counted in 2006. What was Florida's manatee population in 2006?

*Source:* Florida Fish and Wildlife Conservation Commission

***This problem appears as Exercise 11 in Exercise Set 11.5.***

# 11.1

# Introduction to Algebra

Many types of problems require the use of equations in order to be solved efficiently. The study of algebra involves the use of equations to solve problems. Equations are constructed from algebraic expressions.

## **a** Evaluating Algebraic Expressions

In arithmetic, you have worked with expressions such as

$$37 + 86, \quad 7 \times 8, \quad 19 - 7, \quad \text{and} \quad \frac{3}{8}.$$

In algebra, we can use letters for numbers and work with *algebraic expressions* such as

$$x + 86, \quad 7 \times t, \quad 19 - y, \quad \text{and} \quad \frac{a}{b}.$$

Expressions like these should be familiar from the equation and problem solving that we have already done.

Sometimes a letter can stand for various numbers. In that case, we call the letter a **variable**. Let $a = $ your age. Then $a$ is a variable since $a$ changes from year to year. Sometimes a letter can stand for just one number. In that case, we call the letter a **constant**. Let $b = $ your birth year. Then $b$ is a constant.

An **algebraic expression** consists of variables, constants, numerals, operation signs, and/or grouping symbols. When we replace a variable with a number, we say that we are **substituting** for the variable. When we replace all the variables in an expression with numbers and carry out the operations in the expression, we are **evaluating the expression**.

**EXAMPLE 1** Evaluate $x + y$ for $x = 37$ and $y = 29$.

We substitute 37 for $x$ and 29 for $y$ and carry out the addition:

$$x + y = 37 + 29 = 66.$$

The number 66 is called the **value** of the expression when $x = 37$ and $y = 29$.

Algebraic expressions involving multiplication can be written in several ways. For example, "8 times $a$" can be written as

$$8 \times a, \quad 8 \cdot a, \quad 8(a), \quad \text{or simply} \quad 8a.$$

Two letters written together without an operation sign, such as $ab$, also indicates a multiplication.

**EXAMPLE 2** Evaluate $3y$ for $y = 14$.

$$3y = 3(14) = 42$$

Do Exercises 1–3.

1. Evaluate $a + b$ when $a = 38$ and $b = 26$.

2. Evaluate $x - y$ when $x = 57$ and $y = 29$.

3. Evaluate $4t$ when $t = 15$ and when $t = -6.8$.

*Answers*

**1.** 64    **2.** 28    **3.** 60; $-27.2$

**EXAMPLE 3** *Area of a Rectangle.* The area $A$ of a rectangle of length $l$ and width $w$ is given by the formula $A = lw$. Find the area when $l$ is 24.5 in. and $w$ is 16 in.

We substitute 24.5 in. for $l$ and 16 in. for $w$ and carry out the multiplication:

$$A = lw = (24.5 \text{ in.})(16 \text{ in.})$$
$$= (24.5)(16)(\text{in.})(\text{in.})$$
$$= 392 \text{ in}^2, \text{ or } 392 \text{ square inches.}$$

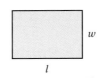

$w$

$l$

Do Exercise 4.

**4.** Find the area of a rectangle when $l$ is 24 ft and $w$ is 8 ft.

Algebraic expressions involving division can also be written in several ways. For example, "8 divided by $t$" can be written as

$$8 \div t, \quad \frac{8}{t}, \quad 8/t, \quad \text{or} \quad 8 \cdot \frac{1}{t},$$

where the fraction bar is a division symbol.

**EXAMPLE 4** Evaluate $\dfrac{a}{b}$ for $a = 63$ and $b = 9$.

We substitute 63 for $a$ and 9 for $b$ and carry out the division:

$$\frac{a}{b} = \frac{63}{9} = 7.$$

---

## Calculator Corner

**Evaluating Algebraic Expressions** We can use a calculator to evaluate algebraic expressions. To evaluate $x - y$ when $x = 48$ and $y = 19$, for example, we press $\boxed{4}\,\boxed{8}\,\boxed{-}\,\boxed{1}\,\boxed{9}\,\boxed{=}$. The calculator displays the result, 29. To evaluate $3x$ when $x = -27$, we press $\boxed{3}\,\boxed{\times}\,\boxed{2}\,\boxed{7}\,\boxed{+/-}\,\boxed{=}$. The result is $-81$. When we evaluate an expression like $\dfrac{x + y}{4}$, we must enclose the numerator in parentheses. To evaluate this expression when $x = 29$ and $y = 15$, we press $\boxed{(}\,\boxed{2}\,\boxed{9}\,\boxed{+}\,\boxed{1}\,\boxed{5}\,\boxed{)}\,\boxed{\div}\,\boxed{4}\,\boxed{=}$. The calculator displays the result, 11.

**Exercises:** Evaluate.

**1.** $x + y$, when $x = 35$ and $y = 16$

**2.** $7t$, when $t = 14$

**3.** $\dfrac{a}{b}$, when $a = 54$ and $b = -9$

**4.** $\dfrac{2m}{n}$, when $m = 38$ and $n = -4$

**5.** $\dfrac{x - y}{7}$, when $x = 94$ and $y = 31$

**6.** $\dfrac{p + q}{12}$, when $p = 47$ and $q = 97$

---

*Answer*

**4.** 192 ft$^2$

**5.** Evaluate $\dfrac{a}{b}$ when $a = -200$ and $b = 8$.

**6.** Evaluate $\dfrac{p + q}{13}$ when $p = 40$ and $q = 25$.

Complete each table by evaluating each expression for the given values.

**7.**

| VALUE OF $x$ | $1 \cdot x$ | $x$ |
|---|---|---|
| $x = 3$ | | |
| $x = -6$ | | |
| $x = 4.8$ | | |

**8.**

| VALUE OF $x$ | $2x$ | $5x$ |
|---|---|---|
| $x = 2$ | | |
| $x = -6$ | | |
| $x = 4.8$ | | |

**EXAMPLE 5** Evaluate $\dfrac{m + n}{12}$ for $m = 8$ and $n = 16$.

$$\frac{m + n}{12} = \frac{8 + 16}{12} = \frac{24}{12} = 2$$

Do Exercises 5 and 6.

## (b) Equivalent Expressions and the Distributive Laws

In solving equations and doing other kinds of work in algebra, we manipulate expressions in various ways. To see how to do this, we consider some examples in which we evaluate expressions.

**EXAMPLE 6** Evaluate $1 \cdot x$ when $x = 5$ and when $x = -8$ and compare the results to $x$.

We substitute 5 for $x$:

$$1 \cdot x = 1 \cdot 5 = 5.$$

Then we substitute $-8$ for $x$:

$$1 \cdot x = 1 \cdot (-8) = -8.$$

We see that $1 \cdot x$ and $x$ represent the same number.

Do Exercises 7 and 8.

We see in Example 6 and in Margin Exercise 7 that the expressions $1 \cdot x$ and $x$ represent the same number for any allowable replacement of $x$. In that sense, the expressions $1 \cdot x$ and $x$ are **equivalent**.

---
**EQUIVALENT EXPRESSIONS**

Two expressions that have the same value for all allowable replacements are called **equivalent**.

---

In the expression $3/x$, the number 0 is not allowable because $3/0$ is not defined. Even so, the expressions $6/(2x)$ and $3/x$ are *equivalent* because they represent the same number for any allowable (not 0) replacement of $x$. For example, when $x = 5$,

$$\frac{6}{2x} = \frac{6}{2 \cdot 5} = \frac{6}{10} = \frac{3}{5} \quad \text{and} \quad \frac{3}{x} = \frac{3}{5}.$$

We see in Margin Exercise 8 that the expressions $2x$ and $5x$ are *not* equivalent.

The fact that $1 \cdot x$ and $x$ are equivalent is a law called the **identity property of 1**. We often refer to the use of the identity property of 1 as "multiplying by 1." We have used multiplying by 1 many times in this text.

---
**THE IDENTITY PROPERTY OF 1 (MULTIPLICATIVE IDENTITY)**

For any number $a$,

$$a \cdot 1 = 1 \cdot a = a.$$

---

*Answers*

**5.** $-25$   **6.** 5   **7.** 3, 3; $-6$, $-6$; 4.8, 4.8
**8.** 4, 10; $-12$, $-30$; 9.6, 24

We now consider two other laws of real numbers called the **distributive laws**. They are the basis of many procedures in both arithmetic and algebra and are probably the most important laws that we use to manipulate algebraic expressions. The first distributive law involves two operations: addition and multiplication.

Let's begin by considering a multiplication problem from arithmetic:

$$\begin{array}{r} 4\ 5 \\ \times\quad 7 \\ \hline 3\ 5 \\ 2\ 8\ 0 \\ \hline 3\ 1\ 5 \end{array}$$

$3\ 5 \leftarrow$ This is $7 \cdot 5$.
$2\ 8\ 0 \leftarrow$ This is $7 \cdot 40$.
$3\ 1\ 5 \leftarrow$ This is the sum $7 \cdot 40 + 7 \cdot 5$.

To carry out the multiplication, we actually added two products. That is,

$$7 \cdot 45 = 7(40 + 5) = 7 \cdot 40 + 7 \cdot 5.$$

Let's examine this further. If we wish to multiply a sum of several numbers by a factor, we can either add and then multiply or multiply and then add.

**EXAMPLE 7** Evaluate $5(x + y)$ and $5x + 5y$ when $x = 2$ and $y = 8$, and compare the results.

We substitute 2 for $x$ and 8 for $y$ in each expression. Then we use the rules for order of operations to calculate.

**a)** $5(x + y) = 5(2 + 8)$
$\qquad\ \ = 5(10)$     Adding within parentheses first and then multiplying
$\qquad\ \ = 50$

**b)** $5x + 5y = 5 \cdot 2 + 5 \cdot 8$
$\qquad\qquad\ = 10 + 40$     Multiplying first and then adding
$\qquad\qquad\ = 50$

The results of (a) and (b) are the same.

Do Exercises 9–11.

The expressions $5(x + y)$ and $5x + 5y$, in Example 7 and in Margin Exercise 9, are equivalent. They illustrate the distributive law of multiplication over addition. Margin Exercises 10 and 11 also illustrate this distributive law.

> ### THE DISTRIBUTIVE LAW OF MULTIPLICATION OVER ADDITION
>
> For any numbers $a$, $b$, and $c$,
>
> $$a(b + c) = ab + ac.$$

In the statement of the distributive law, we know that in an expression such as $ab + ac$, the multiplications are to be done first according to the rules for order of operations. So, instead of writing $(4 \cdot 5) + (4 \cdot 7)$, we can write $4 \cdot 5 + 4 \cdot 7$. However, in $a(b + c)$, we cannot omit the parentheses. If we did, we would have $ab + c$, which means $(ab) + c$. For example, $3(4 + 2) = 18$, but $3 \cdot 4 + 2 = 14$.

**9.** Complete this table.

| VALUES OF $x$ AND $y$ | $5(x + y)$ | $5x + 5y$ |
|---|---|---|
| $x = 6, y = 7$ | | |
| $x = -3, y = 4$ | | |
| $x = -10, y = 5$ | | |

**10.** Evaluate $6x + 6y$ and $6(x + y)$ when $x = 10$ and $y = 5$.

**11.** Evaluate $4(x + y)$ and $4x + 4y$ when $x = 11$ and $y = 5$.

*Answers*

**9.** 65, 65; 5, 5; −25, −25    **10.** 90; 90
**11.** 64; 64

The second distributive law relates multiplication and subtraction. This law says that to multiply by a difference, we can either subtract and then multiply or multiply and then subtract.

> ### THE DISTRIBUTIVE LAW OF MULTIPLICATION OVER SUBTRACTION
>
> For any numbers $a$, $b$, and $c$,
>
> $$a(b - c) = ab - ac.$$

We often refer to "*the* distributive law" when we mean *either or both* of these laws.

Do Exercises 12–14.

**12.** Evaluate $7(x - y)$ and $7x - 7y$ when $x = 9$ and $y = 7$.

**13.** Evaluate $6x - 6y$ and $6(x - y)$ when $x = 10$ and $y = 5$.

**14.** Evaluate $2(x - y)$ and $2x - 2y$ when $x = 11$ and $y = 5$.

What do we mean by the *terms* of an expression? **Terms** are separated by addition signs. If there are subtraction signs, we can find an equivalent expression that uses addition signs.

**EXAMPLE 8** What are the terms of $3x - 4y + 2z$?

$$3x - 4y + 2z = 3x + (-4y) + 2z \qquad \text{Separating parts with } + \text{ signs}$$

The terms are $3x$, $-4y$, and $2z$.

Do Exercises 15 and 16.

What are the terms of each expression?

**15.** $5x - 4y + 3$

**16.** $-4y - 2x + 3z$

The distributive laws are the basis for a procedure in algebra called **multiplying**. In an expression such as $8(a + 2b - 7)$, we multiply each term inside the parentheses by 8:

$$8(a + 2b - 7) = 8 \cdot a + 8 \cdot 2b - 8 \cdot 7 = 8a + 16b - 56.$$

**EXAMPLES** Multiply.

**9.** $9(x - 5) = 9x - 9(5) \qquad$ Using the distributive law of multiplication over subtraction

$$= 9x - 45$$

**10.** $\dfrac{2}{3}(w + 1) = \dfrac{2}{3} \cdot w + \dfrac{2}{3} \cdot 1 \qquad$ Using the distributive law of multiplication over addition

$$= \dfrac{2}{3}w + \dfrac{2}{3}$$

*Answers*

**12.** 14; 14   **13.** 30; 30   **14.** 12; 12
**15.** $5x$; $-4y$; 3   **16.** $-4y$; $-2x$; $3z$

**EXAMPLE 11** Multiply: $-4(x - 2y + 3z)$.

$$-4(x - 2y + 3z) = -4 \cdot x - (-4)(2y) + (-4)(3z) \quad \text{Using both distributive laws}$$

$$= -4x - (-8y) + (-12z) \quad \text{Multiplying}$$

$$= -4x + 8y - 12z$$

We can also do this problem by first finding an equivalent expression with all plus signs and then multiplying:

$$-4(x - 2y + 3z) = -4[x + (-2y) + 3z]$$
$$= -4 \cdot x + (-4)(-2y) + (-4)(3z) = -4x + 8y - 12z.$$

Do Exercises 17–21.

## (c) Factoring

**Factoring** is the reverse of multiplying. To factor, we can use the distributive laws in reverse:

$$ab + ac = a(b + c) \quad \text{and} \quad ab - ac = a(b - c).$$

---

**FACTOR**

To **factor** an expression is to find an equivalent expression that is a product.

---

Look at Example 9. To *factor* $9x - 45$, we find an equivalent expression that is a product, $9(x - 5)$. When all the terms of an expression have a factor in common, we can "factor it out" using the distributive laws. Note the following.

$9x$ has the factors $9, -9, 3, -3, 1, -1, x, -x, 3x, -3x, 9x, -9x$;

$-45$ has the factors $1, -1, 3, -3, 5, -5, 9, -9, 15, -15, 45, -45$.

We remove the *greatest common factor*. In this case, that factor is 9. Thus,

$$9x - 45 = 9 \cdot x - 9 \cdot 5 = 9(x - 5).$$

Remember that an expression is factored when we find an equivalent expression that is a product.

**EXAMPLES** Factor.

**12.** $5x - 10 = 5 \cdot x - 5 \cdot 2 \quad$ Try to do this step mentally.
$$= 5(x - 2) \longleftarrow \text{You can check by multiplying.}$$

**13.** $9x + 27y - 9 = 9 \cdot x + 9 \cdot 3y - 9 \cdot 1 = 9(x + 3y - 1)$

--------- *Caution!* ---------

Note that although $3(3x + 9y - 3)$ is also equivalent to $9x + 27y - 9$, it is *not* the desired form. We can find the desired form by factoring out another factor of 3:

$$9x + 27y - 9 = 3(3x + 9y - 3) = 3 \cdot 3(x + 3y - 1) = 9(x + 3y - 1).$$

Remember to factor out the *greatest common factor*.

**EXAMPLES** Factor. Try to write just the answer, if you can.

**14.** $5x - 5y = 5(x - y)$

**15.** $-3x + 6y - 9z = -3 \cdot x - 3(-2y) - 3(3z) = -3(x - 2y + 3z)$

We usually factor out a negative factor when the first term is negative. The way we factor can depend on the context in which we are working. We might also factor the expression in Example 15 as follows:

$$-3x + 6y - 9z = 3(-x + 2y - 3z).$$

**16.** $18z - 12x - 24 = 6(3z - 2x - 4)$

Remember that you can always check factoring by multiplying. Keep in mind that an expression is factored when it is written as a product.

Do Exercises 22–25.

### Factor.

**22.** $6z - 12$

**23.** $3x - 6y + 9$

**24.** $16a - 36b + 42$

**25.** $-12x + 32y - 16z$

## (d) Collecting Like Terms

Terms such as $5x$ and $-4x$, whose variable factors are exactly the same, are called **like terms**. Similarly, numbers, such as $-7$ and $13$, are like terms. Also, $3y^2$ and $9y^2$ are like terms because the variables have the same exponent. Terms such as $4y$ and $5y^2$ are not like terms, and $7x$ and $2y$ are not like terms.

The distributive laws are used to **collect like terms**. This procedure can also be called **combining like terms**.

**EXAMPLES** Collect like terms. Try to write just the answer, if you can.

**17.** $4x + 2x = (4 + 2)x = 6x$   Factoring out the $x$ using a distributive law

**18.** $2x + 3y - 5x - 2y = 2x - 5x + 3y - 2y$
$$= (2 - 5)x + (3 - 2)y = -3x + y$$

**19.** $3x - x = 3x - 1x = (3 - 1)x = 2x$

**20.** $x - 0.24x = 1 \cdot x - 0.24x = (1 - 0.24)x = 0.76x$

**21.** $x - 6x = 1 \cdot x - 6 \cdot x = (1 - 6)x = -5x$

**22.** $4x - 7y + 9x - 5 + 3y - 8 = 13x - 4y - 13$

**23.** $\dfrac{2}{3}a - b + \dfrac{4}{5}a + \dfrac{1}{4}b - 10 = \dfrac{2}{3}a - 1 \cdot b + \dfrac{4}{5}a + \dfrac{1}{4}b - 10$

$$= \left(\dfrac{2}{3} + \dfrac{4}{5}\right)a + \left(-1 + \dfrac{1}{4}\right)b - 10$$

$$= \left(\dfrac{10}{15} + \dfrac{12}{15}\right)a + \left(-\dfrac{4}{4} + \dfrac{1}{4}\right)b - 10$$

$$= \dfrac{22}{15}a - \dfrac{3}{4}b - 10$$

Do Exercises 26–32.

### Collect like terms.

**26.** $6x - 3x$

**27.** $7x - x$

**28.** $x - 9x$

**29.** $x - 0.41x$

**30.** $5x + 4y - 2x - y$

**31.** $3x - 7x - 11 + 8y + 4 - 13y$

**32.** $-\dfrac{2}{3} - \dfrac{3}{5}x + y + \dfrac{7}{10}x - \dfrac{2}{9}y$

### Answers

**22.** $6(z - 2)$   **23.** $3(x - 2y + 3)$
**24.** $2(8a - 18b + 21)$
**25.** $-4(3x - 8y + 4z)$, or $4(-3x + 8y - 4z)$
**26.** $3x$   **27.** $6x$   **28.** $-8x$   **29.** $0.59x$
**30.** $3x + 3y$   **31.** $-4x - 5y - 7$
**32.** $\dfrac{1}{10}x + \dfrac{7}{9}y - \dfrac{2}{3}$

Evaluate.

**1.** $6x$, when $x = 7$

**2.** $9t$, when $t = 8$

**3.** $\dfrac{x}{y}$, when $x = 9$ and $y = 3$

**4.** $\dfrac{m}{n}$, when $m = 18$ and $n = 3$

**5.** $\dfrac{3p}{q}$, when $p = -2$ and $q = 6$

**6.** $\dfrac{5y}{z}$, when $y = -15$ and $z = -25$

**7.** $\dfrac{x + y}{5}$, when $x = 10$ and $y = 20$

**8.** $\dfrac{p - q}{2}$, when $p = 17$ and $q = 3$

**9.** $ab$, when $a = -5$ and $b = 4$

**10.** $ba$, when $a = -5$ and $b = 4$

 Evaluate.

**11.** $10(x + y)$ and $10x + 10y$, when $x = 20$ and $y = 4$

**12.** $5(a + b)$ and $5a + 5b$, when $a = 16$ and $b = 6$

**13.** $10(x - y)$ and $10x - 10y$, when $x = 20$ and $y = 4$

**14.** $5(a - b)$ and $5a - 5b$, when $a = 16$ and $b = 6$

Multiply.

**15.** $2(b + 5)$

**16.** $4(x + 3)$

**17.** $7(1 - t)$

**18.** $4(1 - y)$

**19.** $6(5x + 2)$

**20.** $9(6m + 7)$

**21.** $7(x + 4 + 6y)$

**22.** $4(5x + 8 + 3p)$

**23.** $-7(y - 2)$

**24.** $-9(y - 7)$

**25.** $-9(-5x - 6y + 8)$

**26.** $-7(-2x - 5y + 9)$

**27.** $\dfrac{3}{4}(x - 3y - 2z)$

**28.** $\dfrac{2}{5}(2x - 5y - 8z)$

**29.** $3.1(-1.2x + 3.2y - 1.1)$

**30.** $-2.1(-4.2x - 4.3y - 2.2)$

**c** Factor. Check by multiplying.

**31.** $2x + 4$

**32.** $5y + 20$

**33.** $30 + 5y$

**34.** $7x + 28$

**35.** $14x + 21y$

**36.** $18a + 24b$

**37.** $5x + 10 + 15y$

**38.** $9a + 27b + 81$

**39.** $8x - 24$

**40.** $10x - 50$

**41.** $32 - 4y$

**42.** $24 - 6m$

**43.** $8x + 10y - 22$  **44.** $9a + 6b - 15$  **45.** $-18x - 12y + 6$  **46.** $-14x + 21y + 7$

**d**  Collect like terms.

**47.** $9a + 10a$

**48.** $14x + 3x$

**49.** $10a - a$

**50.** $-10x + x$

**51.** $2x + 9z + 6x$

**52.** $3a - 5b + 4a$

**53.** $41a + 90 - 60a - 2$

**54.** $42x - 6 - 4x + 20$

**55.** $23 + 5t + 7y - t - y - 27$

**56.** $95 - 90d - 87 - 9d + 3 + 7d$

**57.** $11x - 3x$

**58.** $9t - 13t$

**59.** $6n - n$

**60.** $10t - t$

**61.** $y - 17y$

**62.** $5m - 8m + 4$

**63.** $-8 + 11a - 5b + 6a - 7b + 7$

**64.** $8x - 5x + 6 + 3y - 2y - 4$

**65.** $9x + 2y - 5x$

**66.** $8y - 3z + 4y$

**67.** $\dfrac{11}{4}x + \dfrac{2}{3}y - \dfrac{4}{5}x - \dfrac{1}{6}y + 12$

**68.** $\dfrac{13}{2}a + \dfrac{9}{5}b - \dfrac{2}{3}a - \dfrac{3}{10}b - 42$

**69.** $2.7x + 2.3y - 1.9x - 1.8y$

**70.** $6.7a + 4.3b - 4.1a - 2.9b$

## Skill Maintenance

For a circle with the given radius, find the diameter, the circumference, and the area. Use 3.14 for $\pi$.  [10.3a, b, c]

**71.** $r = 15$ yd

**72.** $r = 8.2$ m

**73.** $r = 9\frac{1}{2}$ mi

**74.** $r = 2400$ cm

For a circle with the given diameter, find the radius, the circumference, and the area. Use 3.14 for $\pi$.  [10.3a, b, c]

**75.** $d = 20$ mm

**76.** $d = 264$ km

**77.** $d = 4.6$ ft

**78.** $d = 10.3$ m

## Synthesis

Collect like terms, if possible, and factor the result, if possible.

**79.** $q + qr + qrs + qrst$

**80.** $21x + 44xy + 15y - 16x - 8y - 38xy + 2y + xy$

# 11.2 Solving Equations: The Addition Principle

## a Using the Addition Principle

Consider the equation $x = 7$. We can easily "see" that the solution of this equation is 7. If we replace $x$ with 7, we get $7 = 7$, which is true. Now consider the equation $x + 6 = 13$. The solution of this equation is also 7, but the fact that 7 is the solution is not as obvious. We now begin to consider principles that allow us to start with an equation like $x + 6 = 13$ and end up with an equation like $x = 7$, in which the variable is alone on one side and for which the solution is easier to find. The equations $x + 6 = 13$ and $x = 7$ are **equivalent**.

> ### EQUIVALENT EQUATIONS
>
> Equations with the same solutions are called **equivalent equations**.

One principle that we use to solve equations is the addition principle, which we have used throughout this text.

> ### THE ADDITION PRINCIPLE
>
> For any numbers $a$, $b$, and $c$,
>
> $$a = b \quad \text{is equivalent to} \quad a + c = b + c.$$

Let's solve $x + 6 = 13$ using the addition principle. We want to get $x$ alone on one side. To do so, we use the addition principle, choosing to add $-6$ on both sides because $6 + (-6) = 0$:

$$x + 6 = 13$$
$$x + 6 + (-6) = 13 + (-6) \qquad \text{Using the addition principle:}$$
$$\text{adding } -6 \text{ on both sides}$$
$$x + 0 = 7 \qquad \text{Simplifying}$$
$$x = 7. \qquad \text{Identity property of 0}$$

The solution of $x + 6 = 13$ is 7.

> Do Margin Exercise 1.

To visualize the addition principle, think of a jeweler's balance. When both sides of the balance hold equal amounts of weight, the balance is level. If weight is added or removed, equally, on both sides, the balance remains level.

1. Solve $x + 2 = 11$ using the addition principle.
   a) First, complete this sentence:
      $2 + \square = 0$.
   b) Then solve the equation.

*Answers*

*Skill to Review:*
1. 4    2. 9

*Margin Exercise:*
1. (a) $2 + -2 = 0$; (b) 9

When we use the addition principle, we sometimes say that we "add the same number on both sides of the equation." This is also true for subtraction, since we can express every subtraction as an addition. That is, since

$$a - c = b - c \quad \text{is equivalent to} \quad a + (-c) = b + (-c),$$

the addition principle tells us that we can "subtract the same number on both sides of an equation."

**EXAMPLE 1**  Solve: $x + 5 = -7$.

We have

$$x + 5 = -7$$
$$x + 5 - 5 = -7 - 5 \qquad \text{Using the addition principle: adding } -5 \text{ on both sides or subtracting 5 on both sides}$$
$$x + 0 = -12 \qquad \text{Simplifying}$$
$$x = -12. \qquad \text{Identity property of 0}$$

To check the answer, we substitute $-12$ in the original equation.

Check:
$$\frac{x + 5 = -7}{-12 + 5 \;?\; -7}$$
$$-7 \;|\qquad \text{TRUE}$$

The solution of the original equation is $-12$.

In Example 1, to get $x$ alone, we used the addition principle and subtracted 5 on both sides. This eliminated the 5 on the left. We started with $x + 5 = -7$, and using the addition principle we found a simpler equation $x = -12$, for which it was easier to "*see*" the solution. The equations $x + 5 = -7$ and $x = -12$ are *equivalent*.

Do Exercises 2 and 3.

Now we solve an equation that involves a subtraction by using the addition principle.

**EXAMPLE 2**  Solve: $-6.5 = y - 8.4$.

We have

$$-6.5 = y - 8.4$$
$$-6.5 + 8.4 = y - 8.4 + 8.4 \qquad \text{Using the addition principle: adding 8.4 to eliminate } -8.4 \text{ on the right}$$
$$1.9 = y.$$

Check:
$$\frac{-6.5 = y - 8.4}{-6.5 \;?\; 1.9 - 8.4}$$
$$-6.5 \;|\qquad \text{TRUE}$$

The solution is 1.9.

Note that equations are reversible. That is, if $a = b$ is true, then $b = a$ is true. Thus, when we solve $-6.5 = y - 8.4$, we can reverse it and solve $y - 8.4 = -6.5$ if we wish.

Do Exercises 4 and 5.

Solve.

**2.** $x + 7 = 2$

**3.** $y + 9 = 13$

Solve.

**4.** $8.7 = n - 4.5$

**5.** $x - 6 = -9$

**a**  Solve using the addition principle. Don't forget to check!

**1.** $x + 2 = 6$

Check:  $x + 2 = 6$
$$\overline{\phantom{x + 2 = 6}}$$
$$\overset{?}{\phantom{|}}$$

**2.** $y + 4 = 11$

Check:  $y + 4 = 11$
$$\overline{\phantom{y + 4 = 11}}$$
$$\overset{?}{\phantom{|}}$$

**3.** $x + 15 = -5$

Check:  $x + 15 = -5$
$$\overline{\phantom{x + 15 = -5}}$$
$$\overset{?}{\phantom{|}}$$

**4.** $t + 10 = -4$

Check:  $t + 10 = -4$
$$\overline{\phantom{t + 10 = -4}}$$
$$\overset{?}{\phantom{|}}$$

**5.** $x + 6 = 8$

**6.** $y + 8 = 37$

**7.** $x + 5 = 12$

**8.** $x + 3 = 7$

**9.** $11 = y + 7$

**10.** $14 = y + 8$

**11.** $-22 = t + 4$

**12.** $-14 = t + 8$

**13.** $x + 16 = -2$

**14.** $y + 34 = -8$

**15.** $y + 9 = -9$

**16.** $x + 13 = -13$

**17.** $x - 9 = 6$

**18.** $x - 9 = 2$

**19.** $t - 3 = 16$

**20.** $t - 5 = 12$

**21.** $y - 8 = -9$

**22.** $y - 6 = -12$

**23.** $x - 7 = -21$

**24.** $x - 5 = -16$

**25.** $5 + t = 7$

**26.** $6 + y = 22$

**27.** $-7 + y = 13$

**28.** $-8 + z = 16$

**29.** $-3 + t = -9$

**30.** $-8 + y = -23$

**31.** $r + \dfrac{1}{3} = \dfrac{8}{3}$

**32.** $t + \dfrac{3}{8} = \dfrac{5}{8}$

**33.** $m + \dfrac{5}{6} = -\dfrac{11}{12}$

**34.** $x + \dfrac{2}{3} = -\dfrac{5}{6}$

**35.** $x - \dfrac{5}{6} = \dfrac{7}{8}$

**36.** $y - \dfrac{3}{4} = \dfrac{5}{6}$

**37.** $-\dfrac{1}{5} + z = -\dfrac{1}{4}$

**38.** $-\dfrac{1}{8} + y = -\dfrac{3}{4}$

**39.** $7.4 = x + 2.3$

**40.** $9.3 = 4.6 + x$

**41.** $7.6 = x - 4.8$

**42.** $9.5 = y - 8.3$

**43.** $-9.7 = -4.7 + y$

**44.** $-7.8 = 2.8 + x$

**45.** $5\dfrac{1}{6} + x = 7$

**46.** $5\dfrac{1}{4} = 4\dfrac{2}{3} + x$

**47.** $q + \dfrac{1}{3} = -\dfrac{1}{7}$

**48.** $47\dfrac{1}{8} = -76 + z$

## Skill Maintenance

Add.   [2.2a], [4.2a], [5.2c]

**49.** $-3 + (-8)$

**50.** $-\dfrac{2}{3} + \dfrac{5}{8}$

**51.** $-14.3 + (-19.8)$

**52.** $3.2 + (-4.9)$

Subtract.   [2.3a], [4.3a], [5.2c]

**53.** $-3 - (-8)$

**54.** $-\dfrac{2}{3} - \dfrac{5}{8}$

**55.** $-14.3 - (-19.8)$

**56.** $3.2 - (-4.9)$

Multiply.   [2.4a], [3.6a], [5.3a]

**57.** $-3(-8)$

**58.** $-\dfrac{2}{3} \cdot \dfrac{5}{8}$

**59.** $-14.3 \times (-19.8)$

**60.** $3.2(-4.9)$

Divide.   [2.5a], [3.7b], [5.4a]

**61.** $\dfrac{-24}{-3}$

**62.** $-\dfrac{2}{3} \div \dfrac{5}{8}$

**63.** $\dfrac{283.14}{-19.8}$

**64.** $\dfrac{-15.68}{3.2}$

## Synthesis

Solve.

**65.** 🖩 $-356.788 = -699.034 + t$

**66.** $-\dfrac{4}{5} + \dfrac{7}{10} = x - \dfrac{3}{4}$

**67.** $x + \dfrac{4}{5} = -\dfrac{2}{3} - \dfrac{4}{15}$

**68.** $8 - 25 = 8 + x - 21$

**69.** $16 + x - 22 = -16$

**70.** $x + x = x$

**71.** $-\dfrac{3}{2} + x = -\dfrac{5}{17} - \dfrac{3}{2}$

**72.** $|x| = 5$

# 11.3 Solving Equations: The Multiplication Principle

## a Using the Multiplication Principle

**OBJECTIVE**

**a** Solve equations using the multiplication principle.

Suppose that $a = b$ is true, and we multiply $a$ by some number $c$. We get the same answer if we multiply $b$ by $c$, because $a$ and $b$ are the same number.

> **THE MULTIPLICATION PRINCIPLE**
>
> For any numbers $a$, $b$, and $c$, $c \neq 0$,
>
> $a = b$  is equivalent to  $a \cdot c = b \cdot c$.

**SKILL TO REVIEW**

Objective 1.7b: Solve equations like $28 \cdot x = 168$.

Solve.

**1.** $8x = 32$

**2.** $3x = 48$

When using the multiplication principle, we sometimes say that we "multiply on both sides of the equation by the same number."

**EXAMPLE 1**  Solve: $5x = 70$.

To get $x$ alone, we multiply by the *multiplicative inverse*, or *reciprocal*, of 5. Then we get the *multiplicative identity* 1 times $x$, or $1 \cdot x$, which simplifies to $x$. This allows us to eliminate 5 on the left.

$$5x = 70 \qquad \text{The reciprocal of 5 is } \tfrac{1}{5}.$$

$$\frac{1}{5} \cdot 5x = \frac{1}{5} \cdot 70 \qquad \text{Multiplying by } \tfrac{1}{5} \text{ to get } 1 \cdot x \text{ and eliminate 5 on the left}$$

$$1 \cdot x = 14 \qquad \text{Simplifying}$$

$$x = 14 \qquad \text{Identity property of 1: } 1 \cdot x = x$$

Check:

$$\begin{array}{c} 5x = 70 \\ \hline 5 \cdot 14 \ \overset{?}{\ } \ 70 \\ 70 \ \Big| \quad \text{TRUE} \end{array}$$

The solution is 14.

The multiplication principle also tells us that we can "divide on both sides of the equation by the same nonzero number." This is because dividing is the same as multiplying by a reciprocal. That is,

$$\frac{a}{c} = \frac{b}{c} \quad \text{is equivalent to} \quad a \cdot \frac{1}{c} = b \cdot \frac{1}{c}, \quad \text{when } c \neq 0.$$

In an expression like $5x$ in Example 1, the number 5 is called the **coefficient**. Example 1 could be done as follows, dividing on both sides by 5, the coefficient of $x$.

**EXAMPLE 2**  Solve: $5x = 70$.

$$5x = 70$$

$$\frac{5x}{5} = \frac{70}{5} \qquad \text{Dividing by 5 on both sides}$$

$$1 \cdot x = 14 \qquad \text{Simplifying}$$

$$x = 14 \qquad \text{Identity property of 1.}$$

The solution is 14.

**1.** Solve. Multiply on both sides.

$$6x = 90$$

**2.** Solve. Divide on both sides.

$$4x = -7$$

Do Margin Exercises 1 and 2.

**Answers**

*Skill to Review:*
**1.** 4  **2.** 16

*Margin Exercises:*
**1.** 15  **2.** $-\dfrac{7}{4}$

**EXAMPLE 3** Solve: $-4x = 92$.

$$-4x = 92$$

$$\frac{-4x}{-4} = \frac{92}{-4}$$  Using the multiplication principle. Dividing by $-4$ on both sides is the same as multiplying by $-\frac{1}{4}$.

$$1 \cdot x = -23$$  Simplifying

$$x = -23$$  Identity property of 1

Check: 
$$\frac{-4x = 92}{-4(-23) \overset{?}{\underset{}{\,}} 92}$$
$$92 \mid \quad \text{TRUE}$$

The solution is $-23$.

**3.** Solve: $-6x = 108$.

Do Exercise 3.

**EXAMPLE 4** Solve: $-x = 9$.

$$-x = 9$$

$$-1 \cdot (-x) = -1 \cdot 9$$  Multiplying by $-1$ on both sides

$$-1 \cdot (-1) \cdot x = -9$$  $\quad -x = (-1) \cdot x$

$$1 \cdot x = -9$$

$$x = -9$$

Check:
$$\frac{-x = 9}{-(-9) \overset{?}{\underset{}{\,}} 9}$$
$$9 \mid \quad \text{TRUE}$$

The solution is $-9$.

**4.** Solve: $-x = -10$.

Do Exercise 4.

In practice, it is generally more convenient to divide on both sides of the equation if the coefficient of the variable is in decimal notation or is an integer. If the coefficient is in fraction notation, it is more convenient to multiply by a reciprocal.

**EXAMPLE 5** Solve: $\dfrac{3}{8} = -\dfrac{5}{4}x$.

$$\frac{3}{8} = -\frac{5}{4}x$$

The reciprocal of $-\frac{5}{4}$ is $-\frac{4}{5}$. There is no sign change.

$$-\frac{4}{5} \cdot \frac{3}{8} = -\frac{4}{5} \cdot \left(-\frac{5}{4}x\right)$$  Multiplying by $-\frac{4}{5}$ to get $1 \cdot x$ and eliminate $-\frac{5}{4}$ on the right

$$-\frac{12}{40} = 1 \cdot x$$

$$-\frac{3}{10} = 1 \cdot x$$  Simplifying

$$-\frac{3}{10} = x$$  Identity property of 1

Check:

$$\frac{3}{8} = -\frac{5}{4}x$$

$$\frac{3}{8} \overset{?}{\;|\;} -\frac{5}{4}\left(-\frac{3}{10}\right)$$

$$\frac{3}{8} \quad \text{TRUE}$$

The solution is $-\dfrac{3}{10}$.

As noted in Section 11.2, if $a = b$ is true, then $b = a$ is true. Thus we can reverse the equation $\frac{3}{8} = -\frac{5}{4}x$ and solve $-\frac{5}{4}x = \frac{3}{8}$ if we wish.

Do Exercise 5.

**5.** Solve: $\dfrac{2}{3} = -\dfrac{5}{6}y$.

**EXAMPLE 6**  Solve: $1.16y = 9744$.

$$\frac{1.16y}{1.16} = \frac{9744}{1.16} \qquad \text{Dividing by 1.16 on both sides}$$

$$y = \frac{9744}{1.16}$$

$$y = 8400 \qquad \text{Simplifying}$$

Check:

$$1.16y = 9744$$

$$1.16(8400) \overset{?}{\;|\;} 9744$$

$$9744 \quad \text{TRUE}$$

The solution is 8400.

Do Exercises 6 and 7.

Solve.

**6.** $1.12x = 8736$

**7.** $6.3 = -2.1y$

Now we use the multiplication principle to solve an equation that involves division.

**EXAMPLE 7**  Solve: $\dfrac{-y}{9} = 14$.

$$9 \cdot \frac{-y}{9} = 9 \cdot 14 \qquad \text{Multiplying by 9 on both sides}$$

$$-y = 126$$

$$-1 \cdot (-y) = -1 \cdot 126 \qquad \text{Multiplying by } -1 \text{ on both sides}$$

$$y = -126$$

Check:

$$\frac{-y}{9} = 14$$

$$\frac{-(-126)}{9} \overset{?}{\;|\;} 14$$

$$\frac{126}{9}$$

$$14 \quad \text{TRUE}$$

The solution is $-126$.

Do Exercise 8.

**8.** Solve: $-14 = \dfrac{-y}{2}$.

*Answers*

**5.** $-\dfrac{4}{5}$  **6.** 7800  **7.** $-3$  **8.** 28

# Exercise Set

**a**  Solve using the multiplication principle. Don't forget to check!

**1.** $6x = 36$

**2.** $4x = 52$

**3.** $5x = 45$

**4.** $8x = 56$

**5.** $84 = 7x$

**6.** $63 = 7x$

**7.** $-x = 40$

**8.** $50 = -x$

**9.** $6x = -42$

**10.** $8x = -72$

**11.** $7x = -49$

**12.** $9x = -54$

**13.** $-12x = 72$

**14.** $-15x = 105$

**15.** $-9x = 45$

**16.** $-7x = 56$

**17.** $-21x = -126$

**18.** $-13x = -104$

**19.** $-2x = -10$

**20.** $-78 = -39p$

**21.** $\dfrac{1}{7}t = -9$

**22.** $-\dfrac{1}{8}y = 11$

**23.** $\dfrac{3}{4}x = 27$

**24.** $\dfrac{4}{5}x = 16$

**25.** $-\dfrac{1}{3}t = 7$

**26.** $-\dfrac{1}{6}x = 9$

**27.** $-\dfrac{1}{3}m = \dfrac{1}{5}$

**28.** $\dfrac{1}{5} = -\dfrac{1}{8}z$

**29.** $-\dfrac{3}{5}r = \dfrac{9}{10}$

**30.** $\dfrac{2}{5}y = -\dfrac{4}{15}$

**31.** $-\dfrac{3}{2}r = -\dfrac{27}{4}$

**32.** $-\dfrac{5}{7}x = -\dfrac{10}{14}$

**33.** $6.3x = 44.1$

**34.** $2.7y = -54$

**35.** $-3.1y = 21.7$

**36.** $-3.3y = 6.6$

**37.** $-38.7m = 309.6$

**38.** $29.4m = 235.2$

**39.** $-\dfrac{2}{3}y = -10.6$

**40.** $-\dfrac{9}{7}y = 12.06$

**41.** $\dfrac{-x}{5} = 10$

**42.** $\dfrac{-x}{8} = -16$

**43.** $\dfrac{t}{-2} = 7$

**44.** $\dfrac{m}{-3} = 10$

## Skill Maintenance

**45.** Find the diameter, the circumference, and the area of a circle whose radius is 10 ft. Use 3.14 for $\pi$.  [10.3a, b, c]

**46.** Find the radius, the circumference, and the area of a circle whose diameter is 24 cm. Use 3.14 for $\pi$.  [10.3a, b, c]

**47.** Find the volume of a rectangular block of granite of length 25 ft, width 10 ft, and height 32 ft.  [10.4a]

**48.** Find the volume of a rectangular solid of length 1.3 cm, width 10 cm, and height 2.4 cm.  [10.4a]

Find the area of each figure.  [10.2b]

**49.**
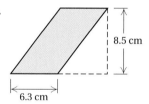
8.5 cm
6.3 cm

**50.**
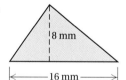
8 mm
16 mm

**51.**
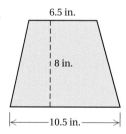
6.5 in.
8 in.
10.5 in.

**52.**
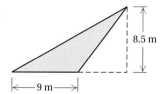
8.5 m
9 m

## Synthesis

Solve.

**53.** $-0.2344m = 2028.732$

**54.** $0 \cdot x = 0$

**55.** $0 \cdot x = 9$

**56.** $4|x| = 48$

**57.** $2|x| = -12$

**58.** A student makes a calculation and gets an answer of 22.5. On the last step, the student multiplies by 0.3 when she should have divided by 0.3. What should the correct answer be?

# Mid-Chapter Review

## Concept Reinforcement

Determine whether each statement is true or false.

_____ **1.** The expression $2(x + 3)$ is equivalent to the expression $2 \cdot x + 3$.   [11.1b]

_____ **2.** To factor an expression is to find an equivalent expression that is a product.   [11.1c]

_____ **3.** Collecting like terms is based on the distributive laws.   [11.1d]

_____ **4.** $3 - x = 4x$ and $5x = -3$ are equivalent equations.   [11.2a]

## Guided Solutions

Fill in each blank with the number, variable, or expression that creates a correct statement or solution.

**5.** Factor:  $6x - 3y + 18$.   [11.1c]

$$6x - 3y + 18 = 3 \cdot \square - 3 \cdot \square + 3 \cdot \square = \square(\square - \square + 6)$$

Solve.   [11.2a], [11.3a]

**6.**  $x + 5 = -3$

$x + 5 - 5 = -3 - \square$

$x + \square = -8$

$x = \square$

**7.**  $-6x = 42$

$\dfrac{-6x}{-6} = \dfrac{42}{\square}$

$\square \cdot x = -7$

$x = \square$

## Mixed Review

Evaluate.   [11.1a]

**8.** $4x$, when $x = -7$

**9.** $\dfrac{a}{b}$, when $a = 56$ and $b = 8$

**10.** $\dfrac{m - n}{3}$, when $m = 17$ and $n = 2$

Multiply.   [11.1b]

**11.** $3(x + 5)$

**12.** $4(2y - 7)$

**13.** $6(3x + 2y - 1)$

**14.** $-2(-3x - y + 8)$

Factor.   [11.1c]

**15.** $3y + 21$

**16.** $5z + 45$

**17.** $9x - 36$

**18.** $24a - 8$

**19.** $4x + 6y - 2$

**20.** $12x - 9y + 3$

**21.** $4a - 12b + 32$

**22.** $30a - 18b - 24$

Collect like terms. [11.1d]

**23.** $7x + 8x$

**24.** $3y - y$

**25.** $5x - 2y + 6 - 3x + y - 9$

Solve. [11.2a], [11.3a]

**26.** $x + 5 = 11$

**27.** $x + 9 = -3$

**28.** $8 = t + 1$

**29.** $-7 = y + 3$

**30.** $x - 6 = 14$

**31.** $y - 7 = -2$

**32.** $3 + t = 10$

**33.** $-5 + x = 5$

**34.** $y + \dfrac{1}{3} = -\dfrac{1}{2}$

**35.** $-\dfrac{3}{2} + z = -\dfrac{3}{4}$

**36.** $4.6 = x + 3.9$

**37.** $-3.3 = -1.9 + t$

**38.** $7x = 42$

**39.** $144 = 12y$

**40.** $17 = -t$

**41.** $6x = -54$

**42.** $-5y = -85$

**43.** $-8x = 48$

**44.** $\dfrac{2}{3}x = 12$

**45.** $-\dfrac{1}{5}t = 3$

**46.** $\dfrac{3}{4}x = -\dfrac{9}{8}$

**47.** $-\dfrac{5}{6}t = -\dfrac{25}{18}$

**48.** $1.8y = -5.4$

**49.** $\dfrac{-y}{7} = 5$

# Understanding Through Discussion and Writing

**50.** Determine whether $(a + b)^2$ and $a^2 + b^2$ are equivalent for all real numbers. Explain. [11.1a]

**51.** The distributive law is introduced before the material on collecting like terms. Why do you think this is? [11.1d]

**52.** Explain the following mistake made by a fellow student. [11.2a]

$$x + \frac{1}{3} = -\frac{5}{3}$$
$$x = -\frac{4}{3}$$

**53.** Explain the following mistake made by a fellow student. [11.3a]

$$\frac{2}{3}x = -\frac{5}{3}$$
$$x = \frac{10}{9}$$

# 11.4

# Using the Principles Together

## OBJECTIVES

**a** Solve equations using both the addition principle and the multiplication principle.

**b** Solve equations in which like terms may need to be collected.

**c** Solve equations by first removing parentheses and collecting like terms.

## a Applying Both Principles

Consider the equation $3x + 4 = 13$. It is more complicated than those in the two preceding sections. In order to solve such an equation, we first isolate the $x$-term, $3x$, using the addition principle. Then we apply the multiplication principle to get $x$ by itself.

**EXAMPLE 1** Solve: $3x + 4 = 13$.

$$3x + 4 = 13$$

$$3x + 4 - 4 = 13 - 4 \qquad \text{Using the addition principle: adding } -4 \text{ or subtracting 4 on both sides}$$

$$3x = 9 \qquad \text{Simplifying}$$

$$\frac{3x}{3} = \frac{9}{3} \qquad \text{Using the multiplication principle: multiplying by } \frac{1}{3} \text{ or dividing by 3 on both sides}$$

$$x = 3 \qquad \text{Simplifying}$$

Check: 
$$\begin{array}{c|c} \hline 3x + 4 = 13 \\ \hline 3 \cdot 3 + 4 \; ? \; 13 \\ 9 + 4 \\ 13 \; | \quad \text{TRUE} \end{array}$$

The solution is 3.

**1.** Solve: $9x + 6 = 51$.

> Do Exercise 1.

**EXAMPLE 2** Solve: $-5x - 6 = 16$.

$$-5x - 6 = 16$$

$$-5x - 6 + 6 = 16 + 6 \qquad \text{Adding 6 on both sides}$$

$$-5x = 22$$

$$\frac{-5x}{-5} = \frac{22}{-5} \qquad \text{Dividing by } -5 \text{ on both sides}$$

$$x = -\frac{22}{5}, \text{ or } -4\frac{2}{5} \qquad \text{Simplifying}$$

Check: 
$$\begin{array}{c|c} \hline -5x - 6 = 16 \\ \hline -5\left(-\dfrac{22}{5}\right) - 6 \; ? \; 16 \\ 22 - 6 \\ 16 \; | \quad \text{TRUE} \end{array}$$

Solve.

**2.** $8x - 4 = 28$

**3.** $-\dfrac{1}{2}x + 3 = 1$

The solution is $-\dfrac{22}{5}$.

> Do Exercises 2 and 3.

*Answers*

1. 5    2. 4    3. 4

**EXAMPLE 3**   Solve: $45 - x = 13$.

$$45 - x = 13$$
$$-45 + 45 - x = -45 + 13 \qquad \text{Adding } -45 \text{ on both sides}$$
$$-x = -32$$
$$-1 \cdot (-x) = -1 \cdot (-32) \qquad \text{Multiplying by } -1 \text{ on both sides}$$
$$-1 \cdot (-1) \cdot x = 32$$
$$x = 32$$

Check:
$$\frac{45 - x = 13}{45 - 32 \; ? \; 13}$$
$$13 \; | \qquad \text{TRUE}$$

The solution is 32.

Do Exercise 4.

**4.** Solve: $-18 - x = -57$.

**EXAMPLE 4**   Solve: $16.3 - 7.2y = -8.18$.

$$16.3 - 7.2y = -8.18$$
$$-16.3 + 16.3 - 7.2y = -16.3 + (-8.18) \qquad \text{Adding } -16.3 \text{ on both sides}$$
$$-7.2y = -24.48$$
$$\frac{-7.2y}{-7.2} = \frac{-24.48}{-7.2} \qquad \text{Dividing by } -7.2 \text{ on both sides}$$
$$y = 3.4$$

Check:
$$\frac{16.3 - 7.2y = -8.18}{16.3 - 7.2(3.4) \; ? \; -8.18}$$
$$16.3 - 24.48 \; |$$
$$-8.18 \; | \qquad \text{TRUE}$$

The solution is 3.4.

Do Exercises 5 and 6.

Solve.

**5.** $-4 - 8x = 8$

**6.** $41.68 = 4.7 - 8.6y$

## (b) Collecting Like Terms

If there are like terms on one side of the equation, we collect them before using the addition principle or the multiplication principle.

**EXAMPLE 5**   Solve: $3x + 4x = -14$.

$$3x + 4x = -14$$
$$7x = -14 \qquad \text{Collecting like terms}$$
$$\frac{7x}{7} = \frac{-14}{7} \qquad \text{Dividing by 7 on both sides}$$
$$x = -2$$

The number $-2$ checks, so the solution is $-2$.

Do Exercises 7 and 8.

Solve.

**7.** $4x + 3x = -21$

**8.** $x - 0.09x = 728$

*Answers*

**4.** 39    **5.** $-\dfrac{3}{2}$    **6.** $-4.3$    **7.** $-3$    **8.** 800

If there are like terms on opposite sides of the equation, we get them on the same side by using the addition principle. Then we collect them. In other words, we get all terms with a variable on one side of the equation and all terms without a variable on the other side.

**EXAMPLE 6** Solve: $2x - 2 = -3x + 3$.

$$2x - 2 = -3x + 3$$

$$2x - 2 + 2 = -3x + 3 + 2 \qquad \text{Adding 2}$$

$$2x = -3x + 5 \qquad \text{Collecting like terms}$$

$$2x + 3x = -3x + 5 + 3x \qquad \text{Adding } 3x$$

$$5x = 5 \qquad \text{Simplifying}$$

$$\frac{5x}{5} = \frac{5}{5} \qquad \text{Dividing by 5}$$

$$x = 1 \qquad \text{Simplifying}$$

Check:

$$\begin{array}{c|c} \multicolumn{2}{c}{2x - 2 = -3x + 3} \\ \hline 2 \cdot 1 - 2 \;?\; -3 \cdot 1 + 3 & \text{Substituting in the original equation} \\ 2 - 2 \;\big|\; -3 + 3 & \\ 0 \;\big|\; 0 & \text{TRUE} \end{array}$$

The solution is 1.

Do Exercises 9 and 10.

In Example 6, we used the addition principle to get all terms with a variable on one side of the equation and all terms without a variable on the other side. Then we collected like terms and proceeded as before. If there are like terms on one side at the outset, they should be collected first.

**EXAMPLE 7** Solve: $6x + 5 - 7x = 10 - 4x + 3$.

$$6x + 5 - 7x = 10 - 4x + 3$$

$$-x + 5 = 13 - 4x \qquad \text{Collecting like terms}$$

$$4x - x + 5 = 13 - 4x + 4x \qquad \text{Adding } 4x$$

$$3x + 5 = 13 \qquad \text{Simplifying}$$

$$3x + 5 - 5 = 13 - 5 \qquad \text{Subtracting 5}$$

$$3x = 8 \qquad \text{Simplifying}$$

$$\frac{3x}{3} = \frac{8}{3} \qquad \text{Dividing by 3}$$

$$x = \frac{8}{3} \qquad \text{Simplifying}$$

The number $\frac{8}{3}$ checks, so $\frac{8}{3}$ is the solution.

Do Exercises 11 and 12.

Solve.

**9.** $7y + 5 = 2y + 10$

**10.** $5 - 2y = 3y - 5$

Solve.

**11.** $7x - 17 + 2x = 2 - 8x + 15$

**12.** $3x - 15 = 5x + 2 - 4x$

*Answers*

**9.** 1  **10.** 2  **11.** 2  **12.** $\frac{17}{2}$

## Clearing Fractions and Decimals

For the equations considered thus far, we generally use the addition principle first. There are, however, some situations in which it is to our advantage to use the multiplication principle first. Consider, for example,

$$\frac{1}{2}x = \frac{3}{4}.$$

The LCM of the denominators is 4. If we multiply by 4 on both sides, we get $2x = 3$, which has no fractions. We have "cleared fractions." Now consider

$$2.3x = 4.78.$$

If we multiply by 100 on both sides, we get $230x = 478$, which has no decimal points. We have "cleared decimals." The equations are then easier to solve. It is your choice whether to clear fractions or decimals, but doing so often eases computations.

In what follows, we use the multiplication principle first to "clear," or "eliminate," fractions or decimals. For fractions, the number by which we multiply is the **least common multiple of all the denominators**.

**EXAMPLE 8** Solve:

$$\frac{2}{3}x - \frac{1}{6} + \frac{1}{2}x = \frac{7}{6} + 2x.$$

The denominators are 3, 6, and 2. The number 6 is the least common multiple of all the denominators. We multiply by 6 on both sides of the equation:

$$6\left(\frac{2}{3}x - \frac{1}{6} + \frac{1}{2}x\right) = 6\left(\frac{7}{6} + 2x\right) \quad \text{Multiplying by 6 on both sides}$$

$$6 \cdot \frac{2}{3}x - 6 \cdot \frac{1}{6} + 6 \cdot \frac{1}{2}x = 6 \cdot \frac{7}{6} + 6 \cdot 2x \quad \text{Using the distributive laws. (Caution! Be sure to multiply all terms by 6.)}$$

$$4x - 1 + 3x = 7 + 12x \quad \text{Simplifying. Note that the fractions are cleared.}$$

$$7x - 1 = 7 + 12x \quad \text{Collecting like terms}$$

$$7x - 1 - 12x = 7 + 12x - 12x \quad \text{Subtracting } 12x$$

$$-5x - 1 = 7 \quad \text{Simplifying}$$

$$-5x - 1 + 1 = 7 + 1 \quad \text{Adding 1}$$

$$-5x = 8 \quad \text{Collecting like terms}$$

$$\frac{-5x}{-5} = \frac{8}{-5} \quad \text{Dividing by } -5$$

$$x = -\frac{8}{5}.$$

**STUDY TIPS**

**BEGINNING TO STUDY FOR THE FINAL EXAM: THREE DAYS TO TWO WEEKS OF STUDY TIME**

1. **Begin by browsing through each chapter, reviewing the high-lighted or boxed information regarding important formulas in both the text and the Summary and Review.** There may be some formulas that you will need to memorize.

2. **Retake the chapter tests that you took in class, assuming your instructor has returned them. You can also use the chapter tests in the text.** Restudy the objectives in the text that correspond to each question you missed.

3. **Work through the Cumulative Review for Chapters 1–11 at the end of this chapter during the last couple of days before the final.** Be careful to avoid any questions corresponding to objectives not covered. Again, restudy the objectives in the text that correspond to each question you missed.

4. **For remaining difficulties, see your instructor, go to a tutoring session, or participate in a study group.**

**Check:**

$$\frac{2}{3}x - \frac{1}{6} + \frac{1}{2}x = \frac{7}{6} + 2x$$

$$\frac{2}{3}\left(-\frac{8}{5}\right) - \frac{1}{6} + \frac{1}{2}\left(-\frac{8}{5}\right) \;?\; \frac{7}{6} + 2\left(-\frac{8}{5}\right)$$

$$-\frac{16}{15} - \frac{1}{6} - \frac{8}{10} \;\Big|\; \frac{7}{6} - \frac{16}{5}$$

$$-\frac{32}{30} - \frac{5}{30} - \frac{24}{30} \;\Big|\; \frac{35}{30} - \frac{96}{30}$$

$$\frac{-32 - 5 - 24}{30} \;\Big|\; \frac{35 - 96}{30}$$

$$-\frac{61}{30} \;\Big|\; -\frac{61}{30} \qquad \text{TRUE}$$

The solution is $-\dfrac{8}{5}$.

**Do Exercise 13.**

**13.** Solve: $\dfrac{7}{8}x - \dfrac{1}{4} + \dfrac{1}{2}x = \dfrac{3}{4} + x.$

To illustrate clearing decimals, we repeat Example 4, but this time we clear the decimals first.

**EXAMPLE 9**  Solve: $16.3 - 7.2y = -8.18.$

$$\underset{\substack{\text{1 decimal}\\\text{place}}}{16.3} \;-\; \underset{\substack{\text{1 decimal}\\\text{place}}}{7.2y} \;=\; \underset{\substack{\text{2 decimal}\\\text{places}}}{-8.18}$$

The greatest number of decimal places in any one number is *two*. Multiplying by 100, which has *two* 0's, will clear the decimals.

| | |
|---|---|
| $100(16.3 - 7.2y) = 100(-8.18)$ | Multiplying by 100 on both sides |
| $100(16.3) - 100(7.2y) = 100(-8.18)$ | Using a distributive law |
| $1630 - 720y = -818$ | Simplifying |
| $1630 - 720y - 1630 = -818 - 1630$ | Subtracting 1630 on both sides |
| $-720y = -2448$ | Collecting like terms |
| $\dfrac{-720y}{-720} = \dfrac{-2448}{-720}$ | Dividing by $-720$ on both sides |
| $y = \dfrac{17}{5}, \text{ or } 3.4$ | |

The number $\dfrac{17}{5}$, or 3.4, checks, as shown in Example 4, so it is the solution.

**14.** Solve: $41.68 = 4.7 - 8.6y.$

**Do Exercise 14.**

*Answers*

**13.** $\dfrac{8}{3}$   **14.** $-\dfrac{43}{10}$, or $-4.3$

## (c) Equations Containing Parentheses

To solve certain kinds of equations that contain parentheses, we first use the distributive laws to remove the parentheses. Then we proceed as before.

**EXAMPLE 10**  Solve: $8x = 2(12 - 2x)$.

$$8x = 2(12 - 2x)$$

$$8x = 24 - 4x \qquad \text{Using the distributive law to multiply and remove parentheses}$$

$$8x + 4x = 24 - 4x + 4x \qquad \text{Adding } 4x \text{ to get all } x\text{-terms on one side}$$

$$12x = 24 \qquad \text{Collecting like terms}$$

$$\frac{12x}{12} = \frac{24}{12} \qquad \text{Dividing by 12}$$

$$x = 2$$

Check:
$$8x = 2(12 - 2x)$$

$$8 \cdot 2 \; ? \; 2(12 - 2 \cdot 2) \qquad \text{We use the rules for order of operations to}$$
$$16 \quad | \quad 2(12 - 4) \qquad \text{carry out the calculations on each side of}$$
$$| \quad 2 \cdot 8 \qquad \text{the equation.}$$
$$| \quad 16 \qquad \text{TRUE}$$

The solution is 2.

Do Exercises 15 and 16.

Here is a procedure for solving the types of equations discussed in this section.

---

### AN EQUATION-SOLVING PROCEDURE

1. Multiply on both sides to clear the equation of fractions or decimals. (This is optional, but it can ease computations.)
2. If parentheses occur, multiply using the distributive laws to remove them.
3. Collect like terms on each side, if necessary.
4. Get all terms with variables on one side and all constant terms on the other side, using the *addition principle.*
5. Collect like terms again, if necessary.
6. Multiply or divide to solve for the variable, using the *multiplication principle.*
7. Check all possible solutions in the original equation.

---

Solve.

**15.** $2(2y + 3) = 14$

**16.** $5(3x - 2) = 35$

*Answers*

**15.** 2   **16.** 3

**EXAMPLE 11**  Solve: $2 - 5(x + 5) = 3(x - 2) - 1$.

$$2 - 5(x + 5) = 3(x - 2) - 1$$

$$2 - 5x - 25 = 3x - 6 - 1 \qquad \text{Using the distributive laws to multiply and remove parentheses}$$

$$-5x - 23 = 3x - 7 \qquad \text{Collecting like terms}$$

$$-5x - 23 + 5x = 3x - 7 + 5x \qquad \text{Adding } 5x$$

$$-23 = 8x - 7 \qquad \text{Collecting like terms}$$

$$-23 + 7 = 8x - 7 + 7 \qquad \text{Adding } 7$$

$$-16 = 8x \qquad \text{Collecting like terms}$$

$$\frac{-16}{8} = \frac{8x}{8} \qquad \text{Dividing by 8}$$

$$-2 = x$$

Check:
$$\begin{array}{c|c} \multicolumn{2}{c}{2 - 5(x + 5) = 3(x - 2) - 1} \\ \hline 2 - 5(-2 + 5) & 3(-2 - 2) - 1 \\ 2 - 5(3) & 3(-4) - 1 \\ 2 - 15 & -12 - 1 \\ -13 & -13 \qquad \text{TRUE} \end{array}$$

The solution is $-2$.

Note that the solution of $-2 = x$ is $-2$, which is also the solution of $x = -2$.

Do Exercises 17 and 18.

**Solve.**

**17.** $3(7 + 2x) = 30 + 7(x - 1)$

**18.** $4(3 + 5x) - 4 = 3 + 2(x - 2)$

---

## Calculator Corner

**Checking Solutions of Equations**   We can use a calculator to check solutions of equations by substituting the possible solution for each occurrence of the variable on the left side of the equation and then on the right side of the equation. If both sides have the same value, then the number that was substituted for the variable is the solution of the equation. In Example 5, for instance, to check if $-2$ is the solution of the equation $3x + 4x = -14$, we substitute $-2$ for $x$ on the left side of the equation. To do this, we press $\boxed{3}$ $\boxed{\times}$ $\boxed{2}$ $\boxed{+/-}$ $\boxed{+}$ $\boxed{4}$ $\boxed{\times}$ $\boxed{2}$ $\boxed{+/-}$ $\boxed{=}$. The result is $-14$. Note that the right side of the equation does not contain a variable. It is $-14$, the value that was obtained when $-2$ was substituted for $x$ on the left side. Thus the number $-2$ is the solution of the equation.

In Example 6, we can check if 1 is the solution of the equation $2x - 2 = -3x + 3$. First, we substitute 1 for $x$ on the left side of the equation by pressing $\boxed{2}$ $\boxed{\times}$ $\boxed{1}$ $\boxed{-}$ $\boxed{2}$ $\boxed{=}$. We get 0. Next, we substitute 1 for $x$ on the right side of the equation by pressing $\boxed{3}$ $\boxed{+/-}$ $\boxed{\times}$ $\boxed{1}$ $\boxed{+}$ $\boxed{3}$ $\boxed{=}$. We get 0 again. Since the same value was obtained on both sides of the equation, the number 1 is the solution.

**Exercises:**

**1.** Use a calculator to check the solutions of the equations in Examples 1–4 and 7–11.

**2.** Use a calculator to check the solutions of the equations in Margin Exercises 1–18.

---

*Answers*

**17.** $-2$   **18.** $-\dfrac{1}{2}$

For Extra Help

*MyMathLab*  Math XL PRACTICE  WATCH  DOWNLOAD  READ  REVIEW

**a**   Solve. Don't forget to check!

**1.** $5x + 6 = 31$

**2.** $8x + 6 = 30$

**3.** $8x + 4 = 68$

**4.** $8z + 7 = 79$

**5.** $4x - 6 = 34$

**6.** $4x - 11 = 21$

**7.** $3x - 9 = 33$

**8.** $6x - 9 = 57$

**9.** $7x + 2 = -54$

**10.** $5x + 4 = -41$

**11.** $-45 = 3 + 6y$

**12.** $-91 = 9t + 8$

**13.** $-4x + 7 = 35$

**14.** $-5x - 7 = 108$

**15.** $-7x - 24 = -129$

**16.** $-6z - 18 = -132$

**b**   Solve.

**17.** $5x + 7x = 72$

**18.** $4x + 5x = 45$

**19.** $8x + 7x = 60$

**20.** $3x + 9x = 96$

**21.** $4x + 3x = 42$

**22.** $6x + 19x = 100$

**23.** $-6y - 3y = 27$

**24.** $-4y - 8y = 48$

**25.** $-7y - 8y = -15$

**26.** $-10y - 3y = -39$

**27.** $10.2y - 7.3y = -58$

**28.** $6.8y - 2.4y = -88$

**29.** $x + \dfrac{1}{3}x = 8$

**30.** $x + \dfrac{1}{4}x = 10$

**31.** $8y - 35 = 3y$

**32.** $4x - 6 = 6x$

**33.** $8x - 1 = 23 - 4x$

**34.** $5y - 2 = 28 - y$

**35.** $2x - 1 = 4 + x$

**36.** $5x - 2 = 6 + x$

**37.** $6x + 3 = 2x + 11$

**38.** $5y + 3 = 2y + 15$

**39.** $5 - 2x = 3x - 7x + 25$

**40.** $10 - 3x = 2x - 8x + 40$

**41.** $4 + 3x - 6 = 3x + 2 - x$

**42.** $5 + 4x - 7 = 4x - 2 - x$

**43.** $4y - 4 + y + 24 = 6y + 20 - 4y$

**44.** $5y - 7 + y = 7y + 21 - 5y$

Solve. Clear fractions or decimals first.

**45.** $\dfrac{7}{2}x + \dfrac{1}{2}x = 3x + \dfrac{3}{2} + \dfrac{5}{2}x$

**46.** $\dfrac{7}{8}x - \dfrac{1}{4} + \dfrac{3}{4}x = \dfrac{1}{16} + x$

**47.** $\dfrac{2}{3} + \dfrac{1}{4}t = \dfrac{1}{3}$

**48.** $-\dfrac{3}{2} + x = -\dfrac{5}{6} - \dfrac{4}{3}$

**49.** $\dfrac{2}{3} + 3y = 5y - \dfrac{2}{15}$

**50.** $\dfrac{1}{2} + 4m = 3m - \dfrac{5}{2}$

**51.** $\dfrac{5}{3} + \dfrac{2}{3}x = \dfrac{25}{12} + \dfrac{5}{4}x + \dfrac{3}{4}$

**52.** $1 - \dfrac{2}{3}y = \dfrac{9}{5} - \dfrac{y}{5} + \dfrac{3}{5}$

**53.** $2.1x + 45.2 = 3.2 - 8.4x$

**54.** $0.96y - 0.79 = 0.21y + 0.46$

**55.** $1.03 - 0.62x = 0.71 - 0.22x$

**56.** $1.7t + 8 - 1.62t = 0.4t - 0.32 + 8$

**57.** $\frac{2}{7}x - \frac{1}{2}x = \frac{3}{4}x + 1$

**58.** $\frac{5}{16}y + \frac{3}{8}y = 2 + \frac{1}{4}y$

**C** Solve.

**59.** $3(2y - 3) = 27$

**60.** $4(2y - 3) = 28$

**61.** $40 = 5(3x + 2)$

**62.** $9 = 3(5x - 2)$

**63.** $2(3 + 4m) - 9 = 45$

**64.** $3(5 + 3m) - 8 = 88$

**65.** $5r - (2r + 8) = 16$

**66.** $6b - (3b + 8) = 16$

**67.** $6 - 2(3x - 1) = 2$

**68.** $10 - 3(2x - 1) = 1$

**69.** $5(d + 4) = 7(d - 2)$

**70.** $3(t - 2) = 9(t + 2)$

**71.** $8(2t + 1) = 4(7t + 7)$

**72.** $7(5x - 2) = 6(6x - 1)$

**73.** $3(r - 6) + 2 = 4(r + 2) - 21$

**74.** $5(t + 3) + 9 = 3(t - 2) + 6$

**75.** $19 - (2x + 3) = 2(x + 3) + x$

**76.** $13 - (2c + 2) = 2(c + 2) + 3c$

**77.** $0.7(3x + 6) = 1.1 - (x + 2)$

**78.** $0.9(2x + 8) = 20 - (x + 5)$

**79.** $a + (a - 3) = (a + 2) - (a + 1)$

**80.** $0.8 - 4(b - 1) = 0.2 + 3(4 - b)$

## Skill Maintenance

In each of Exercises 81–88, fill in the blank with the correct term from the given list. Some of the choices may not be used.

81. The _____ numbers consist of all numbers that can be named in the form $\frac{a}{b}$, where $a$ and $b$ are integers and $b$ is not _____. [3.3a]

82. The symbol $>$ means _____. [1.6c]

83. Two numbers whose product is _____ are called reciprocals of each other. [3.7a]

84. The _____ of a number is its distance from zero on the number line. [2.1d]

85. A(n) _____ angle is an angle whose measure is greater than 90° and less than 180°. [10.5b]

86. A(n) _____ triangle is a triangle in which all sides are the same length. [10.5d]

87. The basic unit of length in the metric system is the _____. [9.2a]

88. To _____ an expression is to find an equivalent expression that is a product. [11.1c]

is greater than
is less than
acute
obtuse
negative
one
zero
factor
multiply
absolute value
opposite
equilateral
scalene
meter
gram
rational
irrational

## Synthesis

Solve.

89. $\dfrac{y - 2}{3} = \dfrac{2 - y}{5}$

90. $3x = 4x$

91. $\dfrac{5 + 2y}{3} = \dfrac{25}{12} + \dfrac{5y + 3}{4}$

92. ▦ $0.05y - 1.82 = 0.708y - 0.504$

93. $\dfrac{2}{3}(2x - 1) = 10$

94. $\dfrac{2}{3}\left(\dfrac{7}{8} - 4x\right) - \dfrac{5}{8} = \dfrac{3}{8}$

95. The perimeter of the figure shown is 15 cm. Solve for $x$.

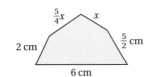

# 11.5
## Applications and Problem Solving

## a) Translating to Algebraic Expressions

In algebra, we translate problems to equations. The different parts of an equation are translations of word phrases to algebraic expressions. To translate, it helps to learn which words translate to certain operation symbols.

**OBJECTIVES**

**a)** Translate phrases to algebraic expressions.

**b)** Solve applied problems by translating to equations.

**KEY WORDS**

| ADDITION ( + ) | SUBTRACTION ( − ) | MULTIPLICATION ( · ) | DIVISION ( ÷ ) |
|---|---|---|---|
| add | subtract | multiply | divide |
| added to | subtracted from | multiplied by | quotient |
| sum | difference | product | divided by |
| total | minus | times | |
| plus | less than | of | |
| more than | decreased by | | |
| increased by | take away | | |

**EXAMPLE 1**  Translate to an algebraic expression:

Twice (or two times) some number.

Think of some number—say, 8. We can write 2 times 8 as $2 \times 8$, or $2 \cdot 8$. We multiplied by 2. To translate to an algebraic expression, we do the same thing using a variable. We can use any variable we wish, such as $x$, $y$, $m$, or $n$. Let's use $y$ to stand for some number. If we multiply by 2, we get an expression

$$y \times 2, \quad 2 \times y, \quad 2 \cdot y, \quad \text{or} \quad 2y.$$

In algebra, $2y$ is the expression used most often.

**EXAMPLE 2**  Translate to an algebraic expression:

Seven less than some number.

We let

$$x = \text{the number.}$$

If the number were 10, then 7 less than 10 would be $10 - 7$, or 3. If we knew the number to be 34, then 7 less than the number would be $34 - 7$. Thus if the number is $x$, then the translation is

$$x - 7.$$

---
*Caution!*
---

Note that $7 - x$ is *not* a correct translation of the expression in Example 2. The expression $7 - x$ is a translation of "seven minus some number" or "some number less than seven."

---

**EXAMPLE 3** Translate to an algebraic expression:

Eighteen more than a number.

We let

$t$ = the number.

Now if the number were 26, then the translation would be $18 + 26$, or $26 + 18$. If we knew the number to be 174, then the translation would be $18 + 174$, or $174 + 18$. The translation we want is

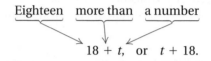

Eighteen   more than   a number

$$18 + t, \quad \text{or} \quad t + 18.$$

**EXAMPLE 4** Translate to an algebraic expression:

A number divided by 5.

We let

$m$ = the number.

If the number were 8, then the translation would be $8 \div 5$, or $8/5$, or $\frac{8}{5}$. If the number were 213, then the translation would be $213 \div 5$, or $213/5$, or $\frac{213}{5}$. The translation is found as follows:

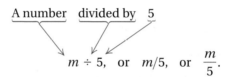

A number   divided by   5

$$m \div 5, \quad \text{or} \quad m/5, \quad \text{or} \quad \frac{m}{5}.$$

**EXAMPLE 5** Translate to an algebraic expression.

| PHRASE | ALGEBRAIC EXPRESSION |
|---|---|
| Five more than some number | $5 + n$, or $n + 5$ |
| Half of a number | $\frac{1}{2} t$, or $\frac{t}{2}$ |
| Five more than three times some number | $5 + 3p$, or $3p + 5$ |
| The difference of two numbers | $x - y$ |
| Six less than the product of two numbers | $mn - 6$ |
| Seventy-six percent of some number | $76\%z$, or $0.76z$ |

Do Exercises 1–9.

Translate to an algebraic expression.

1. Twelve less than some number

2. Twelve more than some number

3. Four less than some number

4. Half of some number

5. Six more than eight times some number

6. The difference of two numbers

7. Fifty-nine percent of some number

8. Two hundred less than the product of two numbers

9. The sum of two numbers

*Answers*

1. $x - 12$   2. $y + 12$, or $12 + y$
3. $m - 4$   4. $\frac{1}{2}p$, or $\frac{p}{2}$
5. $6 + 8x$, or $8x + 6$   6. $a - b$
7. $59\%x$, or $0.59x$   8. $xy - 200$
9. $p + q$

## (b) Five Steps for Solving Problems

We have introduced many new equation-solving tools in this chapter. We now apply them to problem solving. We have purposely used the following strategy throughout this text in order to introduce you to algebra.

> **FIVE STEPS FOR PROBLEM SOLVING IN ALGEBRA**
>
> 1. *Familiarize* yourself with the problem situation.
> 2. *Translate* to an equation.
> 3. *Solve* the equation.
> 4. *Check* your possible answer in the original problem.
> 5. *State* the answer clearly.

Of the five steps, the most important is probably the first one: becoming familiar with the problem situation. The box below lists some hints for familiarization.

> To familiarize yourself with a problem:
>
> - If a problem is given in words, read it carefully. Reread the problem, perhaps aloud. Try to verbalize the problem as if you were explaining it to someone else.
> - Choose a variable (or variables) to represent the unknown and clearly state what the variable represents. Be descriptive! For example, let $L$ = the length in feet, $d$ = the distance in miles, and so on.
> - Make a drawing and label it with known information, using specific units if given. Also, indicate unknown information.
> - Find further information. Look up formulas or definitions with which you are not familiar. (Geometric formulas appear on the inside back cover of this text.) Consult a reference librarian or an expert in the field. You might also find information on the Internet.
> - Create a table that lists all the information you have available. Look for patterns that may help in the translation to an equation.
> - Guess what the answer might be and check the guess. Observe the manner in which the guess is checked. This can be very helpful in determining how to translate the problem to an equation.

**EXAMPLE 6**  *Knitted Scarf.*  Lilly knitted a scarf in three shades of blue, starting with a light-blue section, then a medium-blue section, and finally a dark-blue section. The medium-blue section is one-half the length of the light-blue section. The dark-blue section is one-fourth the length of the light-blue section. The scarf is 7 ft long. Find the length of each section of the scarf.

1. **Familiarize.**  Because the lengths of the medium-blue section and the dark-blue section are expressed in terms of the length of the light-blue section, we let

$$x = \text{the length of the light-blue section.}$$

Then   $\dfrac{1}{2}x$ = the length of the medium-blue section

and   $\dfrac{1}{4}x$ = the length of the dark-blue section.

We make a drawing and label it.

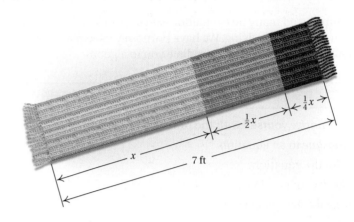

2. **Translate.** From the statement of the problem and the drawing, we know that the lengths add up to 7 ft. This gives us our translation:

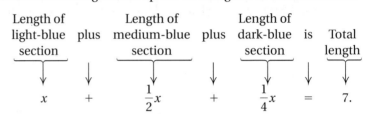

| Length of light-blue section | plus | Length of medium-blue section | plus | Length of dark-blue section | is | Total length |
|---|---|---|---|---|---|---|
| $x$ | $+$ | $\frac{1}{2}x$ | $+$ | $\frac{1}{4}x$ | $=$ | $7.$ |

3. **Solve.** First, we clear fractions and then carry out the solution as follows:

$$x + \frac{1}{2}x + \frac{1}{4}x = 7 \qquad \text{The LCM of the denominators is 4.}$$

$$4\left(x + \frac{1}{2}x + \frac{1}{4}x\right) = 4 \cdot 7 \qquad \text{Multiplying by the LCM, 4}$$

$$4 \cdot x + 4 \cdot \frac{1}{2}x + 4 \cdot \frac{1}{4}x = 4 \cdot 7 \qquad \text{Using the distributive law}$$

$$4x + 2x + x = 28 \qquad \text{Simplifying}$$

$$7x = 28 \qquad \text{Collecting like terms}$$

$$\frac{7x}{7} = \frac{28}{7} \qquad \text{Dividing by 7}$$

$$x = 4.$$

4. **Check.** Do we have an answer to the *original problem*? If the length of the light-blue section is 4 ft, then the length of the medium-blue section is $\frac{1}{2} \cdot 4$ ft, or 2 ft, and the length of the dark-blue section is $\frac{1}{4} \cdot 4$ ft, or 1 ft. The sum of these lengths is 7 ft, so the answer checks.

5. **State.** The length of the light-blue section is 4 ft, the length of the medium-blue section is 2 ft, and the length of the dark-blue section is 1 ft. (Note that we must include the unit, feet, in the answer.)

Do Exercise 10.

**EXAMPLE 7** *Hiking.* At age 79, Earl Shaffer became the oldest person to through-hike all 2100 miles of the Appalachian Trail—from Springer Mountain, Georgia, to Mount Katahdin, Maine. Shaffer through-hiked the trail three times, in 1948 (Georgia to Maine), in 1965 (Maine to Georgia), and in 1998 (Georgia to Maine) near the 50th anniversary of his first hike. At one

10. **Sharing a Sandwich.** A sandwich shop specializes in sandwiches prepared in buns of length 18 in. Suppose Jenny, Emma, and Sarah buy one of these sandwiches and take it back to their apartment. Since they have different appetites, Jenny cuts the sandwich in such a way that Emma gets one-half of what Jenny gets and Sarah gets three-fourths of what Jenny gets. Find the length of each person's sandwich.

*Answer*

10. Jenny: 8 in.; Emma: 4 in.; Sarah: 6 in.

point in 1998, Shaffer stood atop Big Walker Mountain, Virginia, which is three times as far from the northern end of the trail as from the southern end. How far was Shaffer from each end of the trail?

**Source:** Appalachian Trail Conference; Earl Shaffer Foundation

**1. Familiarize.** Let's consider a drawing.

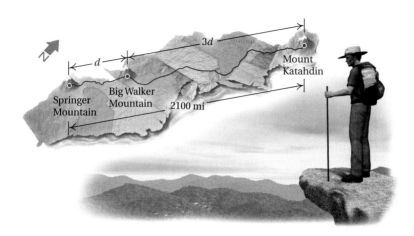

To become familiar with the problem, let's guess a possible distance that Shaffer stood from Springer Mountain—say, 600 mi. Three times 600 mi is 1800 mi. Since 600 mi + 1800 mi = 2400 mi and 2400 mi is greater than 2100 mi, we see that our guess is too large. Rather than guess again, let's use the equation-solving skills we have learned in this chapter. We let

$d$ = the distance, in miles, to the southern end,  and

$3d$ = the distance, in miles, to the northern end.

(We could also let $x$ = the distance to the northern end and $\frac{1}{3}x$ = the distance to the southern end.)

**2. Translate.** From the drawing, we see that the lengths of the two parts of the trail must add up to 2100 mi. This leads to our translation:

$$
\underbrace{\text{Distance to}}_{d} \quad \underbrace{\text{plus}}_{+} \quad \underbrace{\text{Distance to}}_{3d} \quad \underbrace{\text{is}}_{=} \quad \underbrace{\text{2100 mi}}_{2100.}
$$

| Distance to southern end | plus | Distance to northern end | is | 2100 mi |
|---|---|---|---|---|
| $d$ | $+$ | $3d$ | $=$ | $2100.$ |

**3. Solve.** We solve the equation:

$d + 3d = 2100$

$4d = 2100$     Collecting like terms

$\dfrac{4d}{4} = \dfrac{2100}{4}$     Dividing by 4 on both sides

$d = 525.$

**4. Check.** As expected, $d$ is less than 600 mi. If $d = 525$ mi, then $3d = 1575$ mi. Since 525 mi + 1575 mi = 2100 mi, we have a check.

**5. State.** Atop Big Walker Mountain, Shaffer stood 525 mi from Springer Mountain and 1575 mi from Mount Katahdin.

Do Exercise 11.

**11. Running.** In 1997, Yiannis Kouros of Australia set the record for the greatest distance run in 24 hr by running 188 mi. After 8 hr, he was approximately twice as far from the finish line as he was from the start. How far had he run?

**Source:** Australian Ultra Runners Association

*Answer*

**11.** $62\frac{2}{3}$ mi

**EXAMPLE 8** *Truck Rental.* Truck-Rite Rentals rents 16-ft trucks at a daily rate of $29.99 plus 79¢ per mile. Concert Productions has budgeted $400 per day for renting a truck to haul equipment to an upcoming concert. How many miles can a rental truck be driven on a $400 daily budget?

1. **Familiarize.** Let's begin by guessing the answer. Suppose that Concert Productions drives 75 mi. Then the cost is

| Daily charge | plus | | Mileage charge | |
|---|---|---|---|---|
| ($29.99) | plus | (Cost per mile) | times | (Number of miles driven) |
| $29.99 | + | $0.79 | · | 75, |

which is $29.99 + $59.25, or $89.24. Since $89.24 is less than $400, we know that the guess is too low. Nevertheless, checking the guess familiarized us with the way in which the calculation of the cost is made. (Note that we converted 79¢ to $0.79 so that we are using the same unit, dollars, throughout our calculations.)

Now we will translate the problem to an equation. Let $m$ = the number of miles that can be driven for $400.

2. **Translate.** We reword the problem and translate as follows:

| Daily rate | plus | Cost per mile | times | Number of miles driven | is | Total cost |
|---|---|---|---|---|---|---|
| 29.99 | + | 0.79 | · | $m$ | = | 400. |

3. **Solve.** We solve the equation:

$$29.99 + 0.79 \cdot m = 400$$

$$100(29.99 + 0.79m) = 100(400) \qquad \text{Multiplying by 100 on both sides to clear decimals}$$

$$100(29.99) + 100(0.79m) = 40{,}000 \qquad \text{Using the distributive law}$$

$$2999 + 79m = 40{,}000$$

$$79m = 37{,}001 \qquad \text{Subtracting 2999}$$

$$\frac{79m}{79} = \frac{37{,}001}{79} \qquad \text{Dividing by 79}$$

$$m \approx 468. \qquad \text{Rounding down}$$

4. **Check.** We check in the original problem. We multiply 468 by $0.79, getting $369.72. Then we add $369.72 to $29.99 and get $399.71, which is just about the $400 allotted.

5. **State.** The truck can be driven about 468 mi per day on the rental allotment of $400.

Do Exercise 12.

**12. Van Rental.** Truck-Rite also rents cargo vans at a daily rate of $19.99 plus 67¢ per mile. What mileage will allow a salesperson to stay within a daily budget of $100?

*Answer*

**12.** 119 mi per day

**EXAMPLE 9** *Angles of a Triangle.* The second angle of a triangle is twice as large as the first. The measure of the third angle is 20° greater than that of the first angle. How large are the angles?

1. **Familiarize.** We first make a drawing. Since the second and third angles are described in terms of the first angle, we let

$$\text{the measure of the first angle} = x.$$

Then the measure of the second angle = $2x$

and the measure of the third angle = $x + 20$.

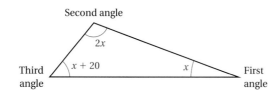

2. **Translate.** To translate, we recall from Section 10.5 that the sum of the measures of the angles of a triangle is 180°.

| Measure of first angle | + | Measure of second angle | + | Measure of third angle | = | 180° |
|---|---|---|---|---|---|---|
| $x$ | + | $2x$ | + | $(x + 20)$ | = | 180 |

3. **Solve.** We solve the equation:

$$x + 2x + (x + 20) = 180$$
$$4x + 20 = 180$$
$$4x + 20 - 20 = 180 - 20$$
$$4x = 160$$
$$\frac{4x}{4} = \frac{160}{4}$$
$$x = 40.$$

Possible measures for the angles are as follows:

First angle: $x = 40°$;
Second angle: $2x = 2(40) = 80°$;
Third angle: $x + 20 = 40 + 20 = 60°$.

4. **Check.** Consider 40°, 80°, and 60°. The second is twice the first, and the third is 20° greater than the first. The sum is 180°. These numbers check.

5. **State.** The measures of the angles are 40°, 80°, and 60°.

---
*Caution!*
---

Units are important in answers. Remember to include them, where appropriate.

---

Do Exercise 13.

**13. Angles of a Triangle.** The second angle of a triangle is three times as large as the first. The third angle measures 30° more than the first angle. Find the measures of the angles.

*Answer*

**13.** 30°, 90°, 60°

**EXAMPLE 10** *Fastest Roller Coasters.* The average top speed of the three fastest steel roller coasters in the United States is 116 mph. The third-fastest roller coaster, Superman: The Escape (located at Six Flags Magic Mountain, Valencia, California), reaches a top speed of 28 mph less than the fastest roller coaster, Kingda Ka (located at Six Flags Great Adventure, Jackson, New Jersey). The second-fastest roller coaster, Top Thrill Dragster (located at Cedar Point, Sandusky, Ohio), has a top speed of 120 mph. What is the top speed of the fastest steel roller coaster?

**Source:** Coaster Grotto

1. **Familiarize.** The **average** of a set of numbers is the sum of the numbers divided by the number of addends.

   We are given that the second-fastest speed is 120 mph. Suppose the three top speeds are 109, 120, and 125. The average is then

   $$\frac{109 + 120 + 125}{3} = \frac{354}{3} = 118,$$

   which is too high. Instead of continuing to guess, let's use the equation-solving skills we have learned in this chapter. We let $x$ represent the top speed of the fastest roller coaster. Then $x - 28$ is the top speed of the third-fastest roller coaster.

2. **Translate.** We reword the problem and translate as follows:

   $$\frac{\begin{array}{c}\text{Speed of} \\ \text{fastest} \\ \text{coaster}\end{array} + \begin{array}{c}\text{Speed of} \\ \text{second-} \\ \text{fastest} \\ \text{coaster}\end{array} + \begin{array}{c}\text{Speed of} \\ \text{third-} \\ \text{fastest} \\ \text{coaster}\end{array}}{\text{Number of roller coasters}} = \begin{array}{c}\text{Average speed} \\ \text{of three} \\ \text{fastest roller} \\ \text{coasters}\end{array}$$

   $$\frac{x + 120 + (x - 28)}{3} = 116.$$

3. **Solve.** We solve as follows:

   $$\frac{x + 120 + (x - 28)}{3} = 116$$

   $$3 \cdot \frac{x + 120 + (x - 28)}{3} = 3 \cdot 116 \qquad \text{Multiplying by 3 on both sides to clear the fraction}$$

   $$x + 120 + (x - 28) = 348$$

   $$2x + 92 = 348 \qquad \text{Collecting like terms}$$

   $$2x = 256 \qquad \text{Subtracting 92}$$

   $$x = 128. \qquad \text{Dividing by 2}$$

4. **Check.** If the top speed of the fastest roller coaster is 128 mph, then the top speed of the third-fastest is $128 - 28$, or 100 mph. The average of the top speeds of the three fastest is $(128 + 120 + 100) \div 3 = 348 \div 3$, or 116 mph. The answer checks.

5. **State.** The top speed of the fastest steel roller coaster in the United States is 128 mph.

Do Exercise 14.

**14. Average Test Score.** Sam's average score on his first three math tests is 77. He scored 62 on the first test. On the third test, he scored nine points more than he scored on his second test. What did he score on the second and third tests?

**EXAMPLE 11** *Simple Interest.* An investment is made at 6% simple interest for 1 year. It grows to $768.50. How much was originally invested (the principal)?

1. **Familiarize.** Suppose that $100 was invested. Recalling the formula for simple interest, $I = Prt$, we know that the interest for 1 year on $100 at 6% simple interest is given by $I = \$100 \cdot 0.06 \cdot 1 = \$6$. Then, at the end of the year, the amount in the account is found by adding the principal and the interest:

$$
\begin{array}{ccccc}
\text{Principal} & + & \text{Interest} & = & \text{Amount} \\
\downarrow & \downarrow & \downarrow & \downarrow & \downarrow \\
\$100 & + & \$6 & = & \$106.
\end{array}
$$

In this problem, we are working backward. We are trying to find the principal, which is the original investment. We let $x =$ the principal. Then the interest earned is 6%$x$.

2. **Translate.** We reword the problem and then translate:

$$
\begin{array}{ccccc}
\text{Principal} & + & \text{Interest} & = & \text{Amount} \\
\downarrow & \downarrow & \downarrow & \downarrow & \downarrow \\
x & + & 6\%x & = & 768.50
\end{array}
$$

Interest is 6% of the principal.

3. **Solve.** We solve the equation:

$$
\begin{aligned}
x + 6\%x &= 768.50 \\
x + 0.06x &= 768.50 \qquad \text{Converting to decimal notation} \\
1x + 0.06x &= 768.50 \qquad \text{Identity property of 1} \\
1.06x &= 768.50 \qquad \text{Collecting like terms} \\
\frac{1.06x}{1.06} &= \frac{768.50}{1.06} \qquad \text{Dividing by 1.06} \\
x &= 725.
\end{aligned}
$$

4. **Check.** We check by taking 6% of $725 and then adding it to $725:

$$6\% \times \$725 = 0.06 \times 725 = \$43.50.$$

Then $\$725 + \$43.50 = \$768.50$, so $725 checks.

5. **State.** The original investment was $725.

Do Exercise 15.

**15. Simple Interest.** An investment is made at 7% simple interest for 1 year. It grows to $8988. How much was originally invested (the principal)?

**STUDY TIPS**

**FINAL STUDY TIP**

You are arriving at the end of your course in Basic College Mathematics. If you have not begun to prepare for the final examination, be sure to read the comments in the Study Tips on pp. 585, 657, and 672.

*Answers*

**14.** Second: 80; third: 89  **15.** $8400

# Translating for Success

1. *Angle Measures.* The measure of the second angle of a triangle is 51° more than that of the first angle. The measure of the third angle is 3° less than twice that of the first angle. Find the measures of the angles.

2. *Sales Tax.* Tina paid $3976 for a used car. This amount included 5% for sales tax. How much did the car cost before tax?

3. *Perimeter.* The perimeter of a rectangle is 2347 ft. The length is 28 ft greater than the width. Find the length and the width.

4. *Fraternity or Sorority Membership.* At Arches Tech University, 3976 students belong to a fraternity or a sorority. This is 35% of the total enrollment. What is the total enrollment at Arches Tech?

5. *Fraternity or Sorority Membership.* At Moab Tech University, 35% of the students belong to a fraternity or a sorority. The total enrollment of the university is 11,360 students. How many students belong to either a fraternity or a sorority?

The goal of these matching questions is to practice step (2), *Translate,* of the five-step problem-solving process. Translate each problem to an equation and select a correct translation from equations A–O.

**A.** $x + (x - 3) + \dfrac{4}{5}x = 384$

**B.** $x + (x + 51) + (2x - 3) = 180$

**C.** $x + (x + 96{,}000) = 180{,}000$

**D.** $2 \cdot 96 + 2x = 3976$

**E.** $x + (x + 1) + (x + 2) = 384$

**F.** $3976 = x \cdot 11{,}360$

**G.** $2x + 2(x + 28) = 2347$

**H.** $3976 = x + 5\%x$

**I.** $x + (x + 28) = 2347$

**J.** $x = 35\% \cdot 11{,}360$

**K.** $x + 96 = 3976$

**L.** $x + (x + 3) + \dfrac{4}{5}x = 384$

**M.** $x + (x + 2) + (x + 4) = 384$

**N.** $35\% \cdot x = 3976$

**O.** $2x + (x + 28) = 2347$

*Answers on page A-19*

6. *Island Population.* There are 180,000 people living on a small Caribbean island. The women out-number the men by 96,000. How many men live on the island?

7. *Wire Cutting.* A 384-m wire is cut into three pieces. The second piece is 3 m longer than the first. The third is four-fifths as long as the first. How long is each piece?

8. *Locker Numbers.* The numbers on three adjoining lockers are consecutive integers whose sum is 384. Find the integers.

9. *Fraternity or Sorority Membership.* The total enrollment at Canyon-lands Tech University is 11,360 students. Of these, 3976 students belong to a fraternity or a sorority. What percent of the students belong to a fraternity or a sorority?

10. *Width of a Rectangle.* The length of a rectangle is 96 ft. The perimeter of the rectangle is 3976 ft. Find the width.

| 11.5 | **Exercise Set** | For Extra Help | Math PRACTICE |  WATCH |  DOWNLOAD |  READ |  REVIEW |
|---|---|---|---|---|---|---|---|

*MyMathLab*

**a**    Translate to an algebraic expression.

**1.** Three less than twice a number

**2.** Three times a number divided by *a*

**3.** The product of 97% and some number

**4.** 43% of some number

**5.** Four more than five times some number

**6.** Seventy-five less than eight times a number

**7.** *Pipe Cutting.*   A 240-in. pipe is cut into two pieces. One piece is three times the length of the other. Let $x =$ the length of the longer piece. Write an expression for the length of the shorter piece.

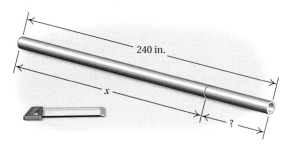

**8.** The price of a book is decreased by 30% during a sale. Let $b =$ the price of the book before the reduction. Write an expression for the sale price.

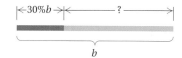

**b**    Solve.

**9.** What number added to 85 is 117?

**10.** Eight times what number is 2552?

**11.** *Manatee Population.*   The manatee, Florida's state marine mammal, is an endangered species. An aerial wintertime manatee census counted 2817 of these animals in 2007. This was 296 fewer than the number counted in 2006. What was Florida's manatee population in 2006?

**Source:** Florida Fish and Wildlife Conservation Commission

**12.** *Mass Transit Boom.*   Americans took 2.8 billion rides on public transit from April through June in 2008. This was the highest ridership for that period in 50 yr and represented an increase of 0.7 billion rides over the same period in 1998. How many rides were taken from April through June in 1998?

**Source:** American Public Transportation Association

**13.** *Statue of Liberty.* The height of the Eiffel Tower is 974 ft, which is about 669 ft higher than the Statue of Liberty. What is the height of the Statue of Liberty?

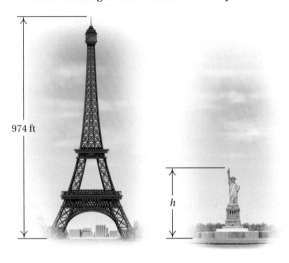

974 ft

$h$

**14.** *Area of Lake Ontario.* The area of Lake Superior is about four times the area of Lake Ontario. The area of Lake Superior is 30,172 $mi^2$. What is the area of Lake Ontario?

**15.** When 17 is subtracted from four times a certain number, the result is 211. What is the number?

**16.** When 36 is subtracted from five times a certain number, the result is 374. What is the number?

**17.** If you double a number and then add 16, you get $\frac{2}{3}$ of the original number. What is the original number?

**18.** If you double a number and then add 85, you get $\frac{3}{4}$ of the original number. What is the original number?

**19.** *Iditarod Race.* The Iditarod sled dog race in Alaska extends for 1049 mi from Anchorage to Nome. If a musher is twice as far from Anchorage as from Nome, how many miles has the musher traveled?

**Source:** Iditarod Trail Commission

**20.** *Cost of Movie Tickets.* The average cost of movie tickets for a family of four was $28.32 in 2008. This was $11.76 more than the cost in 1993. What was the average cost of movie tickets for a family of four in 1993? (These prices include senior discounts and children's prices.)

**Source:** Motion Picture Association of America

**21.** *Pipe Cutting.* A 480-m pipe is cut into three pieces. The second piece is three times as long as the first. The third piece is four times as long as the second. How long is each piece?

**22.** *Rope Cutting.* A 180-ft rope is cut into three pieces. The second piece is twice as long as the first. The third piece is three times as long as the second. How long is each piece of rope?

**23.** *Perimeter of NBA Court.* The perimeter of an NBA basketball court is 288 ft. The length is 44 ft longer than the width. Find the dimensions of the court.

**Source:** National Basketball Association

**24.** *Perimeter of High School Court.* The perimeter of a standard high school basketball court is 268 ft. The length is 34 ft longer than the width. Find the dimensions of the court.

**Source:** Indiana High School Athletic Association

**25.** *Hancock Building Dimensions.* The ground floor of the John Hancock Building in Chicago is a rectangle whose length is 100 ft more than the width. The perimeter is 860 ft. Find the length, the width, and the area of the ground floor.

**26.** *Hancock Building Dimensions.* The top floor of the John Hancock Building in Chicago is in the shape of a rectangle whose length is 60 ft more than the width. The perimeter is 520 ft. Find the length, the width, and the area of the top floor.

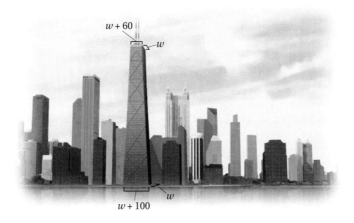

**27.** *Car Rental.* Value Rent-A-Car rents a family-sized car at a daily rate of $69.95 plus 40¢ per mile. Rick is allotted a daily budget of $200. How many miles can he drive per day and stay within his budget?

**28.** *Van Rental.* Value Rent-A-Car rents a van at a daily rate of $84.95 plus 60¢ per mile. Molly rents a van to deliver electrical parts to her customers. She is allotted a daily budget of $250. How many miles can she drive per day and stay within her budget?

**29.** *Angles of a Triangle.* The second angle of a triangular parking lot is four times as large as the first. The third angle is 45° less than the sum of the other two angles. How large are the angles?

**30.** *Angles of a Triangle.* The second angle of a triangular field is three times as large as the first. The third angle is 40° greater than the first. How large are the angles?

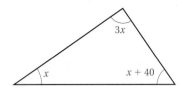

**31.** *Stock Prices.* Sarah's investment in AOL/Time Warner stock grew 28% to $448. How much did she invest?

**32.** *Savings Interest.* Sharon invested money in a savings account at a rate of 6% simple interest. After 1 year, she has $6996 in the account. How much did Sharon originally invest?

**33.** *Credit Cards.* The balance in Will's Mastercard® account grew 2%, to $870, in one month. What was his balance at the beginning of the month?

**34.** *Loan Interest.* Alvin borrowed money from a cousin at a rate of 10% simple interest. After 1 year, $7194 paid off the loan. How much did Alvin borrow?

**35.** *Taxi Fares.* In Beniford, taxis charge $3 plus 75¢ per mile for an airport pickup. How far from the airport can Courtney travel for $12?

**36.** *Taxi Fares.* In Cranston, taxis charge $4 plus 90¢ per mile for an airport pickup. How far from the airport can Ralph travel for $17.50?

**37.** *Tipping.* Leon left a 15% tip for a meal. The total cost of the meal, including the tip, was $41.40. What was the cost of the meal before the tip was added?

**38.** *Tipping.* Selena left an 18% tip for a meal. The total cost of the meal, including the tip, was $40.71. What was the cost of the meal before the tip was added?

**39.** *Standard Billboard Sign.*  A standard rectangular highway billboard sign has a perimeter of 124 ft. The length is 6 ft more than three times the width. Find the dimensions.

**40.** *Two-by-Four.*  The perimeter of a cross section of a "two-by-four" piece of lumber is 10 in. The length is 2 in. more than the width. Find the actual dimensions of the cross section of a two-by-four.

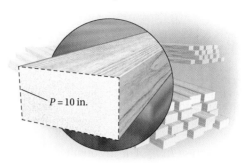

## Skill Maintenance

Calculate.

**41.** $-\dfrac{4}{5} - \dfrac{3}{8}$  [4.3a]

**42.** $-\dfrac{4}{5} + \dfrac{3}{8}$  [4.2a]

**43.** $-\dfrac{4}{5} \cdot \dfrac{3}{8}$  [3.6a]

**44.** $-\dfrac{4}{5} \div \dfrac{3}{8}$  [3.7b]

**45.** $-25.6 \div (-16)$  [5.4a]

**46.** $-25.6(-16)$  [5.3a]

**47.** $-25.6 - (-16)$  [5.2c]

**48.** $-25.6 + (-16)$  [5.2c]

## Synthesis

**49.** Abraham Lincoln's 1863 Gettysburg Address refers to the year 1776 as "Four *score* and seven years ago." Write an equation to find out what a score is.

**50.** *Test Questions.*  A student scored 78 on a test that had 4 seven-point fill-ins and 24 three-point multiple-choice questions. The student answered 1 fill-in incorrectly. How many multiple-choice questions did the student answer correctly?

**51.** The width of a rectangle is $\frac{3}{4}$ of the length. The perimeter of the rectangle becomes 50 cm when the length and the width are each increased by 2 cm. Find the original length and the original width.

**52.** Cookies are set out on a tray for six people to take home. One-third, one-fourth, one-eighth, and one-fifth are given to four people, respectively. The fifth person is given ten cookies, leaving one cookie remaining for the sixth person. Find the original number of cookies on the tray.

**53.** Susanne went to the bank to get $20 in quarters, dimes, and nickels to use to make change at her yard sale. She got twice as many quarters as dimes and 10 more nickels than dimes. How many of each type of coin did she get?

**54.** A student has an average score of 82 on three tests. His average score on the first two tests is 85. What was the score on the third test?

# Summary and Review

---

## Key Terms and Properties

variable, p. 634
constant, p. 634
algebraic expression, p. 634
substituting, p. 634

equivalent expressions, p. 636
terms, p. 638
factoring, p. 639
like terms, p. 640

equivalent equations, p. 643
coefficient, p. 647
clearing fractions, p. 657
clearing decimals, p. 657

| | |
|---|---|
| *Identity Property of 1:* | For any number $a$, $a \cdot 1 = 1 \cdot a = a$. |
| *Distributive Laws:* | For any numbers $a$, $b$, and $c$, $a(b + c) = ab + ac$ and $a(b - c) = ab - ac$. |
| *Addition Principle:* | For any numbers $a$, $b$, and $c$, $a = b$ is equivalent to $a + c = b + c$. |
| *Multiplication Principle:* | For any numbers $a$, $b$, and $c$, $c \neq 0$, $a = b$ is equivalent to $a \cdot c = b \cdot c$. |

---

## Concept Reinforcement

Determine whether each statement is true or false.

_____ **1.** The expression $x - 7$ is not equivalent to the expression $7 - x$.   [11.1b]

_____ **2.** $3y$ and $3y^2$ are like terms.   [11.1d]

_____ **3.** The equations $y + 8 = 6$ and $y - 2$ are equivalent.   [11.2a]

_____ **4.** We can use the multiplication principle to divide on both sides of an equation by the same nonzero number.   [11.3a]

---

## Important Concepts

---

**Objective 11.1b**   Use the distributive laws to multiply expressions like 8 and $x - y$.

| **Example**   Multiply: $3(5x - 2y + 4)$. | **Practice Exercise** |
|---|---|
| $3(5x - 2y + 4) = 3 \cdot 5x - 3 \cdot 2y + 3 \cdot 4$ $\qquad\qquad = 15x - 6y + 12$ | **1.** Multiply: $4(x + 5y - 7)$. |

---

**Objective 11.1c**   Use the distributive laws to factor expressions like $4x - 12 + 24y$.

| **Example**   Factor: $12x - 6y + 9$. | **Practice Exercise** |
|---|---|
| $12x - 6y + 9 = 3 \cdot 4x - 3 \cdot 2y + 3 \cdot 3$ $\qquad\qquad\quad = 3(4x - 2y + 3)$ | **2.** Factor: $24a - 8b + 16$. |

---

**Objective 11.1d**   Collect like terms.

| **Example**   Collect like terms: $4a - 2b - 2a + b$. | **Practice Exercise** |
|---|---|
| $4a - 2b - 2a + b = 4a - 2a - 2b + b$ $\qquad\qquad\qquad\quad = 4a - 2a - 2b + 1 \cdot b$ $\qquad\qquad\qquad\quad = (4 - 2)a + (-2 + 1)b$ $\qquad\qquad\qquad\quad = 2a - b$ | **3.** Collect like terms: $7x + 3y - x - 6y$. |

---

**Objective 11.2a**   Solve equations using the addition principle.

**Example**   Solve: $x + 6 = 8$.

$$x + 6 = 8$$
$$x + 6 - 6 = 8 - 6$$
$$x + 0 = 2$$
$$x = 2$$

The solution is 2.

**Practice Exercise**

**4.** Solve: $y - 4 = -2$.

---

**Objective 11.3a**   Solve equations using the multiplication principle.

**Example**   Solve: $45 = -5y$.

$$45 = -5y$$
$$\frac{45}{-5} = \frac{-5y}{-5}$$
$$-9 = 1 \cdot y$$
$$-9 = y$$

The solution is $-9$.

**Practice Exercise**

**5.** Solve: $9x = -72$.

---

**Objective 11.4a**   Solve equations using both the addition principle and the multiplication principle.

**Example**   Solve: $3x - 2 = 7$.

$$3x - 2 = 7$$
$$3x - 2 + 2 = 7 + 2$$
$$3x = 9$$
$$\frac{3x}{3} = \frac{9}{3}$$
$$x = 3$$

The solution is 3.

**Practice Exercise**

**6.** Solve: $5y + 1 = 6$.

---

**Objective 11.4b**   Solve equations in which like terms may need to be collected.

**Example**   Solve: $2y - 1 = -3y - 8 + 2$.

$$2y - 1 = -3y - 8 + 2$$
$$2y - 1 = -3y - 6$$
$$2y - 1 + 1 = -3y - 6 + 1$$
$$2y = -3y - 5$$
$$2y + 3y = -3y - 5 + 3y$$
$$5y = -5$$
$$\frac{5y}{5} = \frac{-5}{5}$$
$$y = -1$$

The solution is $-1$.

**Practice Exercise**

**7.** Solve: $6x - 4 - x = 2x - 10$.

**Objective 11.4c**   Solve equations by first removing parentheses and collecting like terms.

**Example**   Solve: $8b - 2(3b + 1) = 10$.

$$8b - 2(3b + 1) = 10$$
$$8b - 6b - 2 = 10$$
$$2b - 2 = 10$$
$$2b - 2 + 2 = 10 + 2$$
$$2b = 12$$
$$\frac{2b}{2} = \frac{12}{2}$$
$$b = 6$$

The solution is 6.

**Practice Exercise**

**8.** Solve: $2(y - 1) = 5(y - 4)$.

---

**Objective 11.5a**   Translate phrases to algebraic expressions.

**Example**   Translate to an algebraic expression:  Three less than some number.

We let $n = $ the number. Now if the number were 5, then the translation would be $5 - 3$. Similarly, if the number were 35, then the translation would be $35 - 3$. Thus we see from these numerical examples that if the number were $n$, the translation would be

$$n - 3.$$

**Practice Exercise**

**9.** Translate to an algebraic expression:  Five more than some number.

---

# Review Exercises

**1.** Evaluate $\dfrac{x - y}{3}$ when $x = 17$ and $y = 5$.   [11.1a]

Multiply.   [11.1b]

**2.** $5(3x - 7)$

**3.** $-2(4x - 5)$

**4.** $10(0.4x + 1.5)$

**5.** $-8(3 - 6x + 2y)$

Factor.   [11.1c]

**6.** $2x - 14$

**7.** $6x - 6$

**8.** $5x + 10$

**9.** $12 - 3x + 6z$

Collect like terms.   [11.1d]

**10.** $11a + 2b - 4a - 5b$

**11.** $7x - 3y - 9x + 8y$

**12.** $6x + 3y - x - 4y$

**13.** $-3a + 9b + 2a - b$

Solve. [11.2a], [11.3a]

**14.** $x + 5 = -17$

**15.** $-8x = -56$

**16.** $-\dfrac{x}{4} = 48$

**17.** $n - 7 = -6$

**18.** $15x = -35$

**19.** $x - 11 = 14$

**20.** $-\dfrac{2}{3} + x = -\dfrac{1}{6}$

**21.** $\dfrac{4}{5}y = -\dfrac{3}{16}$

**22.** $y - 0.9 = 9.09$

**23.** $5 - x = 13$

Solve. [11.4a, b, c]

**24.** $5t + 9 = 3t - 1$

**25.** $7x - 6 = 25x$

**26.** $\dfrac{1}{4}x - \dfrac{5}{8} = \dfrac{3}{8}$

**27.** $14y = 23y - 17 - 10$

**28.** $0.22y - 0.6 = 0.12y + 3 - 0.8y$

**29.** $\dfrac{1}{4}x - \dfrac{1}{8}x = 3 - \dfrac{1}{16}x$

**30.** $4(x + 3) = 36$

**31.** $3(5x - 7) = -66$

**32.** $8(x - 2) - 5(x + 4) = 20x + x$

**33.** $-5x + 3(x + 8) = 16$

**34.** Translate to an algebraic expression: [11.5a]
Nineteen percent of some number.

Solve. [11.5b]

**35.** *Dimensions of Wyoming.* The state of Wyoming is roughly in the shape of a rectangle whose perimeter is 1280 mi. The length is 90 mi more than the width. Find the dimensions.

**36.** An entertainment center sold for $2449 in June. This was $332 more than the cost in February. Find the cost in February.

**37.** Ty is paid a commission of $8 for each small appliance he sells. One week, he received $216 in commissions. How many appliances did he sell?

**38.** The measure of the second angle of a triangle is 50° more than that of the first angle. The measure of the third angle is 10° less than twice that of the first angle. Find the measures of the angles.

**39.** After a 30% reduction, a bread maker is on sale for $154. What was the marked price (the price before the reduction)?

**40.** A tax-exempt organization received a bill of $145.90 for office supplies. The bill incorrectly included sales tax of 5%. How much does the organization actually owe?

**41.** A hotel manager's salary is $78,000, which is a 4% increase over the previous year's salary. What was the previous salary?

**42.** *HDTV Price.* An HDTV television sold for $829 in May. This was $38 less than the cost in January. What was the cost in January?

**43.** *Writing Pad.* The perimeter of a rectangular writing pad is 56 cm. The width is 6 cm less than the length. Find the width and the length.

**44.** *Nile and Amazon Rivers.* The total length of the Nile and Amazon Rivers is 13,108 km. If the Amazon were 234 km longer, it would be as long as the Nile. Find the length of each river.

**Source:** *The Handy Geography Answer Book*

**45.** Factor: $6a - 30b + 3$. [11.1c]

**A.** $6(a - 5b)$      **B.** $3(2a - 10b)$
**C.** $3(2a - 10b + 1)$      **D.** $3(2a - b + 1)$

**46.** Collect like terms: $3x - 2y + x - 5y$. [11.1d]

**A.** $4x - 7y$      **B.** $x - 4y$
**C.** $4x + 3y$      **D.** $4x + 7y$

## Synthesis

Solve. [2.1d], [11.4a, b]

**47.** $2|n| + 4 = 50$

**48.** $|3n| = 60$

# Understanding Through Discussion and Writing

**1.** Explain at least three uses of the distributive laws considered in this chapter. [11.1b, c, d]

**2.** Explain the role of the opposite of a number when using the addition principle. [11.2a]

**3.** Explain the role of the reciprocal of a number when using the multiplication principle. [11.3a]

**4.** Describe a procedure that a classmate could use to solve the equation $ax + b = c$ for $x$. [11.4a]

Test

For Extra Help

Step-by-step test solutions are found on the Chapter Test Prep Videos available via the Video Resources on DVD, in *MyMathLab* ▌ , and on You Tube (search "BittingerBasicMathEI" and click on "Channels").

**1.** Evaluate $\dfrac{3x}{y}$ when $x = 10$ and $y = 5$.

Multiply.

**2.** $3(6 - x)$

**3.** $-5(y - 1)$

Factor.

**4.** $12 - 22x$

**5.** $7x + 21 + 14y$

Collect like terms.

**6.** $9x - 2y - 14x + y$

**7.** $-a + 6b + 5a - b$

Solve.

**8.** $x + 7 = 15$

**9.** $t - 9 = 17$

**10.** $3x = -18$

**11.** $-\dfrac{4}{7}x = -28$

**12.** $3t + 7 = 2t - 5$

**13.** $\dfrac{1}{2}x - \dfrac{3}{5} = \dfrac{2}{5}$

**14.** $8 - y = 16$

**15.** $-\dfrac{2}{5} + x = -\dfrac{3}{4}$

**16.** $0.4p + 0.2 = 4.2p - 7.8 - 0.6p$

**17.** $3(x + 2) = 27$

**18.** $-3x - 6(x - 4) = 9$

**19.** Translate to an algebraic expression:
Nine less than some number.

Solve.

**20.** *Perimeter of a Photograph.* The perimeter of a rectangular photograph is 36 cm. The length is 4 cm greater than the width. Find the width and the length.

**21.** *Charitable Contributions.* About $102.3 billion was given to religious organizations in 2007. This represents 33% of all charitable donations that year. How much was donated to all charities?

Sources: Giving USA Foundation; Center on Philanthropy at Indiana University

**22.** *Board Cutting.* An 8-m board is cut into two pieces. One piece is 2 m longer than the other. How long are the pieces?

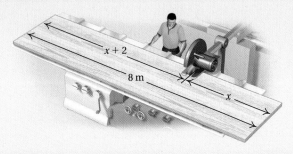

x + 2

8 m

x

**23.** *Savings Account.* Money is invested in a savings account at 5% simple interest. After 1 year, there is $924 in the account. How much was originally invested?

**24.** If you triple a number and then subtract 14, you get $\frac{2}{3}$ of the original number. What is the original number?

**25.** The second angle of a triangle is three times as large as the first. The third angle is 25° less than the sum of the other two angles. Find the measure of the first angle.

**26.** Solve: $5y - 1 = 3y + 7$.
  **A.** −4  **B.** 1
  **C.** 3   **D.** 4

# Synthesis

**27.** Solve: $3|w| - 8 = 37$.

**28.** A movie theater had a certain number of tickets to give away. Five people got the tickets. The first got $\frac{1}{3}$ of the tickets, the second got $\frac{1}{4}$ of the tickets, and the third got $\frac{1}{5}$ of the tickets. The fourth person got 8 tickets, and there were 5 tickets left for the fifth person. Find the total number of tickets given away.

This cumulative review also serves as a review for a final examination covering the entire book. A question that may occur at this point is what notation to use for a particular problem or exercise. Although there is no particular rule, especially as you use mathematics outside the classroom, here is the guideline that we follow: Use the notation given in the problem. That is, if the problem is given using mixed numerals, give the answer in mixed numerals. If the problem is given in decimal notation, give the answer in decimal notation.

**1.** In 47,201, what digit tells the number of thousands?

**2.** Write expanded notation for 7405.

**3.** Write a word name for 7.463.

Add and simplify, if appropriate.

**4.**
```
  7 4 1
+ 2 7 1
```

**5.**
```
  4 9 0 3
  5 2 7 8
  6 3 9 1
+ 4 5 1 3
```

**6.** $-\dfrac{2}{13} + \dfrac{1}{26}$

**7.**
$$2\frac{4}{9}$$
$$+ 3\frac{1}{3}$$

**8.**
```
    2.0 4 8
    6 3.9 1 4
+ 4 2 8.0 0 9
```

**9.** $34.56 + 2.783 + 0.433 + 765.1$

Subtract and simplify, if possible.

**10.**
```
  6 7 4
- 5 2 2
```

**11.**
```
  9 4 6 5
- 8 7 9 1
```

**12.** $\dfrac{7}{8} - \dfrac{2}{3}$

**13.**
$$4\frac{1}{3}$$
$$- 1\frac{5}{8}$$

**14.**
```
  2 0.0
-   0.0 0 2 7
```

**15.** $40.03 - (-5.789)$

Simplify.

**16.** $\dfrac{21}{30}$

**17.** $\dfrac{275}{5}$

Multiply and simplify, if possible.

**18.**
```
  2 9 7
×   1 6
```

**19.**
```
  3 4 9
× 7 6 3
```

**20.** $1\dfrac{3}{4} \cdot 2\dfrac{1}{3}$

**21.** $\dfrac{9}{7} \cdot \dfrac{14}{15}$

**22.** $-12 \cdot \dfrac{5}{6}$

**23.**
```
  3 4.0 9
×     7.6
```

**24.** Convert to a mixed numeral: $\dfrac{18}{5}$.

**37.** Use = or ≠ for ☐ to write a true sentence:

$\dfrac{4}{7}$ ☐ $\dfrac{3}{5}$.

Divide and simplify. State the answer using a whole-number quotient and a remainder.

**25.** $6 \overline{)\ 3\ 4\ 3\ 8}$        **26.** $3\ 4 \overline{)\ 1\ 9\ 1\ 4}$

**38.** Use < or > for ☐ to write a true sentence:

$\dfrac{4}{7}$ ☐ $\dfrac{3}{5}$.

**27.** Write a mixed numeral for the quotient in Exercise 26.

**39.** Which number is greater, $-1.001$ or $-0.9976$?

Divide and simplify, if possible.

**28.** $\dfrac{4}{5} \div \dfrac{8}{15}$

**40.** *Pears.* Find the unit price of each brand listed in the table below. Then determine which package has the lowest unit price.

| BRAND | SIZE | PRICE | UNIT PRICE |
|-------|------|-------|------------|
| A | $8\frac{1}{2}$ oz | $0.95 | |
| B | 15 oz | $1.66 | |
| C | $15\frac{1}{4}$ oz | $1.86 | |
| D | 24 oz | $2.54 | |
| E | 29 oz | $3.07 | |

**29.** $2\dfrac{1}{3} \div (-30)$        **30.** $2.7 \overline{)1\ 0\ 5.3}$

**31.** Round 68,489 to the nearest thousand.

**41.** *Pluto.* The dwarf planet Pluto has a diameter of 1400 mi. Use $\frac{22}{7}$ for $\pi$.

a) Find the circumference of Pluto.
b) Find the volume of Pluto.

**32.** Round 0.4275 to the nearest thousandth.

**33.** Round $21.\overline{83}$ to the nearest hundredth.

*Kitchen Remodeling.* The Reisters spent $26,888 to remodel their kitchen. Complete the table below, which relates percents and costs.

**34.** Determine whether 1368 is divisible by 6.

| | ITEM | PERCENT OF COST | COST |
|---|------|-----------------|------|
| **42.** | Cabinets | 40% | |
| **43.** | Countertops | | $4033.20 |
| **44.** | Appliances | 13% | |
| **45.** | Fixtures | | $8066.40 |
| **46.** | Flooring | 2% | |

**35.** Find all the factors of 15.

**36.** Find the LCM of 16, 25, and 32.

**47.** Use < or > for ☐ to write a true sentence:

987 ☐ 879.

**48.** What part is shaded?

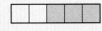

Convert to decimal notation.

**49.** $\dfrac{37}{1000}$

**50.** $-\dfrac{13}{25}$

**51.** $\dfrac{8}{9}$

**52.** 7%

Convert to fraction notation.

**53.** 4.63

**54.** $-7\dfrac{1}{4}$

**55.** 40%

Convert to percent notation.

**56.** $\dfrac{17}{20}$

**57.** 1.5

Solve.

**58.** $234 + y = 789$

**59.** $-3.9 \times y = 249.6$

**60.** $\dfrac{2}{3} \cdot t = \dfrac{5}{6}$

**61.** $\dfrac{8}{17} = \dfrac{36}{x}$

*Source of Donations.* Americans donated a total of $306.4 billion to charitable organizations in 2007. The table below shows the percent of this total donated by each type of source. Use this table to do Exercises 62 and 63.

| SOURCE OF DONATIONS | PERCENT OF TOTAL |
|---|---|
| Individuals | 75% |
| Corporations | 5% |
| Bequests | 7.5% |
| Foundations | 12.5% |

SOURCE: Giving USA Foundation

**62.** Make a circle graph of the data.

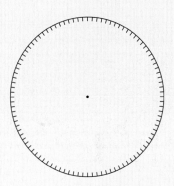

**63.** Make a vertical bar graph of the data, showing the percent of the total contributions corresponding to each source.

**64.** Find the missing angle measure.

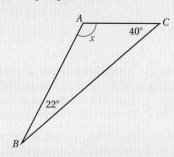

**65.** Classify the triangle in Exercise 64 as right, obtuse, or acute.

Solve.

**66.** *Donations.*   Lorenzo made donations of $627 and $48 to a charity. What was the total donation?

**67.** *Candy Bars.*   A machine wraps 134 candy bars per minute. How long does it take this machine to wrap 8710 bars?

**68.** *Stock Prices.*   A share of stock bought for $29.63 dropped $3.88 before it was resold. What was the price when it was resold?

**69.** *Length of Trip.*   At the start of a trip, a car's odometer read 27,428.6 mi, and at the end of the trip, the reading was 27,914.5 mi. How long was the trip?

**70.** *Taxes.*   From an income of $12,000, amounts of $2300 and $1600 are paid for federal and state taxes. How much remains after these taxes have been paid?

**71.** *Teacher Salary.*   A substitute teacher was paid $87 per day for 9 days. How much was she paid altogether?

**72.** *Walking Distance.*   Celeste walks $\frac{3}{5}$ km per hour. At this rate, how far would she walk in $\frac{1}{2}$ hr?

**73.** *Sweater Costs.*   Eight identical sweaters cost a total of $679.68. What is the cost of each sweater?

**74.** *Paint Needs.*   Eight gallons of exterior paint covers 400 ft$^2$. How much paint is needed to cover 650 ft$^2$?

**75.** *Simple Interest.*   What is the simple interest on $4000 principal at 5% for $\frac{3}{4}$ years?

**76.** *Commission Rate.*   A real estate agent received $5880 commission on the sale of an $84,000 home. What was the rate of commission?

**77.** *Population Growth.*   The population of Waterville is 29,000 this year and is increasing at 4% per year. What will the population be next year?

**78.** *Student Ages.*   The ages of students in a math class at a community college are as follows:

18, 21, 26, 31, 32, 18, 50.

Find the average, the median, and the mode of their ages.

Evaluate.

**79.** $18^2$

**80.** $7^3$

Simplify.

**81.** $\sqrt{9}$

**82.** $\sqrt{121}$

**83.** Approximate to three decimal places: $\sqrt{20}$.

Complete.

**84.** $\frac{1}{3}$ yd = _____ in.

**85.** 4280 mm = _____ cm

**86.** 3 days = _____ hr

**87.** 20,000 g = _____ kg

**88.** 5 lb = _____ oz

**89.** 0.008 cg = _____ mg

**90.** 8190 mL = _____ L

**91.** 20 qt = _____ gal

**92.** Find the length of the third side of this right triangle. Give an exact answer and an approximation to three decimal places.

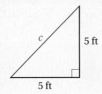

**93.** Find the diameter, the circumference, and the area of this circle. Use 3.14 for $\pi$.

**94.** Find the perimeter and the area.

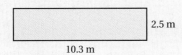

Find the area.

**95.**

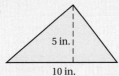

**96.**

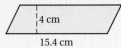

**97.**

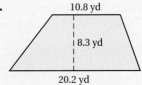

**98.** Find the volume.

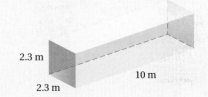

Find the volume. Use 3.14 for $\pi$.

**99.**

**100.**

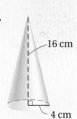

Solve.

**101.** $7 - x = 12$

**102.** $-4.3x = -17.2$

**103.** $5x + 7 = 3x - 9$

**104.** $5(x - 2) - 8(x - 4) = 20$

Compute and simplify.

**105.** $12 \times 20 - 10 \div 5$

**106.** $4^3 - 5^2 + (16 \cdot 4 + 23 \cdot 3)$

**107.** $|(-1) \cdot 3|$

**108.** $17 + (-3)$

**109.** $-\dfrac{1}{3} - \left(-\dfrac{2}{3}\right)$

**110.** $(-6) \cdot (-5)$

**111.** $-\dfrac{5}{7} \cdot \dfrac{14}{35}$

**112.** $\dfrac{48}{-6}$

Translate to an algebraic expression.

**113.** 17 more than some number

**114.** 38 percent of some number

Solve.

**115.** *Rollerblade Costs.* Sam and Susan purchased rollerblades for a total of $107. Sam paid $17 more for his rollerblades than Susan did. What did Susan pay?

**116.** *Savings Investment.* Money is invested in a savings account at 8% simple interest. After 1 year, there is $1134 in the account. How much was originally invested?

**117.** *Wire Cutting.* A 143-m wire is cut into three pieces. The second piece is 3 m longer than the first. The third is four-fifths as long as the first. How long is each piece?

Solve.

**118.** $\dfrac{2}{3}x + \dfrac{1}{6} - \dfrac{1}{2}x = \dfrac{1}{6} - 3x$

**119.** $29.966 - 8.673y = -8.18 + 10.4y$

**120.** Collect like terms: $\dfrac{1}{4}x - \dfrac{3}{4}y + \dfrac{1}{4}x - \dfrac{3}{4}y.$

   **A.** 0             **B.** $-\dfrac{2}{3}y$

   **C.** $\dfrac{1}{2}x - \dfrac{3}{2}y$     **D.** $-\dfrac{1}{2}x - \dfrac{1}{2}y$

**121.** Factor out the greatest common factor: $8x + 4y - 12z.$

   **A.** $2(4x - 2y - 6z)$     **B.** $4(2x + y - 3z)$

   **C.** $4(2x - 3z)$          **D.** $2(4x + 4y - 12z)$

**122.** Divide: $-\dfrac{13}{25} \div \left(-\dfrac{13}{5}\right).$

   **A.** $\dfrac{169}{125}$           **B.** 5

   **C.** $\dfrac{125}{169}$          **D.** $\dfrac{1}{5}$

**123.** Add: $-27 + (-11).$

   **A.** $-38$           **B.** $-16$

   **C.** 16            **D.** 38

# Synthesis

**124.** The sum of two numbers is 430. The difference is 40. Find the numbers.

# Photo Credits

# Answers

## CHAPTER 1

### Exercise Set 1.1, p. 6

**1.** 5 thousands   **3.** 5 hundreds   **5.** 9   **7.** 7
**9.** 5 thousands + 7 hundreds + 0 tens + 2 ones, or
5 thousands + 7 hundreds + 2 ones
**11.** 9 ten thousands + 3 thousands + 9 hundreds + 8 tens +
6 ones   **13.** 2 thousands + 0 hundreds + 5 tens + 8 ones, or
2 thousands + 5 tens + 8 ones   **15.** 1 thousand +
5 hundreds + 7 tens + 6 ones   **17.** 1 billion +
4 hundred millions + 2 ten millions + 4 millions +
1 hundred thousand + 6 ten thousands + 1 thousand +
9 hundreds + 4 tens + 8 ones   **19.** 9 ten millions +
9 millions + 8 hundred thousands + 8 ten thousands +
6 thousands + 5 hundreds + 6 tens + 8 ones
**21.** 6 hundred thousands + 1 ten thousand + 7 thousands +
2 hundreds + 4 tens + 9 ones   **23.** Eighty-five
**25.** Eighty-eight thousand   **27.** One hundred twenty-three
thousand, seven hundred sixty-five   **29.** Seven billion, seven
hundred fifty-four million, two hundred eleven thousand, five
hundred seventy-seven   **31.** Seven hundred thousand, six
hundred thirty-four   **33.** Three million, forty-eight thousand,
five   **35.** 2,233,812   **37.** 8,000,000,000
**39.** 50,324   **41.** 632,896   **43.** 1,600,000,000
**45.** 64,186,000   **47.** 138

### Calculator Corner, p. 10

**1.** 121   **2.** 1602   **3.** 1932   **4.** 864

### Exercise Set 1.2, p. 12

**1.** 387   **3.** 164   **5.** 5198   **7.** 100   **9.** 8503   **11.** 5266
**13.** 4466   **15.** 6608   **17.** 34,432   **19.** 101,310   **21.** 230
**23.** 18,424   **25.** 31,685   **27.** 132 yd   **29.** 1661 ft
**31.** 570 ft   **33.** 8 ten thousands   **34.** Nine billion, three
hundred forty-six million, three hundred ninety-nine
thousand, four hundred sixty-eight
**35.** $1 + 99 = 100, 2 + 98 = 100, \ldots, 49 + 51 = 100$. Then
49 100's = 4900 and $4900 + 50 + 100 = 5050$.

### Calculator Corner, p. 15

**1.** 28   **2.** 47   **3.** 67   **4.** 119   **5.** 2128   **6.** 2593

### Exercise Set 1.3, p. 17

**1.** 44   **3.** 533   **5.** 39   **7.** 14   **9.** 369   **11.** 26   **13.** 234
**15.** 417   **17.** 5382   **19.** 2778   **21.** 3069   **23.** 1089
**25.** 7748   **27.** 4144   **29.** 56   **31.** 454   **33.** 3831
**35.** 3749   **37.** 2191   **39.** 4418   **41.** 43,028   **43.** 95,974
**45.** 4206   **47.** 1305   **49.** 9989   **51.** 48,017   **53.** 1024

**54.** 12,732   **55.** 90,283   **56.** 29,364   **57.** 1345   **58.** 924
**59.** 22,692   **60.** 10,920   **61.** Six million, three hundred
seventy-five thousand, six hundred two   **62.** 7 ten thousands
**63.** 3; 4

### Calculator Corner, p. 23

**1.** 448   **2.** 21,970   **3.** 6380   **4.** 39,564   **5.** 180,480
**6.** 2,363,754

### Exercise Set 1.4, p. 24

**1.** 520   **3.** 564   **5.** 1527   **7.** 64,603   **9.** 4770   **11.** 3995
**13.** 870   **15.** 1920   **17.** 46,296   **19.** 14,652   **21.** 258,312
**23.** 798,408   **25.** 20,723,872   **27.** 362,128   **29.** 302,220
**31.** 49,101,136   **33.** 25,236,000   **35.** 20,064,048
**37.** 529,984 sq mi   **39.** 8100 sq ft   **41.** 12,685   **42.** 10,834
**43.** 427,477   **44.** 111,110   **45.** 1241   **46.** 8889
**47.** 254,119   **48.** 66,444   **49.** 247,464 sq ft

### Calculator Corner, p. 30

**1.** 28   **2.** 123   **3.** 323   **4.** 36

### Exercise Set 1.5, p. 32

**1.** 12   **3.** 1   **5.** 22   **7.** 0   **9.** Not defined   **11.** 6
**13.** 55 R 2   **15.** 108   **17.** 307   **19.** 753 R 3   **21.** 74 R 1
**23.** 92 R 2   **25.** 1703   **27.** 987 R 5   **29.** 12,700   **31.** 127
**33.** 52 R 52   **35.** 29 R 5   **37.** 40 R 12   **39.** 90 R 22
**41.** 29   **43.** 105 R 3   **45.** 1609 R 2   **47.** 1007 R 1   **49.** 23
**51.** 107 R 1   **53.** 370   **55.** 609 R 15   **57.** 304
**59.** 3508 R 219   **61.** 8070   **63.** Perimeter   **64.** Minuend
**65.** Digits; periods   **66.** Dividend   **67.** Factors; product
**68.** Additive   **69.** Associative   **70.** Divisor; remainder;
dividend   **71.** 54, 122; 33, 2772; 4, 8   **73.** 30 buses

### Mid-Chapter Review: Chapter 1, p. 35

**1.** False   **2.** True   **3.** True   **4.** False   **5.** True   **6.** False
**7.**                                           95,406,237

Ninety-five million,

four hundred six thousand,

two hundred thirty-seven
**8.**   $\overset{5\ 9\ 14}{\cancel{6\ 0\ 4}}$   **9.** 6 hundreds   **10.** 6 ten thousands

   $-\ 4\ 9\ 7$
   $\overline{\phantom{0}1\ 0\ 7}$
**11.** 6 thousands   **12.** 6 ones   **13.** 2   **14.** 6   **15.** 5   **16.** 1
**17.** 5 thousands + 6 hundreds + 0 tens + 2 ones, or
   5 thousands + 6 hundreds + 2 ones

**18.** 6 ten thousands + 9 thousands + 3 hundreds + 4 tens + 5 ones   **19.** One hundred thirty-six   **20.** Sixty-four thousand, three hundred twenty-five   **21.** 308,716   **22.** 4,567,216   **23.** 798   **24.** 1030   **25.** 7922   **26.** 7534   **27.** 465   **28.** 339   **29.** 1854   **30.** 4328   **31.** 216   **32.** 15,876   **33.** 132,275   **34.** 5,679,870   **35.** 253   **36.** 112 R 5   **37.** 23 R 19   **38.** 144 R 31   **39.** 25 m   **40.** 8 sq in.   **41.** When numbers are being added, it does not matter how they are grouped.   **42.** Subtraction is not commutative. For example, $5 - 2 = 3$, but $2 - 5 \neq 3$.   **43.** Answers will vary. Suppose one coat costs $150. Then the multiplication $4 \cdot \$150$ gives the cost of four coats. Or, suppose one ream of copy paper costs $4. Then the multiplication $\$4 \cdot 150$ gives the cost of 150 reams.   **44.** If we use the definition of division, $0 \div 0 = a$ such that $a \cdot 0 = 0$. We see that $a$ could be *any* number since $a \cdot 0 = 0$ for any number $a$. Thus we cannot say that $0 \div 0 = 0$. This is why we agree not to allow division by 0.

### Exercise Set 1.6, p. 43

**1.** 50   **3.** 460   **5.** 730   **7.** 900   **9.** 100   **11.** 1000   **13.** 9100   **15.** 32,800   **17.** 6000   **19.** 8000   **21.** 45,000   **23.** 373,000   **25.** $80 + 90 = 170$   **27.** $8070 - 2350 = 5720$   **29.** 220; incorrect   **31.** 890; incorrect   **33.** $7300 + 9200 = 16,500$   **35.** $6900 - 1700 = 5200$   **37.** 1600; correct   **39.** 1500; correct   **41.** $10,000 + 5000 + 9000 + 7000 = 31,000$   **43.** $92,000 - 23,000 = 69,000$   **45.** $50 \cdot 70 = 3500$   **47.** $30 \cdot 30 = 900$   **49.** $900 \cdot 300 = 270,000$   **51.** $400 \cdot 200 = 80,000$   **53.** $350 \div 70 = 5$   **55.** $8450 \div 50 = 169$   **57.** $1200 \div 200 = 6$   **59.** $8400 \div 300 = 28$   **61.** $11,200   **63.** $18,900; no   **65.** Answers will vary depending on the options chosen.   **67.** (a) $2,190,000; (b) $2,170,000   **69.** 90 people   **71.** <   **73.** >   **75.** <   **77.** >   **79.** >   **81.** >   **83.** $190,078 > 172,000$, or $172,000 < 190,078$   **85.** $1694 < 5249$, or $5249 > 1694$   **87.** 86,754   **88.** 13,589   **89.** 48,824   **90.** 4415   **91.** 1702   **92.** 17,748   **93.** 54 R 4   **94.** 208   **95.** Left to the student   **97.** Left to the student

### Exercise Set 1.7, p. 52

**1.** 14   **3.** 0   **5.** 90,900   **7.** 450   **9.** 352   **11.** 25   **13.** 29   **15.** 0   **17.** 79   **19.** 45   **21.** 8   **23.** 14   **25.** 32   **27.** 143   **29.** 17,603   **31.** 37   **33.** 1035   **35.** 66   **37.** 324   **39.** 743   **41.** 175   **43.** 335   **45.** 18,252   **47.** 104   **49.** 45   **51.** 4056   **53.** 2847   **55.** 15   **57.** 205   **59.** 457   **61.** 142 R 5   **62.** 142   **63.** 334   **64.** 334 R 11   **65.** <   **66.** >   **67.** >   **68.** <   **69.** 6,376,000   **70.** 6,375,600   **71.** 347

### Translating for Success, p. 63

**1.** E   **2.** M   **3.** D   **4.** G   **5.** A   **6.** O   **7.** F   **8.** K   **9.** J   **10.** H

### Exercise Set 1.8, p. 64

**1.** 550 ft   **3.** 1250 ft   **5.** 95 milligrams   **7.** 18 rows   **9.** 1502 hr   **11.** 2054 mi   **13.** 2,073,600 pixels   **15.** 268,000 men   **17.** 168 hr   **19.** $23 per month   **21.** $233   **23.** $9276   **25.** 151,500   **27.** 1,190,000 motorcycles   **29.** $78   **31.** $40 per month   **33.** $24,456   **35.** 35 weeks; 2 episodes   **37.** 236 gal   **39.** 21 columns   **41.** (a) 4200 sq ft; (b) 268 ft   **43.** $247   **45.** 645 mi; 5 in.   **47.** 56 cartons   **49.** 32 $10 bills   **51.** $400   **53.** 525 min, or 8 hr 45 min   **55.** 700 min, or 11 hr 40 min   **57.** 104 seats   **59.** 106 bones   **61.** 234,600   **62.** 234,560   **63.** 235,000   **64.** 22,000   **65.** 16,000   **66.** 4000   **67.** 8000   **68.** 320,000   **69.** 720,000   **70.** 46,800,000   **71.** 792,000 mi; 1,386,000 mi

### Calculator Corner, p. 72

**1.** 243   **2.** 15,625   **3.** 20,736   **4.** 2048

### Calculator Corner, p. 74

**1.** 49   **2.** 85   **3.** 36   **4.** 0   **5.** 73   **6.** 49

### Exercise Set 1.9, p. 77

**1.** $3^4$   **3.** $5^2$   **5.** $7^5$   **7.** $10^3$   **9.** 49   **11.** 729   **13.** 20,736   **15.** 243   **17.** 22   **19.** 20   **21.** 100   **23.** 1   **25.** 49   **27.** 5   **29.** 434   **31.** 41   **33.** 88   **35.** 4   **37.** 303   **39.** 20   **41.** 70   **43.** 295   **45.** 32   **47.** 906   **49.** 62   **51.** 102   **53.** 32   **55.** $94   **57.** 401   **59.** 110   **61.** 7   **63.** 544   **65.** 708   **67.** 27   **69.** 452   **70.** 835   **71.** 13   **72.** 37   **73.** 2342   **74.** 4898   **75.** 25   **76.** 100   **77.** 104,286 sq mi   **78.** 98 gal   **79.** $24; 1 + 5 \cdot (4 + 3) = 36$   **81.** $7; 12 \div (4 + 2) \cdot 3 - 2 = 4$

### Summary and Review: Chapter 1, p. 80

#### Concept Reinforcement

**1.** True   **2.** True   **3.** False   **4.** False   **5.** True   **6.** False

#### Important Concepts

**1.** 2 thousands   **2.** 65,302   **3.** 3237   **4.** 225,036   **5.** 315 R 14   **6.** 36,500   **7.** 36,000   **8.** <   **9.** 36   **10.** 216

#### Review Exercises

**1.** 8 thousands   **2.** 3   **3.** 2 thousands + 7 hundreds + 9 tens + 3 ones   **4.** 5 ten thousands + 6 thousands + 0 hundreds + 7 tens + 8 ones, or 5 ten thousands + 6 thousands + 7 tens + 8 ones   **5.** 4 millions + 0 hundred thousands + 0 ten thousands + 7 thousands + 1 hundred + 0 tens + 1 one, or 4 millions + 7 thousands + 1 hundred + 1 one   **6.** Sixty-seven thousand, eight hundred nineteen   **7.** Two million, seven hundred eighty-one thousand, four hundred twenty-seven   **8.** Four million, eight hundred seventeen thousand, nine hundred forty-one   **9.** 476,588   **10.** 1,563,000,000   **11.** 14,272   **12.** 66,024   **13.** 21,788   **14.** 98,921   **15.** 5148   **16.** 1689   **17.** 2274   **18.** 17,757   **19.** 5,100,000   **20.** 6,276,800   **21.** 506,748   **22.** 27,589   **23.** 5,331,810   **24.** 12 R 3   **25.** 5   **26.** 913 R 3   **27.** 384 R 1   **28.** 4 R 46   **29.** 54   **30.** 452   **31.** 5008   **32.** 4389   **33.** 345,800   **34.** 345,760   **35.** 346,000   **36.** 300,000   **37.** >   **38.** <   **39.** $41,300 + 19,700 = 61,000$   **40.** $38,700 - 24,500 = 14,200$   **41.** $400 \cdot 700 = 280,000$   **42.** 8   **43.** 45   **44.** 58   **45.** 0   **46.** $4^3$   **47.** 10,000   **48.** 36   **49.** 65   **50.** 233   **51.** 260   **52.** 165   **53.** $502   **54.** $484   **55.** 1982   **56.** 19 cartons   **57.** $13,585   **58.** 14 beehives   **59.** 98 sq ft; 42 ft   **60.** 137 beakers filled; 13 mL left over   **61.** $27,598   **62.** B   **63.** A   **64.** D   **65.** 8   **66.** $a = 8, b = 4$   **67.** 6 days

#### Understanding Through Discussion and Writing

**1.** No; if subtraction were associative, then $a - (b - c) = (a - b) - c$ for any $a$, $b$, and $c$. But, for example,

$$12 - (8 - 4) = 12 - 4 = 8,$$

whereas

$$(12 - 8) - 4 = 4 - 4 = 0.$$

Since $8 \neq 0$, this example shows that subtraction is not associative.   **2.** By rounding prices and estimating their sum, a shopper can estimate the total grocery bill while shopping. This is particularly useful if the shopper wants to spend no more than a certain amount.   **3.** Answers will vary. Anthony is driving from Kansas City to Minneapolis, a distance of 512 mi. He stops for gas after driving 183 mi. How much farther must he drive?

**4.** The parentheses are not necessary in the expression $9 - (4 \cdot 2)$. Using the rules for order of operations, the multiplication would be performed before the subtraction even if the parentheses were not present. The parentheses are necessary in the expression $(3 \cdot 4)^2$; $(3 \cdot 4)^2 = 12^2 = 144$, but $3 \cdot 4^2 = 3 \cdot 16 = 48$.

**Test: Chapter 1, p. 85**

**1.** [1.1a] 5   **2.** [1.1b] 8 thousands + 8 hundreds + 4 tens + 3 ones   **3.** [1.1c] Thirty-eight million, four hundred three thousand, two hundred seventy-seven
**4.** [1.2a] 9989   **5.** [1.2a] 63,791   **6.** [1.2a] 3165
**7.** [1.2a] 10,515   **8.** [1.3a] 3630   **9.** [1.3a] 1039
**10.** [1.3a] 6848   **11.** [1.3a] 5175   **12.** [1.4a] 41,112
**13.** [1.4a] 5,325,600   **14.** [1.4a] 2405   **15.** [1.4a] 534,264
**16.** [1.5a] 3 R 3   **17.** [1.5a] 70   **18.** [1.5a] 97
**19.** [1.5a] 805 R 8   **20.** [1.6a] 35,000   **21.** [1.6a] 34,530
**22.** [1.6a] 34,500   **23.** [1.6b] 23,600 + 54,700 = 78,300
**24.** [1.6b] 54,800 − 23,600 = 31,200
**25.** [1.6b] 800 · 500 = 400,000   **26.** [1.6c] >
**27.** [1.6c] <   **28.** [1.7b] 46   **29.** [1.7b] 13   **30.** [1.7b] 14
**31.** [1.7b] 381   **32.** [1.8a] 83 calories   **33.** [1.8a] 20 staplers
**34.** [1.8a] 1,256,615 sq mi   **35. (a)** [1.2b], [1.4b] 300 in., 5000 sq in.; 264 in., 3872 sq in.; 228 in., 2888 sq in.;
**(b)** [1.8a] 2112 sq in.   **36.** [1.8a] 1852 12-packs; 7 cakes left over   **37.** [1.8a] $95   **38.** [1.9a] $12^4$   **39.** [1.9b] 343
**40.** [1.9b] 100,000   **41.** [1.9c] 31   **42.** [1.9c] 98
**43.** [1.9c] 2   **44.** [1.9c] 18   **45.** [1.9d] 216   **46.** [1.9c] A
**47.** [1.4b], [1.8a] 336 sq in.   **48.** [1.9c] 9   **49.** [1.8a] 80 payments

## CHAPTER 2

**Exercise Set 2.1, p. 92**

**1.** −282   **3.** 24; −2   **5.** 950,000,000; −460
**7.** −34; 15   **9.** −39,868
**11.**
**13.**
**15.** >   **17.** <   **19.** >   **21.** <   **23.** >   **25.** <
**27.** <   **29.** >   **31.** 3   **33.** 18   **35.** 325   **37.** 29
**39.** 300   **41.** 53   **43.** 13,549   **44.** 1107   **45.** 50,609
**46.** 219   **47.** 259   **48.** 8778   **49.** >   **51.** =

**Calculator Corner, p. 97**

**1.** 13   **2.** −8   **3.** −12

**Exercise Set 2.2, p. 98**

**1.** −7   **3.** −4   **5.** 0   **7.** −8   **9.** −7   **11.** −27   **13.** 0
**15.** −42   **17.** 0   **19.** 0   **21.** 3   **23.** −9   **25.** 7   **27.** 0
**29.** 45   **31.** −2   **33.** −7   **35.** −19   **37.** −6   **39.** −1
**41.** 39   **43.** 43   **45.** −1093   **47.** −24   **49.** 26   **51.** −103
**53.** −46   **55.** −9   **57.** 14   **59.** −65   **61.** 5   **63.** 14
**65.** −10   **67.** 3456   **68.** 137,016   **69.** 4,533,324
**70.** 408 R 3   **71.** 127 R 46   **72.** 221 R 331   **73.** 641,540
**74.** 641,500   **75.** 642,000   **77.** All positive   **79.** −640,979
**81.** Negative

**Exercise Set 2.3, p. 103**

**1.** −4   **3.** −7   **5.** −6   **7.** 0   **9.** −4   **11.** −7   **13.** −6
**15.** 0   **17.** 11   **19.** −14   **21.** 5   **23.** −7   **25.** −5
**27.** −3   **29.** −23   **31.** −68   **33.** −73   **35.** 116   **37.** −2
**39.** −5   **41.** −1   **43.** −5   **45.** 22   **47.** 19   **49.** −20
**51.** −8   **53.** −10   **55.** −3   **57.** −7   **59.** −30   **61.** 0
**63.** −1   **65.** 37   **67.** −62   **69.** 6   **71.** 107   **73.** 219

**75.** 20,602 ft   **77.** −3°   **79.** Down 3 ft   **81.** −$85
**83.** 100°F   **85.** 5676 ft   **87.** 64   **88.** 125   **89.** 29
**90.** 41   **91.** 66   **92.** 3   **93.** 8 cans   **94.** 288 oz
**95.** True   **97.** True   **99.** True   **101.** True

**Mid-Chapter Review: Chapter 2, p. 106**

**1.** True   **2.** False   **3.** True
**4.** $-x = -(-4) = 4$;
$-(-x) = -(-(-4)) = -(4) = -4$
**5.** $5 - 13 = 5 + (-13) = -8$   **6.** $-6 - (-7) = -6 + 7 = 1$
**7.** 450; −79   **8.** 20; −23
**9.**
**10.**   **11.** <   **12.** <
**13.** >   **14.** >   **15.** 15   **16.** 18   **17.** 0   **18.** 12
**19.** 5   **20.** −7   **21.** 0   **22.** 49   **23.** 19   **24.** 2   **25.** −2
**26.** −2   **27.** 0   **28.** −17   **29.** −10   **30.** −7   **31.** −9
**32.** −2   **33.** −10   **34.** 16   **35.** 2   **36.** −12   **37.** −4
**38.** −6   **39.** −2   **40.** 13   **41.** 9   **42.** −23   **43.** 75
**44.** 14   **45.** 33°C   **46.** $48   **47.** Answers will vary.
**48.** The absolute value of a number is its distance from 0, and distance is always nonnegative.   **49.** Answers may vary. If we think of the addition on the number line, we start at a negative number and move to the left. This always brings us to a point on the negative portion of the number line.   **50.** Yes; consider $m - (-n)$, where both $m$ and $n$ are positive. Then, $m - (-n) = m + n$. Now $m + n$, the sum of two positive numbers, is positive.

**Exercise Set 2.4, p. 110**

**1.** −16   **3.** −24   **5.** −72   **7.** 16   **9.** 42   **11.** −120
**13.** −238   **15.** 1200   **17.** 84   **19.** −12   **21.** 24   **23.** 21
**25.** −69   **27.** 27   **29.** −18   **31.** −45   **33.** 420   **35.** 36
**37.** −60   **39.** 150   **41.** −90   **43.** 1911   **45.** 56   **47.** 30
**49.** −960   **51.** 1764   **53.** −240   **55.** −756   **57.** −720
**59.** −30,240   **61.** Average   **62.** <   **63.** Commutative
**64.** Divisor   **65.** Absolute value   **66.** Opposites
**67.** Difference   **68.** Quotient   **69. (a)** One must be negative and one must be positive. **(b)** Either or both must be zero.
**(c)** Both must be negative or both must be positive.

**Translating for Success, p. 116**

**1.** I   **2.** C   **3.** G   **4.** L   **5.** B   **6.** O   **7.** A   **8.** F
**9.** D   **10.** M

**Exercise Set 2.5, p. 117**

**1.** −6   **3.** −13   **5.** −2   **7.** 4   **9.** −8   **11.** 2   **13.** −12
**15.** −8   **17.** Not defined   **19.** −9   **21.** −54°C   **23.** $26
**25.** 32 m below sea level   **27.** −7   **29.** −7   **31.** −334
**33.** 14   **35.** 1880   **37.** 12   **39.** 8   **41.** −86   **43.** 37
**45.** −1   **47.** −10   **49.** −67   **51.** −7988   **53.** −3000
**55.** 60   **57.** 1   **59.** 10   **61.** −2   **63.** −4
**65.** 8 thousands   **66.** 8 millions   **67.** 8 ones
**68.** 8 hundreds   **69.** 2203   **70.** 848   **71.** 37,239
**72.** 11,851   **73.** 4992 sq ft; 284 ft   **74.** $928   **75.** −2
**77.** Positive   **79.** Negative   **81.** Positive

**Summary and Review: Chapter 2, p. 120**

Concept Reinforcement

**1.** False   **2.** True   **3.** False   **4.** True

Important Concepts

**1.**   **2.** <   **3. (a)** 17; **(b)** 14
**4.** −3   **5.** −8   **6.** 14   **7.** 72   **8.** −90   **9.** 4   **10.** −4
**11.** −12

## Review Exercises

**1.** $-45; 72$    **2.** 38    **3.** 7    **4.** 0    **5.** $-2$    **6.** $<$    **7.** $>$
**8.** $>$    **9.** $<$    **10.**
**11.**    **12.** $-8$    **13.** 14    **14.** 34
**15.** 5    **16.** $-3$    **17.** $-7$    **18.** $-4$    **19.** $-5$    **20.** 4
**21.** $-14$    **22.** $-8$    **23.** 54    **24.** $-39$    **25.** $-56$
**26.** $-210$    **27.** $-7$    **28.** $-3$    **29.** 6    **30.** 41    **31.** $-62$
**32.** $-5$    **33.** 2    **34.** $-180$    **35.** 8-yd gain    **36.** $-\$130$
**37.** $10 per share    **38.** $19    **39.** C    **40.** B
**41. (a)** $-7 + (-6) + (-5) + (-4) + (-3) + (-2) +$
$(-1) + 0 + 1 + 2 + 3 + 4 + 5 + 6 + 7 + 8 = 8;$ **(b)** 0
**42.** $9 - (3 - 4) + 5 = 15$    **43.** $-5$    **44.** $-5$

### Understanding Through Discussion and Writing

**1.** If the negative integer has the greater absolute value, the answer is negative.    **2.** We know that $a + (-a) = 0$, so the opposite of $-a$ is $a$. That is, $-(-a) = a$.    **3.** Answers will vary. At 4 P.M., the temperature in Circle City was 23°F. By 11 P.M., the temperature had dropped 32°F. What was the temperature at 11 P.M.?    **4.** We know that the product of an even number of negative numbers is positive, and the product of an odd number of negative numbers is negative. Since $(-7)^8$ is equivalent to the product of eight negative numbers, it will be a positive number. Similarly, since $(-7)^{11}$ is equivalent to the product of eleven negative numbers, it will be a negative number.    **5.** Jake is expecting the multiplication to be performed before the division.    **6.** Consider $\frac{a}{b} = q$, where $a$ and $b$ are both negative integers. Then $q \cdot b = a$, so $q$ must be positive.

### Test: Chapter 2, p. 125

**1.** [2.1c] $<$    **2.** [2.1c] $>$    **3.** [2.1c] $>$    **4.** [2.1c] $<$
**5.** [2.1d] 7    **6.** [2.1d] 94    **7.** [2.1d] $-27$    **8.** [2.2b] $-23$
**9.** [2.2b] 14    **10.** [2.2b] 8
**11.** [2.1b]    **12.** [2.3a] 78
**13.** [2.2a] $-8$    **14.** [2.2a] 2    **15.** [2.3a] 10    **16.** [2.3a] $-25$
**17.** [2.2a] 15    **18.** [2.4a] $-48$    **19.** [2.4a] 24    **20.** [2.5a] $-9$
**21.** [2.5a] 9    **22.** [2.5a] $-4$    **23.** [2.5c] 109
**24.** [2.3b] 14°F higher    **25.** [2.3b] Up 15 pts
**26.** [2.5b] 16,080    **27.** [2.5b] 2°C each minute    **28.** [2.2b] D
**29.** [2.1d], [2.3a] 15    **30.** [2.3b] 2385 m
**31. (a)** [2.3a] $-4, -9, -15$; **(b)** [2.3a] $-2, -6, -10$;
**(c)** [2.3a] $-18, -24, -31$; **(d)** [2.5a] $4, -2, 1$

# CHAPTER 3

### Calculator Corner, p. 130

**1.** No    **2.** Yes    **3.** Yes    **4.** No

### Exercise Set 3.1, p. 133

**1.** No    **3.** Yes    **5.** 1, 2, 3, 6, 9, 18    **7.** 1, 2, 3, 6, 9, 18, 27, 54
**9.** 1, 2, 4    **11.** 1    **13.** 1, 2, 7, 14, 49, 98    **15.** 1, 3, 5, 15, 17, 51, 85, 255    **17.** 4, 8, 12, 16, 20, 24, 28, 32, 36, 40    **19.** 20, 40, 60, 80, 100, 120, 140, 160, 180, 200    **21.** 3, 6, 9, 12, 15, 18, 21, 24, 27, 30    **23.** 12, 24, 36, 48, 60, 72, 84, 96, 108, 120    **25.** 10, 20, 30, 40, 50, 60, 70, 80, 90, 100    **27.** 9, 18, 27, 36, 45, 54, 63, 72, 81, 90    **29.** No    **31.** Yes    **33.** Yes    **35.** No    **37.** No
**39.** Neither    **41.** Composite    **43.** Prime    **45.** Prime
**47.** $2 \cdot 2 \cdot 2$    **49.** $2 \cdot 7$    **51.** $2 \cdot 3 \cdot 7$    **53.** $5 \cdot 5$
**55.** $2 \cdot 5 \cdot 5$    **57.** $13 \cdot 13$    **59.** $2 \cdot 2 \cdot 5 \cdot 5$    **61.** $5 \cdot 7$
**63.** $2 \cdot 2 \cdot 2 \cdot 3 \cdot 3$    **65.** $7 \cdot 11$    **67.** $2 \cdot 2 \cdot 7 \cdot 103$
**69.** $3 \cdot 17$    **71.** $2 \cdot 2 \cdot 2 \cdot 2 \cdot 3 \cdot 5 \cdot 5$    **73.** $3 \cdot 7 \cdot 13$
**75.** $2 \cdot 3 \cdot 11 \cdot 17$    **77.** 26    **78.** 256    **79.** 425    **80.** 4200
**81.** 0    **82.** 22    **83.** 1    **84.** 3    **85.** $946    **86.** 201 min, or 3 hr 21 min    **87.** Row 1: 48, 90, 432, 63; row 2: 7, 2, 2, 10, 8, 6, 21, 10; row 3: 9, 18, 36, 14, 12, 11, 21; row 4: 29, 19, 42

### Exercise Set 3.2, p. 139

**1.** 46, 224, 300, 36, 45,270, 4444, 256, 8064, 21,568    **3.** 224, 300, 36, 4444, 256, 8064, 21,568    **5.** 300, 36, 45,270, 8064
**7.** 36, 45,270, 711, 8064    **9.** 324, 42, 501, 3009, 75, 2001, 402, 111,111, 1005    **11.** 55,555, 200, 75, 2345, 35, 1005    **13.** 56, 784, 200    **15.** 200    **17.** 313,332, 7624, 111,126, 876, 1110, 5128, 64,000, 9990    **19.** 313,332, 111,126, 876, 1110, 9990
**21.** 9990    **23.** 1110, 64,000, 9990    **25.** 138    **26.** 139
**27.** 874    **28.** 56    **29.** 26    **30.** 13    **31.** 234    **32.** 4003
**33.** 45 gal    **34.** 4320 min    **35.** $2 \cdot 2 \cdot 2 \cdot 3 \cdot 5 \cdot 5 \cdot 13$
**37.** $2 \cdot 2 \cdot 3 \cdot 3 \cdot 7 \cdot 11$    **39.** 95,238

### Exercise Set 3.3, p. 146

**1.** Numerator: 3; denominator: 4    **3.** Numerator: 11; denominator: 2    **5.** Numerator: 0; denominator: 7
**7.** $\frac{6}{12}$    **9.** $\frac{1}{8}$    **11.** $\frac{4}{8}$    **13.** $\frac{9}{8}$    **15.** $\frac{7}{6}$    **17.** $\frac{3}{4}$    **19.** $\frac{12}{12}$    **21.** $\frac{4}{8}$
**23.** $\frac{5}{8}$    **25.** $\frac{4}{7}$    **27.** $\frac{12}{16}$    **29.** $\frac{38}{16}$    **31. (a)** $\frac{2}{8}$; **(b)** $\frac{6}{8}$
**33. (a)** $\frac{3}{8}$; **(b)** $\frac{5}{8}$    **35. (a)** $\frac{5}{8}$; **(b)** $\frac{5}{3}$; **(c)** $\frac{3}{8}$; **(d)** $\frac{3}{5}$    **37. (a)** $\frac{1014}{100,000}$;
**(b)** $\frac{750}{100,000}$; **(c)** $\frac{812}{100,000}$; **(d)** $\frac{922}{100,000}$; **(e)** $\frac{865}{100,000}$; **(f)** $\frac{1379}{100,000}$
**39. (a)** $\frac{4}{15}$; **(b)** $\frac{4}{11}$; **(c)** $\frac{11}{15}$    **41. (a)** $\frac{16}{100}$; **(b)** $\frac{18.5}{100}$    **43.** 0    **45.** 7
**47.** 1    **49.** 1    **51.** 0    **53.** 1    **55.** 1    **57.** 0    **59.** 1
**61.** $-18$    **63.** Not defined    **65.** Not defined    **67.** 34,560
**68.** 34,600    **69.** 35,000    **70.** 30,000    **71.** 24,465 members
**72.** 71 gal    **73.** 1666    **74.** 5361    **75.** $-7$    **76.** $-15$
**77.** $\frac{1}{6}$    **79.** $\frac{2}{16}$, or $\frac{1}{8}$    **81.**    **83.**

### Exercise Set 3.4, p. 155

**1.** $\frac{4}{15}$    **3.** $\frac{70}{9}$    **5.** $-\frac{2}{15}$    **7.** $-\frac{8}{11}$    **9.** $\frac{49}{64}$    **11.** $\frac{5}{8}$    **13.** $-\frac{14}{39}$
**15.** $\frac{1}{6}$    **17.** $-\frac{6}{5}$    **19.** $\frac{3}{5}$    **21.** $\frac{21}{4}$    **23.** $\frac{9}{16}$    **25.** $-\frac{1}{1000}$    **27.** $\frac{2}{5}$
**29.** $\frac{36}{25}$    **31.** $\frac{66}{5}$    **33.** $-\frac{7}{100}$    **35.** $\frac{40}{3}$ yd    **37.** $\frac{1}{2625}$    **39.** $\frac{1}{24}$
**41.** $\frac{9}{20}$    **43.** 3001    **44.** 204 R 8    **45.** $-204$    **46.** $-700$
**47.** 8 thousands    **48.** 8 millions    **49.** 8 ones
**50.** 8 hundreds    **51.** 1    **52.** 169    **53.** 3    **54.** 48
**55.** $\frac{71,269}{180,433}$    **57.** $-\frac{56}{1125}$

### Calculator Corner, p. 160

**1.** $\frac{14}{15}$    **2.** $\frac{7}{8}$    **3.** $\frac{138}{167}$    **4.** $\frac{7}{25}$

### Exercise Set 3.5, p. 162

**1.** $\frac{5}{10}$    **3.** $\frac{20}{32}$    **5.** $-\frac{27}{30}$    **7.** $\frac{28}{32}$    **9.** $\frac{20}{48}$    **11.** $\frac{-51}{-54}$    **13.** $\frac{75}{45}$
**15.** $\frac{42}{132}$    **17.** $\frac{1}{2}$    **19.** $-\frac{3}{4}$    **21.** $\frac{1}{5}$    **23.** $-3$    **25.** $\frac{3}{4}$    **27.** $\frac{7}{8}$
**29.** $\frac{6}{5}$    **31.** $\frac{1}{3}$    **33.** $-6$    **35.** $-\frac{1}{3}$    **37.** $\frac{13}{47}$    **39.** $-\frac{45}{112}$
**41.** $=$    **43.** $\neq$    **45.** $=$    **47.** $\neq$    **49.** $=$    **51.** $\neq$    **53.** $=$
**55.** $\neq$    **57.** $\frac{7}{25}$    **59.** $\frac{1}{4}$    **61.** 526,761 yd$^2$; 3020 yd    **62.** $163
**63.** 186    **64.** $-3$    **65.** 3520    **66.** 89
**67.** No; $\frac{63}{82} \neq \frac{77}{100}$ because $63 \cdot 100 \neq 82 \cdot 77$.

### Mid-Chapter Review: Chapter 3, p. 164

**1.** True    **2.** False    **3.** False    **4.** True    **5.** $\frac{25}{25} = 1$
**6.** $\frac{0}{-9} = 0$    **7.** $\frac{-8}{1} = -8$    **8.** $\frac{6}{13} = \frac{18}{39}$
**9.** $\frac{70}{225} = \frac{2 \cdot 5 \cdot 7}{3 \cdot 3 \cdot 5 \cdot 5} = \frac{5}{5} \cdot \frac{2 \cdot 7}{3 \cdot 3 \cdot 5} = 1 \cdot \frac{14}{45} = \frac{14}{45}$
**10.** 84, 17,576, 224, 132, 594, 504, 1632
**11.** 84, 300, 132, 500, 180    **12.** 17,576, 224, 500
**13.** 84, 300, 132, 120, 1632    **14.** 300, 180, 120    **15.** Prime
**16.** Prime    **17.** Composite    **18.** Neither    **19.** 1, 2, 4, 5, 8, 10, 16, 20, 32, 40, 80, 160; $2 \cdot 2 \cdot 2 \cdot 2 \cdot 2 \cdot 5$
**20.** 1, 2, 3, 6, 37, 74, 111, 222; $2 \cdot 3 \cdot 37$
**21.** 1, 2, 7, 14, 49, 98; $2 \cdot 7 \cdot 7$

22. 1, 3, 5, 7, 9, 15, 21, 35, 45, 63, 105, 315; 3 · 3 · 5 · 7   23. $\frac{8}{24}$, or $\frac{1}{3}$   24. $\frac{8}{6}$, or $\frac{4}{3}$   25. $\frac{7}{9}$   26. $\frac{8}{45}$   27. $-\frac{40}{11}$   28. $-\frac{8}{45}$
29. $\frac{21}{32}$   30. $-\frac{10}{3}$   31. $\frac{5}{16}$   32. $-\frac{27}{28}$   33. $\frac{2}{5}$   34. $\frac{11}{3}$   35. 1
36. 0   37. $\frac{21}{29}$   38. Not defined   39. =   40. ≠
41. $\frac{60}{500}$, or $\frac{3}{25}$   42. $\frac{21}{10,000}$ mi²   43. Find the product of two prime numbers.   44. If we use the divisibility tests, it is quickly clear that none of the even-numbered years is a prime number. In addition, the divisibility tests for 5 and 3 show that 2001, 2005, 2007, 2013, 2015, and 2019 are not prime numbers. Then the years 2003, 2009, 2011, and 2017 can be divided by prime numbers to determine whether they are prime. When we do this, we find that 2003, 2011, and 2017 are prime numbers. If the divisibility tests are not used, each of the numbers from 2000 to 2020 can be divided by prime numbers to determine if it is prime.   45. It is possible to cancel only when identical *factors* appear in the numerator and the denominator of a fraction. Situations in which it is not possible to cancel include the occurrence of identical *addends* or *digits* in the numerator and the denominator.   46. No; since the only factors of a prime number are the number itself and 1, two different prime numbers cannot contain a common factor (other than 1).

### Exercise Set 3.6, p. 168

1. $\frac{1}{3}$   3. $\frac{1}{8}$   5. $-\frac{1}{10}$   7. $\frac{1}{6}$   9. $\frac{27}{10}$   11. $-\frac{14}{9}$   13. 1   15. 1
17. 1   19. 1   21. 2   23. −4   25. 9   27. −9   29. $\frac{26}{5}$
31. $\frac{98}{5}$   33. 60   35. −30   37. $\frac{1}{5}$   39. $-\frac{9}{25}$   41. $\frac{11}{40}$
43. $\frac{5}{14}$   45. $\frac{5}{8}$ in.   47. $48,280   49. 625 addresses
51. $\frac{1}{3}$ cup   53. $115,500   55. 160 mi   57. Food: $8400; housing: $10,500; clothing: $4200; savings: $3000; taxes: $8400; other expenses: $7500   59. 35   60. 85   61. 125   62. 120
63. 4989   64. 8546   65. 6498   66. 6407   67. −16
68. 11   69. −4   70. 0   71. $\frac{129}{485}$   73. $\frac{1}{12}$   75. $\frac{1}{168}$

### Translating for Success, p. 176

1. C   2. H   3. A   4. N   5. O   6. F   7. I   8. L
9. D   10. M

### Exercise Set 3.7, p. 177

1. $\frac{6}{5}$   3. $\frac{1}{6}$   5. 6   7. $-\frac{3}{10}$   9. $\frac{4}{5}$   11. $-\frac{4}{15}$   13. 4   15. 2
17. $\frac{1}{8}$   19. $\frac{3}{7}$   21. −8   23. 35   25. −1   27. $-\frac{2}{3}$   29. $\frac{9}{4}$
31. 144   33. 75   35. −2   37. $\frac{3}{5}$   39. −315   41. 960 extension cords   43. 32 pairs   45. 24 bowls   47. 16 L
49. 288 km; 108 km   51. $\frac{1}{16}$ in.   53. Associative
54. Factors   55. Prime   56. Denominator   57. Additive
58. Reciprocals   59. Whole   60. Equation   61. $\frac{9}{19}$
63. 36   65. $\frac{3}{8}$

### Summary and Review: Chapter 3, p. 180

#### Concept Reinforcement

1. True   2. False   3. True   4. True

#### Important Concepts

1. 1, 2, 4, 8, 13, 26, 52, 104   2. 2 · 2 · 2 · 13   3. 0, 1, 18
4. $\frac{56}{96}$   5. $\frac{5}{14}$   6. ≠   7. $-\frac{70}{9}$   8. $\frac{7}{10}$   9. $\frac{7}{3}$ cups

#### Review Exercises

1. 1, 2, 3, 4, 5, 6, 10, 12, 15, 20, 30, 60   2. 1, 2, 4, 8, 11, 16, 22, 44, 88, 176   3. 8, 16, 24, 32, 40, 48, 56, 64, 72, 80   4. Yes
5. No   6. Prime   7. Neither   8. Composite   9. 2 · 5 · 7
10. 2 · 3 · 5   11. 3 · 3 · 5   12. 2 · 3 · 5 · 5
13. 2 · 2 · 2 · 3 · 3 · 3 · 3   14. 2 · 3 · 5 · 5 · 5 · 7   15. 4344, 600, 93, 330, 255,555, 780, 2802, 711   16. 140, 182, 716, 2432, 4344, 600, 330, 780   17. 140, 716, 2432, 4344, 600, 780
18. 2432, 4344, 600   19. 140, 95, 475, 600, 330, 255,555, 780

20. 4344, 600, 330, 780, 2802   21. 255,555, 711
22. 140, 600, 330, 780   23. Numerator: 2; denominator: 7
24. $\frac{3}{5}$   25. $\frac{7}{6}$   26. $\frac{2}{7}$   27. (a) $\frac{3}{5}$; (b) $\frac{5}{3}$; (c) $\frac{3}{8}$   28. $\frac{2}{5}$   29. $\frac{1}{4}$
30. 1   31. 0   32. $\frac{39}{7}$   33. 18   34. $-\frac{1}{3}$   35. $-\frac{11}{23}$
36. Not defined   37. 6   38. $\frac{2}{7}$   39. $-\frac{32}{225}$   40. $\frac{15}{100} = \frac{3}{20}$; $\frac{38}{100} = \frac{19}{50}$; $\frac{23}{100} = \frac{23}{100}$; $\frac{24}{100} = \frac{6}{25}$   41. ≠   42. =   43. ≠
44. =   45. $\frac{3}{2}$   46. 56   47. $-\frac{5}{2}$   48. −24   49. $\frac{2}{3}$   50. $\frac{1}{14}$
51. $-\frac{2}{3}$   52. $\frac{1}{22}$   53. $\frac{3}{20}$   54. $\frac{10}{7}$   55. $\frac{5}{4}$   56. $-\frac{1}{3}$   57. 9
58. $-\frac{36}{47}$   59. $\frac{9}{2}$   60. −2   61. $\frac{11}{6}$   62. $-\frac{1}{4}$   63. $\frac{9}{4}$
64. 300   65. 1   66. $\frac{4}{9}$   67. $\frac{3}{10}$   68. −240   69. 9 days
70. About 22,800,000 metric tons   71. 1000 km   72. $\frac{1}{3}$ cup; 2 cups   73. $15   74. 60 bags   75. D   76. B
77. $a = 11,176$; $b = 9887$   78. 13, 11, 101, 37

### Understanding Through Discussion and Writing

1. The student is probably multiplying the divisor by the reciprocal of the dividend rather than multiplying the dividend by the reciprocal of the divisor.
2. $9432 = 9 \cdot 1000 + 4 \cdot 100 + 3 \cdot 10 + 2 \cdot 1 = 9(999 + 1) + 4(99 + 1) + 3(9 + 1) + 2 \cdot 1 = 9 \cdot 999 + 9 \cdot 1 + 4 \cdot 99 + 4 \cdot 1 + 3 \cdot 9 + 3 \cdot 1 + 2 \cdot 1$. Since 999, 99, and 9 are each a multiple of 9, $9 \cdot 999$, $4 \cdot 99$, and $3 \cdot 9$ are multiples of 9. This leaves $9 \cdot 1 + 4 \cdot 1 + 3 \cdot 1 + 2 \cdot 1$, or $9 + 4 + 3 + 2$. If $9 + 4 + 3 + 2$, the sum of the digits, is divisible by 9, then 9432 is divisible by 9.   3. Taking $\frac{1}{2}$ of a number is equivalent to multiplying the number by $\frac{1}{2}$. Dividing by $\frac{1}{2}$ is equivalent to multiplying by the reciprocal of $\frac{1}{2}$, or 2. Thus taking $\frac{1}{2}$ of a number is not the same as dividing by $\frac{1}{2}$.
4. We first consider an object, and take $\frac{4}{7}$ of it. We divide the object into 7 equal parts and take 4 of them, as shown by the shading below.

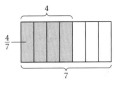

Next, we take $\frac{2}{3}$ of the shaded area above. We divide it into 3 equal parts and take two of them, as shown below.

The entire object has been divided into 21 parts, 8 of which have been shaded twice. Thus, $\frac{2}{3} \cdot \frac{4}{7} = \frac{8}{21}$.   5. Since $\frac{1}{7}$ is a smaller number than $\frac{2}{3}$, there are more $\frac{1}{7}$'s in 5 than $\frac{2}{3}$'s. Thus, $5 \div \frac{1}{7}$ is a larger number than $5 \div \frac{2}{3}$.   6. No; in order to simplify a fraction, we must be able to remove a factor of the type $\frac{n}{n}$, $n \neq 0$, where $n$ is a factor that the numerator and the denominator have in common.

### Test: Chapter 3, p. 185

1. [3.1a] 1, 2, 3, 4, 5, 6, 10, 12, 15, 20, 25, 30, 50, 60, 75, 100, 150, 300   2. [3.1c] Prime   3. [3.1c] Composite
4. [3.1d] 2 · 3 · 3   5. [3.1d] 2 · 2 · 3 · 5   6. [3.2a] Yes
7. [3.2a] No   8. [3.2a] No   9. [3.2a] Yes
10. [3.3a] Numerator: 4; denominator: 5   11. [3.3a] $\frac{3}{4}$
12. [3.3a] $\frac{3}{7}$   13. [3.3a] (a) $\frac{371}{555}$; (b) $\frac{184}{555}$   14. [3.3b] 26
15. [3.3b] 1   16. [3.3b] 0   17. [3.5b] $-\frac{1}{2}$   18. [3.5b] 6
19. [3.3b] Not defined   20. [3.3b] Not defined   21. [3.5b] $-\frac{2}{3}$
22. [3.5c] =   23. [3.5c] ≠   24. [3.6a] 32   25. [3.6a] $-\frac{3}{2}$
26. [3.6a] $\frac{5}{2}$   27. [3.6a] $\frac{2}{9}$   28. [3.7a] $\frac{8}{5}$   29. [3.7a] −4

**30.** [3.7a] $\frac{1}{18}$  **31.** [3.7b] $\frac{8}{5}$  **32.** [3.7b] $-18$  **33.** [3.7b] $\frac{18}{7}$
**34.** [3.7c] $-64$  **35.** [3.7c] $\frac{7}{4}$  **36.** [3.6b] 4375 students
**37.** [3.7d] $\frac{3}{40}$ m  **38.** [3.7d] 5 qt  **39.** [3.6b] $\frac{3}{4}$ in.
**40.** [3.3a] C  **41.** [3.6b] $\frac{7}{48}$ acre  **42.** [3.6a], [3.7b] $-\frac{7}{960}$

## Cumulative Review: Chapters 1-3, p. 187

**1.** [1.1c] Seven million, four hundred fifty-three thousand, sixty-two  **2.** [1.1a] 4 ten thousands  **3.** [3.3b] 5
**4.** [3.3b] 1  **5.** [3.5b] $-\frac{4}{3}$  **6.** [3.3b] Not defined  **7.** [3.7a] $\frac{1}{8}$
**8.** [3.7a] $-\frac{2}{3}$  **9.** [1.2a] 5387  **10.** [2.2a] $-12$  **11.** [1.3a] 1384
**12.** [2.3a] $-16$  **13.** [1.4a] 16,452  **14.** [2.4a] $-96$
**15.** [3.6a] $-\frac{25}{2}$  **16.** [3.6a] $\frac{2}{3}$  **17.** [1.5a] 451 R 22  **18.** [2.5a] 8
**19.** [3.7b] $-\frac{1}{49}$  **20.** [3.7b] $\frac{4}{3}$  **21.** [1.9c] 0  **22.** [1.9c] 3
**23.** [2.5c] $-14$  **24.** [2.5c] $-4$  **25.** [1.9c] 86
**26.** [1.6a] 166,000  **27.** [1.9b] 81  **28.** [1.9b] 125
**29.** [1.9b] 16  **30.** [2.1c] $<$  **31.** [1.6c] $>$  **32.** [2.1d] 33
**33.** [2.1d] 86  **34.** [2.1d] 0  **35.** [2.2b] $-29$  **36.** [2.2b] 144
**37.** [2.2b] $-7$  **38.** [2.1b] ←+—+—+—+—+—+—+—+—+—+—+→
$-2\ -1\ \ 0\ \ 1\ \ 2\ \ 3\ \ 4\ \ 5\ \ 6\ \ 7\ \ 8\ \ 9\ \ 10$
**39.** [3.1c] Prime  **40.** [3.1c] Neither  **41.** [3.1c] Composite
**42.** [3.1d] $2 \cdot 2 \cdot 3 \cdot 3$  **43.** [3.1d] $3 \cdot 3 \cdot 3 \cdot 5$
**44.** [3.1d] $2 \cdot 3 \cdot 3 \cdot 3 \cdot 5 \cdot 5$  **45.** [3.2a] Divisible by 2, 3, 5, 6, 10; not divisible by 4, 8, 9  **46.** [3.5c] $\neq$  **47.** [3.5c] $=$
**48.** [1.7b] 44  **49.** [3.7c] $-\frac{3}{5}$  **50.** [1.8a] 2062
**51.** [1.8a] \$36  **52.** [2.3b], [2.5b] $-17°C$
**53.** [3.6b] 14 students  **54.** **(a)** [1.2a], [1.4a] 5, 20;
**(b)** [2.2a], [2.4a] $-4$, $-25$; **(c)** [1.2a], [1.4a], [2.2a], [2.4a] 10, 10; $-10$, $-10$

# CHAPTER 4

## Exercise Set 4.1, p. 195

**1.** 4  **3.** 50  **5.** 40  **7.** 54  **9.** 150  **11.** 120  **13.** 72
**15.** 420  **17.** 144  **19.** 288  **21.** 30  **23.** 90  **25.** 72
**27.** 60  **29.** 36  **31.** 900  **33.** 48  **35.** 50  **37.** 143
**39.** 420  **41.** 378  **43.** 810  **45.** 2160  **47.** 9828
**49.** 6000  **51.** Every 60 yr  **53.** Every 420 yr
**55.** 659 tornadoes  **56.** 940,000,000  **57.** $-\frac{8}{7}$  **58.** 33,135
**59.** $\frac{2}{3}$  **60.** 77,699  **61.** 5 in. by 24 in.

## Exercise Set 4.2, p. 201

**1.** 1  **3.** $\frac{3}{4}$  **5.** $-\frac{1}{6}$  **7.** $\frac{7}{24}$  **9.** $-\frac{1}{10}$  **11.** $\frac{19}{24}$  **13.** $\frac{9}{10}$
**15.** $\frac{1}{18}$  **17.** $\frac{31}{100}$  **19.** $\frac{41}{60}$  **21.** $\frac{9}{100}$  **23.** $\frac{7}{8}$  **25.** $\frac{13}{24}$  **27.** $\frac{17}{24}$
**29.** $\frac{3}{4}$  **31.** $\frac{437}{500}$  **33.** $\frac{13}{40}$  **35.** $\frac{391}{144}$  **37.** $\frac{37}{12}$ mi  **39.** $\frac{13}{16}$ lb
**41.** $\frac{4}{5}$ qt; $\frac{8}{5}$ qt; $\frac{2}{5}$ qt  **43.** $\frac{27}{32}$  **45.** $\frac{51}{32}$ in.  **47.** 210,528
**48.** 4,194,000  **49.** $-56$  **50.** 108  **51.** 544,683 votes
**52.** 5 votes  **53.** 84 votes  **54.** 510,314 votes
**55.** 21,484,549 votes  **56.** 33,128,217 votes
**57.** 6,140,000 lb  **58.** $\frac{3}{64}$ acre  **59.** $\frac{4}{15}$; \$320

## Translating for Success, p. 208

**1.** J  **2.** E  **3.** D  **4.** B  **5.** I  **6.** N  **7.** A  **8.** C
**9.** L  **10.** F

## Exercise Set 4.3, p. 209

**1.** $\frac{2}{3}$  **3.** $\frac{3}{4}$  **5.** $-\frac{5}{8}$  **7.** $\frac{1}{24}$  **9.** $-\frac{1}{2}$  **11.** $\frac{9}{14}$  **13.** $\frac{3}{5}$
**15.** $-\frac{7}{10}$  **17.** $\frac{17}{60}$  **19.** $\frac{53}{100}$  **21.** $\frac{26}{75}$  **23.** $\frac{9}{100}$  **25.** $-\frac{19}{24}$
**27.** $\frac{1}{10}$  **29.** $-\frac{1}{24}$  **31.** $\frac{13}{16}$  **33.** $-\frac{31}{75}$  **35.** $\frac{13}{75}$  **37.** $<$
**39.** $>$  **41.** $<$  **43.** $<$  **45.** $>$  **47.** $>$  **49.** $<$  **51.** $\frac{1}{15}$
**53.** $-\frac{22}{15}$  **55.** $\frac{1}{2}$  **57.** $\frac{5}{12}$ hr  **59.** $\frac{1}{32}$ in.  **61.** $\frac{1}{4}$ tub
**63.** $\frac{11}{20}$ lb  **65.** 1  **66.** Not defined  **67.** Not defined
**68.** 4  **69.** $-\frac{4}{21}$  **70.** $\frac{3}{2}$  **71.** 21  **72.** $\frac{1}{32}$
**73.** About 12.8 billion, or 12,800,000,000, crayons
**74.** 9 cups  **75.** $\frac{14}{3553}$  **77.** $\frac{21}{40}$ km  **79.** $\frac{19}{24}$  **81.** $-\frac{17}{144}$

**83.** $>$  **85.** *Day 1:* Cut off $\frac{1}{7}$ of bar and pay him. *Day 2:* Cut off $\frac{2}{7}$ of bar. Trade him for the $\frac{1}{7}$. *Day 3:* Give him back the $\frac{1}{7}$. *Day 4:* Trade him the $\frac{4}{7}$ for his $\frac{3}{7}$. *Day 5:* Give him the $\frac{1}{7}$ again. *Day 6:* Trade him the $\frac{2}{7}$ for the $\frac{1}{7}$. *Day 7:* Give him the $\frac{1}{7}$ again. This assumes that he does not spend parts of the gold bar immediately.

## Exercise Set 4.4, p. 215

**1.** $\frac{65}{2}, \frac{125}{6}, \frac{47}{4}$  **3.** $4\frac{1}{4}, 3\frac{1}{3}, 1\frac{1}{8}$  **5.** $\frac{17}{4}$  **7.** $\frac{13}{4}$  **9.** $-\frac{81}{8}$  **11.** $\frac{51}{10}$
**13.** $\frac{103}{5}$  **15.** $-\frac{59}{6}$  **17.** $\frac{73}{10}$  **19.** $\frac{13}{8}$  **21.** $-\frac{51}{4}$  **23.** $-\frac{43}{10}$
**25.** $\frac{203}{100}$  **27.** $\frac{200}{3}$  **29.** $-\frac{279}{50}$  **31.** $\frac{1621}{16}$  **33.** $3\frac{3}{5}$  **35.** $4\frac{2}{3}$
**37.** $-4\frac{1}{2}$  **39.** $5\frac{7}{10}$  **41.** $-7\frac{4}{7}$  **43.** $7\frac{1}{2}$  **45.** $11\frac{1}{2}$  **47.** $-1\frac{1}{2}$
**49.** $7\frac{57}{100}$  **51.** $-43\frac{1}{8}$  **53.** $108\frac{5}{8}$  **55.** $618\frac{1}{5}$  **57.** $40\frac{4}{7}$
**59.** $55\frac{1}{51}$  **61.** $2292\frac{23}{35}$  **63.** 45,800  **64.** 45,770  **65.** $\frac{8}{15}$
**66.** $\frac{21}{25}$  **67.** $-\frac{16}{27}$  **68.** $-\frac{583}{669}$  **69.** 18  **70.** $-\frac{5}{2}$  **71.** $\frac{1}{4}$
**72.** $\frac{2}{5}$  **73.** $-24$  **74.** 49  **75.** $\frac{2560}{3}$  **76.** $-\frac{4}{3}$  **77.** $237\frac{19}{541}$
**79.** $8\frac{2}{3}$  **81.** $52\frac{2}{7}$

## Mid-Chapter Review: Chapter 4, p. 218

**1.** True  **2.** True  **3.** False  **4.** False
**5.**
$$\frac{11}{42} - \frac{3}{35} = \frac{11}{2 \cdot 3 \cdot 7} - \frac{3}{5 \cdot 7}$$
$$= \frac{11}{2 \cdot 3 \cdot 7} \cdot \left(\frac{5}{5}\right) - \frac{3}{5 \cdot 7} \cdot \left(\frac{2 \cdot 3}{2 \cdot 3}\right)$$
$$= \frac{11 \cdot 5}{2 \cdot 3 \cdot 7 \cdot 5} - \frac{3 \cdot 2 \cdot 3}{5 \cdot 7 \cdot 2 \cdot 3}$$
$$= \frac{55}{2 \cdot 3 \cdot 5 \cdot 7} - \frac{18}{2 \cdot 3 \cdot 5 \cdot 7}$$
$$= \frac{55 - 18}{2 \cdot 3 \cdot 5 \cdot 7} = \frac{37}{210}$$

**6.**
$$x + \frac{1}{8} = \frac{2}{3}$$
$$x + \frac{1}{8} - \frac{1}{8} = \frac{2}{3} - \frac{1}{8}$$
$$x + 0 = \frac{2}{3} \cdot \frac{8}{8} - \frac{1}{8} \cdot \frac{3}{3}$$
$$x = \frac{16}{24} - \frac{3}{24}$$
$$x = \frac{13}{24}$$

**7.**
45 and 50 ⟶ 120
50 and 80 ⟶ 720
30 and 24 ⟶ 400
18, 24, and 80 ⟶ 450
30, 45, and 50

**8.** $\frac{16}{45}$  **9.** $\frac{25}{12}$  **10.** $-\frac{1}{18}$  **11.** $-\frac{19}{90}$  **12.** $\frac{7}{240}$  **13.** $\frac{156}{119}$
**14.** $\frac{79}{720}$  **15.** $-\frac{6}{91}$  **16.** $\frac{22}{45}$ mi  **17.** $\frac{101}{20}$ hr  **18.** $\frac{1}{15}, \frac{2}{7}, \frac{3}{10}, \frac{4}{9}$
**19.** $\frac{13}{80}$  **20.** $17\frac{8}{15}$  **21.** C  **22.** C  **23.** No; if one number is a multiple of the other, for example, the LCM is the larger of the numbers.  **24.** We multiply by 1, using the notation $n/n$, to express each fraction in terms of the least common denominator.
**25.** Write $\frac{8}{5}$ as $\frac{16}{10}$ and $\frac{8}{2}$ as $\frac{40}{10}$ and since taking 40 tenths away from 16 tenths would give a result less than 0, it cannot possibly be $\frac{8}{3}$. You could also find the sum $\frac{8}{3} + \frac{8}{2}$ and show that it is not $\frac{8}{5}$.
**26.** No; $2\frac{1}{3} = \frac{7}{3}$ but $2 \cdot \frac{1}{3} = \frac{2}{3}$.

## Exercise Set 4.5, p. 225

**1.** $28\frac{3}{4}$  **3.** $185\frac{7}{8}$  **5.** $6\frac{1}{2}$  **7.** $2\frac{11}{12}$  **9.** $14\frac{7}{12}$  **11.** $12\frac{1}{10}$
**13.** $16\frac{5}{24}$  **15.** $21\frac{1}{2}$  **17.** $27\frac{7}{8}$  **19.** $27\frac{13}{24}$  **21.** $1\frac{3}{5}$
**23.** $4\frac{1}{10}$  **25.** $21\frac{17}{24}$  **27.** $12\frac{1}{4}$  **29.** $15\frac{3}{8}$  **31.** $7\frac{5}{12}$
**33.** $13\frac{3}{8}$  **35.** $11\frac{5}{18}$  **37.** $14\frac{13}{24}$ flats  **39.** $7\frac{5}{16}$ lb  **41.** $6\frac{5}{12}$ in.
**43.** $20\frac{1}{12}$ yd  **45.** $5\frac{7}{8}$ in.  **47.** $95\frac{1}{2}$ mi  **49.** $134\frac{1}{4}''$
**51.** $20\frac{1}{8}$ in.  **53.** $78\frac{1}{12}$ in.  **55.** $3\frac{4}{5}$ hr  **57.** $28\frac{3}{4}$ yd
**59.** $7\frac{3}{8}$ ft  **61.** $5\frac{3}{8}$ yd  **63.** $4\frac{5}{6}$ ft $\times 10\frac{7}{12}$ ft  **65.** 16 packages
**66.** 286 cartons; 2 oz left over  **67.** Yes  **68.** No  **69.** No
**70.** Yes  **71.** No  **72.** Yes  **73.** Yes  **74.** Yes  **75.** $\frac{10}{13}$
**76.** $-\frac{1}{10}$  **77.** $-8\frac{5}{8}$  **79.** $-11\frac{1}{24}$

## Calculator Corner, p. 234

**1.** $\frac{7}{12}$  **2.** $\frac{11}{10}$  **3.** $\frac{35}{16}$  **4.** $\frac{3}{10}$  **5.** $10\frac{2}{15}$  **6.** $1\frac{1}{28}$  **7.** $10\frac{11}{15}$
**8.** $2\frac{91}{115}$

## Translating for Success, p. 235

**1.** O **2.** K **3.** F **4.** D **5.** H **6.** G **7.** L **8.** E
**9.** M **10.** J

## Exercise Set 4.6, p. 236

**1.** $22\frac{2}{3}$ **3.** $-2\frac{5}{12}$ **5.** $8\frac{1}{6}$ **7.** $9\frac{31}{40}$ **9.** $-24\frac{91}{100}$ **11.** $95\frac{2}{3}$
**13.** $6\frac{1}{4}$ **15.** $1\frac{1}{5}$ **17.** $-3\frac{9}{16}$ **19.** $1\frac{1}{8}$ **21.** $1\frac{8}{43}$ **23.** $\frac{9}{40}$
**25.** 45,000 beagles **27.** About 690,000 **29.** $62\frac{1}{2}$ ft$^2$
**31.** $13\frac{1}{3}$ tsp **33.** $343\frac{3}{4}$ lb **35.** 68°F **37.** About 1,800,750
**39.** About 301,360,000 **41.** $5\frac{1}{2}$ cups of flour, $2\frac{2}{3}$ cups of sugar
**43.** 15 mpg **45.** 400 cu ft **47.** $16\frac{1}{2}$ servings **49.** $35\frac{115}{256}$ sq in.
**51.** $59,538\frac{1}{8}$ sq ft **53.** Divisor; quotient; dividend
**54.** Common **55.** Composite **56.** Divisible; divisible
**57.** Multiplications; divisions; additions; subtractions
**58.** Addends **59.** Numerator **60.** Reciprocal **61.** $360\frac{60}{473}$
**63.** $-2\frac{25}{64}$ **65.** $\frac{4}{9}$ **67.** $1\frac{4}{5}$

## Exercise Set 4.7, p. 244

**1.** $\frac{1}{24}$ **3.** $\frac{2}{5}$ **5.** $-\frac{4}{7}$ **7.** $\frac{59}{30}$, or $1\frac{29}{30}$ **9.** $\frac{3}{20}$ **11.** $\frac{211}{8}$, or $26\frac{3}{8}$
**13.** $\frac{7}{16}$ **15.** $\frac{1}{36}$ **17.** $-\frac{5}{8}$ **19.** $\frac{37}{48}$ **21.** $\frac{25}{72}$ **23.** $\frac{103}{16}$, or $6\frac{7}{16}$
**25.** $16\frac{7}{96}$ mi **27.** $9\frac{19}{40}$ lb **29.** 0 **31.** 1 **33.** $\frac{1}{2}$ **35.** $\frac{1}{2}$
**37.** 0 **39.** 1 **41.** 3 **43.** 13 **45.** 2 **47.** $1\frac{1}{2}$
**49.** $\frac{1}{2}$ **51.** $271\frac{1}{2}$ **53.** 3 **55.** 100 **57.** $29\frac{1}{2}$ **59.** $\frac{8}{3}$
**60.** $\frac{3}{8}$ **61.** Prime: 5, 7, 23, 43; composite; 9, 14; neither: 1
**62.** 59 R 77 **63.** 16 people **64.** 43 mg **65.** $a = 2, b = 8$
**67.** The largest is $\frac{4}{3} + \frac{5}{2} = \frac{23}{6}$.

## Summary and Review: Chapter 4, p. 247

### Concept Reinforcement

**1.** True **2.** True **3.** False **4.** True

### Important Concepts

**1.** 156 **2.** $\frac{28}{45}$ **3.** $\frac{4}{35}$ **4.** < **5.** $-\frac{103}{99}$ **6.** $\frac{26}{3}$ **7.** $-7\frac{5}{6}$
**8.** $7\frac{27}{28}$ **9.** $-14\frac{14}{25}$ **10.** About 4,500,000 **11.** $\frac{9}{2}$, or $4\frac{1}{2}$
**12.** $4\frac{1}{2}$

### Review Exercises

**1.** 36 **2.** 90 **3.** 30 **4.** 1404 **5.** $\frac{63}{40}$ **6.** $-\frac{11}{48}$ **7.** $\frac{25}{12}$
**8.** $\frac{891}{1000}$ **9.** $\frac{1}{3}$ **10.** $\frac{1}{8}$ **11.** $-\frac{5}{27}$ **12.** $-\frac{19}{18}$ **13.** > **14.** >
**15.** $\frac{19}{40}$ **16.** $-\frac{7}{5}$ **17.** $\frac{15}{2}$ **18.** $\frac{67}{8}$ **19.** $\frac{13}{9}$ **20.** $-\frac{75}{7}$
**21.** $2\frac{1}{3}$ **22.** $6\frac{3}{4}$ **23.** $-12\frac{3}{5}$ **24.** $3\frac{1}{2}$ **25.** $877\frac{1}{3}$ **26.** $456\frac{5}{23}$
**27.** $10\frac{2}{5}$ **28.** $11\frac{11}{15}$ **29.** $10\frac{2}{3}$ **30.** $8\frac{1}{4}$ **31.** $7\frac{7}{9}$
**32.** $4\frac{11}{15}$ **33.** $4\frac{3}{20}$ **34.** $13\frac{3}{8}$ **35.** 16 **36.** $-3\frac{1}{2}$ **37.** $2\frac{21}{50}$
**38.** $-6$ **39.** $-12$ **40.** $1\frac{1}{17}$ **41.** $\frac{1}{8}$ **42.** $\frac{9}{10}$ **43.** $4\frac{1}{4}$ yd
**44.** $177\frac{3}{4}$ in$^2$ **45.** $50\frac{1}{4}$ in$^2$ **46.** $1\frac{73}{100}$ in. **47.** 24 lb
**48.** $850 **49.** $8\frac{3}{8}$ cups **50.** $13\frac{1}{4}$ in. $\times$ $13\frac{1}{4}$ in.:
perimeter = 53 in., area = $175\frac{9}{16}$ sq in.; $13\frac{1}{4}$ in. $\times$ $3\frac{1}{4}$ in.:
perimeter = 33 in., area = $43\frac{1}{16}$ sq in. **51.** $63\frac{2}{3}$ pies; $19\frac{1}{3}$ pies
**52.** 1 **53.** $\frac{7}{40}$ **54.** 3 **55.** $\frac{77}{240}$ **56.** $\frac{1}{2}$ **57.** 0 **58.** 1
**59.** 7 **60.** 10 **61.** $5\frac{1}{2}$ **62.** 0 **63.** $2\frac{1}{2}$ **64.** $28\frac{1}{2}$ **65.** A
**66.** D **67.** 12 min **68.** $\frac{6}{3} + \frac{5}{4} = 3\frac{1}{4}$

### Understanding Through Discussion and Writing

**1.** No; if the sum of the fractional parts of the mixed numerals is $n/n$, then the sum of the mixed numerals is an integer. For example, $1\frac{1}{5} + 6\frac{4}{5} = 7\frac{5}{5} = 8$. **2.** Answers may vary. A wheel makes $33\frac{1}{3}$ revolutions per minute. It rotates for $4\frac{1}{2}$ min. How many revolutions does it make? **3.** The student is multiplying the whole numbers to get the whole-number portion of the answer and multiplying fractions to get the fraction part of the answer. The student should have converted each mixed numeral to fraction notation, multiplied, simplified, and then converted back to a mixed numeral. The correct answer is $4\frac{6}{7}$.

**4.** It might be necessary to find the least common denominator before adding or subtracting. The least common denominator is the least common multiple of the denominators. **5.** Answers may vary. Suppose that a room has dimensions $15\frac{3}{4}$ ft by $28\frac{5}{8}$ ft. The equation $2 \cdot 15\frac{3}{4} + 2 \cdot 28\frac{5}{8} = 88\frac{3}{4}$ gives the perimeter of the room, in feet. **6.** Note that $5 \cdot 3\frac{2}{7} = 5\left(3 + \frac{2}{7}\right) = 5 \cdot 3 + 5 \cdot \frac{2}{7}$. The products $5 \cdot 3$ and $5 \cdot \frac{2}{7}$ should be added rather than multiplied together. The student could also have converted $3\frac{2}{7}$ to fraction notation, multiplied, simplified, and converted back to a mixed numeral. The correct answer is $16\frac{3}{7}$.

## Test: Chapter 4, p. 253

**1.** [4.1a] 48 **2.** [4.1a] 600 **3.** [4.2a] 3 **4.** [4.2a] $-\frac{5}{24}$
**5.** [4.2a] $\frac{921}{1000}$ **6.** [4.3a] $\frac{1}{3}$ **7.** [4.3a] $\frac{1}{12}$ **8.** [4.3a] $-\frac{31}{40}$
**9.** [4.3c] $\frac{15}{4}$ **10.** [4.3c] $\frac{1}{4}$ **11.** [4.3b] > **12.** [4.4a] $\frac{7}{2}$
**13.** [4.4a] $-\frac{79}{8}$ **14.** [4.4a] $4\frac{1}{2}$ **15.** [4.4a] $-8\frac{2}{9}$
**16.** [4.4b] $162\frac{7}{11}$ **17.** [4.5a] $14\frac{1}{5}$ **18.** [4.5a] $14\frac{5}{6}$
**19.** [4.5b] $4\frac{7}{24}$ **20.** [4.5b] $6\frac{1}{6}$ **21.** [4.6a] 39 **22.** [4.6a] $-4\frac{1}{2}$
**23.** [4.6b] 2 **24.** [4.6b] $-\frac{1}{36}$ **25.** [4.6c] About 105 kg
**26.** [4.6c] 80 books **27.** [4.5c] (a) 3 in.; (b) $4\frac{1}{2}$ in.
**28.** [4.3d] $\frac{1}{16}$ in. **29.** [4.7a] $6\frac{11}{36}$ ft **30.** [4.7a] $3\frac{1}{2}$
**31.** [4.7a] $\frac{3}{4}$ **32.** [4.7b] 0 **33.** [4.7b] 1 **34.** [4.7b] 16
**35.** [4.7b] $1214\frac{1}{2}$ **36.** [4.1a] D **37.** [4.1a] (a) 24, 48, 72; (b) 24
**38.** [4.3b], [4.5c] Rebecca walks $\frac{17}{56}$ mi farther.

## Cumulative Review: Chapters 1-4, p. 255

**1.** (a) [4.5c] $14\frac{13}{24}$ mi; (b) [4.7a] $4\frac{61}{72}$ mi **2.** [1.8a] 31 people
**3.** [3.6b] $\frac{2}{5}$ tsp; 4 tsp **4.** [4.6c] 16 pieces **5.** [1.8a] $108
**6.** [4.2b] $\frac{33}{20}$ mi **7.** [3.3a] $\frac{5}{6}$ **8.** [3.3a] $\frac{4}{3}$ **9.** [1.2a] 8982
**10.** [1.3a] 4518 **11.** [2.2a] $-15$ **12.** [4.2a] $\frac{5}{12}$
**13.** [2.3a] $-14$ **14.** [4.3a] $-\frac{13}{12}$ **15.** [4.5a] $8\frac{1}{4}$ **16.** [4.5b] $1\frac{1}{6}$
**17.** [2.4a] $-75$ **18.** [3.6a] $\frac{3}{2}$ **19.** [3.6a] 15 **20.** [4.6a] $7\frac{1}{3}$
**21.** [1.5a] 715 **22.** [1.5a] 56 R 11 **23.** [4.4b] $56\frac{11}{45}$
**24.** [1.1a] 5 **25.** (a) [4.6c] $142\frac{1}{4}$ ft$^2$; (b) [4.5c] 54 ft
**26.** [1.6a] 38,500 **27.** [4.1a] 72 **28.** [4.7a] $\frac{1377}{100}$, or $13\frac{77}{100}$
**29.** [4.3b] > **30.** [3.5c] = **31.** [4.3b] < **32.** [4.7b] 1
**33.** [4.7b] $\frac{1}{2}$ **34.** [4.7b] 0 **35.** [3.5b] $\frac{4}{5}$ **36.** [3.3b] 0
**37.** [3.5b] 32 **38.** [4.4a] $\frac{37}{9}$ **39.** [4.4a] $5\frac{2}{3}$ **40.** [1.7b] 93
**41.** [4.3c] $\frac{5}{9}$ **42.** [3.7c] $-\frac{12}{7}$ **43.** [1.7b] 905
**44.** [3.1a, c, d], [3.2a]
Factors of 68: 1, 2, 4, 17, 34, 68
Factorization of 68: $2 \cdot 2 \cdot 17$, or $2 \cdot 34$
Prime factorization of 68: $2 \cdot 2 \cdot 17$
Numbers divisible by 6: 12, 54, 72, 300
Numbers divisible by 8: 8, 16, 24, 32, 40, 48, 64, 864
Numbers divisible by 5: 70, 95, 215
Prime numbers: 2, 3, 17, 19, 23, 31, 47, 101
**45.** [3.1c] 2003

# CHAPTER 5

## Exercise Set 5.1, p. 265

**1.** Four hundred eighty-six and thirty-four hundredths
**3.** One hundred forty-six thousandths **5.** Two hundred forty-nine and eighty-nine hundredths **7.** Three and seven hundred eighty-five thousandths **9.** Negative thirty-four and eight hundred ninety-one thousandths **11.** $\frac{83}{10}$ **13.** $\frac{356}{100}$
**15.** $\frac{4603}{100}$ **17.** $-\frac{13}{100,000}$ **19.** $-\frac{10,008}{10,000}$ **21.** $\frac{20,003}{1000}$ **23.** 0.8
**25.** 8.89 **27.** $-3.798$ **29.** 0.0078 **31.** 0.00019
**33.** $-0.376193$ **35.** 99.44 **37.** 3.798 **39.** $-2.1739$
**41.** 8.953073 **43.** 0.58 **45.** 0.91 **47.** $-0.0009$
**49.** 235.07 **51.** $\frac{4}{100}$ **53.** $-0.432$ **55.** 0.1 **57.** $-0.5$

**59.** 2.7  **61.** −123.7  **63.** 0.89  **65.** 0.67  **67.** −1.00
**69.** −0.09  **71.** 0.325  **73.** −17.002  **75.** 10.101
**77.** −9.999  **79.** 800  **81.** 809.473  **83.** 809  **85.** 34.5439
**87.** 34.54  **89.** 35  **91.** 6170  **92.** 6200  **93.** 6000
**94.** $2 \cdot 2 \cdot 2 \cdot 2 \cdot 5 \cdot 5 \cdot 5$, or $2^4 \cdot 5^3$
**95.** $2 \cdot 3 \cdot 3 \cdot 5 \cdot 17$, or $2 \cdot 3^2 \cdot 5 \cdot 17$  **96.** $2 \cdot 7 \cdot 11 \cdot 13$
**97.** $2 \cdot 2 \cdot 2 \cdot 7 \cdot 7 \cdot 11$, or $2^3 \cdot 7^2 \cdot 11$  **99.** −2.109, −2.108,
−2.1, −2.0302, −2.018, −2.0119, −2.000001  **101.** 6.78346
**103.** 0.03030

### Calculator Corner, p. 270

**1.** 317.645  **2.** 506.553  **3.** 17.15  **4.** 49.08  **5.** 4.4
**6.** 33.83  **7.** 454.74  **8.** 0.99

### Exercise Set 5.2, p. 274

**1.** 334.37  **3.** 1576.215  **5.** 132.56  **7.** 50.0248
**9.** 40.007  **11.** 977.955  **13.** 771.967  **15.** 8754.8221
**17.** 49.02  **19.** 85.921  **21.** 2.4975  **23.** 3.397  **25.** 8.85
**27.** 3.37  **29.** 1.045  **31.** 3.703  **33.** 0.9902  **35.** 99.66
**37.** 4.88  **39.** 0.994  **41.** 17.802  **43.** 51.13  **45.** 32.7386
**47.** 4.0622  **49.** −3.29  **51.** −2.5  **53.** −7.2  **55.** 3.379
**57.** −16.6  **59.** 2.5  **61.** −3.519  **63.** 9.601  **65.** 75.5
**67.** 3.8  **69.** −10.292  **71.** −8.8  **73.** 11.65  **75.** 384.68
**77.** 582.97  **79.** −20,033.3
**81.** The "Balance forward" column should read:
   $ 9704.56
   9677.12
   10,677.12
   10,553.17
   10,429.15
   10,416.72
   12,916.72
   12,778.94
   12,797.82
   9997.82
**83.** 35,000  **84.** 34,000  **85.** $\frac{1}{6}$  **86.** $\frac{34}{45}$  **87.** $\frac{8}{15}$  **88.** $-\frac{31}{12}$
**89.** $16\frac{1}{2}$ servings  **90.** $60\frac{1}{5}$ mi  **91.** 345.8

### Calculator Corner, p. 279

**1.** 48.6  **2.** 6930.5  **3.** 142.803  **4.** 0.5076  **5.** 7916.4
**6.** 20.4153

### Exercise Set 5.3, p. 283

**1.** 60.2  **3.** 6.72  **5.** 0.252  **7.** 0.522  **9.** 237.6
**11.** −583,686.852  **13.** −780  **15.** 8.923  **17.** 0.09768
**19.** −0.782  **21.** 521.6  **23.** 3.2472  **25.** −897.6
**27.** 322.07  **29.** 55.68  **31.** 3487.5  **33.** 50.0004
**35.** 114.42902  **37.** 13.284  **39.** 90.72  **41.** 0.0028728
**43.** 0.72523  **45.** −1.872115  **47.** 99.84  **49.** 45,678
**51.** 2888¢  **53.** 66¢  **55.** $0.34  **57.** $34.45
**59.** 47,300,000,000  **61.** 9,300,000  **63.** 23,400,000,000
**65.** $11\frac{1}{5}$  **66.** $-\frac{35}{72}$  **67.** $2\frac{7}{15}$  **68.** $7\frac{2}{15}$  **69.** 342  **70.** 87
**71.** 4566  **72.** 1257  **73.** −7  **74.** −14
**75.** $10^{21} = 1$ sextillion  **77.** $10^{24} = 1$ septillion

### Calculator Corner, p. 287

**1.** 14.3  **2.** 2.56  **3.** 0.064  **4.** 75.8

### Exercise Set 5.4, p. 293

**1.** 2.99  **3.** 23.78  **5.** 7.48  **7.** 7.2  **9.** −1.143
**11.** −4.041  **13.** 0.07  **15.** 70  **17.** 20  **19.** 0.4
**21.** 0.41  **23.** 8.5  **25.** 9.3  **27.** −0.625  **29.** 0.26
**31.** 15.625  **33.** 2.34  **35.** 0.47  **37.** 0.2134567
**39.** −2.359  **41.** 4.26487  **43.** −169.4  **45.** 1023.7
**47.** −4256.1  **49.** 9.3  **51.** −0.0090678  **53.** 45.6
**55.** 2107  **57.** −302.997  **59.** 446.208  **61.** 24.14
**63.** −5.0072  **65.** 19.3204  **67.** 473.188278  **69.** −9.51
**71.** 911.13  **73.** 205  **75.** $1288.36  **77.** 5.42 million stays

**79.** $\frac{6}{7}$  **80.** $\frac{7}{8}$  **81.** $-\frac{19}{73}$  **82.** $-\frac{23}{31}$
**83.** $2 \cdot 2 \cdot 3 \cdot 3 \cdot 19$, or $2^2 \cdot 3^2 \cdot 19$
**84.** $2 \cdot 3 \cdot 3 \cdot 3 \cdot 3$, or $2 \cdot 3^4$  **85.** $3 \cdot 3 \cdot 223$, or $3^2 \cdot 223$
**86.** $5 \cdot 401$  **87.** $15\frac{1}{8}$  **88.** $5\frac{7}{8}$  **89.** 6.254194585  **91.** 1000
**93.** 100

### Mid-Chapter Review: Chapter 5, p. 297

**1.** False  **2.** True  **3.** True
**4.**     $y + 12.8 = 23.35$
$y + 12.8 - 12.8 = 23.35 - 12.8$
$y + 0 = 10.55$
$y = 10.55$
**5.** $5.6 + 4.3 \times (6.5 - 0.25)^2 = 5.6 + 4.3 \times (6.25)^2$
$= 5.6 + 4.3 \times 39.0625$
$= 5.6 + 167.96875$
$= 173.56875$
**6.** Nine and sixty-nine hundredths  **7.** 1,050,000
**8.** $\frac{453}{100}$  **9.** $-\frac{287}{1000}$  **10.** 0.13  **11.** −5.09  **12.** 0.7
**13.** −6.39  **14.** 35.67  **15.** 8.002  **16.** 28.462  **17.** 28.46
**18.** 28.5  **19.** 28  **20.** 50.095  **21.** 1214.862  **22.** −3.772
**23.** 18.24  **24.** 272.19  **25.** 5.593  **26.** 15.55  **27.** −58.77
**28.** 4.14  **29.** 92.871  **30.** −8123.6  **31.** 2.937  **32.** 5.06
**33.** −3.2  **34.** 763.4  **35.** 0.914036  **36.** 2045¢  **37.** $1.47
**38.** −105.3  **39.** 8.4  **40.** 59.774  **41.** 33.83  **42.** The
student probably rounded over successively from the
thousandths place as follows: 236.448 ≈ 236.45 ≈ 236.5 ≈ 237.
The student should have considered only the tenths place and
rounded down.  **43.** The decimal points were not lined up
before the subtraction was carried out.
**44.** $10 \div 0.2 = \frac{10}{0.2} = \frac{10}{0.2} \cdot \frac{10}{10} = \frac{100}{2} = 100 \div 2.$
**45.** $0.247 \div 0.1 = \frac{247}{1000} \div \frac{1}{10} = \frac{247}{1000} \cdot \frac{10}{1} = \frac{247 \cdot 10}{10 \cdot 100} = \frac{247}{100} =$
$2.47 \neq 0.0247; 0.247 \div 10 = \frac{247}{1000} \div 10 = \frac{247}{1000} \cdot \frac{1}{10} = \frac{247}{10,000} =$
$0.0247 \neq 2.47$

### Exercise Set 5.5, p. 304

**1.** 0.23  **3.** 0.6  **5.** −0.325  **7.** 0.2  **9.** −0.85
**11.** 0.375  **13.** −0.975  **15.** 0.52  **17.** −20.016
**19.** 0.25  **21.** −1.16  **23.** 1.1875  **25.** $0.2\overline{6}$  **27.** $0.\overline{3}$
**29.** $-1.\overline{3}$  **31.** $-1.\overline{16}$  **33.** $0.\overline{571428}$  **35.** $-0.91\overline{6}$
**37.** 0.3; 0.27; 0.267  **39.** 0.3; 0.33; 0.333
**41.** −1.3; −1.33; −1.333  **43.** −1.2; −1.17; −1.167  **45.** 0.6;
0.57; 0.571  **47.** −0.9; −0.92; −0.917  **49.** 0.2; 0.18; 0.182
**51.** −0.3; −0.28; −0.278  **53.** (a) 0.429; (b) 0.75; (c) 0.571;
(d) 1.333  **55.** 15.8 mpg  **57.** 17.8 mpg  **59.** 15.2 mph
**61.** $29.5625; $29.56  **63.** $27.875; $27.88  **65.** $31.484375;
$31.48  **67.** 11.06  **69.** 8.4  **71.** $-417.51\overline{6}$  **73.** 0
**75.** 2.8125  **77.** −0.07675  **79.** 317.14  **81.** 0.1825
**83.** 18  **85.** −2.736  **87.** 21  **88.** $238\frac{7}{8}$  **89.** −10
**90.** $\frac{43}{52}$  **91.** $50\frac{5}{24}$  **92.** $30\frac{7}{10}$  **93.** $1\frac{1}{2}$  **94.** $14\frac{13}{24}$
**95.** $\frac{25}{24}$ cups, or $1\frac{1}{24}$ cups  **96.** $\frac{133}{100}$ in., or $1\frac{33}{100}$ in.  **97.** $0.\overline{142857}$
**99.** $0.\overline{428571}$  **101.** $0.\overline{714285}$  **103.** $0.\overline{1}$  **105.** $0.\overline{001}$

### Exercise Set 5.6, p. 311

**1.** (d)  **3.** (c)  **5.** (a)  **7.** (c)  **9.** 1.6  **11.** 6  **13.** 60
**15.** 2.3  **17.** 180  **19.** (a)  **21.** (c)  **23.** (b)  **25.** (b)
**27.** 1800 ÷ 9 = 200 posts; answers may vary
**29.** $2 · 12 = $24; answers may vary  **31.** Repeating
**32.** Multiple  **33.** Distributive  **34.** Solution
**35.** Multiplicative  **36.** Commutative  **37.** Denominator;
multiple  **38.** Divisible; divisible  **39.** Yes  **41.** No
**43.** (a) +, ×; (b) +, ×, −

### Translating for Success, p. 321

**1.** I  **2.** C  **3.** N  **4.** A  **5.** G  **6.** B  **7.** D  **8.** O
**9.** F  **10.** M

**Exercise Set 5.7, p. 322**

**1.** \$43.1 billion     **3.** 6.29 million passengers
**5.** \$151.1 million     **7.** \$0.51     **9.** 102.8°F
**11.** \$64,333,333.33     **13.** Area: 8.125 sq cm; perimeter: 11.5 cm
**15.** 22,691.5 mi     **17.** 20.2 mpg     **19.** 11.9752 cu ft
**21.** 78.1 cm     **23.** 28.5 cm     **25.** \$24.33     **27.** 2.31 cm
**29.** 876 calories     **31.** \$1171.74     **33.** 227.75 sq ft
**35.** 0.364     **37.** 2152.56 sq yd     **39.** 10.8¢     **41.** \$906.50
**43.** 6.052 billion     **45.** 1.4°F     **47.** \$262,153     **49.** \$83,782
**51.** \$1,401,429     **53.** \$29,133     **55.** \$53.04     **57.** \$3745.41
**59.** \$1406.75     **61.** 6335     **62.** $\frac{31}{24}$     **63.** $6\frac{5}{6}$     **64.** $-\frac{2}{15}$
**65.** $\frac{1}{24}$     **66.** 2803     **67.** $-\frac{23}{15}$     **68.** $1\frac{5}{6}$     **69.** $\frac{129}{251}$     **70.** $\frac{5}{16}$
**71.** $-\frac{13}{25}$     **72.** $-\frac{25}{19}$     **73.** 28 min     **74.** $7\frac{1}{5}$ min
**75.** 186 calories     **76.** 30 calories     **77.** \$17.28

**Summary and Review: Chapter 5, p. 329**

Concept Reinforcement

**1.** True     **2.** False     **3.** True     **4.** False     **5.** True

Important Concepts

**1.** $\frac{5093}{100}$     **2.** $-81.7$     **3.** 42.159     **4.** 153.35     **5.** 38.611
**6.** 207.848     **7.** 19.11     **8.** $-0.176$     **9.** 60,437     **10.** 7.4
**11.** 0.047     **12.** $-15,690$

Review Exercises

**1.** 6,590,000     **2.** 3,100,000,000     **3.** Three and forty-seven hundredths     **4.** Thirty-one thousandths     **5.** Negative thirteen and four tenths     **6.** Seven ten-thousandths     **7.** $\frac{9}{100}$
**8.** $\frac{4561}{1000}$     **9.** $-\frac{89}{1000}$     **10.** $-\frac{30,227}{10,000}$     **11.** 0.034     **12.** $-4.2603$
**13.** 27.91     **14.** $-867.006$     **15.** 0.034     **16.** 0.91     **17.** 0.741
**18.** $-1.038$     **19.** 17.4     **20.** 17.43     **21.** 17.429     **22.** 17
**23.** 574.519     **24.** 0.6838     **25.** $-209.5$     **26.** 45.551
**27.** 29.2092     **28.** 790.29     **29.** 46.142     **30.** $-70.8109$
**31.** 12.96     **32.** 0.14442     **33.** 4.3     **34.** $-0.02468$     **35.** 0.45
**36.** 45.2     **37.** 1.022     **38.** $-7.5$     **39.** 0.2763     **40.** $-1389.2$
**41.** 496.2795     **42.** $-6.95$     **43.** 42.54     **44.** 4.9911
**45.** \$15.52     **46.** 1.9 lb     **47.** \$784.47     **48.** \$171.24
**49.** 14.5 mpg     **50. (a)** 106.2 lb; **(b)** 15.2 lb     **51.** 272
**52.** 216     **53.** \$125     **54.** 2.6     **55.** $-1.28$     **56.** 2.75
**57.** $-3.25$     **58.** $1.1\overline{6}$     **59.** $1.\overline{54}$     **60.** 1.5     **61.** 1.55
**62.** 1.545     **63.** \$82.73     **64.** \$4.87     **65.** 2493¢     **66.** 986¢
**67.** $-1.5805$     **68.** 57.1449     **69.** $-15.6375$     **70.** D
**71.** B     **72. (a)** $2.56 \times 6.4 \div 51.2 - 17.4 + 89.7 = 72.62$;
**(b)** $(11.12 - 0.29) \times 3^4 = 877.23$
**73.** $1 = 3 \cdot \frac{1}{3} = 3(0.33333333\ldots) = 0.99999999\ldots$, or $0.\overline{9}$

Understanding Through Discussion and Writing

**1.** Count the number of decimal places. Move the decimal point that many places to the right and write the result over a denominator of 1 followed by that many zeros.
**2.** $346.708 \times 0.1 = \frac{346,708}{1000} \times \frac{1}{10} = \frac{346,708}{10,000} = 34.6708 \neq 3467.08$
**3.** When the denominator of a fraction is a multiple of 10, long division is not the fastest way to convert the fraction to decimal notation. Many times when the denominator is a factor of some multiple of 10, this is also the case. The latter situation occurs when the denominator has only 2's or 5's or both as factors.
**4.** Multiply by 1 to get a denominator that is a power of 10:
$$\frac{44}{125} = \frac{44}{125} \cdot \frac{8}{8} = \frac{352}{1000} = 0.352.$$
We can also divide to find that $\frac{44}{125} = 0.352$.

**Test: Chapter 5, p. 334**

**1.** [5.3b] 8,900,000,000     **2.** [5.3b] 3,756,000     **3.** [5.1a] Two and thirty-four hundredths     **4.** [5.1a] One hundred five and five ten-thousandths     **5.** [5.1b] $\frac{91}{100}$     **6.** [5.1b] $-\frac{2769}{1000}$
**7.** [5.1b] 0.074     **8.** [5.1b] $-3.7047$     **9.** [5.1b] 756.09

**10.** [5.1b] $-91.703$     **11.** [5.1c] 0.162     **12.** [5.1c] $-0.06$
**13.** [5.1c] 0.9     **14.** [5.1d] 6     **15.** [5.1d] 5.68
**16.** [5.1d] 5.678     **17.** [5.1d] 5.7     **18.** [5.2a] 0.7902
**19.** [5.2c] $-18.3$     **20.** [5.2a] 1033.23     **21.** [5.2b] 48.357
**22.** [5.2b] 19.0901     **23.** [5.2c] $-316.4642$     **24.** [5.3a] 0.03
**25.** [5.3a] $-0.21345$     **26.** [5.3a] 73,962     **27.** [5.4a] $-4.75$
**28.** [5.4a] 30.4     **29.** [5.4a] 0.19     **30.** [5.4a] $-0.34689$
**31.** [5.4a] 34,689     **32.** [5.4b] $-84.26$     **33.** [5.2d] 8.982
**34.** [5.7a] \$133.99     **35.** [5.7a] 28.3 mpg     **36.** [5.7a] \$592.45
**37.** [5.7a] \$293.93     **38.** [5.7a] 67.44 million passengers
**39.** [5.6a] 198     **40.** [5.6a] 4     **41.** [5.5a] 1.6     **42.** [5.2a] 0.88
**43.** [5.5a] $-5.25$     **44.** [5.5a] 0.75     **45.** [5.5a] $-1.\overline{2}$
**46.** [5.5a] $2.\overline{142857}$     **47.** [5.5b] 2.1     **48.** [5.5b] 2.14
**49.** [5.5b] 2.143     **50.** [5.4c] 40.0065     **51.** [5.4c] 384.8464
**52.** [5.5c] 302.4     **53.** [5.5c] $52.339\overline{4}$     **54.** [5.3b] B
**55.** [5.7a] \$35     **56.** [5.1b, c] $-\frac{13}{15}, -\frac{17}{20}, -\frac{11}{13}, -\frac{15}{19}, -\frac{5}{7}, -\frac{2}{3}$

**Cumulative Review: Chapters 1–5, p. 337**

**1.** [4.4a] $\frac{20}{9}$     **2.** [5.1b] $-\frac{3051}{1000}$     **3.** [5.5a] $-1.\overline{4}$     **4.** [5.5a] $0.\overline{54}$
**5.** [3.1c] Prime     **6.** [3.2a] Yes     **7.** [1.9c] 1754
**8.** [5.4c] $-0.136$     **9.** [5.1d] 584.97     **10.** [5.5b] 218.56
**11.** [5.6a] 160     **12.** [5.6a] 4     **13.** [1.6b] 12,800,000
**14.** [5.6a] 6     **15.** [4.5a] $6\frac{1}{20}$     **16.** [1.2a] 139,116
**17.** [4.2a] $\frac{31}{18}$     **18.** [5.2c] $-141.847$     **19.** [1.3a] 710,137
**20.** [5.2b] 13.097     **21.** [4.5b] $\frac{5}{7}$     **22.** [4.3a] $-1\frac{1}{10}$
**23.** [3.6a] $-\frac{1}{6}$     **24.** [1.4a] 5,317,200     **25.** [5.3a] 4.78
**26.** [5.3a] 0.0279431     **27.** [5.4a] 2.122     **28.** [1.5a] 1843
**29.** [5.4a] 13,862.1     **30.** [3.7b] $\frac{5}{6}$     **31.** [5.2d] 0.78
**32.** [1.7b] 28     **33.** [5.4b] $-8.62$     **34.** [1.7b] 367,251
**35.** [4.3c] $\frac{1}{18}$     **36.** [3.7c] $-\frac{1}{2}$     **37.** [1.8a] 24,929 transplants
**38.** [1.8a] \$177 billion     **39.** [3.7d] \$1500     **40.** [3.6b] \$2400
**41.** [5.7a] \$258.77     **42.** [4.5c] $6\frac{1}{2}$ lb     **43.** [4.2b] 2 lb
**44.** [5.7a] 467.28 sq ft     **45.** [4.7a] $\frac{9}{32}$     **46.** [5.4c] $-527.04$
**47.** [5.7a] \$2.39     **48.** [4.6c] 144 packages

**CHAPTER 6**

**Exercise Set 6.1, p. 344**

**1.** $\frac{4}{5}$     **3.** $\frac{178}{572}$     **5.** $\frac{0.4}{12}$     **7.** $\frac{3.8}{7.4}$     **9.** $\frac{56.78}{98.35}$     **11.** $\frac{8\frac{3}{4}}{9\frac{5}{6}}$     **13.** $\frac{21}{4}, \frac{4}{21}$
**15.** $\frac{100}{84.7}, \frac{84.7}{100}$     **17.** $\frac{113}{366}$     **19.** $\frac{1000}{126}, \frac{126}{1000}$     **21.** $\frac{60}{100}, \frac{100}{60}$     **23.** $\frac{2}{3}$
**25.** $\frac{3}{4}$     **27.** $\frac{12}{25}$     **29.** $\frac{7}{9}$     **31.** $\frac{2}{3}$     **33.** $\frac{14}{15}$     **35.** $\frac{1}{2}$     **37.** $\frac{3}{4}$
**39.** $\frac{478}{213}, \frac{213}{478}$     **41.** $=$     **42.** $\neq$     **43.** $\neq$     **44.** $=$     **45.** 500
**46.** 9.5     **47.** $-14.5$     **48.** $-152$     **49.** $6\frac{7}{20}$ cm     **50.** $17\frac{11}{20}$ cm
**51.** $\frac{30}{47}$     **53.** $1:2:3$

**Exercise Set 6.2, p. 350**

**1.** 40 km/h     **3.** 7.48 mi/sec     **5.** 25 mpg     **7.** 32 mpg
**9.** 43,728 people/sq mi     **11.** 25 mph; 0.04 hr/mi
**13.** About 24.6 points/game     **15.** 0.623 gal/ft$^2$
**17.** 186,000 mi/sec     **19.** 124 km/h     **21.** 25 beats/min
**23.** 26.188¢/oz; 26.450¢/oz; 16 oz     **25.** 8.867¢/oz; 7.320¢/oz;
75 oz     **27.** 13.889¢/oz; 17.464¢/oz; 18 oz     **29.** 12.991¢/oz;
10.346¢/oz; 26 oz     **31.** B: 18.719¢/oz; E: 14.563¢/oz;  Brand E
**33.** A: 10.375¢/oz; B: 9.139¢/oz; H: 8.022¢/oz;  Brand H
**35.** 4063 theaters     **36.** $37\frac{1}{2}$ servings     **37.** \$9.1 billion
**38.** 281.4 million tracks     **39.** 109.608     **40.** 67,819
**41.** $-5833.56$     **42.** 466,190.4     **43.** 6-oz: 10.833¢/oz;
5.5-oz: 10.909¢/oz

**Exercise Set 6.3, p. 358**

**1.** No     **3.** Yes     **5.** Yes     **7.** No     **9.** 45     **11.** 12     **13.** 10
**15.** 20     **17.** 5     **19.** 18     **21.** 22     **23.** 28     **25.** $\frac{28}{3}$, or $9\frac{1}{3}$
**27.** $\frac{26}{9}$, or $2\frac{8}{9}$     **29.** 5     **31.** 5     **33.** 0     **35.** 14     **37.** 2.7
**39.** 1.8     **41.** 0.06     **43.** 0.7     **45.** 12.5725     **47.** 1     **49.** $\frac{1}{20}$
**51.** $\frac{3}{8}$     **53.** $\frac{16}{75}$     **55.** $\frac{51}{16}$, or $3\frac{3}{16}$     **57.** $\frac{546}{185}$, or $2\frac{176}{185}$     **59.** Quotient

**60.** Sum **61.** Average **62.** Dollars; cents **63.** Subtrahend
**64.** Terminating **65.** Commutative **66.** Cross products
**67.** Approximately 2731.4 **69.** Ruth: 1.863 strikeouts per
home run; Schmidt: 3.436 strikeouts per home run

### Mid-Chapter Review: Chapter 6, p. 361

**1.** True **2.** True **3.** False **4.** False
**5.** $\frac{120 \text{ mi}}{2 \text{ hr}} = \frac{120}{2} \frac{\text{mi}}{\text{hr}} = 60 \text{ mi/hr}$
**6.** $\quad \frac{x}{4} = \frac{3}{6}$
$\quad x \cdot 6 = 4 \cdot 3$
$\quad \frac{x \cdot 6}{6} = \frac{4 \cdot 3}{6}$
$\quad\quad x = 2$
**7.** $\frac{4}{7}$ **8.** $\frac{313}{199}$ **9.** $\frac{35}{17}$ **10.** $\frac{59}{101}$ **11.** $\frac{2}{3}$ **12.** $\frac{1}{3}$ **13.** $\frac{8}{7}$
**14.** $\frac{25}{19}$ **15.** $\frac{2}{1}$ **16.** $\frac{5}{1}$ **17.** $\frac{2}{7}$ **18.** $\frac{3}{5}$ **19.** 60.75 mi/hr, or
60.75 mph **20.** 48.67 km/h **21.** 13 m/sec **22.** 16.17 ft/sec
**23.** 27 in./day **24.** About 0.929 free throw made/attempt
**25.** 11.611¢/oz **26.** 49.917¢/oz **27.** Yes **28.** No
**29.** No **30.** Yes **31.** 12 **32.** 40 **33.** 9 **34.** 35
**35.** 2.2 **36.** 4.32 **37.** $\frac{1}{2}$ **38.** $\frac{65}{4}$, or $16\frac{1}{4}$ **39.** Yes;
every ratio $\frac{a}{b}$ can be written as $\frac{\frac{a}{b}}{1}$. **40.** By making some
sketches, we see that the rectangle's length must be twice
the width. **41.** The student's approach will work. However,
when we use the approach of equating cross products, we
eliminate the need to find the least common denominator.
**42.** The instructor thinks that the longer a student studies, the
higher his or her grade will be. An example is the situation in
which one student gets a test grade of 96 after studying for 8 hr
while another student gets a score of 78 after studying for $6\frac{1}{2}$ hr.
This is represented by the proportion $\dfrac{96}{8} = \dfrac{78}{6\frac{1}{2}}$.

### Translating for Success, p. 368

**1.** N **2.** I **3.** A **4.** K **5.** J **6.** F **7.** M **8.** B
**9.** G **10.** E

### Exercise Set 6.4, p. 369

**1.** 11.04 hr **3.** 60 students **5.** (a) About 122 gal;
(b) 3080 mi **7.** 212.52 million, or 212,520,000 **9.** 175 bulbs
**11.** 2975 ft$^2$ **13.** 450 pages **15.** (a) 28.065 British pounds;
(b) \$15,392.84 **17.** (a) 13,399.625 Japanese yen; (b) \$47.76
**19.** 13,500 mi **21.** 880 calories **23.** 120 lb **25.** 64 gal
**27.** 100 oz **29.** 954 deer **31.** 58.1 mi **33.** 9.75 gal
**35.** (a) 171 games; (b) about 47 home runs **37.** Neither
**38.** Prime **39.** Composite **40.** Composite **41.** Prime
**42.** $2 \cdot 2 \cdot 2 \cdot 101$, or $2^3 \cdot 101$ **43.** $2 \cdot 2 \cdot 2 \cdot 7$, or $2^3 \cdot 7$
**44.** $2 \cdot 433$ **45.** $3 \cdot 31$ **46.** $2 \cdot 2 \cdot 5 \cdot 101$, or $2^2 \cdot 5 \cdot 101$
**47.** 17 positions **49.** 2150 earned runs **51.** CD player:
\$150; receiver: \$450; speakers: \$300

### Exercise Set 6.5, p. 378

**1.** 25 **3.** $\frac{4}{3}$, or $1\frac{1}{3}$ **5.** $x = \frac{27}{4}$, or $6\frac{3}{4}$; $y = 9$ **7.** $x = 7.5$;
$y = 7.2$ **9.** 1.25 m **11.** 36 ft **13.** 7 ft **15.** 100 ft
**17.** 4 **19.** $10\frac{1}{2}$ **21.** $x = 6$; $y = 5.25$; $z = 3$ **23.** $x = 5\frac{1}{3}$,
or $5.\overline{3}$; $y = 4\frac{2}{3}$, or $4.\overline{6}$; $z = 5\frac{1}{3}$, or $5.\overline{3}$ **25.** 20 ft **27.** 152 ft
**29.** \$59.81 **30.** 9.63 **31.** 679.4928 **32.** $-2.74568$
**33.** 27,456.8 **34.** $-0.549136$ **35.** 0.85 **36.** 1.825
**37.** 0.909 **38.** $-0.843$ **39.** 13.75 ft **41.** 1.25 cm
**43.** 3681.437 **45.** $x = 0.4$; $y \approx 0.35$

### Summary and Review: Chapter 6, p. 382

#### Concept Reinforcement

**1.** True **2.** True **3.** False **4.** True

#### Important Concepts

**1.** $\frac{17}{3}$ **2.** $\frac{8}{7}$ **3.** \$7.50/hr **4.** A: 9.964¢/oz; B: 10.281¢/oz;
Brand A **5.** Yes **6.** $\frac{27}{8}$ **7.** 175 mi **8.** 21

#### Review Exercises

**1.** $\frac{47}{84}$ **2.** $\frac{46}{1.27}$ **3.** $\frac{83}{100}$ **4.** $\frac{0.72}{197}$ **5.** (a) $\frac{12,480}{16,640}$, or $\frac{3}{4}$; (b) $\frac{16,640}{29,120}$,
or $\frac{4}{7}$ **6.** $\frac{3}{4}$ **7.** $\frac{9}{16}$ **8.** 26 mpg **9.** 6300 revolutions/min
**10.** 0.638 gal/ft$^2$ **11.** 6.33¢/tablet **12.** 11.208¢/oz
**13.** 14.969¢/oz; 12.479¢/oz; 15.609¢/oz; 48 oz **14.** Yes
**15.** No **16.** 32 **17.** 7 **18.** $\frac{1}{40}$ **19.** 24 **20.** 27 circuits
**21.** (a) 267 Canadian dollars; (b) 46.82 U.S. dollars
**22.** 832 mi **23.** 27 acres **24.** About 3,418,140 lb
**25.** 6 in. **26.** About 13,644 lawyers **27.** $x = \frac{14}{3}$, or $4\frac{2}{3}$
**28.** $x = \frac{56}{5}$, or $11\frac{1}{5}$; $y = \frac{63}{5}$, or $12\frac{3}{5}$ **29.** 40 ft
**30.** $x = 3$; $y = \frac{21}{2}$, or $10\frac{1}{2}$; $z = \frac{15}{2}$, or $7\frac{1}{2}$ **31.** B **32.** C
**33.** 1.329¢/sheet; 1.554¢/sheet; 1.110¢/sheet; 6 big rolls
**34.** $x = 4258.5$; $z \approx 10,094.3$ **35.** Finishing paint: 11 gal;
primer: 16.5 gal

### Understanding Through Discussion and Writing

**1.** In terms of cost, a low faculty-to-student ratio is less expensive
than a high faculty-to-student ratio. In terms of quality of
education and student satisfaction, a high faculty-to-student
ratio is more desirable. A college president must balance the cost
and quality issues. **2.** Yes; unit prices can be used to solve
proportions involving money. In Example 3 of Section 6.4, for
instance, we would have divided \$90 by the unit price, or the
price per ticket, to find the number of tickets that could be
purchased for \$90. **3.** Leslie used 4 gal of gasoline to drive
92 mi. At the same rate, how many gallons would be needed to
travel 368 mi? **4.** Yes; consider the following pair of triangles.

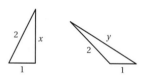

Two pairs of sides are proportional, but we can see that $x$ is
shorter than $y$ so the ratio of $x$ to $y$ is clearly not the same as the
ratio of 1 to 1 (or 2 to 2).

### Test: Chapter 6, p. 387

**1.** [6.1a] $\frac{85}{97}$ **2.** [6.1a] $\frac{0.34}{124}$ **3.** [6.1b] $\frac{9}{10}$ **4.** [6.1b] $\frac{25}{32}$
**5.** [6.2a] 0.625 ft/sec **6.** [6.2a] $1\frac{1}{3}$ servings/lb
**7.** [6.2a] 32 mpg **8.** [6.2b] About 15.563¢/oz
**9.** [6.2b] 16.475¢/oz; 13.980¢/oz; 11.490¢/oz; 16.660¢/oz;
100 oz **10.** [6.3a] Yes **11.** [6.3a] No **12.** [6.3b] 12
**13.** [6.3b] 360 **14.** [6.3b] 42.1875 **15.** [6.3b] 100
**16.** [6.4a] 1512 km **17.** [6.4a] 4.8 min **18.** [6.4a] 525 mi
**19.** [6.5a] 66 m **20.** [6.4a] (a) 3501.45 Hong Kong dollars;
(b) \$102.17 **21.** [6.4a] About \$59.17 **22.** [6.5a] $x = 8$;
$y = 8.8$ **23.** [6.5b] $x = \frac{24}{5}$, or 4.8; $y = \frac{32}{5}$, or 6.4; $z = 12$
**24.** [6.2a] C **25.** [6.4a] 5888 marbles

### Cumulative Review: Chapters 1–6, p. 389

**1.** [5.2a] 513.996 **2.** [4.5a] $6\frac{3}{4}$ **3.** [4.2a] $-\frac{7}{20}$
**4.** [5.2b] 30.491 **5.** [5.2b] 72.912 **6.** [4.3a] $-\frac{5}{12}$
**7.** [5.3a] 222.076 **8.** [5.3a] 567.8 **9.** [4.6a] $-3$ **10.** [5.4a] 43
**11.** [1.5a] 899 **12.** [3.7b] $-\frac{3}{2}$ **13.** [1.1b] 3 ten thousands +
0 thousands + 0 hundreds + 7 tens + 4 ones, or 3 ten
thousands + 7 tens + 4 ones **14.** [5.1a] One hundred twenty
and seven hundredths **15.** [5.1c] 0.7 **16.** [5.1c] $-0.799$
**17.** [3.1d] $2 \cdot 2 \cdot 2 \cdot 2 \cdot 3 \cdot 3$, or $2^4 \cdot 3^2$ **18.** [4.1a] 90
**19.** [3.3a] $\frac{5}{8}$ **20.** [3.5b] $\frac{5}{8}$ **21.** [5.5c] 5.718 **22.** [5.5c] 0.179
**23.** [6.1a] $\frac{0.3}{15}$ **24.** [6.3a] Yes **25.** [6.2a] 55 m/sec
**26.** [6.2b] 8-oz can **27.** [6.3b] 30.24 **28.** [5.4b] 26.4375
**29.** [3.7c] $\frac{8}{9}$ **30.** [6.3b] 128 **31.** [5.2d] $-102.46$ **32.** [4.3c] $\frac{76}{175}$
**33.** [3.6b] 390 cal **34.** [6.4a] 7 min **35.** [4.5c] $2\frac{1}{4}$ cups
**36.** [3.7d] 12 doors **37.** [5.7a], [6.4a] 42.2025 mi
**38.** [5.7a] 132 orbits **39.** [3.1c] D **40.** [6.2a] B
**41.** [6.5a] $10\frac{1}{2}$ ft

## CHAPTER 7

### Calculator Corner, p. 394
**1.** 0.14  **2.** 0.00069  **3.** 0.438  **4.** 1.25

### Exercise Set 7.1, p. 396
**1.** $\frac{90}{100}$; $90 \times \frac{1}{100}$; $90 \times 0.01$  **3.** $\frac{12.5}{100}$; $12.5 \times \frac{1}{100}$; $12.5 \times 0.01$
**5.** 0.67  **7.** 0.456  **9.** 0.5901  **11.** 0.1  **13.** 0.01  **15.** 2
**17.** 0.001  **19.** 0.0009  **21.** 0.0018  **23.** 0.2319
**25.** 0.14875  **27.** 0.565  **29.** 0.97  **31.** 0.07; 0.08
**33.** 0.548  **35.** 47%  **37.** 3%  **39.** 870%  **41.** 33.4%
**43.** 75%  **45.** 40%  **47.** 0.6%  **49.** 1.7%  **51.** 27.18%
**53.** 2.39%  **55.** 27%  **57.** 5.7%; 17.6%  **59.** 90.6%; 88%
**61.** 0.64; 0.3; 0.04; 0.02  **63.** $33\frac{1}{3}$  **64.** $37\frac{1}{2}$  **65.** $9\frac{3}{8}$
**66.** $18\frac{9}{16}$  **67.** $-5\frac{11}{14}$  **68.** $-111\frac{2}{3}$  **69.** $0.\overline{6}$  **70.** $0.\overline{3}$
**71.** $-0.8\overline{3}$  **72.** $1.41\overline{6}$  **73.** $2.\overline{6}$  **74.** $-0.9375$  **75.** 50%
**77.** 70%  **79.** 20%

### Calculator Corner, p. 399
**1.** 52%  **2.** 38.46%  **3.** 110.26%  **4.** 171.43%  **5.** 59.62%
**6.** 28.31%

### Exercise Set 7.2, p. 403
**1.** 41%  **3.** 5%  **5.** 20%  **7.** 30%  **9.** 50%  **11.** 87.5%,
or $87\frac{1}{2}$%  **13.** 80%  **15.** $66.\overline{6}$%, or $66\frac{2}{3}$%  **17.** $16.\overline{6}$%, or
$16\frac{2}{3}$%  **19.** 18.75%, or $18\frac{3}{4}$%  **21.** 81.25%, or $81\frac{1}{4}$%
**23.** 16%  **25.** 5%  **27.** 34%  **29.** 8%; 59%  **31.** 22%
**33.** 12%  **35.** 15%  **37.** $\frac{17}{20}$  **39.** $\frac{5}{8}$  **41.** $\frac{1}{3}$  **43.** $\frac{1}{6}$  **45.** $\frac{29}{400}$
**47.** $\frac{1}{125}$  **49.** $\frac{203}{800}$  **51.** $\frac{176}{225}$  **53.** $\frac{711}{1100}$  **55.** $\frac{3}{2}$  **57.** $\frac{13}{40,000}$
**59.** $\frac{1}{3}$  **61.** $\frac{3}{50}$  **63.** $\frac{3}{25}$  **65.** $\frac{3}{4}$  **67.** $\frac{3}{20}$  **69.** $\frac{209}{1000}$

**71.**

| Fraction Notation | Decimal Notation | Percent Notation |
|---|---|---|
| $\frac{1}{8}$ | 0.125 | 12.5%, or $12\frac{1}{2}$% |
| $\frac{1}{6}$ | $0.1\overline{6}$ | $16.\overline{6}$%, or $16\frac{2}{3}$% |
| $\frac{1}{5}$ | 0.2 | 20% |
| $\frac{1}{4}$ | 0.25 | 25% |
| $\frac{1}{3}$ | $0.\overline{3}$ | $33.\overline{3}$%, or $33\frac{1}{3}$% |
| $\frac{3}{8}$ | 0.375 | 37.5%, or $37\frac{1}{2}$% |
| $\frac{2}{5}$ | 0.4 | 40% |
| $\frac{1}{2}$ | 0.5 | 50% |

**73.**

| Fraction Notation | Decimal Notation | Percent Notation |
|---|---|---|
| $\frac{1}{2}$ | 0.5 | 50% |
| $\frac{1}{3}$ | $0.\overline{3}$ | $33.\overline{3}$%, or $33\frac{1}{3}$% |
| $\frac{1}{4}$ | 0.25 | 25% |
| $\frac{1}{6}$ | $0.1\overline{6}$ | $16.\overline{6}$%, or $16\frac{2}{3}$% |
| $\frac{1}{8}$ | 0.125 | 12.5%, or $12\frac{1}{2}$% |
| $\frac{3}{4}$ | 0.75 | 75% |
| $\frac{5}{6}$ | $0.8\overline{3}$ | $83.\overline{3}$%, or $83\frac{1}{3}$% |
| $\frac{3}{8}$ | 0.375 | 37.5%, or $37\frac{1}{2}$% |

**75.** 70  **76.** 5  **77.** $-400$  **78.** 18.75  **79.** 23.125
**80.** 25.5  **81.** 4.5  **82.** 8.75  **83.** $33\frac{1}{3}$  **84.** $37\frac{1}{2}$  **85.** $83\frac{1}{3}$
**86.** $20\frac{1}{2}$  **87.** $-43\frac{1}{8}$  **88.** $-62\frac{1}{6}$  **89.** $18\frac{3}{4}$  **90.** $7\frac{4}{9}$  **91.** $\frac{18}{17}$
**92.** $\frac{209}{10}$  **93.** $-\frac{203}{2}$  **94.** $-\frac{259}{8}$  **95.** $11.\overline{1}$%  **97.** $257.\overline{46317}$%
**99.** $0.01\overline{5}$  **101.** $1.04\overline{142857}$  **103.** $\frac{1}{6}$%, $\frac{2}{7}$%, 0.5%, $1\frac{1}{6}$%, 1.6%,
$16\frac{1}{6}$%, 0.2, $\frac{1}{2}$, $0.\overline{54}$, 1.6

### Calculator Corner, p. 410
**1.** $5.04  **2.** 0.0112  **3.** 450  **4.** $1000  **5.** 2.5%  **6.** 12%

### Exercise Set 7.3, p. 411
**1.** $a = 32\% \times 78$  **3.** $89 = p \times 99$  **5.** $13 = 25\% \times b$
**7.** 234.6  **9.** 45  **11.** $18  **13.** 1.9  **15.** 78%  **17.** 200%
**19.** 50%  **21.** 125%  **23.** 40  **25.** $40  **27.** 88  **29.** 20
**31.** 6.25  **33.** $846.60  **35.** 1216  **37.** $\frac{9}{100}$  **38.** $\frac{179}{100}$
**39.** $\frac{875}{1000}$, or $\frac{7}{8}$  **40.** $\frac{125}{1000}$, or $\frac{1}{8}$  **41.** $-\frac{9375}{10,000}$, or $-\frac{15}{16}$
**42.** $-\frac{6875}{10,000}$, or $-\frac{11}{16}$  **43.** 0.89  **44.** 0.07  **45.** $-0.3$
**46.** $-0.017$  **47.** $800 (can vary); $843.20  **49.** $10,000 (can
vary); $10,400  **51.** $1875

### Exercise Set 7.4, p. 417
**1.** $\frac{37}{100} = \frac{a}{74}$  **3.** $\frac{N}{100} = \frac{4.3}{5.9}$  **5.** $\frac{25}{100} = \frac{14}{b}$  **7.** 68.4  **9.** 462
**11.** 40  **13.** 2.88  **15.** 25%  **17.** 102%  **19.** 25%
**21.** 93.75%, or $93\frac{3}{4}$%  **23.** $72  **25.** 90  **27.** 88  **29.** 20
**31.** 25  **33.** $780.20  **35.** 200  **37.** 8  **38.** 4000  **39.** 8
**40.** 2074  **41.** 100  **42.** 15  **43.** $8.0\overline{4}$  **44.** $\frac{3}{16}$, or 0.1875
**45.** $\frac{43}{48}$ qt  **46.** $\frac{1}{8}$ T  **47.** $1170 (can vary); $1118.64

### Mid-Chapter Review: Chapter 7, p. 419
**1.** True  **2.** False  **3.** True  **4.** $\frac{1}{2}\% = \frac{1}{2} \cdot \frac{1}{100} = \frac{1}{200}$
**5.** $\frac{80}{1000} = \frac{8}{100} = 8\%$  **6.** $5.5\% = \frac{5.5}{100} = \frac{55}{1000} = \frac{11}{200}$
**7.** $0.375 = \frac{375}{1000} = \frac{37.5}{100} = 37.5\%$

**8.** $15 = p \times 80$
$\frac{15}{80} = \frac{p \times 80}{80}$
$\frac{15}{80} = p$
$0.1875 = p$
$18.75\% = p$
**9.** 0.28 **10.** 0.0015
**11.** 0.05375 **12.** 2.4 **13.** 71% **14.** 9% **15.** 38.91%
**16.** 18.75%, or $18\frac{3}{4}\%$ **17.** 0.5% **18.** 74% **19.** 600%
**20.** $83.\overline{3}\%$, or $83\frac{1}{3}\%$ **21.** $\frac{17}{20}$ **22.** $\frac{3}{6250}$ **23.** $\frac{91}{400}$ **24.** $\frac{1}{6}$
**25.** 62.5%, or $62\frac{1}{2}\%$ **26.** 45% **27.** 58 **28.** $16.\overline{6}\%$, or $16\frac{2}{3}\%$
**29.** 2560 **30.** $50 **31.** 20% **32.** $455 **33.** 0.05%, 0.1%,
$\frac{1}{2}\%$, 1%, 5%, 10%, $\frac{13}{100}$, 0.275, $\frac{3}{10}$, $\frac{7}{20}$ **34.** D **35.** B **36.** Some
will say that the conversion will be done most accurately by first
finding decimal notation. Others will say that it is more efficient to
become familiar with some or all of the fraction and percent
equivalents that appear inside the back cover and to make the
conversion by going directly from fraction notation to percent
notation. **37.** Since $40\% \div 10 = 4\%$, we can divide 36.8 by 10,
obtaining 3.68. Since $400\% = 40\% \times 10$, we can multiply 36.8 by
10, obtaining 368. **38.** Answers may vary. Some will say this is a
good idea since it makes the computations in the solution easier.
Others will say it is a poor idea since it adds an extra step to the
solution. **39.** They all represent the same number.

## Translating for Success, p. 428

**1.** J **2.** M **3.** N **4.** E **5.** G **6.** H **7.** O **8.** C
**9.** D **10.** B

## Exercise Set 7.5, p. 429

**1.** South Korea: 62,381 students; Japan: 35,272 students
**3.** $46,656 **5.** 140 items **7.** 940,000,000 acres **9.** $36,400
**11.** 76 items correct; 4 items incorrect **13.** Egypt: 24,596,374;
United States: 62,315,472 **15.** About 17.3%
**17.** $230.10 **19.** About 612,000 fast-food cooks **21.** Alcohol:
43.2 mL; water: 496.8 mL **23.** Air Force: 25.2%; Army: 36.5%;
Navy: 25.3%; Marines: 13.0% **25.** 5% **27.** 15% **29.** About
34.7% **31.** $33\frac{1}{3}\%$ **33.** About 49.5% **35.** About 59%
**37.** About 9.5% **39.** 34.375%, or $34\frac{3}{8}\%$ **41.** 15,081; 2.5%
**43.** 5,130,632; 20.2% **45.** 1,466,465; 13.3% **47.** 1.0%
**49.** $2.\overline{27}$ **50.** $-0.44$ **51.** $-3.375$ **52.** $4.\overline{7}$ **53.** 0.92
**54.** $0.8\overline{3}$ **55.** 0.4375 **56.** 2.317 **57.** $-3.4809$
**58.** $-0.675$ **59.** $42

## Exercise Set 7.6, p. 440

**1.** $9.56 **3.** $1.62 **5.** $11.59; $171.39 **7.** 4% **9.** 2%
**11.** $5600 **13.** $116.72 **15.** $276.28 **17.** $194.08
**19.** $2625 **21.** 12% **23.** $185,000 **25.** $5440 **27.** 15%
**29.** $355 **31.** $30; $270 **33.** $2.55; $14.45 **35.** $125;
$112.50 **37.** 40%; $360 **39.** $50; 30% **41.** $849; 21.2%
**43.** 18 **44.** $\frac{22}{7}$ **45.** 265.625 **46.** $-1.5$ **47.** $0.\overline{5}$ **48.** $2.\overline{09}$
**49.** $0.91\overline{6}$ **50.** $1.\overline{857142}$ **51.** $-2.\overline{142857}$ **52.** $-1.58\overline{3}$
**53.** 4,030,000,000,000 **54.** 5,800,000 **55.** 42,700,000
**56.** 6,090,000,000,000 **57.** $4.99 **59.** $17,700

## Calculator Corner, p. 447

**1.** $16,357.18 **2.** $12,764.72

## Exercise Set 7.7, p. 450

**1.** $8 **3.** $113.52 **5.** $925 **7.** $671.88 **9. (a)** $147.95;
**(b)** $10,147.95 **11. (a)** $84.14; **(b)** $6584.14
**13. (a)** $46.03; **(b)** $5646.03 **15.** $441 **17.** $2802.50
**19.** $7853.38 **21.** $99,427.40 **23.** $4243.60
**25.** $28,225.00 **27.** $9270.87 **29.** $129,871.09
**31.** $4101.01 **33.** $1324.58 **35.** Interest: $20.88; amount
applied to principal: $4.69; balance after the payment: $1273.87
**37. (a)** $98; **(b)** interest: $86.56; amount applied to principal:
$11.44; **(c)** interest: $51.20; amount applied to principal: $46.80;

**(d)** At 12.6%, the principal is reduced by $35.36 more than at the
21.3% rate. The interest at 12.6% is $35.36 less than
at 21.3%. **39.** Reciprocals **40.** Divisible by 6 **41.** Additive
**42.** Unit price **43.** Perimeter **44.** Divisible by 3
**45.** Prime **46.** Proportional **47.** 9.38%

## Summary and Review: Chapter 7, p. 453

### Concept Reinforcement

**1.** True **2.** False **3.** True

### Important Concepts

**1.** 0.62625 **2.** $63.\overline{63}\%$, or $63\frac{7}{11}\%$ **3.** $\frac{17}{250}$ **4.** $4.1\overline{6}\%$, or $4\frac{1}{6}\%$
**5.** 10,000 **6.** About 15.3% **7.** 6% **8.** $185,000
**9.** Simple interest: $22.60; total amount due: $2522.60
**10.** $6594.26

### Review Exercises

**1.** 0.04; 0.144 **2.** 0.621; 0.842 **3.** 37.5%, or $37\frac{1}{2}\%$
**4.** $33.\overline{3}\%$, or $33\frac{1}{3}\%$ **5.** 170% **6.** 6.5% **7.** $\frac{6}{25}$ **8.** $\frac{63}{1000}$
**9.** $30.6 = p \times 90$; 34% **10.** $63 = 84\% \times b$; 75
**11.** $a = 38\frac{1}{2}\% \times 168$; 64.68 **12.** $\frac{24}{100} = \frac{16.8}{b}$; 70
**13.** $\frac{42}{30} = \frac{N}{100}$; 140% **14.** $\frac{10.5}{100} = \frac{a}{84}$; 8.82 **15.** 178 students;
84 students **16.** 46% **17.** 2500 mL **18.** 12% **19.** 92
**20.** $24 **21.** 6% **22.** 11% **23.** $42; $308 **24.** 14%
**25.** $2940 **26.** About 18.3% **27.** $36 **28. (a)** $394.52;
**(b)** $24,394.52 **29.** $121 **30.** $7575.25 **31.** $9504.80
**32. (a)** $129; **(b)** interest: $100.18; amount applied to principal:
$28.82; **(c)** interest: $70.72; amount applied to principal: $58.28;
**(d)** At 13.2%, the principal is decreased by $29.46 more than at the
18.7% rate. The interest at 13.2% is $29.46 less than at 18.7%.
**33.** A **34.** C **35.** $66.\overline{6}\%$, or $66\frac{2}{3}\%$ **36.** About 19%

### Understanding Through Discussion and Writing

**1.** A 40% discount is better. When successive discounts are
taken, each is based on the previous discounted price rather
than on the original price. A 20% discount followed by a 22%
discount is the same as a 37.6% discount off the original price.
**2.** Let $S = $ the original salary. After both raises have been given,
the two situations yield the same salary: $1.05 \cdot 1.1S = 1.1 \cdot 1.05S$.
However, the first situation is better for the wage earner, because
$1.1S$ is earned the first year when a 10% raise is given while in
the second situation $1.05S$ is earned that year. **3.** No; the 10%
discount was based on the original price rather than on the sale
price. **4.** For a number $n$, 40% of 50% of $n$ is $0.4(0.5n)$, or
$0.2n$, or 20% of $n$. Thus taking 40% of 50% of a number is the
same as taking 20% of the number. **5.** The interest due on the
30-day loan will be $41.10 while that due on the 60-day loan will
be $131.51. This could be an argument in favor of the 30-day
loan. On the other hand, the 60-day loan puts twice as much
cash at the firm's disposal for twice as long as the 30-day loan
does. This could be an argument in favor of the 60-day loan.
**6.** Answers will vary.

### Test: Chapter 7, p. 459

**1.** [7.1b] 0.147 **2.** [7.1b] 38% **3.** [7.2a] 137.5% **4.** [7.2b] $\frac{13}{20}$
**5.** [7.3a, b] $a = 40\% \cdot 55$; 22 **6.** [7.4a, b] $\frac{N}{100} = \frac{65}{80}$; 81.25%
**7.** [7.5a] 16,692 kidney transplants; 6224 liver transplants; 2263
heart transplants **8.** [7.5a] About 611 at-bats
**9.** [7.5b] 6.7% **10.** [7.5a] 60.6% **11.** [7.6a] $25.20; $585.20
**12.** [7.6b] $630 **13.** [7.6c] $40; $160 **14.** [7.7a] $8.52
**15.** [7.7a] $5356 **16.** [7.7b] $1110.39 **17.** [7.7b] $11,580.07
**18.** [7.5b] Plumber: 757,000, 7.4%; veterinary assistant: 29,000,
40.8%; motorcycle repair technician: 21,000, 14.3%; fitness
professional: 235,000, 63,000 **19.** [7.6c] $50; about 14.3%
**20.** [7.7c] Interest: $36.73; amount applied to the principal:
$17.27; balance after payment: $2687 **21.** [7.3a, b], [7.4a, b] B
**22.** [7.6b] $194,600 **23.** [7.6b], [7.7b] $2546.16

**Cumulative Review: Chapters 1-7, p. 461**

**1.** [7.1b] 0.44; 0.3  **2.** [7.1b] 26.9%  **3.** [7.2a] 112.5%
**4.** [5.5a] $-2.1\overline{6}$  **5.** [6.1a] $\frac{5}{0.5}$, or $\frac{10}{1}$  **6.** [6.2a] $\frac{70\,km}{3\,hr}$, or
$23.\overline{3}$ km/hr, or $23\frac{1}{3}$ km/hr  **7.** [3.5c], [4.3b] <  **8.** [3.5c], [4.3b] <
**9.** [1.6b], [5.6a] 296,200  **10.** [1.6b] 50,000
**11.** [1.9d] 13  **12.** [5.4c] $-1.5$  **13.** [4.2a], [4.5a] $\frac{91}{30}$, or $3\frac{1}{30}$
**14.** [5.2c] $-44.06$  **15.** [1.2a] 515,150  **16.** [5.2b] 0.02
**17.** [4.5b] $\frac{2}{3}$  **18.** [4.3a] $-\frac{110}{63}$  **19.** [3.6a] $\frac{1}{6}$  **20.** [1.4a] 853,142,400
**21.** [5.3a] 1.38036  **22.** [4.6b] $-\frac{3}{2}$, or $-1\frac{1}{2}$  **23.** [5.4a] 12.25
**24.** [1.5a] 123 R 5  **25.** [1.7b] 95  **26.** [5.2d] $-293.87$
**27.** [3.7c] $-9$  **28.** [4.3c] $\frac{1}{12}$  **29.** [6.3b] 40  **30.** [6.3b] $\frac{176}{21}$
**31.** [7.5a] 462,259 visitors  **32.** [1.8a] $177,500
**33.** [6.4a] About 65 games  **34.** [5.7a] $84.95
**35.** [6.2b] 7.495 cents/oz  **36.** [4.2b], [4.4a] $\frac{3}{2}$ mi, or $1\frac{1}{2}$ mi
**37.** [6.4a] 60 mi  **38.** [7.7b] $12,663.69  **39.** [4.6c] 5 pieces
**40.** [7.5b] About 14.2%; 883,000 technicians  **41.** [4.2a] C
**42.** [7.5b] A  **43.** [7.3b], [7.4b] 200%

## CHAPTER 8

**Exercise Set 8.1, p. 471**

**1.** Average: $36.\overline{5}$; median: 21; mode: 2  **3.** Average: 21; median:
18.5; mode: 29  **5.** Average: 21; median: 20; modes: 5, 20
**7.** Average: 5.38; median: 5.7; no mode exists  **9.** Average:
239.5; median: 234; mode: 234  **11.** 23 mpg  **13.** 2.7
**15.** Average: $4.19; median: $3.99; mode: $3.99  **17.** 90
**19.** 263 days  **21.** Bulb A: average time $= 1171.25$ hr; bulb B:
average time $\approx 1251.58$ hr; bulb B is better  **23.** 225.05
**24.** 126.0516  **25.** $\frac{3}{35}$  **26.** $-\frac{14}{15}$  **27.** 118.75%  **28.** 68.75%
**29.** 51.2%  **30.** 97.81%  **31.** $a = 30$; $b = 58$  **33.** $3475

**Exercise Set 8.2, p. 477**

**1.** 100 calories  **2.** Boca All American Flame Grilled Meatless;
Franklin Farms Portabella Fresh; Gardenburger Portabella
**5.** Average: 3.5 g; median: 3 g; mode: 3 g  **7.** Most expensive:
Lightlife Meatless Light; least expensive: Boca All American
Flame Grilled Meatless  **9.** Greatest fat: Veggie Patch Garlic
Portabella; least fat: Franklin Farms Portabella Fresh and
Lightlife Meatless Light  **11.** 92°  **13.** 108°  **15.** 85°, 60%;
90°, 40%; 100°, 10%  **17.** 90° and higher  **19.** 30% and
higher  **21.** $90\% - 40\% = 50\%$  **23.** 483,612,200 mi
**25.** Neptune  **27.** All  **29.** 11 Earth diameters
**31.** Average: 31,191.75 mi; median: 19,627.5 mi; no mode exists
**33.** White rhino  **35.** About 1350 rhinos
**37.** About 4100 rhinos  **39.** Cabinets: $13,444; countertops:
$4033.20; appliances: $2151.04; fixtures: $806.64
**40.** Cabinets: $3597.55; countertops: $1276.55; labor: $2901.25;
flooring: $696.30

**Mid-Chapter Review: Chapter 8, p. 481**

**1.** True  **2.** True  **3.** False
**4.** $\frac{60 + 45 + 115 + 15 + 35}{5} = \frac{270}{5} = 54$  **5.** 2.1, 4.8, 6.3, 8.7, 11.3,
14.5; 6.3 and 8.7; $\frac{6.3 + 8.7}{2} = \frac{15}{2} = 7.5$; the median is 7.5.
**6.** Average: 83; median: 45; mode: 29  **7.** Average: 18.45;
median: 13.895; no mode  **8.** Average: $\frac{1}{2}$; median: $\frac{1}{2}$; no mode
**9.** Average: 126; median: 116; no mode  **10.** Average: $6.09;
median: $5.24; modes: $4.96 and $5.24  **11.** Average: $\frac{27}{32}$;
median: $\frac{13}{16}$; no mode  **12.** Average: 6; median: 7; modes:
5 and 7  **13.** Average: 38.2; median: 38.2; no mode
**14.** 8 oz  **15.** 6%  **16.** Hershey's Special Dark chocolate bar
**17.** 7 oz  **18.** Nabisco Chips Ahoy cookies  **19.** 90 guns per
100 civilians  **20.** 60 more guns per 100 civilians
**21.** 45 guns per 100 civilians; 55 guns per 100 civilians
**22.** India  **23.** 47.5 guns per 100 civilians  **24.** 50 guns per
100 civilians  **25.** At an average speed of 20 mph, the trip would
take $1\frac{1}{2}$ hr (30 mi $\div$ 20 mph $= 1\frac{1}{2}$ hr). But the driver could have
driven at a speed of 75 mph for a brief period during that time.

**26.** Answers may vary. Some would ask for the average salary
since it is a center point that places equal emphasis on all the
salaries in the firm. Some would ask for the median salary
since it is a center point that deemphasizes the extremely high
and extremely low salaries. Some would ask for the mode of
the salaries since it might indicate the salary you are most
likely to earn.

**Exercise Set 8.3, p. 489**

**1.** Miniature tall bearded  **3.** About 23.2 in.  **5.** 16 in. to 26 in.
**7.** Tall bearded  **9.** 25 in.  **11.** 190 calories  **13.** 1 slice of
chocolate cake with fudge frosting  **15.** 1 cup of premium
chocolate ice cream  **17.** About 125 calories  **19.** 1950 and
1970  **21.** About 175,000 bachelor's degrees
**23.**

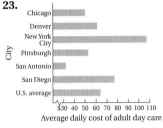

**25.** Chicago, Denver,
Pittsburgh, and
San Antonio
**27.** $43 higher
**29.** New York City
**31.** 27 min
**33.** 3.2 million tourists;
4.3 million tourists
**35.** 2007 to 2008

**37.** 9–11 A.M.  **39.** About 300 bank crimes
**41.**

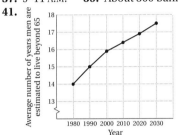

**43.** 25%  **45.** 10.1%
**46.** Natural  **47.** Add;
divide  **48.** Simple
**49.** Marked price; rate of
discount; discount; sale
price  **50.** Compound
**51.** Repeating
**52.** Distributive
**53.** Commutative

**Translating for Success, p. 496**

**1.** D  **2.** B  **3.** J  **4.** K  **5.** I  **6.** F  **7.** N  **8.** E
**9.** L  **10.** M

**Exercise Set 8.4, p. 497**

**1.** 11%  **3.** 93,750 students  **5.** Japan  **7.** 18%
**9.** 38%, or about 390 people

**11.**

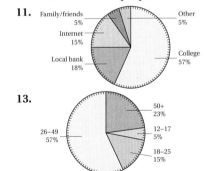

**13.**

**Summary and Review: Chapter 8, p. 499**

**Concept Reinforcement**

**1.** False  **2.** True  **3.** True

**Important Concepts**

**1.** Average: 8; median: 8; mode: 8  **2.** Quaker Organic Maple &
Brown Sugar; $0.54 per serving  **3.** 12 g  **4.** Arrowhead
Stadium and Candlestick Park  **5.** About $110 million more
**6.** 80 yr and older  **7.** 27%

## Review Exercises

**1.** 38.5 **2.** 13.4 **3.** 1.55 **4.** 1840 **5.** $16.\overline{6} **6.** 321.\overline{6}
**7.** 96 **8.** 28 mpg **9.** 3.1 **10.** 38.5 **11.** 14 **12.** 1.8
**13.** 1900 **14.** $17 **15.** 375 **16.** Average: $260; median:
$228 **17.** 26 **18.** 11 and 17 **19.** 0.2 **20.** 700 and 800
**21.** $17 **22.** 20 **23.** Battery A: average ≈ 43.04 hr;
battery B: average = 41.55 hr; battery A is better. **24.** $30.37
**25.** $10.85 **26.** $3.76 **27.** 18 champions **28.** 30–34 years
**29.** 19 more champions **30.** 2004, 2007, 2008 **31.** 60%
**32.** About 52% **33.** 2006 **34.** 2001 **35.** 2007
**36.** 2001, 2002, 2005 **37.** 2002, 2005 **38.** 19 higher
**39.** 284 **40.** About 283 **41.** 36% **42.** Marines
**43.** 350,000 **44.** 10% more **45.** D **46.** A
**47.**

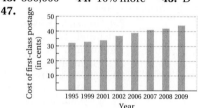

**48.**

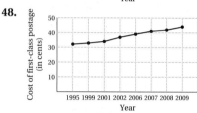

**49.**

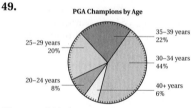

**50.** $a = 316, b = 349$

## Understanding Through Discussion and Writing

**1.** The equation could represent a person's average income
during a 4-yr period. Answers may vary. **2.** Bar graphs that
show change over time can be successfully converted to line
graphs. Other bar graphs cannot be successfully converted to
line graphs. **3.** We can use circle graphs to visualize how
the numbers of items in various categories compare in size.
**4.** A bar graph is convenient for showing comparisons. A line
graph is convenient for showing a change over time as well as to
indicate patterns or trends. The choice of which to use to graph
a particular set of data would probably depend on the type of
data analysis desired. **5.** The average, the median, and the
mode are "center points" that characterize a set of data. You
might use the average to find a center point that is midway
between the extreme values of the data. The median is a center
point that is in the middle of all the data. That is, there are as
many values less than the median as there are values greater
than the median. The mode is a center point that represents the
value or values that occur most frequently. **6.** Circle graphs
are similar to bar graphs in that both allow us to tell at a glance
how items in various categories compare in size. They differ in
that circle graphs show percents whereas bar graphs show
actual numbers of items in a given category.

## Test: Chapter 8, p. 505

**1.** [8.1a] 49.5 **2.** [8.1a] 2.6 **3.** [8.1a] 15.5
**4.** [8.1b, c] 50.5; no mode exists **5.** [8.1b, c] 3; 1 and 3
**6.** [8.1b, c] 17.5; 17 and 18 **7.** [8.1a] 33 mpg **8.** [8.1a] 76

**9.** [8.1a] 2.9 **10.** [8.1d] Bar A: average ≈ 8.417; bar B:
average ≈ 8.417; equal quality **11.** [8.2a] 179 lb
**12.** [8.2a] 5 ft 3 in.; medium frame **13.** [8.2a] 9 lb
**14.** [8.2a] 32 lb **15.** [8.2b] Spain **16.** [8.2b] Norway and
the United States **17.** [8.2b] 900 lb **18.** [8.2b] 1000 lb
**19.** [8.3b]

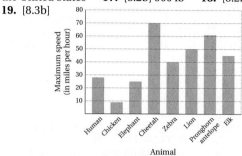

Animal

**20.** [8.2a], [8.3a] 61 mph **21.** [8.2a], [8.3a] No; the zebra can
run 12 mph faster than a human.
**22.** [8.1a], [8.2a], [8.3a] 41 mph
**23.** [8.1b], [8.2a], [8.3a] 42.5 mph **24.** [8.3c] 53%
**25.** [8.3c] 41% **26.** [8.3c] 1967 **27.** [8.3c] 2006
**28.** [8.3b]

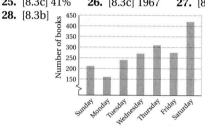

Day

**29.** [8.3d]

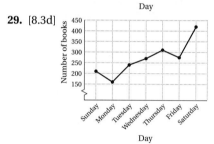

Day

**30.** [8.4b]

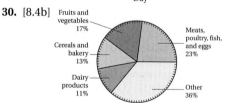

**31.** [8.4a] C **32.** [8.1a, b] $a = 74, b = 111$

## Cumulative Review: Chapters 1-8, p. 509

**1.** [5.3b] 62,100,000,000 **2.** [8.1a] 24 mpg
**3.** [1.1a] 5 hundreds **4.** [1.9c] 128 **5.** [3.1a] 1, 2, 3, 4, 5, 6,
10, 12, 15, 20, 30, 60 **6.** [5.1d] 52.0 **7.** [4.4a] $\frac{33}{10}$
**8.** [5.3b] $2.10 **9.** [7.2a] 35% **10.** [6.3a] No **11.** [4.5a] $6\frac{7}{10}$
**12.** [5.2c] −2.5091 **13.** [4.3a] $-\frac{4}{3}$ **14.** [5.2b] 325.43
**15.** [4.6a] 15 **16.** [1.4a] 2,740,320 **17.** [3.7b] $\frac{9}{10}$
**18.** [4.4b] $4361\frac{1}{2}$ **19.** [6.3b] $9\frac{3}{5}$ **20.** [3.7c] $\frac{3}{4}$
**21.** [5.4b] −6.8 **22.** [1.7b] 15,312 **23.** [6.2b] 20.6¢/oz
**24.** [7.5a] 3324 students **25.** [4.6c] $\frac{1}{4}$ yd **26.** [3.6b] $\frac{3}{8}$ cup
**27.** [5.7a] 6.2 lb **28.** [1.8a] 2572 billion kWh
**29.** [7.5a] 2.5%; 0.2% **30.** [7.5b] About 4.1%
**31.** [4.2b], [4.3d] $\frac{1}{4}$ **32.** [6.4a] 1122 defective valves
**33.** [5.7a] $9.55 **34.** [7.6b] 7% **35.** [8.1a, b] $56.52; $56.76

**36.** [8.3b]

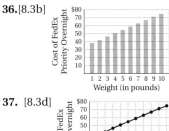

**37.** [8.3d]

$80
70
60
50
40
30
20
10

Cost of FedEx Priority Overnight

1 2 3 4 5 6 7 8 9 10
Weight (in pounds)

**38.** [7.5b], [8.1a] 12% decrease

## CHAPTER 9

### Exercise Set 9.1, p. 516

**1.** 12 **3.** $\frac{1}{12}$ **5.** 5280 **7.** 108 **9.** 7 **11.** $1\frac{1}{2}$, or 1.5
**13.** 26,400 **15.** $5\frac{1}{4}$, or 5.25 **17.** $3\frac{1}{3}$ **19.** 37,488
**21.** $1\frac{1}{2}$, or 1.5 **23.** $1\frac{1}{4}$, or 1.25 **25.** 110 **27.** 2 **29.** 300
**31.** 30 **33.** $\frac{1}{36}$ **35.** 126,720 **37.** 2988 **39.** $\frac{37}{400}$ **40.** $\frac{7}{8}$
**41.** $\frac{11}{40}$ **42.** $\frac{57}{100}$ **43.** 137.5%, or $137\frac{1}{2}$% **44.** $66.\overline{6}$%, or $66\frac{2}{3}$%
**45.** 25% **46.** 43.75%, or $43\frac{3}{4}$% **47.** $\frac{49.2}{368.7}$, or $\frac{492}{3687}$, or $\frac{164}{1229}$, $\frac{368.7}{49.2}$,
or $\frac{3687}{492}$, or $\frac{1229}{164}$ **48.** $\frac{96,385}{9,870,000}$, or $\frac{19,277}{1,974,000}$; $\frac{9,870,000}{96,385}$, or $\frac{1,974,000}{19,277}$
**49.** Length: 5400 in., or 450 ft; breadth: 900 in., or 75 ft;
height: 540 in., or 45 ft

### Exercise Set 9.2, p. 523

**1.** (a) 1000; (b) 0.001 **3.** (a) 10; (b) 0.1 **5.** (a) 0.01;
(b) 100 **7.** 6700 **9.** 0.98 **11.** 8.921 **13.** 0.05666
**15.** 566,600 **17.** 4.77 **19.** 688 **21.** 0.1 **23.** 100,000
**25.** 142 **27.** 0.82 **29.** 450 **31.** 0.000024 **33.** 0.688
**35.** 230 **37.** 3.92 **39.** 180; 0.18 **41.** 278; 27.8
**43.** 48,440; 48.44 **45.** 400; 400 **47.** 0.027; 0.00027
**49.** 442,000; 44,200 **51.** 0.234 **52.** 0.108 **53.** 13.8474
**54.** $-80\frac{1}{2}$ **55.** 37.5% **56.** 62.5% **57.** $66.\overline{6}$%, or $66\frac{2}{3}$%
**58.** 20% **59.** 0.9 **60.** 0.108 **61.** $\frac{47}{75}$ **62.** $\frac{59}{72}$
**63.** 1.0 m **65.** 1.4 cm

### Exercise Set 9.3, p. 527

**1.** 100.65 **3.** 727.4394 **5.** 104.585 **7.** 289.62
**9.** 112.63 **11.** 9.14 **13.** 81.8896 **15.** 1250.061
**17.** 6.18109 **19.** 1376.136 **21.** 0.51181

| | yd | cm | in. | m | mm |
|---|---|---|---|---|---|
| **23.** | 0.2361 | 21.59 | $8\frac{1}{2}$ | 0.2159 | 215.9 |
| **25.** | 52.9934 | 4844 | 1907.0828 | 48.44 | 48,440 |
| **27.** | 4 | 365.6 | 144 | 3.656 | 3656 |
| **29.** | 0.000295 | 0.027 | 0.0106299 | 0.00027 | 0.27 |
| **31.** | 483.548 | 44,200 | 17,401.54 | 442 | 442,000 |

**33.** 28,700,000 **34.** 763,000,000,000 **35.** 1 in. = 25.4 mm
**37.** 21.3 mph

### Mid-Chapter Review: Chapter 9, p. 529

**1.** False **2.** True **3.** False **4.** True **5.** True
**6.** $16\frac{2}{3}$ yd $= 16\frac{2}{3} \times 1$ yd $= \frac{50}{3} \times 3$ ft $= 50$ ft

**7.** 13,200 ft = 13,200 ft $\times \frac{1\text{ mi}}{5280\text{ ft}} = 2.5$ mi
**8.** 520 mm = 520 mm $\times \frac{1\text{ m}}{1000\text{ mm}} = 0.52$ m $\times \frac{1\text{ km}}{1000\text{ m}} =$
0.00052 km **9.** 10,200 mm = 10,200 mm $\times \frac{1\text{ m}}{1000\text{ mm}} \approx$
10.2 m $\times \frac{3.281\text{ ft}}{1\text{ m}} = 33.4662$ ft **10.** 9680 **11.** 70
**12.** 2.405 **13.** 0.00015 **14.** 22,680 **15.** 1200
**16.** 60,000 **17.** 850 **18.** 5 **19.** 1251 **20.** 52,800
**21.** 180,000 **22.** 84,000 **23.** 700 **24.** 0.4
**25.** 0.8009 **26.** 10 **27.** 19,200 **28.** 100
**29.** 0.000000001 **30.** 400,000 **31.** 760,320 **32.** 118.116
**33.** 0.8 **34.** 2.285 **35.** 60 **36.** 0.0025
**37.** $\frac{1}{4}$ yd — 24,000 dm
144 in. — 1320 yd
2400 m — 2400 mm
0.75 mi — 9 in.
24 m — 0.024 km
240 cm — 12 ft
**38.** 3.5 ft, 2 yd, 100 in., $\frac{1}{100}$ mi, 1000 in., 430 ft, 6000 ft
**39.** 150 hm, 13 km, 310 dam, 300 m, 33,000 mm, 3240 cm,
250 dm **40.** $\frac{1}{2}$ ft, 65 cm, 2 yd, 2.5 m, 1.5 mi, 3 km
**41.** The student should have multiplied by $\frac{1}{12}$ (or divided by 12)
to convert inches to feet. The correct procedure is as follows:
$$23\text{ in.} = 23\text{ in.} \times \frac{1\text{ ft}}{12\text{ in.}} = \frac{23\text{ in.}}{12\text{ in.}} \times 1\text{ ft}$$
$$= \frac{23}{12} \times \frac{\text{in.}}{\text{in.}} \times 1\text{ ft} = \frac{23}{12} \times 1\text{ ft} = \frac{23}{12}\text{ ft.}$$
**42.** Metric units are based on tens, so computations and
conversions with metric units can be done by moving a decimal
point. American units, which are not based on tens, require
more complicated arithmetic in computations and conversions.
**43.** A larger unit can be expressed as an equivalent number of
smaller units. Thus when converting from a larger unit to a
smaller unit, we can express the quantity as a number times
1 unit and then substitute the equivalent number of smaller
units for the larger unit. When converting from a smaller unit
to a larger unit, we multiply by 1 using one larger unit in the
numerator and the equivalent number of smaller units in the
denominator. This allows us to "cancel" the smaller units,
leaving the larger units. **44.** Answers may vary.

### Exercise Set 9.4, p. 535

**1.** 2000 **3.** 3 **5.** 64 **7.** 12,640 **9.** 0.1 **11.** 5
**13.** 26,000,000,000 lb **15.** 1000 **17.** 10 **19.** $\frac{1}{100}$, or 0.01
**21.** 1000 **23.** 10 **25.** 234,000 **27.** 5.2 **29.** 6.7
**31.** 0.0502 **33.** 8.492 **35.** 58.5 **37.** 800,000 **39.** 1000
**41.** 0.0034 **43.** 0.0603 **45.** 1000 **47.** 0.325 **49.** 210,600
**51.** 0.0049 **53.** 125 mcg **55.** 0.875 mg; 875 mcg
**57.** 4 tablets **59.** 8 cc **61.** $\frac{7}{20}$ **62.** $\frac{99}{100}$ **63.** $\frac{171}{200}$
**64.** $\frac{171}{500}$ **65.** $\frac{3}{8}$ **66.** $\frac{2}{3}$ **67.** $\frac{5}{6}$ **68.** $\frac{1}{6}$ **69.** 187,200
**70.** About 324 students **71.** \$24.30 **72.** 12%
**73.** 7500 sheets **74.** 46 test tubes; 2 cc left over
**75.** 144 packages **77.** (a) 109.134 g; (b) 9.104 g; (c) Golden
Jubilee: 3.85 oz; Hope: 0.321 oz

### Exercise Set 9.5, p. 542

**1.** 1000; 1000 **3.** 87,000 **5.** 0.049 **7.** 0.000401
**9.** 78,100 **11.** 320 **13.** 10 **15.** 32 **17.** 20
**19.** 14 **21.** 88

| | gal | qt | pt | cups | oz |
|---|---|---|---|---|---|
| **23.** | 1.125 | 4.5 | 9 | 18 | 144 |
| **25.** | 16 | 64 | 128 | 256 | 2048 |
| **27.** | 0.25 | 1 | 2 | 4 | 32 |
| **29.** | 0.3984375 | 1.59375 | 3.1875 | 6.375 | 51 |

| | L | mL | cc | cm$^3$ |
|---|---|---|---|---|
| **31.** | 2 | 2000 | 2000 | 2000 |
| **33.** | 64 | 64,000 | 64,000 | 64,000 |
| **35.** | 0.355 | 355 | 355 | 355 |

**37.** 2000 mL   **39.** 0.32 L   **41.** 59.14 mL   **43.** 500 mL
**45.** 125 mL/hr   **47.** 9   **49.** $\frac{1}{5}$   **51.** 6   **53.** 15   **55.** 45.2%
**56.** 99.9%   **57.** 33.$\overline{3}$%, or $33\frac{1}{3}$%   **58.** 66.$\overline{6}$%, or $66\frac{2}{3}$%
**59.** 55%   **60.** 105%   **61.** 88%   **62.** 8%
**63.** $\frac{10.6}{71.3}$, or $\frac{106}{713}$; $\frac{60.7}{10.6}$, or $\frac{607}{106}$   **64.** $\frac{1}{4}$; $\frac{3}{4}$; $\frac{1}{3}$   **65.** 0.21$\overline{3}$ oz
**67.** $0.787/L

**Calculator Corner, p. 547**

**1.** 41°F   **2.** 122°F   **3.** 20°C   **4.** 45°C

**Exercise Set 9.6, p. 548**

**1.** 24   **3.** 60   **5.** $365\frac{1}{4}$   **7.** 0.05   **9.** 8.2   **11.** 6.5
**13.** 10.75   **15.** 336   **17.** 4.5   **19.** 56   **21.** 86,164.2 sec
**23.** 77°F   **25.** 104°F   **27.** 186.8°F   **29.** −4°F
**31.** 35.6°F   **33.** −11.2°F   **35.** 5432°F   **37.** 30°C
**39.** −25°C   **41.** 81.$\overline{1}$°C   **43.** 60°C   **45.** 20°C   **47.** −12.$\overline{2}$°C
**49.** 37°C   **51.** (a) 136°F = 57.$\overline{7}$°C, $56\frac{2}{3}$°C = 134°F; (b) 2°F
**53.** −11.$\overline{1}$°C   **55.** Compound   **56.** Ratio   **57.** Median
**58.** Commission   **59.** Similar   **60.** Mean   **61.** Composite
**62.** Subtrahend   **63.** About 0.03 yr   **65.** About 31,688 yr
**67.** 0.25

**Translating for Success, p. 553**

**1.** E   **2.** H   **3.** O   **4.** A   **5.** G   **6.** C   **7.** M   **8.** I
**9.** K   **10.** N

**Exercise Set 9.7, p. 554**

**1.** 144   **3.** 640   **5.** $\frac{1}{144}$   **7.** 198   **9.** 396   **11.** 12,800
**13.** 27,878,400   **15.** 5   **17.** 1   **19.** $\frac{1}{640}$, or 0.0015625
**21.** 25,792   **23.** 5,210,000   **25.** 140   **27.** 23.456
**29.** 0.085214   **31.** 2500   **33.** 0.4728   **35.** $240   **36.** $212
**37.** (a) $484.11; (b) $15,984.11   **38.** (a) $209.59; (b) $8709.59
**39.** (a) $220.93; (b) $6620.93   **40.** (a) $37.97; (b) $4237.97
**41.** 10.76   **43.** 1.67   **45.** About 14,233 m$^2$

**Summary and Review: Chapter 9, p. 556**

**Concept Reinforcement**

**1.** True   **2.** False   **3.** True   **4.** False   **5.** False   **6.** False

**Important Concepts**

**1.** $\frac{7}{3}$, or $2\frac{1}{3}$ or 2.$\overline{3}$   **2.** 13,200   **3.** 1200   **4.** 0.000046
**5.** 10.94   **6.** 5.14   **7.** 0.00978   **8.** 64   **9.** 42.67   **10.** 1
**11.** 154.4°F   **12.** 40°C   **13.** 9   **14.** 5240

**Review Exercises**

**1.** $2\frac{2}{3}$   **2.** 30   **3.** 0.03   **4.** 0.004   **5.** 72   **6.** 400,000
**7.** $1\frac{1}{6}$   **8.** 0.15   **9.** 218.8   **10.** 32.18   **11.** 10; 0.01
**12.** 305,000; 30,500   **13.** 112   **14.** 0.004   **15.** $\frac{4}{15}$, or 0.2$\overline{6}$
**16.** 0.464   **17.** 180   **18.** 4700   **19.** 16,140   **20.** 830
**21.** $\frac{1}{4}$, or 0.25   **22.** 0.04   **23.** 200   **24.** 30   **25.** 0.0007
**26.** 0.06   **27.** 1600   **28.** 400   **29.** 1.25   **30.** 50
**31.** 160   **32.** 7.5   **33.** 13.5   **34.** 60   **35.** 0.0302
**36.** 5.25   **37.** 6 mL   **38.** 3000 mL   **39.** 250 mcg
**40.** 21.2°F   **41.** 113°F   **42.** 20°C   **43.** −28.$\overline{8}$°C   **44.** 36
**45.** 300,000   **46.** 14.375   **47.** 0.06   **48.** C   **49.** B
**50.** 17.64 sec   **51.** 1 gal = 128 oz, so 1 oz of water (as capacity)
weighs $\frac{8.3453}{128}$ lb, or about 0.0652 lb. An ounce of pennies weighs
$\frac{1}{16}$ lb, or 0.0625 lb. Thus an ounce of water (as capacity) weighs
more than an ounce of pennies.

## Understanding Through Discussion and Writing

**1.** Grams are more easily converted to other units of mass than
ounces. Since 1 gram is much smaller than 1 ounce, masses that
might be expressed using fractional or decimal parts of ounces
can often be expressed by whole numbers when grams are used.
**2.** A single container is all that is required for both types of
measure.   **3.** Consider the table on p. 521 in the text. Moving
one place in the table corresponds to moving the decimal point
one place. To convert units of length, we determine the
corresponding number of moves in the diagram and then move
the decimal point that number of places. Since area involves
square units, we multiply the number of moves by 2 and move
the decimal point that number of places when converting units
of area.   **4.** Since metric units are based on 10, they are more
easily converted than American units.
**5.** (a) 23°C = 73.4°F, so you would want to play golf.
(b) 10°C = 50°F, so you would not want to take a bath.
(c) Since 0°C = 32°F, the freezing point of water, the lake would
certainly be frozen at the lower temperature of −10°C and it
would be safe to go ice skating.   **6.** 1 m ≈ 3.281 ft, so
1 square meter ≈ 3.281 ft × 3.281 ft = 10.764961 square feet.
1 yd = 3 ft, so 1 square yard = 3 ft × 3 ft = 9 square feet. Thus
one square meter is larger than 1 square yard.

**Test: Chapter 9, p. 561**

**1.** [9.1a] 48   **2.** [9.1a] $\frac{1}{3}$   **3.** [9.2a] 6000   **4.** [9.2a] 0.87
**5.** [9.3a] 182.8   **6.** [9.3a] 1490.4   **7.** [9.2a] 5; 0.005
**8.** [9.2a] 1854.2; 185.42   **9.** [9.5a] 3.08   **10.** [9.5a] 240
**11.** [9.4a] 64   **12.** [9.4a] 8220   **13.** [9.4b] 3800
**14.** [9.4b] 0.4325   **15.** [9.4b] 2.2   **16.** [9.6a] 300
**17.** [9.6a] 360   **18.** [9.5a] 32   **19.** [9.5a] 1280   **20.** [9.5a] 40
**21.** [9.4c] 370   **22.** [9.6b] 35°C   **23.** [9.6b] 138.2°F
**24.** [9.3a] 1.094; 100; 39.370; 1000   **25.** [9.3a] 36,377.2; 14,328;
363.772; 363,772   **26.** [9.5b] 2500 mL   **27.** [9.4c] 1500 mcg
**28.** [9.5b] About 118.28 mL   **29.** [9.7a] 1728   **30.** [9.7b] 0.0003
**31.** [9.6b] D   **32.** [9.3a] About 39.47 sec

**Cumulative Review: Chapters 1–9, p. 563**

**1.** [5.3b] 37,300,000; 54,600,000   **2.** [6.2a] 33 mpg
**3.** [7.5b] About 26.3%   **4.** [7.5b] About 1.9%
**5.** [1.4a] 50,854,100   **6.** [3.6a] $-\frac{1}{12}$   **7.** [5.5c] −13.8
**8.** [4.6b] $\frac{2}{3}$   **9.** [5.4a] 35.6   **10.** [4.5b] $\frac{5}{6}$   **11.** [3.2a] Yes
**12.** [3.2a] No   **13.** [3.1d] 3 · 3 · 11   **14.** [4.1a] 245
**15.** [5.5b] 35.8   **16.** [5.1a] One hundred three and sixty-four
thousandths   **17.** [8.1a, b] 17.8$\overline{3}$; 17.5   **18.** [7.1b] 8%
**19.** [7.2a] 60%   **20.** [9.1a] 6   **21.** [9.4a] 0.375, or $\frac{3}{8}$
**22.** [9.6b] 59°F   **23.** [9.5a] 87   **24.** [9.6a] 0.15, or $\frac{3}{20}$
**25.** [9.2a] 0.17   **26.** [9.3a] 3539.8   **27.** [9.5a] 2
**28.** [9.4c] 230   **29.** [9.7a] 108   **30.** [5.4b] −150.5
**31.** [1.7b] 19,248   **32.** [3.7c] $\frac{15}{2}$   **33.** [4.3c] $-\frac{58}{35}$
**34.** [4.2b], [4.4a] $1\frac{1}{4}$ hr   **35.** [5.3b], [5.7a] 181,784,400,000 lb
**36.** [5.7a], [6.2a] 255.8 mi; 15.9875 mpg   **37.** [7.6b] $16,125
**38.** [7.7b] $2388.10   **39.** [6.4a] $8\frac{3}{4}$ lb   **40.** [3.6b] $13,200
**41.** [7.5a] Yes   **42.** [9.4a] 240 oz   **43.** [1.8a] 60 dips
**44.** [4.5c], [4.6c] About 25 or 26 dips   **45.** [5.7a] $119.40
**46.** [1.8a], [5.7a] $89.70   **47.** [3.7c] $\frac{5}{2}$

## CHAPTER 10

**Exercise Set 10.1, p. 569**

**1.** 17 mm   **3.** 15.25 in.   **5.** 18 km   **7.** 30 ft   **9.** 16 yd
**11.** 88 ft   **13.** 182 mm   **15.** 27 ft   **17.** 122 cm
**19.** 172 in., or 14 ft 4 in.   **21.** (a) 228 ft; (b) $1046.52
**23.** $19.20   **24.** $96   **25.** 1000   **26.** 1331   **27.** 225
**28.** 484   **29.** 49   **30.** 64   **31.** 5%   **32.** 11%   **33.** 64 in.

**Exercise Set 10.2, p. 577**

**1.** $15 \text{ km}^2$ **3.** $1.4 \text{ in}^2$ **5.** $6\frac{1}{4} \text{ yd}^2$ **7.** $8100 \text{ ft}^2$ **9.** $50 \text{ ft}^2$
**11.** $169.883 \text{ cm}^2$ **13.** $41\frac{2}{9} \text{ in}^2$ **15.** $484 \text{ ft}^2$
**17.** $3237.61 \text{ km}^2$ **19.** $28\frac{57}{64} \text{ yd}^2$ **21.** $32 \text{ cm}^2$ **23.** $60 \text{ in}^2$
**25.** $104 \text{ ft}^2$ **27.** $45.5 \text{ in}^2$ **29.** $8.05 \text{ cm}^2$ **31.** $297 \text{ cm}^2$
**33.** $7 \text{ m}^2$ **35.** $1197 \text{ m}^2$ **37.** **(a)** About $8473 \text{ ft}^2$; **(b)** about
$102 **39.** $630.36 \text{ ft}^2$ **41.** **(a)** $819.75 \text{ ft}^2$; **(b)** 3 gal; **(c)** $74.85
**43.** $80 \text{ cm}^2$ **45.** $675 \text{ cm}^2$ **47.** $21 \text{ cm}^2$ **49.** $144 \text{ ft}^2$
**51.** 234 **52.** 230 **53.** 336 **54.** 24 **55.** 0.724
**56.** 0.0724 **57.** 2520 **58.** 6 **59.** 28 **60.** $22\frac{1}{2}$ **61.** 12
**62.** 46,252.8 **63.** $30,474.86 **64.** $413,458.31
**65.** $641,566.26 **66.** $429,610.21 **67.** $16,914 \text{ in}^2$

**Calculator Corner, p. 584**

**1.** Left to the student **2.** Left to the student

**Exercise Set 10.3, p. 587**

**1.** 14 cm; 44 cm; $154 \text{ cm}^2$ **3.** $1\frac{1}{2}$ in.; $4\frac{5}{7}$ in.; $1\frac{43}{56} \text{ in}^2$
**5.** 16 ft; 100.48 ft; $803.84 \text{ ft}^2$ **7.** 0.7 cm; 4.396 cm; $1.5386 \text{ cm}^2$
**9.** Diameter: 12.74 ft; circumference: about 40 ft; area: about
$127.41 \text{ ft}^2$; about $46.67 \text{ ft}^2$ larger than the medium net and about
$19.76 \text{ ft}^2$ larger than the large net **11.** About $30.96 \text{ in}^2$ larger
**13.** About 24,889 mi **15.** Maximum circumference of barrel:
$8\frac{9}{14}$ in.; minimum circumference of handle: $2\frac{86}{133}$ in.
**17.** $65.94 \text{ yd}^2$ **19.** 45.68 ft **21.** 26.84 yd **23.** 45.7 yd
**25.** $100.48 \text{ m}^2$ **27.** $6.9972 \text{ cm}^2$ **29.** $64.4214 \text{ in}^2$ **31.** 543
**32.** 0.00543 **33.** 5 lb **34.** $730 **35.** $43,560 \text{ ft}^2$; 1311.6 ft;
$599.96 **37.** $43,595.47395 \text{ ft}^2$; 739.9724 ft; $449.97
**39.** $43,560 \text{ ft}^2$; 844 ft; $449.97

**Mid-Chapter Review: Chapter 10, p. 591**

**1.** False **2.** True **3.** True **4.** False **5.** True
**6.** $p = 2 \cdot (10 \text{ ft} + 3 \text{ ft})$ **7.** $A = \frac{1}{2} \cdot 12 \text{ cm} \cdot 8 \text{ cm}$
$\quad p = 2 \cdot (13 \text{ ft})$ $\quad A = \frac{12 \cdot 8}{2} \text{ cm}^2$
$\quad p = 26 \text{ ft};$ $\quad A = \frac{96}{2} \text{ cm}^2$, or $48 \text{ cm}^2$
$\quad A = 10 \text{ ft} \cdot 3 \text{ ft}$
$\quad A = 10 \cdot 3 \cdot \text{ft} \cdot \text{ft}$
$\quad A = 30 \text{ ft}^2$
**8.** $C \approx 3.14 \cdot 10.2 \text{ in.}$ **9.** 76 mm
$\quad C = 32.028 \text{ in.};$

$\quad A \approx 3.14 \cdot 5.1 \text{ in.} \cdot 5.1 \text{ in.}$
$\quad A = 81.6714 \text{ in}^2$
**10.** $P = 50\frac{2}{3} \text{ ft}; A = 160\frac{4}{9} \text{ ft}^2$ **11.** $800 \text{ in}^2$ **12.** $\frac{9}{16} \text{ yd}^2$
**13.** $66 \text{ km}^2$ **14.** $C = 43.96 \text{ in.}; A = 153.86 \text{ in}^2$
**15.** $C = 27.004 \text{ cm}; A = 58.0586 \text{ cm}^2$ **16.** Area of a circle with
radius 4 ft: $16 \cdot \pi \text{ ft}^2$; Area of a square with side 4 ft: $16 \text{ ft}^2$;
Circumference of a circle with radius 4 ft: $8 \cdot \pi$ ft; Area of a
rectangle with length 8 ft and width 4 ft: $32 \text{ ft}^2$; Area of a triangle
with base 4 ft and height 8 ft: $16 \text{ ft}^2$; Perimeter of a square with
side 4 ft: 16 ft; Perimeter of a rectangle with length 8 ft and width
4 ft: 24 ft **17.** The area of a 16-in.-diameter pizza is
approximately $3.14 \cdot 8 \text{ in.} \cdot 8 \text{ in.}$, or $200.96 \text{ in}^2$. At $16.25, its unit
price is $\dfrac{\$16.25}{200.96 \text{ in}^2}$, or about $0.08 \text{ in}^2$. The area of 10-in.-diameter
pizza is approximately $3.14 \cdot 5 \text{ in.} \cdot 5 \text{ in.}$, or $78.5 \text{ in}^2$. At $7.85, its
unit price is $\dfrac{\$7.85}{78.5 \text{ in}^2}$, or $0.10/\text{in}^2$. Since the 16-in.-diameter
pizza has the lower unit price, it is a better buy.
**18.** No; let $l$ and $w$ represent the length and the width of the
smaller rectangle. Then $3 \cdot l$ and $3 \cdot w$ represent the length and the
width of the larger rectangle. The area of the first rectangle is $l \cdot w$,
but the area of the second is $3 \cdot l \cdot 3 \cdot w = 3 \cdot 3 \cdot l \cdot w = 9 \cdot l \cdot w$,
or 9 times the area of the smaller rectangle. **19.** Yes; let $s$
represent the length of a side of the larger square. Then $\frac{1}{2}s$

represents the length of a side of the smaller square. The perimeter
of the larger square is $4 \cdot s$, and the perimeter of the smaller square
is $4 \cdot \frac{1}{2}s = \frac{1}{2} \cdot 4s$, or $\frac{1}{2}$ the perimeter of the larger square.
**20.** For a rectangle with length $l$ and width $w$,
$\quad P = l + w + l + w$
$\quad\quad = (l + w) + (l + w)$
$\quad\quad = 2 \cdot (l + w).$
We also have
$\quad P = l + w + l + w$
$\quad\quad = (l + l) + (w + w)$
$\quad\quad = 2 \cdot l + 2 \cdot w.$
**21.** See p. 573 of the text. **22.** No; let $r$ = radius of the smaller
circle. Then its area is $\pi \cdot r \cdot r$, or $\pi r^2$. The radius of the larger
circle is $2r$, and its area is $\pi \cdot 2r \cdot 2r$, or $4\pi r^2$, or $4 \cdot \pi r^2$. Thus the
area of the larger circle is 4 times the area of the smaller circle.

**Exercise Set 10.4, p. 598**

**1.** $768 \text{ cm}^3$ **3.** $45 \text{ in}^3$ **5.** $75 \text{ m}^3$ **7.** $357\frac{1}{2} \text{ yd}^3$
**9.** $803.84 \text{ in}^3$ **11.** $353.25 \text{ cm}^3$ **13.** $41,580,000 \text{ yd}^3$
**15.** $4,186,666.67 \text{ in}^3$ **17.** $124.72 \text{ m}^3$ **19.** $1950\frac{101}{168} \text{ ft}^3$
**21.** $113,982 \text{ ft}^3$ **23.** $24.64 \text{ cm}^3$ **25.** $\frac{33}{40} \text{ yd}^3$
**27.** $4747.68 \text{ cm}^3$ **29.** About $904 \text{ ft}^3$ **31.** $143.72 \text{ cm}^3$
**33.** $32,993,440,000 \text{ mi}^3$ **35.** About $77.7 \text{ in}^3$ **37.** $61,600 \text{ m}^3$
**39.** $5832 \text{ yd}^3$ **41.** $646.74 \text{ cm}^3$ **43.** 396 **44.** 3 **45.** 14
**46.** 253,440 **47.** 12 **48.** 335,808 **49.** 24 **50.** 14
**51.** 0.566 **52.** 13,400 **53.** $1064 \text{ mi}^3$ **54.** $24,360 \text{ mi}^2$
**55.** 260.4 ft **56.** $1087.8 \text{ mi}^3$ **57.** $3540.68 \text{ km}^3$
**59.** $0.477 \text{ m}^3$

**Exercise Set 10.5, p. 609**

**1.** Angle $GHI$, angle $IHG$, $\angle GHI$, $\angle IHG$, or $\angle H$ **3.** $10°$
**5.** $180°$ **7.** $130°$ **9.** Obtuse **11.** Acute **13.** Straight
**15.** Obtuse **17.** Acute **19.** Obtuse **21.** $79°$ **23.** $23°$
**25.** $32°$ **27.** $61°$ **29.** $177°$ **31.** $41°$ **33.** $95°$ **35.** $78°$
**37.** Scalene; obtuse **39.** Scalene; right **41.** Equilateral;
acute **43.** Scalene; obtuse **45.** $46°$ **47.** $120°$ **49.** $58°$
**51.** Three **52.** Parallel **53.** Gram **54.** Prime
**55.** Angle **56.** Cents; dollars **57.** Perimeter
**58.** Multiplicative; additive **59.** $m\angle 2 = 67.13°$;
$m\angle 3 = 33.07°$; $m\angle 4 = 79.8°$; $m\angle 5 = 67.13°$
**61.** $m\angle ACB = 50°$; $m\angle CAB = 40°$; $m\angle EBC = 50°$;
$m\angle EBA = 40°$; $m\angle AEB = 100°$; $m\angle ADB = 50°$

**Calculator Corner, p. 613**

**1.** 6.6 **2.** 9.7 **3.** 19.8 **4.** 17.3 **5.** 24.9 **6.** 24.5
**7.** 121.2 **8.** 115.6 **9.** 85.4

**Translating for Success, p. 616**

**1.** K **2.** G **3.** B **4.** H **5.** O **6.** M **7.** E **8.** A
**9.** D **10.** I

**Exercise Set 10.6, p. 617**

**1.** 10 **3.** 21 **5.** 25 **7.** 19 **9.** 23 **11.** 100 **13.** 6.928
**15.** 2.828 **17.** 4.243 **19.** 2.449 **21.** 3.162 **23.** 8.660
**25.** 14 **27.** 13.528 **29.** $c = \sqrt{34}; c \approx 5.831$
**31.** $c = \sqrt{98}; c \approx 9.899$ **33.** $a = 5$ **35.** $b = 8$
**37.** $c = 26$ **39.** $b = 12$ **41.** $b = \sqrt{1023}; b \approx 31.984$
**43.** $c = 5$ **45.** $\sqrt{250} \text{ m} \approx 15.8 \text{ m}$ **47.** $\sqrt{8450} \text{ ft} \approx 91.9 \text{ ft}$
**49.** $h = \sqrt{500} \text{ ft} \approx 22.4 \text{ ft}$ **51.** $\sqrt{211,200.00} \text{ ft} \approx 14,532.7 \text{ ft}$
**53.** $\sqrt{832} \text{ ft} \approx 28.8 \text{ ft}$ **55.** 1000 **56.** 100 **57.** 100,000
**58.** 10,000 **59.** $994,274.04 **60.** $1,686,435.46
**61.** $87,087.71 **62.** $90,450.48 **63.** The areas are the same.

**Summary and Review: Chapter 10, p. 620**

**Concept Reinforcement**

**1.** True **2.** False **3.** False **4.** True **5.** True **6.** True

## Important Concepts

**1.** 27.8 ft; 46.74 ft$^2$  **2.** 15.5 m$^2$  **3.** 8.75 ft$^2$  **4.** 80 m$^2$
**5.** 37.68 in.  **6.** 616 cm$^2$  **7.** 1683.3 m$^3$  **8.** $30\frac{6}{35}$ ft$^3$
**9.** 1696.537813 cm$^3$  **10.** 26.49375 ft$^3$  **11.** Complement:
52°; supplement: 142°  **12.** 87°  **13.** $\sqrt{208} \approx 14.422$

## Review Exercises

**1.** 23 ft  **2.** 4.4 m  **3.** 228 ft; 2808 ft$^2$  **4.** 36 ft; 81 ft$^2$
**5.** 17.6 cm; 12.6 cm$^2$  **6.** 22.5 m$^2$  **7.** 29.64 yd$^2$  **8.** 88 m$^2$
**9.** $145\frac{5}{9}$ in$^2$  **10.** 840 ft$^2$  **11.** 8 m  **12.** $\frac{14}{11}$ in., or $1\frac{3}{11}$ in.
**13.** 14 ft  **14.** 20 cm  **15.** 50.24 m  **16.** 8 in.  **17.** 200.96 m$^2$
**18.** $5\frac{1}{11}$ in$^2$  **19.** 1038.555 ft$^2$  **20.** 93.6 yd$^3$  **21.** 193.2 cm$^3$
**22.** 31,400 ft$^3$  **23.** 33.49$\overline{3}$ cm$^3$  **24.** 4.71 in$^3$
**25.** 942 cm$^3$  **26.** 26.28 ft$^2$; 20.28 ft  **27.** 54°  **28.** 180°
**29.** 140°  **30.** 90°  **31.** Acute  **32.** Straight  **33.** Obtuse
**34.** Right  **35.** 49°  **36.** 136°  **37.** 60°  **38.** Scalene
**39.** Right  **40.** 8  **41.** 9.110  **42.** $c = \sqrt{850}$; $c \approx 29.155$
**43.** $b = \sqrt{51}$; $b \approx 7.141$  **44.** $c = \sqrt{89}$ ft; $c \approx 9.434$ ft
**45.** $a = \sqrt{76}$ cm; $a \approx 8.718$ cm  **46.** About 28.8 ft
**47.** About 79 m  **48.** About 85.9 ft  **49.** B  **50.** B
**51.** 100 ft$^2$  **52.** 7.83998704 m$^2$  **53.** 47.25 cm$^2$

## Understanding Through Discussion and Writing

**1.** Add 90° to the measure of the angle's complement.
**2.** This could be done using the technique in Example 7 of
Section 10.4. We could also approximate the volume with the
volume of a similarly shaped rectangular solid. Another method
is to break the egg and measure the capacity of its contents.
**3.** Show that the sum of the squares of the lengths of the legs is
the same as the square of the length of the hypotenuse.
**4.** Divide the figure into 3 triangles.

The sum of the measures of the angles of each triangle is 180°,
so the sum of the measures of the angles of the figure is 3 · 180°,
or 540°.  **5.** The volume of the cone is half the volume of the
dome. It can be argued that a cone-cap is more energy-efficient
since there is less air under it to be heated and cooled.
**6.** Volume of two spheres, each with radius $r$: $2\left(\frac{4}{3}\pi r^3\right) = \frac{8}{3}\pi r^3$;
volume of one sphere with radius $2r$: $\frac{4}{3}\pi(2r)^3 = \frac{32}{3}\pi r^3$. The
volume of the sphere with radius $2r$ is four times the volume of
the two spheres, each with radius $r$: $\frac{32}{3}\pi r^3 = 4 \cdot \frac{8}{3}\pi r^3$.

## Test: Chapter 10, p. 628

**1.** [10.1a], [10.2a] 32.82 cm; 65.894 cm$^2$
**2.** [10.1a], [10.2a] $19\frac{1}{2}$ in.; $23\frac{49}{64}$ in$^2$  **3.** [10.2b] 25 cm$^2$
**4.** [10.2b] 12 m$^2$  **5.** [10.2b] 18 ft$^2$  **6.** [10.3a] $\frac{1}{4}$ in.
**7.** [10.3a] 9 cm  **8.** [10.3b] $\frac{11}{14}$ in.  **9.** [10.3c] 254.34 cm$^2$
**10.** [10.2c], [10.3d] 65.46 km; 103.815 km$^2$  **11.** [10.4a] 84 cm$^3$
**12.** [10.4e] 420 in$^3$  **13.** [10.4b] 1177.5 ft$^3$
**14.** [10.4c] 4186.$\overline{6}$ yd$^3$  **15.** [10.4d] 113.04 cm$^3$
**16.** [10.5a] 90°  **17.** [10.5a] 35°  **18.** [10.5a] 180°
**19.** [10.5a] 113°  **20.** [10.5b] Right  **21.** [10.5b] Acute
**22.** [10.5b] Straight  **23.** [10.5b] Obtuse  **24.** [10.5e] 35°
**25.** [10.5d] Isosceles  **26.** [10.5d] Obtuse
**27.** [10.5c] Complement: 25°; supplement: 115°  **28.** [10.6a] 15
**29.** [10.6b] 9.327  **30.** [10.6c] $c = 40$
**31.** [10.6c] $b = \sqrt{60}$; $b \approx 7.746$  **32.** [10.6c] $c = \sqrt{2}$; $c \approx 1.414$
**33.** [10.6c] $b = \sqrt{51}$; $b \approx 7.141$  **34.** [10.6d] About 15.8 m
**35.** [10.4c] D  **36.** [9.1a], [10.2a] 2 ft$^2$
**37.** [9.1a], [10.2b] 1.875 ft$^2$  **38.** [9.1a], [10.4a] 0.65 ft$^3$
**39.** [9.1a], [10.4d] 0.033 ft$^3$  **40.** [9.1a], [10.4b] 0.055 ft$^3$

## Cumulative Review: Chapters 1–10, p. 631

**1.** [5.3b] 79,200,000  **2.** [10.4e] 93,750 lb  **3.** [4.5a] $4\frac{1}{6}$
**4.** [5.2b] 87.52  **5.** [1.5a] 1234  **6.** [2.5c] −595
**7.** [5.5c] −33.2  **8.** [1.9d] 2  **9.** [5.1b] $\frac{1209}{1000}$  **10.** [7.2b] $\frac{17}{100}$
**11.** [4.3b] <  **12.** [3.5c] =  **13.** [9.1a] 3.5  **14.** [9.6a] $\frac{3}{20}$
**15.** [9.6b] 59  **16.** [9.5a] 87  **17.** [9.7a] 27  **18.** [9.2a] 0.17
**19.** [4.3c] $-\frac{13}{8}$  **20.** [6.3b] $4\frac{2}{7}$  **21.** [5.4b] 113.4
**22.** [6.3b] $14\frac{2}{5}$  **23.** [10.1a], [10.2b] 380 cm; 5500 cm$^2$
**24.** [10.3a, b, c] 70 in.; 220 in; 3850 in$^2$  **25.** [10.4c] $179,666\frac{2}{3}$ in$^3$
**26.** [8.1a] 99  **27.** [7.7a] $84  **28.** [7.7b] $22,376.03
**29.** [10.6d] 17 m  **30.** [7.6a] 6%  **31.** [4.5c] $2\frac{1}{8}$ yd
**32.** [5.7a] $37.42  **33.** [6.2b] The 8-qt box  **34.** [3.6b] $\frac{7}{20}$ km
**35.** [10.5e] 30°  **36.** [10.5d] Isosceles  **37.** [10.5d] Obtuse
**38.** [9.1a], [10.4b] 94,200 ft$^3$  **39.** [9.1a], [10.4a] 1.342 ft$^3$

# CHAPTER 11

## Calculator Corner, p. 635

**1.** 51  **2.** 98  **3.** −6  **4.** −19  **5.** 9  **6.** 12

## Exercise Set 11.1, p. 641

**1.** 42  **3.** 3  **5.** −1  **7.** 6  **9.** −20  **11.** 240; 240
**13.** 160; 160  **15.** $2b + 10$  **17.** $7 - 7t$  **19.** $30x + 12$
**21.** $7x + 28 + 42y$  **23.** $-7y + 14$  **25.** $45x + 54y - 72$
**27.** $\frac{3}{4}x - \frac{9}{4}y - \frac{3}{2}z$  **29.** $-3.72x + 9.92y - 3.41$  **31.** $2(x + 2)$
**33.** $5(6 + y)$  **35.** $7(2x + 3y)$  **37.** $5(x + 2 + 3y)$
**39.** $8(x - 3)$  **41.** $4(8 - y)$  **43.** $2(4x + 5y - 11)$
**45.** $-6(3x + 2y - 1)$, or $6(-3x - 2y + 1)$  **47.** $19a$  **49.** $9a$
**51.** $8x + 9z$  **53.** $-19a + 88$  **55.** $4t + 6y - 4$  **57.** $8x$
**59.** $5n$  **61.** $-16y$  **63.** $17a - 12b - 1$  **65.** $4x + 2y$
**67.** $\frac{39}{20}x + \frac{1}{2}y + 12$  **69.** $0.8x + 0.5y$  **71.** 30 yd; 94.2 yd;
706.5 yd$^2$  **72.** 16.4 m; 51.496 m; 211.1336 m$^2$  **73.** 19 mi;
59.66 mi; 283.385 mi$^2$  **74.** 4800 cm; 15,072 cm;
18,086,400 cm$^2$  **75.** 10 mm; 62.8 mm; 314 mm$^2$  **76.** 132 km;
828.96 km; 54,711.36 km$^2$  **77.** 2.3 ft; 14.444 ft; 16.6106 ft$^2$
**78.** 5.15 m; 32.342 m; 83.28065 m$^2$  **79.** $q(1 + r + rs + rst)$

## Exercise Set 11.2, p. 645

**1.** 4  **3.** −20  **5.** 2  **7.** 7  **9.** 4  **11.** −26  **13.** −18
**15.** −18  **17.** 15  **19.** 19  **21.** −1  **23.** −14
**25.** 2  **27.** 20  **29.** −6  **31.** $\frac{7}{3}$  **33.** $-\frac{7}{4}$  **35.** $\frac{41}{24}$
**37.** $-\frac{1}{20}$  **39.** 5.1  **41.** 12.4  **43.** −5  **45.** $1\frac{5}{6}$  **47.** $-\frac{10}{21}$
**49.** −11  **50.** $-\frac{1}{24}$  **51.** −34.1  **52.** −1.7  **53.** 5
**54.** $-\frac{31}{24}$  **55.** 5.5  **56.** 8.1  **57.** 24  **58.** $-\frac{5}{12}$  **59.** 283.14
**60.** −15.68  **61.** 8  **62.** $-\frac{16}{15}$  **63.** −14.3  **64.** −4.9
**65.** 342.246  **67.** $-\frac{26}{15}$  **69.** −10  **71.** $-\frac{5}{17}$

## Exercise Set 11.3, p. 650

**1.** 6  **3.** 9  **5.** 12  **7.** −40  **9.** −7  **11.** −7  **13.** −6
**15.** −5  **17.** 6  **19.** 5  **21.** −63  **23.** 36  **25.** −21
**27.** $-\frac{3}{5}$  **29.** $-\frac{3}{2}$  **31.** $\frac{9}{2}$  **33.** 7  **35.** −7  **37.** −8
**39.** 15.9  **41.** −50  **43.** −14  **45.** 20 ft; 62.8 ft; 314 ft$^2$
**46.** 12 cm; 75.36 cm; 452.16 cm$^2$  **47.** 8000 ft$^3$  **48.** 31.2 cm$^3$
**49.** 53.55 cm$^2$  **50.** 64 mm$^2$  **51.** 68 in$^2$  **52.** 38.25 m$^2$
**53.** −8655  **55.** No solution  **57.** No solution

## Mid-Chapter Review: Chapter 11, p. 652

**1.** False  **2.** True  **3.** True  **4.** False
**5.** $6x - 3y + 18 = 3 \cdot 2x - 3 \cdot y + 3 \cdot 6 = 3(2x - y + 6)$
**6.**
$$x + 5 = -3$$
$$x + 5 - 5 = -3 - 5$$
$$x + 0 = -8$$
$$x = -8$$
**7.**
$$-6x = 42$$
$$\frac{-6x}{-6} = \frac{42}{-6}$$
$$1 \cdot x = -7$$
$$x = -7$$

**8.** $-28$ **9.** 7 **10.** 5 **11.** $3x + 15$ **12.** $8y - 28$
**13.** $18x + 12y - 6$ **14.** $6x + 2y - 16$ **15.** $3(y + 7)$
**16.** $5(z + 9)$ **17.** $9(x - 4)$ **18.** $8(3a - 1)$
**19.** $2(2x + 3y - 1)$ **20.** $3(4x - 3y + 1)$
**21.** $4(a - 3b + 8)$ **22.** $6(5a - 3b - 4)$ **23.** $15x$
**24.** $2y$ **25.** $2x - y - 3$ **26.** 6 **27.** $-12$ **28.** 7
**29.** $-10$ **30.** 20 **31.** 5 **32.** 7 **33.** 10 **34.** $-\frac{5}{6}$ **35.** $\frac{3}{4}$
**36.** 0.7 **37.** $-1.4$ **38.** 6 **39.** 12 **40.** $-17$ **41.** $-9$
**42.** 17 **43.** $-6$ **44.** 18 **45.** $-15$ **46.** $-\frac{3}{2}$ **47.** $\frac{5}{3}$
**48.** $-3$ **49.** $-35$ **50.** They are not equivalent. For example, let $a = 2$ and $b = 3$. Then $(a + b)^2 = (2 + 3)^2 = 5^2 = 25$, but $a^2 + b^2 = 2^2 + 3^2 = 4 + 9 = 13$. **51.** We use the distributive law when we collect like terms even though we might not always write this step. **52.** The student probably added $\frac{1}{3}$ on the right side of the equation rather than adding $-\frac{1}{3}$ (or subtracting $\frac{1}{3}$). The correct solution is $-2$. **53.** The student apparently multiplied by $-\frac{2}{3}$ on both sides rather than dividing by $\frac{2}{3}$ on both sides. The correct solution is $-\frac{5}{2}$.

### Calculator Corner, p. 660

**1.** Left to the student **2.** Left to the student

### Exercise Set 11.4, p. 661

**1.** 5 **3.** 8 **5.** 10 **7.** 14 **9.** $-8$ **11.** $-8$ **13.** $-7$
**15.** 15 **17.** 6 **19.** 4 **21.** 6 **23.** $-3$ **25.** 1 **27.** $-20$
**29.** 6 **31.** 7 **33.** 2 **35.** 5 **37.** 2 **39.** 10 **41.** 4
**43.** 0 **45.** $-1$ **47.** $-\frac{4}{3}$ **49.** $\frac{2}{5}$ **51.** $-2$ **53.** $-4$ **55.** $\frac{4}{5}$
**57.** $-\frac{28}{27}$ **59.** 6 **61.** 2 **63.** 6 **65.** 8 **67.** 1 **69.** 17
**71.** $-\frac{5}{3}$ **73.** $-3$ **75.** 2 **77.** $-\frac{51}{31}$ **79.** 2 **81.** Rational; zero **82.** Is greater than **83.** One **84.** Absolute value
**85.** Obtuse **86.** Equilateral **87.** Meter **88.** Factor
**89.** 2 **91.** $-2$ **93.** 8 **95.** 2 cm

### Translating for Success, p. 674

**1.** B **2.** H **3.** G **4.** N **5.** J **6.** C **7.** L **8.** E
**9.** F **10.** D

### Exercise Set 11.5, p. 675

**1.** $2x - 3$ **3.** $97\%y$, or $0.97y$ **5.** $5x + 4$, or $4 + 5x$
**7.** $\frac{1}{3}x$, or $240 - x$ **9.** 32 **11.** 3113 manatees **13.** 305 ft
**15.** 57 **17.** $-12$ **19.** $699\frac{1}{3}$ mi **21.** First: 30 m; second: 90 m; third: 360 m **23.** Length: 94 ft; width: 50 ft
**25.** Length: 265 ft; width: 165 ft; area: 43,725 ft$^2$ **27.** 325 mi
**29.** First: 22.5°; second: 90°; third: 67.5° **31.** $350
**33.** $852.94 **35.** 12 mi **37.** $36 **39.** Length: 48 ft; width: 14 ft **41.** $-\frac{47}{40}$ **42.** $-\frac{17}{40}$ **43.** $-\frac{3}{10}$ **44.** $-\frac{32}{15}$
**45.** 1.6 **46.** 409.6 **47.** $-9.6$ **48.** $-41.6$
**49.** $1863 = 1776 + (4s + 7)$, or $1863 - 1776 = 4s + 7$
**51.** Length: 12 cm; width: 9 cm **53.** Quarters: 60; dimes: 30; nickels: 40

### Summary and Review: Chapter 11, p. 680

#### Concept Reinforcement

**1.** True **2.** False **3.** False **4.** True

#### Important Concepts

**1.** $4x + 20y - 28$ **2.** $8(3a - b + 2)$ **3.** $6x - 3y$ **4.** 2
**5.** $-8$ **6.** 1 **7.** $-2$ **8.** 6 **9.** $n + 5$, or $5 + n$

#### Review Exercises

**1.** 4 **2.** $15x - 35$ **3.** $-8x + 10$ **4.** $4x + 15$
**5.** $-24 + 48x - 16y$ **6.** $2(x - 7)$ **7.** $6(x - 1)$
**8.** $5(x + 2)$ **9.** $3(4 - x + 2z)$ **10.** $7a - 3b$
**11.** $-2x + 5y$ **12.** $5x - y$ **13.** $-a + 8b$ **14.** $-22$
**15.** 7 **16.** $-192$ **17.** 1 **18.** $-\frac{7}{3}$ **19.** 25 **20.** $\frac{1}{2}$
**21.** $-\frac{15}{64}$ **22.** 9.99 **23.** $-8$ **24.** $-5$ **25.** $-\frac{1}{3}$ **26.** 4
**27.** 3 **28.** 4 **29.** 16 **30.** 6 **31.** $-3$ **32.** $-2$ **33.** 4
**34.** $19\%x$, or $0.19x$ **35.** Length: 365 mi; width: 275 mi

**36.** $2117 **37.** 27 appliances **38.** 35°, 85°, 60° **39.** $220
**40.** $138.95 **41.** $75,000 **42.** $867 **43.** Width: 11 cm; length: 17 cm **44.** Amazon: 6437 km; Nile: 6671 km
**45.** C **46.** A **47.** $23, -23$ **48.** $20, -20$

### Understanding Through Discussion and Writing

**1.** The distributive laws are used to multiply, factor, and collect like terms in this chapter. **2.** For an equation $x + a = b$, we add the opposite of $a$ on both sides of the equation to get $x$ alone. **3.** For an equation $ax = b$, we multiply by the reciprocal of $a$ on both sides of the equation to get $x$ alone. **4.** We add $-b$ (or subtract $b$) on both sides and simplify. Then we multiply by the reciprocal of $a$ (or divide by $a$) on both sides and simplify.

### Test: Chapter 11, p. 685

**1.** [11.1a] 6 **2.** [11.1b] $18 - 3x$ **3.** [11.1b] $-5y + 5$
**4.** [11.1c] $2(6 - 11x)$ **5.** [11.1c] $7(x + 3 + 2y)$
**6.** [11.1d] $-5x - y$ **7.** [11.1d] $4a + 5b$ **8.** [11.2a] 8
**9.** [11.2a] 26 **10.** [11.3a] $-6$ **11.** [11.3a] 49
**12.** [11.4b] $-12$ **13.** [11.4a] 2 **14.** [11.4a] $-8$
**15.** [11.2a] $-\frac{7}{20}$ **16.** [11.4b] 2.5 **17.** [11.4c] 7
**18.** [11.4c] $\frac{5}{3}$ **19.** [11.5a] $x - 9$ **20.** [11.5b] Width: 7 cm; length: 11 cm **21.** [11.5b] About $310 billion
**22.** [11.5b] 3 m, 5 m **23.** [11.5b] $880 **24.** [11.5b] 6
**25.** [11.5b] 25.625° **26.** [11.4b] D
**27.** [2.1d], [11.4a] $15, -15$ **28.** [11.5b] 60 tickets

### Cumulative Review: Chapters 1-11, p. 687

**1.** [1.1a] 7 **2.** [1.1b] 7 thousands + 4 hundreds + 0 tens + 5 ones, or 7 thousands + 4 hundreds + 5 ones
**3.** [5.1a] Seven and four hundred sixty-three thousandths
**4.** [1.2a] 1012 **5.** [1.2a] 21,085 **6.** [4.2a] $-\frac{3}{26}$ **7.** [4.5a] $5\frac{7}{9}$
**8.** [5.2a] 493.971 **9.** [5.2a] 802.876 **10.** [1.3a] 152
**11.** [1.3a] 674 **12.** [4.3a] $\frac{5}{24}$ **13.** [4.5b] $2\frac{17}{24}$
**14.** [5.2b] 19.9973 **15.** [5.2c] 45.819 **16.** [3.5b] $\frac{7}{10}$
**17.** [3.5b] 55 **18.** [1.4a] 4752 **19.** [1.4a] 266,287
**20.** [4.6a] $4\frac{1}{12}$ **21.** [3.6a] $\frac{6}{5}$ **22.** [3.4a] $-10$
**23.** [5.3a] 259.084 **24.** [4.4a] $3\frac{3}{5}$ **25.** [1.5a] 573
**26.** [1.5a] 56 R 10 **27.** [4.4b] $56\frac{5}{17}$ **28.** [3.7b] $\frac{3}{2}$
**29.** [4.6b] $-\frac{7}{90}$ **30.** [5.4a] 39 **31.** [1.6a] 68,000
**32.** [5.1d] 0.428 **33.** [5.5b] 21.84 **34.** [3.2a] Yes
**35.** [3.1a] 1, 3, 5, 15 **36.** [4.1a] 800 **37.** [3.5c] $\neq$
**38.** [4.3b] $<$ **39.** [5.1c] $-0.9976$ **40.** [6.2b] 11.176¢/oz, 11.067¢/oz, 12.197¢/oz, 10.583¢/oz, 10.586¢/oz; brand D has the lowest unit price. **41.** [10.3b], [10.4c] **(a)** 4400 mi; **(b)** about 1,437,333,333 mi$^3$ **42.** [7.5a] $10,755.20
**43.** [7.5a] 15% **44.** [7.5a] $3495.44 **45.** [7.5a] 30%
**46.** [7.5a] $537.76 **47.** [1.6c] $>$ **48.** [3.3a] $\frac{3}{5}$
**49.** [5.1b] 0.037 **50.** [5.5a] $-0.52$ **51.** [5.5a] $0.\overline{8}$
**52.** [7.1b] 0.07 **53.** [5.1b] $\frac{463}{100}$ **54.** [4.4a] $-\frac{29}{4}$ **55.** [7.2b] $\frac{2}{5}$
**56.** [7.2a] 85% **57.** [7.1b] 150% **58.** [1.7b] 555
**59.** [5.4b] $-64$ **60.** [3.7c] $\frac{5}{4}$ **61.** [6.3b] $\frac{153}{2}$, or $76\frac{1}{2}$, or 76.5
**62.** [8.4b]

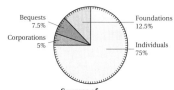

Sources of Charitable Donations

**63.** [8.3b]

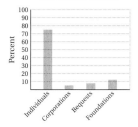

Sources of Charitable Donations

**64.** [10.5e] 118°    **65.** [10.5d] Obtuse    **66.** [1.8a] $675
**67.** [1.8a] 65 min    **68.** [5.7a] $25.75    **69.** [5.7a] 485.9 mi
**70.** [1.8a] $8100    **71.** [1.8a] $783    **72.** [3.4b] $\frac{3}{10}$ km
**73.** [5.7a] $84.96    **74.** [6.4a] 13 gal    **75.** [7.7a] $150
**76.** [7.6b] 7%    **77.** [7.5b] 30,160    **78.** [8.1a, b, c] 28; 26; 18
**79.** [1.9b] 324    **80.** [1.9b] 343    **81.** [10.6a] 3    **82.** [10.6a] 11
**83.** [10.6b] 4.472    **84.** [9.1a] 12    **85.** [9.2a] 428    **86.** [9.6a] 72
**87.** [9.4b] 20    **88.** [9.4a] 80    **89.** [9.4b] 0.08    **90.** [9.5a] 8.19
**91.** [9.5a] 5    **92.** [10.6c] $c = \sqrt{50}$ ft; $c \approx 7.071$ ft
**93.** [10.3a, b, c] 20.8 in.; 65.312 in.; 339.6224 in²
**94.** [10.1a], [10.2a] 25.6 m; 25.75 m²    **95.** [10.2b] 25 in²

**96.** [10.2b] 61.6 cm²    **97.** [10.2b] 128.65 yd²
**98.** [10.4a] 52.9 m³    **99.** [10.4b] 803.84 ft³
**100.** [10.4d] 267.94$\overline{6}$ cm³    **101.** [11.4a] −5    **102.** [11.3a] 4
**103.** [11.4b] −8    **104.** [11.4c] $\frac{2}{3}$    **105.** [1.9c] 238
**106.** [1.9c] 172    **107.** [2.1d], [2.4a] 3    **108.** [2.2a] 14
**109.** [4.3a] $\frac{1}{3}$    **110.** [2.4a] 30    **111.** [3.6a] $-\frac{2}{7}$    **112.** [2.5a] −8
**113.** [11.5a] $y + 17$, or $17 + y$    **114.** [11.5a] 38%$x$, or 0.38$x$
**115.** [11.5b] $45    **116.** [11.5b] $1050
**117.** [11.5b] 50 m; 53 m; 40 m    **118.** [11.4b] 0    **119.** [11.4b] 2
**120.** [11.1d] C    **121.** [11.1c] B    **122.** [3.7b] D    **123.** [2.2a] A
**124.** [11.5b] 235 and 195

# Glossary

## A

**Absolute value**   The distance of a number from 0 on the number line

**Acute angle**   An angle whose measure is greater than 0° and less than 90°

**Acute triangle**   A triangle in which all three angles are acute

**Addend**   In addition, a number being added

**Additive identity**   The number 0

**Additive inverse**   Two numbers are additive inverses if their sum is 0. Also called the opposite.

**Algebraic expression**   An expression consisting of variables, constants, numerals, operation signs, and/or grouping symbols

**Angle**   A set of points consisting of two rays (half-lines) with a common endpoint (vertex)

**Area**   The number of square units that fill a plane region

**Arithmetic mean**   A center point of a set of numbers found by adding the numbers and dividing by the number of items of data; also called mean or average

**Associative law of addition**   The statement that when three numbers are added, regrouping the addends gives the same sum

**Associative law of multiplication**   The statement that when three numbers are multiplied, regrouping the factors gives the same product

**Average**   A center point of a set of numbers found by adding the numbers and dividing by the number of items of data; also called the arithmetic mean or mean

## B

**Bar graph**   A graphic display of data using bars proportional in length to the numbers represented

**Base**   In exponential notation, the number being raised to a power

## C

**Celsius**   A temperature scale for metric measure

**Circle graph**   A graphic means of displaying data using sectors of a circle; often used to show the percent of a quantity in different categories or to show visually the ratio of one category to another; also called a pie chart

**Circumference**   The distance around a circle

**Coefficient**   The numerical multiplier of a variable

**Commission**   A percent of total sales paid to a salesperson

**Commutative law of addition**   The statement that when two numbers are added, changing the order in which the numbers are added does not affect the sum

**Commutative law of multiplication**   The statement that when two numbers are multiplied, changing the order in which the numbers are multiplied does not affect the product

**Complementary angles**   Two angles for which the sum of their measures is 90°

**Composite number**   A natural number, other than 1, that is not prime

**Compound interest**   Interest paid on interest

**Constant**   A known number

**Cross products**   Given an equation with a single fraction on each side, the products formed by multiplying the left numerator and the right denominator, and the left denominator and the right numerator

## D

**Decimal notation**   A representation of a number containing a decimal point

**Denominator**   The bottom number in a fraction

**Diameter**   A segment that passes through the center of a circle and has its endpoints on the circle

**Difference**   The result of subtracting one number from another

**Digit**   A number 0, 1, 2, 3, 4, 5, 6, 7, 8, or 9 that names a place-value location

**Discount**   The amount subtracted from the original price of an item to find the sale price

**Distributive law of multiplication over addition**   The statement that multiplying a factor by the sum of two numbers gives the same result as multiplying the factor by each of the two numbers and then adding

**Distributive law of multiplication over subtraction** The statement that multiplying a factor by the difference of two numbers gives the same result as multiplying the factor by each of the two numbers and then subtracting

**Dividend** In division, the number being divided

**Divisible** The number $a$ is divisible by another number $b$ if there exists a number $c$ such that $a = b \cdot c$

**Divisor** In division, the number dividing another number

### E

**Equation** A number sentence that says that the expressions on either side of the equals sign, $=$, represent the same number

**Equilateral triangle** A triangle in which all sides are the same length

**Equivalent equations** Equations with the same solutions

**Equivalent expressions** Expressions that have the same value for all allowable replacements

**Equivalent fractions** Fractions that name the same number

**Even number** A number that is divisible by 2; that is, it has an even ones digit

**Exponent** In expressions of the form $a^n$, the number $n$; for $n$ a natural number, $a^n$ represents $n$ factors of $a$

**Exponential notation** A representation of a number using a base raised to a power

### F

**Factor** *Verb:* to write an equivalent expression that is a product. *Noun:* a multiplier

**Factorization** A number expressed as a product of natural numbers

**Fahrenheit** A temperature scale for American measure

**Fraction notation** A number written using a numerator and a denominator

### H

**Hypotenuse** In a right triangle, the side opposite the right angle (also the longest side)

### I

**Identity property of 1** The statement that the product of a number and 1 is always the original number

**Identity property of 0** The statement that the sum of a number and 0 is always the original number

**Inequality** A mathematical sentence using $<$, $>$, $\leq$, $\geq$, or $\neq$

**Integers** The whole numbers and their opposites

**Interest rate** A percentage of an amount invested or borrowed

**Isosceles triangle** A triangle in which two or more sides are the same length

### L

**Least common denominator (LCD)** The least common multiple of the denominators of two or more fractions

**Least common multiple (LCM)** For two or more natural numbers, the smallest number that is a multiple of all the numbers

**Legs** In a right triangle, the two sides that form the right angle

**Like terms** Terms that have exactly the same variable factors

**Line graph** A graphic means of displaying data by connecting adjacent data points with line segments

### M

**Marked price** The original price of an item

**Mean** A center point of a set of numbers found by adding the numbers and dividing by the number of items of data; also called the arithmetic mean or average

**Median** In a set of data listed in order from smallest to largest, the middle number if there is an odd number of data items, or the average of the two middle numbers if there is an even number of data items

**Minuend** The number from which another number is being subtracted

**Mixed numeral** A number represented by an integer and a fraction less than 1

**Mode** The number or numbers that occur most often in a set of data. If each number occurs the same number of times, there is no mode.

**Multiple** A product of a number and some natural number

**Multiplicative identity** The number 1

### N

**Natural numbers** The numbers 1, 2, 3, 4, 5, ...

**Negative integers** The integers to the left of zero on the number line

**Numerator** The top number in a fraction

### O

**Obtuse angle** An angle whose measure is greater than $90°$ and less than $180°$

**Obtuse triangle** A triangle in which one angle is an obtuse angle

**Opposite** Two numbers are opposites if their sum is 0. Also called the additive inverse.

**Original price** The price of an item before a discount is deducted

### P

**Parallelogram** A four-sided polygon with two pairs of parallel sides

**Percent notation** A representation of a number as parts per 100

**Perimeter** The distance around an object, or the sum of the lengths of its sides

# Index